1D and Multi-D Modeling Techniques for IC Engine Simulation

1D and Multi-D Modeling Techniques for IC Engine Simulation

A. ONORATI AND G. MONTENEGRO

Warrendale, Pennsylvania, USA

400 Commonwealth Drive
Warrendale, PA 15096-0001 USA
E-mail: CustomerService@sae.org
Phone: 877-606-7323 (inside USA and Canada)
724-776-4970 (outside USA)
FAX: 724-776-0790

Library of Congress Catalog Number 2019948855
http://dx.doi.org/10.4271/9780768099522

ISBN-Print 978-0-7680-9352-0

To purchase bulk quantities, please contact: SAE Customer Service

E-mail: CustomerService@sae.org
Phone: 877-606-7323 (inside USA and Canada)
724-776-4970 (outside USA)
Fax: 724-776-0790

Visit the SAE International Bookstore at books.sae.org

Chief Product Officer
Frank Menchaca

Publisher
Sherry Dickinson Nigam

Director of Content Management
Kelli Zilko

Production Associate
Erin Mendicino

Manufacturing Associate
Adam Goebel

contents

CHAPTER 5

3D-CFD Combustion Models for SI and CI Engines 181

Tommaso Lucchini and Yuri Wright

CHAPTER 6

Control-Oriented Gas Dynamic Simulation via Model Order Reduction 221

Stephanie Stockar and Marcello Canova

CHAPTER 7

Modeling of EGR Systems 257

José Galindo, Héctor Climent, and Roberto Navarro

CHAPTER 12

3D Simulation Models for After-Treatment Systems 417

O. Haralampous and G. Koltsakis

CHAPTER 13

Modeling of IC Engine Silencers and Tailpipe Noise: 1D and 3D Approaches 457

Mats Åbom, Francisco D. Denia, and Antonio J. Torregrosa

preface

Nowadays the IC engine is the most common thermal machine in the world, playing a key role in several application fields: passenger cars, motorcycles and motorbikes, on-road commercial transport, boats and ships, off-road for industrial use, and power generation. In particular, the automotive sector is currently experiencing a significant evolution, with a range of alternative powertrains available to achieve a clean mobility. However, undoubtedly, the IC engine will represent the powertrain solution for the majority of vehicles worldwide during the next decades, of course considering hybrid and electrified architectures. An optimized IC engine will be fundamental to guarantee low fuel consumption and near-zero pollutant emissions, minimizing CO_2 production as well. Innovative combustion modes, synthetic fuels, turbocharging, and advanced after-treatment systems will allow to achieve this challenging target.

In this scenario, in order to investigate and compare different IC engine technologies, modeling techniques and simulation codes will continue to be fundamental in the design chain. A "virtual" IC engine, thanks to advanced modeling approaches, gives the opportunity to test quickly the effect of different solutions on performances and emissions.

During the last two decades, a significant research effort in the field of 1D/3D engine modeling has been carried out in universities, software companies, and research centers, providing notable improvements and enhancements. This progress has allowed a further spread of simulation tools among industries for the daily engine design process.

A lot of interesting conference and journal papers have been published by several research groups about 1D and 3D thermo-fluid dynamic modeling of engines and components, with a wide range of innovative contributions, as reported in the reference list of the following chapters. This book is an attempt to provide a description of the most significant and recent achievements in the field of 1D engine simulation models and coupled 1D-3D modeling techniques, including 0D combustion models, quasi-3D methods, and some 3D model applications.

The organization of the book is the following:

Chapter 1 is dedicated to outline the state of the art of 1D thermo-fluid dynamic simulation models, briefly describing the history of wave dynamics simulation and the most significant contribution in terms of numerical methods and simulation tools.

Chapter 2 is focused on the importance of 1D and 3D CFD simulation tools to achieve a robust design by means of a virtual engine environment. This chapter describes when and how 1D and 3D simulation tools, as well as a combination of both, can be used, pointing out which characteristics are required from these models to establish a fast, reliable, and effective virtual engine design.

Chapter 3 is dedicated to 0D and quasi-D thermodynamic combustion models for SI and CI engines, highlighting the recent developments in this field to improve 1D engine simulation codes in terms of combustion and emission prediction.

Chapter 4 deals with the topic of turbine and compressor modeling by means of advanced boundary conditions and techniques in the framework of 1D simulation models; a notable portion of IC engines is turbocharged today, so the accuracy of turbocharger sub-models is really important.

Chapter 5 is a useful summary of 3D CFD combustion models for SI and CI engines, intended to provide an overview of the most significant approaches to engineers and experts typically more familiar with the 1D simulation world.

Chapter 6 describes the numerical methods available to reduce the complexity and computational demand of 1D gas dynamic models in order to achieve a control-oriented air path system modeling that takes into account the wave motion. This approach allows to develop a high-fidelity 1D fast engine simulator.

Chapter 7 deals with the 1D simulation of high pressure and low pressure EGR systems for NOx control, pointing out the recommended approaches to model the EGR valve and EGR coolers; the issues of water condensation and fouling are also discussed, due to their influence on the thermal effectiveness and pressure loss of EGR coolers.

Chapter 8 introduces the importance of co-simulation and X-in-the-loop testing approaches resorting to 1D engine models. The application of real-time capable, 1D crank-angle-resolved engine models can reduce the effort for test and validation under complex boundary conditions.

Chapter 9 describes the numerical techniques developed to achieve a robust coupling between 1D and 3D modeling tools, allowing a hybrid simulation code; applications and results are discussed to highlight the advantages of this approach for some specific cases.

Chapter 10 explains how to extend the capability of 1D simulation tools by means of a quasi-3D approach, useful for the modeling of complex shape components. The numerical techniques adopted in this case, to achieve an intermediate tool between simplified 1D network models and coupled 1D-3D models, are discussed.

Chapter 11 is dedicated to the principles and governing equations for the 1D modeling of after-treatment components; the state of the art of 1D simulation in this field is discussed, highlighting the methods to achieve a robust solution of detailed chemical kinetics, gas phase species diffusion, as well as pore diffusion within a multi-layered washcoat.

Chapter 12 describes typical examples of exhaust after-treatment modeling via the 3D approach; applications to 3-way catalysts, including 3D heat transfer problems, to SCR systems, with simulation of the injected urea-water solution, and to diesel particulate filters during regeneration are discussed.

Chapter 13 is focused on the modeling of IC engine silencers and tailpipe noise by means of 1D and 3D approaches; different techniques available for silencer modeling and intake and exhaust noise prediction are described, including linear (frequency domain) and nonlinear (time domain) methods, comprising 1D and 3D techniques.

In order to cope with this challenging work, the editors had the opportunity to rely on the precious collaboration of several colleagues from different well-known research groups worldwide, who accepted to contribute to the preparation of this book with a chapter related to their area of expertise. The editors are truly grateful to all the authors for the patient, valuable, and careful work, indispensable to make the content useful for engineers, researchers, and students working in these fields.

Angelo Onorati and Gianluca Montenegro

about the editors

Angelo Onorati graduated in 1989 in Mechanical Engineering at the Politecnico di Milano, then he achieved a Ph.D. in Energy Engineering in 1993. He became a lecturer at the Department of Energy of Politecnico di Milano in 1993 (in the field of "Fluid Machines"). From 1998 to 2003, he was an associate professor in the same university, and then since the beginning of 2004 he become a full professor. His main research subjects are 1D and CFD modeling of unsteady reacting flows in IC engine duct systems and after-treatment devices, modeling of SI and CI engine combustion process and prediction of emissions, and prediction of tailpipe noise and silencer modeling. He is the author of more than 120 publications.

He coordinates the activity of the ICE Group at the Department of Energy of PoliMi, which is active in the field of IC engine modeling and simulation, for the development and application of advanced 3D (LibICE/OpenFOAM) and 1D (GASDYN) CFD codes.

He is involved in the editorial board of the *IJER* (*International Journal of Engine Research*) and of the *SAE International Journal of Engines*. In April 2011, he received the Lloyd L. Withrow distinguished speaker SAE award. He is currently involved as co-organizer of the session "0-D and 1-D Modeling and Numerics" for the WCX SAE World Congress.

Gianluca Montenegro achieved the MSc degree in mechanical engineering at the Politecnico di Milano in 1999 and the PhD in energy engineering in 2002, in the same university. He became permanent staff in 2006 working as assistant professor, and he is associate professor at Politecnico di Milano, Department of Energy, since 2015.

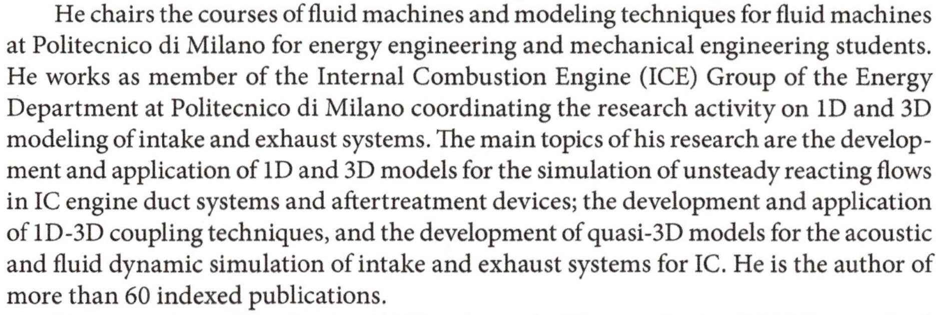

He chairs the courses of fluid machines and modeling techniques for fluid machines at Politecnico di Milano for energy engineering and mechanical engineering students. He works as member of the Internal Combustion Engine (ICE) Group of the Energy Department at Politecnico di Milano coordinating the research activity on 1D and 3D modeling of intake and exhaust systems. The main topics of his research are the development and application of 1D and 3D models for the simulation of unsteady reacting flows in IC engine duct systems and aftertreatment devices; the development and application of 1D-3D coupling techniques, and the development of quasi-3D models for the acoustic and fluid dynamic simulation of intake and exhaust systems for IC. He is the author of more than 60 indexed publications.

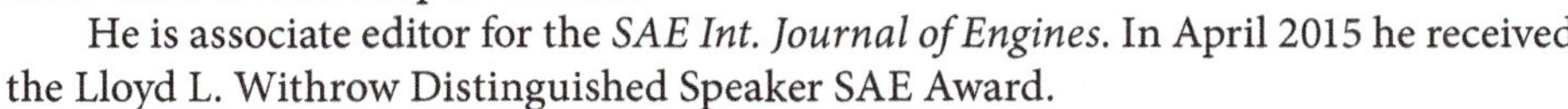

He is associate editor for the *SAE Int. Journal of Engines*. In April 2015 he received the Lloyd L. Withrow Distinguished Speaker SAE Award.

CHAPTER 1

State of the Art of 1D Thermo-Fluid Dynamic Simulation Models

Angelo Onorati and Gianluca Montenegro
Politecnico di Milano

1.1 Recent Advances in IC Engines and Future Perspectives

The internal combustion (IC) engine is currently the most common thermal machine in the world, with roughly 230 million new engines built every year, considering all the application fields: passenger cars, motorcycles and motorbikes, on-road commercial transport, boats and ships, off-road for industrial use, and power generation.

The IC engine is a volumetric thermal machine, invented at the end of the nineteenth century (Eugenio Barsanti and Felice Matteucci, 1854; Jean Etienne Lenoir, 1860; Nikolaus Otto, 1867; Rudolf Diesel, 1892). After an initial smooth penetration in the industrial sector for mechanical and electric power generation, it started to be massively applied in the transport sector, nowadays reaching around 90 million passenger cars produced every year. A few forecasts about future trends in the automotive industry predict a growth of world production to roughly 150 million cars per year in 2030, most of which (around 80%) still powered by an IC engine. See, for example, Figure 1.1, which highlights a possible future scenario regarding passenger car powertrains.

At present, due to the fast and challenging development of alternative powertrains, mainly in the automotive field (i.e., electric motors with batteries or fuel cells), the IC engine is sometimes considered as an "old technology," destined to disappear during the next years. Is it really an old technology? By this point of view, the fuel cell was invented in 1839 (William R. Grove, UK); the battery was invented in 1799 by Alessandro Volta, Italy; the term "photovoltaics" or "photovoltaic effect" was coined later from his name. Considering other important prime movers designed today for renewable energy production, also wind turbines are certainly much older (the windmill was invented in the seventh century AC in eastern Persia), as well as hydraulic turbines, invented and

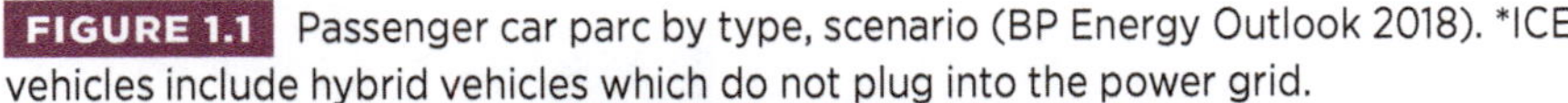

FIGURE 1.1 Passenger car parc by type, scenario (BP Energy Outlook 2018). *ICE vehicles include hybrid vehicles which do not plug into the power grid.

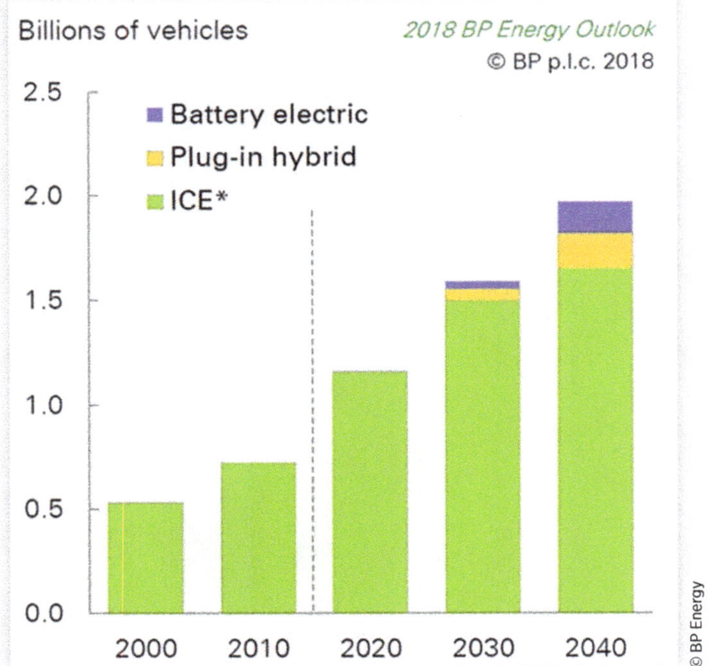

applied in different forms during the Roman age (around first century BC). In general, the concept of most prime movers was invented centuries ago, but the continuous research and development has allowed a deep evolution of each of them.

This is the case of the IC engine, which continues to experience significant technology advancements and innovations, supported by the intense research activities in this field. Important efforts from research centers, universities, and industries are currently made to achieve improvements in performance, fuel consumption, and pollutant and CO_2 emissions. A strong driver in this sense is certainly the compliance with stringent emission legislations implemented in several countries (current Euro 6d in Europe, Tier 3 in US, etc.), with the introduction of the real-driving emission (RDE) procedure, an additional test to achieve new vehicle homologation [1, 2].

As a matter of fact, the increasing stringency of pollutant emission regulations for on-road and off-road engines is pushing the research and development process toward advanced and complex solutions, once unpractical or too expensive. The requirement of extremely low CO_2 production is another key driver, which is guiding the development of several new technologies, including the "electrification" of vehicles for most applications.

At a glance, several best available technologies are currently applied to achieve the required performances from a modern IC engine. For example, the adoption of a Miller/Atkinson cycle [3, 4, 5, 6, 7] is quite common on new engines: it allows to reduce the effective engine compression ratio, by means of a shorter compression stroke, followed by a longer expansion stroke, resulting in higher thermodynamic efficiency. Moreover, lower charge temperatures are reached in the cylinder, with a corresponding reduction of NOx and knock propensity. The effective compression ratio can be reduced by implementing either a late intake valve closure (LIVC) or an early intake valve closure (EIVC) [5, 6]. A flexible adjustment of intake valve lifts and timings is nowadays feasible, thanks to the introduction of variable valve actuation (VVA) and variable valve timing (VVT) mainly for the intake valves, sometimes with a fully variable, flexible technology [8, 9, 10]. This is a key technology for both SI and CI engines, as a practical method not only to improve volumetric efficiency but also to achieve a smart management of internal EGR, or to increase after-treatment system temperatures during warm up after a cold start.

In the field of SI engines, cylinder deactivation [11, 12] can be useful to increase engine efficiency at part load, by a significant reduction of gas exchange work, thanks to the deactivation of a few cylinders; in this way the active cylinders can be operated at higher loads, reaching higher efficiency.

Concerning the charge ignition, a significant improvement for SI engines can be achieved by the adoption of new ignition systems such as the "corona discharge" [13], which allows a better start of combustion even with lean or diluted mixtures. One major critical issue of the turbocharged SI engine is the onset of knock, which can be currently mitigated (at high loads) by the adoption of gasoline direct injection, water injection, cooled EGR (exhaust gas recirculation) [14, 15, 16, 17], variable compression ratio [18, 19], and Miller cycle as well [5].

With regard to the CI engine, a notable improvement of combustion quality has been achieved by the continuous increase of injection pressure into the cylinder, achieved by common rail systems, with the adoption of piezo-injectors to manage multiple injections with close and short injection events, up to five injections or more. New combustion modes, such as HCCI (*homogeneous charge compression ignition*), LTC *low-temperature combustion*), and RCCI (*reactivity controlled compression ignition*), are under investigation to provide better fuel consumptions and emission reduction [20]. Recent advances in the innovative *spark-assisted compression ignition* (SACI) [21, 22, 23] appear very promising in terms of global efficiency and emission reduction, resorting to a jet ignitor [24] to operate at very high boost, high compression ratio and lean mixtures.

In most cases turbocharging is extensively adopted as a method to downsize the engine, keeping or increasing performances, while reducing emissions and fuel consumption. In this field, variable geometry turbines (VGT) for the CI engine and fixed geometry turbine (FGT) with a waste-gate valve for the SI engines are the most typical solutions. Dual stage and sequential turbocharging are advanced solutions which render the matching of the two turbochargers to the engine even more complex, with control issues to cope with [25, 26]. Moreover, most engines are characterized by the adoption of EGR, with short route (high pressure) and long route (low pressure) systems and EGR cooler [27, 28, 29], to achieve low NOx emission and mitigate knock tendency in some conditions.

Another innovative, promising field is represented by the future alternative fuels which will be available for the IC engine: biofuels (bioethanol, biodiesel, biogas) [30] or synthetic liquid/gaseous fuels, also named *e-fuels* when produced from renewable electricity [31]. E-fuels can significantly help the transport sector to meet the CO_2 targets and reduce engine-out soot emission, thanks to their characteristic properties. IC engines are extremely flexible and can be designed to exploit the positive features of new fuels, becoming gradually available in the international scenario.

Significant developments are continuously achieved in the field of after-treatment systems, one of the best measure to guarantee a notable reduction of pollutant emissions in the exhaust gas and provide clean mobility solutions. Research efforts are focused on the improvement of the next generation of three-way and oxidation catalysts, selective catalytic reduction (SCR) catalysts with urea injection, NOx traps, Diesel particulate filters (DPF) and gasoline particulate filters (GPF), hybrid solutions to achieve a compact particulate filter with SCR (SCR on filter) [32, 33].

Moreover, new IC engines will have to comply with the recently implemented "RDE" test, as an additional requirement for vehicle homologation, which allows to cover the entire speed and load range of the IC engine [34]. In this way all pollutant emissions will be measured more reliably onboard during the RDE test, including a wide range of operating conditions. This aspect has significant implications on the optimization and calibration of future IC engines [35].

Finally, the evident trend to the gradual electrification of vehicles is changing the requirements to the IC engine, in terms of performances, efficiency, and emissions, depending on the architecture considered: hybrid electric vehicle (HEV) or plug-in hybrid electric vehicle (PHEV). In any case, with hybrid powertrain solutions, the IC engine is generally characterized by a complex architecture, with high-boosting EGR circuits and advanced after-treatment technologies, and is coupled to several mechanical and electrical components of the driveline [36]. The optimization of the whole vehicle system is a challenging objective, involving the IC engine, electric motor and power electronics, battery, cooling system, vehicle, etc., aiming at the best solution in terms of fuel consumption, emissions, CO_2 production, and performances. Different operating conditions, including cold start and transients, can be studied and possibly optimized [37].

As a matter of fact, the ongoing competition between the IC engine and alternative powertrains, such as the electric motor powered with batteries (full electric vehicle) [38] or the fuel cell (fed with hydrogen or hydrocarbons) [39], as a possible solution in the mid-long term, is clearly beneficial to promote the further development and evolution of the IC engine, which can certainly continue to offer a sustainable solution to mobility, contributing to reduce the environmental impact of means of transport during the next decades.

1.2 The Key Role of IC Engine Simulation Models

The complex and challenging scenario described above requires an advanced approach to analyze different technologies and integrated solutions (ranging from new combustion modes to turbocharging and after-treatment systems). Several different configurations for a new IC engine need to be tested and compared, to finally select the most promising solutions.

One key aspect in this context is the massive adoption of simulation codes, to investigate the influence of new technical solutions and guide the design of new engines. In this way engineers can analyze a "virtual engine" [40], instead of relying only on the test bench (which is an expensive and slow process) to check the effects of new designs. During the last 20 years, a general, evident improvement of the predictiveness of thermo-fluid dynamic simulation models has been achieved thanks to the continuous research and development work of universities, software companies, and research centers. Engineers are confident that the simulation tool can provide useful, reliable results to guide the design process.

It is nowadays evident that the extensive application of simulation models, to optimize different technologies on a virtual engine, allows a cost-effective development process and a significant reduction of the time duration of the whole design process. However, the success degree of this design activity, mainly based on simulation tools, strongly depends on the accuracy and reliability of modeling approaches, which is fundamental to achieve a trustable analysis and optimization of the powertrain architectures at a very early stage of the concept development. In particular, in the field of thermo-fluid dynamic modeling, both 0D/1D (zero-dimensional/one-dimensional) simulation codes and 3D computational fluid dynamic (CFD) codes are currently applied to support the design process of new IC engines. The former are able to achieve a quick modeling of the whole engine, including the complete intake and exhaust systems with air boxes, turbochargers, EGR circuits, after-treatment, and silencers [27, 41], providing a reliable prediction of global performances and emissions. The latter requires a significant computational effort, due to the high number of computational cells to describe the fluid dynamic system; hence the computer run times are around hours or days. Hence, CFD codes are applied to carry out a deep investigation of IC engine components, such as the combustion chamber, head ducts, turbocharger, and exhaust after-treatment devices [42, 43].

In general, 3D and 1D thermo-fluid dynamic simulations are carried out separately, even if the predicted unsteady flows in the intake/exhaust pipes, provided by 1D codes, are used as boundary conditions for the 3D modeling of the cylinder gas exchange process. On the other hand, during the last decade, the use of coupled 1D-3D simulation codes has become a common practice [44, 45], in order to apply the 1D approach to the whole engine pipe system, apart from some components (typically duct junctions, plenums, and silencers) in which the unsteady flows are significantly multidimensional, requiring a detailed 3D modeling.

In this way the 3D simulation can achieve an improved local accuracy, capturing the details of the fluid flow and wave propagation, while retaining the realistic engine operating conditions provided by the IC engine 1D simulation. The 1D and the 3D calculations in different domains are coupled by means of suitable numerical procedures at the interface region [46, 47]. A wide description of these topics can be found in the following Chapters 2 and 9.

1.3 Brief History of Wave Dynamics

The importance of wave motion in the pipes of IC engine is well known and will be described in details in this book. However, it is useful to consider the studies on wave motion in engine ducts as part of a much wider research activity on wave motion in fluids.

The interest of scientists and engineers for waves travelling in fluids dates back to the antiquity: Pythagoras realized the relationship between a vibrating string of a musical instrument and the propagation of sound waves in air (550 BC) [48, 49]. During the Roman period, the generation and propagation of waves in aqueducts was observed and studied [50, 51] to improve the characteristics of those imposing structures. At the end of the fifteenth century, Leonardo da Vinci carried out several detailed studies about waves in water flows [52, 53], observing the motion of waves and currents and investigating how to control the water stream (see Figure 1.2).

With regard to sound wave propagation in different media, Isaac Newton conducted his pioneer work on the calculation of the speed of sound in the second half of the seventeenth century, as reported in his famous work *Principia Mathematica* [54]. His approximated approach was based on the isothermal compression assumption, which led to an error around 15-20%; it was studied and corrected one century later by Pierre-Simon Laplace, who introduced the assumption of adiabatic compression and proposed the current formulation of the Newton-Laplace equation for the calculation of the speed of sound [55].

FIGURE 1.2 Leonardo, *Study of Water Passing Obstacles*, c. 1508-1509 (Codex Atlanticus, 185v).

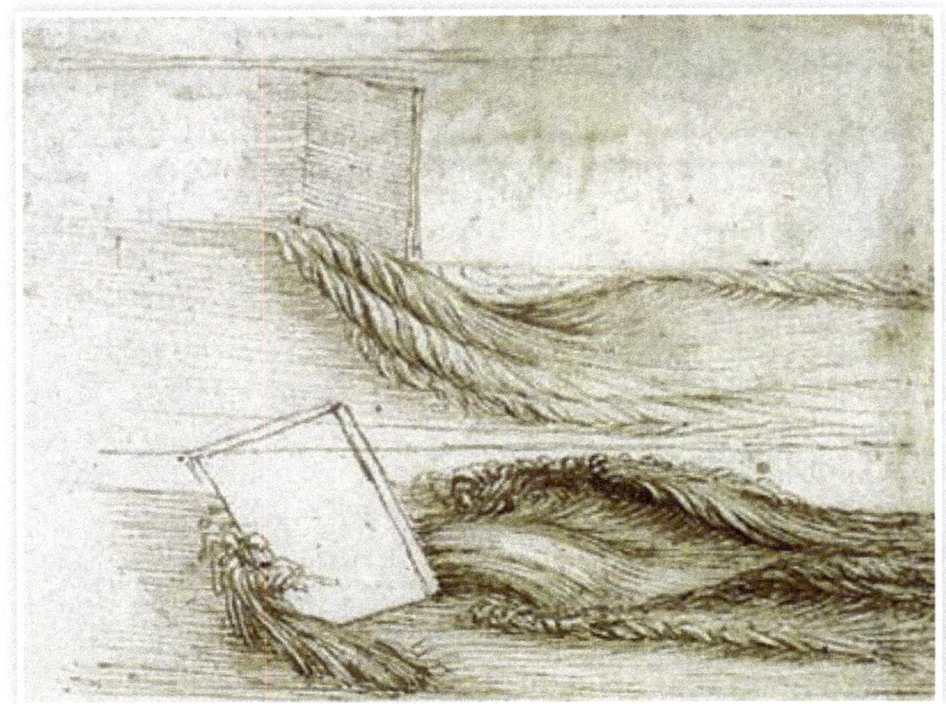

During the eighteenth century, there was a clear focus on linear wave propagation and acoustics, as documented by the work of D'Alembert [56] (the well-known "wave equation" or "D'Alembert equation," 1747) and Euler (1766) [56]. With regard to nonlinear waves, representative of the propagation of finite amplitude perturbations in a medium, Euler was already aware of the difference between linear and nonlinear wave laws and proposed a nonlinear equation for plane wave propagation in air. At the beginning of the nineteenth century, Poisson (1808) [57] gave a contribution to the understanding of nonlinear wave propagation, highlighting that finite amplitude waves propagate at a velocity which is the sum of local fluid velocity and the speed of sound. Later, Stokes (1848) [58] studied the shape change of finite amplitude waves during propagation in a fluid without losses and the formation of discontinuities or shocks. A few years later, Rankine (1870) and Hugoniot (1887) gave an important contribution to the theory of shock waves, formulating the correct equations for normal shocks, including the influence of dissipative effects (Rankine-Hugoniot equations) [59]. Then, one century later Earnshow (1860) [60] was the first to formulate an equation for nonlinear plane waves, relating the velocity of wave propagation to the amplitude of pressure perturbations. In the same period, Riemann (1858) [61] discovered how to transform the partial differential equations describing wave motion into ordinary differential equations, resorting to the concept of characteristic lines and Riemann variables; this allowed the development of the famous *method of characteristics* (MOC) [62]. Moreover, at the end of the nineteenth century, Rayleigh investigated the propagation of axial waves of finite amplitude [63]. Since then, several studies were conducted on finite plane wave dynamics in pipes.

1.4 IC Engine Gas Dynamics

In the field of IC engine gas dynamics, the previous studies of Earnshow and Riemann were very useful to allow the investigation of wave dynamics in IC engine pipe systems. In particular, in 1923 Capetti [64] was among the first to investigate finite pressure wave propagation in IC engine pipes, on the basis of Earnshow's nonlinear wave theory; he included friction and area variation in the Earnshow's equation, to predict wave dynamics effects. Similarly, Bannister and Mucklow (1948) applied the finite wave theory of Earnshow to the sudden discharge of exhaust gas from a cylinder to a pipe [65].

However, most of the studies were conducted by means of the graphical solution of the MOC, following Riemann's work. De Haller (1945) [66] applied this technique to analyze the wave dynamics in the exhaust system of an IC engine. A few years later, Jenny (1950) [67] extended the graphical solution method taking account of friction and heat transfer effects, to study wave action in exhaust systems, including the turbocharger. Shapiro (1954) [68] and Rudinger (1955) [69] described extensively the application of the graphical MOC to IC engine pipe systems. Similarly, Wallace (1956) [70] and Benson (1958) [71] investigated the wave action in exhaust systems for two-stroke engines by means of graphical methods of characteristics. However, at that time it was already evident that the complex construction of wave diagrams, necessary for the graphical solution method, represented a major drawback of this approach. This was the reason why Rudinger [69] and Shapiro [68] began to propose numerical methods to address the solution of nonlinear hyperbolic system of conservation equations, overcoming the limitations of the graphical MOC. On the other hand, Benson, Garg, and Woollatt (1964) in their seminal paper [72] were the first to propose a numerical solution of the MOC by means of a computer. Since then, during the 1970s and the 1980s, the *mesh method of characteristics* became the most common and successful numerical method in the field of IC engine gas dynamics. Benson and his group at the UMIST (University of Manchester) gave a fundamental contribution to the growth of this research field, developing numerical simulation models and boundary conditions for most of the typical engine components (valves, open ends, sudden area changes, junctions, turbochargers, etc.). In 1982 Benson's work was described in the famous book *The Thermodynamics and Gas Dynamics of Internal Combustion Engines* [73], which represented (and still represents today) a fundamental textbook for generations of researchers, engineers, and students working in this field. Since then, several researchers followed Benson's approach and models, giving significant contributions on engine modeling by the mesh method of characteristics. It is impossible to cite most of them, see, for example, Winterbone [74, 75], Baruah [76], Blair [77, 78], Bingham [79], Sierens [80], Payri [81], Jones [82], Corberan [83], Onorati [84], and many others, as reported in Winterbone and Pearson's book [85] and in Weaving's book [86].

As a matter of fact, the MOC represented the most common approach for IC engine gas dynamics simulation until the 1990s, due to its simplicity and physical insight to predict plane wave propagation in pipe systems. However, it suffered from some evident limitations related to the low numerical accuracy (only first order in space and time), the non-conservative form of the fundamental equations required to develop the method, and the inability to cope with the calculation of shock formation in the pipes (in fact the method is not *shock capturing*) [85]. Due to these drawbacks, the MOC was gradually superseded in the 1980s and 1990s by more advanced and robust techniques. However, still today the MOC is commonly adopted in many simulation models as a sound method for the treatment of boundary conditions, due to the comprehensive work published and the intuitive approach for the modeling of every kind of flow boundary regions [73, 85].

Starting from the 1980s, a new generation of numerical techniques, belonging to the wide family of "shock capturing" methods, was gradually introduced to solve nonlinear hyperbolic systems of governing equations. An exhaustive description of different shock-capturing numerical methods and their evolution for the application to IC engine gas dynamics is available in the well-known book *Theory of Engine Manifold Design* [85] written by Winterbone and Pearson (2000), that, together with the companion book *Design Techniques for Engine Manifolds* [87] (1999) by the same authors, still today represents the most complete reference work for researchers and engineers working in this field. A wide description of numerical methods for computational fluid dynamics is also reported in [88, 89].

In what follows only a summary of the most important shock-capturing numerical methods is reported, pointing out their impact on the evolution of IC engine fluid dynamic modeling.

In general, it is possible to distinguish two important families of numerical methods: **symmetric** and **upwind** shock-capturing schemes, to solve the hyperbolic problem of conservation equations. Symmetric schemes apply in every node the same space-centered, finite difference discretization, which does not depend on the flow characteristics; conversely, upwind schemes apply a biased space discretization in each node, depending on the direction of propagation of the information in a flow field [85, 87, 88, 89].

The first significant developments of numerical methods for the solution of unsteady flow equations were due to Courant, Friedrichs, and Lewy (1928) [90] and later to von Neumann (1944) [91]. In the 1950s, several scientists worked on the development of robust shock-capturing methods with high accuracy: von Neumann and Richtmyer (1950) [92] investigated numerical methods to calculate shocks; Hartree (1953) [93] formulated the mesh method of characteristics that was so successful a few years later in the field of IC engine modeling; Courant, Isaacson, and Rees (1952) [94] proposed the so-called CIR method (from their initials) for the solution of governing equations in non-conservative form; the CIR method (as well as the MOC) can be considered as the first example of upwind scheme.

A major contribution in the field of symmetric shock-capturing methods was provided by Lax (1954) [95] and later by Lax and Wendroff (1960 and 1964) [96, 97], who finally devised numerical techniques with second-order accuracy (in space and time) to solve the governing equations in conservative form. This was fundamental to overtake the typical limitations of the MOC. The second-order symmetric scheme proposed by Lax and Wendroff [85, 97] can be regarded as the classic reference symmetric scheme. The disadvantage of this method is related to the evaluation of the Jacobian matrix, which is computationally demanding. To simplify the numerical calculation, Richtmyer and Morton (1967) [98] suggested a two-step variant of the Lax and Wendroff method, which was successfully applied in several simulation codes. Their work was followed by MacCormack (1969), who proposed a renowned alternative symmetric numerical method (with second-order accuracy as well) [99], adopted in some numerical models for IC engine calculations [100, 101]. Similarly, Lerat and Peyret (1974) [102] proposed a general, non-centered numerical technique with equivalent accuracy. However, it was soon evident that every symmetric finite difference scheme with second-order accuracy was characterized by numerical overshoots in the proximity of discontinuous solutions, such as shock waves and contact discontinuities [85, 103]. These spurious oscillations, known as Gibbs phenomenon, could be eliminated or reduced by means of different techniques. Boris and Book [104, 105] devised the flux-corrected transport (FCT) technique for symmetric shock-capturing schemes, whose variants were described by Niessner and Bulaty [106].

On the other hand, **upwind** shock-capturing schemes were already investigated by Godunov (1959) [103], who proposed a first-order accurate method to calculate intercell fluxes by solving a series of local Riemann problems (discontinuous neighboring states) arising from a piecewise-constant reconstruction of the solution along the cells. The exact Riemann solver adopted required a significant computational effort, so that approximate Riemann solvers were proposed later by Roe (1981) [107, 108], Steger, and Warming (1981, flux vector splitting methods) [109] and van Leer [110].

Thanks to Godunov's work, it was possible to clearly understand the numerical behavior of high-order upwind schemes; it was evident that a piecewise-linear reconstruction of the solution allowed a second-order accurate method, however affected by spurious oscillations (Gibbs phenomenon) in the case of linear schemes with constant coefficients. The so-called Godunov's theorem inspired other scientists to develop second-order (or higher) numerical schemes with nonlinear coefficients, to prevent the appearance of numerical overshoots in the proximity of discontinuous solutions. Harten (1983) [111] described high-resolution schemes and introduced the total variation diminishing (TVD) criterion, an innovative approach to guarantee the absence of spurious oscillations in the solution. Van Leer (1974) [112], Sod (1978) [113], Sweby (1984) [114], and others [85, 89] extensively studied these numerical methods, devising advanced flux-limiting techniques compatible with the TVD criterion, able to prevent the appearance of numerical overshoots in the solution.

1.5 Overview of IC Engine 1D Simulation Codes

As previously outlined, a first example of computer simulation model for IC engine wave dynamics was proposed by Benson, Garg, and Woollatt (1964) [72]. Their numerical code was based on the mesh method of characteristics, a significant development step with respect to the previous studies regarding the graphical solution of the MOC by Jenny (1950) [67], Shapiro (1954) [68], and Rudinger (1955) [69]. Since then, during the 1970s and the 1980s the *mesh method of characteristics* became the most popular numerical technique for the simulation of IC engine gas dynamics, with several contributions from different research groups. Certainly a fundamental contribution was provided by Benson and his group of researchers at the University of Manchester (UMIST, Manchester, UK), developing successful numerical models for IC engine simulations named Mk11, Mk12, and successive updates. Their work contributed to spread the simulation tools among industries and universities [73, 115]. In the same period, other IC engine modeling tools were developed on the basis of the mesh method of characteristics. An important contribution was given by Blair and his group at the Queen's University of Belfast [77, 78, 79, 116, 117]. In general, during the 1970s there was a notable proliferation of 1D fluid dynamic models for IC engine simulation, which contributed to increase the range of applications and validations, with a resulting improvement of robustness of the numerical codes.

Since the 1970s, but mainly during the 1980s and the 1990s, the MOC was gradually replaced by second-order shock-capturing methods, whereas the characteristic-based approach was frequently confirmed for the treatment of boundary conditions. This was mainly due to its simplicity and physical insight to represent plane wave reflection and transmission at boundaries, as well as to the comprehensive work published for the modeling of every kind of flow boundary region [73, 85]. Finite difference shock-capturing numerical techniques were already adopted in the late 1970s in the PROMO code, developed in Germany at the Ruhr University Bochum by Seifert and his co-workers [118, 119], who described the adoption of the Lax-Wendroff technique. A similar pioneering development was achieved by Dwyer et al. [120], by Takizawa et al. [121], and later by Liu et al. [122].

Similarly, Poloni and Winterbone (1987) [123] described the advantages of the second-order Lax-Wendroff method with respect to the first-order MOC, introducing finite difference schemes in the UMIST's codes. A few years later, Pearson and Winterbone et al. [124, 125] described the application of flux-limiting techniques like the FCT, to prevent spurious oscillations in the solution. Most of their work is described in detail in the previously cited books *Theory of Engine Manifold Design* [85] and *Design Techniques for Engine Manifolds* [87].

In the early 1980s, a different shock-capturing numerical method named FRAM (filtering remedy and methodology) was developed in the USA by Chapman et al. [126, 127] to solve the conservation equations of 1D unsteady flows. This technique was based on a staggered grid, adopting a local, nonlinear artificial viscosity term to eliminate nonphysical oscillations in the solution. This approach was implemented in the MANDY engine simulation code, developed and applied at Ford Motor Company [128] for many years. A few years later, Morel et al. (1988) developed the IRIS 1D code for IC engine simulation [129], following a second-order accurate, explicit, finite volume method, based on a staggered mesh. A similar approach was applied in the commercial code WAVE, developed by Morel et al. (1990) [130, 131, 132], exploiting numerical techniques originally developed for multidimensional CFD Navier-Stokes codes [133] and then modified for application to 1D flows. In the 1990s [134, 135, 136], the same class of numerical methods was adopted by Morel, Silvestri, and co-workers in the commercial code GT-Power, which is nowadays the most common 1D simulation tool for IC engine modeling. In the early 1990s, high-order accurate numerical methods were adopted in the commercial code BOOST [137, 138], which makes use of ENO schemes [139, 140]. In the late 1990s, the research 1D code GASDYN was developed at Politecnico di Milano by Onorati, D'Errico, Ferrari et al. [141], exploiting both classic finite difference shock-capturing schemes and high-order numerical methods such as the CE-SE (conservation element-solution element) method and the discontinuous Galerkin FEM [141, 142, 143]. The successive developments of the code were focused on the transport of chemical species in the exhaust system, including the catalytic converter, and the modeling of combustion and pollutant emissions [144, 145, 146].

Another example of research and development in the field of 1D engine simulation is represented by the WAM code, developed by CMT-Motores Termicos at the University of Valencia (Spain). Their work started in the 1980s, following typical approaches based on the mesh method of characteristics [81], as proposed by Benson, and then continued during the last three decades with the introduction of both classical [100] and advanced numerical methods [147, 148], as well as with the development of boundary conditions. The source code is currently available as OpenWAM [149]. To conclude, three significant examples are represented by: the 1D engine model LESoft, developed at Lotus Engineering (United Kingdom) during the last 20 years, which makes use of classical numerical methods [150, 151]; the 1DIME code, developed during the same time framework at the University of Naples (Italy) by Bozza and coworkers, making use of finite volume, TVD numerical techniques [152, 153]; and finally the AMESim code including the CFD-1D library, under LMS Imagine.Lab platform [154].

From the analysis above, it is evident that the last 30 years have been very fruitful and stimulating in this field, in terms of research and development of advanced 1D numerical tools for engine simulation. This contributed to spread the application of simulation models from universities and research centers to automotive industry, with a robust trend during the 1990s and the following years. This was supported by the growth of commercial software tools, as discussed above, which catalyzed the research activity of several research groups from both universities and industries.

Considering only the last 20 years of activity, the recent advanced developments have been mainly focused on:

1. Innovative numerical methods for the solution of the 1D conservation equations
2. The modeling of combustion process and emissions
3. Turbines and compressors
4. Complex intake and exhaust systems and junctions, quasi-3D models
5. Numerical techniques to achieve a coupled 1D-3D simulation
6. Control-oriented simulation models
7. EGR circuits
8. After-treatment systems
9. Silencers and tailpipe noise

The modeling approach has been mainly 1D, but frequently also hybrid 1D-3D and quasi-3D methods have been developed and applied [44, 155]. The areas above will be the topics of the following chapters in this book.

1.6 Conservation Equations

From a general point of view, a 1D thermo-fluid dynamic model for the simulation of the whole engine system is able to carry out a prediction of volumetric efficiency, torque, power, fuel consumption, pollutant emissions, and tailpipe noise. Most codes can simulate both SI and CI engines, multicylinder and multivalve, naturally aspirated or turbocharged, including intake systems with complex geometry, air filters, pressure losses, EGR circuits, exhaust systems with multi-pipe junctions, after-treatment components, and silencers. Moreover, the calculation model can be applied to evaluate the intake and exhaust gas dynamic noise emitted downhill the silencers with a complex shape, in terms of sound pressure level spectra and overall noise level.

The core of the simulation model is represented by the solver for the numerical integration of the 1D conservation equations, by means of a variety of numerical methods, as discussed previously in detail [72, 73, 74, 75, 76, 77, 78, 79, 80, 81, 82, 83, 84, 85, 86, 87, 88, 89, 90, 91, 92, 93, 94, 95, 96, 97, 98, 99, 100, 101, 102, 103, 104, 105, 106, 107, 108, 109, 110, 111, 112, 113, 114, 115, 116, 117, 118, 119, 120, 121, 122, 123, 124, 125, 126, 127, 128, 129, 130, 131, 132, 133, 134, 135, 136, 137, 138, 139, 140, 141, 142, 143, 144, 145, 146, 147, 148]. With the same level of importance, the treatment of discontinuities for the flow as boundary conditions is represented typically by poppet valves, sudden area changes, multi-pipe junctions, concentrated pressure losses, turbines, compressors, and many others. These can be generally modeled resorting to the classical mesh method of characteristics [73, 85] or to other approaches [126, 127, 130].

In what follows, only the fundamental conservation equations for a 1D, unsteady, compressible flow are described, considering both the cases of perfect gas and transport of chemical species. This brief discussion may be useful to recall the fundamental equations of any 1D fluid dynamic engine model.

The governing equations for 1D, unsteady, compressible flows in ducts with variable cross section (Figure 1.3) can be written in strong conservative form, suitable for the

FIGURE 1.3 The one-dimensional differential conservation equations are written for a duct with infinitesimal length and variable cross section.

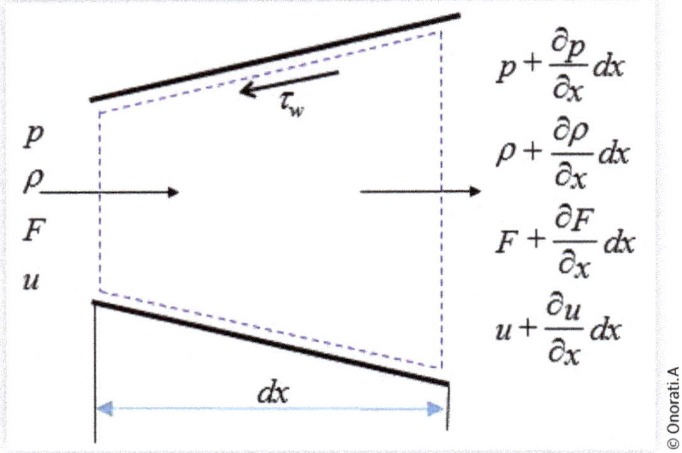

accurate numerical integration [85]. In general, two possible approaches may be highlighted:

1. A simple one, based on the perfect gas assumption, without transport of species or chemical reactions along the intake and exhaust ducts
2. A complex one, allowing chemical reactions in the unsteady flows, with transport of chemical species along the ducts [141]

1.6.1 Perfect Gas Assumption

In this case the conservation equations of mass, momentum, and energy are written as follows:

$$\text{Mass:} \quad \frac{\partial(\rho F)}{\partial t}+\frac{\partial(\rho u F)}{\partial x}=0 \tag{1.1}$$

$$\text{Momentum:} \quad \frac{\partial(\rho u F)}{\partial t}+\frac{\partial(\rho u^2 F+pF)}{\partial x}-\frac{pdF}{dx}+\rho GF=0 \tag{1.2}$$

$$\text{Energy:} \quad \frac{\partial(\rho e_0 F)}{\partial t}+\frac{\partial(\rho u h_0 F)}{\partial x}-\rho qF=0 \tag{1.3}$$

where

P, u, ρ, e_0, h_0 are pressure, flow velocity, density, stagnation specific internal energy, and enthalpy, respectively
F is the cross-sectional area
$G=4f_w u|u|/2d$
d is the duct diameter
f_w is the friction factor at the duct-wall
q is the heat transferred per unit mass per unit time

The presence of friction and heat transfer in the "*source*" terms of Equations (1.2) and (1.3) actually arises from the one-dimensional flow assumption, consisting in a plug-flow velocity distribution in each pipe cross-section. As a consequence, 1D models cannot describe any thermal or fluid-dynamic boundary layer, and need some theoretical-experimental correlation for the empirical estimation of f_w and q.

A further equation must be added to solve the problem, describing the fluid behavior on the basis of different assumptions. The classical approach, useful for many applications, considers a perfect gas state equation:

$$p = \rho R^{*} T \tag{1.4}$$

where R^* is a fixed specific gas constant. Different R^* values need to be specified for intake and exhaust pipes, which limit the 1D analysis to engines where intake and exhaust systems are completely decoupled (no port fuel injection (PFI), no short-circuiting, or exhaust gas recirculation (EGR)).

The above formulation of the fundamental equations provides a good conservation of mass along tapered pipes in case of both steady and unsteady flows, since the duct cross-section area F appears in the partial derivative terms of the Equations (1.1-1.3) [122, 141].

Finally, the conservation equations can be written in compact, vector form as follows:

$$\frac{\partial \mathbf{W}(x,t)}{\partial t} + \frac{\partial \mathbf{F}(\mathbf{W})}{\partial x} + \mathbf{B}(\mathbf{W}) + \mathbf{C}(\mathbf{W}) = 0 \tag{1.5}$$

where

$$\mathbf{W}(x,t) = \begin{bmatrix} \rho F \\ \rho u F \\ \rho e_0 F \end{bmatrix}, \quad \mathbf{F}(\mathbf{W}) = \begin{bmatrix} \rho u F \\ \rho u^2 F + pF \\ \rho u h_0 F \end{bmatrix}, \tag{1.6}$$

$$\mathbf{B}(\mathbf{W}) = \begin{bmatrix} 0 \\ -p\dfrac{dF}{dx} \\ 0 \end{bmatrix}, \quad \mathbf{C}(\mathbf{W}) = \begin{bmatrix} 0 \\ \rho G F \\ -\rho q F \end{bmatrix}, \tag{1.7}$$

1.6.2 Transport of Chemical Species with Reactions

In this second case, the conservation equations of mass, momentum, energy, and convective transport of species can be written as follows:

$$\text{Mass}: \quad \frac{\partial(\rho F)}{\partial t} + \frac{\partial(\rho u F)}{\partial x} = 0 \tag{1.8}$$

$$\text{Momentum}: \quad \frac{\partial(\rho u F)}{\partial t} + \frac{\partial\left(\rho u^2 F + pF\right)}{\partial x} - \frac{p dF}{dx} + \rho G F = 0 \tag{1.9}$$

$$\text{Energy}: \quad \frac{\partial(\rho e_0 F)}{\partial t} + \frac{\partial(\rho u h_0 F)}{\partial x} - \rho(q + q_{re}) F = 0 \tag{1.10}$$

$$\text{Specie continuity}: \quad \frac{\partial(\rho Y_j F)}{\partial t} + \frac{\partial(\rho u Y_j F)}{\partial x} - \rho \dot{Y}_j F = 0, \quad j = 1,2,\ldots,N-1. \tag{1.11}$$

In addition to the previous set of variables, here in Equation (1.10) q_{re} is the heat released by chemical reactions in gas phase per unit mass per unit time; in Equation (1.11), $Y_j = m_j/m$ is the mass fraction in the control volume for the specie j, whereas $\dot{Y}_j$ is the rate of variation of mass fraction due to chemical reactions involving the jth specie [85, 144].

In particular, the transport of chemical species along the duct system is accounted for by means of the $N-1$ specie continuity in Equation (1.11), being N the number of species advected. These equations are based on the hypothesis of no diffusion in the flow, so that the species are simply advected, with source terms to represent chemical reactions (unsteady reacting flows) [145]. Only $N-1$ equations of type (1.11) are required for N species, since the Nth equation is simply:

$$\sum_{j=1}^{N} Y_j = 1 \Rightarrow Y_N = 1 - \sum_{j=1}^{N-1} Y_j. \tag{1.12}$$

A further equation must be added to solve the problem, describing the fluid behavior on the basis of different assumptions. The most general approach considers a mixture of N ideal gaseous species, obeying the ideal gas state equation:

$$p = \frac{\rho RT}{\sum_{j=1}^{N} X_j M_j} \tag{1.13}$$

where

X_j and M_j are the mole fraction and molar mass of the jth specie, respectively
R is the universal gas constant

The specific heats depend on both gas temperature and chemical composition; the internal energy of the jth specie of the mixture can be expressed by means of polynomial relationships, in which the coefficients for each chemical specie have been determined on the basis of the JANAF and NASA data [141].

This second, more complex approach can be suitable when the 1D modeling of the after-treatment system is addressed, so that the change in gas composition due to the specie conversion through a three-way catalyst can be accounted for, as well as the heat released in the gas phase. This approach is also useful for tracking the gas composition even in the absence of any species conversion and heat release, and has to be utilized when PFI and/or EGR are concerned.

The above conservation equations can be written in compact, vector form as follows:

$$\frac{\partial \mathbf{W}(x,t)}{\partial t} + \frac{\partial \mathbf{F}(\mathbf{W})}{\partial x} + \mathbf{B}(\mathbf{W}) + \mathbf{C}(\mathbf{W}) = 0 \tag{1.14}$$

$$\text{where:}\quad \mathbf{W}(x,t) = \begin{bmatrix} \rho F \\ \rho u F \\ \rho e_0 F \\ \rho \mathbf{Y} F \end{bmatrix}, \quad \mathbf{F}(\mathbf{W}) = \begin{bmatrix} \rho u F \\ \rho u^2 F + pF \\ \rho u h_0 F \\ \rho u \mathbf{Y} F \end{bmatrix}, \tag{1.15}$$

$$\mathbf{B}(\mathbf{W}) = \begin{bmatrix} 0 \\ -p\dfrac{dF}{dx} \\ 0 \\ 0 \end{bmatrix}, \quad \mathbf{C}(\mathbf{W}) = \begin{bmatrix} 0 \\ \rho G F \\ -\rho(q + q_{re})F \\ -\rho \dot{\mathbf{Y}} F \end{bmatrix}, \quad \mathbf{Y} = \begin{bmatrix} Y_1 \\ \vdots \\ Y_{N-1} \end{bmatrix}. \tag{1.16}$$

FIGURE 1.4 Meshes along a duct for the numerical solution of 1D differential conservation equations. The numerical method can provide the solution only in the interior nodes; to complete the calculation, boundary conditions, based on Riemann variables, are required on the left and right faces of the duct (cell 1-2 and cell $I - I + 1$).

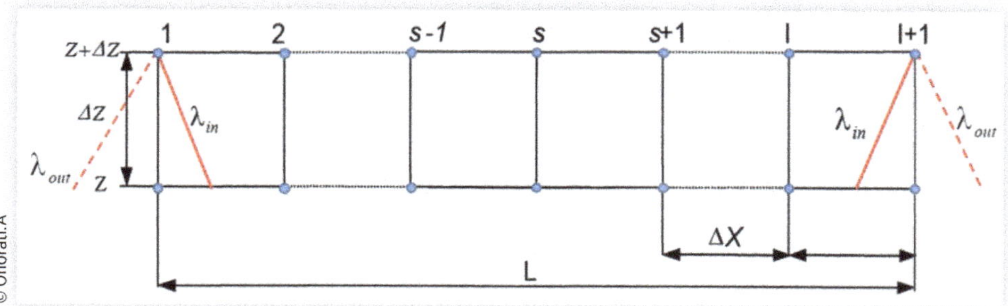

Both in the case of perfect gas and ideal gas with transport of chemical species, the set of conservation laws represent a nonlinear hyperbolic system of partial differential equations, to be solved numerically by advanced shock-capturing numerical schemes [85, 86, 87, 88, 89, 90, 91, 92, 93, 94, 95, 96, 97, 98, 99, 100, 101, 102, 103, 104, 105, 106, 107, 108, 109, 110, 111, 112, 113, 114, 115, 116, 117, 118, 119, 120, 121, 122, 123, 124, 125, 126, 127, 128, 129, 130, 131, 132, 133, 134, 135, 136, 137, 138, 139, 140, 141, 142, 143, 144, 145, 146, 147, 148].

The conservation equations discussed above must be solved node by node along the pipes of the intake and exhaust system. In general, a pipe is characterized by a certain number of meshes, depending of the mesh length (Δx) (see Figure 1.4).

The numerical method is applied only in the internal nodes, whereas at the boundaries a specific treatment is required, to correctly account for the interaction of the unsteady flow with the boundary condition and determine the consequent transmission and reflection of waves. In several simulation codes, a characteristic-based approach is adopted, to solve the boundary problem by means of Riemann variables λ_{in} and λ_{out}, as described in detail, for example, in [73, 79, 81, 85, 101, 148].

Sometimes the shape of an intake or exhaust duct can be complex, made of several sub-pipes as represented in Figure 1.5, to account for tapered ducts, curved ducts with different friction coefficient, or parts with a different wall temperature. This complex pipe can be managed as a continuous duct by the numerical method, with possible smooth variations of diameter (divergent or convergent pipes) and changes in wall temperatures or friction coefficients from one sub-element to another. Abrupt area changes are not admitted by this approach (these must be represented by specific

FIGURE 1.5 Example of a complex pipe made of several sub-elements connected together. The pipe is treated as a continuous duct by the numerical method.

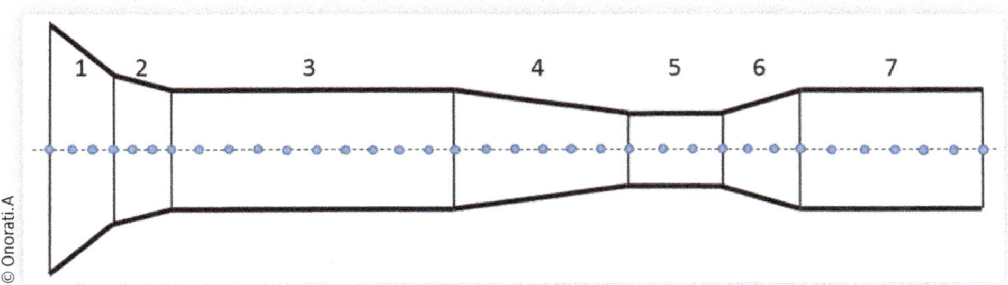

boundary conditions for sudden enlargements and contractions). To clarify this approach, in Figure 1.5 a complex pipe is reported: sub-elements with variations of diameter, friction, and wall temperature are connected to form a continuous pipe. For example, a bent sub-element along the duct can have a higher friction coefficient at the walls, depending on the geometrical parameters of the curve. Moreover, each sub-element can have a distinct mesh size, if required.

Definitions, Acronyms, and Abbreviations

Symbols

P - Pressure
u - Flow velocity
ρ - Density
T - Gas temperature
e_0 - Stagnation specific internal energy
h_0 - Stagnation specific internal enthalpy
F - Cross-sectional area of duct
d - Duct diameter
f_w - Friction factor at the duct-wall
$G = -4f_w u|u|/2d$
q - Heat transferred per unit mass per unit time
q_{re} - Heat released by chemical reactions in gas phase (per unit mass per unit time)
Y_j - Mass fraction of specie j rate of variation of mass fraction due to chemical reactions
X_j - Mole fraction of the jth specie
M_j - Molar mass of the jth specie
R^* - Specific gas constant
R - Universal gas constant
D_x - Mesh length
D_Z - Time step
$\lambda_{in}, \lambda_{out}$ - Riemann variables at boundaries

Abbreviations

0D - Zero-dimensional
1D, 3D - One-dimensional, three-dimensional
CFD - Computational fluid dynamics
CI - Compression ignition
DPF - Diesel particulate filter
EGR - Exhaust gas recirculation
EIVC - Early intake valve closure
ENO - Essentially nonoscillatory
FCT - Flux corrected transport
FEM - Finite element method
FGT - Fixed geometry turbine
GPF - Gasoline particulate filter
HCCI - Homogeneous charge compression ignition
HEV - Hybrid electric vehicle

LIVC - Late intake valve closure
LTC - Low temperature combustion
MOC - Method of characteristics
PHEV - Plug-in hybrid electric vehicle
quasiD - Quasi-dimensional
RCCI - Reactivity controlled compression ignition
RDE - Real drive emission
SACI - Spark-assisted compression ignition
SCR - Selective catalytic reduction
SI - Spark ignition
TVD - Total variation diminishing
VVA - Variable valve actuation
VVT - Variable valve timing
VGT - Variable geometry turbine

References

1. Martin Weiss, P.B., Hummel, R., and Steininger, N., "A Complementary Emissions Test for Light-Duty Vehicles: Assessing the Technical Feasibility of Candidate Procedures," European Commission, Joint Research Centre, Institute for Energy and Transport (IET), 2013.
2. Maschmeyer, H., Kluin, M., and Beidl, C., "Real Driving Emissions - A Paradigm Change for Development," *MTZ Worldwide* 76, no. 2 (2015): 16-21.
3. Li, T., Gao, Y., Wang, J., and Chen, Z., "The Miller Cycle Effects on Improvement of Fuel Economy in a Highly Boosted, High Compression Ratio, Direct-Injection Gasoline Engine: EIVC vs. LIVC," *Energy Convers. Manag.* 79 (2014): 59-65, doi:10.1016/j.enconman.2013.12.022.
4. Okamoto, K., Zhang, F.-R., Shimogata, S., and Shoji, F., "Development of a Late Intake-Valve Closing (LIVC) Miller Cycle for Stationary Natural Gas Engines - Effect of EGR Utilization," SAE Technical Paper 972948, 1997, doi:10.4271/972948.
5. Luisi, S., Doria, V., Stroppiana, A., Millo, F. et al., "Experimental Investigation on Early and Late Intake Valve Closures for Knock Mitigation through Miller Cycle in a Downsized Turbocharged Engine," SAE Technical Paper 2015-01-0760, 2015, doi:10.4271/2015-01-0760.
6. Luo, X., Teng, H., Lin, Y., Li, B. et al., "A Comparative Study on Influence of EIVC and LIVC on Fuel Economy of a TGDI Engine Part II: Influences of Intake Event and Intake Valve Closing Timing on the Cylinder Charge Motion," SAE Technical Paper 2017-01-2246, 2017, doi:10.4271/2017-01-2246.
7. Langwiesner, M., Krüger, C., Donath, S., and Bargende, M., "Combustion Engine Concepts with Extended Expansion in Hybrid Powertrains," *18th Stuttgart International Symposium on Automotive and Engine Technology*, Stuttgart, Germany, 2018.
8. Flierl, R., Paulov, M., Knecht, A., and Hannibal, W., "Investigations with a Mechanically Fully Variable Valve Train on a 2.0l Turbo Charged Four Cylinder Engine," SAE Technical Paper 2008-01-1352, 2008, doi:10.4271/2008-01-1352.
9. De Simio, L., Gambino, M., Iannaccone, S., Borrelli, L. et al., "Experimental Analysis of a Natural Gas Fueled Engine and 1-D Simulation of VVT and VVA Strategies," SAE Technical Paper 2013-24-0111, 2013, doi:10.4271/2013-24-0111.

10. Millo, F., Luisi, S., Stroppiana, A., and Borean, F., "Effects of Different Geometries of the Cylinder Head on the Combustion Characteristics of a VVA Gasoline Engine," SAE Technical Paper 2013-24-0057, 2013, doi:10.4271/2013-24-0057.
11. Zammit, J.P., McGhee, M.J., Shayler, P.J., and Pegg, I., "Benefits of Cylinder Deactivation on a Diesel Engine and Restrictions due to Low Boost," in: *Internal Combustion Engines: Performance, Fuel Economy Emission* (Elsevier, 2013), 95-108, doi:10.1533/9781782421849.3.95.
12. Flierl, R., Lauer, F., Breuer, M., and Hannibal, W., "Cylinder Deactivation with Mechanically Fully Variable Valve Train," *SAE Int. J. Engines* 5, no. 2 (2012): 207-215, doi:10.4271/2012-01-0160.
13. Pineda, D., Wolk, B., Chen, J., and Dibble, R., "Application of Corona Discharge Ignition in a Boosted Direct-Injection Single Cylinder Gasoline Engine: Effects on Combustion Phasing, Fuel Consumption, and Emissions," *SAE Int. J. Engines* 9, no. 3 (2016): 1970-1988, doi:10.4271/2016-01-9045.
14. Potteau, S., Lutz, P., Leroux, S., Moroz, S. et al., "Cooled EGR for a Turbo SI Engine to Reduce Knocking and Fuel Consumption," SAE Technical Paper 2007-01-3978, 2007, doi:10.4271/2007-01-3978.
15. Roth, D.B., Keller, P., and Becker, M., "Requirements of External EGR Systems for Dual Cam Phaser Turbo GDI Engines," SAE Technical Paper 2010-01-0588, 2010, doi:10.4271/2010-01-0588.
16. Hoppe, F., Thewes, M., Baumgarten, H., and Dohmen, J., "Water Injection for Gasoline Engines: Potentials, Challenges, and Solutions," *Int. J. Engine Res.* 17 (2016): 86-96, doi:10.1177/1468087415599867.
17. Bozza, F., De Bellis, V., and Teodosio, L., "Potentials of Cooled EGR and Water Injection for Knock Resistance and Fuel Consumption Improvements of Gasoline Engines," *Appl. Energy* 169 (2016): 112-125, doi:10.1016/j.apenergy.2016.01.129.
18. Wittek, K., "CO2-Potential of a Two-Stage VCR System in Combination with Future Gasoline Powertrains," *33rd International Vienna Motor Symposium*, Vienna, Austria, 2012.
19. Ferrey, P., Miehe, Y., Constensou, C., and Collee, V., "Potential of a Variable Compression Ratio Gasoline SI Engine with Very High Expansion Ratio and Variable Valve Actuation," *SAE Int. J. Engines* 7, no. 1 (2014): 468-487, doi:10.4271/2014-01-1201.
20. Roberts, G., Rousselle, C., Musculus, M., Wissink, M. et al., "RCCI Combustion Regime Transitions in a Single-Cylinder Optical Engine and a Multi-Cylinder Metal Engine," *SAE Int. J. Engines* 10, no. 5 (2017): 2392-2413, doi:10.4271/2017-24-0088.
21. Takazawa, M., Komura, K., and Kitamura, T., "Transient Control Technology of Spark Assisted HCCI," SAE Technical Paper 2015-01-0880, 2015, doi:10.4271/2015-01-0880.
22. Li, L., Xie, H., Chen, T., Yu, W. et al., "Experimental Study on Spark Assisted Compression Ignition (SACI) Combustion with Positive Valve Overlap in a HCCI Gasoline Engine," SAE Technical Paper 2012-01-1126, 2012, doi:10.4271/2012-01-1126.
23. Chiodi, M., Kachele, A., Bargende, M., Kock, D. et al., "Development of an Innovative Combustion Process: Spark-Assisted Compression Ignition," *18th Stuttgart International Symposium on Automotive and Engine Technology*, Stuttgart, Germany, 2018.

24. Attard, W., Blaxill, H., Anderson, E., and Litke, P., "Knock Limit Extension with a Gasoline Fueled Pre-Chamber Jet Igniter in a Modern Vehicle Powertrain," *SAE Int. J. Engines* 5, no. 3 (2012): 1201-1215, doi:10.4271/2012-01-1143.

25. Zhang, Q., Brace, C., Akehurst, S., Burke, R. et al., "Simulation Study of the Series Sequential Turbocharging for Engine Downsizing and Fuel Efficiency," SAE Technical Paper 2013-01-0935, 2013, doi:10.4271/2013-01-0935.

26. Zhang, H., Shi, L., Cui, Y., Deng, K. et al., "Study on the Optimal Control Strategy of Transient Process for Diesel Engine with Sequential Turbocharging System," SAE Technical Paper 2016-01-2157, 2016, doi:10.4271/2016-01-2157.

27. Cornolti, L., Onorati, A., Cerri, T., Montenegro, G. et al., "1D Simulation of a Turbocharged Diesel Engine with Comparison of Short and Long EGR Route Solutions," *Applied Energy*, 111 (2013): 1-15.

28. Körfer, T. et al., "Emission-Based EGR Strategies in Diesel Engines for RDE Requirements," *MTZ Worldwide* 75, no. 9 (2014): 10-17.

29. Hoepke, B., Jannsen, S., Kasseris, E., and Cheng, W.K., "EGR Effects on Boosted SI Engine Operation and Knock Integral Correlation," *SAE Int. J. Engines*, 5, no. 2 (2012): 547-559, doi:10.4271/2012-01-0707.

30. Cheng, W.K. and Onorati, A., "Editorial: Biofuels in Internal Combustion Engines," *International Journal of Engine Research* 16, no. 5 (2015): 609-609, doi:10.1177/1468087415597824.

31. Siegemund, S., Schmidt, P., Trommler, M., Weindorf, W. et al., "The Potential of Electricity-Based Fuels for Low-Emission Transport in the EU. An Expertise by LBST and dena," Deutsche Energie-Agentur GmbH (dena), German Energy Agency, Berlin, 2017.

32. Johnson, T., "Review of Vehicular Emissions Trends," *SAE Int. J. Engines* 8, no. 3 (2015): 1152-1167, doi:10.4271/2015-01-0993.

33. Johnson, T., "Vehicular Emissions in Review," *SAE Int. J. Engines* 9, no. 2 (2016): 1258-1275, doi:10.4271/2016-01-0919.

34. Bosteels, D., "Real Driving Emissions and Test Cycle Data from 4 Modern European Vehicles," *IQPC 2nd International Conference Real Driving Emissions*, Düsseldorf, 2014.

35. Naber, D., Bareiss, S., Kufferath, A., Krüger, M. et al., Measures to Fulfill "Real Driving Emission (RDE)" with Diesel Passenger Cars," *18th Internationales Stuttgarter Symposium*, March 2018.

36. Wurzenberger, J., Prah, I., Tominc, P., and Katrasnik, T., "Optimization of Hybrid Power Trains-Physical Based Modeling for Concept Design," SAE Technical Paper 2012-01-0359, 2012, doi:10.4271/2012-01-0359.

37. Hamza, K. and Laberteaux, K., "A Study on Optimal Powertrain Sizing of Plugin Hybrid Vehicles for Minimizing Criteria Emissions Associated with Cold Starts," SAE Technical Paper 2018-01-0406, 2018, doi:10.4271/2018-01-0406.

38. Safari, M., "Battery Electric Vehicles: Looking behind to Move Forward," *Energy Policy* 115 (2018): 54-65, doi:10.1016/j.enpol.2017.12.053.

39. Walters, M., Wick, M., Pischinger, S., Ogrzewalla, J. et al., "Fuel Cell System Development: A Strong Influence on FCEV Performance," SAE Technical Paper 2018-01-1305, 2018, doi:10.4271/2018-01-1305.

40. Martin, J., Arnau, F., Piqueras, P., and Auñon, A., "Development of an Integrated Virtual Engine Model to Simulate New Standard Testing Cycles," SAE Technical Paper 2018-01-1413, 2018, doi:10.4271/2018-01-1413.

41. Deppenkemper, K., Özyalcin, C., Ehrly, M., Schoenen, M. et al., "1D Engine Simulation Approach for Optimizing Engine and Exhaust Aftertreatment Thermal Management for Passenger Car Diesel Engines by Means of Variable Valve Train (VVT) Applications," SAE Technical Paper 2018-01-0163, 2018, doi:10.4271/2018-01-0163.

42. Stiesch, G., *Modeling Engine Spray and Combustion Processes* (Springer, 2003), ISBN:978-3-540-00682-4.

43. D'Adamo, A., Breda, S., Fontanesi, S., and Cantore, G., "A RANS-Based CFD Model to Predict the Statistical Occurrence of Knock in Spark-Ignition Engines," *SAE Int. J. Engines* 9, no. 1 (2016): 618-630, doi:10.4271/2016-01-0581.

44. da Silva Trindade, W., "Use of 1D-3D Coupled Simulation to Develop an Intake Manifold System," SAE Technical Paper 2010-01-1534, 2010, doi:10.4271/2010-01-1534.

45. Bozza, F., Gimelli, A., Piazzesi, R., Fortunato, F. et al., "The Prediction of the Performance and Gasdynamic Noise Emitted by a Medium-Size Spark-Ignition Engine by Means of 1D and 3D Analyses," SAE Technical Paper 2007-01-0380, 2007, doi:10.4271/2007-01-0380.

46. Montenegro, G. and Onorati, A., "Modeling of Silencers for I.C. Engine Intake and Exhaust Systems by Means of an Integrated 1D-multiD Approach," *SAE Int. J. Engines* 1, no. 1 (2009): 466-479, doi:10.4271/2008-01-0677.

47. Montenegro, G., Onorati, A., Fairbrother, R., and Liu, S., "Modeling of Silencers for Internal Combustion Engines: 1D-3D Coupling, Network of 1D Elements and a Generic 3D Cell Approach," SAE Technical Paper 2009-24-0133, 2009, doi:10.4271/2009-24-0133.

48. Marcovici, M., *The 50 Most Influential People in History* (Books on Demand, 2014), ISBN-13:9783735721921.

49. Ferguson, H., *The Music of Pythagoras* (Pub. Walker, 2008), ISBN:978-0-8027-1631-6.

50. Chanson, H., "Hydraulics of Roman Aqueducts: Steep Chutes, Cascades and Dropshafts," *American Journal of Archaeology* 104, no. 1 (January 2000): 47-72.

51. Chanson, H., "An Experimental Study of Roman Dropshaft Hydraulics," *J. Hyd. Res., IAHR* 40, no. 1 (2002): 3-12.

52. Macagno, E., "Some Remarkable Experiments of Leonardo da Vinci," *Houille Blanche* 46, no. 6 (1991): 463-471.

53. https://www.royalcollection.org.uk/collection/912660/recto-studies-of-flowing-water-with-notes-verso-studies-of-flowing-water-with.

54. Newton, I., *Principia* (Cambridge University Press, 1687).

55. Finn, B.S., "Laplace and the Speed of Sound," *Isis* 55, no. 1 (1964): 7-19, doi:10.1086/349791.

56. Enflo, B.O. and Hedberg, C.M., *Theory of Nonlinear Acoustics in Fluids* (Kluwer Academic Publishers, 2002), ISBN-10:1402005725.

57. Poisson, S.D., "Memoire sur la theorie du son," *J. de l'Ecole Polytechnic* 7 (1808): 319-392.

58. Stokes, G.G., "On a Difficulty in the Theory of Sound," *Philos. Mag.* 23 (1848): 349-356.

59. Enderson, L., "Chapter 2: General Laws for Propagation of Shock Waves through Matter," in: *Handbook of Shock Waves*, vol. 1 (Academic Press, 2000), ISBN:9780120864300.

60. Earnshow, S., "On the Mathematical Theory of Sound," *Philos. Trans. R. Soc.* 150 (1860): 133-148.

61. Riemann, B., *Gesammelte Mathematische Werke* (Leipzig: Teubner, 1892).

62. Jeffrey, A. and Taniuti, T., "Chapter 2: The Method of Characteristics," in: *Non-Linear Wave Propagation with Applications to Physics and Magneto-Hydrodynamics*, Mathematics in Science and Engineering, vol. 9 (1964), 65-110, ISBN:9780123749178.

63. Rayleigh, J., "Aerial Plane Waves of Finite Amplitude," *Proc. R. Soc. Lond., Ser. A* 84 (1910): 247-284.

64. Capetti, A., "Contributo allo studio del flusso nei cilindri dei motori veloci," *Ingegneria* 2 (1923): 206-210.

65. Bannister, F.K. and Mucklow, G.F., "Wave Action Following Sudden Release of Compressed Gas from a Cylinder," *Proc. Inst. Mech. E* 159 (1948): 269.

66. De Haller, R., "The Application of a Graphic Method to Some Dynamic Problems in Gases," *Sulzer Technical Review* 1, no. 6 (1945).

67. Jenny, E., "Unidimensional Transient Flow with Consideration of Friction, Heat Transfer and Change of Section," *Brown Boveri Review* 37, no. 11 (1950): 447.

68. Shapiro, A.H., *The Dynamics and Thermodynamics of Compressible Fluid Flow*, vol. II (The Ronald Press, 1954).

69. Rudinger, G., *Wave Diagrams for Non-steady Flows in Ducts* (D. Van Nostrand Co. Inc., 1955).

70. Wallace, F.J. and Boxer, G., "Wave Action in Diffusers for Exhaust Pipe Systems with Special Reference to the Scavenging of Two-Stroke Engines," *Proc. Inst. Mech. E* 170 (1956).

71. Benson, R.S., "The Discharge Process in a Two-Stroke Cycle Engine Cylinder and Exhaust System," Ph.D. thesis, University of Liverpool, 1958.

72. Benson, R.S., Garg, R.D., and Woollatt, D. "A Numerical Solution of Unsteady Flow Problems," *Int. J. Mech. Sci.* 6, no. 1 (1964): 117-144.

73. Benson, R.S., *The Thermodynamics and Gas Dynamics of Internal Combustion Engines*, vol. 1 (Clarendon Press - Oxford, 1982).

74. Low, S.C., Benson, R.S., and Winterbone, D.E., "Computer Aided Design Package for Diesel Engine Manifold System," SAE Technical Papers 790277, 1979, doi:10.4271/790277.

75. Winterbone, D.E. and Yoshitomi, M., "The Accuracy of Calculating Wave Action in Engine Intake Manifolds," SAE Technical Paper 900677, 1990, doi:10.4271/900677.

76. Benson, R.S., Annand, W.J.D., Baruah, P.C., A Simulation Model Including Intake and Exhaust Systems for a Single Cylinder Four-Stroke Cycle Spark Ignition Engine," *International Journal of Mechanical Sciences* 17, no. 2 (1975): 97-124.

77. Coates, S.W. and Blair, G.P., "Further Studies of Noise Characteristics of Internal Combustion Engine Exhaust Systems," SAE Technical Paper 740713, 1974, doi:10.4271/740713.

78. Kirkpatrick, S.J., Blair, G.P., Fleck, R., and McMullan, R.K., "Experimental Evaluation of 1-D Computer Codes for the Simulation of Unsteady Gas Flow Through Engines - A First Phase," SAE Technical Paper 941685, 1994, doi:10.4271/941685.

79. Bingham, J.F. and Blair, G.P., "An Improved Branched Pipe Model for Multi-Cylinder Automotive Engine Calculations," *Proc. Instn. Mech. Eng., Part D, Journal of Automobile Engineering* 199, no. D1 (1985): 65-77.

80. Sierens, R., Van Hove, W., and Snauwaert, P., "Comparison of Measured and Calculated Gas Velocities in the Inlet Channel of a Single Cylinder Reciprocating Engine," *Proceedings of the Institution of Mechanical Engineers, Part A: Journal of Power and Energy* 198 no. 1 (1984): 61-69.

81. Payri, F., Corberan, J.M., and Boada, F., "Modifications to the Method of Characteristics for the Analysis of the Gas Exchange Process in Internal Combustion Engines," *Proceedings of the Institution of Mechanical Engineers, Part D: Transport Engineering* 200, no. D4 (1986): 259-266.

82. Jones, A.D. and Brown, G.L., "Determination of Two-Stroke Engine Exhaust Noise by the Method of Characteristics," *Journal of Sound and Vibration* 82, no. 3 (1982): 305-327.

83. Corberan, J.M., "New Constant Pressure Model for N-Branch Junctions," *Proceedings of the Institution of Mechanical Engineers, Part D: Journal of Automobile Engineering* 206, no. 2 (1992): 117-123.

84. Onorati, A., "Prediction of the Acoustical Performances of Muffling Pipe Systems by the Method of Characteristics," *Journal of Sound and Vibration* 171, no. 3 (1994): 369-395.

85. Winterbone, D.E. and Pearson, R.J., *Theory of Engine Manifold Design: Wave Action Methods for IC Engines* (Professional Engineering Publishing, 2000).

86. Weaving, P.M., *Internal Combustion Engineering: Science & Technology* (Elsevier Applied Science, 1990).

87. Winterbone, D.E. and Pearson, R.J., *Design Techniques for Engine Manifolds: Wave Action Methods for IC Engines* (Professional Engineering Pub. Limited, 1999).

88. Hirsch, C., *Numerical Computation of Internal and External Flows* (Elsevier, 2007).

89. Toro, E.F., *Riemann Solvers and Numerical Methods for Fluid Dynamics* (Springer, 1997).

90. Courant, R., Friedrichs, K.O., and Lewy, H., "Über die partiellen Differenzengleichungen der mathematischen Physik," *Mathematische Annalen* 100 (1928): 32-74.

91. von Neumann, J., "Proposal and Analysis of a New Numerical Method for the Treatment of Hydrodynamical Shock Problems," National Defence and Research Committee Report AM551, 1944.

92. von Neumann, J. and Richtmyer, R.D., "Method for the Numerical Calculation of Hydrodynamic Shocks," *Journal of Applied Physics* 21 (1950): 232-237.

93. Hartree, D.R., "Some Practical Methods of Using Characteristics in the Calculation of Non-steady Compressible Flow," US Atomic Energy Commission Report AECU-2713, 1953.

94. Courant, R., Isaacson, E., and Rees, M., "On the Solution of Nonlinear Hyperbolic Differential Equations by Finite Differences," *Communications on Pure and Applied Mathematics* 5 (1952): 243-249.

95. Lax, P.D., "Weak Solutions of Nonlinear Hyperbolic Equations and Their Numerical Computation," *Communications on Pure and Applied Mathematics* 7 (1954): 159-193.

96. Lax, P.D. and Wendroff, B., "Systems of Conservation Laws," *Communications on Pure and Applied Mathematics* 13 (1960): 217-237.

97. Lax, P.D. and Wendroff, B., "Difference Schemes for Hyperbolic Equations with High Order of Accuracy," *Communications on Pure and Applied Mathematics* 17 (1964): 381-398.

98. Richtmyer, R.D. and Morton, K.W., *Difference Methods for Initial Value Problems* (New York: Interscience, 1967).

99. MacCormack, R.W., "The Effect of Viscosity in Hypervelocity Impact Cratering," *AIAA Paper 69-354*, 1969.

100. Payri, F., Torregrosa, A.J., and Chust, M.D., "Application of MacCormack Schemes to IC Engine Exhaust Noise Prediction," *Journal of Sound and Vibration* 195, no. 5 (1996): 757-773.

101. Onorati, A., "A White Noise Approach for Rapid Gas Dynamic Modeling of IC Engine Silencers," *3rd International Conference on Computers in Reciprocating Engines and Gas Turbines, IMechE*, C499/0521996, 219-228, 1996.

102. Lerat, A. and Peyret, R., "Non-centred Schemes and Shock Propagation Problems," *Computer Fluids* 2 (1974): 35-52.

103. Godunov, S.K., "A Difference Scheme for Numerical Computation of Discontinuous Solutions of Hydrodynamics Equations," *Mat. Sb.* 47 (1959): 271-306; English translation in US Joint Publication Research Service, JPRS 7226, 1960.

104. Boris, J.P. and Book, D.L., "Flux-Corrected Transport. I SHASTA, a Fluid Transport Algorithm That Works," *Journal of Computational Physics* 11 (1973): 38-69.

105. Book, D.L. and Boris, J.P., "Flux-Corrected Transport. II Generalizations of the Method," *Journal of Computational Physics* 18 (1975): 248-283.

106. Niessner, H. and Bulaty, T., "A Family of Flux-Correction Methods to Avoid Overshoot Occurring with Solutions of Unsteady Flow Problems," *Proceedings of the GAMM Conference of Numerical Methods of Fluid Mechanics*, 241-250, 1981.

107. Roe, P.L., "Approximate Riemann Solvers, Parameter Vectors and Difference Schemes," *J. Comp. Phys.* 43 (1981): 357-372.

108. Roe, P.L., "Characteristic-Based Schemes for the Euler Equations," *Annual Review of Fluid Mechanics* 18 (1986): 337-365.

109. Steger, J.L. and Warming, R.F., "Flux Vector Splitting of the Inviscid Gas Dynamic Equations with Application to Finite-Difference Methods," *Journal of Computational Physics* 40 (1981): 236-293.

110. van Leer, B., "Flux-Vector Splitting for the Euler Equations," *Eighth International Conference on Numerical Methods in Fluid Dynamics*, 507-512, 1982.

111. Harten, A., "High Resolution Schemes Using Flux Limiters for Hyperbolic Conservation Laws," *Journal of Computational Physics* 49 (1983): 357-393.

112. van Leer, B., "Towards Ultimate Conservative Difference Scheme: 2. Monotonicity and Conservation Combined in a Second-Order Scheme," *Journal of Computational Physics* 14 (1974): 361-370.

113. Sod, G.A., "Survey of Several Finite-Difference Methods for Systems of Non-linear Hyperbolic Conservation Laws," *J. Comput. Phys.* 27 (1978): 1-31.

114. Sweby, P.K., "High Resolution Schemes Using Flux Limiters for Hyperbolic Conservation Laws," *SIAM J. Numer. Anal.* 21 (1984): 995-1011.

115. Horlock, J.H. and Winterbone, D.E., *The Thermodynamics and Gas Dynamics of Internal-Combustion Engines*, vol. 2 (Clarendon Press - Oxford, 1986).

116. Blair, G.P., Goulburn, J.R., "An Unsteady Flow Analysis of Exhaust Systems for Multi-Cylinder Automobile Engines," SAE Technical Paper 690469, 1969, doi:10.4271/690469.

117. Blair, G.P., "Computer-Aided Design of Small Two-Stroke Engines for Both Performance Characteristics and Noise Levels," IMechE Conference Publ., 1978-5, 1978.

118. Seifert, H., "Analysis of Unsteady Flows in Intake Manifold of a Multi-Cylinder Gasoline Engine," *MTZ Motortechnische Zeitschrift* 39, no. 1 (1978): 25-30.

119. Seifert, H., "A Mathematical Model for Simulation of Processes in an Internal Combustion Engine," *Acta Astronaut.* 6 (1979): 1361-1376.

120. Dwyer, H., Allen, R., Ward, M., Karnopp, D. et al., "Shock Capturing Finite Difference Methods for Unsteady Gas Transfer," *7th Fluid and Plasma Dynamics Conference*, AIAA Paper 74-521, 1974.

121. Takizawa, M., Uno, T., Oue, T., and Yura, T., "A Study of Gas Exchange Process Simulation of an Automotive Multi-Cylinder Internal Combustion Engine," SAE Technical Paper 820410, 1982, doi:10.4271/820410.

122. Liu, J., Schorn, N., Schernus, C., and Peng, L., "Comparison Studies on the Method of Characteristics and Finite Difference Methods for One-Dimensional Gas Flow through IC Engine Manifold". SAE Technical Paper 960078, 1996, doi:10.4271/960078.

123. Poloni, M., Winterbone, D.E., and Nichols, J.R., "Comparison of Unsteady Flow Calculations in a Pipe by the Method of Characteristics and the Two-Step Differential Lax-Wendroff Method," *International Journal of Mechanical Science* 29, no. 5 (1987): 367-378.

124. Winterbone, D.E., Pearson, R.J., and Zhao, Y., "Numerical Simulation of Intake and Exhaust Flows in a High Speed Multi-Cylinder Petrol Engine Using the Lax-Wendroff Method," *IMechE International Conference on Computers in Engine Technology*, Cambridge, U.K., 1991.

125. Pearson, R.J. and Winterbone, D.E., "The Simulation of Gas Dynamics in Engine Manifolds Using Non-linear Symmetric Difference Scheme," *Proceedings of the Institution of Mechanical Engineers, Part C: Journal of Mechanical Engineering Science* 211 (1997): 601-616.

126. Chapman, M., "FRAM - Nonlinear Damping Algorithms for the Continuity Equation," *J. Comp. Phys.*, 44 (1981): 84-103.

127. Chapman, M., Novak, J.M., and Stein, R.A., "Numerical Modeling of Inlet and Exhaust Flows in Multi-cylinder Internal Combustion Engines," *Flows in Internal Combustion Engines*, Uzkan, T., editor (Austin, TX: ASME WAM, 1982).

128. Norman, K.R., Selamet, A., and Novak, J.M., "Perforated Muffler Manifold Catalyst," *Journal of Sound and Vibration* 218, no. 4 (1998): 711-734.

129. Morel, T., Keribar, R., and Blumberg, P.N., "A New Approach to Integrating Engine Performance and Component Design Analysis through Simulation," SAE Technical Paper 880131, 1988, doi:10.4271/880131.

130. Morel, T., Flemming, M., and LaPointe, L., "Characterization of Manifold Dynamics in the Chrysler 2.2 S.I. Engine by Measurements and Simulation," SAE Technical Paper 900697, 1990, doi:10.4271/900679.

131. Morel, T., Morel, J., and Blaser, D., "Fluid Dynamic and Acoustic Modeling of Concentric-Tube Resonators/Silencers," SAE Technical Paper 910072, 1991, doi:10.4271/910072.

132. Sapsford, S., Richards, V., Amlee, D., Morel, T. et al., "Exhaust System Evaluation and Design by Non-Linear Modeling," SAE Technical Paper 920686, 1992, doi:10.4271/920686.

133. Gosman, A., "Multidimensional Modeling of Cold Flows and Turbulence in Reciprocating Engines," SAE Technical Paper 850344, 1985, doi:10.4271/850344.

134. Stobart, R.K., May, A., Challen, B., and Morel, T., "Engine and Control System Modelling to Reduce Powertrain Development Risk," *IEE Colloquium (Digest)* 388 (1997): 2/1-2/4.

135. Stobart, R., May, A., Challen, B., and Morel, T., "Modeling for Diesel Engine Control: The CPower Environment". SAE Technical Paper 980794, 1998, doi:10.4271/980794.

136. Morel, T., Silvestri, J., Goerg, K., and Jebasinski, R., "Modeling of Engine Exhaust Acoustics," SAE Technical Paper 1999-01-1665, 1999, doi:10.4271/1999-01-1665.

137. Laimboeck, F., Glanz, R., Modre, E., and Rothbauer, R., "AVL Approach for Small 4-Stroke Cylinderhead, Port and Combustion Chamber Layout," SAE Technical Paper 1999-01-3344, 1999, doi:10.4271/1999-01-3344.

138. Fairbrother, R. and Tonsa, M., "Acoustic Simulation of after Treatment Devices Using Linear and Nonlinear Methods," *Proceedings of the Tenth International Congress on Sound and Vibration*, 3195-3202, 2003.

139. Harten, A., Engquist, B., Osher, S., and Chakravarthy, S.R., "Uniformly High Order Accurate Essentially Non-oscillatory Schemes, III," *Journal of Computational Physics* 71, no. 2 (1987): 231-303.

140. Giannattasio, P. and Dadone, A., "Applications of a High Resolution Shock Capturing Scheme to the Unsteady Flow Computation in Engine Ducts," *Inst. Mech. Eng. C*, C430/055, 1991.

141. Onorati, A. and Ferrari, G., "Modeling of 1-D Unsteady Flows in I.C. Engine Pipe Systems: Numerical Methods and Transport of Chemical Species," SAE Technical Paper 980782, 1998, doi:10.4271/980782.

142. Chang, S.C. and To, W., "A Brief Description of a New Numerical Framework for Solving Conservation Laws, The Method of Space-Time Conservation Element and Solution Element," NASA Technical Memorandum 105757, 1999.

143. Briz, G. and Giannattasio, P., "Applicazione dello schema numerico Conservation Element - Solution Element al calcolo del flusso intazionario nei condotti dei motori a c.i.," *Proceedings of the 48th ATI National Congress*, 233-247 1993 (in Italian).

144. Onorati, A., D'Errico, G., and Ferrari, G., "1D Fluid Dynamic Modeling of Unsteady Reacting Flows in the Exhaust System with Catalytic Converter for S.I. Engines," SAE Technical Paper 2000-01-0210, 2000, doi:10.4271/2000-01-0210.

145. Onorati, A., Ferrari, G., D'Errico, G., and Montenegro, G., "The Prediction of 1D Unsteady Flows in the Exhaust System of a S.I. Engine Including Chemical Reactions in the Gas and Solid Phase," SAE Technical Paper 2002-01-0003, 2002, doi:10.4271/2002-01-0003.

146. D'Errico, G., Ferrari, G., Onorati, A., and Cerri, T., "Modeling the Pollutant Emissions from a S.I. Engine," SAE Technical Paper 2002-01-0006, 2002, doi:10.4271/2002-01-0006.

147. Gascón, L. and Corberán, J.M., "Construction of Second-Order TVD Schemes for Non-homogeneous Hyperbolic Conservation Laws," *Journal of Computational Physics* 172 (2001): 261-297.

148. Payri, F., Galindo, J., Serrano, J.R., and Arnau, F.J., "Analysis of Numerical Methods to Solve One-Dimensional Fluid-Dynamic Governing Equations under Impulsive Flow in Tapered Ducts," *International Journal of Mechanical Science* 46 (2004): 981-1004.

149. http://openwam.webs.upv.es/docs/.

150. http://www.lotuscars.com/engineering/engineering-software/.

151. Chan, K., Ordys, A., Volkov, K., and Olga Duran, O., "Comparison of Engine Simulation Software for Development of Control System," *Modelling and Simulation in Engineering* 2013 (2013), Article ID 401643.

152. Bozza, F., Gimelli, A., and Tuccillo, R., "The Control of a VVA-Equipped SI Engine Operation by Means of 1D Simulation and Mathematical Optimization," SAE Technical Paper 2002-01-1107, 2002, doi:10.4271/2002-01-1107.

153. Bozza, F. and Torella, E., "The Employment of a 1D Simulation Model for A/F Ratio Control in a VVT Engine," SAE Technical Paper 2003-01-0027, 2003, doi:10.4271/2003-01-0027.

154. Alix, G., Pera, C., Bohbot, J., and Baldari, A., "Comparison of 0D and 1D Duct System Modeling for Naturally Aspirated Spark Ignition Engines," SAE Technical Paper 2011-01-1898, 2011, doi:10.4271/2011-01-1898.

155. Montenegro, G., Onorati, A., and Della Torre, A., "The Prediction of Silencer Acoustical Performances by 1D, 1D-3D and Quasi-3D Non-linear Approaches," *Computers & Fluids* 71 (2013): 208-223.

CHAPTER 2

Virtual Engine Development: 1D- and 3D-CFD up to Full Engine Simulation

Mahir Tim Keskin, Michael Grill, Marco Chiodi, and Michael Bargende
FKFS & IVK University of Stuttgart

2.1 Introduction

More than 140 years after the invention of the internal combustion engine, there is still remarkable room for efficiency enhancement and emissions reduction. Facing this constant challenge of improvement—as the only option in order to ensure the future success of this thermodynamic machine—the development process of modern internal combustion engines must be as efficient as possible. In order to exploit this room for improvement, the following aspects and modern development trends must be intensively investigated and analyzed:

- Complex injection systems and strategies (from homogeneous to stratified mixtures, dual fuels, etc.)
- Innovative combustion concepts (spark ignition (SI), Diesel, HCCI, spark-assisted compression ignition, pre-chamber spark plugs, lean combustion, high exhaust gas recirculation (EGR) rates, post-oxidation, etc.) suitable for high compression ratios
- Innovative and CO_2-neutral fuels (bio and synthetic fuel, gaseous and liquid, etc.)
- Variable valve trains (Miller and Atkinson cycle, scavenging concepts, etc.)
- Different charging concepts (turbocharging, supercharging, e-booster, etc.)
- Variable compression ratio (VCR) mechanisms
- Different hybridization concepts

This list clearly shows that the internal combustion engine of the future is becoming very complex and it must fulfill more and more stringent emission regulations. Moreover, it has to remain affordable and reliable and ensure drivability and emotion. Accordingly,

FIGURE 2.1 The analysis and simulation tools available for IC-engine development.

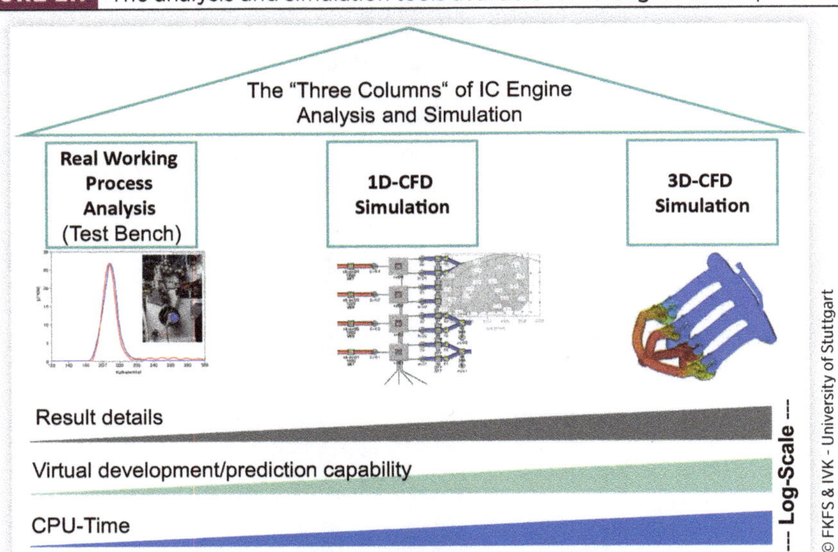

in order to manage this complexity, a proper combination of different development tools is required, so that a more comprehensive investigation of numerous concept ideas within the limited financial and short temporal framework can be performed. This includes combined one dimensional (1D)- and three-dimensional computational fluid dynamic (3D-CFD) simulations as well as a single-cylinder test bench (equipped with indicating system for experiment and data evaluation toward a reliable analysis of the real working-process operation, optical access, etc.). These tools have ideally been implemented before the installation of a more expensive full engine device. This is required since each of these development tools shows different advantages and drawbacks for the different applications. As shown in Figure 2.1, the degree of result details and the prediction capability for a reliable virtual engine development—the main advantages of the 3D-CFD approach—are unfortunately linked to very high CPU time and complexity that easily can become unaffordable. However, the improvement process of these simulation tools will influence the best practice combination toward a successful engine development in the future.

A good understanding of the processes occurring within the engine—as well as influential geometrical and control-related factors—plays a key role for a successful development. However, the investigation of multiple engine variants on the test bench, e.g., different channel and combustion chamber geometries, is extremely expensive and time-consuming.

Supported by continuously growing computing power and further development in simulation software, numerical simulations are increasingly being incorporated into the development process. 3D-CFD simulations enable a comprehensive analysis of the processes occurring within the engine. They deliver information which would be impossible or very difficult to acquire via measurement, e.g., via optical access (turbulence, mixture formation, etc.). In particular, 3D-CFD can be used to investigate flow-dependent processes with high temporal and spatial resolution. For example, the mixture formation, supported by the charge motion, or the combustion process, is significantly influenced by the turbulence. Furthermore, the influence of adjustments to geometry, valve timing, etc. can be evaluated, and changes regarding the specific power, fuel consumption, and pollutant emissions can be assessed. As described above, 3D-CFD simulation accomplishes a dual role: as a measurement (a virtual eye within an engine

FIGURE 2.2 Main capabilities of measurement and simulation tools.

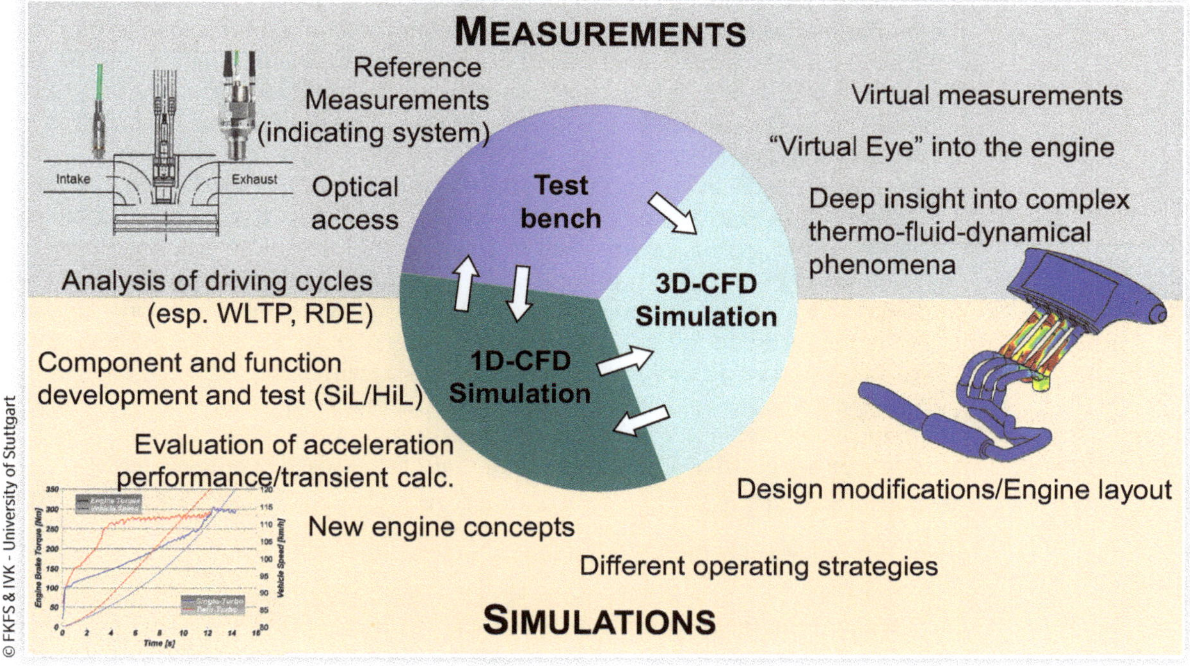

that allows an extension of the experimental measurements) and simulation device, respectively (see Figure 2.2).

1D simulations are particularly useful both in concept studies and in the transient calculation of engine processes, particularly under consideration of further drivetrain components. Here the 3D-CFD results can be used for operating point-dependent validation of zero-dimensional (0D)/1D models (turbulence, residual gas content, etc.). In addition, experimental investigations are an important basis for validation and of course will never be completely replaced by simulations. However, the development process of a combustion engine can be improved by reordering the areas of focus.

A combination of quantitatively fewer but more targeted and high-quality investigations on the test bench, with comprehensive numerical simulations, offers various advantages (see Figure 2.3). Within the same development time, significantly more engine variants can be virtually analyzed and compared, sometimes even simultaneously. Without major financial risk, also unusual or novel concepts, which can usually not be implemented on the test bench, will be taken into account. An extensive matrix of virtually tested engine variants (virtual engine development) thereby increases the confidence in engine design decisions, which are also less dependent on the long-term, individual experience of particular development engineers.

As introduced above, designing a new engine is a complex and challenging task, which requires to use both a cleverly worked out, universally applicable procedure and powerful, reliable tools. The first question to be asked is thus which overall method is to be applied in an engine development process. Secondly, it has to be decided which tool category is best suited for the single steps

FIGURE 2.3 Future target in the implementation of measurement and simulation tools.

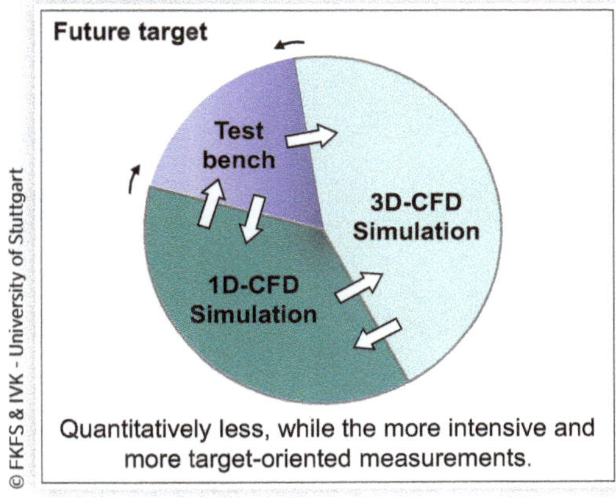

along the development path: where should 1D simulation be used, when is 3D-CFD beneficial, and where do we need the test bench results? Finally, it is necessary to look at the available models and sub-models in the single tool category to find out if they have the characteristics that are needed in order to establish a fast, reliable, and effective virtual engine design. This last step requires the largest amount of know-how and technical knowledge and will be discussed in detail in Sections 2.2, 2.3, and 2.4. The first two questions, however, can be answered in brief within this introduction.

Regarding the overall procedure, the so-called V-model is widely used in systems engineering in general and within the automotive industry in particular. It generally comprises two streams: a top-down-oriented "design stream," which starts with the overall concept and architecture based on the identified requirements and then goes down more and more into the detailed design of sub-systems and components, and a bottom-up-oriented "testing stream," which starts at component level and ends at the validation and verification of the final product; see Figure 2.4.

Applied to the engine development process, that means that first of all concept studies have to be done based on legal requirements (emissions, fleet consumption) and customer needs (drivability, fuel consumption). This leads to questions such as:

- Which displacement and number of cylinders should the engine have?
- Which level of hybridization is required to meet targets?
- What kind of combustion process has to be used, and what are the consequences for the charging system and the exhaust gas aftertreatment?

which can be answered best by 1D simulation. This is the case not only because there are no alternatives available—there are no test bench engines or detailed geometries for 3D-CFD simulations at this early stage—but also because 1D simulation combines a good predictive accuracy with very low computational times, which is a prerequisite for a high number of investigated configurations.

Once the basic concept is settled, there is still a large number of issues to be investigated, such as setting operating strategies or possible benefits from a variable valve train. For such topics, it is again vital to investigate a large number of different configurations over a variety of operating conditions. Even relatively simple simulation tasks, for instance, an engine speed sweep at full load for a turbocharged Diesel engine, will quickly lead to a considerable amount of required case simulations if the aim is to find the optimal variable turbine geometry (VTG) position for certain peak pressure and

FIGURE 2.4 V-model for a typical engine development process.

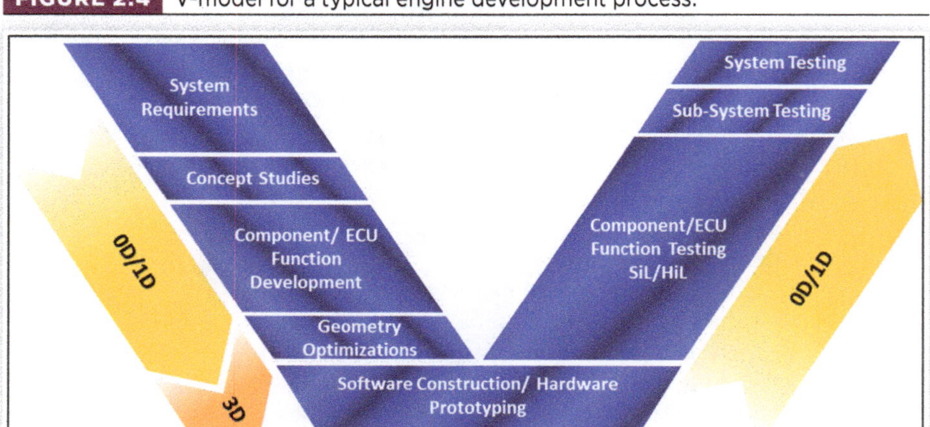

exhaust temperature limitations (see Section 2.6). Evidently the computational demands will be much higher even for transient investigations, which become increasingly important given the change from rather static type certifying driving cycles to more dynamic ones or even real driving cycles. Thus, a multitude of simulations has to be run, requiring very fast computational times while maintaining the capacity to make reliable predictions. Again, these demands can almost exclusively be met by 1D simulation, coupled with 0D models for the processes in the combustion chamber.

The use of 1D simulation extends even lower down the hierarchy, for instance when boundary conditions for the design of single components are needed, and concerns also the thermal management of the engine or the oil circuit. At some point, however, 3D-CFD becomes the more powerful tool, for instance when trying to find the optimal geometry for the inlet pipes and valves in order to combine high tumble generation and low coefficients of discharge for a gasoline engine. This represents also a good example of how the whole process can have iterations and how 1D and 3D departments need to be in close collaborations: the achievable tumble number effects the burn duration which in turn decides many details of the engine design and operation.

Regarding the testing stream, 1D simulation can be a very helpful tool, too. While testing functions and components, hardware-in-the-loop (HiL) or software-in-the-loop (SiL) simulations are frequently used. The focus is here on very short computation times close to real time, which can be achieved again best by 1D simulation (albeit with special adaptions).

Given the importance of 1D simulation along all the steps the engine development process, it makes sense to have first a closer look at the sub-models it relies on. Basically, it can be split in the flow simulation (the actual 1D-CFD) and the simulation of the processes in the combustion chamber, which is 0D from a thermodynamic point of view (0D or quasidimensional). Whereas 1D-CFD flow simulation usually requires most of the computational time in such a simulation setup, it is rather straightforward regarding the agreement between measured and predicted values. However, the opposite is the case for the mentioned 0D models—often labelled as quasidimensional models due to the fact that they try to describe a complex, 3D behavior within the framework of comparably simple mathematical equations—they have manageable computation times in general but are crucial for the overall predictive quality of the data obtained from engine simulation. It is thus important to identify the requirements such models should be able to meet and how that can be assessed. This will be discussed in the following sections.

2.2 Model Requirements

The question what defines a "good" model can be answered relatively quickly from a general point of view. First and foremost, to fulfil its intended purpose, the model should have a high predictive power, i.e., each sub-model needs to predict the impact of changed variables on the target variable reliably and accurately. Obviously a model that fails to do so is of little use for any investigation, but that alone is not enough to ensure that the user can actually work with the model in an effective way within the overall engine development process. The importance of these "practical" requirements must not be underestimated. This is particularly apparent for the computational time, which has to be sufficiently short to ensure that all the required simulations for the investigation can be completed within a reasonable time frame. However, "soft" requirements are often equally essential. This includes model tuning as well as support from the software provider. The model calibration process in particular is a sticking point in the assessment of models: How extensive is the needed measurement database? How many tuning parameters are there? Can they be tuned independently or do they influence each other? Are they associated with a clear, physical effect, or do they just represent a number and the model is more or less a black box for the user? Ideally, a model should contain only very few tuning parameters that can be set at constant values over the whole operating range of the engine, such that a single operating point is enough for calibration, resulting in a tuning process that can be easily automated via an optimizer tool. This will guarantee optimal results with the highest possible user-friendliness while saving valuable time. In this context, support from the model developer is really helpful as well. It is particularly effective if the model developer is also model user at the same time, as experience and know-how are pivotal for dealing with support requests.

2.3 Assessment of Quasidimensional Models

As already explained, the scientifically most challenging and result-wise most critical part of 1D simulation are the quasidimensional models for the processes in the combustion chamber. In order to get meaningful results from the simulation model, the user has to make sure he uses all the necessary sub-models relevant to the investigation target, and above all he has to make sure they work as they should. This should be checked before the virtual engine design starts, as flaws in the simulation tools will inevitably lead to flaws in the engine design. This chapter will thus try to give a guideline of how to assess the quality of quasidimensional models, outlining the general requirements first before giving a wide range of practical examples.

2.3.1 General Assessment Guidelines

The most difficult (but also the most "beautiful") part of developing quasidimensional models arises from the fact that the most important mechanisms have to be identified out of a myriad of physical and chemical influences on engine behavior. The model developer thus has to find the right "deepness" in modeling: on the one hand, he or she has to make sure all the relevant effect mechanisms are included; on the other hand, complexity should be reduced; quasidimensional models should be as simple as possible, but not *simpler*—they still need to make the right predictions.

From a user's point of view, it is often hard to decide if this balancing act has been performed in a satisfying manner. It is necessary to have a good understanding and knowledge of the relevant processes to judge a model from its description alone and to answer the question if it makes sense to include this or that mechanism—as it is well possible that it only seemingly increases the model's precision but in fact leads to higher complexity, more difficult model tuning, and, at worse, even worse predictive accuracy. That applies even more to "black box models," e.g., neuronal network models, where the user has no chance to see inside the model at all. Either way, however, there are two very simple tests that can always be performed by the user to quickly get an idea of a model's quality:

- **Sensitivity test**: Ideally the simulation model should react to any change of operating parameters in exactly the same way the real engine would do. This can be tested by means of a simple sensitivity analysis: changing an input parameter and monitoring the model's reaction to it. For instance, a Wiebe burn rate model will not react to a change in engine speed, which is contradictory to the observations on the test bench, and indicates that a better burn rate model is needed. This is, of course, a very basic example, but the same method can be used to judge any simulation model, with the focus set on the quantities that are particularly relevant for the investigation target (see practical examples).

- **Continuity test**: Within the scope of engine investigation, system responses have a continuous nature generally—gradually changing one parameter leads to a gradual change of the monitored result quantities. This can easily be checked by the user and may seem like a matter of course at first; however, it is not uncommon to see models that fail the continuity test, and users only become aware of it when they see "something strange" in the results. It is thus recommended to test some commonly used input parameters for continuity before starting the actual simulations if a new model is used (see practical examples).
- **User-friendliness test**: This is probably not a single test per se, but users should not underestimate the importance of simple and straightforward model handling. Questions users should ask themselves after trying their hand on a new model include, among others, the following: Do I know what the various tuning parameters mean? Do I have automated assist for calibration (optimization tool)? Is there support from experts I can contact when in doubt? Again, practical examples where the user-friendliness becomes crucial are given in the following sections.

2.3.2 Practical Examples for SI Engines

The following list illustrates how the three tests from the previous section can be applied to the most important sub-models for SI engines:

2.3.2.1 BURN RATE MODEL

The burn rate model is the single most important sub-model for an engine model as it determines the engine performance. However, it is important to keep in mind that a good burn rate model needs to be coupled with further sub-models (see below) before performing a sensitivity test. The predicted burn rate should then react correctly to changes of any parameter that influences the combustion in the real engine, most noticeably air-fuel ratio, residual gas content, ignition timing, and valve timings.

The ignition timing is also a good parameter to test the continuity of the model behavior. If the ignition timing is varied in small steps, the predicted burn rates should show a similarly continuous change regarding the start of combustion; see Figure 2.5. It is not acceptable that different ignition timings lead to the same burn rate.

Apart from the general remarks regarding the tuning process, another aspect that is important to judge a burn rate model's user-friendliness is the question if internal controllers are available. For instance, it is generally recommendable to use the MFB50 point rather than the ignition timing, either to reproduce burn rates from a pressure trace analysis (PTA) or to make sure that the ignition timing is optimal regarding the engine efficiency. In such cases it is a huge advantage if an internal controller takes care of this task as it saves a lot of computational time and is much more comfortable for the user. The same applies to other quantities that often need to be controlled, such as the peak pressure or the maximum rate of pressure rise; see Figure 2.6.

If gasoline combustion processes other than the conventional homogeneous, spark-ignited one are to be investigated, it will be generally necessary to have additional, specific burn rate models, e.g., for stratified combustion [2] or HCCI [3]. It is useful if these models can be offered by the same developer to avoid inconsistencies from differing sub-models. The additional models themselves can be assessed similarly to the procedure that has already been discussed.

FIGURE 2.5 Reaction of a burn rate model (here expressed via peak pressure and timings of 5%/50% mass fraction burned) to a continuous change of an input parameter (here: ignition timing from 10° CA BTDC in steps of 0.1°CA), see model [1].

MFB5
MFB50
Peak Pressure
Crank Angle [°CA ATDC]
Cylinder Pressure [bar]
Spark Timing [°CA BTDC]
0
5
10
15
20
25
0
10
20
30
40
50
60
70
80
5
5.5
6
6.5
7
7.5
8
8.5
9
9.5
10

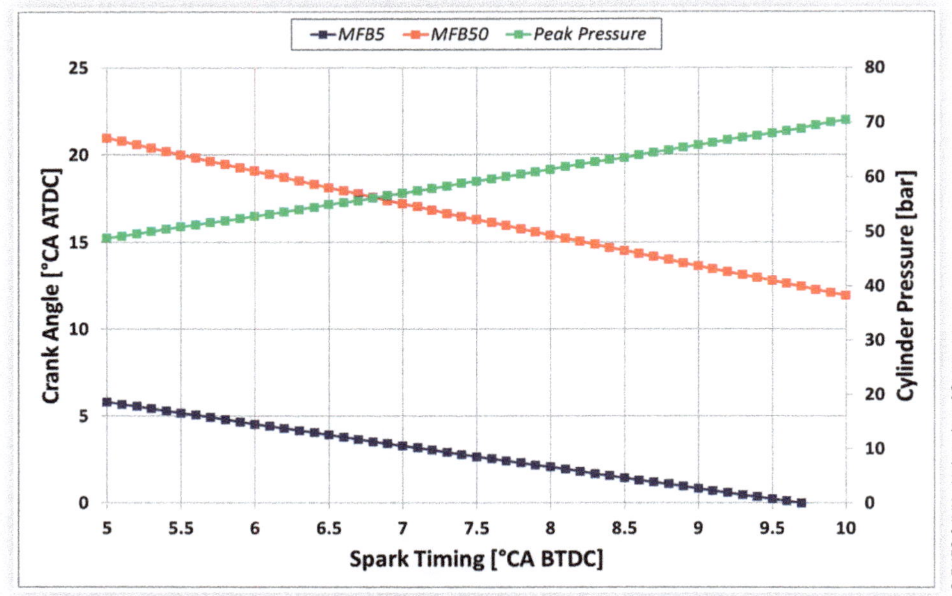

FIGURE 2.6 GUI of a burn rate model allowing the use of internal controllers for MFB50, peak pressure and rate of maximum pressure rise.

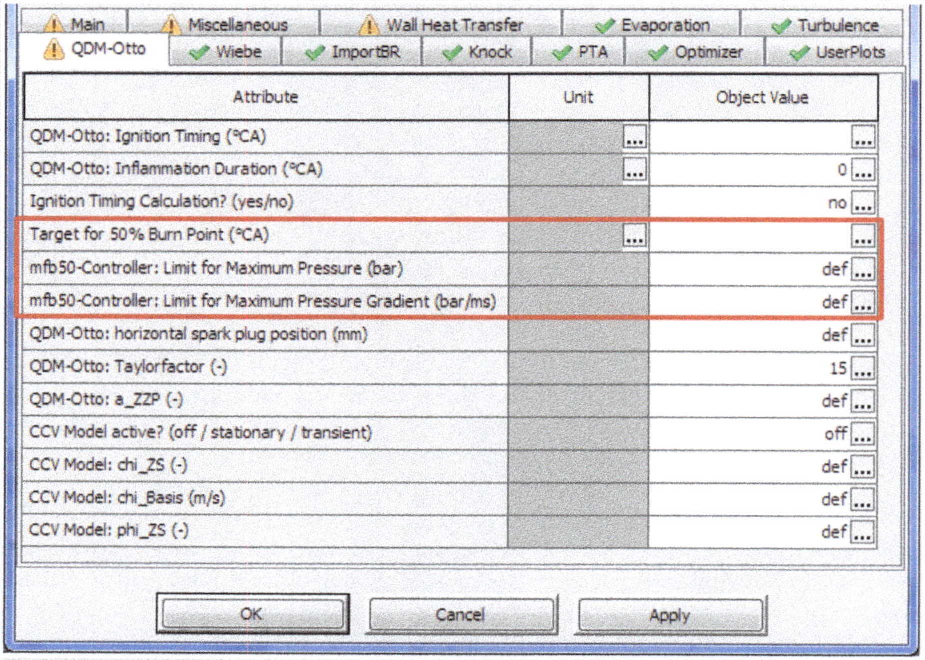

Main
Miscellaneous
Wall Heat Transfer
Evaporation
Turbulence
QDM-Otto
Wiebe
ImportBR
Knock
PTA
Optimizer
UserPlots
Attribute
Unit
Object Value
QDM-Otto: Ignition Timing (°CA)
QDM-Otto: Inflammation Duration (°CA)
0
Ignition Timing Calculation? (yes/no)
no
Target for 50% Burn Point (°CA)
mfb50-Controller: Limit for Maximum Pressure (bar)
def
mfb50-Controller: Limit for Maximum Pressure Gradient (bar/ms)
def
QDM-Otto: horizontal spark plug position (mm)
def
QDM-Otto: Taylorfactor (-)
15
QDM-Otto: a_ZZP (-)
def
CCV Model active? (off / stationary / transient)
off
CCV Model: chi_ZS (-)
def
CCV Model: chi_Basis (m/s)
def
CCV Model: phi_ZS (-)
def
OK
Cancel
Apply

2.3.2.2 TURBULENCE/CHARGE MOTION MODEL

As turbulence is one of the most important quantities influencing the combustion in SI engines, the burn rate model needs to be coupled with a turbulence model. Ideally, the turbulence model is based on the calculation of charge motion during gas exchange. Such a combined turbulence/charge motion model is then able to predict the impact of different cam contours on turbulence and hence on the burn rate. The model depicted in Figure 2.7 predicts correctly the reduced turbulence and the slower combustion for early intake valve closing (IVC) timings. This should be one of the aspects that should be checked when performing a sensitivity test for turbulence models. Otherwise, any investigation analyzing the benefits of variable valve timings will yield flawed results, as only the benefits for the gas exchange cycle, but not the impact on the high pressure part will be predicted.

2.3.2.3 LAMINAR FLAME SPEED MODEL

Another sub-model that needs to be coupled with the burn rate model is the laminar flame speed model.

For many years, correlations based on measurement data were used predominantly; recently, however, new correlations based on reaction kinetics have been developed (e.g., [6]), which avoid the shortcomings of the previous ones. The necessity of such reaction-kinetics-based correlations becomes immediately obvious when comparing the range of measurement data with a typical temperature/pressure trace during combustion; see Figure 2.8. It is evident that extrapolation from measurement data by a factor of approx. 50 is inevitably linked with considerable uncertainties and inaccuracies.

Another example for the shortcomings of measurement-based approaches is how the influence of mixture dilution (via excess air or EGR) is modelled—a particularly important aspect considering the trend to high EGR rates or lean mixtures. Whereas simple, partly even linear correction terms have been used in old models, reaction-kinetic-based models depict a rather complex behavior where not only the influence of EGR and excess air is distinctly different but also varying depending on the temperature; see Figure 2.9. This behavior results in considerable changes of the burn rate when

FIGURE 2.7 Reaction of a turbulence/burn rate model to different cam contours, see model [4].

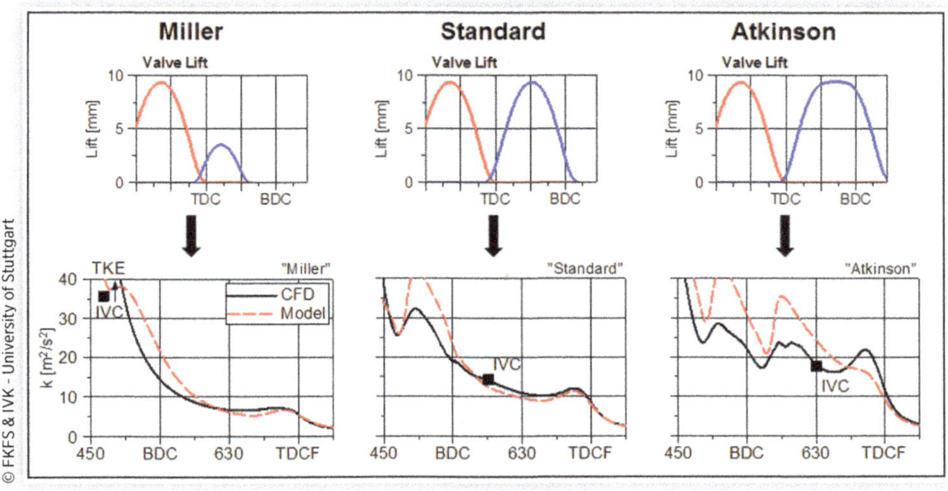

FIGURE 2.8 Comparison of available laminar flame speed measurement data with a typical temperature/pressure trace during combustion [5].

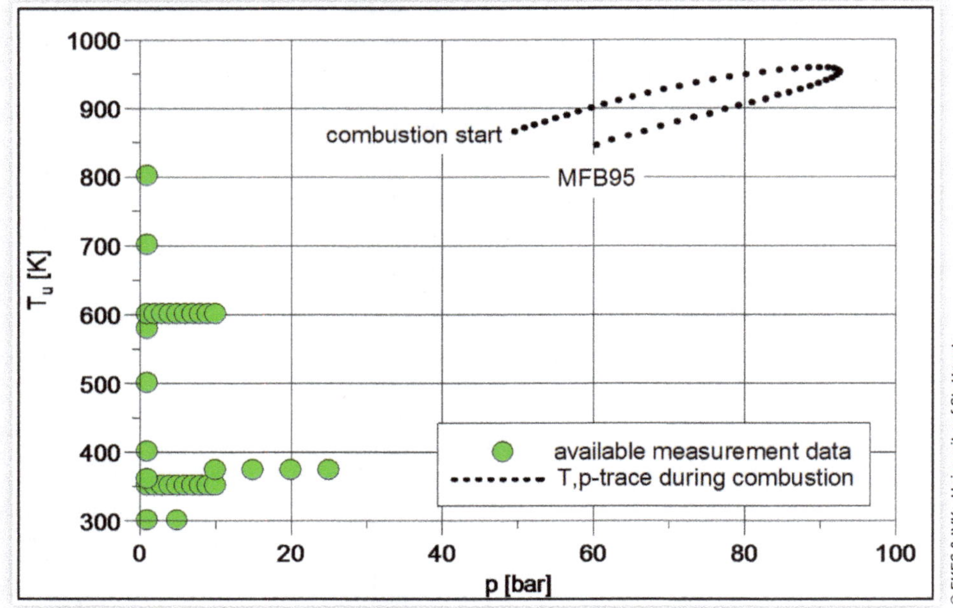

FIGURE 2.9 Influence of mixture dilution on laminar flame speed for a measurement-based model (Heywood) [7] and a reaction-kinetic-based model (Hann) [6].

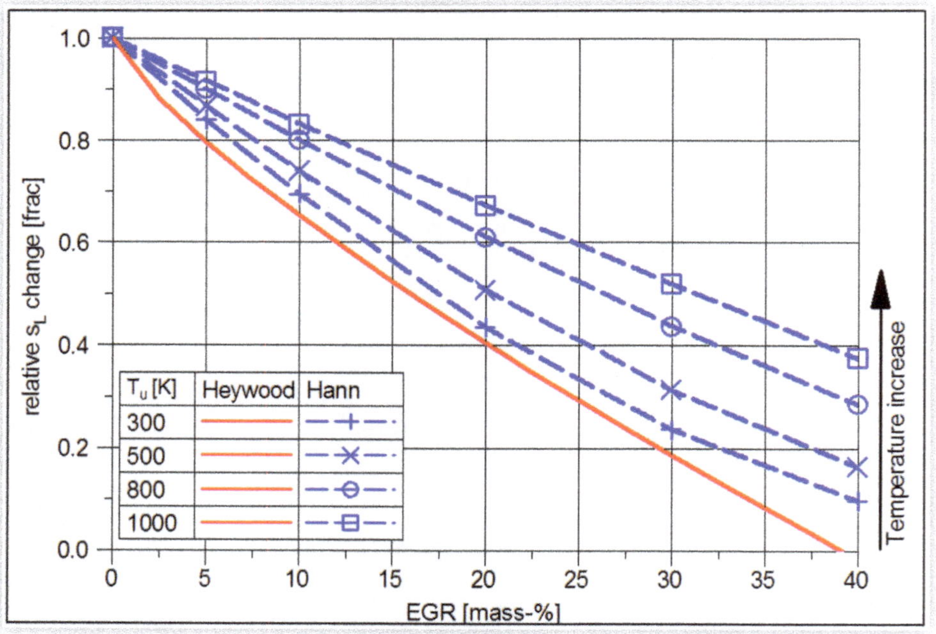

FIGURE 2.10 Predictive quality of the models from Figure 2.9 (Heywood, Hann) for an EGR sweep in comparison to burn rates from pressure trace analysis (PTA).

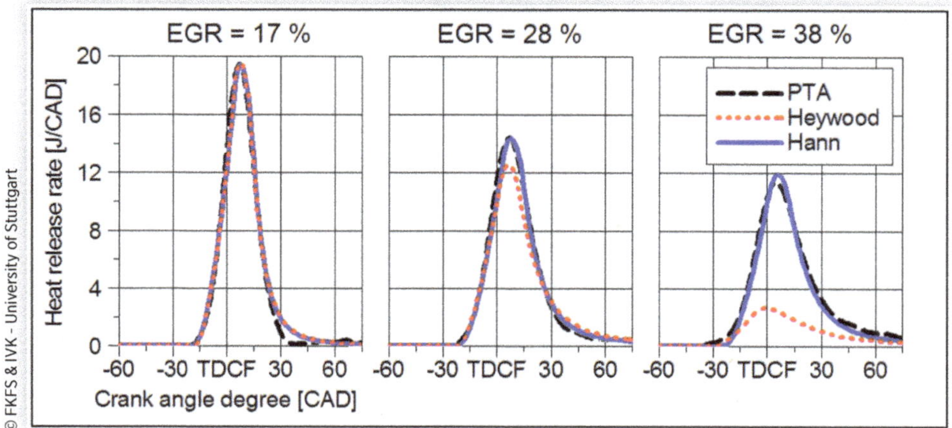

simulating an EGR sweep, for example; see Figure 2.10. If measurement data is available, this can be used for an enhanced sensitivity test that not only checks if the model reacts correctly from a qualitative point of view but also from a quantitative one.

A further aspect that needs to be considered when assessing laminar flame speed models is their ability to predict the behavior of different fuels. It is beneficial to have a uniform approach for the most important fuels—such as gasoline, methane, and methane-based fuels like natural gas—to make sure that no inconsistencies arise when the influence of a new fuel on the combustion in an existing engine is to be investigated. This condition is met by the model in [6] for instance.

2.3.2.4 FLAME GEOMETRY MODEL

The flame geometry model is another sub-model that is needed for the burn rate prediction. As it is based on geometrical calculations on the assumption of spherical flame propagation, there is not much that can go wrong here, but still it is a good example to discuss the question of the right "deepness" in modeling. For instance, the combustion chamber can be modelled to be simply disk-shaped, or much more detailed descriptions of the shape may be used. The latter may seem to be more precise at first, but in actual fact such an approach does not make much sense in a quasidimensional model: the accuracy will not be improved in this way, as the basic assumption of spherical flame propagation is a much bigger simplification anyway. The only consequence will be increased computational times and a much more complicated handling of the model from a user's point of view, given that more input data has to be delivered to the model. This example also shows nicely that the question of the right modeling deepness and user-friendliness are often closely correlated.

2.3.2.5 CCV MODEL

Models that can predict the cyclic variations of gasoline engines—or the coefficient of variance to be more precise—have not been available in 0D/1D simulation for a long time. One of the first models allowing to do so was presented in [8]. Predicting cyclic variations is not only important regarding the lean limit; it is also crucial for a correct brake-specific fuel consumption (BSFC) prediction. The fuel consumption of an average cycle does not equal the average fuel consumption of the single working cycle due to the fact that there is no linear relation between indicated mean effective pressure (IMEP)

FIGURE 2.11 Impact of cyclic variations on BSFC [5]; see model [8].

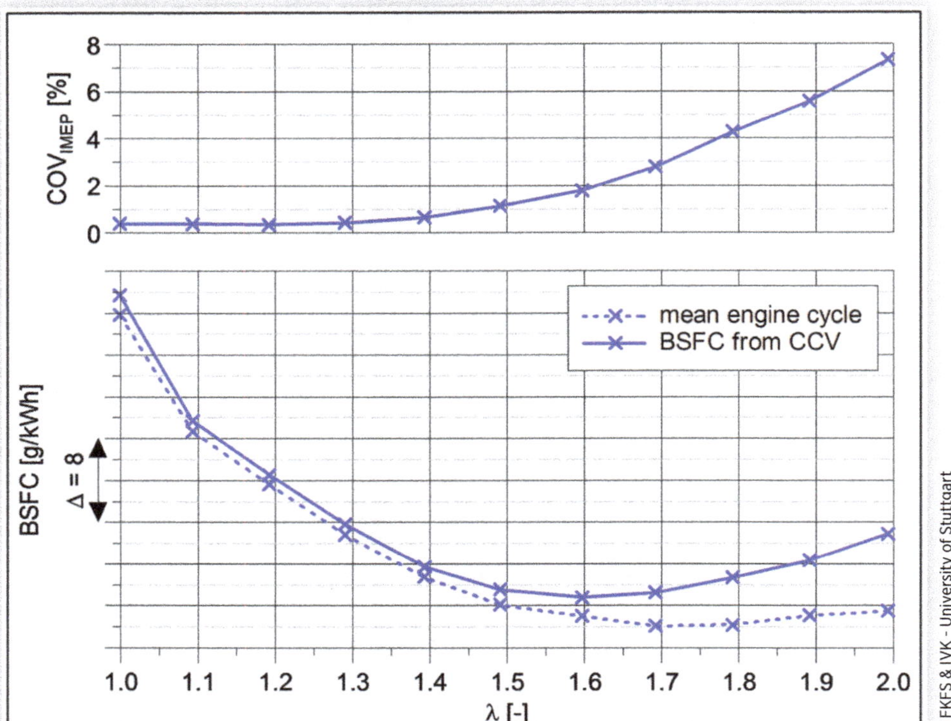

and BSFC: faster burning working cycles generally only offer a small efficiency benefit over the average cycle, whereas particularly slow burning cycles (misfiring in an extreme case) show a considerable worse efficiency.

Thus a good sensitivity test for a CCV model should check if it is able to predict the impact of cyclic variation on fuel consumption. As shown in Figure 2.11, the effect becomes increasingly apparent with higher cyclic variations and becomes significant particularly for cases with very lean mixture or high EGR rates—both features that will be used more and more often in modern gasoline engines. Investigations with a CCV model that cannot predict a BSFC correction factor will inevitably lead to flawed results.

2.3.2.6 KNOCK MODEL

Such as a CCV model is needed to predict the lean limit, a knock model is needed to predict the knock limit. Together, they can be used to determine the operating limits of a gasoline engine that have to be considered for almost all types of investigations.

Knock models are again a good example to discuss the right modeling deepness. For many years, simple Arrhenius-based approaches have been considered sufficient to model the autoignition behavior of the air-fuel mixture. However, newer investigations, such as [9, 10] were able to show that for a significant range of operating conditions, two-stage ignition occurs in gasoline engines; see Figure 2.12.

A knock model that is not designed to account for two-stage ignition will thus inevitably fail under such operating conditions; it would be too simple. The benefit of modeling two-stage ignition is clearly visible in Figure 2.13. However, it is important to note that for a correct prediction, not only the correct shape of the ignition delay curve has to

FIGURE 2.12 Operating conditions under which two-stage ignition occurs [9].

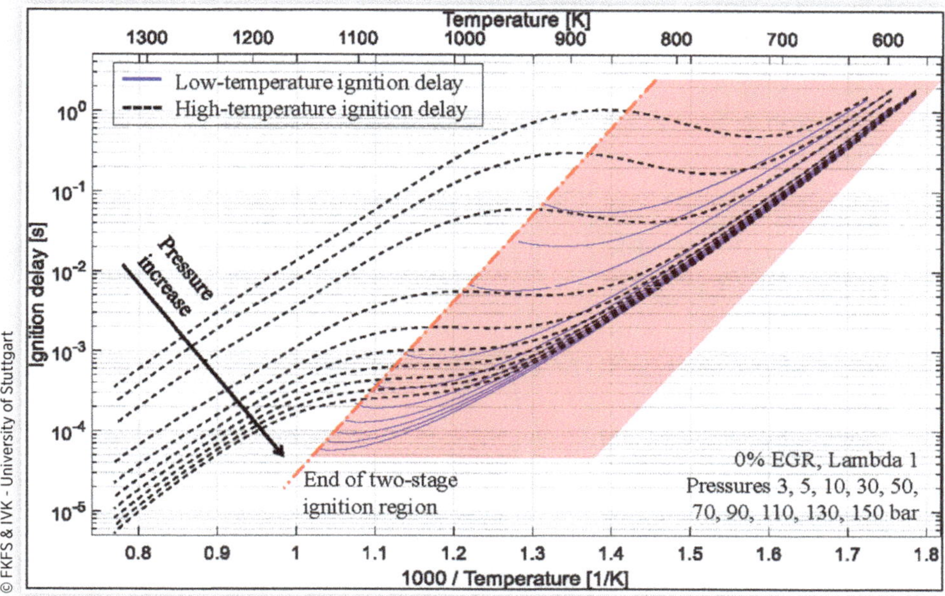

FIGURE 2.13 Correlation between measured (Y axis) and simulated knock onset (X axis) for knock models with single-stage and two-stage approach; red squares indicate that the model did not predict autoignition at all.

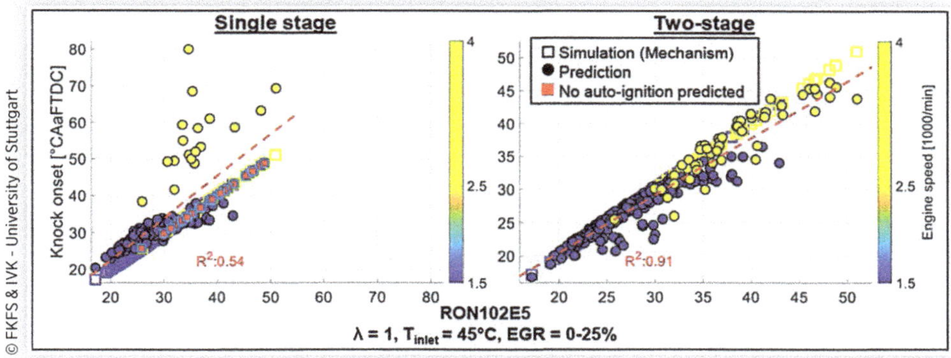

be modelled but also the heat release from the low-temperature ignition. This is often omitted in models as the first ignition stage occurs in the unburnt mixture during combustion and is thus masked by the heat release from combustion. Albeit, as shown in Figure 2.14, the temperature increase resulting from the low-temperature ignition can be quite substantial (up to 100 K depending on the boundary conditions—increasing engine speeds and high EGR rates cause the temperature rise resulting from the low-temperature ignition to decline), which obviously influences the following chemical reactions and thus the overall ignition delay of the mixture significantly.

Figure 2.15 gives a concluding overview over the achievable accuracy with a state-of-the-art knock model that takes two-stage ignition with heat release from the first stage into account. The prediction error is basically always below 2°CA, whereas single-stage knock models show unacceptable discrepancies.

FIGURE 2.14 Simulated temperature profiles of different working cycles with autoignition in two stages depending on boundary conditions [5].

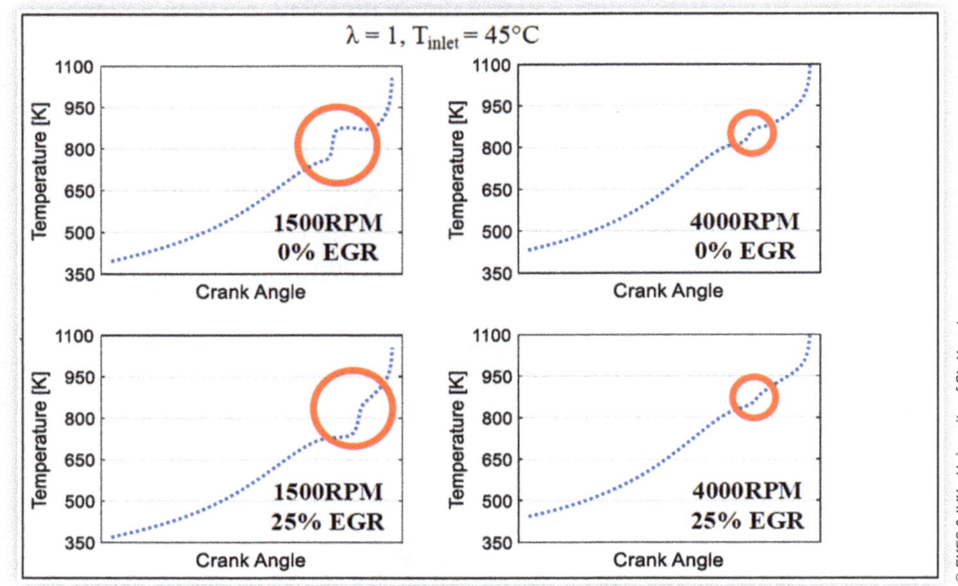

FIGURE 2.15 Knock boundary prediction with a single-stage knock model ("OLD," [12]) and a two-stage knock model ("NEW," [11]), from [11].

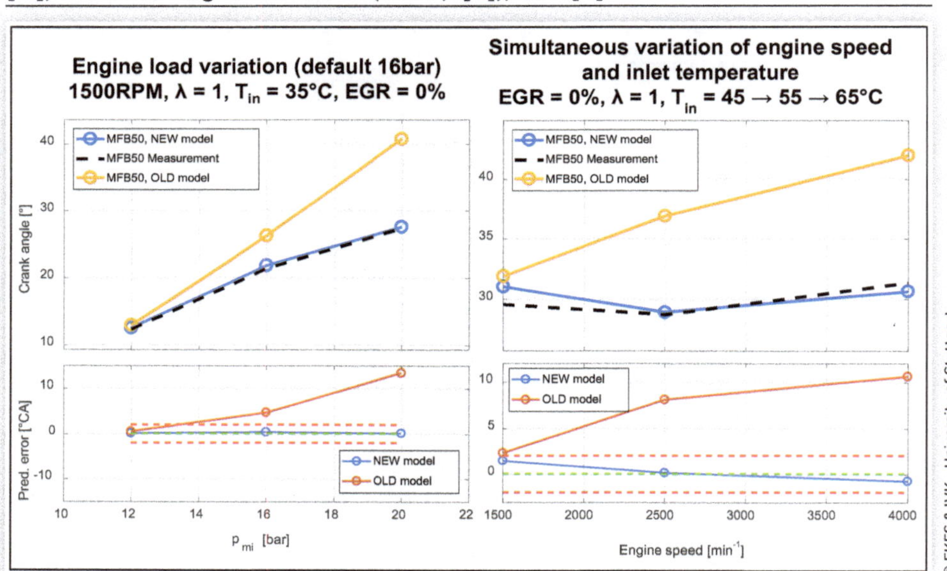

2.3.2.7 NOx MODEL

The ability to predict NOx emissions of gasoline engines has been considered of secondary importance for a long time, and such an assessment may still be valid for stoichiometric operation where the three-way catalyst will take care of the NOx emissions anyway. However, once lean operation is considered, engine-out emissions (and NOx emissions in particular) become an important quantity, answering for instance the question how lean the engine has to be operated to comply with legislation limits without complex aftertreatment.

Most available models are based on the Zeldovich mechanism and yield good results from a qualitative point of view. Matching results also quantitatively and probably requires some kind of modeling of inhomogeneities in the combustion chamber, or in the simplest case at least a scaling factor for model tuning. Even the latter is something many NOx models do not feature, although they should with regard to user-friendliness.

To sum up, all the sub-models available for SI engines combined allow for simulations with very high predictive power. The impact of changes in control parameters such as spark timing, air-to-fuel ratio, engine speed, load, or EGR rate on quantities of interest like BSFC or NO_x emissions should be predicted accurately and reliably without any necessary input by the user, i.e., there should not be any calibration parameters for the influence of EGR rate, engine speed, spark timing, etc. as the underlying physical and chemical effect mechanisms should all be embedded in the respective sub-models.

2.3.3 Practical Examples for CI Engines

2.3.3.1 BURN RATE/INJECTION MODEL

Regarding its importance, what has been said for gasoline engines applies as well for Diesel engines. It is thus important to conduct several sensitivity tests, including Diesel-specific input parameters like rail pressure or injection profiles for instance. Figure 2.16 shows how a good model can predict the burn rate correctly for a wide range of operating conditions with a single set of tuning parameters. Similarly, the model is able to predict the impact of injection rate modulation on the burn rate correctly (see [14] for instance).

FIGURE 2.16 Comparison of measured and calculated burn rates for a passenger car Diesel engine at different operating points (top, IMAP 4 bar; bottom, IMEP 10 bar), from [13].

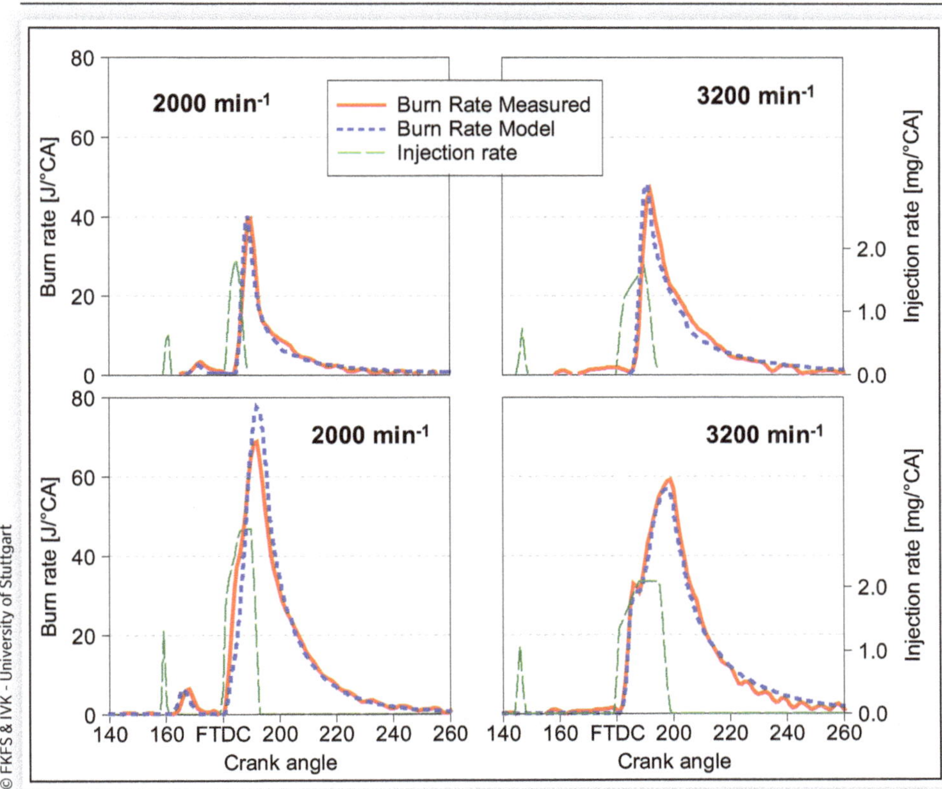

FIGURE 2.17 Input mask of a user-friendly injection model.

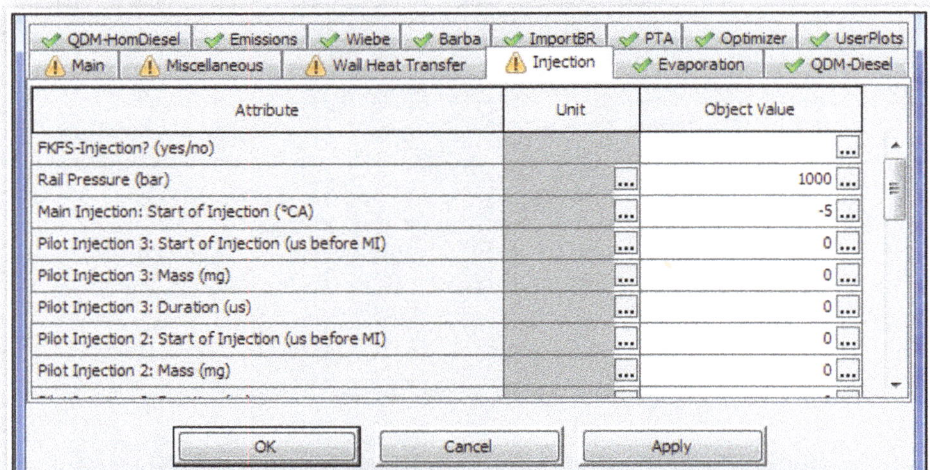

This is a prerequisite for virtual optimization of Diesel combustion by means of digital rate shaping for instance.

Diesel burn rate models are also a very good example to highlight the importance of user-friendliness. As combustion is closely linked to injection in Diesel engines, the handling of the injection models becomes crucial from a user's point of view, especially when considering multiple injections. There are considerable differences regarding such injection models. Some of them try to model the injection rate in a very detailed way—requiring thus quite a lot of input parameters and hydraulic measurements for reference—whereas other ones refrain to a simple approach where the user can stick to some basic quantities directly accessible from the engine's ECU. Given that experience shows that in the phenomenological model class the latter approach is perfectly sufficient in terms of accuracy, it ideally combines the right modeling "deepness" with highest user-friendliness. Figure 2.17 shows an example of the input mask of an injection model featuring these qualities.

2.3.3.2 WALL HEAT MODEL

Wall heat models are obviously not only required for Diesel engines, but the following example is intended to show why a good wall heat model is needed in 0D/1D simulation, or rather several of them, as the question which one is best may well change depending on investigation targets and the overall simulation setup.

Investigating a variation of the start of injection (SOI), important results such as the fuel consumption or the NOx emissions will depend on the used wall heat model. This is due to the fact that even with the same integral wall heat losses over a working cycles, the relative losses during compression on the one hand and combustion on the other hand will differ for different wall heat models, changing thus the benefit or penalty for earlier or later combustion timings.

As shown in Figure 2.18, the prediction from the Bargende model for the optimal SOI timing differs considerably from the Woschni model's prediction, with the former being very close to measurement data (not shown in the diagram). Although that might not be the case generally and in all conceivable cases, it can be reasonably argued that having more than just one wall heat model at one's disposal can be very important when tuning a model to measurement data to get reliable predictions. In particular, the Bargende

FIGURE 2.18 Prediction of optimal SOI timing depending on the used wall heat model.

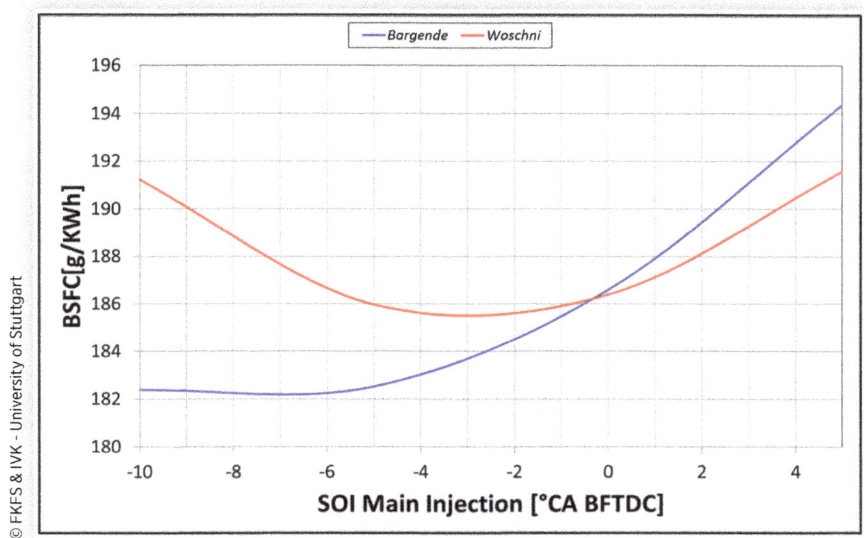

model has proved to make accurate predictions for a wide range of different combustion processes [15], also in cases where the widely used Woschni model fails to do so.

2.3.3.3 EMISSION MODELS

For obvious reasons, the prediction of NOx emissions (and, to a lesser degree, also of soot emissions) is an important part of Diesel engine simulation. A sensitivity test should thus be conducted before using an emissions model. Two aspects should be checked in particular: the model's ability to predict emissions during load steps and its ability to predict emissions in operating points close to stoichiometric conditions. Figure 2.19

FIGURE 2.19 Predicted NOx emissions for a variation of the air-to-fuel ratio; base model is described in [16], improved model in [17].

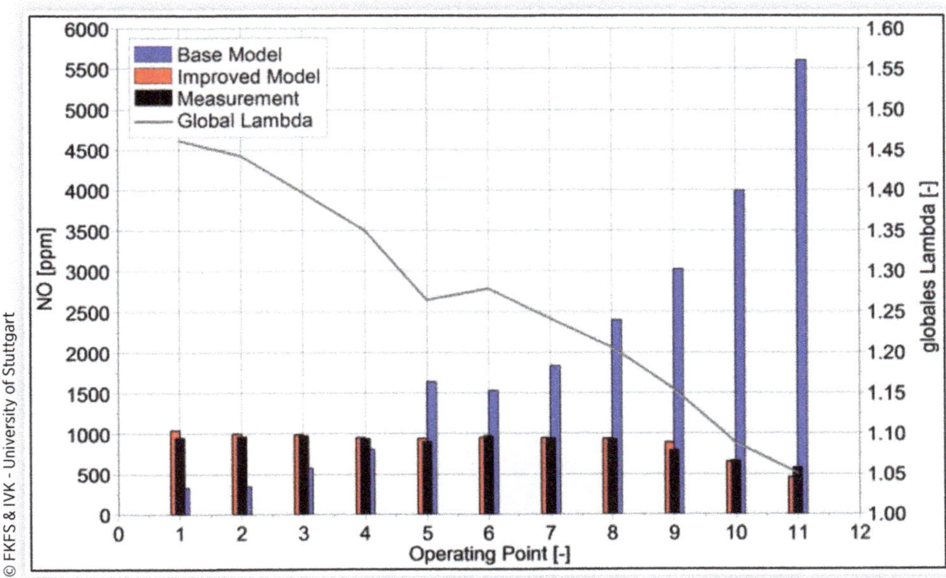

shows a comparison between two different NOx models, highlighting the huge difference in predictive power between the two models.

Last but not least, it should be stated that it is highly beneficial for the user if all of the discussed 0D combustion chamber sub-models are available in a single tool that can be easily coupled with 1D flow simulation, because in the end he will typically purchase a tool or package rather than individual sub-models.

2.4 Assessment of 3D Models (for Fast-Response 3D-CFD-Simulations)

The processes occurring in the working fluid of an internal combustion engine are numerous and very complex. Therefore, an accurate numerical modeling of the processes is required. For this task, two basic approaches have been developed. These can be mainly categorized as thermodynamic or fluid dynamic in nature, depending on whether the implemented equations are based only on energy conservation (0D) or on a full analysis of the fluid motion (multidimensional) [18, 19, 20].

An engine simulation tool independently from its approach is actually a collection of various engine process models (physical, chemical, and thermodynamic phenomena or just control/actuation models like injector model and SI model that in a successive step generate a phenomenon). Here the conservation equations in the simulation tool (see Figure 2.20) are the "webs" for all the information transfers among the engine process models that are needed for calculating the final results.

Although the conservation equations that set the balance of the thermodynamic engine processes are well known and evident like in any other thermodynamic system, the procedure toward a reliable and appropriate process modeling does not follow a unique way and still represents a most complicated and controversial task. Due to the complexity of engine processes, the insufficient understanding at fundamental level

FIGURE 2.20 Engine process modeling in the 3D-CFD-simulation.

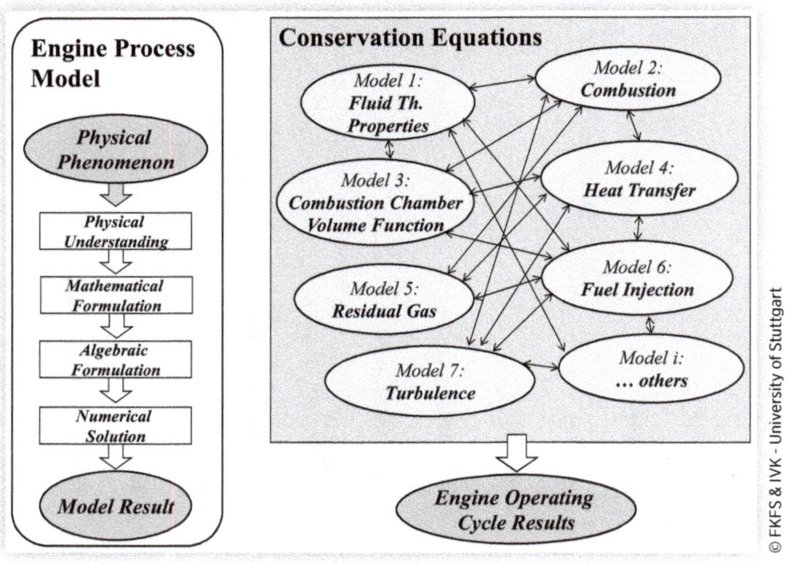

and very often still the limited computational resources, it is not possible to model engine processes that describe all important aspects starting from the basic governing equations alone. First of all, in order to govern the complexity, each modelled process must be limited to its relevant effects on engine behavior that have to be analyzed, then the formulations of the critical features of the processes have to be based on a keen combination of assumptions, approximations, and phenomenological and eventually empirical relations. This procedure permits both to bridge gaps in our phenomena understanding and to lessen the computational time by reducing the number of required equations [18, 19, 20].

For any simulation approach, depending on the context, the formulations of engine processes that have stood the test of time show different level of sophistication. Each of them is able to predict with varying degrees of completeness, versatility, and accuracy the predominant structure of the investigated process.

The formulation of engine processes is an active practice that continues to develop as soon as our understanding of the physics and chemistry of the phenomena expands and as soon as our ability to properly convert the process understanding first into a mathematical formulation, then into an algebraic formulation and finally into a numerical implementation, increases (see Figure 2.20). Any of these steps between the physical phenomenon and the resulting model is responsible for the general accuracy of the model. For example, a particular emphasis only on the numerical discretization would for sure not lead to a more accurate analysis. Similarly, any engine process model in the complex information exchange of the simulation tools is not more accurate than its weakest link. The weakest links are often not only represented by the mathematical formulation of the model or by the problem related to the numerical implementation but also by the accuracy of the input variables from other models. No model can provide reliable results until their inputs are reliable and, in case of a 3D-CFD-simulation, mesh-independent. In conclusion each critical phenomenon should always be described by engine models at comparable levels of sophistication in order to establish a balance of complexity and details among the engine process models.

The development and validation steps of a 3D-CFD model are a particular challenge. Due to its spatial and temporal resolution, a 3D-CFD model implemented into an existing 3D-CFD code and applied to an engine mesh is able to locally reproduce an engine process (local implementation), but its results, if at all, can be measured and validated only globally (see Figure 2.21). For this reason, due to the absence of reliable local

FIGURE 2.21 Development, validation, and application of engine process models for the 3D-CFD simulation.

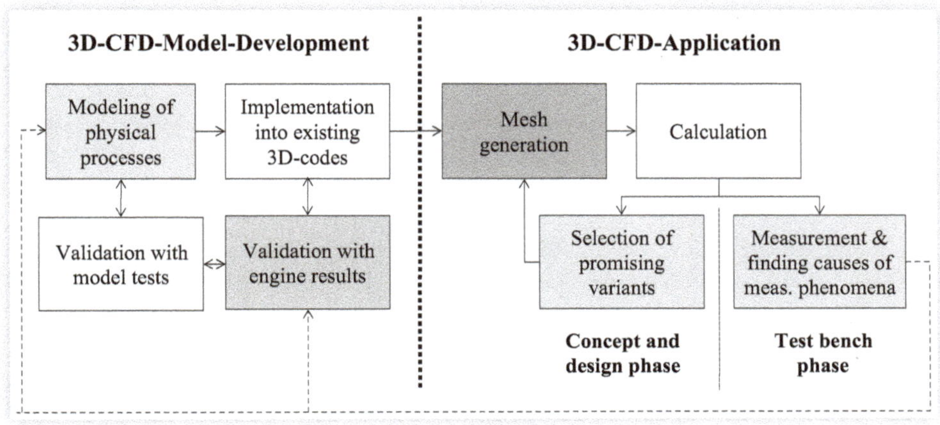

measurements of the investigated process on a real "unmodified" engine, the validation of the 3D-engine models still represents a critical and often "subjective" step.

A simpler validation way (validation with model tests) is the application of the 3D-CFD model to a mesh reproducing a laboratory device (e.g., a pressure chamber) that allows a more accurate comparison with experimental measurements. Unfortunately, the simulated and measured processes are, under laboratory conditions, far away, e.g., from the real behavior of the process in the combustion chamber of an internal combustion engine.

The desire to improve the engine process modeling in the 3D-CFD simulation is very high. First of all, this would allow to visualize and to investigate all relevant phenomena, also that ones that elude recording by measuring devices, and then, after a more comprehensive evaluation of the engine behavior, it would allow a more efficient and rapid selection of promising variants. Especially in the near future, the emphasis on process modeling will remarkably gain in importance; in particular, the necessity to explore new engine concepts toward a drastic reduction of CO_2 and pollutant emissions, respectively, will require robust simulation programs able to manage the rising complexity.

In the following parts of the chapter, a more detailed description of the engine simulation approaches is reported. The focus is on an innovative fast-response 3D-CFD tool (QuickSim) developed at both the FKFS (Research Institute for Automotive and Internal Combustion Engines-Stuttgart) and the IVK-University of Stuttgart. This introduced a new concept in the simulation of internal combustion engines that aims to increase the relevance and reliance of the 3D-CFD-simulation in the engine development process.

2.4.1 Model and Calculation Layout in an Innovative Fast-Response 3D-CFD Tool

The basic idea of the calculation layout of this innovative fast-response 3D-CFD tool is briefly reported here (see Figure 2.22). This layout is based on a modular structure of engine models, databases or neural networks, variable/parameter exchange devices, and control functions. The information/parameter can be linked and assigned to the following three main groups:

2.4.1.1 TEST BENCH AND LABORATORY ENVIRONMENT

Here, if available, measurement data (or eventually results from external simulation tools) is provided for automatic calibration, comparison, and validation.

2.4.1.2 ZERO-DIMENSIONAL ENVIRONMENT

The core of the calculation environment in this 3D-CFD approach is the evaluation tool that collects, averages, extrapolates, etc., the countless variables $a_i(\vec{x}_j,\phi)$ (temperature, pressure, velocity, species concentration, etc.) which are provided by the 3D-CFD simulation for each cell j in the mesh at any time step or crank angle ϕ.

Parts of the outputs of the evaluation tool are similar to the output of a test bench with a modern indicating system, others, e.g., are similar to the results provided by LIF technologies in a pressure chamber for the investigation of complex phenomena like those occurring during the fuel injection (spray penetration, droplet size, droplet velocity, etc.).

In a second step, the outputs of the evaluation tool are progressively at disposal for other kinds of applications like the real working-process analysis (WP) (0D or thermodynamic analysis of the engine operating cycle). In this case the real

FIGURE 2.22 Calculation layout in an innovative fast-response 3D-CFD tool.

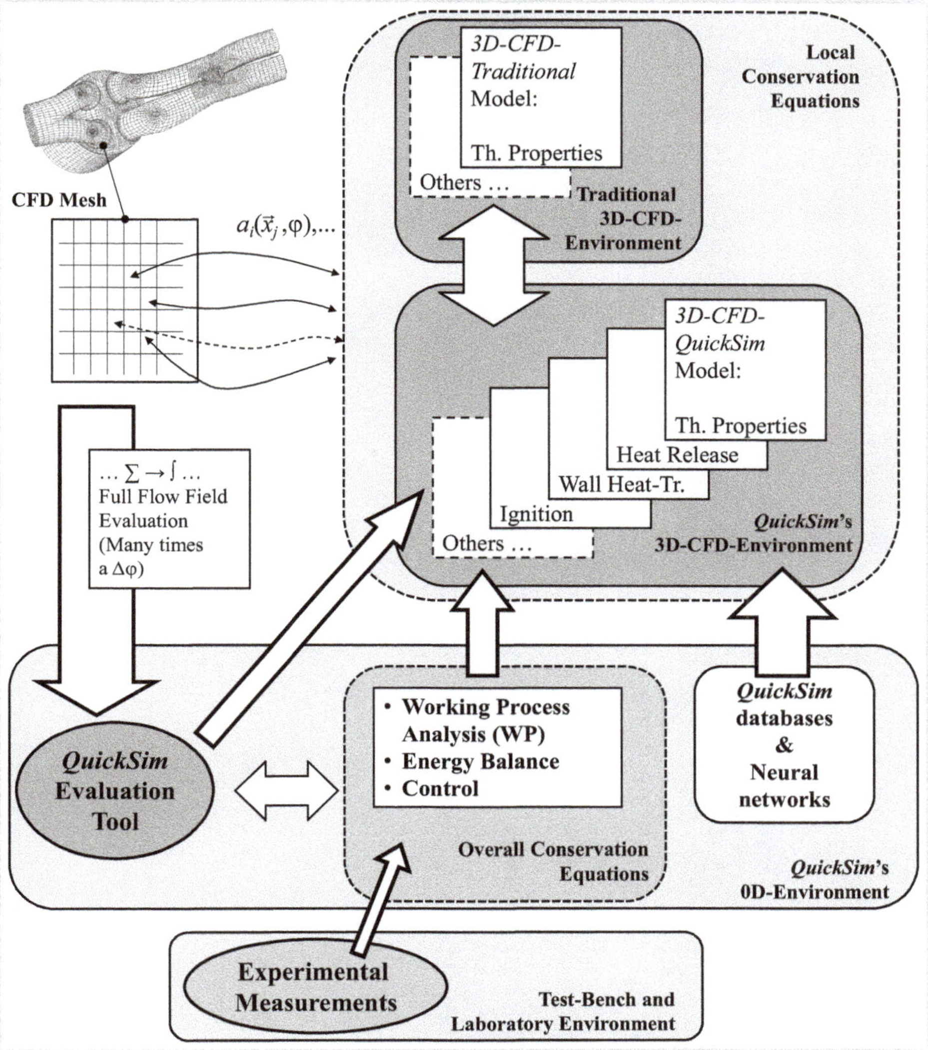

working-process analysis uses the 3D-CFD-simulation as a virtual engine test bench. Another implementation of the evaluation tool consists of establishing a real-time feedback of any variable as an enhanced information exchange process between the evaluation process and the 3D-CFD simulation. This approach allows the 3D-CFD simulation to have both at disposal, local, and global variables in each cell of the mesh. The implementation of both variable types in the 3D models can be properly combined (e.g., when a local variable due to convergence difficulties is not reliable anymore or when a formulation of a phenomenological or quasidimensional model is recommendable).

In a third and last step, all the outputs from the WP are progressively at disposal as a feedback for the 3D-CFD simulation for a comparison between the two approaches and eventually for control. In this way it is possible to establish an internal coupling between the 3D-CFD simulation and the WP at any time step, both based on the same simulation evolution.

The last element of this environment is represented by a collection of libraries (databases and neural networks) which help the local 3D-CFD engine models with fast-response inputs whose direct calculation would be too time expensive (e.g., the thermodynamic properties of the working fluid, ignition delay, etc.)

2.4.1.3 3D-CFD ENVIRONMENT

The background of this environment is the solution of the conservation equations solved by the 3D-CFD code. These conservation equations are the "webs" for all the information transfers among the engine process models that are needed for calculating the final results ensuring the correct mass, momentum, and energy balance at the local level.

A few phenomena or engine processes are calculated using traditional 3D-CFD models (e.g., turbulence, viscosity, spray collision, droplet vaporization, etc.); others have been replaced by new 3D dedicated models (*QuickSim* models). The formulation of these models can be conveniently chosen regarding the peculiarities of the investigated process (physical understanding, mathematical and algebraic description), the numerical solution (in particular the mesh dependence), and the required computational time. The information exchange within the calculation layout (see Figure 2.22) is always able to provide the necessary inputs also for "unconventional" formulations of local 3D-CFD models [18, 19, 20].

2.4.2 New Developed 3D-CFD Models

Here a selection of newly developed 3D-CFD models implemented into this innovative fast-response 3D-CFD tool in the last 20 years is reported:

- Model for the thermophysical properties of the working fluid for any fuel CnHmOrNq
- Wall heat transfer model
- SI model (flame propagation model in the near region of the spark plug)
- Heat release models for SI, HCCI, and Diesel engines
- Self-ignition model
- Model for gas and liquid fuel injection
- Model for the dense spray region
- Model for choking flows (Mach ≥1)
- Model for flow streams through the valve seats at low valve lift
- Model for an improved setting of initial conditions
- Model for an improved setting of boundary conditions

The following case briefly illustrates the concept of one of these newly developed 3D-CFD engine models:

2.4.2.1 3D-CFD ENGINE HEAT TRANSFER

The wall heat transfer represents the thermodynamic boundary condition of the combustion chamber, that, in particular during the working period, is the main source of inaccuracy in the 3D-CFD analysis of the engine operating cycle.

Very often in the 3D-CFD simulation, attention is mainly paid to the modeling of phenomena like combustion (chemical kinetic reactions) and fuel spray formation, instead of ensuring an accurate balance in the solution of all relevant engine processes. This is a fatal error that easily compromises the quality of the simulation results because a simulation running with wrong thermodynamic boundary conditions will never deliver reliable results.

The convective heat-transfer process is governed by the temperature and velocity profile within the near-wall region (boundary layer), which is, in internal combustion engines, locally and temporally drastically under the influence of the mean flow motion, the turbulence, the flame propagation, etc.—*limited physical understanding.* In order to solve the Navier-Stokes conservation equations within the boundary layer (laminar sub-layer and turbulent boundary layer), complex mathematical models for anisotropic turbulence, turbulence separation, impinging jet, flame quenching phenomena, etc. are required—*limited mathematical formulation.* Nowadays such mathematical models have been proposed only for simple wall geometries and flow patterns, respectively [18]. Therefore, not a single validated 3D-CFD model, which permits an accurate investigation of the boundary layer of a combustion chamber, has been successfully implemented in simulation programs.

On the other side, since the end of the 1960s, phenomenological heat-transfer models have been developed and implemented in the real working-process analysis (WP) after comprehensive validation with extensive measurements at the test benches. These models (like Woschni, Hohenberg, and Bargende) are based on the Re-Nu correlation (dimensional analysis), in which the assumption of the Nusselt, Reynolds, and Prandtl number relationships follow that found for turbulent flow in pipes with additional adaptations to the engine case. Starting from this assessment, the basic idea in the development of a new 3D-CFD heat-transfer model is to find a way how to use the above-introduced phenomenological approach (0D global approach) locally in a 3D-CFD mesh. This model is implemented locally, like the wall function, in only a single wall-cell layer, so that computing overheads are not caused. The classic 3D-CFD wall function is still used for assuring the no-slip velocity condition and the temperature condition at the wall.

The modeling of the heat-transfer coefficients chosen for the 3D-CFD-simulation, similarly to the real working-process analysis, is based on the following correlations [18]:

- Bargende's correlation during the compression and expansion stroke (working cycle)
- Hohenberg's correlation during the intake and exhaust stroke (gas exchange period)

The aim of this work is to develop a phenomenological 3D heat transfer model $\alpha_1(\vec{x}_j,\phi)$, which, first of all, ensures a high accuracy in predicting the overall heat transfer rate $(dQ_W/d\phi)_{CFD}$ at every time step $d\phi$. This condition is a prerequisite for a thorough prediction of the engine energy balance. In this manner, the overall heat transfer rate calculated from the real working-process analysis (WP) $(dQ_W/d\phi)_{WP}$ is taken as the representative (leading) value.

Target:

$$\left(\frac{dQ_W}{d\phi}\right)_{CFD} = \left(\frac{dQ_W}{d\phi}\right)_{WP}$$

with

$$\left(\frac{dQ_W}{d\phi}\right)_{CFD} = \sum_j \alpha_1(\vec{x}_j,\phi)\cdot\left[T_C(\vec{x}_j,\phi) - T_W(\vec{x}_j)\right]\cdot A_j\cdot\frac{dt}{d\phi}$$

FIGURE 2.23 The "internal coupling" between the 3D-CFD code and the real working-process analysis (WP).

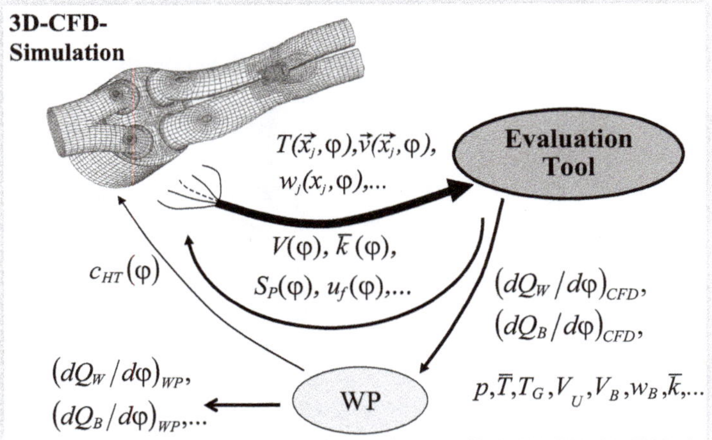

where $T_C(\vec{x}_j,\phi)$ is the temperature at the center node of a cell j adjacent to the wall with the surface A_j and a constant temperature $T_W(\vec{x}_j)$ over the time.

The development of $a_i(\vec{x}_j,\phi)$ have been performed by using both the 3D-CFD evaluation tool (collection and evaluation of the local variables in the mesh and then the calculation of the global/averaged variables) and the real working-process analysis (WP) implemented into the fast-response 3D-CFD tool via user subroutines. This becomes an "internal 3D-CFD-WP coupling" (see Figure 2.23) as the decisive step of this engine model development. This approach ensures not only the same "thermodynamic conditions" between 3D-CFD and WP but also provides, at every time step, the necessary average values of the variables used in the real working-process analysis back to the 3D simulation.

$$c_{HT}(\phi)=\left(\frac{dQ_W}{d\phi}\right)_{CFD}\bigg/\left(\frac{dQ_W}{d\phi}\right)_{WP}$$

Choosing a variant of $a_i(\vec{x}_j,\phi)$, which is mostly a combination of global (averaged overall variables) and local variables, respectively, the heat-transfer rate of the CFD-simulation $(dQ_W/d\phi)_{CFD}$ has been calculated and compared to that of the working-process analysis $(dQ_W/d\phi)_{WP}$. If they differ, at every time step, an overall correction variable $c_{HT}(\phi)$ is determined and set in the 3D-CFD simulation in order to correct the wall heat flux. In this manner a reliable and representative overall heat-transfer rate can been imposed as a boundary condition.

Concluding, the 3D phenomenological heat transfer $a_i(\vec{x}_j,\phi)$ during the working period is then given by:

$$\alpha_1(\vec{x},\phi)=253.5\cdot V(\phi)^{-0.073}\cdot p(\vec{x}_j,\phi)^{0.78}\cdot \bar{T}(\vec{x}_j,\phi)^{-0.477}\cdot w(\vec{x}_j,\phi)^{0.78}$$

where

$V(\phi)$ is the global cylinder volume.

$p(\vec{x}_j,\phi)$ is the local pressure term.

$T(\vec{x}_j,\phi)$ is the local temperature term.

$w(\vec{x}_j,\phi)$ is a local velocity term.

2.5 Basics of Engine Design

As explained in the introduction, one of the most important tasks that can be performed with 0D/1D simulation are concept studies for engine design in early development stages. In particular the engine behavior at full load and the resulting questions of how to choose design parameters for the engine are a crucial aspect to be investigated and are thus discussed in some detail in this section. It should be kept in mind, however, that there are many more important aspects that need to be considered for basic engine design that cannot be discussed in the scope of this chapter, but that can be investigated in a satisfying manner with 0D/1D simulations in general. For instance, the thermal engine behavior, which is important for issues such as catalyst light-off can be predicted very reliably within this model class, provided that the engine model has all the necessary input data regarding thermal masses and conductivities.

2.5.1 Full Load Design for SI Engines

In general the maximum achievable power output of a SI engine is limited by the following events:

- Exceeding of the maximum permissible exhaust temperature (in case of turbocharged SI engines)
- Exceeding of the maximum permissible peak cylinder pressure (or maximum rate of pressure rise)
- Occurrence of knock

All of these events will lead to mechanical failure of the engine or one of its components and thus have to be avoided in any case. Whereas concrete numerical values can be tagged to the respective limits depending on the structural design of the engine block and the engineering material of the turbine, they can be considered to be "soft" in as far as the power output can still be increased after they have been hit first by very simple countermeasures: shifting the ignition spark to later timings will reduce both peak pressure and knock tendency but increase exhaust temperature—which will be typically met by fuel enrichment. Obviously, however, apart from the negative effect on fuel consumption, these measures cannot be pursued ad libitum, but only as long as the "hard" limits are hit: there is both a maximum permissible degree of fuel enrichment (in actual fact, the limit might very well be set at stoichiometric conditions already due to current emission legislation) and a latest permissible ignition timing. Beyond these points, extremely slow combustion or misfires will occur, making a further power increase impossible. Furthermore, when considering turbocharged SI engines, the limitations of the boosting system have to be taken into account as well; the compressor must be operated between its surge and choke line while still leaving some altitude and heat reserve at rated power, resulting in a maximum achievable boost pressure for maximum permissible exhaust back pressure.

It is important to keep all these interrelationships into mind when designing a new engine concept. When doing so, the following design parameters, among others, can be chosen relatively freely:

- Engine displacement/boosting system
  Depending on the engine's target power output, a decision has to be taken whether it is to be reached with a relatively high displacement and a relatively low mean effective

pressure or vice versa. Throughout the last decades, a tendency to engines with higher specific power output can be observed, known as downsizing. This is mainly due to the load point shift to regions of higher thermal efficiency and the possibility to make use of the exhaust energy in a turbocharger. However, there is a certain limit regarding the question which degree of downsizing is optimal, as higher mean effective pressures lead to higher knock tendencies and higher demands on the boosting systems. This might become particularly relevant if lean or ultra-lean combustion processes are considered, where higher engine displacements might make it possible to stick to a single-stage boosting system whereas downsized engines might require more sophisticated ones.

- Engine speed at rated power
  Obviously a certain target power can be reached by different engine speed/torque combinations. Choosing a higher engine speed at rated power will lower the necessary mean effective pressure, bringing potential benefits regarding the choice of compression ratio for instance. A lower engine speed at rated power, on the other hand, has the advantage of reduced friction losses and allows for an easier turbocharger matching as the engine speed spread in general will be lower.
- Compression ratio
  Given a certain combustion process, the theoretically optimal compression ratio changes over engine speed and load, with the highest values reached at low loads and high speeds. When using a fixed compression ratio—still very much state of the art—higher values thus mean better fuel consumption at lower loads and possibly higher fuel consumption at full load (due to knock). Usually the choice of the compression ratio is dictated by knock at full knock, which means that there is a direct impact of the necessary mean effective pressure on the maximum permissible compression ratio. However, it can be easily assessed in simulation how big the possible benefit of a VCR can be, allowing a solid base for decisions whether or not to incorporate VCR systems.
- Bore/stroke ratio
  The bore/stroke ratio is another geometrical quantity that can be optimized in engine concept simulations. Longer strokes induce higher turbulence levels, which is generally beneficial for higher efficiencies and lower knock tendencies, but cause higher friction losses and higher mechanical stress due to the increase in mean piston speed at constant engine speed. For a meaningful investigation, it is indispensable to have turbulence models at one's disposal that can predict the impact of longer strokes on turbulence.
- Intake channel design
  Although it is very difficult if not impossible to derive concrete channel design recommendations from 0D/1D simulations by the very nature of this kind of simulation (3D-CFD would be the measure of choice for such investigations), certain known correlations between tumble numbers and coefficient of discharge can be used to account for the trade-off between low flow losses on the one hand and high turbulence generation on the other hand. This can then help to decide which turbulence level should be aimed at for a certain engine concept.
- Valve train
  Valve lift profiles and the respective opening and closing timings can be optimized for rated power to ensure the highest possible amount of air reaches the cylinder, which is particularly relevant in naturally aspirated engines. In turbocharged engines it might sometimes make sense to use higher boost pressures in conjunction with a

Miller cycle to reduce the effective compression ratio while making use of the intercooler to mitigate knock, making variable valve trains interesting not only in part load but in fact also for full load design.

- Combustion process
  The question whether a conventional, stoichiometric combustion process is to be used or something more innovative—e.g., homogeneous lean combustion—has wide implications on all the other parameters discussed so far for apparent reasons. Lean or ultra-lean combustion will fundamentally change the demands on the boosting system for instance. A further example would be the use of external EGR at full load, which can be used to mitigate knock while reducing exhaust temperatures. This in turn makes fuel enrichment unnecessary and allows for a high fuel benefit at full load.

Already from these very basic considerations, it is obvious that given the complexity and multitude of causal interrelationships between design parameters and engine behavior, it is indispensable to use 0D/1D simulation models as it would be incredibly time and cost expensive to do such investigation on the engine test bench. Of course, however, the simulation model used for concept decision has to include all the necessary sub-models in high quality as discussed in the previous section. Otherwise, there will be inevitably wrong concept decisions and thus flawed engine designs.

2.5.2 Full Load Design for CI Engines

Similarly to SI engines, three general limitations for CI engines can be named:

- Exhaust temperature
- Peak cylinder pressure
- Air-to-fuel ratio (lean mixtures due to soot formation)

For practical implications, there is often also an additional torque limit to smooth out the torque curve which is only of secondary importance mostly for design considerations. Again the limit value for exhaust temperature results from the engineering material used for the turbine, and the maximum peak cylinder pressure depends on the structural design of the engine block. Both values can be usually seen as fixed for a certain design task. Figure 2.24 shows an example of a typical torque curve of a turbocharged CI engine with the respective limitations.

Similarly to SI engines, engine control parameters can be used to make sure the engine is operated within these limits. In case of CI engines, the timing and the amount of fuel of the main injection are the most important parameters. Later timings are used to reduce the peak pressure, whereas lower fuel masses reduce the exhaust temperature. This determines the maximum achievable torque at a certain engine speed for given limit values and thus also the achievable power output. Generally, the highest torque is reached when both limits are hit at the same time; see Figure 2.25.

In modern turbocharged CI engines, turbines with VTG are used generally. The rack position then represents an additional degree of freedom. For each possible rack position, a different combination of SOI and injected fuel mass can be found where both peak pressure and exhaust temperature limits are reached—and the resulting torque will differ as well. There is no general rule where this optimal rack position is situated (it will also change depending on the boundary conditions), which means that the rack position has to be varied

FIGURE 2.24 Typical torque curve of a turbocharged CI engines and relevant limitations.

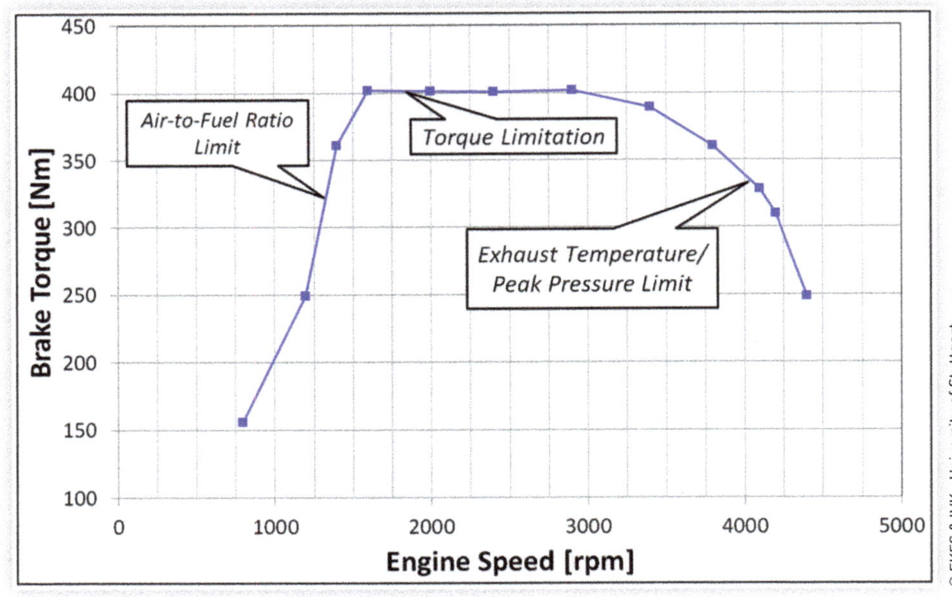

FIGURE 2.25 Torque map (isolines in [Nm]) at full load for a CI engine with a line of constant peak pressure (blue) and a line of constant exhaust temperature (magenta), fixed VTG position.

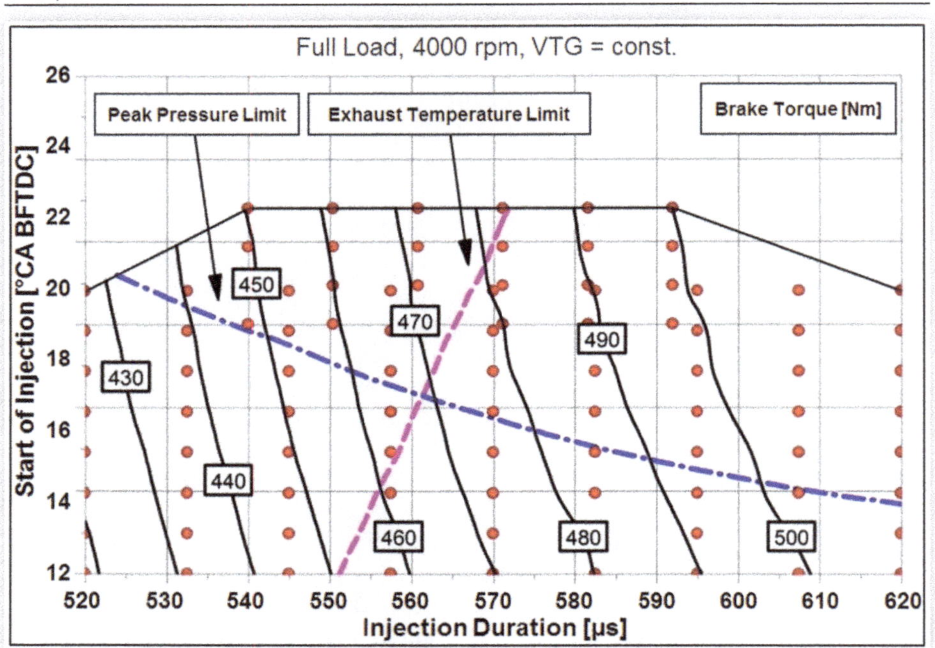

FIGURE 2.26 Influence of rack position on start of injection, injected fuel mass, torque and boost pressure; dashed lines indicate peak pressure (top left) and exhaust temperature (top right).

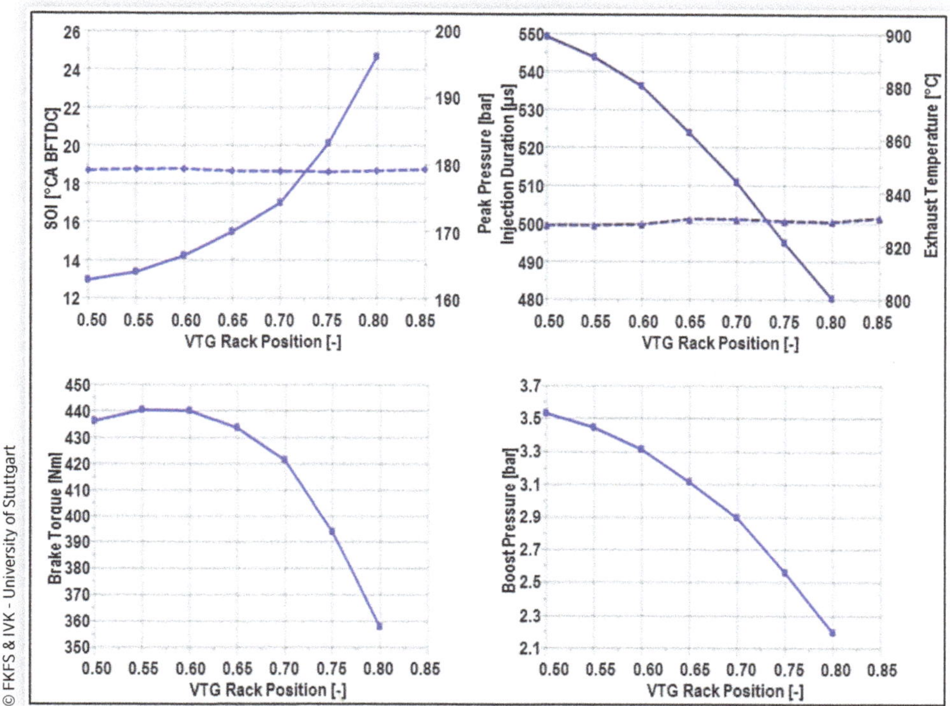

in a set of simulations to find the optimal set of control parameters; see Figure 2.26. It is interesting to note that in particular the highest torque and the highest boost pressure are not reached at the same rack position.

When it comes to design parameters influencing the power output, it can be clearly stated that the boosting system has the single most important impact. Again, 0D/1D simulation is a very valuable tool here to compare different concepts. Further aspects include the rail pressure—a faster combustion will help to reduce exhaust temperatures—and the compression ratio. Here, the optimal choice depends largely on the concrete values regarding the peak pressure limit and the exhaust temperature limit. For instance, relatively low peak pressure values will require relatively low compression ratios, as otherwise very late MFB50 timings with the associated disadvantages regarding thermal efficiency might occur. This might even go so far that additional boost pressure lowers the power output, as the additional fuel that can be injected then cannot compensate the efficiency loss due to the then even later MFB50 timings.

2.6 Application Examples

This section is intended to present a selection from a wide range of application examples in order to give an idea of the many and varied possible applications of virtual engine development. Single aspects such as acceleration simulation for instance as well as full virtual engine development from scratch to prototype will be discussed.

2.6.1 Example 1: Vehicle Acceleration Simulation and Cross-Comparison of Different Engine Concepts

As already mentioned, the investigation of different engine concepts is one of the most important tasks of 0D/1D simulation. Having toolboxes at one's disposal with modular virtual engines containing all the necessary sub-systems is thus often a very valuable starting point for all kinds of investigation (something that can be offered by experienced engineering service providers). In this example, two different boosting systems—a single-turbo and a twin-turbo concept, both coupled with a 3-cylinder, 1.2 L engine with 130 kW rated power—are compared by means of an acceleration test, in which the vehicle is accelerated at full load in a fixed gear from a certain starting speed.

Figure 2.27 shows the results when accelerating from 30 to 50 km/h for both concepts. It is clearly noticeable that the engine brake torque of the single-turbo concept is lower during almost the entire acceleration process. In order to ensure the same vehicle acceleration nonetheless, a higher transmission ratio (for both gear ratio and final drive ratio) had been chosen. As can be seen in Figure 2.28, this leads to engine operation at higher engine speeds and thus to a similar power output. Figure 2.28 also shows the reason for the higher torque of the twin turbo: the boost pressure is much higher over the entire time range. Furthermore, the trapping ratio depicted in the same diagram allows an assessment of how good scavenging works. For the acceleration process under discussion, only slight scavenging effects can be observed for the twin turbo due to the fast increase in engine speed during acceleration.

In Figures 2.29 and 2.30, the same diagrams are depicted for an acceleration process from 80 to 120 km/h in sixth gear. For both concepts, the gear ratios are identical (unlike it was the case for fourth gear in the previous example). For a fair comparison, the final drive ratio for the respective concepts was chosen in such a way that the same top speed is reached. Under these boundary conditions, the acceleration of the single-turbo-powered vehicle is considerably slower.

FIGURE 2.27 Engine torque and vehicle speed for two different turbocharging concepts when accelerating from 30 to 50 km/h in fourth gear [21].

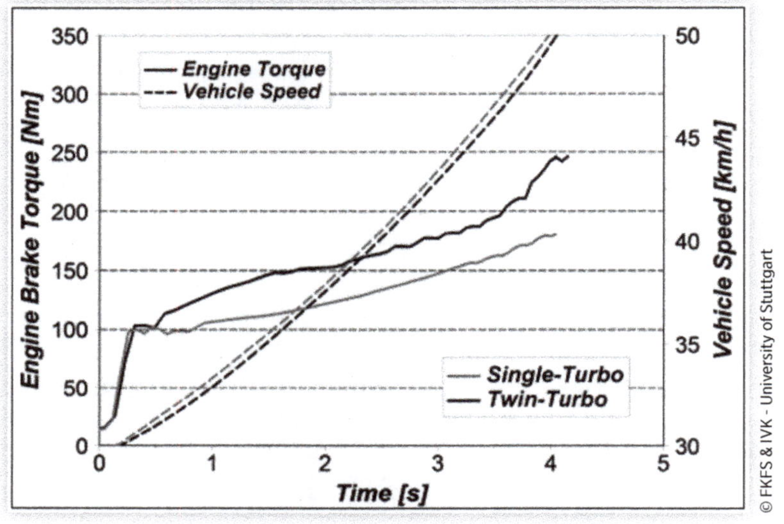

FIGURE 2.28 Boost pressure, engine speed, and trapping ratio for two different turbocharging concepts when accelerating from 30 to 50 km/h in fourth gear [21].

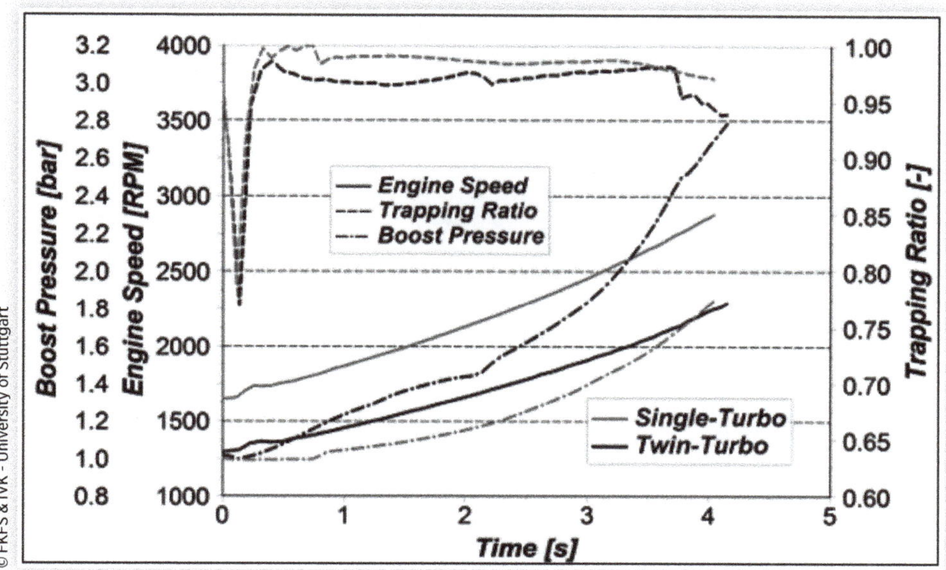

FIGURE 2.29 Engine torque and vehicle speed for two different turbocharging concepts when accelerating from 80 to 120 km/h in sixth gear [21].

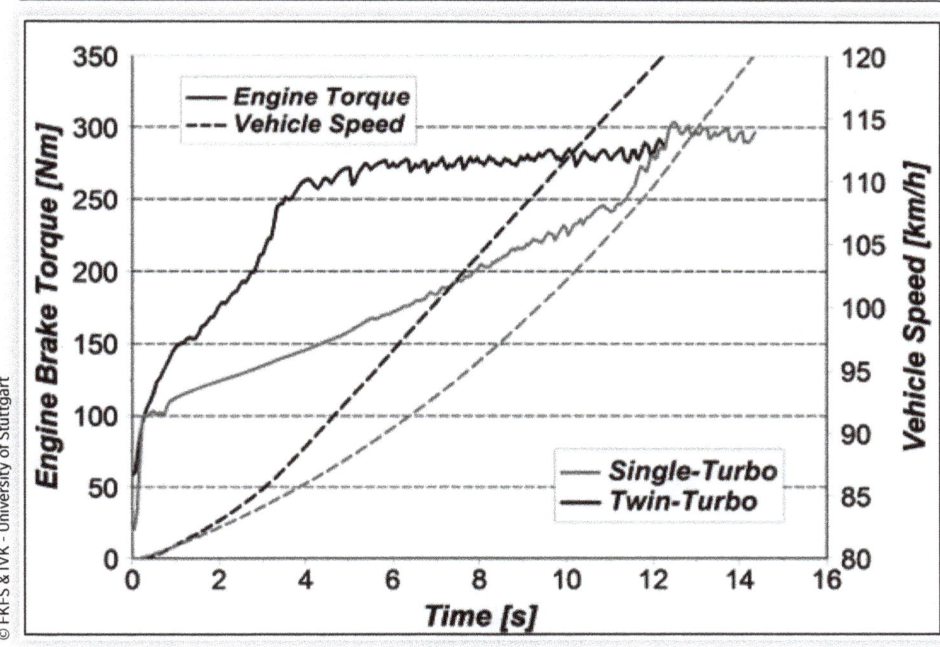

FIGURE 2.30 Boost pressure, engine speed, and trapping ratio for two different turbocharging concepts when accelerating from 80 to 120 km/h in sixth gear [21].

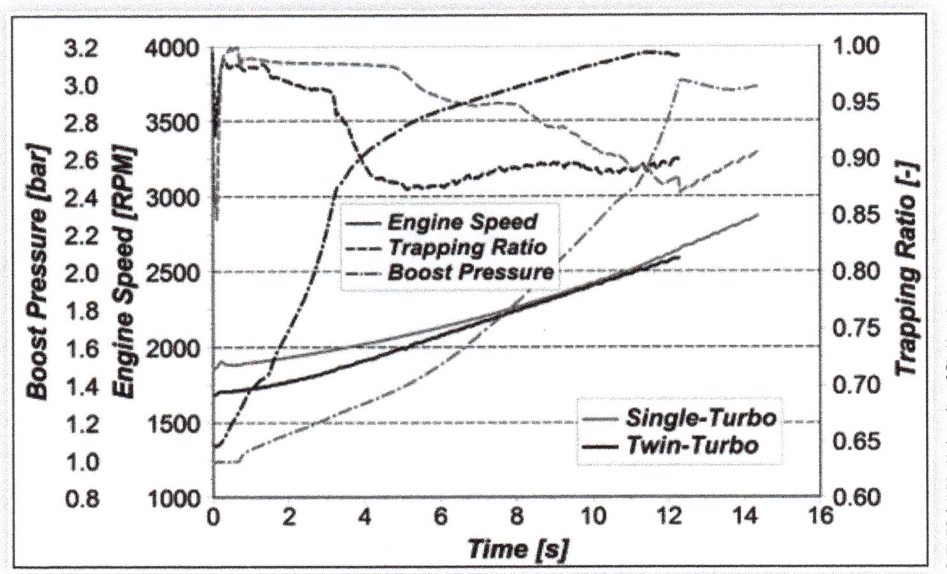

As apparent in Figure 2.30, both concepts make abundant use of scavenging, with the twin-turbo concept doing so very early on and thus ensuring a very fast torque buildup. For such calculations, a certain adjustment speed of the variable valve train has to be assumed. Such adjustment speeds can be derived for instance from dynamic oil circuit simulations. If a controlled oil pump is used there, various simulations under different boundary conditions (e.g., oil pressure at the start of calculation) can be performed at part load, and the resulting adjustment speed can be calculated. This can then be taken into account for vehicle acceleration tests similar to the one under discussion. It might then occur, for instance, that the twin-turbo concept reacts much more sensitive to a reduction of the maximum adjustment speed, whereas the single turbo's response is much more stable. This allows assessing the combination of different technologies within the overall powertrain system, which is particularly important in the context of increased system complexity. As an aside, it should be noted that again for this kind of simulations, very reliable sub-models are needed, e.g., for knock (visible in the engine torque curves in Figure 2.29 starting from 250 Nm onward roughly) and cycle-to-cycle variations. Knock (at the associated later MFB50 timings) will lead to a different position of the turbocharger's operating point, and likewise cyclic variations will have an impact on turbocharger efficiency in the low-end torque region.

2.6.2 Example 2: Tuning of 1D Flow Model

In the previous sections, the focus has mainly been on the 0D or quasidimensional models for in-cylinder processes, but the 1D flow simulation is of course also an important part of a 0D/1D simulation. However, the modeling of flow through tubes is usually very precise and no user input is required. Typically, it is sufficient to model pipe lengths and diameters according to geometry data, with additional fine tuning to replicate pressure losses and temperatures along the flow path correctly. The overall

FIGURE 2.31 Measured and simulated pressure traces in the intake and exhaust path at full load, 1500 rpm in a turbocharged 4-cylinder SI engine, from [22].

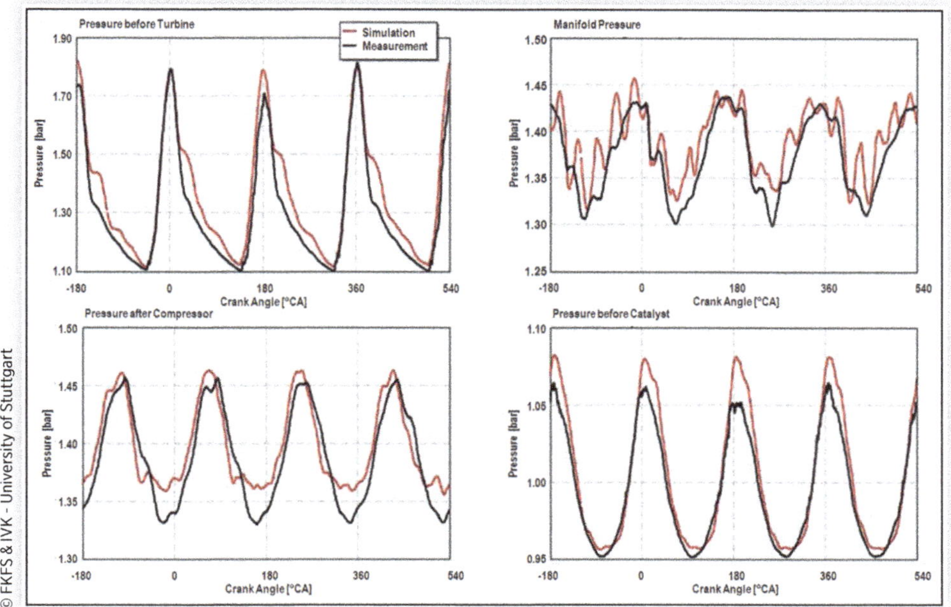

result then replicates the measured pressure and pressure pulsations in the intake and exhaust path quite well. Figure 2.31 shows an example of the accuracy that can be typically reached.

2.6.3 Example 3: Virtual Development of a High-Performance CNG Engine, Full Engine Simulations with a Fast-Response 3D-CFD Tool

The described case here is one of many investigated over the years with the fast-response 3D-CFD tool [18, 19, 20, 23, 24, 25, 26, 27, 28, 29]. Natural gas as fuel represents a very innovative and promising solution toward decreasing of CO_2 emissions. Thanks to a very high knock resistance, natural gas can be successfully used also in high-performance engines (e.g., race engines). On the other hand, in comparison to a gasoline engine, the homogenization of the air-fuel mixture is more critical because even high gas velocities at the injector nozzle cause a very low fuel penetration (low mass density of gaseous fuels). Therefore, mixture homogenization or stratification depends much more on charge motion as usual for liquid fuels. For this reason, the design of the intake system is a crucial step for the optimization of a compressed natural gas (CNG) engine that must be accurately investigated; otherwise high cylinder-to-cylinder and cycle-to-cycle variations compromise the engine output.

The case introduced here refers to a monovalent CNG endurance race engine (24 h of Nuerburgring), which has been derived from a turbocharged 4-cylinder 2.0 L gasoline engine with direct injection. The main modifications are the following: the high knock resistance of natural gas permits to run with a higher compression ratio, i.e., the piston

FIGURE 2.32 3D-CFD domain of the full engine.

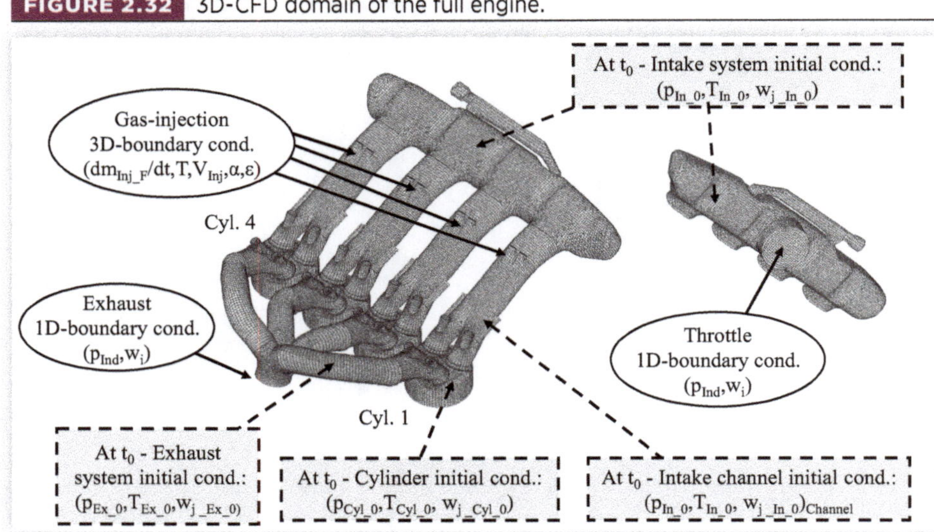

shape has been modified with a top part that raises the compression ratio from $\varepsilon = 10$ of the gasoline version to $\varepsilon = 12$. Some mechanical parts have been reinforced or modified in order to better stand the higher thermal and structural stresses (in particular the crankshaft, the connecting rod, and the valve seats). The most remarkable modification concerns the injection system. Here the CNG engine has been converted from gasoline direct injection to gas manifold injection (the hole left by the injector in the combustion chamber has been closed). Due to the high demand of fuel, even two gas injectors for each cylinder (see Figure 2.32) are required for ensuring the fuel injection within a reasonable period of time.

Investigations at the test bench of the first CNG-engine version (development step 0) equipped with a full indicating system have shown extremely high cycle-to-cycle variations of the cylinders 2 and 3. Since the maximal allowed pressure in cylinder, due to structural design of the cylinder head, is limited to 140 bar, it becomes evident that the cylinders 2 and 3 are not able to fully exploit their thermodynamic potential. The measurements, of course, are able to suggest mixture formation problems as the causes of the behavior of the central cylinders, but an accurate analysis of the engine processes toward improvements cannot be performed. Therefore, like in this case, a fast-response 3D-CFD simulation is able to find within a short time explanations and answers to this problem, so that successive engine development steps can meet improvements. Here briefly a selection of these steps are reported.

2.6.3.1 RESULTS (ENGINE DEVELOPMENT STEP 0)

The goal of a homogeneous fuel-air mixture in CNG engines is a very complex task. The fuel injectors are able of a very low radial penetration, and even an injection velocity of approx. 500 m/s generates a poor axial penetration. Usually after an axial penetration of ca. 25 mm, the fuel does not have a relative velocity to the background fluid anymore. From this distance the fuel is essentially transported by the charge motion. Following some results of the full-engine simulation at 5500 rpm WOT are shown. Due to confidentiality reasons, often the variable profiles are normalized to the maximum value of cylinder 1. Starting from the analysis of the fluid motion, e.g., the mass flows through the intake and exhaust valves (EVs) (see Figure 2.33),

FIGURE 2.33 (Step 0) Mass flow through the intake valves (IVs) and exhaust valves (EVs)-5500 rpm WOT—fifth cycle.

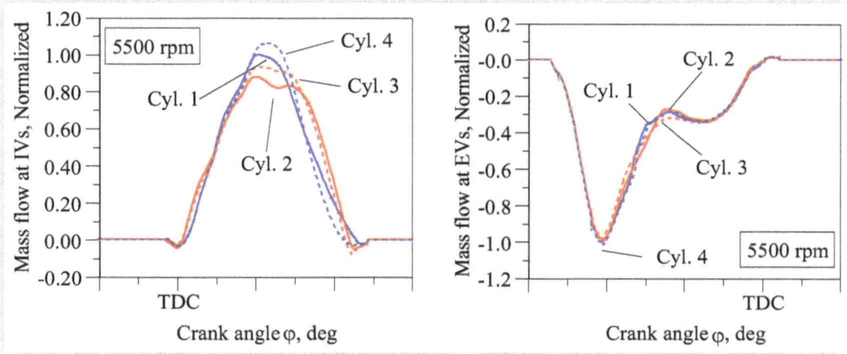

it becomes evident how the airbox design influences the intake phase of each cylinder. The external cylinders 1 and 4 have a relatively sharp peak while the central cylinders 2 and 3 show a flatter maximum. This behavior is mainly due to the small volume of the airbox and the central position of the throttle manifold (see Figure 2.32), that having a relatively constant mass flow reduces the pressure waves in the facing cylinders, i.e., the central ones.

The flow through the throttle manifold causes also a more asymmetric mixture formation in the intake runners of the central cylinders due to a major deflection of the fuel jets (see Figures 2.34 and 2.35). Here it is interesting to notice (see Figure 2.35) that the total mass flows through the intake valves (IVs) (e.g., of the cylinder 3) is minimally higher for IV1, but the differences are remarkably amplified if only the fuel mass flows are evaluated. Therefore, due to the low density of the gas and the wide distance of the gas injectors from the valves, a small asymmetry in the intake process generates a considerable deflection of the fuel jets that increases the fuel concentration in the intake channel of valve 1 in case of cylinder 3 (see Figure 2.34) and the opposite for cylinder 2.

The result of the mixture formation process at the ignition point (IP) is shown in Figure 2.36 (lambda maps). Here it becomes evident that the central cylinders have an inadequate homogenization. In particular cycle-to-cycle variations of lambda at the spark plug (from lean to a rich mixture and back) are responsible for remarkable variations of the combustion duration and consequently remarkable variations of the IMEP (high σ_imep).

The results of the 3D-CFD simulations (see Figure 2.37) show also a reduction of the tumble ratio in the central cylinders. Since the breakdown of the tumble directly influences the turbulence level within the combustion chamber (overall and local at the spark plug—see Figure 2.37), it follows a less beneficial turbulence profile for the combustion process. The turbulence maps at the ignition point of each cylinder (IP), in comparison to the fuel distribution, show a good symmetry of the turbulence in the central cylinders between IV 1 and IV 2 (see Figure 2.38).

Focusing now on the combustion process, the temperature distributions in Figure 2.39 show the flame positions at CA = 10° after FTDC (burned mass fraction ca. 35%). The flames in the external cylinders remain in the central position of the combustion chamber and still have a circular shape. In contrast the centers of the burned zones in the internal cylinders have moved near the EVs and have a less circular shape (indication of a more sensitive combustion). The analysis of the

FIGURE 2.34 (Step 0) Fuel mass fraction and flow field during the intake phase of each cylinder—90° after TDC (cylinder relative)—5500 rpm WOT (full engine mesh, fifth cycle).

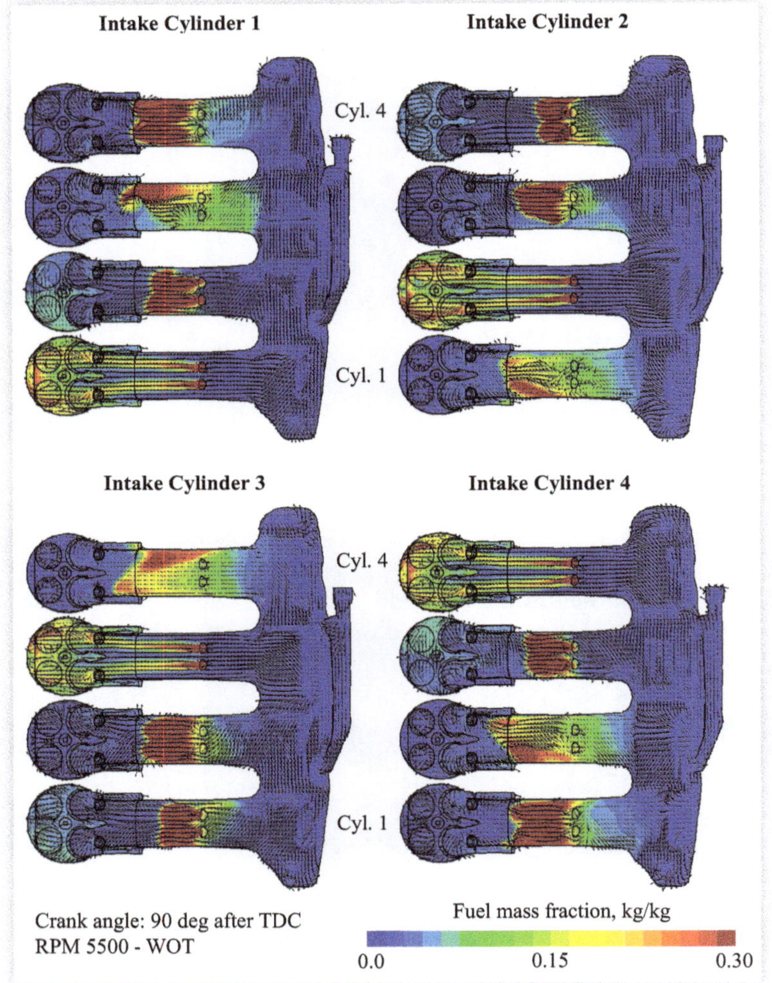

FIGURE 2.35 (Step 0) Fuel mass fraction in the intake channels of cyl. 3 (section 5 mm from valve seat) - 60° after TDC and mass flow through each intake valve of cyl. 3—5500 rpm WOT—fifth cycle.

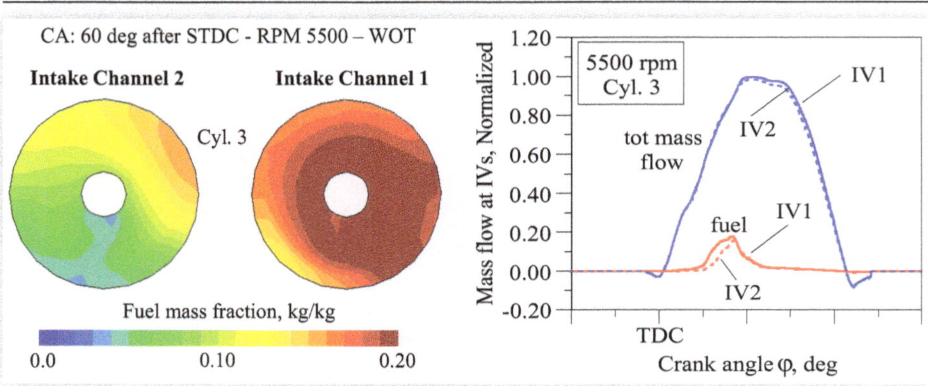

FIGURE 2.36 (Step 0) Lambda distribution at the ignition point of each cylinder—IP = 25° before FTDC (cylinder relative)—5500 rpm WOT (full engine mesh, fifth cycle).

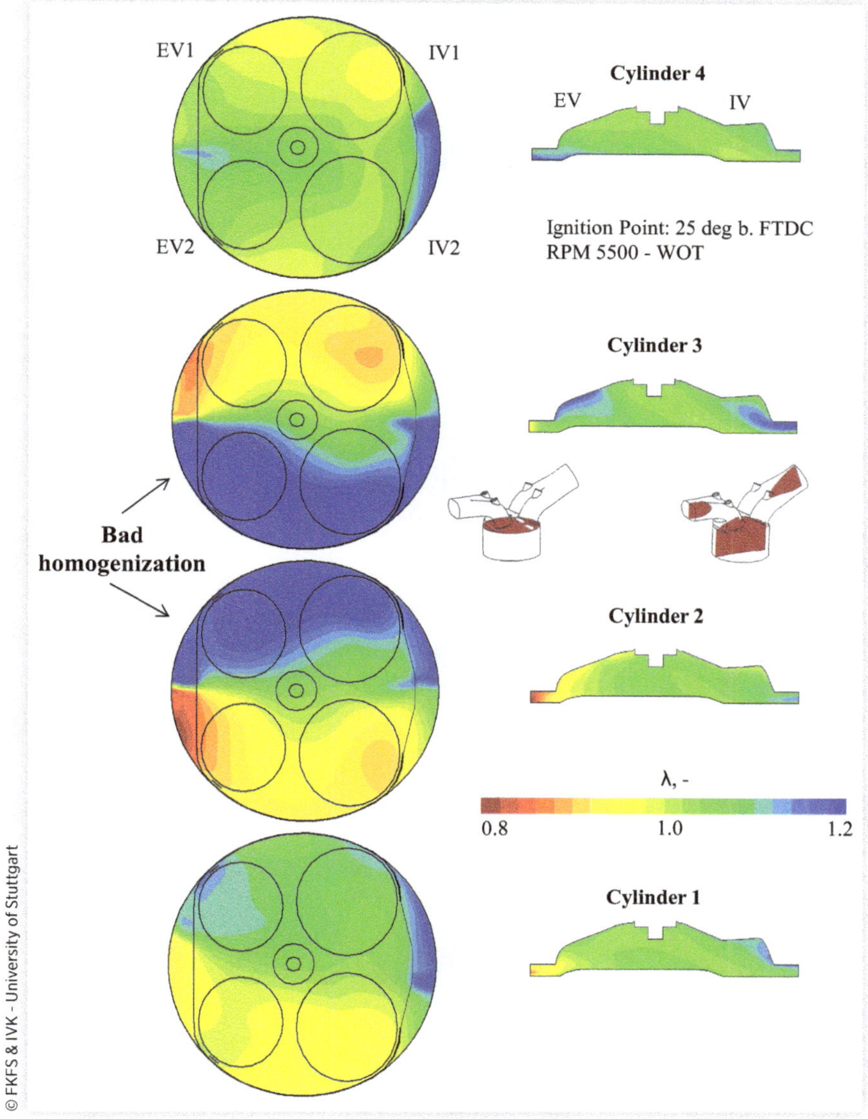

FIGURE 2.37 (Step 0) Tumble ratio and cylinder turbulence and spark plug turbulence—5500 rpm WOT—fifth cycle.

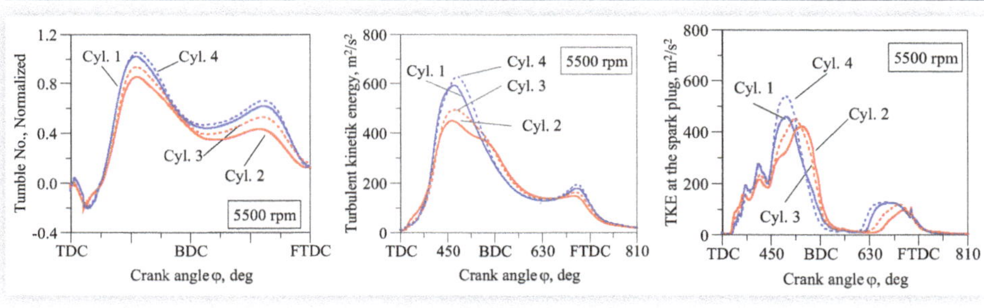

FIGURE 2.38 (Step 0) (TKE) Turbulence at the ignition point of each cylinder—IP = 25° before FTDC (cylinder relative)—5500 rpm WOT (full engine mesh, fifth cycle).

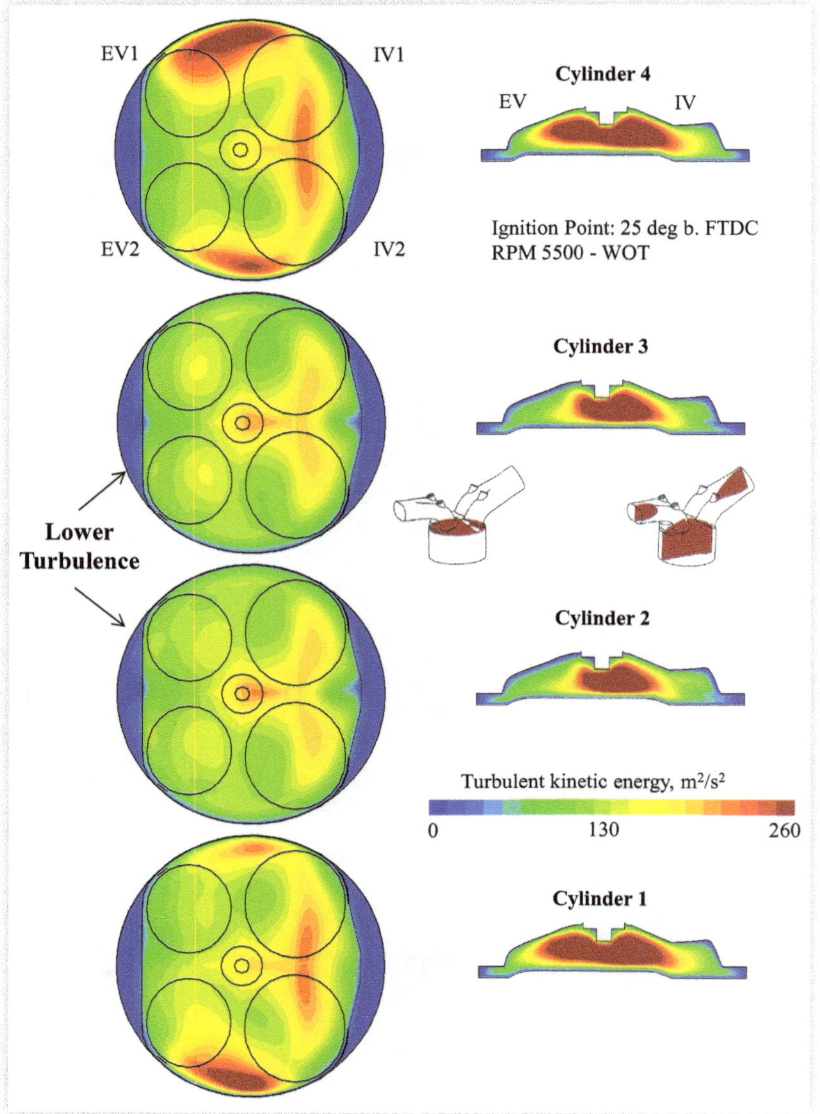

combustion profile reported in Figure 2.40 shows at the end also a longer combustion duration of the central cylinders. The combination of a lower turbulence level and a worse mixture formation in the central cylinders causes definitively a loss of performance of the engine.

Looking in general for improvement of a turbocharged race engine, many factors have to be taken into account. Depending on the race track characteristic, race duration, vehicle characteristics, etc., not only the maximal power output is a relevant parameter but also the torque profile within the engine speed range of interest and the engine drivability (engine performance during transients) are very important. Within the given engine limitations fixed by regulations (airbox pressure, max. air

FIGURE 2.39 (Step 0) Temperature distribution during combustion of each cylinder—10° after FTDC (cylinder relative)—5500 rpm WOT (full engine mesh, fifth cycle).

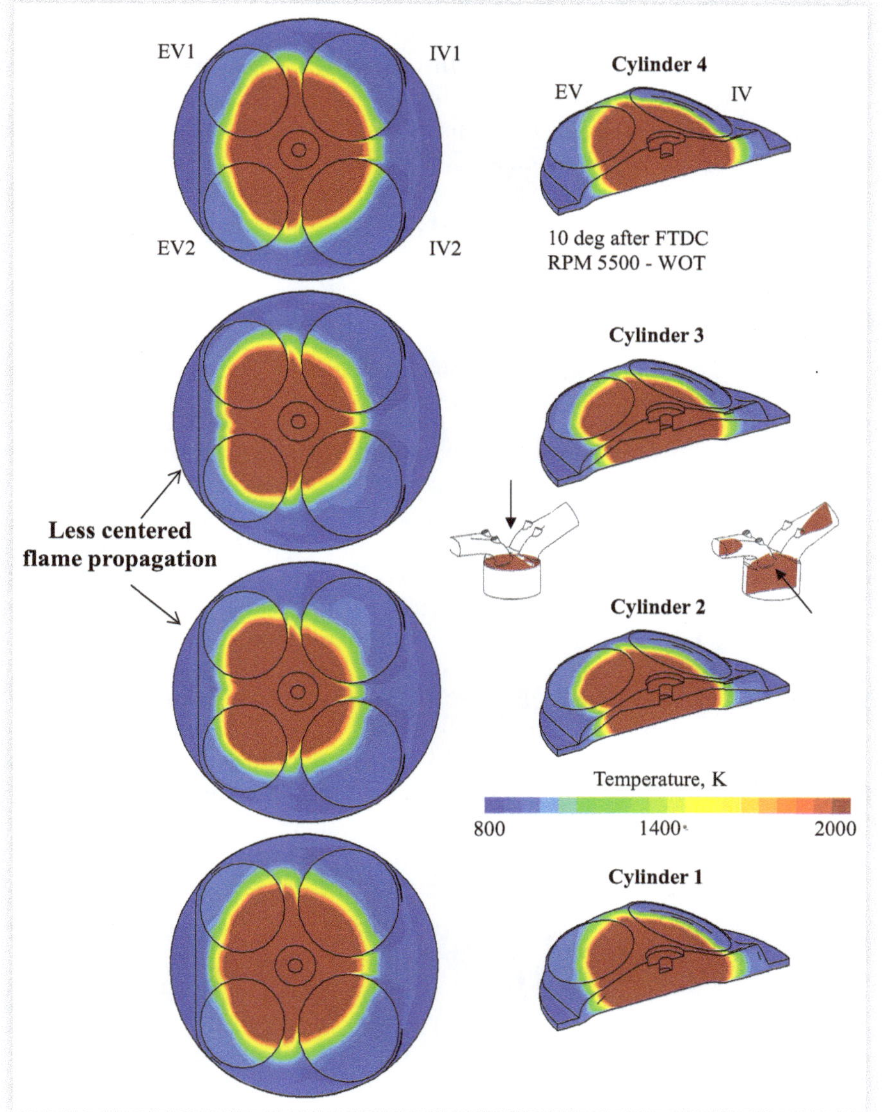

consumption) or structural reasons (max. allowed cylinder pressure, exhaust gas temperature, etc.), the engine must be developed finding the "best" compromise among cylinder filling, turbulence (reduction of combustion duration toward increasing of indicated efficiency), charge exchange optimization (reduction of work losses during the exchange phase), and mixture homogenization within the engine speed range of interest.

2.6.3.2 IMPROVEMENTS (ENGINE DEVELOPMENT STEP 1)

Starting from the analysis of the engine "step 0," different designs of the airbox have been simulated and evaluated using the fast-response 3D-CFD simulation tool

FIGURE 2.40 (Step 0) Burn rate profile—5500 rpm WOT—fifth cycle.

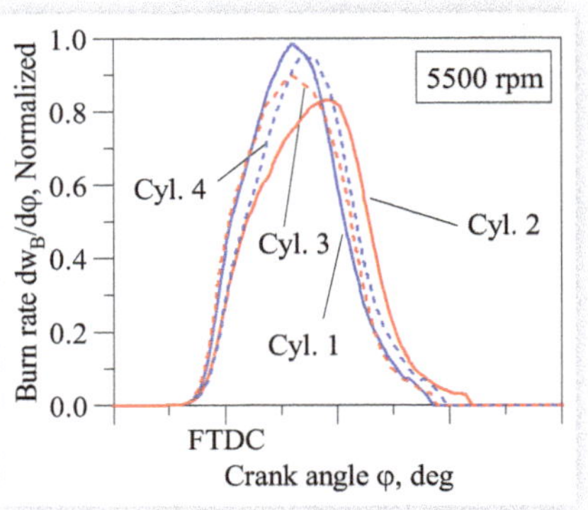

introduced in this chapter. Based on the results of this virtual development process, only the promising solutions have been realized and then tested at the test bench. Thanks to this virtual development, it has been possible to speed up the engine development process by limiting the budget. Here a selection of these investigations is reported. Due to confidential reasons, the following results of full-engine simulations focus mainly on the visualization of the mixture formation and turbulence within the combustion chamber at the IP.

The starting idea here is to increase the volume of the airbox by a factor of ca. 2.5. This aims to reduce the pressure waves within the airbox toward a more "soft" intake process (see Figure 2.41):

- Step 1.0: Greater airbox volume with throttle near the intake runners.
- Step 1.1: Same airbox volume as "step 1.0" but higher distance between throttle and runners.

Analyzing the concentration of the fuel within the intake channels of cylinder 3 (see Figure 2.42), it becomes evident how the position of the throttle is able to remarkably influence the deviation of the fuel jets. In case of "step 1.0," the asymmetry between the two channels increases also in comparison to the "step 0" (smaller airbox volume). In case of "step 1.1," the results show improvements, i.e., the repartition of the fuel between the two intake channels is now better.

The analysis of the mass flows through the IVs (see Figure 2.42) shows in case of "step 1.0" higher differences in the peaks of the total mass flows through each valve, i.e., in this case not only the fuel but also the repartition of the intake air between the two valves becomes worse. This causes an increasing deflection of the fuel jets. Here the shape of the curves of the total mass flow is more "waved" and lightly retarded in the valve closing phase (no backflow before IVC). In case of "step 1.1," the

FIGURE 2.41 (Step 1) Investigation on different airbox designs.

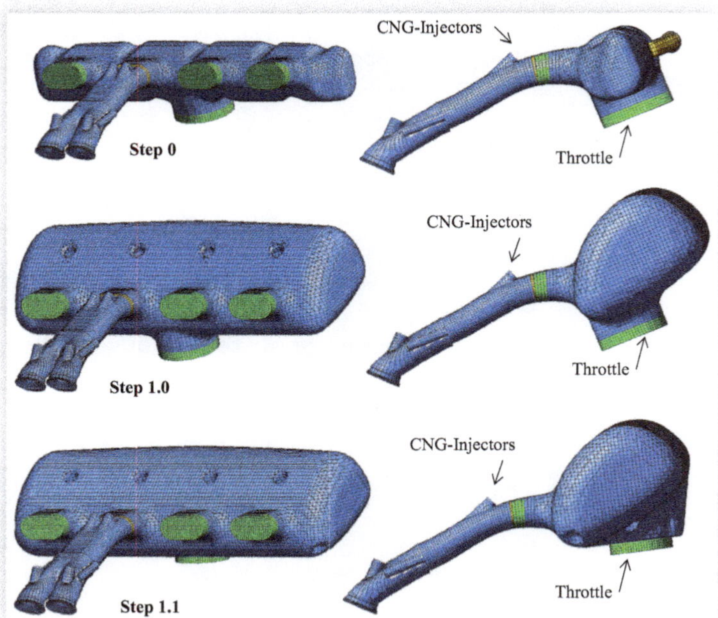

FIGURE 2.42 (Step 1) Fuel mass fraction in the intake channels of cyl. 3 (section 5 mm from valve seats)—60° after TDC (cylinder relative) and mass flows through the valves—5500 rpm WOT (full engine mesh, fifth cycle).

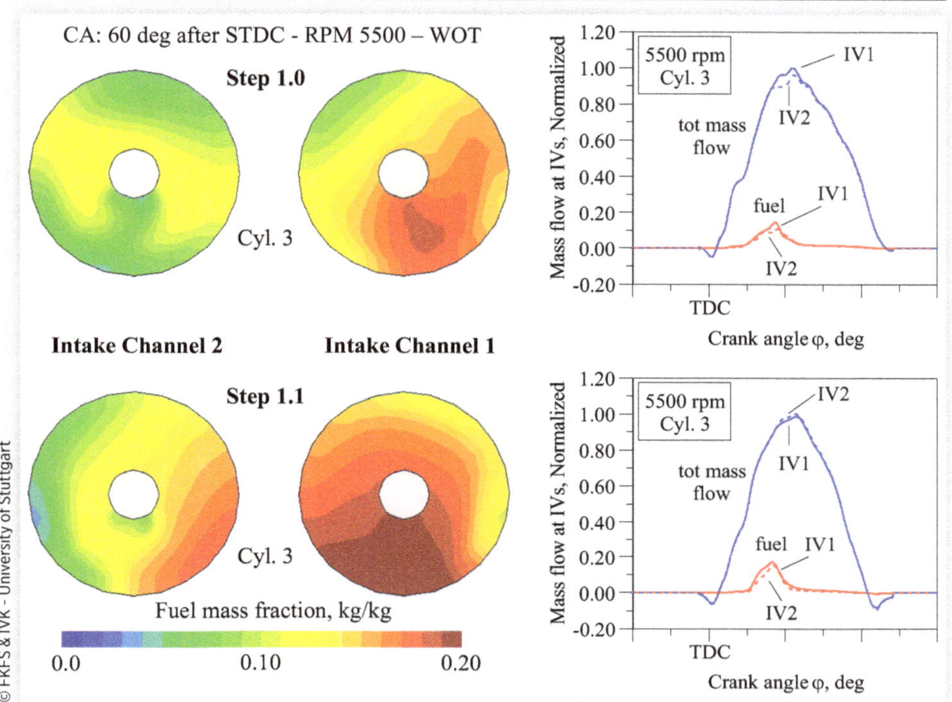

asymmetries in general become lower and the curve profile of the total mass flow remains similar to the case of "step 0."

The results of the simulations (here briefly reported only for the cylinder 3—see Figure 2.43) show for "step 1.0" also a worsening of both the mixture formation and the turbulence level within the combustion chamber at the IP in comparison to "step 0."

This means that only the increasing of the airbox volume does not directly influence the charge motion. In case of "step 1.1," where the throttle is more distant to the intake runners, the mixture formation remarkably improves in comparison to "step 0." Here,

FIGURE 2.43 (Step 1) Lambda and turbulence (TKE) distribution at the ignition point of cylinder 3—IP = 25° before FTDC—5500 rpm WOT (full engine mesh, fifth cycle).

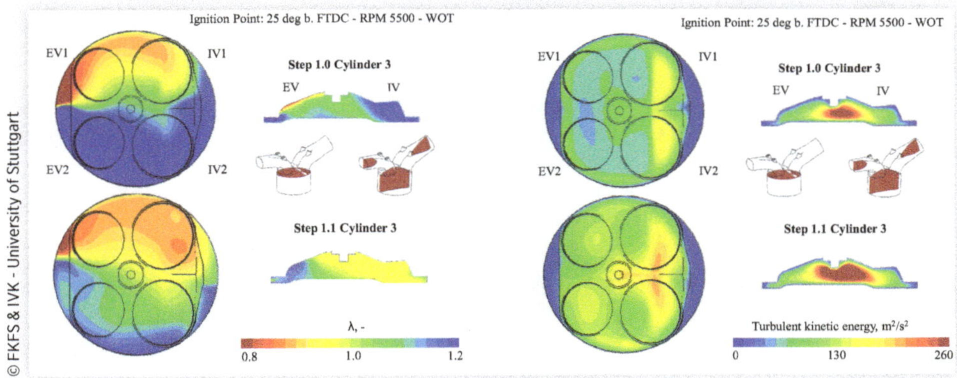

due to the particular design of the airbox, the flow through the throttle manifold can first "calm down" in the airbox, so that the interactions with the flow within the intake runners of each cylinder can be reduced.

2.6.3.3 IMPROVEMENTS (ENGINE DEVELOPMENT STEP 2)

In a second engine development step, some investigations have been focused on the influence of a filter plate within the airbox and of different positions of the fuel injectors on the charge motion (see Figure 2.44).

- Step 2.0: Similar airbox volume and design as "step 1.1" with an internal filter plate.
- Step 2.1: Same airbox as step 2.0 but injector position nearer to the cylinder head.

The analysis of the mass flows through the IVs (see Figure 2.45) shows very low differences in the profiles of the total mass flows between the valves. Also the deviation of the fuel jets becomes much lower than in the previous cases. Due to the near position of the fuel injectors to the IVs in case of "step 2.1," the peak of fuel that flows through the valves is brought forward in comparison to the other cases (the fuel injection profile is the same in all the cases).

The filter plate introduces a separation of the airbox into two chambers. Thanks to this approach, the flow from the throttle can be better distributed over the whole filter-plate surface with a resulting more uniform and lower velocity. It follows a reduction of the influence on the charge motion in the intake runners among the cylinders.

The airbox design with the standard injector position ("step 2.0") shows a surprisingly good homogenization within the combustion chamber (see Figure 2.45). As expected using an airbox with integrated filter (see Figure 2.46), the turbulence level in all combustion chambers shows a moderate reduction of turbulence in comparison to the basis "step 0" or "step 1.0," which sensibly increases the combustion duration (negative influence on the indicated efficiency).

FIGURE 2.44 (Step 2) Investigation on the same airbox design with different injector positions.

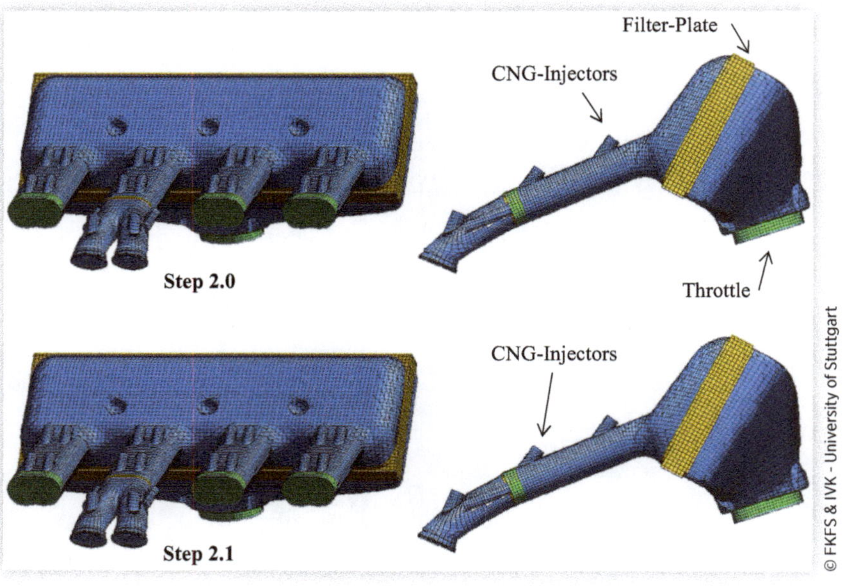

FIGURE 2.45 (Step 2) Lambda distribution at the ignition point of cylinder 3—IP = 25° before FTDC and mass flows through the valves—5500 rpm WOT (full engine mesh, fifth cycle).

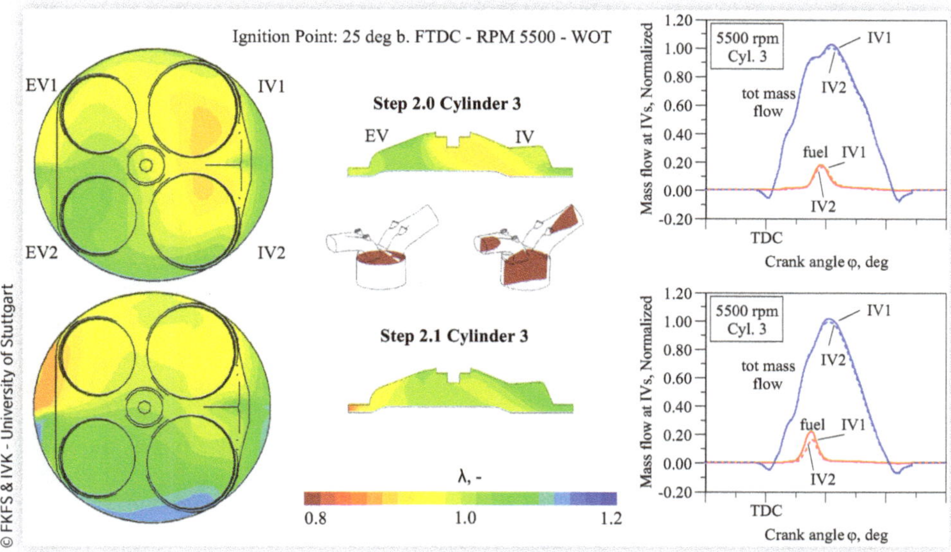

FIGURE 2.46 (Step 2) Turbulence (TKE) at the ignition point of cylinder 3—IP = 25° before FTDC—5500 rpm WOT (full engine mesh, fifth cycle).

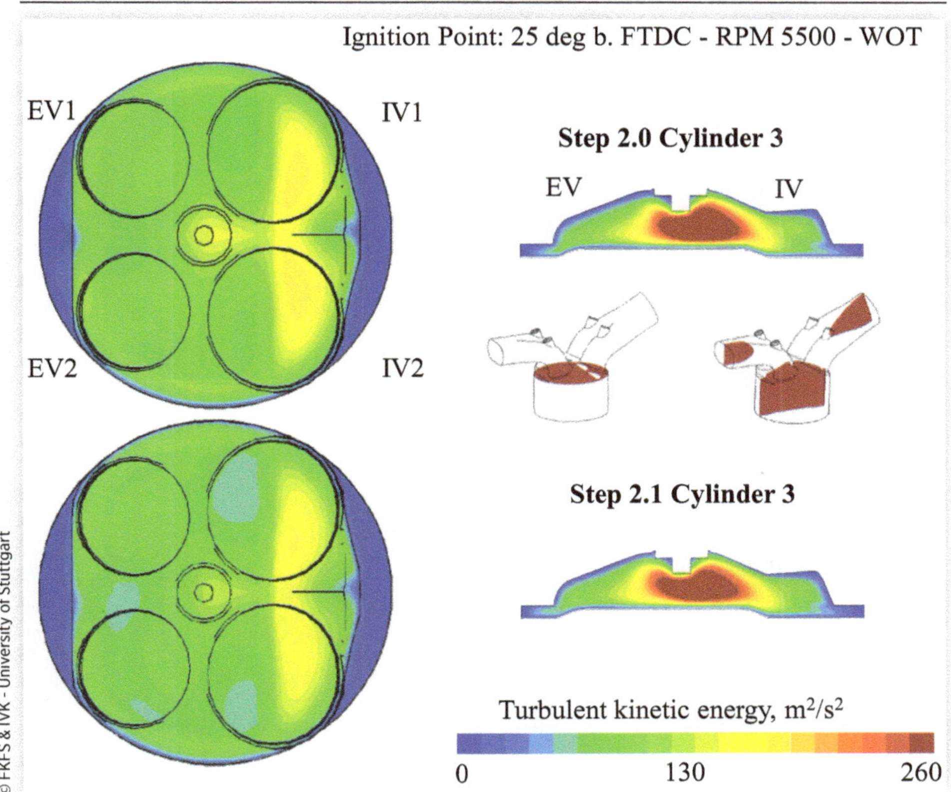

The configuration with fuel injectors nearer to the cylinder head ("step 2.1") shows a light worsening of the mixture formation in comparison to "step 2.0" with the same level of turbulence (as expected different fuel injector positions in the manifolds do not influence the turbulence profile during the whole operating cycle). Conclusively, solution 2.1, in comparison to "step 2.0," thanks to the nearer position of the fuel injectors, does not have advantages in terms of IMEP but only in terms of drivability due to a faster response at any load changes.

Looking for performance enhancement of a race engine (SI) combined with high drivability. there is a conflict among stable combustion (low σ_imep mainly due to a homogeneous mixture formation), high imep (short combustion duration at full load mainly due to high turbulence at the flame front), and short delay between fuel injection in the intake runners and optimal mixture formation in the combustion chambers during transient operating conditions. Considering the development steps reported in this work, the most promising step for successive investigations has been represented by the step "1.1" (best compromise between quality of mixture formation and turbulence level).

Based on these considerations, the development process of any engines requires an accurate analysis and evaluation of the occurring engine processes toward a clear identification of component design configurations, valve motion strategies, injection strategies, etc., which indisputably allow an enhancement of the engine performance.

Abbreviations

0D - Zero-dimensional
1D - One-dimensional
3D - Three-dimensional
ATDC - After top dead center
BSFC - Brake-specific fuel consumption
BTDC - Before top dead center
CA - Crank angle
CCV - Cycle-to-cycle variations
CFD - Computational fluid dynamics
CI - Compression ignition
CNG - Compressed natural gas
ECU - Electronic control unit
EGR - Exhaust gas recirculation
GUI - Graphical user interface
HCCI - Homogeneous charge compression ignition
HiL - Hardware-in-the-loop
IMEP - Indicated mean effective pressure
IC - Internal combustion
IVC - Inlet valve closing
LIF - Laser-induced fluorescence
MFB - Mass fraction burned
PTA - Pressure trace analysis
RDE - Real drive emissions
SI - Spark ignition

SiL - Software-in-the-loop
SOI - Start of injection
VCR - Variable compression ratio
VTG - Variable turbine geometry
WLTP - Worldwide harmonized light vehicle test procedure
WP - Working process

References

1. Grill, M., Billinger, T., and Bargende, M., "Quasi-Dimensional Modeling of Spark Ignition Engine Combustion with Variable Valve Train," SAE Technical Paper 2006-01-1107, 2006, doi:10.4271/2006-01-1107.
2. Schmid, A., Grill, M., Berner, H.-J., and Bargende, M., "Development of a Quasi-Dimensional Combustion Model for Stratified SI-Engines," SAE Technical Paper 2009-01-2659, 2009, doi:10.4271/2009-01-2659.
3. Keskin, M., Grill, M., and Bargende, M., "Development of a Fast, Predictive Burn Rate Model for Gasoline-HCCI," SAE Technical Paper 2016-01-0569, 2016, doi:10.4271/2016-01-0569.
4. Bossung, C., Bargende, M., Dingel, O., and Grill, M., "A Quasi-Dimensional Charge Motion and Turbulence Model for Engine Process Calculations," *15. Internationales Stuttgarter Symposium*, Stuttgart, March 17-18, 2015.
5. Grill, M., Fandakov, A., Hann, S., Keskin, M. et al., "Lean Combustion, EGR or gHCCI at High Load: Challenging Tasks in the 0D/1D Engine Simulation," *Internationaler Motorenkongress*, Baden-Baden, 2018.
6. Hann, S., Grill, M., and Bargende, M., "Reaction Kinetics Calculations and Modeling of the Laminar Flame Speeds of Gasoline Fuels," SAE Technical Paper 2018-01-0857, 2018, doi:10.4271/2018-01-0857.
7. Heywood, J., *Internal Combustion Engine Fundamentals* (McGraw-Hill Series in Mechanical Engineering, 1988), ISBN:978-0070286375.
8. Wenig, M., Grill, M., and Bargende, M., "A New Approach for Modeling Cycle-to-Cycle Variations within the Framework of a Real Working-Process Simulation," SAE Technical Paper 2013-01-1315, 2013, doi:10.4271/2013-01-1315.
9. Fandakov, A., Grill, M., Bargende, M., and Kulzer, A., "Two-Stage Ignition Occurrence in the End Gas and Modeling Its Influence on Engine Knock," *SAE Int. J. Engines* 10, no. 4 (2017): 2109-2128, doi:10.4271/2017-24-0001.
10. Pasternak, M., Netzer, C., Mauss, F., Fischer, M. et al., "Simulation of the Effects of Spark Timing and External EGR on Gasoline Combustion under Knock-Limited Operation at High Speed and Load," in: Guenther, M. and Sens, M., eds., *Knocking in Gasoline Engines* (Berlin: IAV Automotive Engineering, 2017), 121-142.
11. Fandakov, A., Grill, M., Bargende, M., and Kulzer, A, "A Two-Stage Knock Model for the Development of Future SI Engine Concepts," SAE Technical Paper 2018-01-0855, 2018, doi:10.4271/2018-01-0855.
12. Schmid, A., Grill, M., Berner, H.-J., and Bargende, M., "A New Approach for SI-Knock Prediction," *3rd International Conference on Knocking in Gasoline Engines*, Berlin, 2010.

13. Grill, M., Rether, D., Schmid, A., and Bargende, M., "Quasi-Dimensional and Empirical Modeling of Compression-Ignition Engine Combustion and Emissions," SAE Technical Paper 2010-01-0151, 2010, doi:10.4271/2010-01-0151.
14. Kožuch, P., Maderthaner, K., Grill, M., and Schmid, A., "Combustion and Emissions Modelling on Heavy-Duty Engines of Daimler AG," *9th International Symposium on Combustion Diagnostics*, Baden-Baden, Germany, June 8-9, 2010.
15. Bargende, M. and Heinle, M., "Einige Ergänzungen zur Berechnung der Wandwärmeverluste in der Prozessrechnung," *13. Tagung "Der Arbeitsprozess des Verbrennungsmotors"*, Graz, 2011.
16. Kožuch, P., "Ein phänomenologisches Modell zur kombinierten Stickoxid- und Rußberechnung bei direkteinspritzenden Dieselmotoren," Ph.D. thesis, University of Stuttgart, 2004.
17. Kaal, B., Grill, M, and Bargende, M., "Transient Simulation of Nitrogen Oxide Emissions of CI Engines," SAE Technical Paper 2016-01-1002, 2016, doi:10.4271/2016-01-1002.
18. Chiodi, M., *An Innovative 3D-CFD-Approach towards Virtual Development of Internal Combustion Engines* (Vieweg+Teubner Verlag, June 2010), ISBN:978-3-8348-1540-8.
19. Chiodi, M. and Bargende, M., "Improvement of Engine Heat-Transfer Calculation in the Three-Dimensional Simulation Using a Phenomenological Heat-Transfer Model," SAE Technical Paper 2001-01-3601, 2001, doi:10.4271/2001-01-3601.
20. Grill, M., Schmid, A., Chiodi, M., Berner, H.-J. et al., "Calculating the Properties of User-Defined Working Fluids for Real Working-Process Simulations," SAE Technical Paper 2007-01-0936, 2007, doi:10.4271/2007-01-0936.
21. Schmid, A., Grill, M., Berner, H.-J., and Bargende, M., "Virtuelle Optimierung an einem Twin-Turbo Dreizylinder-Ottomotor im FTP75," IAV-Tagung Motorprozessrechnung, Berlin, 2011.
22. Höpke, B., Pischinger, S., Haussmann, F., and Bargende, M., "ATL-Dynamik," *FVV-Frühjahrstagung 2015, Informationstagung Motoren*, Bad Neuenahr, March 26-27, 2015.
23. Seboldt, D., Lejsek, D., Wentsch, M., Chiodi, M. et al., "Numerical and Experimental Studies on Mixture Formation with an Outward-Opening Nozzle in a SI Engine with CNG-DI," SAE Technical Paper 2016-01-0801, 2016, doi:10.4271/2016-01-0801.
24. Koch, D., Wachtmeister, G., Wentsch, M., Chiodi, M. et al., "Investigation of the Mixture Formation Process with Combined Injection Strategies in High-Performance SI-Engines," *16th Stuttgart Symposium*, Stuttgart, Germany, March 2016.
25. Wentsch, M., Chiodi, M., Bargende, M., Poetsch, Ch. et al., "Virtuelle Motorentwicklung als Erfolgsfaktor in der F.I.A. Rallye-Weltmeisterschaft (WRC)," *12th International Symposium on Combustion Diagnostics*, Baden-Baden, Germany, May 2016.
26. Schneider, S., Friedrich, H., Chiodi, M., and Bargende, M., "Development and Experimental Investigation of a Two-Stroke Opposed-Piston Free-Piston Engine," SAE Technical Paper 2016-32-0046, 2016, doi:10.4271/2016-32-0046.
27. Schneider, S., Friedrich, H., Chiodi, M., and Bargende, M., "Analysis of SI and HCCI Combustion in a Two-Stroke Opposed-Piston Free-Piston Engine," SAE Technical Paper 2017-32-0037, 2017, doi:10.4271/2017-32-0037.

28. Wentsch, M., Chiodi, M., and Bargende, M., "Fuel Injection Analysis with a Fast Response 3D-CFD Tool," SAE Technical Paper 2017-24-0103, 2017, doi:10.4271/2017-24-0103.

29. Chiodi, M., Kaechele, A., Bargende, M., Wichelhaus, D. et al., "Development of an Innovative Combustion Process: Spark-Assisted Compression Ignition," *SAE Int. J. Engines* 10, no. 5 (2017): 2486-2499, doi:10.4271/2017-24-0147.

CHAPTER

3

Advanced 0D and QuasiD Thermodynamic Combustion Models for SI and CI Engines

Fabio Bozza and Vincenzo De Bellis
University of Naples "Federico II"

Alessio Dulbecco
Institut Carnot IFPEN Transports Energie

CHAPTER 3

This chapter focuses on advanced 0D and quasiD combustion models for spark ignition (SI) and compression ignition (CI) engines. Numerical simulation is more and more integrated into the engine development process. According to the considered physical contents, each model presents different accuracy and computational time. In this chapter, the most common thermodynamic combustion models are presented. The physical background of each model is described, underlining its suitability for a given application. Specific aspects, such as pollutant emissions formation and abnormal combustions, are also detailed. New requirements and present trends of development are discussed as well. The chapter is structured into three sections. Section 3.1 (Physical Background (Combustion Regimes for SI and CI Engines)) is an introduction to the physics of the combustion in internal combustion engines (ICE). A focus on the combustion modes (auto-ignition, premixed flames, and diffusion flames) and on the interaction of flames with turbulence will be given. Both are essential in combustion simulation: they allow, from one side, to formulate the hypotheses defining the physical bases of the combustion models and, on the other side, to select the most suitable approach for a given application. Section 3.2 (SI Engines Modeling) is dedicated to SI engine combustion modeling. This section is further divided into four. In Section 3.2.1, different approaches to turbulent combustion in SI engines are presented, selected among the ones available in the current literature. The discussed models are representative of the most important modeling families.

They are organized according to two major criteria: the computational time and the level of detail of physical contents. Section 3.2.2 presents the most common turbulence modeling approaches to compute the mean turbulent flow field within the combustion chamber. The characterization of turbulence in a 0D framework is, for sure, one of the most challenging aspects of the modeling activity, because of its stochastic and 3D features. Nevertheless, such information are today fundamental to correctly predict the turbulent combustion, especially in engine architectures equipped with advanced technologies such as variable valve actuation (VVA), variable valve timing (VVT), variable compression ratio (VCR), or direct injection (DI) systems. Section 3.2.3 focuses on the pollutant emission modeling for SI engines and in particular on CO and NO_x emissions. It is well known that pollutants emissions are one of the major concerns of car manufacturers, and their reduction during the combustion process represents an effective solution to comply with related legislations. The importance of tightly coupling pollutant models with the combustion one to predict with accuracy the pollutant kinetics will be pointed out. Section 3.2.4 presents modeling approaches to abnormal combustions and, in particular, to the knock phenomenon. This is a challenging task because of its local thermochemical origin. Two major approaches, available in literature, are detailed: the first, based on ignition delay phenomenological modeling, the second, resorting to the computation of chemical kinetics. Section 3.3 (CI Engines Modeling) is dedicated to CI engine combustion modeling. The section includes four main sections. In Section 3.3.1, the phenomenological aspects of the CI combustion process are presented. The other sections concern the modeling aspects. Section 3.3.2 focuses on spray computation, Section 3.3.3 approaches turbulent combustion description, and Section 3.3.4 regards pollutant emissions, and, in particular, NO_x and soot emissions. The need for a strict information exchange between pollutants and combustion models to accurately describe pollutant kinetics will be underlined.

3.1 Physical Background (Combustion Regimes for SI and CI Engines)

The fuel oxidation in engine combustion chambers is a complex process due to the fact that reactive mixture is not homogeneous in terms of temperature and composition. Furthermore, chemical reactions take place in a turbulent environment, which amplifies the stochastic features of combustion. In the last decades, the researchers developed numerical approaches aiming to understand and describe in details the combustion scenario. Depending on the relative weight of the different concurring phenomena, such as chemical oxidation mechanism, mixture stratification, and turbulence, three fundamental combustion regimes were identified and characterized by means of experiments: auto-ignition, premixed flames, and diffusion flames [1]. These regimes represent limit cases of fuel oxidation processes. In what follows, the physical background of each combustion mode will be recalled, focusing on the specific aspects and features used for formulating the state of the art of the 0D/quasiD combustion models.

3.1.1 Auto-Ignition

Auto-ignition (AI) is a rapid and spontaneous combustion mechanism. It is the result of a set of chemical reactions which release heat and produce unstable chemical species of partial combustion (peroxides, aldehydes, hydrogen peroxides, etc.), which initiate the combustion process without the requirement of an external heat source. This is possible when the heat released by the former reactions is sufficient to increase the mixture temperature and, accordingly, to rise chemical activity. AI of a homogeneous gaseous reactive mixture is controlled by temperature, pressure, composition (equivalence ratio and dilution), and fuel formulation. It is characterized by an AI delay, τ_{AI}. This last does not have a unique definition in literature. Depending on convention or experimental facility, it can be defined as shown in Figure 3.1 (left) as:

- The time interval from the instant when the mixture attains the reference initial thermochemical conditions and the time when temperature gradient is maximum
- The time interval from the instant when the mixture attains the reference initial thermochemical conditions and the time when the molar concentration of a given combustion radical, usually the OH, is maximum

τ_{AI} can be considered as inversely proportional to the oxidation reaction rate. Accordingly, it can be reasonably represented by an inverse Arrhenius-like expression:

$$\tau_{AI} = Ap^{-n}e^{\left(\frac{E_a}{RT}\right)} \tag{3.1}$$

where

A, n and E_a are constants determined experimentally, depending on the studied mixture

R is the mixture gas constant

p and T are initial pressure and temperature of the referring operating conditions, respectively

FIGURE 3.1 Auto-ignition combustion mode schematics: one step AI (left), two step AI with the presence of a cool flame (right).

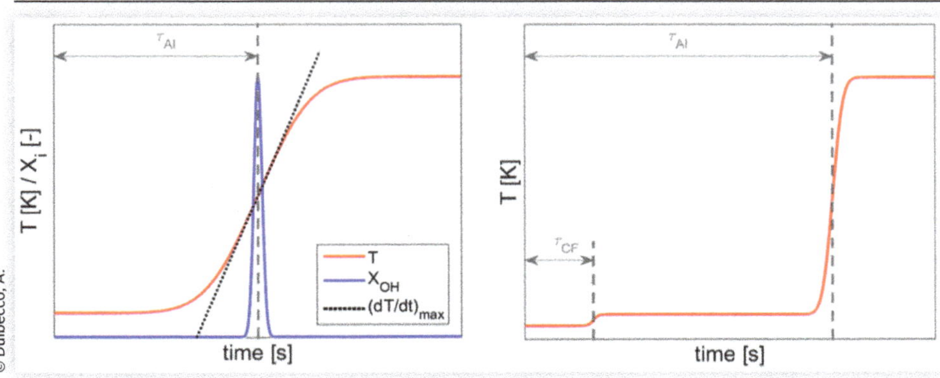

τ_{AI} has an exponential dependence on temperature and an inverse dependence on reactant concentration (dependence on p). From a composition viewpoint, when reactive mixture is diluted with inert gas or air, τ_{AI} usually increases. This is because the diluent absorbs part of the released heat by lowering the overall temperature and, accordingly, the mixture reactivity. However, this is not a general rule, especially for higher fuel concentrations and for fuels with high molar weight. Actually, in these cases, the number of hydrocarbon molecules is higher, and, consequently, molecular collisions between oxidant and oxidizer more likely occur, favoring the first kinetic steps of the oxidation process. Depending on fuel and thermodynamic conditions, AI can take place in one or more steps. For instance, Figure 3.1 (right) shows an AI process taking place into two steps, the first identified by a reduced temperature increase, while the second by a more important one. The first step, occurring after the delay τ_{CF}, is called cool flame (several tenths to several hundred degrees) and is caused by exothermic peroxidation reactions. In case of multistep AI, τ_{AI} can be defined as the sum of the delays, both depending on pressure, temperature, and compositions, in an independent manner, following, sometimes, opposite trends. The AI process characterization becomes more complex when mixture is heterogeneous. In this case, additional variables have to be introduced to account for heterogeneities and mixing. Furthermore, in case of two-phase mixture, proper corrections are required to account for mixture cooling related to fuel evaporation heat subtraction.

3.1.2 Premixed Flames

Premixed combustion relies on the hypothesis that fuel and oxidizer are completely mixed. In premixed flames, the unburnt mixture is ignited by the diffusion of heat and combustion radicals coming from the adjacent combusting region. This thin region, hosting auto-sustaining chemical reactions, separates unburnt gas from burnt gas and propagates toward unburnt gas. It is commonly labelled as flame front. The temperature in the flame can reach values higher than 2000 K. A premixed flame can propagate through a mixture having a laminar or turbulent aerodynamic field; accordingly, it is common to speak about laminar or turbulent premixed flames, respectively.

3.1.2.1 LAMINAR

The displacement speed perpendicular to the flame front relatively to unburnt gas flow is called laminar flame speed, u_L (Figure 3.2). Because of the extremely low flame thickness, δ_L, it can be considered and studied supposing it as a mono-dimensional problem. Within the flame front, two regions are distinguished:

- The preheating region, $\delta_L - \delta_R$, where mixture is subjected to heat and species mass transfer coming from the adjacent reaction zone; no significant heat release due to oxidation reactions is detectable in this zone.
- The reaction region, δ_R, where oxidation reactions take place releasing an amount of heat sufficient to self-sustain the flame.

FIGURE 3.2 Laminar premixed flame structure.

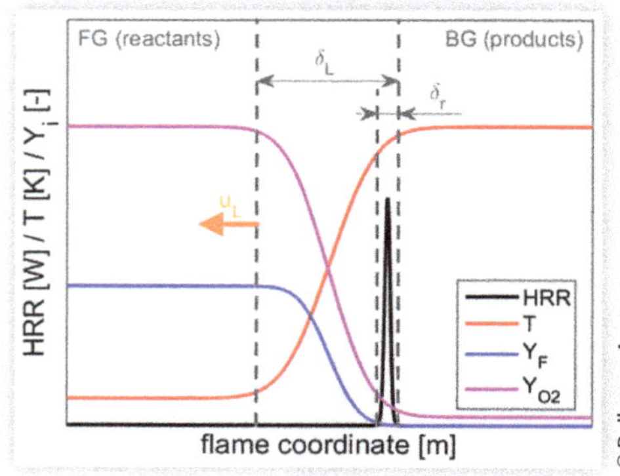

For most premixed flames, the ratio of the preheating region thickness and reaction region thickness is about 10. The laminar flame speed is an intrinsic property of the reactive mixture and depends on heat and mass transfer phenomena. It can be considered as an indicator of the mixture propensity to burn. Starting from mass and energy balance equations, it is possible to write:

$$u_L \propto \sqrt{\frac{\lambda \omega_r}{c_p \rho}} \tag{3.2}$$

where

λ is the thermal conductivity
ω_r is the combustion reaction rate
ρ is the mixture density
c_p is the thermal capacity at constant pressure

In turn, δ_L can be expressed as:

$$\delta_L \propto \frac{\lambda}{c_p \rho u_L} \propto \sqrt{\frac{\lambda}{c_p \rho \omega_r}} \tag{3.3}$$

The flame properties also depend on unburnt gas thermochemical conditions. Fuel formulation and mixture composition (equivalence ratio, ϕ, and dilution) influence the flame thickness and speed, as shown in Figure 3.3.

As shown in the figure, the maximum values of u_L are obtained for slightly reach mixtures, which corresponds to equivalence ratios of highest chemical reactivity (ϕ = 1.05:1.15). Increasing (diminishing) the equivalence ratio reduces the laminar flame speed up to reaching the upper (lower) limit of flammability. In those conditions, the heat released by the oxidation process is not enough to compensate the heat losses from the reaction zone and to sustain chemical kinetics. An increasing of the unburnt gas temperature enhances u_L, because of the increasing of the oxidation reaction rate and of the thermal conductivity, despite the diminishing of the gas density. A pressure increase penalizes u_L, since it induces a reduction of the transport phenomena within the flame front, despite a higher reactant concentration. Regarding the impact of the reactive mixture dilution with an inert gas, its increasing slows down u_L because of the higher thermal capacity of the mixture, which, in turn, reduces the local temperature. Several authors in the literature [3] showed that the laminar flame speed under different operating conditions can be related to the one measured at a reference state, $u_{L,0}$, by means of a power-law relation:

$$u_L = u_{L,0}\left(\frac{T}{T_0}\right)^{\alpha}\left(\frac{p}{p_0}\right)^{\beta} \tag{3.4}$$

with $\alpha = A + B(\phi - 1)$; $\beta = C + D(\phi - 1)$; and $u_{L,0} = E + F(\phi - \phi_m)^2$, where T_0 and p_0 are the reference thermodynamic conditions (T_0 = 288 K, p_0 = 1 atm), ϕ_m is the mixture equivalence ratio of maximum laminar flame speed, and A to F are calibration constants, which depend on the fuel formulation. The laminar premixed flame speed reduction induced by the presence of dilution gases can be described by the following relation:

$$u_L = u_L\left(1 - G X_{dil}^{\gamma}\right) \tag{3.5}$$

where

X_{dil}^{γ} is the dilution gas molar fraction
G and γ are calibration constants

FIGURE 3.3 Laminar burning velocity of several fuels as a function of equivalence ratio, at 0.1 MPa and 300 K. Lines are least squares polynomial fits to experimental data [2].

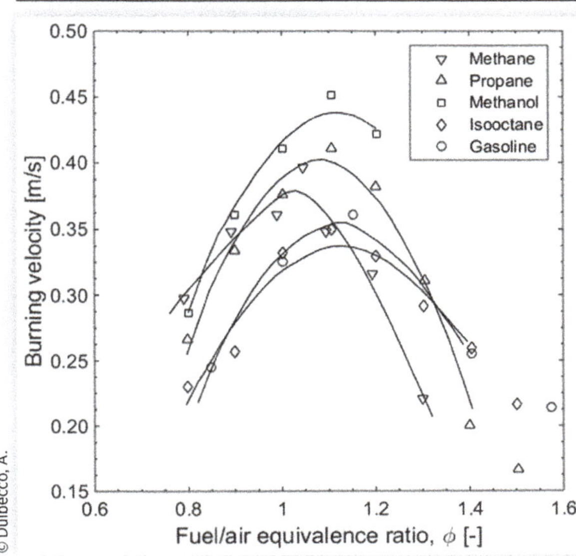

The above correlations do not account for the strain influence which decreases, but sometimes increases, the unstretched laminar flame speed. Its impact highly depends on the mixture Lewis number, *Le*, defined as the ratio of heat and mass diffusivities.

3.1.2.2 TURBULENT

Differently from laminar flames, turbulent flames interact with the aerodynamic flow field. Accordingly, the flame front is influenced by turbulence and vice versa. This interaction may lead to a strong increasing of fuel consumption rate and of flame thickness. The flame-turbulence interaction can be characterized by means of dimensionless numbers; the most important dimensionless numbers used in turbulent combustion theory are introduced below.

3.1.2.2.1 **Turbulence Reynolds Number.** The turbulence Reynolds number based on the integral length scale, Re_t, is the ratio between inertial forces related to turbulent flow motion and viscous forces. It is defined as:

$$Re_t = \frac{u' l_t}{\nu} \tag{3.6}$$

where

u' is the turbulent velocity

l_t is the integral length scale of turbulence

ν is the kinematic viscosity of the mixture

3.1.2.2.2 **Damköhler Number.** The Damköhler number, *Da*, is the ratio between the turbulence characteristic time at the integral length scale level, τ_{l_t}, and the chemical kinetics characteristic time, τ_{ch}. It is defined as:

$$Da = \frac{\tau_{l_t}}{\tau_{ch}} = \frac{l_t}{u'} \frac{u_L}{\delta_L} \tag{3.7}$$

3.1.2.2.3 **Karlovitz Number.** The Karlovitz number, *Ka*, is the ratio between the chemical kinetics characteristic time, τ_{ch}, and the turbulence characteristic time at the scale of the smallest turbulence eddies (Kolmogorov length scale, l_η), τ_η. It is defined as:

$$Ka = \frac{1}{Da(l_\eta)} = \frac{\tau_{ch}}{\tau_\eta} = \frac{\delta_L}{u_L} \frac{u_\eta}{l_\eta} \tag{3.8}$$

where u_η represents the turbulent velocity at the Kolmogorov scale. This last, assuming the turbulence as being homogeneous and isotropic, is computed as:

$$u_\eta = \sqrt[3]{\frac{l_\eta}{l_t}} u' \tag{3.9}$$

The three numbers are related by the following expression:

$$\mathrm{Re}_t = Da^2 Ka^2 \tag{3.10}$$

FIGURE 3.4 Turbulent premixed combustion diagram [4]. Axes variables use log-log scales.

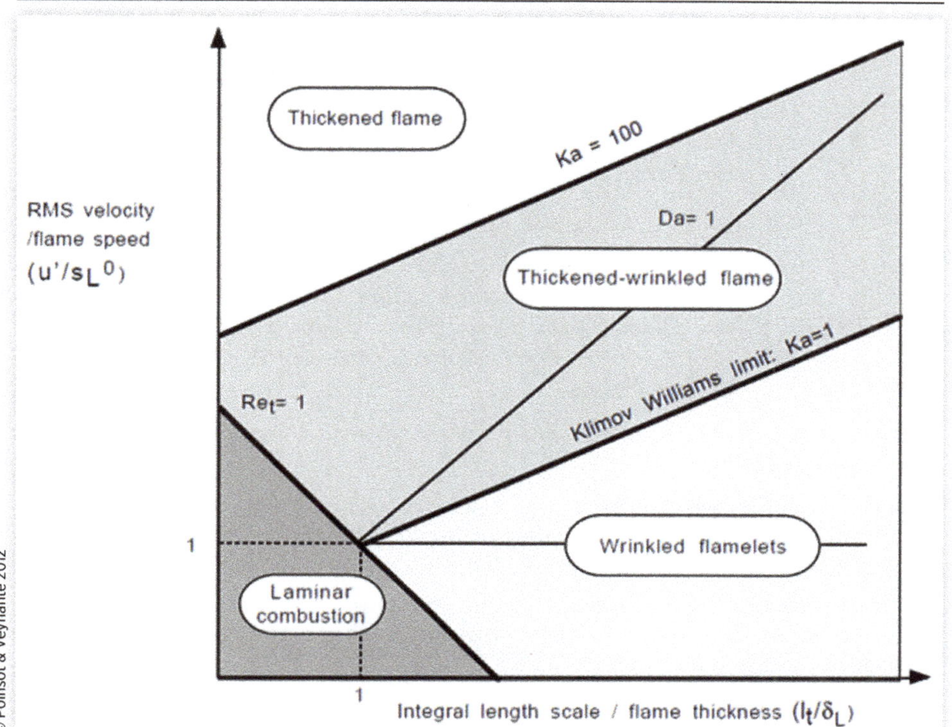

Based on the Damköhler and Karlovitz numbers, various combustion regimes can be identified according to characteristics length (l_t/δ_L) and velocity (u'/u_L) ratios [4], Figure 3.4:

- $Re_t < 1$, laminar flame combustion regime.
- $Ka < 1$, the chemical time scale is shorter than any turbulent scale and the flame thickness is smaller than the lowest turbulent scale. In this regime, the chemistry is fast, the flame has a thin structure corresponding to a laminar flame, and the front is wrinkled by the turbulence (wrinkled flamelet).
- $Ka > 1$ and $Ka < 100$, some of the turbulent eddies, smaller than the flame thickness, are able to modify the inner structure of the flame and in particular to penetrate within the preheating region. The flame is once again wrinkled by the turbulence, but it cannot be considered as having a laminar structure (thickened wrinkled flame). In this regime, the stretch induced by the smaller turbulent vortices are larger than the critical flame stretch, leading to flame quenching.
- $Ka > 100$, for these conditions, the turbulence has shorter characteristic times with respect to chemical ones. The species mixing is so fast that the overall reaction rate is limited by the chemistry. In this regime, the turbulent eddies can penetrate the reaction region of the flame. This last cannot be considered as a laminar structure (thickened flame). This regime includes also the well-stirred reactor combustion regime.

3.1.3 Diffusion Flames

In diffusion flames, fuel and the oxidizer are introduced separately within the combustion region. Mixing is the main issue in diffusion flames as it must assure the blending

FIGURE 3.5 Laminar diffusion flame structure.

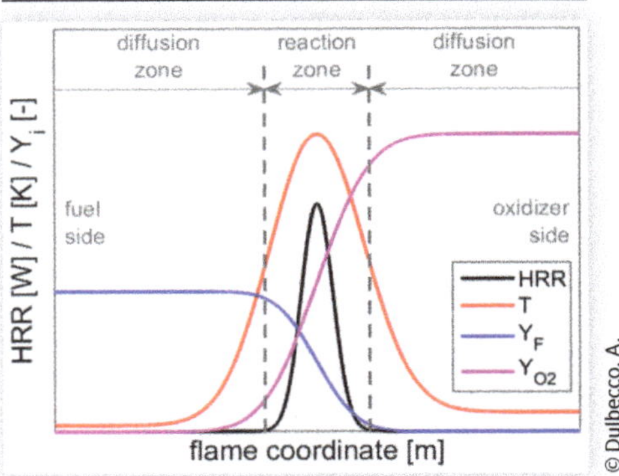

of the reactants fast enough for combustion to proceed. Fuel and oxidizer diffuse toward the reaction zone, where they react by releasing heat and generating burnt gases. The temperature, which is maximum in this zone, gradually reduces away from the reaction zone toward fuel and oxidizer stream temperatures. A diffusion flame can exist in a mixture characterized by both laminar or turbulent aerodynamic field, referring to it as laminar or turbulent diffusion flame, respectively.

3.1.3.1 LAMINAR

In the laminar diffusion flames, two thermochemical states have to be specified, one for each reactant. Figure 3.5 is a schematic of a one-dimensional laminar diffusion flame structure.

Figure 3.5 helps to illustrate some important characteristics of diffusion flames:

- Mixture is highly heterogeneous and its local equivalence ratio can vary ideally from zero to the infinite*.
- Only a reduced range of mixture equivalence ratios is favorable to the establishment of exothermic chemical activity. The most favorable mixture for flame stabilization is characterized by the stoichiometric composition.
- To maintain a steady-state diffusion flame, the opposed flows of fuel and oxidizer have to feed continuously the reaction zone with unburnt gases and generate an adequate strain for mixing.
- Speed and thickness of a diffusion flame, unlike a premixed one, are not intrinsic properties of the reactive mixture. Moreover, they do not depend only on mixture properties but also on fluid dynamics conditions, these last substantially related to the flame stretch [4].

To address diffusion flames and diffusion combustion analysis, some hypotheses and notions have to be introduced. The most important ones are presented below.

3.1.3.1.1 **Decompositions into Mixing and Flame Structure Problems.** In the numerical description of the combustion process, a common practice is to separate the diffusion flame computation into two problems:

- Fuel and oxidizer mixing: the field of fuel/oxidizer mixture is determined based on the boundary and initial conditions, geometry, and flow field.
- Flame structure: the relations between mixture variables (species mass fractions, temperatures) and chemistry assumptions are used to construct all flame variables.

3.1.3.1.2 **Fuel Mixture Fraction.** The fuel mixture fraction, Z, identifies a combustion independent variable which allows to characterize the mixture composition. In 0D and quasiD combustion models, it is common to adopt several assumptions to ease the solution of the mixing problem. The most relevant is that all the species have the same diffusion coefficient. According to the laminar diffusion flame theory proposed in [5], combining the transport equations of fuel mass fraction, Y_F, and of oxidizer mass fraction, Y_O, after some manipulations, it is possible to demonstrate that:

$$Z = Y_{F_r} \tag{3.11}$$

* This is true if the fuel stream does not contain oxygen and if the oxidizer stream does not contain fuel.

FIGURE 3.6 Inner structure of laminar non-premixed flame in the *Z*-space [5].

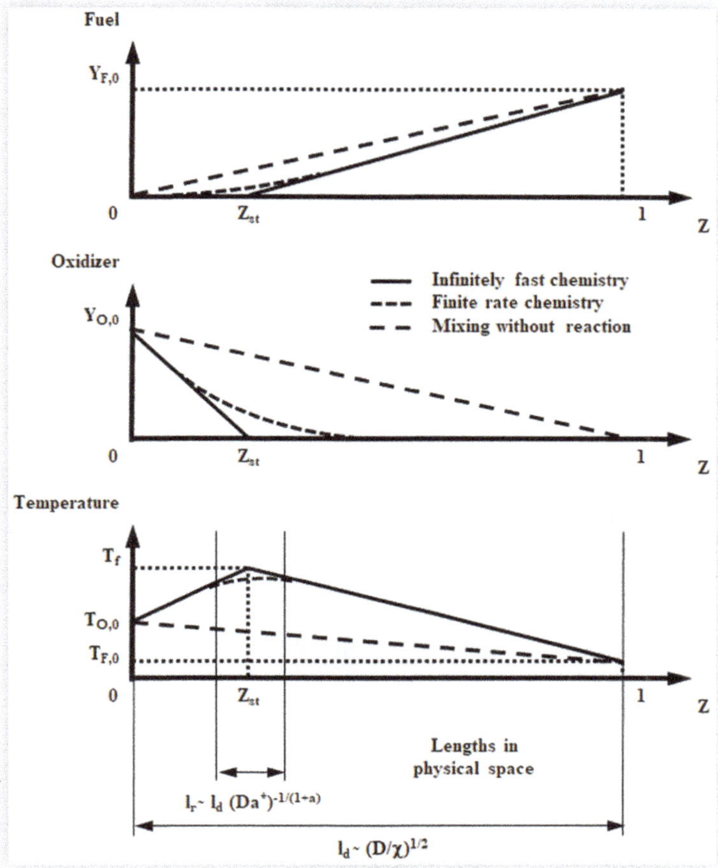

where Y_{F_T}, the fuel-tracer mass fraction, is the local fuel mass fraction of a nonreactive mixture (mixing line in Figure 3.6).

3.1.3.1.3 **Scalar Dissipation Rate.** The flame thickness of a diffusion flame mainly depends on flame stretch, Figure 3.6. To characterize this, it is common to introduce the mixture fraction scalar dissipation rate, χ, defined as:

$$\chi = D\left(\frac{\partial Z}{\partial x_j}\frac{\partial Z}{\partial x_j}\right) = D\left|\nabla Z\right|^2 \tag{3.12}$$

where D is the mass diffusivity. χ measures the inverse of the diffusive time, τ_χ:

$$\tau_\chi = \chi^{-1} \tag{3.13}$$

An increasing of χ is representative of a rising of mass and heat transfers through the stoichiometric surface of the flame.

3.1.3.1.4 **Chemical-Kinetics Time Scale.** In the same manner, a chemical kinetics characteristic time, τ_{ch}, can be defined to characterize the chemical reaction rate. This time depends, generally, on the fuel reaction rate.

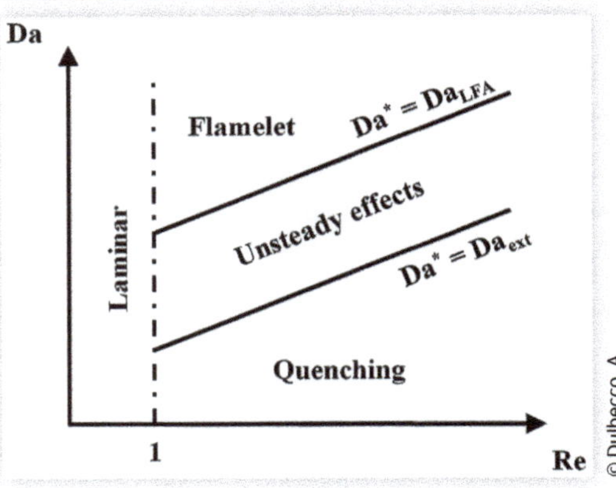

FIGURE 3.7 Non-premixed turbulent combustion regimes [4].

3.1.3.1.5 **Damköhler Number in the Diffusion Flames.** The diffusion flame Damköhler number, Da^*, can be expressed as:

$$Da^* = \frac{\tau_\chi}{\tau_{ch}} \tag{3.14}$$

When chemistry cannot be considered infinitely fast compared to all other time scale, Da^* allows to give a better description of the flame structure, which becomes more complex, Figure 3.7.

3.1.3.1.6 **Flame Structure.** Fuel mixture fraction and scalar dissipation rate are used to describe the inner structure of diffusion flames, Figure 3.6.

Figure 3.6 represents fuel and oxidizer mass fractions and temperature evolutions within a stationary non-premixed flame front, as a function of the mixture composition, expressed in terms of fuel mixture fraction. Three states of the mixture are depicted: pure mixing (no combustion), combustion with infinitely fast chemistry, and finite rate chemistry. As shown on the bottom of the figure, the scalar dissipation rate at the stoichiometric surface and the flame Damköhler number allow to translate the diffusion flame structure given in the Z-space into a diffusion flame structure in the physical space. The expressions to determine the characteristic thicknesses of the flame, the diffusive thickness, l_d, and the reaction zone thickness, l_r, are given in Figure 3.6.*

3.1.3.2 TURBULENT

As diffusion flames do not exhibit intrinsic properties, such as a propagation velocity or a flame thickness, their classification is not trivial. According to [4], two non-dimensional numbers allow to describe a turbulent non-premixed flame, Figure 3.7:

- The Reynolds number, characterizing the flow regime
- The Damköhler number, characterizing the reaction zone

In Figure 3.7, Da_{LFA} and Da_{ext} represent the laminar flame assumption (LFA) transition Damköhler number and the quenching transition Damköhler number, respectively. Four combustion regimes are identified:

- The laminar flame region ($Re < 1$). The flow field is laminar, hence the flame has a laminar structure.
- The flamelet region ($Re > 1$ and $Da^* > Da_{LFA}$). The flame front behaves as a laminar flame characterized by the same scalar dissipation rate: the chemistry is fast enough to follow the flow changes induced by turbulence.
- The unsteady effects region, identified by $Re > 1$ and $Da_{ext} < Da^* < Da_{LFA}$. The unsteady flow effects are not negligible: the chemical time scale has the same magnitude order as the turbulence one.
- The quenching region, $Da^* < Da_{ext}$. The strain induced by the turbulence on the flame is so intense that flame quenching occurs.

* The coefficient a in the l_r formula in Figure 3.6 is the order of a global one-step reaction.

3.2 SI Engines Modeling

3.2.1 Combustion Models for SI Engines

3.2.1.1 SINGLE ZONE

The simplest approach for the combustion modeling is based on a single-zone schematization, where all thermodynamic properties are assumed homogenous within the cylinder. In this case, the energy equation, coupled to the ideal gas law and to the volume variation law, is enough to describe the evolution of the thermodynamic state during the engine period with the valves closed. Energy equation writes:

$$\frac{d(me)}{dt} = -p\frac{dV}{dt} - p\frac{dQ_w}{dt} \tag{3.15}$$

After some manipulations, the temperature derivative can be obtained:

$$\frac{dT}{dt} = \frac{1}{mc_v}\left(-p\frac{dV}{dt} - p\frac{dQ_w}{dt} - \frac{\partial e}{\partial x_b}\frac{dx_b}{dt}\right) \tag{3.16}$$

In this equation, volume variations can be computed based on the engine geometrical characteristics, and the rate of wall heat transfer, Q_w, is estimated by empirical correlations (for instance, the one proposed by Woschni [6]). The variation of the internal energy with the composition depends on the thermochemical properties of the in-cylinder mixture and can be computed by empirical correlations. The most crucial term to be evaluated is the residual mass fraction rate. This is usually described using purely mathematical S-shaped functions. To this aim, the most suitable function was proposed by Wiebe [7]. It has the following formulation:

$$x_b = 1 - \exp\left[-a\left(\frac{\theta - \theta_0}{\Delta\vartheta}\right)^{m+1}\right] \tag{3.17}$$

This simple equation proved to provide a correct reproduction of the experimental pressure trace, but, not being a phenomenological model, it requires a case-dependent tuning of the constants a and m.

3.2.1.2 TWO ZONES

A more advanced approach involves the subdivision of the in-cylinder volume in two zones where thermochemical properties are assumed homogenous. The volume is divided in burnt and unburnt zone, separated by a thin surface which schematizes the flame front. For each zone, the energy balance is written as:

$$\frac{d(m_b e_b)}{dt} = -p\frac{dV_b}{dt} - p\frac{dQ_{w,b}}{dt} + h_u\frac{dm_b}{dt} \tag{3.18}$$

$$\frac{d(m_u e_u)}{dt} = -p\frac{dV_u}{dt} - p\frac{dQ_{w,u}}{dt} - h_u\frac{dm_b}{dt} \tag{3.19}$$

where $h_u dm_b/dt$ is the coupling term. A common assumption is that the thermodynamic state is different between the zones (different internal energy, temperature, density, etc.), but the pressure is the same. The integration of these equations requires the same information as the single-zone formulation (ideal gas and volume variation laws) and the estimation of the coupling term. It can be computed once again by a simple Wiebe function or by a phenomenological model, which attempts to describe on physical bases a complex phenomenon, such as the propagation in a turbulent field of the flame front. This phenomenon is not yet fully understood, involving a superimposition of different aspects, which affect turbulent flame dynamics. As stated above, it is widely recognized that the combustion in an ICE mainly occurs in the so-called "wrinkled or corrugated flamelet" regimes, where the flame front thickness is smaller than the lowest turbulent length scale. In this regime, the turbulence does not modify the inner flame structure, and the complex chemical kinetics occurring within the flame front can be synthesized by the laminar flame speed. The turbulence intensity is indeed usually higher than the laminar speed. Various combustion models have been proposed since many years, trying to physically describe these phenomena. The peculiar aspect of each combustion model consists in the mechanism inducing the turbulence-related burn-rate enhancement [8, 9]. Other approaches were also proposed, where the combustion chamber is described as a stochastic reactor and the combustion is modeled by probability density functions [10]. The most common combustion models are the eddy burn-up, the fractal, and the CFM (coherent flame model) approaches. The eddy burn-up model describes the entrainment of unburnt mixture in flame front and the subsequent combustion by an exponential decay. It proved to properly agree with the experimental burnt mass fraction trends [11], even if its formulation is not based on any experimental evidence. The fractal model and the CFM try to estimate the flame front wrinkling and related effects on the burning rate on the basis of experimental investigations, as reported in several publications [9, 12, 13, 14].

3.2.1.3 EDDY BURN-UP

The eddy burn-up model is based on the original approach proposed by Keck [15], whose conceptual scheme is reported in Figure 3.8. Following the above schematization, the combustion phenomenon is divided in two main steps, the first one being the entrainment of unburnt mixture within in the turbulent flame brush, and of thickness proportional to the Taylor length scale λ [16]. According to Equation (3.20), the entrainment rate depends on the unburnt gas density, ρ_u, on the flame front area, A_L, assumed not corrugated, and on the entrainment speed, that in turn is described as the sum of laminar, u_L, and turbulent, u_T, components. This last is usually assumed proportional the turbulent intensity, u'. In a second stage, according to Equation (3.21), the entrained but still unburnt mass, $(m_e - m_b)$, is oxidized in a characteristic time scale, τ, given by Equation (3.22) [16].

FIGURE 3.8 Eddy burn-up combustion schematization.

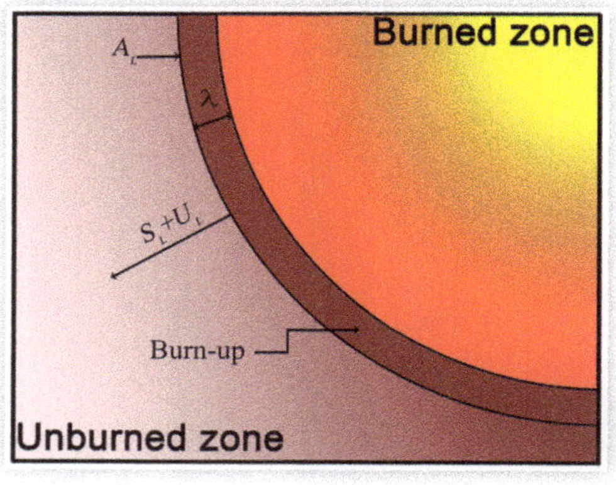

$$\frac{dm_e}{dt} = \rho_u A_L \left(u_L + u_T\right) \tag{3.20}$$

$$\left(\frac{dm_b}{dt}\right)_{eddy\ burn-up} = \frac{m_e - m_b}{\tau} \tag{3.21}$$

$$\tau = \frac{\lambda}{S_L} \tag{3.22}$$

In the various versions of the eddy burn-up model [15, 17], proper corrections are introduced to describe the transition from a laminar combustion to a turbulent one. Usually, those corrections are related to characteristic time and length scales of combustion

FIGURE 3.9 Comparison between eddy burn-up and experimental burn rates, for three different engines and rotational speeds.

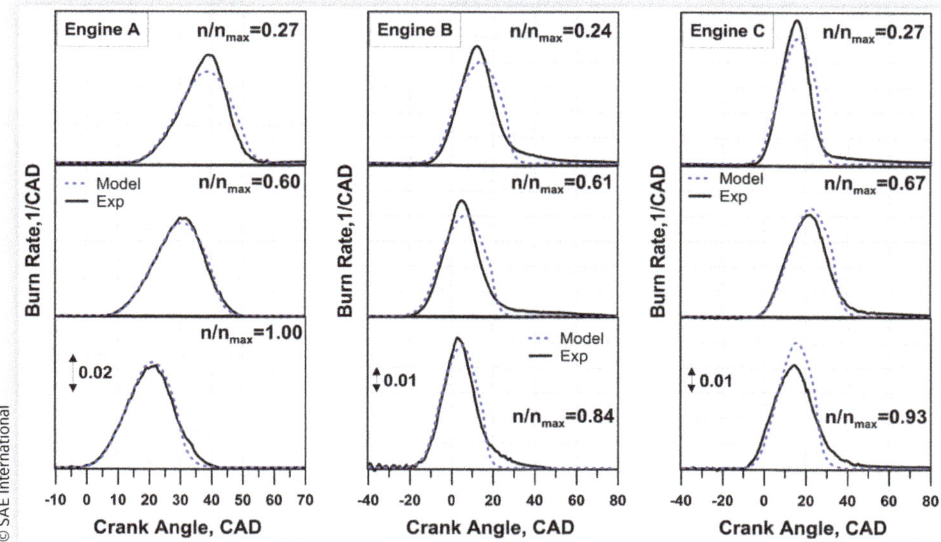

and turbulence. Eddy burn-up model does not require any specific treatment of combustion completion and flame-wall interaction. Equation (3.21), in fact, expresses an exponential burn rate decay, which automatically reproduces the typical burn fraction tail. A more direct control of burn rate slowdown during wall combustion is hence not possible. It must be pointed out that entrainment and burn-up schematization of the combustion process have never been experimentally observed [18]. Experiments indicate a corrugated flame front, sometimes presenting pockets of unburnt gas that burn up in-wards. For these reasons, Equations (3.20-3.22), rather than providing a physical description of combustion process, must be more properly considered as a mathematical modeling of the S-shaped burn fraction profile.

The accuracy and reliability of a combustion model can be assessed by the comparison between the predicted burn rate and the experimental one. Other comparisons, for instance regarding the global engine performance (BMEP, BSFC, mass flow rate, etc.) or the in-cylinder pressure cycles, may be affected by effects not directly related to the combustion process, such as the duct and/or valve fluid dynamic permeability, the heat losses, the mechanical losses, etc. For this reason, the reliability of all the considered combustion models will be discussed with reference to the burn rate profiles. In addition, to demonstrate the consistency of physical background, no model tuning will be applied changing the engine operating condition. As an example, in Figure 3.9, the results from the eddy burn-up approach model are compared with the experimental trends, for three engines, labelled as A, B, and C, and three rotational speeds (normalized by the maximum engine speed). The plots confirm the good model capability in describing the combustion process along its whole development and completion. The eddy burn-up model is able to follow the speed-related modifications in the burn rate peaks, with some major inaccuracies for the engine C. The beginning of the combustion process is correctly captured, while a certain disagreement appears during the completion phase.

3.2.1.4 FRACTAL APPROACH

The original version of the fractal combustion model has been proposed some decades ago [12]. The theoretical background of this approach is more directly founded on the

FIGURE 3.10 Schematic of the fractal combustion model.

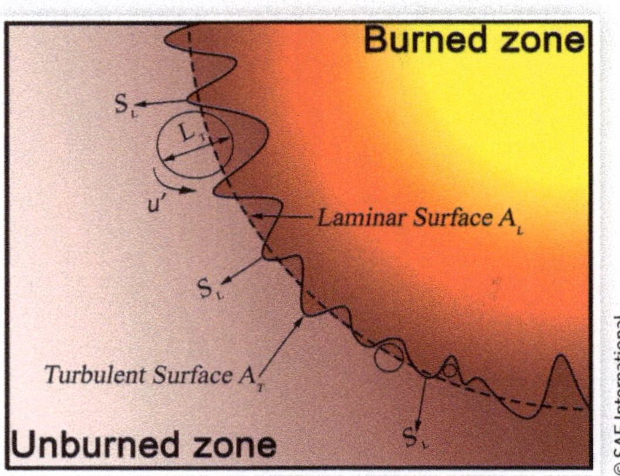

combustion regime occurring in a SI engine, falling in the wrinkled-corrugated flamelet zones reported in the well-known combustion regime diagram of Figure 3.4 [19]. In this combustion regime, the flame front exhibits an increased surface, A_T, due to the interaction with the turbulence (Figure 3.10). Consistently with this physical background, the fractal model directly describes the enhancement of flame front surface, resorting to the concepts of fractal geometry. Experimental observations proved in fact that a wrinkled flame front presents a fractal behavior, resulting in the auto-similarity of its basic structure [13, 14, 20, 21]. This gives the possibility to relate the turbulent flame front extent to the laminar one according to turbulence characteristic speed, time, and length scales [12, 19]. In the fractal formulation, the burn rate is proportional to the wrinkling factor, A_T/A_L:

$$\left(\frac{dm_b}{dt}\right)_{fractal} = \rho_u A_T u_L = \rho_u A_L u_L \left(\frac{A_T}{A_L}\right) = \rho_u A_L u_L \left(\frac{l_{\max}}{l_{\min}}\right)^{D_3-2} \tag{3.23}$$

According to the classical expression reported in [22], the wrinkling ratio is evaluated based on the fractal dimension, D_3, and the maximum and minimum wrinkling scales, $l_{\max}$ and $l_{\min}$. They are also interpreted as the scales of macro and micro vortices of the turbulent flow field. A proper estimation of those parameters has to be provided for the model closure. $l_{\max}$ is related to a macroscopic characteristic dimension of the flame front, such as the flame radius, r_f, the combustion chamber height, or the burnt surface/volume. $l_{\min}$ is commonly taken proportional to the size of the smallest turbulent eddy [12], expressed by the Kolmogorov length scale, l_k. Based on experimental evidences [21], the fractal dimension can be related to the turbulence intensity and the laminar flame speed. The above described fractal model actually applies for a fully developed and freely expanding turbulent flame. At the combustion beginning, the flame front is not effectively corrugated by the turbulence, while, at the end of the process, it is characterized by a progressive interaction with the walls. To handle those phenomena, proper reductions of the burning rate have to be applied [23, 24]. The fractal model reliability is demonstrated in Figure 3.11, with reference to the same engines and operating conditions as eddy burn-up case. The model is quite accurate, also considering that no case-dependent tuning is applied. The burn rate shape and peak are correctly reproduced at changing rotational speed. Looking at the comparison of Figures 3.9 and 3.11, the fractal model accuracy appears slightly higher than the eddy burn-up one, especially in the description of the combustion tail. In addition, the rotational speed variations are better captured.

3.2.1.5 COHERENT FLAME MODEL

The original version of the CFM was proposed in [25]. The CFM belongs to the flame surface density model family. The fuel burning rate in this approach is computed as:

$$\frac{dm_b}{dt} = \rho_u u_L A_T = \rho_u u_L \Xi A_L \tag{3.24}$$

where

ρ_u is the fresh gases density
A_T is the turbulent flame surface
u_L is the laminar flame speed

FIGURE 3.11 Comparison between fractal and experimental burn rates, for three different engines and rotational speeds.

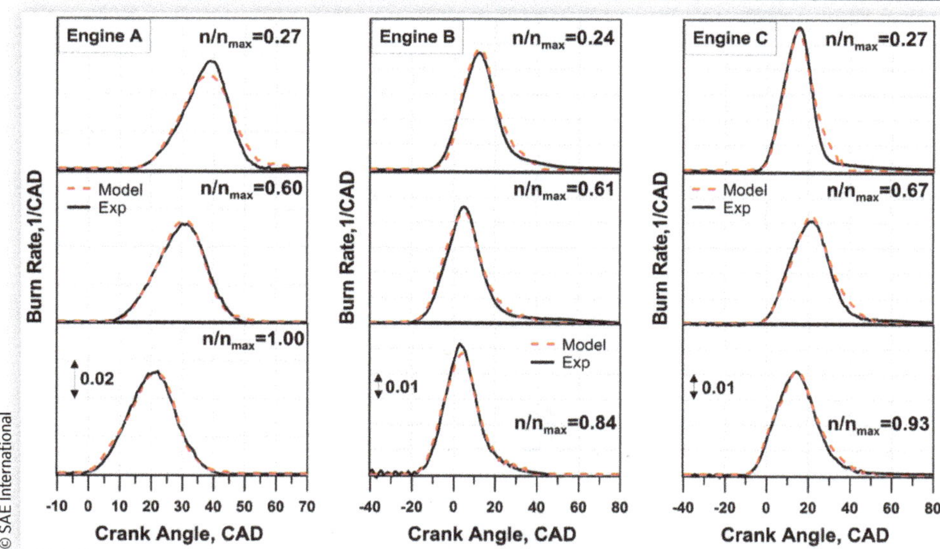

This last is computed by correlations calibrated on experimental results like the one proposed in [3], allowing to explicit the u_L dependence on thermochemical properties of the mixture. The turbulent flame surface A_T is written as the product of a mean flame surface, A_L, and a flame front wrinkling, Ξ, Figure 3.12 (left). The mean evolution of the flame surface over many engine cycles was investigated by [26]. These experiments suggest that the mean flame front grows spherically in the cylinder, at least during the initial phase of combustion. A_L is hence computed assuming a spherical flame propagating from the spark plug. Once flame approaches the combustion chamber walls, the spherical assumption of the burnt gas volume, V_b, is no longer valid. For this reason in [9], A_L was a priori tabulated as a function of piston position and V_b, Figure 3.12 (right). The mean flame surface is then obtained at each time step by interpolating the look-up table using the current piston position and V_b.

The wrinkling factor Ξ is computed by means of a 0D physical equation, obtained by reduction of the 3D computational fluid dynamics (CFD) equation for the flame surface density [27], describing its temporal evolution. This allows to correctly capture

FIGURE 3.12 Schematic of the CFM combustion model (left); principle of mean flame surface tabulation (right) [9].

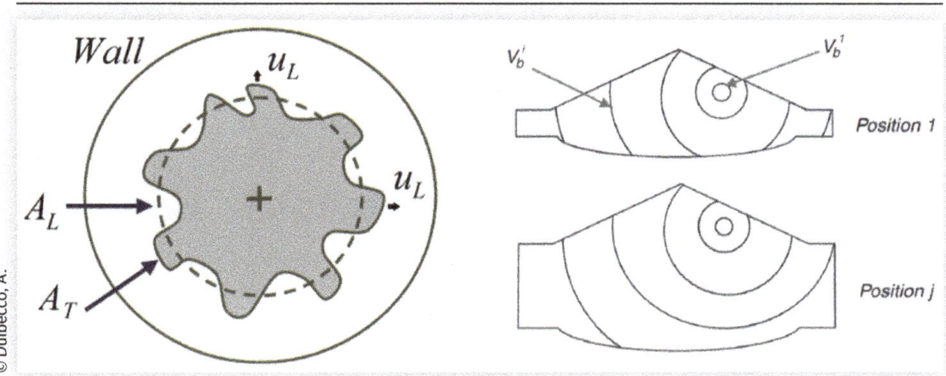

CHAPTER 3

FIGURE 3.13 Evolution of the flame wrinkling Ξ with time for different turbulence levels in a closed vessel for propane (left) and methane (right). Simulations are compared to experiments. Bold lines correspond to the equilibrium assumption [27].

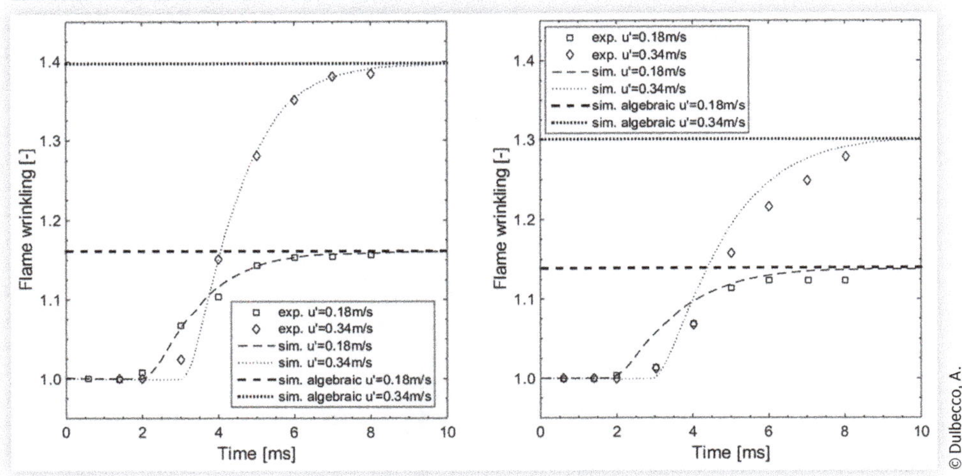

the laminar/turbulent transition of the flame propagation process, Figure 3.13, which depends on current thermodynamic and flow conditions:

$$\frac{1}{\Xi}\frac{d\Xi}{dt} = \Gamma\left(\frac{u'}{u_L}, \frac{l_t}{\delta_L}\right)\frac{u'}{l_t}\left(\frac{\Xi_{eq} - \Xi}{\Xi_{eq} - 1}\right) - \frac{2}{r_b}(1+\tau)(\Xi - 1)u_L \tag{3.25}$$

where

Ξ_{eq} is the wrinkling factor at equilibrium
u' is the instantaneous velocity fluctuation
Γ is the efficiency function of the turbulent flow on the flame strain [28]
l_t is the integral length scale
$\tau = (\rho_u/\rho_b - 1)$ is the thermal expansion rate
$r_b = (3V_b/4\pi)^{1/3}$ is the mean flame radius
δ_L is the laminar flame thickness [29]

The first RHS term of Equation (3.25) corresponds to the flame strain by all turbulent structures, while the second describes the effect of the thermal expansion, which limits the flame front wrinkling by imposing a positive curvature. Ξ_{eq} is computed as:

$$\Xi_{eq} = \left(1 + 2\left(\frac{u'}{u_L}\right)\sqrt{\frac{\Gamma C}{Sc}}\right) \tag{3.26}$$

with C a modeling constant and Sc the Schmidt number, defined as the ratio of momentum diffusivity (kinematic viscosity) and mass diffusivity.

3.2.2 Turbulence Submodels for SI Engines

A reliable and predictive turbulence model has to be able to describe the effects of variations in both operating conditions and the engine architecture. On the other hand, the development of such a 0D model represents a very challenging task, since it aims to synthesize very complex 3D phenomena.

The current literature presents various approaches to face the above issue, most of them classified as k-ε and K-k models. Original proposals of k-ε models are reported in [30] and [31]. In those works, 0D equations of turbulent kinetic energy, k, and dissipation rate, ε, are derived from 3D turbulence models. A first attempt to take into account the contribution to turbulence from ordered flow structures is proposed in [32], where source terms related to the swirl and squish motions are introduced. A more recent version of such approach is proposed in [33]. A different method consists in describing the energy cascade from mean flow kinetic energy, K, into turbulent kinetic energy, k, usually referred as K-k models [34, 35]. Turbulence dissipation is, hence, derived from the integral length scale, l_t, which, in turn, is related to the instantaneous cylinder volume and piston-head distance or reconstructed by mathematical functions [34, 35].

A refinement of a K-k approach involves a more direct description of the ordered flow motions, instead of the overall in-cylinder mean flow velocity. A simplified methodology is proposed in [36], where a linear decrease of the tumble vortex is imposed as a function of the piston distance from the cylinder head. In [37, 38], more physical tumble models are introduced, capable of considering the geometry of the intake system and the valve and the cylinder port inclination. In [39], a model is developed, describing both tumble and swirl motions and their dissipation in turbulence. A critical issue in such methodologies is the description of the flow structure decay during the compression stroke, due to the shear stresses. 3D CFD studies showed that the decay rate well correlates with the piston-head distance. For this reason, the above phenomenon is described by a decay function depending on the piston position normalized by the bore [39]. In recent works [40, 41], K-k and k-ε models are synthesized, leading to the so-called K-k-ε model. Such methodology takes into account the energy cascade mechanism typical of K-k model and directly describes the turbulence dissipation. In addition, a tumble model is provided by a dedicated additional equation. Such refined approaches proved a good accuracy in predicting both tumble ratio and turbulence intensity under various operating conditions, valve strategy, and engines.

Based on the above literature overview, phenomenological models throughout time have reached an improved predictive capability. In the past, models were commonly validated in an indirect way comparing numerical and experimental burn rates [17, 42]. The availability of experimental data about mean flow and/or turbulence fields for model validation purpose was still rather limited. In addition, most of the available experimental data on tumble and turbulence were based on measuring techniques which allowed to observe only a limited portion of the cylinder. These were suitable for validating a 3D model [43], where the velocities could be calculated at different spatial locations inside the cylinder. In phenomenological models, where spatial homogeneity of thermodynamic properties and fluid dynamic variables are assumed, the use of location-specific experimental data for validation purpose is rather questionable. However, even more recent and advanced dynamic flow visualization techniques, allowing for the investigation of the tumble flow field, are limited to a very reduced set of operating conditions [44], because of their high costs and complexity.

The most reliable approach for phenomenological model validation is hence based on the 3D CFD results, which easily include the main fluid dynamic phenomena occurring inside the combustion chamber. In addition, during time, 3D simulations are becoming more and more accurate and reliable, thanks to the adoption of more refined numerical approaches and to the growing computing power. 3D codes, moreover, allow to investigate various operating conditions and engine geometries more easily than experimental techniques. Phenomenological model validation, based on averaged 3D results, has been successfully followed in [34, 37, 39, 41].

The formulation of k-ε equations in a 0D fashion can arise from turbulence model employed in 3D RANS simulations. In this field, the two-equation k-ε RNG model [45] for turbulence description is widely adopted. Starting from these equations, the 0D version of k-ε balances are:

$$\frac{d(mk)}{dt} = (\dot{m}k)_{inc} - (\dot{m}k)_{out} + \frac{2}{3}\frac{\dot{\rho}}{\rho}\left(-m\nu_t\frac{\dot{\rho}}{\rho} + mk\right) + P - m\varepsilon \tag{3.27}$$

$$\frac{d(m\varepsilon)}{dt} = (\dot{m}\varepsilon)_{inc} - (\dot{m}\varepsilon)_{out} + c_{\varepsilon_1}\frac{\varepsilon}{k}\left[P - \frac{2}{3}m\nu_t\left(\frac{\dot{\rho}}{\rho}\right)^2 + \frac{2}{3}mk\frac{\dot{\rho}}{\rho}\right] + \\ - c_{\varepsilon_2}\frac{m\varepsilon^2}{k} - c_{\varepsilon_2}m\varepsilon\frac{\dot{\rho}}{\rho} - \frac{c_\mu\eta^3\left(1-\dfrac{\eta}{\eta_0}\right)m\varepsilon^2}{1+\beta\eta^3}\frac{m\varepsilon^2}{k} \tag{3.28}$$

where $c_{\varepsilon i}$ and c_μ are model constants from 3D formulation. The main assumptions regard the negligibility of the external gravitational forces and of the diffusive effects. According to the 3D definitions, the turbulent viscosity, ν_t, and the parameter, η, are computed by:

$$\nu_t = c_\mu\frac{k^2}{\varepsilon} \quad \eta = \sqrt{\frac{P}{m\nu_t}\frac{k}{\varepsilon}} \tag{3.29}$$

P represents a turbulence kinetic energy production, and it is related to the strain tensor of the mean flow velocity. Finally, the velocity divergence in the 3D equations du_i/dx_i is substituted by the term $\dot{\rho}/\rho$, according to the continuity balance.

Compared to the 3D equations, the 0D variant includes two additional terms to describe the contribution of the convective flows, both incoming and outcoming, through the valves. Some authors integrate the k-ε system only during the engine cycle portion with the valves closed (compression and expansion phases). In this case, convective terms are neglected, while production contribution is modeled in a simplified way [33]. More complex approaches involve the description of the mean flow, both ordered and unordered flow structures. In this case, two additional equations have to be formulated to characterize the time evolution of the specific angular momentum associated to the tumble motion and to the mean flow kinetic energy, defined as:

$$T = U_T r_T \quad K = \frac{U_{fK}^2}{2} \tag{3.30}$$

where

U_{fK} and U_T are the mean flow and tumble vortex velocities
r_T is the related tumble radius

To be more precise, K includes contributions of both unorganized and organized flows developing inside the cylinder. Based on the above definitions, angular momentum and kinetic energy balance equations look like:

$$\frac{d(mT)}{dt} = (\dot{m}T)_{inc} - (\dot{m}T)_{out} - f_d\frac{mT}{t_T} \tag{3.31}$$

$$\frac{d(mK)}{dt} = (\dot{m}K)_{inc} - (\dot{m}K)_{out} - f_d\frac{mK}{t_T} + mK\frac{\dot{\rho}}{\rho} - P \tag{3.32}$$

The first and second terms in both equations describe the incoming and outcoming convective flows through the valves, respectively, while the third one expresses the decay due to the shear stresses with combustion chamber walls. To model this effect, a decay function, f_d, [39] and a characteristic time scale, t_T [40], are commonly introduced.

Coherently with k formulation, K equation includes an additive compressibility term, $\mu K\dot{\rho}/\rho$, and a subtractive turbulent production term, P, expressing the energy cascade mechanism. The latter quantity, related to the strain tensor rate, S_{ij}, requires some modeling assumptions. Various proposals are available in the literature. Most of them [34, 35, 41] assumes that P is proportional directly to the mean flow kinetic energy, K, and inversely to the characteristic time scale, t_T. Other approaches [40] hypothesize that the main driver for turbulence production is only the kinetic energy associated to unordered flow structures, obtained as a difference between the overall kinetic energy K and the one associated to the tumble motion, $K_T = U_T^2/2$. Concerning the convective terms, most common formulations depend on the isentropic velocity through the valves corrected by the proper discharge and/or tumble coefficients [40].

A turbulence model including the equations of both k, K, ε, and T represents the most complex possible approach available in the current literature. A certain simplification can be obtained without considering a direct ε modeling, but an indirect estimation by the following constitutive equation:

$$\varepsilon = c_\mu^{3/4} \frac{k^{3/2}}{l_t} \tag{3.33}$$

Such relation, which strictly speaking only applies at a local level, can be extended in a 0D framework introducing minor inaccuracies. Following along these lines, the global reliability of model is maintained by an accurate description of the integral length scale l_t. In this regard, 3D simulations proved that this parameter only slightly changes with the engine operating conditions. This allows to model l_t by a case-independent mathematical reconstruction which only depends on the in-cylinder geometry (instantaneous cylinder volume, piston-head distance) or is a sequence of simple shape functions (see example in Figure 3.14). This simplified approach demonstrated an accuracy comparable with the most complex one, also including the integration of the dissipation rate equation [40].

FIGURE 3.14 Integral length scale reconstruction and comparison with 3D outcomes under various engine operating conditions.

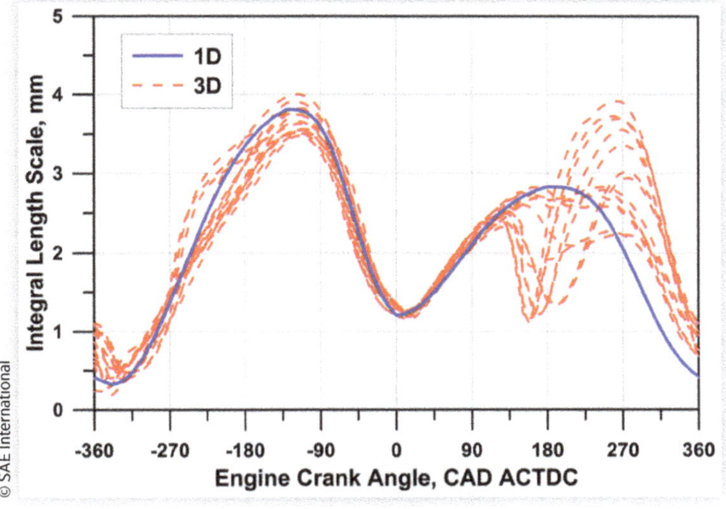

FIGURE 3.15 0D/3D model comparison of mean flow velocity, for various rotational speeds and valve strategies.

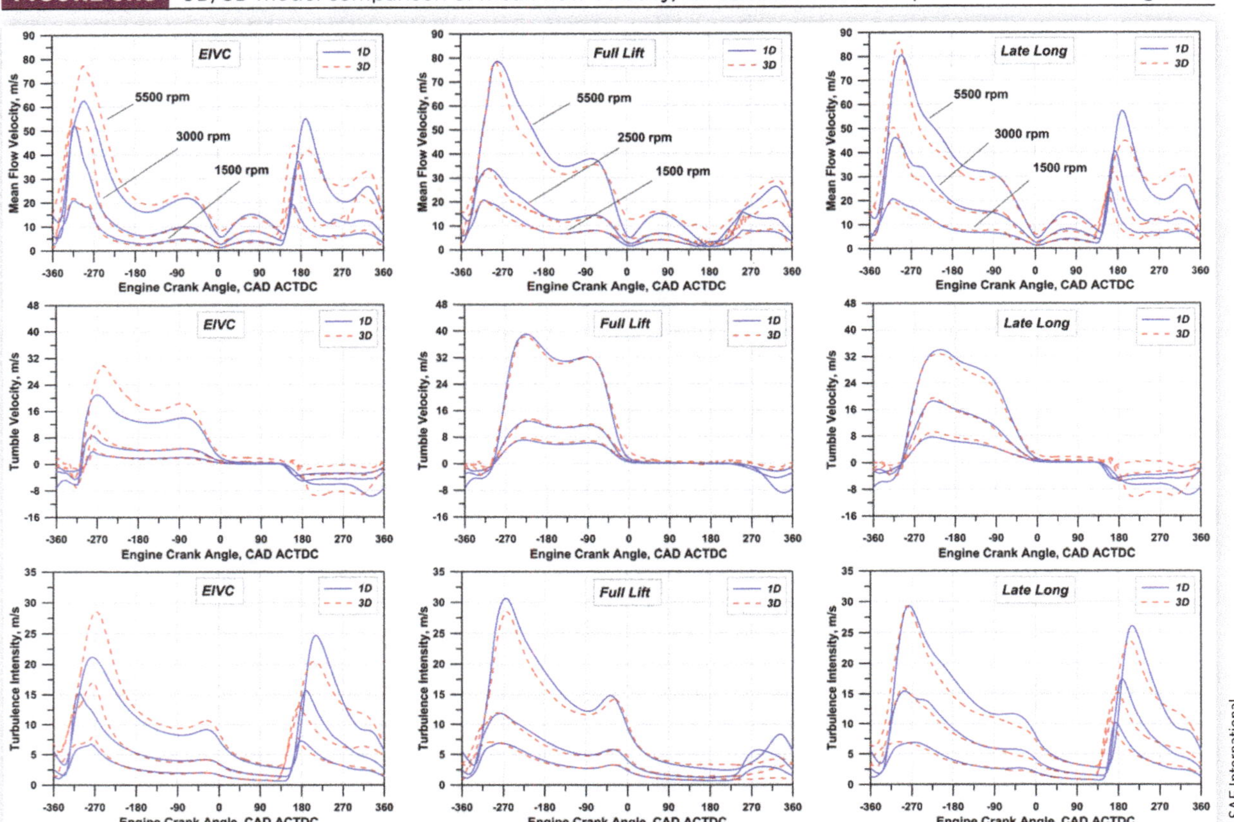

As stated above, the turbulence model reliability is usually demonstrated by the comparison with 3D outcomes. As an example, in Figure 3.15, the results of *K-k-ε* model coupled to the tumble description are compared with the mass-averaged 3D outcomes along the whole engine cycle. The comparisons concern an automotive twin-cylinder engine equipped with a VVA system on the intake side. This last is able to realize both early intake valve closing (EIVC), conventional (here labelled as "full lift"), and late intake valve closing (LIVC) (here labelled as "late long") strategies. The assessment deals with the mean flow velocity, the tumble velocity, and the turbulence intensity. To verify the model consistency of physical background, no model tuning is applied changing the engine operating condition. The overall model behavior is quite satisfactory, since most of the analyzed operating conditions are reproduced with relevant accuracy. Mean flow velocity, tumble velocity, and turbulent intensity denote a very good agreement with 3D profiles during the whole engine cycle.

3.2.3 Emission Models for SI Engines

Most critical pollutants for SI engine applications are carbon monoxide (CO), nitrogen oxides (NO_x), and unburnt hydrocarbons (UHC). While the first two are mainly related to the thermochemical states of unburnt and burnt mixtures, HC have multiple origins, some of them depending on local geometrical features, whose contributions vary significantly depending on the engine design. For this, even if approaches exist to deal with specific sources of HC in 0D and quasiD [2], this topic will not be developed in what follows.

3.2.3.1 CARBON MONOXIDE

In premixed combustion, CO can be produced both through the flame front and in burnt gases [2]. CO formation within the flame front mainly occurs in the presence of burning rich mixtures. To describe this process, instead of using a classical stoichiometric chemical reaction to represent fuel oxidation, in [9] a set of two equations is proposed:

$$\begin{array}{l} C_xH_yO_z + \left(x + \frac{y}{4} - \frac{z}{2}\right)O_2 \rightarrow xCO_2 + \frac{y}{2}H_2O \quad \text{for } \phi < 0.98 \\ \begin{cases} \alpha_1\left(C_xH_yO_z + \left(x + \frac{y}{4} - \frac{z}{2}\right)O_2 \rightarrow xCO_2 + \frac{y}{2}H_2O\right) + \\ +(1-\alpha_1)\left(C_xH_yO_z + \left(\frac{x}{2} - \frac{z}{2}\right)O_2 \rightarrow xCO + \frac{y}{2}H_2\right) \quad \text{for } \phi \geq 0.98 \end{cases} \end{array} \tag{3.34}$$

These reactions are weighted using a parameter α_1, which can be expressed considering the atomic conservation of C, H, and O, as:

$$\alpha_1 = \frac{0.98\dfrac{4x+y-2z}{\phi} - 2(x-z)}{2x+y} \tag{3.35}$$

Accordingly, for an increasing of ϕ, α_1 decreases and the amount of produced CO increases. In burnt gas, due to the high temperature, chemical reactions influencing CO emissions also occur. Indeed, burnt gas composition evolves in time toward chemical equilibrium concentrations, with time scales dictated by temperature. It is assumed that around peak cycle conditions ($T_b \approx 2800$ K; $p > 15$ bar) burnt gas composition is almost at equilibrium. To account for this mechanism, a modeling approach based on a reduced chemical-mechanism was first proposed in [46]. In [9] a different approach was proposed, where an equilibrium scheme, similar to the one proposed in [47], is adopted:

$$\begin{cases} N_2 \leftrightarrow 2N \\ O_2 \leftrightarrow 2O \\ O_2 + 2CO \leftrightarrow 2CO_2 \\ H_2 \leftrightarrow 2H \\ O_2 + H_2 \leftrightarrow 2OH \\ O_2 + 2H_2O \leftrightarrow 4OH \end{cases} \tag{3.36}$$

By solving the above equation system, the equilibrium species concentrations, *[i]*, are determined. To be solved, the set of six equations, based on ten species, must be integrated with four atomic conservation equations for C, O, H, and N. As equilibrium concentrations are not reached instantaneously, but depend on chemical activity, a pseudo kinetics is used as long as temperature is sufficiently high ($T_b > 1700$ K):

$$\frac{dm_i}{dt} = \frac{m_i^{eq} - m_i}{\tau_{chem}} \tag{3.37}$$

where m_i^{eq} and m_i are, respectively, the mass of the ith species in the cylinder at equilibrium, based on the computed concentration *[i]*, and the actual mass and τ_{chem} the

CHAPTER 3

FIGURE 3.16 Comparison of simulated and experimental pollutant engine-out emissions relative to different engine operating conditions [9].

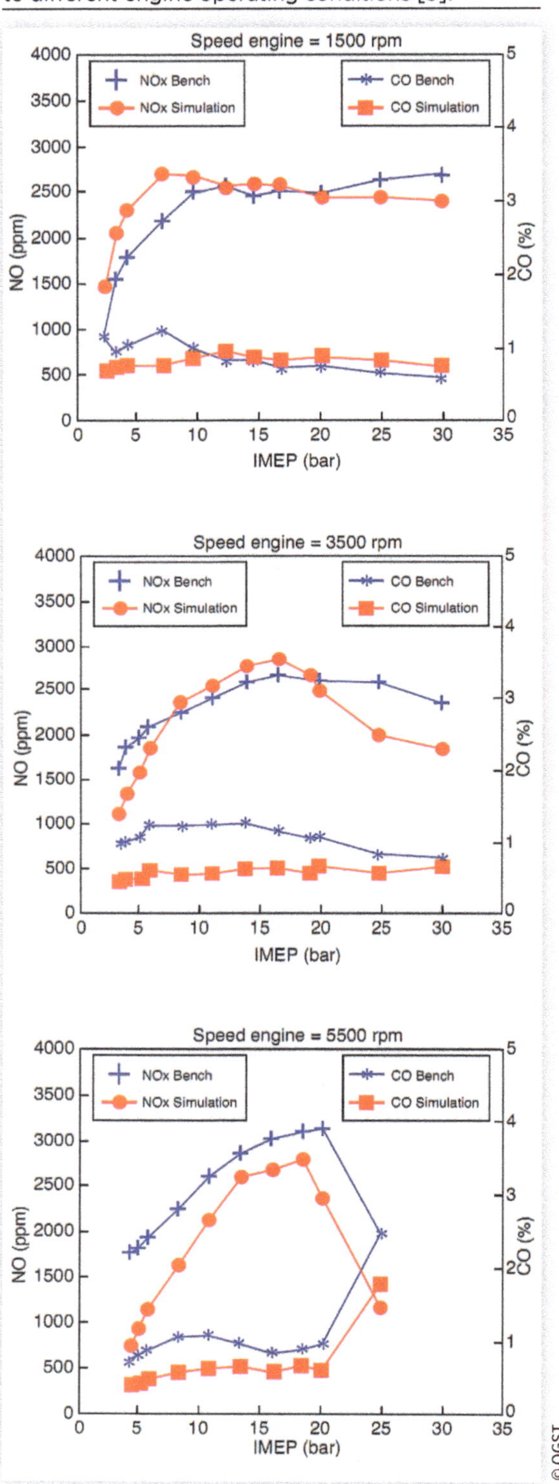

characteristic time to relax the current burnt gas composition toward chemical equilibrium one. Chemical kinetics being strongly dependent on temperature, τ_{chem} is expressed as a linear function of the burnt gas temperature. Figure 3.16 shows some results obtained by using the described approach over engine operating conditions representative of a complete engine map.

3.2.3.2 NITROGEN OXIDES

NO_x emissions, mostly represented by NO and NO_2 species, result from out-of-equilibrium chemical processes which are controlled by the kinetics of relevant reactions. NO_2 emission is generally small compared to NO emission (this last, in average, estimated as 98% of the total), even if for given engine operating point (EOP) can attain 30% of the overall NO_x emissions [2]. Accordingly, in what follows, only NO formation will be detailed, by assuming its concentration enough to obtain reliable estimations of global NO_x emissions. Four paths leading to NO formation are commonly retained in the literature:

- Thermal NO path, the most important in ICE, describes NO production at high temperature through atmosphere-N_2 oxidation.
- Prompt NO path describes NO production at high temperature within the flame front through atmosphere-N_2 oxidation, by means of CH_x radicals. This mechanism is generally neglected. This choice is commonly justified by the fact that flame is too thin to allow a sufficient residence time to produce relevant NO compared to the thermal NO path.
- Nitrous NO path describes the NO production through the action of N_2O as a third body reaction species. According to [48], this mechanism is preponderant at high-pressure to low-temperature conditions ($T_b \approx 1800$ K) where thermal NO is frozen. Those conditions are not the most frequent in common engine architectures.
- Fuel NO path describes the formation of NO due to the presence of nitrogen in the fuel molecule. Relatively to present automotive fuels, this is a negligible source of NO.

In 0D/quasiD simulation, NO_x modeling often reduces to thermal NO modeling. The most common approach consists of the integration within the combustion model of the extended Zeldovich chemical mechanism [49], describing NO formation around stoichiometry. This mechanism is based on a set of three reversible equations:

$$\begin{cases} N_2 + O \leftrightarrow NO + N \\ O_2 + N \leftrightarrow NO + O \\ OH + N \leftrightarrow H + NO \end{cases} \tag{3.38}$$

TABLE 1 Rate constants for the extended Zeldovich NO formation mechanism [2]

Coefficient	Value [cm³/mol/s]	Temperature range [K]
K_1^+	7.6e13exp(-38,000/*T*)	2000-5000
K_1^-	1.6e13	300-5000
K_2^+	6.4e9T_{bg}exp(-3150/*T*)	300-3000
K_2^-	1.5e9T_{bg}exp(-19,500/*T*)	1000-3000
K_3^+	4.1e13	300-2500
K_3^-	2.0e14exp(-23,650/*T*)	2200-4500

This system is solved using the kinetic constants K_i of the different equations i, with indexes + for the forward direction and – for the backward one. The evolution of NO is then computed as:

$$\frac{d[\mathrm{NO}]}{dt} = K_1^+[\mathrm{O}][\mathrm{N_2}] + K_2^+[\mathrm{N}][\mathrm{O_2}] + K_3^+[\mathrm{N}][\mathrm{OH}] + \\ -K_1^-[\mathrm{NO}][\mathrm{N}] - K_2^-[\mathrm{NO}][\mathrm{O}] - K_3^-[\mathrm{NO}][\mathrm{H}] \tag{3.39}$$

The evolution of N_2, O_2, N, O, H, and OH is then computed by exploiting the relations made explicit by the equation system (3.38). K_i values proposed in [2] are recalled in Table 1. In [9], to account for the fact that reactions of NO formation are rapidly frozen during the expansion stroke [2], the extended Zeldovich mechanism is only solved for $T_b > 2500$ K.

Figure 3.16 shows CO and NO_x emissions computational results compared to experiments [9]. Results refer to a four-stroke homogeneous charge GDI single-cylinder engine working under different operating conditions, in terms of engine speed and load. According to the figure, engine-out emission trends and quantitative values can be reasonably predicted by using modeling approaches essentially based on the thermochemical state of burnt gas, as those proposed in Sections 3.2.3.1 and 3.2.3.2.

3.2.4 Knock Models for SI Engines

It is generally accepted that engine knock is the result of auto-ignition in the end gas before it is being reached by the flame front emanating from the spark plug [50, 51, 52]. Starting from this assumption, two families of knock model have been developed during time: ignition-delay model and chemical kinetic model.

3.2.4.1 IGNITION-DELAY MODELS

In a constant pressure-temperature environment, the auto-ignition time, τ_{AI}, can be estimated by empirical correlations, which commonly have an Arrhenius-like formulation [53]:

$$\tau = Ap^{-B}\exp\left(-\frac{E_a}{RT}\right) \tag{3.40}$$

The equation parameters, A, B, and E_a, only depend on the air/fuel mixture composition and quality (octane number of primary reference fuel [PRF], equivalence ratio, residual fraction). The application of such approach to the engine simulation has to take into account the pressure-temperature variability during the cycle induced by the piston motion and by the combustion development. Following the proposal of Livengood and Wu [54], the estimation of the auto-ignition conditions is hence evaluated by the integration over the time of a variable τ_{AI}, according to the current in-cylinder state:

$$\int \frac{dt}{\tau_{AI}} = 1 \tag{3.41}$$

Knock occurrence is hence assumed when the above integral reaches the unity. Such basic approach does not take into account the negative temperature coefficient, typical of various common fuels, and it cannot describe the onset of cool flame. Those issues are addressed by more complex formulations of the auto-ignition time. A modified version of the Livengood and Wu integral was also introduced to describe the activation of differentiated chemistries at different temperatures (two- or three-stage integral) [55, 56]. The intensity of the knock is usually related to some parameters detected at the knock occurrence (in-cylinder volume, unburnt fraction), to the fuel characteristics (octane number), and/or to the mixture composition (equivalence ratio, residual fraction) [57].

3.2.4.2 KINETIC MODELS

A more advanced approach to describe the development of the auto-ignition reactions is based on chemical kinetic models, applied to the unburnt (end gas) zone. Starting from the general formulation of the energy equation in the unburnt zone, and expressing the internal energy, e_u, as a function of the temperature and of the species mass fraction, x_i:

$$e_u(x_i, T_u) = \sum_{i=1}^{N_{spec}} x_i e_i(T_u) \tag{3.42}$$

after some rearrangements, the unburnt temperature equation is obtained:

$$\frac{dT_u}{dt} = \left[\frac{dp}{\rho_u c_{pu}} - \left(\frac{dQ_{w,u}}{dt} + \frac{dQ_{chem}}{dt}\right)\right] \Big/ m_u c_{pu} \tag{3.43}$$

$$\frac{dQ_{chem}}{dt} = m_u \sum_{i=1}^{N_{spec}} e_i \frac{dx_i}{dt} \tag{3.44}$$

In a standard combustion modeling, unburnt gases are characterized by a frozen composition ($dx_i/dt = 0$). In a kinetic model, the above simplification is removed, and a detailed kinetic scheme is solved, together with the mass and energy equations in the unburnt zone. Various kinetic schemes are available in the literature, differing in terms of number of elements, species, and reactions. Since the chemistry solution has to be accomplished at runtime during the engine simulation, simplified or semi-detailed kinetic scheme is usually selected. More suitable scheme for engine simulation are the one proposed by Tanaka and Keck [58] for a PRF (5 elements, 32 species, and 55 reactions) or the one developed by Andrae (5 elements, 137 species, 635 reactions) for a TRF [59]. Those mechanisms handle both low- and high-temperature reactions and are tuned to well reproduce the ignition delay and evolution of most important species.

FIGURE 3.17 Typical pressure and temperatures of numerical trends in the presence of knock.

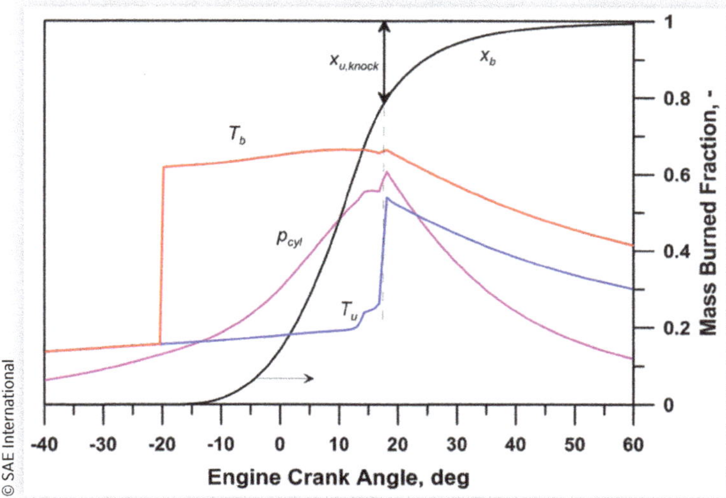

Knock event is assumed as the condition when "high-temperature" reactions activate in the unburnt zone, resulting in a sudden jump in the unburnt temperature profile during the final stage of the combustion process. To better understand the numerical procedure for knock detection, representative traces of in-cylinder pressure, temperature, and mass fraction burnt, p, T_u, T_b, and x_b, are shown in Figure 3.17, for a very severe knocking operation. Looking at the unburnt temperature trend, in a first stage only "low-temperature" reactions activate, while in a second stage, "high-temperature" mechanism starts, as well, causing a sudden jump in the trend. At the same time, a sudden increase in the pressure profile appears. The knock intensity can be quantified by empirical correlations, as the ones introduced in the previous section, or following the alternative approach proposed in [60]. In this case, the knock index, Δp_{knock}, is computed as the pressure increase in an instantaneous isochoric combustion of the end gas still unburnt at knock event:

$$\Delta p_{knock} = \frac{R}{c_v} \frac{x_{u,knock} m_f \mathrm{LHV}}{V_{knock}} \tag{3.45}$$

where

R and c_v are the average gas constant and the constant volume specific heat of the in-cylinder gases

m_f and LHV are the mass and the low heating value of the fuel

V_{knock} and $x_{u,knock}$ are the in-cylinder volume and end-gas unburnt fraction at the knock event (Figure 3.17)

This formulation of the knock index closely resembles the experimental MAPO definition, providing the opportunity to employ the same constraints for knock indices both numerically and experimentally.

Another output of a kinetic approach is the species temporal variation in the unburnt zone. As an example, in Figure 3.18, typical species trends under knock-free (left) and intense knocking (right) operation are shown, respectively. It can be observed that in the first case, only "low-temperature" reactions activate during the last stage of the combustion process (at about 25 CAD after the FTDC). Correspondingly, a partial oxidation of the considered fuel components occurs, inducing the formation of a large amount of CO.

FIGURE 3.18 Species trends in the unburnt zone under knock-free (left) and knocking (right) operation.

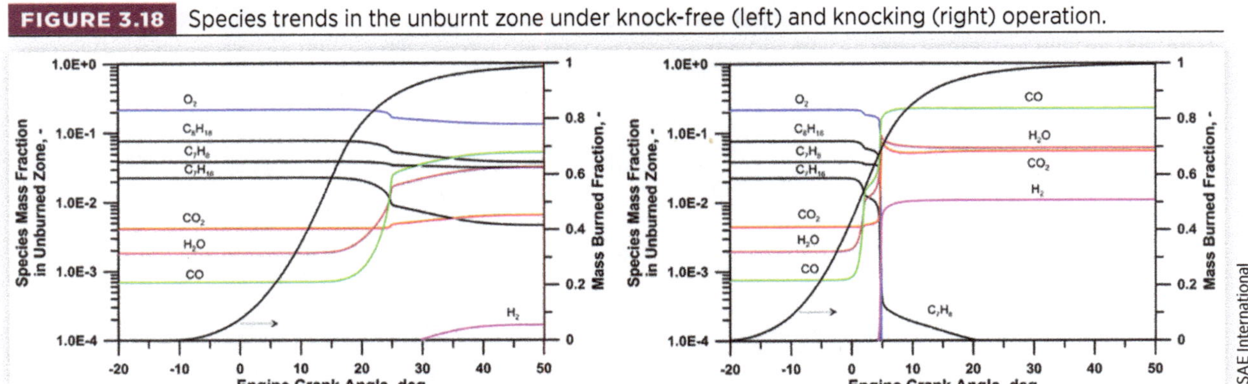

On the other hand, in the second case, the "high-temperature" mechanism (occurring about 5 CAD after the FTDC) closely follows the "low-temperature" mechanism. This happens when the unburnt mass fraction left behind the flame front is very high ($x_{u,knock} \approx 0.3$), corresponding to the presence of very intense knock. The "high-temperature" reactions cause an almost instantaneous consumption of the fuel components, excepting the toluene that is characterized by a slower kinetic. The above oxidation process also induces the formation of incomplete combustion species, such as CO and H_2.

The knock model can be validated through the comparison between the numerical and experimental spark advance at knock borderline (KLSA). To this aim, the maximum knock magnitude (threshold level) has to be specified, and the spark timing has to be properly adjusted so to realize the most advanced possible combustion phasing with an acceptable knock intensity. As an example, in Figure 3.19 the numerical KLSA is compared with the experimental one for a high-performance naturally aspirated engine. The assessment regards full load operation over a wide range of rotational speeds. In the figure, it can be noted that an excellent numerical/experimental agreement confirms the accuracy of the knock model. In this example, an average absolute error of 0.33 CAD is reached, with a maximum difference of 1.1 CAD at 3000 rpm.

FIGURE 3.19 Numerical/experimental KLSA at full load.

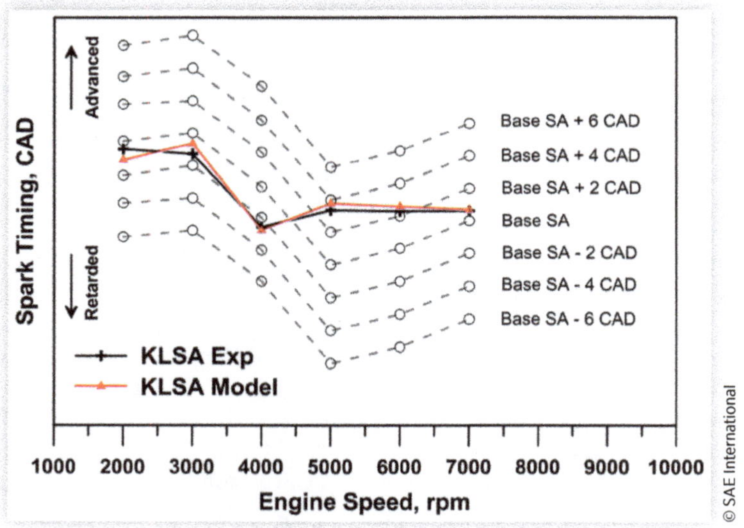

3.3 CI Engines Modeling

CI engines draw great attention due to their high efficiency potential, and this is reflected also by the efforts put in modeling their behavior [2, 61]. The complex physical phenomena taking place in the cylinder during fuel oxidation process make the combustion modeling a challenging task, especially for system simulation approaches for which a reduced computational time is a priority. To develop such modeling approaches, two opposite viewpoints can be envisaged:

- Starting from experimental data,* fuel mass consumption rates are derived and used to identify mathematical formulations, based on the phenomenology of the subjacent physics, able to describe experiments. Typical CI combustion models developed by following this approach are those proposed in [62, 63].
- starting from 3D CFD software, which in last decades knew strong developments and whose performance showed its potential to be used as an accurate virtual test bench, single physical phenomena can be isolated within the complex combustion process scenario. This allows to extract understanding about the phenomena themselves, but also the conditioned interaction between them, and use it to build more detailed physical-based 0D/quasiD combustion models in the perspective to make them more robust in terms of predictivity and suitable to interpolate/extrapolate the engine behavior. Typical CI combustion models developed by following this approach are those proposed in [64, 65].

Different approaches exist in literature, varying from single-zone to quasiD approaches. Their origins are usually dictated by user application constraints, expressed in terms of computational cost and required level of physical contents. This section wants to give an overview of recent 0D/quasiD thermodynamic CI combustion models and make the reader aware about the potential and the limits of a given approach, to help him in choosing the most adapted formulation for a given target; in fact, being models, by definition, an approximation of reality, it does not exist the best model, but the most appropriate for a given application.

3.3.1 CI Combustion Phenomenology

The considered thermodynamic system is an open system delimited by the cylinder head, liner, and piston surface; the system exchanges heat with the ambient through the walls, while exchanges of mass and enthalpy take place through valves seats, injector orifices, and blow-by surfaces. During the intake process, fresh unburnt gas, in the most general case a mixture of air and dilution gas, is admitted within the cylinder with a given kinetic energy and turbulence intensity, due to the high Reynolds number of the flow, and mixes with the residual burnt gas. Depending on the design of the intake port, admitted mass organizes in a coherent macroscopic vortex of swirl motion. During injection, Figure 3.20(A), because of the high injection pressure, liquid fuel, subjected to the presence of cavitation phenomena within the injector nozzle and to hydrodynamic instabilities within the combustion chamber, fragments into droplets and ligaments (primary breakup) which, in turn, atomize into small droplets (secondary breakup). Liquid fuel

* Experiments performed at engine test bench allow to determine in-cylinder mixture conditions at intake valve closing (IVC). Such information together with other boundary conditions (engine speed, injected fuel mass, wall temperatures, etc.) and the measured in-cylinder pressure trace during the close system part of the engine cycle allows to estimate, by applying the energy conservation principle, the heat release rate (HRR) law, and, consequently, at the first order by means of the fuel lower heating value (LHV), the fuel mass consumption rate.

FIGURE 3.20 Combustion scenario in a CI engine. Pressure traces (motored and fired), heat release rate (HRR), and fuel mass fraction burned (FMFB) detail the combustion process. Letters indicate key instants of the process [61].

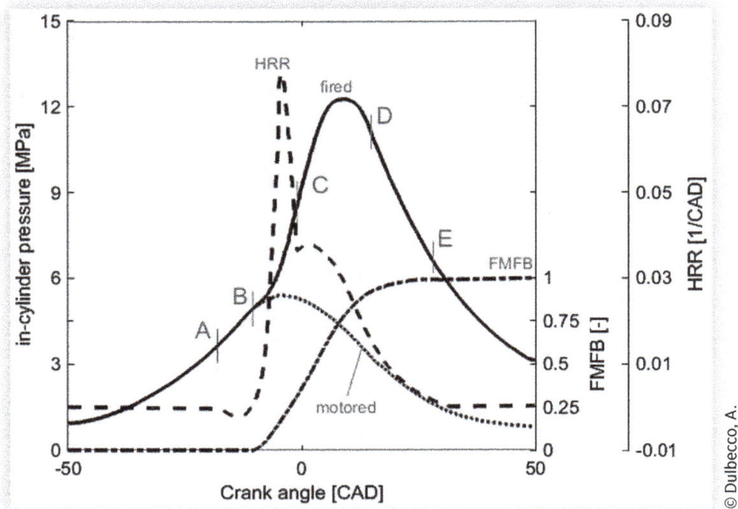

within the cylinder heat-up evaporates and mixes with in-cylinder mixture forming a spray region where fuel and ambient gas coexist.

Liquid penetration attains a maximum value and remains stable up to the end of the fuel injection process, while the spray gaseous phase continues to penetrate. Spray development is mainly headed by the momentum related to fuel injection; nevertheless the internal aerodynamic flow can influence the process as well. The mixture within the spray is an heterogeneous reactive mixture which, because of the severe thermodynamic conditions within the cylinder, mainly generated by the work of the piston on the system during the engine cycle compression stroke, for equivalence ratios close to stoichiometry, auto-ignites after a short ignition delay, Figure 3.20(B). Spatial and temporal localization of first auto-ignition sites are still difficult to be determined with accuracy. Following auto-ignition, two combustion modes characterize the heat release. First, premixed partially homogeneous reactive mixture, formed before auto-ignition, reacts relatively fast producing the peak of heat release due to premixed combustion, Figure 3.20(B→C); this phase is limited by the chemical kinetics. Then, the remaining vaporized fuel burns in a diffusion flame surrounding the spray region, whose heat release rate (HRR) is limited by mixing, Figure 3.20(C→D). The heat released by combustion contributes to increase the in-cylinder pressure and temperature, enhancing the evaporation process of the remaining liquid fuel, the diffusion of the generated gaseous fuel, and acceleration of the oxidation reactions. During the expansion stroke, the ambient gas did not take part in combustion mixes with hot burnt or partially oxidized gases, leading to a completion of the combustion process, Figure 3.20(D→E). Recent experiments [66] showed that spray confinement (spray-wall interaction) and, for multi-hole injector nozzle systems, spray-to-spray interaction impact spray development and, consequently, combustion.

A combustion model aims to reproduce, through computation, the physical processes occurring within the combustion chamber during the engine cycle:

- To relate engine control strategies and combustion performance in terms of heat release law and pollutant emissions
- To get an insight into reactive-fluid-mechanics complex interactions in order to extract useful understanding for engine optimization.

A wide overview of 0D and quasiD CI combustion model exists in literature. Models range from 0D single zone, in which mixture is considered as homogeneous [62, 63, 67, 68, 69], 0D two to N-zones, in which different regions are identified to characterize specific mixture heterogeneities [65, 70, 71, 72, 73, 74, 75, 76], up to quasiD Eulerian [77, 78] or Lagrangian [79, 80] description of spray and mixture formation. These last aim to better account for relevant aspects such as spray to spray and injection to injection interactions, but also spray to wall interactions, the latter depending on combustion chamber geometry. Two or more zones are necessary to acquire relevant information about the physics heading pollutant production: in fact pollutant kinetics is strongly affected by local thermo-chemical conditions of the mixture. This makes single-zone approaches not suitable for reliable pollutant computations. Most common approaches adopt a two-zone formalism: the mixture within the cylinder is considered as made of unburnt gas, gas which still does not have undertaken the combustion process, and burnt gas, hot gas product of combustion. A detailed description of the approaches is out of the scope of this section, whose ambition is to sensitize about the origin of model formulations, as well as about the potential and limitation of a given approach. For this reason, by following the thread of the phenomenological description of the combustion process presented above, the different concepts and relative modeling approaches will be introduced and discussed by paying attention to detail their limits with respect to modern engine architectures. To structure the thought about the continuation of this work, a general synoptic diagram relative to CI engine combustion modeling, describing the interactions of the different submodels building up the complete combustion model, is presented in Figure 3.21.

FIGURE 3.21 Synoptical diagram of a CI combustion chamber model.

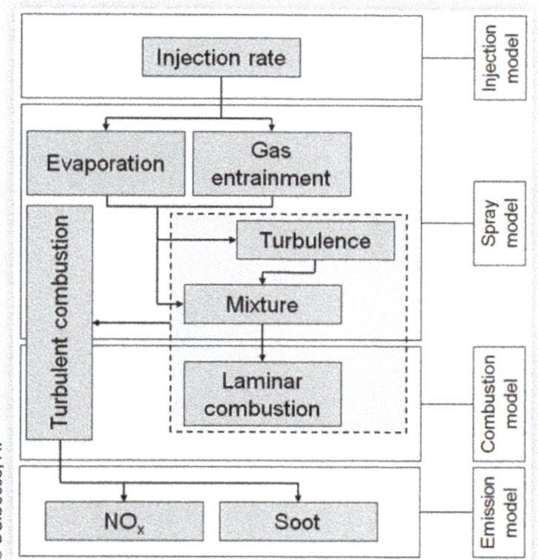

3.3.2 Spray Models for CI Engines

Injector nozzle holes represent the interface between injection system and combustion chamber. Injected liquid fuel leaves the nozzle at high speed due to the high injection pressure. The process bringing to evaporating fuel droplets is complex due to the multitude of involved physical phenomena and the large range of spatial and temporal scales, Section 3.3.1. Accordingly, a detailed numerical description of the process is prohibitive in system simulation approaches, in which a global description of the phenomenology is preferred. The term *spray model* is used here to indicate the different steps of the process through which in-cylinder reactive gaseous-mixture region is obtained. The major steps of the process are:

- Fuel evaporation
- Ambient gas entrainment in the spray region
- Fuel and ambient air mixing

In what follows the abovementioned topics will be detailed.

3.3.2.1 FUEL EVAPORATION

Liquid fuel evaporation is the first and one of the most important aspects to be handled in combustion modeling. Two major evaporation regimes can be encountered in direct injection engines, depending on fuel properties and thermofluid dynamics conditions. If local interphase controlled evaporation regime, transport rates of mass, momentum, and energy limit the overall vaporization rate, and the turbulent mixing must be fast relative to those processes. In other words, sufficient energy is entrained into the spray, but the transfer to the liquid phase is limited, and local velocity and temperature differences exist between the phases in the spray. In local interface-controlled regimes, diameter and velocity droplet distributions must be taken into account. Modeling approaches such as those proposed in [79] discretize the injected mass of fuel into parcels. Each parcel contains a given mass of fuel separated into droplets. The droplet number depends on the quality of the atomization process, described by means of dedicated approaches [81]. Accordingly, each parcel is associated with a fuel drop size distribution (the simplest one is a uniform size distribution), which allows to characterize the evaporation process on the base of the well-known D^2 model [82]. Otherwise, if mixing controlled evaporation regime, mass, momentum, and energy, local interphase transport rates are fast relatively to turbulent mixing rates. In this limit, no local velocity or temperature differences exist between the phases in the spray, which are, locally, in equilibrium. Turbulent mixing controls fuel vaporization by determining the rate at which energy (ambient gas) is entrained into the spray for heating and vaporizing the fuel. Under mixing-controlled regime, it is essential to correctly estimate the ambient gas entrainment rate within the spray region to compute fuel evaporation. Approaches based on this assumption are presented in [64, 70]. According to [83], fuel vaporization in conventional DI CI sprays is mainly controlled by air entrainment (mixing-controlled regime). This observation allows to simplify the process description by focusing on air entrainment rather than on liquid/gaseous phase interface transfer processes. From a modeling viewpoint, in literature, several approaches exist to describe fuel vaporization. Some authors suppose the liquid fuel injection as being instantaneously vaporized [84], others relate the evaporation rate to the instantaneous liquid fuel mass, m_{Fi}, and an evaporation characteristic time, τ_{ev}:

$$\frac{dm_{F_{ev}}}{dt} = C_{ev}\frac{m_{F_l}}{\tau_{ev}} \tag{3.46}$$

where C_{ev} is a calibration constant to adjust the evaporation rate with respect to experiments. Different ways exist to estimate the variable τ_{ev}. Some authors [71] consider it as dependent mainly on the in-cylinder thermodynamic conditions and in particular the temperature T, leading to:

$$\tau_{ev} = C_{\tau_{ev,1}} + \frac{C_{\tau_{ev,2}}}{T} \tag{3.47}$$

in which $C_{\tau_{ev,1}}$ and $C_{\tau_{ev,2}}$ are calibration constants to adjust the evaporation behavior to the fuel characteristics. Other authors [64, 70], based on the observation of [83], elaborated a model for estimating τ_{ev} allowing to account for the impact of injection conditions, fuel properties, and thermochemical states of fluids on its determination, Figure 3.22. For that purpose, τ_{ev} was considered as being proportional to the time required by the injected liquid fuel to fully evaporate, t_{ev}. This evaporation time was then computed as the ratio of the maximum liquid penetration, L, computed according to [83], and the fuel injection velocity, u_{inj}:

$$\tau_{ev} \propto t_{ev} = \frac{L}{u_{inj}} \tag{3.48}$$

FIGURE 3.22 Impact of ambient gas thermodynamic conditions on fuel evaporation characteristic time [85].

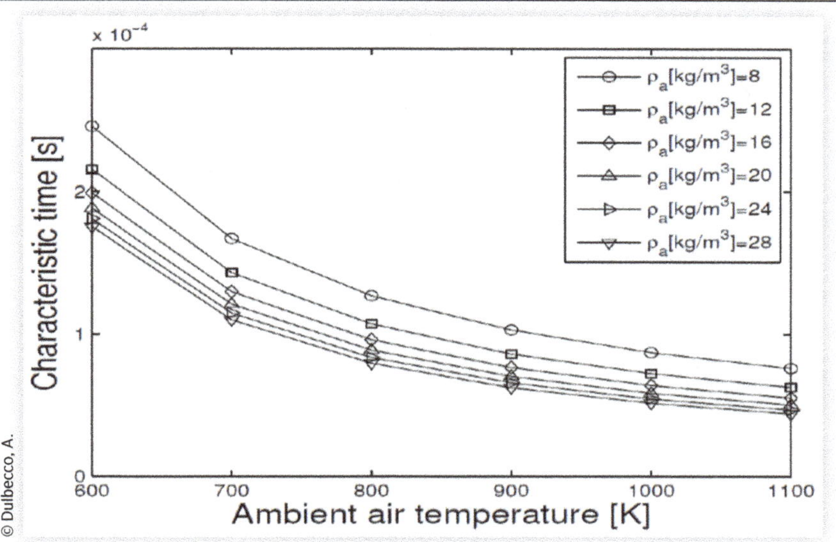

3.3.2.2 AMBIENT GAS ENTRAINMENT IN THE SPRAY REGION

Ambient gas entrainment within the spray region is a key aspect for conventional CI engines as it influences fuel evaporation, Section 3.3.2.1, and, together with it, the mean equivalence ratio within the spray region, but also the combustion process and pollutant emissions. Different approaches exist in literature to describe ambient gas entrainment ranging from empirical to more physical descriptions. In [69], an empirical formulation was introduced, which recognizes two phases:

- Entrainment during fuel injection, characterized by an ambient gas entrainment, $\dot{m}_A$, proportional, through the coefficient Λ_0 (Λ representing the ambient gas/fuel ratio), to fuel evaporation rate (Λ_0 is fixed to guarantee a mean equivalence ratio $\phi = 1.25$ in the spray):

$$\frac{dm_A}{dt} = \Lambda_0 \frac{dm_{F_{ev}}}{dt} \tag{3.49}$$

- Entrainment after the end of injection, the process is supposed to be headed by turbulent mass transport caused by the different fuel concentration between the spray zone and the surrounding gas. Accordingly, first a fuel mass diffusion term is computed as:

$$\left.\frac{dm_F}{dt}\right|_{tt} = cDRe^{0.5} A_Z \frac{\rho_F}{d_Z} \tag{3.50}$$

where ρ_F is the fuel density and d_Z the supposed-spherical spray region diameter (their ratio is representative of the fuel gradient), A_Z is the surface of the spray, $DRe^{0.5}$ is an empirical turbulent transport variable, and c is a model calibration constant. Then, the mass of entrained ambient gas is obtained as:

$$\frac{dm_A}{dt} = \frac{\left.\frac{dm_F}{dt}\right|_{tt}}{m_F - \left.\frac{dm_F}{dt}\right|_{tt}} m_A \tag{3.51}$$

Other authors [64, 70] proposed a physical-based approach to compute ambient gas entrainment through detailed spray computation, based on the results obtained in [86]. For this, the original formulation of the phenomenological spray model given in [86] was first adapted to ICE and then used to compute mass entrainment and spray volume. The scaling law describing the gaseous spray penetration, S, expressed in terms of non-dimensional coordinates t^+ and x^+, respectively for temporal and spatial dimensions, writes:

$$S = \left[\left(\frac{1}{\frac{t_{inj}}{t^+}}\right)^q + \left(\frac{1}{\left(\frac{t_{inj}}{t^+}\right)^{\frac{1}{2}}}\right)\right]^{-\frac{1}{q}} x^+ \tag{3.52}$$

where q is a coefficient fixed to 2.2 by authors, t_{inj} is the relative injection-time coordinate, and:

$$x^+ = \frac{\sqrt{C_a} d_h \sqrt{\frac{\rho_{F_l}}{\rho_A}}}{a \tan\left(\frac{\vartheta}{2}\right)} \qquad t^+ = \frac{x^+}{u_{inj}} \tag{3.53}$$

where

ρ_{F_l} and ρ_A are the densities of liquid fuel and ambient gas, respectively
C_a is the nozzle area contraction coefficient
d_h is the injector holes diameter
a is a coefficient to account for the spray cross-section velocity distribution ($a = 0.75$ for uniform distribution assumption [87])
θ is the spray opening angle

This last is calculated by the following equation:

$$\tan\left(\frac{\vartheta}{2}\right) = C_\vartheta\left[\left(\left|\frac{\rho_A}{\rho_{F_l}}\right|\right)^{0.19} - 0.0043\left(\frac{\rho_{F_l}}{\rho_A}\right)^{0.5}\right] \tag{3.54}$$

where C_θ is a fitting coefficient determined experimentally ($C_\theta = 0.26$). Figure 3.23 (right) shows the response of the gaseous spray penetration model to variations of ambient

FIGURE 3.23 Scheme of the spray model from (left); gaseous spray penetration for different ambient gas conditions (right) [85].

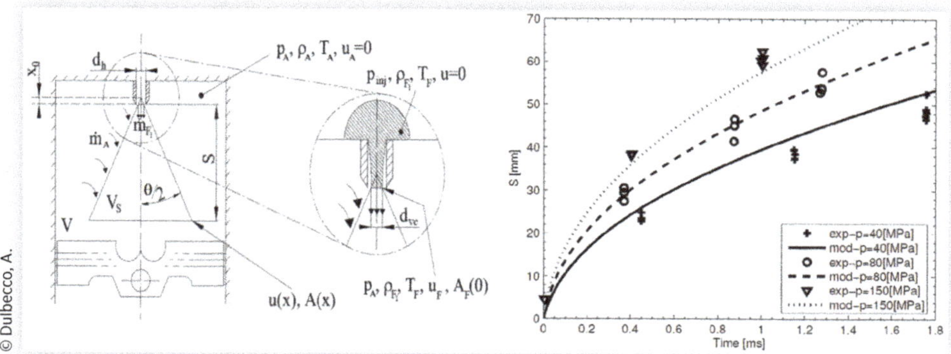

gas density. By assuming a conical shape of the spray region, Figure 3.23 (left), as supposed in [86], the spray volume derivative is given by:

$$\frac{dV_S}{dt} = \frac{d}{dt}\left(\frac{\pi}{3}\tan^2\left(\frac{\vartheta}{2}\right)\left[\left(S + x_0^3\right) - x_0^3\right]\right) \tag{3.55}$$

with x_0 defined as in Figure 3.23. The ambient gas entrainment is then computed as:

$$\frac{dm_A}{dt} = \rho_A \frac{dV_S}{dt} \tag{3.56}$$

3.3.2.3 FUEL AND AMBIENT AIR MIXING

In DI engines, to obtain a valuable reactive mixture is crucial for engine performance (heat release law and pollutant formation) [2]. Mixing is so determinant that, depending on the application of the conceived engine, it constraints engine design.* This implies that mixing is a key point of combustion process and its modeling is crucial. Once identified a given spray volume, mass, and mean composition, which is already a first step in characterizing the overall in-cylinder heterogeneities, Sections 3.3.2.1 and 3.3.2.2, still remains to describe in-spray mixture inhomogeneity. For this in literature several approaches exist. The simplest approach consists of considering spray mixture as homogeneous [69]. On the other side, quasiD models as the Lagrangian approaches proposed in [79], which detail axial and radial discretization of the spray volume, intrinsically give access to such information. In the same fashion, the Eulerian approach proposed by [77] computes for an axial discretization of spray fuel distribution based on momentum conservation laws. An a priori self-similar radial distribution is then introduced at each spray cross section to characterize the radial distribution.

In [88], the fuel/ambient gas mixing is described by means of a discrete PDF model. This last is based on the interaction of fixed-equivalence ratio classes mixing, according

* CI engines cover a wide range of powers. Generally, the smaller the bore, the higher is the engine speed, so to increase the specific power. Higher speed reduces the time available to combustion and, accordingly, to mixing, what constraints the engine design.

FIGURE 3.24 Representation of a β-PDF evolution (top). At the beginning representing a perfectly unmixed mixture (a), intermediary mixing stages (b-c) and perfectly stirred mixture (d). Comparison of 0D results with 3D results relative to a fuel mixture fraction distribution in a CI engine combustion chamber for different crank angles (bottom) [85]; SOI occurs at −6 CAD ATDC.

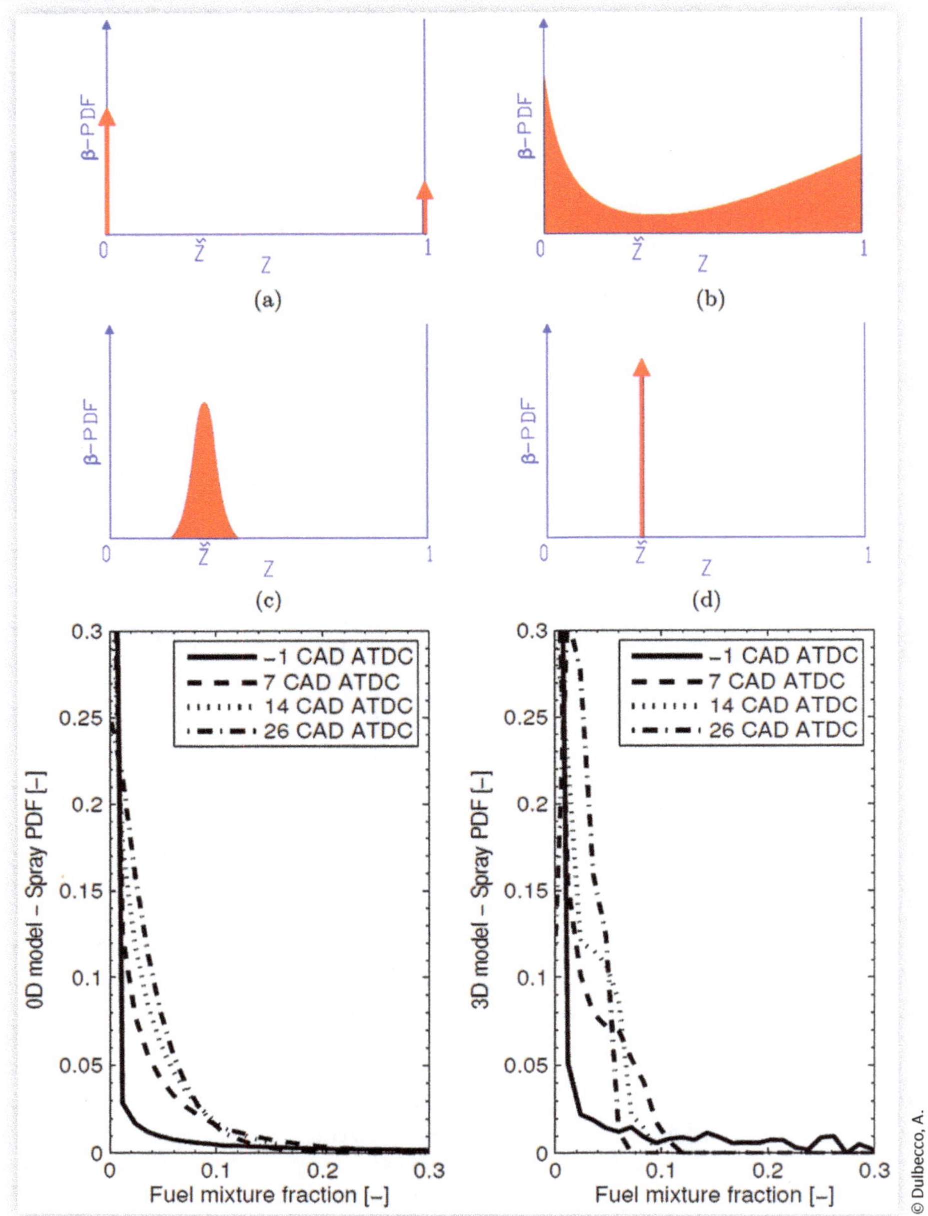

to stochastic theories. In the same fashion, a phenomenological approach was proposed in [70], which is grounded on the use of an a priori PDF distribution, namely, a β-PDF, to characterize fuel/ambient gas mixture stratification. Such an approach only requires one additional ODE to compute the fuel mixture fraction variance, $\widetilde{Z''^2}$, which allows to characterize spray heterogeneities with an interesting level of detail, Figure 3.24, and a limited computational effort. The differential equation for the fuel mixture fraction variance, obtained by integrating the 3D equation over the domain, holds:

$$\frac{d\tilde{Z}''^2}{dt} = -2C_{ds}\frac{\tilde{Z}''^2}{\frac{\tilde{k}}{\tilde{\varepsilon}}} + \frac{1}{m_S}\left(\tilde{Z}^2 - \tilde{Z}''^2\right)\frac{dm_A}{dt} + $$

$$+\frac{1}{m_S}\left[\tilde{Z}^2\left(1-\frac{1}{Z_S}\right)+\left(Z_S-\tilde{Z}\right)^2\frac{1}{Z_S}-\tilde{Z}''^2\right]\frac{dm_F}{dt} \quad (3.57)$$

where the first RHS term represents the variance dissipation due to turbulence, with C_{ds} a calibration coefficient, and the second and third terms represent variance sources due to ambient gas entrainment and fuel evaporation in the spray region, respectively. The mean fuel mixture fraction of the spray, $\tilde{Z}$, is computed as $\tilde{Z} = m_F/m_S$, where m_S is the total gaseous mass within the considered spray. Z_S is the mixture fraction saturation value at the liquid gas interface of the evaporating fuel.

3.3.2.4 TURBULENCE MODELS FOR CI ENGINES

In CI engines, turbulence is very important as it holds the overall mixing process, which is directly related to combustion heat release and pollutant emissions. Most of CI modeling approaches suppose that turbulence is mainly generated by high-pressure fuel injection [69, 70, 71, 84], what allows to neglect other sources of kinetic energy, such as intake and exhaust flow contributions. In [69, 71] the evolution of the turbulent kinetic energy, k, is dictated by the following ODE:

$$\frac{dk}{dt} = -C_{diss}\frac{1}{d_h}k^{\frac{3}{2}} + C_{turb}\left.\frac{dk}{dt}\right|_{spray} = -C_{diss}\frac{1}{d_h}k^{\frac{3}{2}} + C_{turb}\frac{1}{m_S}\frac{dm_{F_{inj}}}{dt}u_{inj}^2 \quad (3.58)$$

with d_h the nozzle hole diameter, supposed to be proportional to the turbulence integral length scale, $m_{F_{inj}}$ the mass of injected fuel, and u_{inj} the fuel injection velocity. C_{diss} and C_{turb} are constant calibration parameters to adjust the relative weights of turbulence dissipation and production terms, respectively [69]. Next generation of CI engine architectures, integrating VVA, VVT, or VCR systems, operating Miller or Atkinson thermodynamic cycles, seems to require more detailed turbulence approaches to better represent its impact on combustion and pollutant emissions. In this context, the improvement of the turbulence models would consider the integration of additional turbulence sources other than spray: intake and exhaust flows generating swirl and tumble vortices, squish, and dissipation contributions due to bowl geometry [89].

3.3.3 Combustion Models for CI Engines

One of the difficulties in modeling CI engine combustion process is that, in the most general case, both auto-ignition, premixed, and diffusion combustion modes are present. According to the phenomenology of a CI combustion process, Section 3.3.1, before AI, fuel and ambient gas mix with each other to form a reactive mixture. Then, spots of AI appear where thermochemical conditions are the most favorable and premixed flame

FIGURE 3.25 Illustration of a conceptual Diesel spray steady-state combustion model [90].

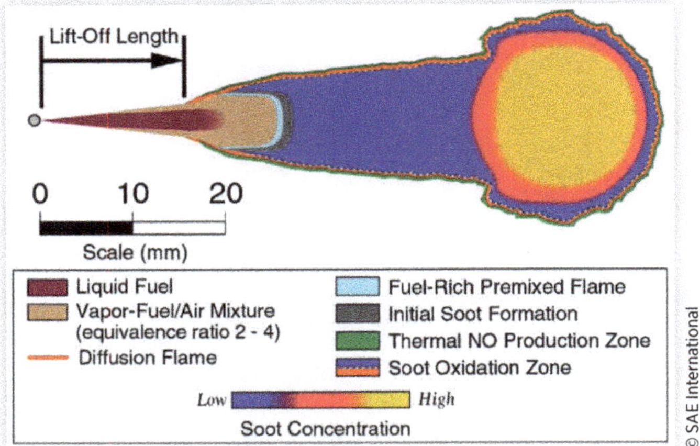

fronts propagate through the spray burning reactive mixture. Different conditions can prevent the propagation of the flame front:

- The presence of regions where mixture is outside the suitable flammability range: too lean (periphery of the spray) or too rich (spray core)
- The interaction of two or more flame fronts propagating against each other
- The interaction of the flame front with walls: mixture in proximity of walls is usually colder and less turbulent than elsewhere. This reduces the flame propagation speed*.

According to Figure 3.25, premixed flame stabilizes near the inner rich mixture region of the spray, where liquid fuel droplets evaporate and gaseous fuel diffuses into ambient gas. The presence of the flame enhances the evaporation process, while the presence of turbulent flow enhances the mixing. It was shown experimentally that an equivalence ratio profile going from rich to lean mixture is obtained moving along the radial coordinate from the axis of the spray. In this scenario, combustion reactions localize at the interface between rich and lean mixture regions, where mixture conditions are the most favorable. The fuel oxidation process is mainly headed by turbulence, which mixes the two reactant streams: evaporating fuel and ambient gas entrainment. Accordingly, this combustion regime is featured as fuel burning in diffusion flame, whose reaction rate depends on jet characteristics, mixture properties, and turbulence.

As premixed and diffusion combustion coexist along the combustion process, the global fuel consumption is generally expressed as the sum of the two contributions:

$$\frac{dm_F}{dt} = \frac{dm_{F_{pre}}}{dt} + \frac{dm_{F_{diff}}}{dt} \tag{3.59}$$

Below, modeling approaches for the three combustion modes are presented and discussed.

* This aspect, important in SI combustion modeling, is usually neglected in CI combustion modeling.

3.3.3.1 AUTO-IGNITION PROCESS

Ignition delay is a crucial variable in CI engines, both since it is hardly predictable and because it is followed by extremely rapid heat release, Figure 3.20, usually identified as premixed combustion phase. This phase, even if it favors the thermodynamic cycle efficiency, if not correctly phased on the engine cycle, penalizes engine durability and noise emissions. According to Section 3.2.4, the chemical processes leading to AI of reactive mixtures are strongly related to fuel formulation and to the complex fuel molecule decomposition mechanisms. These last involve low-temperature reaction kinetics and are responsible for the formation of unstable intermediate species such as peroxides, aldehydes, etc., bringing to fast chain reaction oxidation initiation [91], Section 3.1.1. In this context, AI process, being essentially related to chemistry, is commonly modeled by means of an Arrhenius-like equation:

$$\tau_{AI} = \tau_{ch} = Ap^{-n}e^{\left(\frac{E_a}{RT}\right)} \tag{3.60}$$

Nevertheless, this equation applies to homogeneous mixtures at specific thermodynamic conditions. To adapt this approach to DI CI engines, several authors [74, 69] proposed to describe the AI delay as a combination of physical, τ_{ph}, and chemical, τ_{ch}, delays:

$$\tau_{AI} = \tau_{ph} + \tau_{ch} \tag{3.61}$$

τ_{ph}, commonly known as Magnussen term [92]. This depicts the influence of the injection system and mixture preparation (breakup, evaporation, and mixing) on AI delay. To account for the continuous variation of mixture pressure and temperature during the engine cycle, it is common to assume that AI takes place when the following identity is verified:

$$\int_{t_{SOI}}^{t_{SOC}} \frac{1}{\tau_{AI}} dt = 1 \tag{3.62}$$

where t_{SOI} and t_{SOC} represent the start of injection time and start of combustion time, respectively.

3.3.3.2 PREMIXED COMBUSTION

Premixed combustion in CI combustion modeling usually concerns the injected fuel mass evaporated before AI. This represents the mass of fuel which had the time to mix with ambient gas. Accordingly, premixed combustion in CI engine combustion represents the fast HRR corresponding to a peak of fuel consumption, Figure 3.20. Phenomenologically, to model this process, several approaches exist. In [69], the premixed spray region is viewed as a mixture cloud which leans out after the injection, Section 3.3.2.2. In this scenario, premixed combustion is modeled as a two-step process. First, from one AI location, a premixed flame, similar to those presented in Section 3.2.1, propagates, spherically, through the reactive mixture, considered as homogeneous (ascending (a) phase of the peak), then, during flame propagation, other AI spots appear and flames interact with each other, reducing the overall HRR (descending (d) phase of the peak):

$$\left.\frac{dm_{F_{pre}}}{dt}\right|_a = \rho_u Y_{F_u} A_L\left(u_L + u_T\right) \tag{3.63}$$

CHAPTER 3

$$\left.\frac{dm_{F_{pre}}}{dt}\right|_d = \frac{1}{3} cg \frac{(u_L + u_T)}{r_z} m_{F_{pre}} \tag{3.64}$$

where

Y_{F_u} is the fuel mass fraction in unburnt gas
c is a calibration constant
r_Z is the radius of the premixed zone
$m_{F_{pre}}$ is the remaining mass of fuel, the limiting factor
g is an empirical function to account for the lowering of the HRR as a function of entrained ambient gas:

$$g = \frac{1}{\Lambda^2} \quad \text{with } \Lambda = \frac{m_A}{m_F} \tag{3.65}$$

The premixed combustion HRR rate is then computed by taking the minimum of the two-phase contributions, according to the equation below:

$$\frac{dm_{F_{pre}}}{dt} = \min\left(\left.\frac{dm_{F_{pre}}}{dt}\right|_a, \left.\frac{dm_{F_{pre}}}{dt}\right|_d\right) \tag{3.66}$$

It is interesting to note that turbulent flame speed depends on laminar flame speed, which, according to the turbulent combustion theory, Section 3.1.2.1, accounts for the thermochemical state of the burning mixture. This allows to describe flame extinction phenomena due to an excessive mixture lean out, Section 3.3.2.2, or HRR reduction due to the presence of dilution gases. In [68], another approach to compute premix combustion HRR is proposed. It assumes that premixed combustion reaction mechanism is similar to the one holding AI delay. Accordingly, an Arrhenius-like equation is used to characterize it:

$$\frac{dm_{F_{pre}}}{dt} = k_1 \lambda \text{AFR}_{stoich} e^{-\frac{k_2 T_i}{T}} \frac{m_{F_{pre}}^2}{V_{mix}} (t - t_{SOC})^2 \tag{3.67}$$

where

k_1 and k_2 are calibration constants
λ is the air/fuel excess ratio
T_i is the activation temperature
V_{mix} is the premixed mixture volume

Naturally, this formulation allows to depict the typical premixed combustion HRR peak.

3.3.3.3 DIFFUSION COMBUSTION

Diffusion combustion in CI combustion modeling usually concerns the injected fuel mass evaporated after AI. Turbulent diffusion flame combustion process is generally characterized with a frequency approach [84, 69, 71]. The fuel burning rate in diffusion mode can then be written as:

$$\frac{dm_{F_{diff}}}{dt} = fm_{F_{diff}} \tag{3.68}$$

where f is a mixing frequency function, computed as the ratio of a mixing velocity, u_{mix}, and a mixing length scale, l_{mix}. The function f in literature can assume different formulations. In [69, 71] the mixing length was considered as depending on engine geometry and mixture composition:

$$l_{mix} = \sqrt[3]{\frac{\phi V_{cyl}}{n_h}} \tag{3.69}$$

with ϕ the fuel/air equivalence ratio, V_{cyl} the cylinder volume, and n_h the number of nozzle holes, while mixing velocity was computed as the composition of two contributions: the first depending on engine design and the second related to turbulence intensity:

$$u_{mix} = \sqrt{C_g v_{pist}^2 + C_k k} \tag{3.70}$$

with

v_{pist} is the mean piston velocity
C_g and C_k are the adjustment parameters

These last allow to calibrate, for a given CI engine, the HRR to fit experimental data. To account for the influence of burnt gases on the diffusion reaction rate, a correction factor is generally introduced [71]. The final expression for the diffusion flame burning rate holds:

$$\frac{dm_{F_{diff}}}{dt} = \left(1 - X_{bg}\right)^{\alpha_{res}} \frac{u_{mix}}{l_{mix}} m_{F_{diff}} \tag{3.71}$$

with

X_{bg} is the burnt gases concentration in unburnt gas mixture
α_{res} is the dedicated calibration parameter

Differently in [68], the mixing length, once again considered as depending on engine geometry and mixture composition, was expressed as:

$$l_{mix} = \sqrt[3]{V_{cyl}} \frac{\mathrm{AFR}_{stoic_{dil}}}{\mathrm{AFR}_{stoic}} \tag{3.72}$$

with $\mathrm{AFR}_{stoic_{dil}}$ and AFR_{stoic}, respectively, the air/fuel ratio at stoichiometry with and without ambient gas dilution. The mixing velocity was computed as directly related to the turbulence intensity:

$$u_{mix} = \sqrt{k} \tag{3.73}$$

CHAPTER 3

The final expression for the diffusion flame burning rate holds:

$$\frac{dm_{Fdiff}}{dt} = C_{mo}\frac{u_{mix}}{l_{mix}} m_{Fdiff} \tag{3.74}$$

The last formulation for diffusion flame burning rate naturally accounts for dilution gas impact on diffusion flame HRR.

3.3.3.4 COMBUSTION PROCESS COMPUTATION

This section presents combustion process computation results obtained with a two-zone DI CI combustion model [71] integrating AI, premixed, and diffusion modeling approaches detailed in Section 3.3.3. Results refers to a four-cylinder CR injection system 2.3 L displacement engine, the operating conditions varying from 1000 to 2750 rpm, 1 to 18 bar of IMEP, and 0% to 50% of EGR rate. Ninety-nine EOPs were measured at the engine test bed to characterize the whole engine operating domain, Figure 3.26(a). The calibration of the model was done once and all parameters were kept constant for all simulations. Figure 3.26 shows combustion HRR computation details relative to four engine operating conditions, depicted by filled square markers in Figure 3.27(a), representative of very different engine speed and engine load operating condition. EGR rate varies significantly from one point to another.

As shown in Figure 3.26, AI delay model captures well the beginning of the spray AI, both for pilot of main injections, although in-cylinder mixture composition and thermodynamic conditions highly differ from one EOP to another. In the same fashion, it can be observed that shares between premixed and diffusion combustion are fairly represented. While pilot injections burn essentially in premix combustion regime, main injections generally mix the two modes in proportions dictated by the operating conditions: while in (a) the contribution of premixed combustion is evident both from analysis of experiments and simulation results, in (c) its relative contribution is much less evident. Combustion intensity also is globally well reproduced, witnessing that the adopted modeling approaches are accurate enough for the studied engine architecture. In Figure 3.27, plots (b) to (d) present summary plots relative to combustion computation over the complete engine operating domain (a): computed IMEP and peak and peak location of in-cylinder pressure are compared to experiments. As shown, model performance is relatively stable over the complete database and allows to reproduce experimental results with an interesting level of accuracy.

3.3.4 Emission Models for CI Engines

Four major pollutants are generated during a CI combustion process: NO_x, particulate matter (soot), CO, and UHC. In the following, modeling approaches to compute and predict their evolution during the combustion process are discussed. From a general viewpoint, pollutant formation/consumption processes depend on local thermochemical properties of the mixture, in most cases through exponential relationships. This means that a minimum level of physical details in describing the combustion scenario is required to compute reliable and consistent emissions. This minimum level varies depending on the considered pollutant species. The most critical pollutants in conventional CI combustion are NO_x and soot. Their dependence on local mixture thermochemical properties

FIGURE 3.26 Comparison of combustion process computational results with experiments relative to four engine operating conditions (from a to d). EOP specificities are given in the upper-left corner of the corresponding plot. For each EOP, in-cylinder pressure and HRR curves are shown; SOI and fuel injection rate trace are also plotted to better illustrate the process leading to combustion.

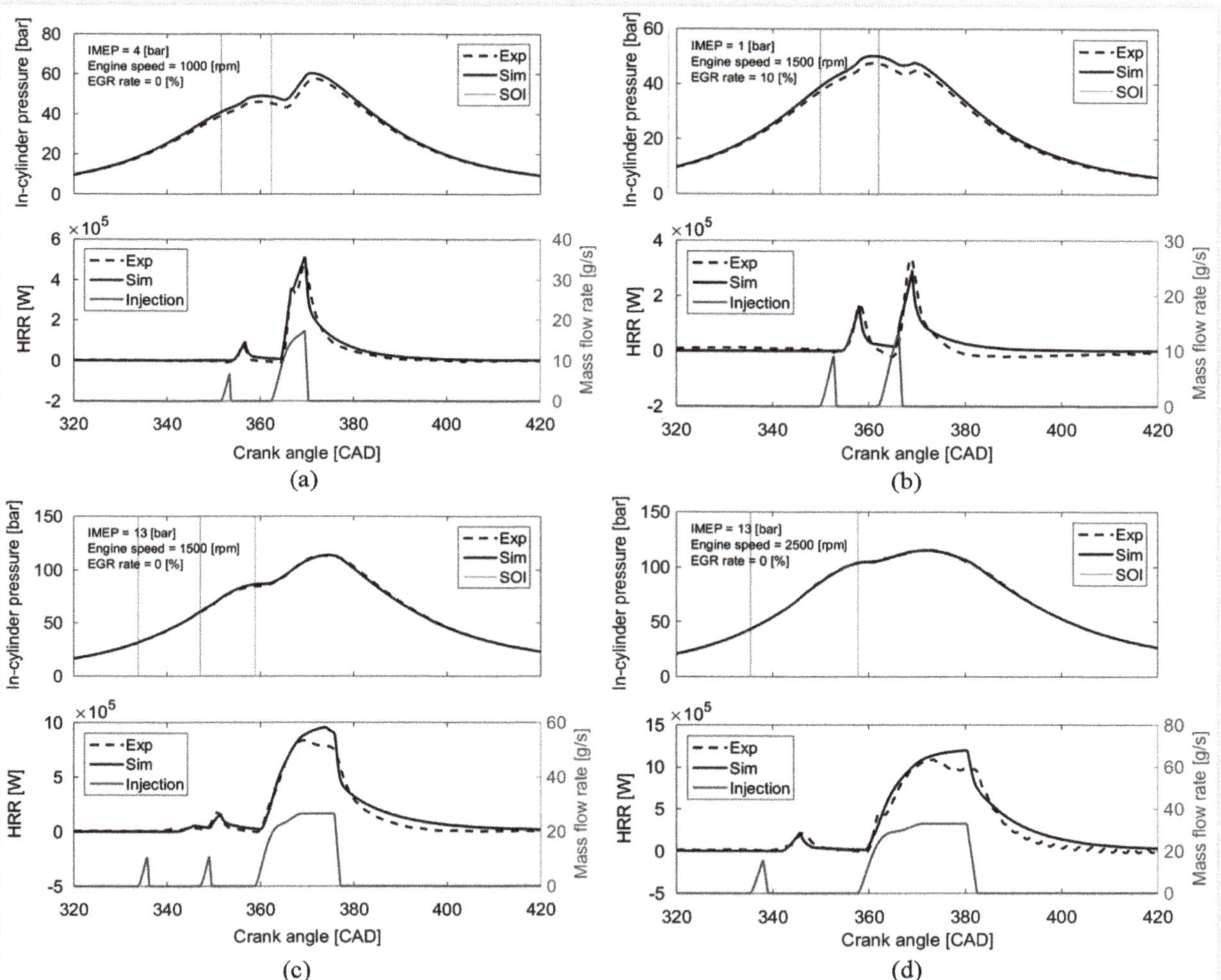

is shown in Figure 3.28. In the figure, conventional Diesel combustion is compared to other combustion modes, highlighting the benefits of such innovative concepts in terms of pollutant emissions.

In what follows, only NO_x and soot emission modeling will be approached.

3.3.4.1 NITROGEN OXIDES

Due to the complexity of the combustion scenario, NO_x emission computation in CI engines is more challenging than in SI engines [2, 94, 95]. Even if thermal NO mechanism remains in most cases the major source of NO_x production, because of mixture heterogeneities, the presence and in-cylinder recirculation of burnt gases interacting with flames and spray-to-spray and injection-to-injection iterations, the use of the extended Zeldovich model, Section 3.2.3.2, is not straightforward. Putting that in perspective,

FIGURE 3.27 Global engine database: engine map (a); comparison of IMEP simulation results with experiments (±0.5 bar error dashed lines) (b); comparison of peak in-cylinder pressure simulation results with experiments (±4 bar error dashed lines) (c); comparison of peak in-cylinder pressure location simulation results with experiments (±2 CAD error dashed lines) (d). Additional information about error type and statistics are given in the figure.

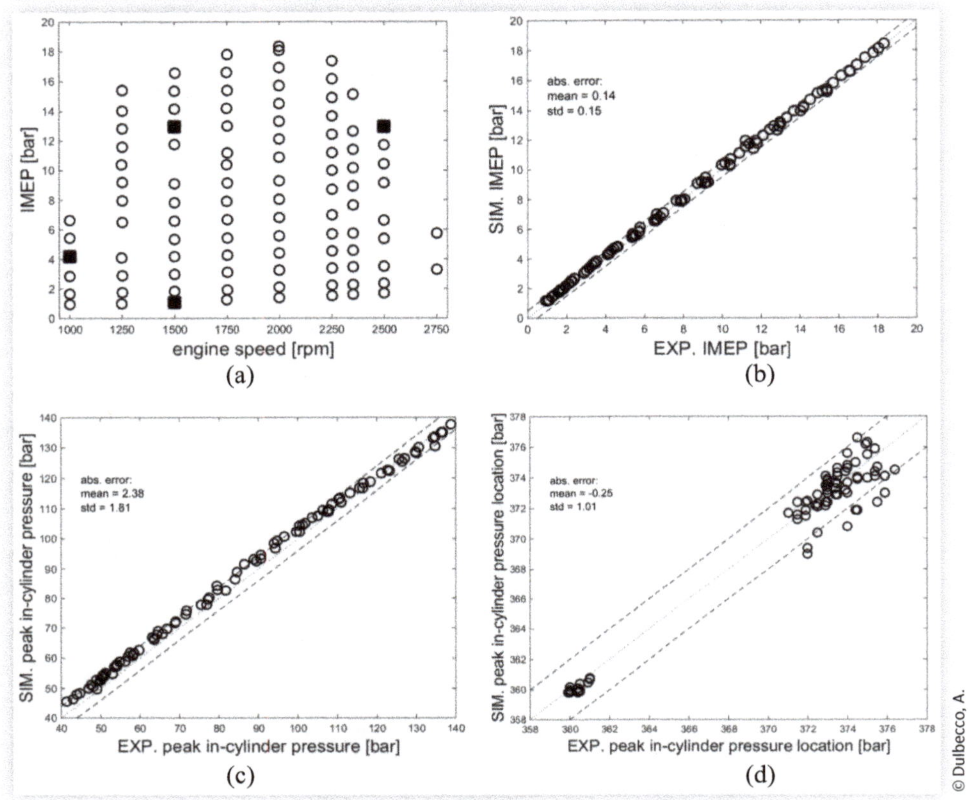

FIGURE 3.28 Conventional Diesel, low temperature combustion (LTC), homogeneous charge compression ignition (HCCI), and premixed controlled compression ignition (PCCI) concept on ϕ-T map [93].

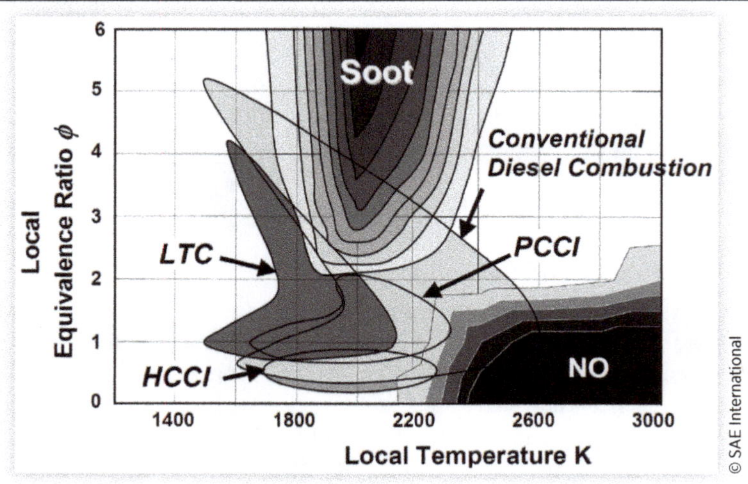

[71] proposed a simpler approach, suitable for a two-zone combustion model formalism, based on a single step chemistry approach expressed as:

$$O_2 + N_2 \leftrightarrow 2NO \tag{3.75}$$

in which chemical kinetics, in terms of forward (F) and backward (B) reaction rates, is depicted on the base of Arrhenius-like equations, expressed in terms of burnt gas mean descriptors (subscript b):

$$\left.\frac{d[NO]_b}{dt}\right|_F = A_F[O_2]_b[N_2]_b e^{\left(\frac{E_{aF}}{RT_b}\right)} \tag{3.76}$$

$$\left.\frac{d[NO]_b}{dt}\right|_B = A_B[O_2]_b[N_2]_b e^{\left(\frac{E_{aB}}{RT_b}\right)} \tag{3.77}$$

with A_i and E_{a_i} pre-exponential factor and activation energy parameters used to calibrate the model and globally integrate and account for all NO production/destruction sources. Because of its simplicity, such an approach requires several EOPs (of the order of tens), well distributed over the engine operating domain, to optimize the calibration. Figure 3.29 shows the comparison of simulation results with experiments. Experimental data refers to the engine database presented in Section 3.3.3.4. As shown, the correlation of simulated with experimental emissions is relatively good, showing that this modeling approach, despite the simple formulation, captures major contributions to NO production over a wide (two magnitude orders) emission scale, Figure 3.29 (left). For sure, limitations of the approach appear when a high level of accuracy is required to estimate low NO_x emission levels, Figure 3.29(right). Other approaches to NO_x emissions, aiming to integrate local phenomena impacting the formation/consumption processes, exist in literature [94]. Such approaches integrate more and

FIGURE 3.29 Comparison of NO_x emission simulation results with experiments. On the left, results refer to the whole engine database presented in Section 3.3.3.4; on the right a zoomed plot shows in more details the model accuracy relatively to low NO_x emission EOPs. Dashed lines represent ±20% error bars; error type and statistics are given in the figure. Results refer to a 2.3 L four-cylinder DI Diesel engine [71].

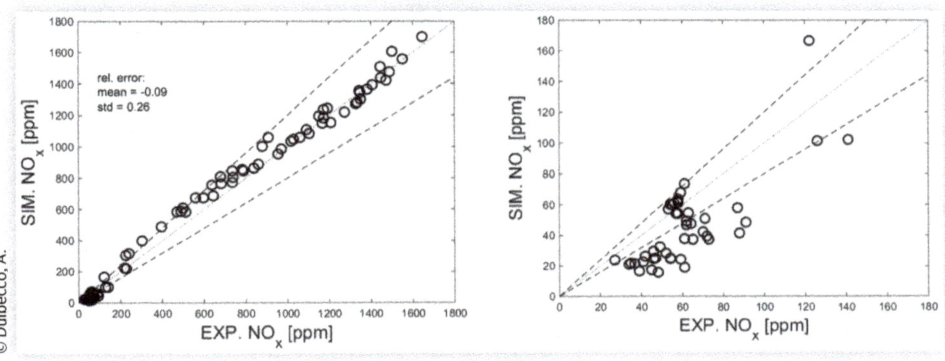

more detailed information about the combustion process into NO_x process description, to account for the different sources of NO_x but also introduce corrective factors to better account for the influence of engine design on emissions. A detailed description of such approaches would lead to digressions out of the perimeter of this chapter.

3.3.4.2 SOOT

Soot production depends on complex physical mechanisms, also including multi-phase phenomena. An additional difficulty to deal with in 0D-quasiD soot modeling is the need to account for local mixture thermochemical properties, requiring the computation of local information. A complete description of soot mechanics is out of the scope of this section. Nevertheless, a reminder of its main steps is useful to understand the modeling choices done by model developers. Fuel molecules undertaking a pyrolysis process originate unsaturated hydrocarbons (e.g., acetylene), which react with radicals to form aromatic compounds. Those compounds add with each other forming soot precursors and soot nuclei. Finally, soot nuclei become soot particles through processes of surface growing, coagulation, and aggregation [2]. Many factors influence soot mechanics such as temperature, pressure, and equivalence ratio, but also fuel molecule structure and composition. Because of this complexity, soot models based on mean burnt gas thermochemical variables only can give access to qualitative emission trends depending on engine operating conditions. Accordingly, to describe soot production, authors commonly adopt multizone approaches, allowing to better estimate local conditions [73]. Recently, based on experimental evidences [87], models to compute soot emission on the base of local thermochemical mixture conditions adapted to a 0D formalism were developed [96, 97]. The main idea was to exploit the understanding used to formulate the conceptual Diesel spray combustion model proposed in [90], to relate soot formation and oxidation processes to the most relevant processes taking place within the combustion chamber: the rich combustion process at lift-off length (*lol*), responsible for soot precursors formation, and soot oxidation due to favorable conditions encountered at the diffusion flame surrounding the spray region, Figure 3.25. Depending on the operating conditions, ambient gas thermochemical state is known. On the base of the phenomenological approaches in the literature, *lol* can be estimated quite precisely [87] as well as the corresponding mixture composition of burning gas, which directly influences soot production. In the same fashion, flame temperature and composition of the diffusion combustion are computed by assuming that flame stabilizes at stoichiometry. Then, the soot kinetics is described by means of the following equation:

$$\frac{dm_{soot}}{dt} = \left.\frac{dm_{soot}}{dt}\right|_p - \left.\frac{dm_{soot}}{dt}\right|_o \tag{3.78}$$

where the first RHS term represents soot production (p) contribution, while the second represents soot oxidation (o) contribution. Soot production is modeled as:

$$\left.\frac{dm_{soot}}{dt}\right|_p = C_{sp}\eta_\phi \phi_{lol}^0 m_{F_{dif}}\, p^{0.5} e^{\frac{-E_{sp}}{RT_{sp}}} \tag{3.79}$$

where

C_{sp} is a calibration constant to adjust the model behavior to different fuels and engines
η_ϕ is an efficiency factor enabling soot production varying in the interval [0,1]*

* η_ϕ passes from one to zero by means of a smoothed transition function when values pass below the value of two, according to experimental results shown in [87].

$m_{F_{dif}}$ represents the mass of fuel within the combustion chamber available to burn in diffusion combustion
p is the in-cylinder pressure
E_{sp} is the activation energy of the soot production reaction (E_{sp} = 52,335 J/mol)
R is the perfect gas constant

T_{sp} represents the adiabatic temperature of combustion at *lol*, relative to an unburnt gas having fuel/ambient gas composition and a temperature equal to that of vaporization (temperature at the liquid/gas interface).

Soot oxidation is modeled as:

$$\left.\frac{dm_{soot}}{dt}\right|_o = C_{so} m_s Y_{O2_{st}} p^{1.8} e^{\frac{-E_{so}}{RT_{so}}} \tag{3.80}$$

where
C_{so} is a calibration constant to adjust the model behavior to different fuels and engines
m_s represents the mass of soot within the combustion chamber
$Y_{O2_{st}}$ is the oxygen mass fraction at stoichiometric conditions
E_{so} is the activation energy of the soot oxidation reaction (E_{so} = 58,615 J/mol)

T_{so} represents the adiabatic temperature of combustion in the region where soot is oxidized, that is, the temperature relative to a stoichiometric diffusion flame, whose unburnt gas temperature was assumed to be the same as the temperature of the soot formation region. Once all of the fuel is consumed, soot oxidation depends on the bulk mean temperature and oxidation quenches when the bulk mean temperature drops below 1000 K [2]. Figure 3.30 shows the comparison of computed soot emissions with experiments; experimental data refers to the engine database presented in Section 3.3.3.4. As shown, the model allows to correctly localize and qualitatively describe soot emission variations over the engine map; nevertheless quantitative results still present deviations compared to experiments.

FIGURE 3.30 Comparison of soot emission simulation results with experiments over a complete engine map. Results refer to a 2.3 L four-cylinder DI Diesel engine [97].

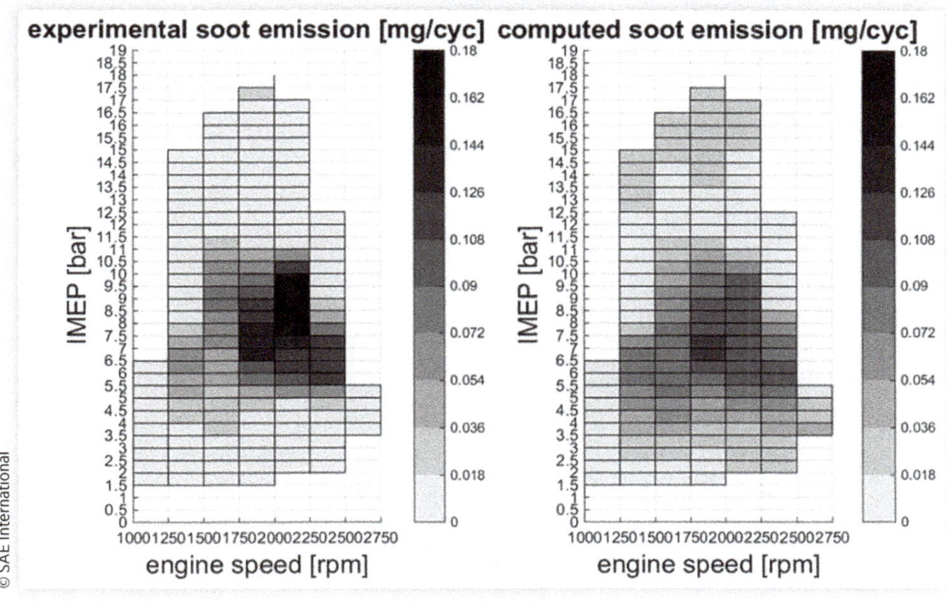

Definitions, Acronyms, and Abbreviations

Acronyms

0D-1D-3D - Zero-one-three-dimensional
ACTDC - After compression top dead center
AFR - Air/fuel ratio
AI - Autoignition
BG - Burnt gas
BMEP - Brake mean effective pressure
BSFC - Brake-specific fuel consumption
CAD - Crank angle degree
CI - Compression ignition
CFM - Coherent flame model
CFD - Computational fluid dynamics
CR - Common rail
DI - Direct injection
EIVC - Early intake valve closing
EOP - Engine operating point
EGR - Exhaust gas recirculation
FG - Fresh gas
FMFB - Fuel mass fraction burned
GDI - Gasoline direct injection
HCCI - Homogeneous charge compression ignition
HRR - Heat release rate
ICE - Internal combustion engine
IMEP - Indicated mean effective pressure
IVC - Intake valve closing
KLSA - Knock limited spark advance
LFA - Laminar flame assumption
LHV - Low heating value
LIVC - Late intake valve closing
lol - lift-off length
LTC - Low temperature combustion
FTDC - Firing top dead center
MAPO - Maximum amplitude of pressure oscillation
ODE - Ordinary differential equation
PCCI - Premixed controlled compression ignition
PDF - Probability density function
PRF - Primary reference fuel
RANS - Reynolds averaged Navier-Stokes
RHS - Right-hand side
RNG - Re-normalization group
rpm - Round per minute
SI - Spark ignition
SOC - Start of combustion
SOI - Start of injection
TDC - Top dead center
TRF - Toluene reference fuel

VCR - Variable compression ratio
VVA - Variable valve actuation
VVT - Variable valve timing

Symbols

A - Pre-exponential factor
A_L - Laminar flame area
A_T - Turbulent flame area
A_Z - Spray surface area
c_p, c_v - Specific heat at constant pressure/volume
C_a - Nozzle area contraction coefficient
CO - Carbon monoxide
d_h - Injector hole diameter
d_z - Supposed-spherical spray region diameter
D - Mass diffusivity, diameter
D_3 - Flame front fractal dimension
Da - Damköhler number
e - Specific internal energy
E_a - Activation energy
f_d - Tumble decay function
h - Specific enthalpy
HC (UHC) - Hydrocarbons (unburnt hydrocarbons)
k - Turbulent kinetic energy
K - Mean flow kinetic energy, reaction rate constant
Ka - Karlovitz number
l - Length scale
l_d - Diffusive thickness
l_k - Kolmogorov length scale
l_{min}, l_{max} - Minimum/maximum flame front wrinkling scale
l_{mix} - Mixing length scale
l_r - Reaction zone thickness
l_t - Turbulence integral length scale
L - Maximum liquid penetration
Le - Lewis number
m - Mass
n - Engine rotational speed
NO_x - Nitrogen oxides
p - Pressure
P - Mean flow kinetic energy dissipation rate
Q - Heat
r_b, r_f - Flame radius
r_T - Tumble radius
r_Z - Radius of premixed zone
R - Gas constant
Re, Re_t - Reynolds number, turbulent Reynolds number
S - Gaseous spray penetration
Sc - Schmidt number
t - Time
T - Temperature, tumble momentum
u - Velocity

CHAPTER 3

u' - Turbulence intensity
u_L - Laminar flame speed
u_T - Turbulent flame speed
u_{inj} - Fuel injection velocity
u_{mix} - Mixing velocity
U_{fK} - Mean flow velocity
U_T - Tumble velocity
v_{pist} - Mean piston velocity
V - Volume
x - spatial dimension
x, Y - Mass fraction
X - Concentration, molar fraction
Z - Fuel mixture fraction

Greeks

δ_L - Laminar flame thickness
δ_R - Flame reaction zone thickness
ε - Turbulence dissipation rate
θ - Engine crank angle, spray opening angle
λ - Thermal conductivity, Taylor length scale, air/fuel excess ratio
μ - Dynamic viscosity
ν - Kinematic viscosity
ν_t - Turbulent viscosity
Ξ - Flame front wrinkling factor
ρ - Density
τ - Time scale, thermal expansion rate
τ_{AI} - Auto ignition time
τ_{CF} - Cool flame time
ϕ - Air/fuel equivalence ratio
χ - Mixture fraction scalar dissipation rate
ω_r - Combustion reaction rate

Subscripts

0 - Reference condition
a - Air
A - Ambient gas
b - Burnt
bg - Burnt gas
ch, chem - Referring to chemistry
cyl - Cylinder
dil - Dilution gas
e - Entrained
eq - Referring to equilibrium condition
ev - Evaporation
ext - Referring to quenching transition
f, F - Fuel
$\mathbf{F_{dif}}$ - Diffusive burned liquid fuel
$\mathbf{F_{ev}}$ - Fuel evaporation
$\mathbf{F_{inj}}$ - Injected fuel

F_l - Liquid fuel
F_{pre} - Premixed burned liquid fuel
F_u - Fuel in unburnt gas
i - Referring to *i*th species, *i*th dimension
inj - Injection
l - Liquid
mix - Premixed mixture
O - Oxidizer
ph - Physical
s - Soot
so - Soot oxidation
sp - Soot production
st, stoich - Stoichiometric
S - Spray, saturation
T - Tumble
u - Unburnt
w - Wall
η - Referring to Kolmogorov scale

Superscripts

. - Time derivative
~ - Mean
” - Variance

References

1. Kuo, K.K., *Principles of Combustion* (Wiley, 2005).
2. Heywood, J.B., *Internal Combustion Engine Fundamentals* (McGraw-Hill, 1988).
3. Metghalchi, M. and Keck, J., “Burning Velocities of Mixtures of Air with Methanol, Isooctane, and Indolene at High Pressure and Temperature,” *Combustion and Flame* 48 (1982): 191-210.
4. Poinsot, T. and Veynante, D., *Theoretical and Numerical Combustion* (Edwards, 2005).
5. Veynante, D. and Vervisch, L., “Turbulent Combustion Modeling,” *Progress in Energy and Combustion Science* 28 (2002): 193-266.
6. Woschni, G., “A Universally Applicable Equation for the Instantaneous Heat Transfer Coefficient in the Internal Combustion Engine,” SAE Technical Paper 670931, 1967, doi:10.4271/670931.
7. Vibe, I.I., *Brennverlauf und Kreisprocess von Verbennungsmotoren* (Berlin: VEB Verlag Technik, 1970).
8. Demesoukas, S., Caillol, C., Higelin, P., and Boiarciuc, A., “Zero-Dimensional Spark Ignition Combustion Modeling - A Comparison of Different Approaches,” SAE Technical Paper 2013-24-0022, 2013, doi:10.4271/2013-24-0022.
9. Richard, S., Bougrine, S., Font, G., Lafossas, F.A. et al., “On the Reduction of a 3D CFD Combustion Model to Build a Physical 0D Model for Simulating Heat Release, Knock and Pollutants in SI Engines,” *Oil & Gas Science and Technology* 64, no. 3 (2009): 223-242, doi:10.2516/ogst/2008055.

10. Perlman, C., Frojd, K., Seidel, L., Tuner, M. et al., "A Fast Tool for Predictive IC Engine In-Cylinder Modelling with Detailed Chemistry," SAE Technical Paper 2012-01-1074, 2012, doi:10.4271/2012-01-1074.
11. Verhelst, S. and Sheppard, C.G.W., "Multi-Zone Thermodynamic Modelling of Spark-Ignition Engine Combustion – An Overview," *Energy Conversion and Management* 50, no. 5 (2009): 1326-1335, doi:https://doi.org/10.1016/j.enconman.2009.01.002.
12. Gouldin, F., "An Application of Fractals to Modeling Premixed Turbulent Flames," *Combustion and Flame* 68, no. 3 (1987): 249-266, doi:10.1016/0010-2180(87)90003-4.
13. Franke, C., Wirth, A., and Peters, N., "New Aspects of the Fractal Behaviour of Turbulent Flames," *23rd Symposium (International) on Combustion*, Orleans, 1990.
14. Gatowsky, J.A. and Heywood, J.B., "Flame Photographs in a Spark-Ignition Engine," *Combustion and Flame* 56 (1984): 71-81.
15. Keck, J., Heywood, J., and Noske, G., "Early Flame Development and Burning Rates in Spark Ignition Engines and Their Cyclic Variability," SAE Technical Paper 870164, 1987, doi:10.4271/870164.
16. Tabaczynski, R.J., "Turbulence and Turbulent Combustion in Spark-Ignition Engines," *Progress in Energy and Combustion Science* 2, no. 3 (1976): 143-165, doi:https://doi.org/10.1016/0360-1285(76)90010-1.
17. Morel, T., Rackmil, C., Keribar, R., and Jennings, M., "Model for Heat Transfer and Combustion in Spark Ignited Engines and Its Comparison with Experiments," SAE Technical Paper 880198, 1988, doi:10.4271/880198.
18. Hicks, R., Lawes, M., Sheppard, C., and Whitaker, B., "Multiple Laser Sheet Imaging Investigation of Turbulent Flame Structure in a Spark Ignition Engine," SAE Technical Paper 941992, 1994, doi:10.4271/941992.
19. Peters, N., "Laminar Flamelet Concepts in Turbulent Combustion," *Symposium (International) on Combustion* 21, no. 1 (1988): 1231-1250, doi:http://dx.doi.org/10.1016/S0082-0784(88)80355-2.
20. Bates, C.S., "Flame Imaging Studies of Combustion Completion in a SI Four-Stroke Engine," TvU Thoughtventions Unlimited LLC, http://www.tvu.com/.
21. North, G.L. and Santavicca, D.A., "The Fractal Nature of Premixed Turbulent Flames," *Combustion Science and Technology* 72, no. 4-6 (1990): 215-232, doi:10.1080/00102209008951648.
22. Matthews, R. and Chin, Y., "A Fractal-Based SI Engine Model: Comparisons of Predictions with Experimental Data," SAE Technical Paper 910075, 1991, doi:10.4271/910079.
23. Rivas, M., Higelin, P., Caillol, C., Sename, O. et al., "Validation and Application of a New 0D Flame/Wall Interaction Sub Model for SI Engines," *SAE Int. J. Engines* 5, no. 3 (2012): 718-733, doi:10.4271/2011-01-1893.
24. De Bellis, V., Bozza, F., and Tufano, D., "A Comparison Between Two Phenomenological Combustion Models Applied to Different SI Engines," SAE Technical Paper 2017-01-2184, 2017, doi:10.4271/2017-01-2184.
25. Candel, S. and Poinsot, T., "Flame Stretch and the Balance Equation for the Flame Area," *Combustion Science and Technology* 70 (1990): 1-15.
26. Baritaud, T., "Combustion and Fluid Dynamic Measurements in a Spark Ignition Engine: Effects of Thermochemistry and Velocity Field, Turbulent Flame Speeds," SAE Technical Paper 892098, 1989, doi:10.4271/892098.

27. Richard, S. and Veynante, D., "A 0-D Flame Wrinkling Equation to Describe the Turbulent Flame Surface Evolution in SI Engines," *C.R. Mecanique* 343 (2015): 219-231.

28. Charlette, F., Meneveau, C., and Veynante, D., "A Power Law Flame Wrinkling Model for LES of Premixed Turbulent Combustion," *Combust. Flame* 131 (2002): 159-180.

29. Blint, R.J., "The Relationship of the Laminar Flame Width to Flame Speed," *Combust. Sci. Technol.* 49 (1986): 79-92.

30. Borgnakke, C., Arpaci, V., and Tabaczynski, R., "A Model for the Instantaneous Heat Transfer and Turbulence in a Spark Ignition Engine," SAE Technical Paper 800287, 1980, doi:10.4271/800287.

31. Morel, T. and Mansour, N., "Modeling of Turbulence in Internal Combustion Engines," SAE Technical Paper 820040, 1982, doi:10.4271/820040.

32. Morel, T. and Keribar, R., "A Model for Predicting Spatially and Time Resolved Convective Heat Transfer in Bowl-in-Piston Combustion Chambers," SAE Technical Paper 850204, 1985, doi:10.4271/850204.

33. Sjeric, M., Kozarac, D., and Bogensperger, M., "Implementation of a Single Zone k-ε Turbulence Model in a Multi Zone Combustion Model," SAE Technical Paper 2012-01-0130, 2012, doi:10.4271/2012-01-0130.

34. Dulbecco, A., Richard, S., Laget, O., and Aubret, P., "Development of a Quasi-Dimensional K-k Turbulence Model for Direct Injection Spark Ignition (DISI) Engines Based on the Formal Reduction of a 3D CFD Approach," SAE Technical Paper 2016-01-2229, 2016, doi:10.4271/2016-01-2229.

35. Kim, N., Kim, J., Ko, I., Choi, H. et al., "A Study on the Refinement of Turbulence Intensity Prediction for the Estimation of In-Cylinder Pressure in a Spark-Ignited Engine," SAE Technical Paper 2017-01-0525, 2017, doi:10.4271/2017-01-0525.

36. Lafossas, F., Colin, O., Le Berr, F., and Menegazzi, P., "Application of a New 1D Combustion Model to Gasoline Transient Engine Operation," SAE Technical Paper 2005-01-2107, 2005, doi:10.4271/2005-01-2107.

37. Ramajo, D., Zanotti, A., and Nigro, N., "Assessment of a Zero-Dimensional Model of Tumble in Four-Valve High Performance Engine," *International Journal of Numerical Methods for Heat & Fluid Flow* 17, no. 8 (2007): 770-787, doi:10.1108/09615530710825765.

38. Achuth, M. and Mehta, P.S. "Predictions of Tumble and Turbulence in Four-Valve Pentroof Spark Ignition Engines," *International Journal of Engine Research* 2, no. 3 (2001): 209-227, doi:10.1243/1468087011545442.

39. Grasreiner, S., Neumann, J., Luttermann, C., Wensing, M. et al., "A Quasi-Dimensional Model of Turbulence and Global Charge Motion for Spark Ignition Engines with Fully Variable Valvetrains," *International Journal of Engine Research* 15, no. 7 (2014): 805-816, doi:10.1177/1468087414521615.

40. Bozza, F., Teodosio, L., De Bellis, V. et al., "Refinement of a 0D Turbulence Model to Predict Tumble and Turbulent Intensity in SI Engines. Part II: Model Concept, Validation and Discussion," SAE Technical Paper 2018-01-0856, 2018, doi:10.4271/2018-01-0856.

41. Fogla, N., Bybee, M., Mirzaeian, M., Millo, F. et al., "Development of a K-k-ε Phenomenological Model to Predict In-Cylinder Turbulence," *SAE Int. J. Engines* 10, no. 2 (2017): 562-575, doi:10.4271/2017-01-0542.

42. Dai, W., Newman, C., and Davis, G., "Predictions of In-Cylinder Tumble Flow and Combustion in SI Engines with a Quasi-Dimensional Model," SAE Technical Paper 961962, 1996, doi:10.4271/961962.

43. Jones, P. and Junday, J., "Full Cycle Computational Fluid Dynamics Calculations in a Motored Four Valve Pent Roof Combustion Chamber and Comparison with Experiment," SAE Technical Paper 950286, 1995, doi:10.4271/950286.

44. Khalighi, B., El Tahry, S., Haworth, D., and Huebler, M., "Computation and Measurement of Flow and Combustion in a Four-Valve Engine with Intake Variations," SAE Technical Paper 950287, 1995, doi:10.4271/950287.

45. Yakhot, V. and Orszag, S.A., "Renormalization Group Analysis of Turbulence. I. Basic Theory," *J. Sci. Comput.* 1 (1986): 3-51, doi:10.1007/BF01061452.

46. Newhall, H.K., "Kinetics of Engine Generated Nitrogen Oxides and Carbon Monoxide," *Proceedings of the 20th International Symposium on Combustion*, 603-613, 1968.

47. Meintjes, K. and Morgan, A.P., "GMR-5827," General Motors Research Publications, 1987.

48. Mellor, A.M., Mellor, J.P., Duffy, K.P., Easley, W.L. et al., "Skeletal Mechanism for NOx Chemistry in Diesel Engines," SAE Technical Paper 981450, 1998, doi:10.4271/981450.

49. Lavoie, G.A., Heywood, J.B., and Keck, J.C., "Experimental and Theoretical investigation of Nitric Oxide formation in Internal Combustion Engines," *Combust. Sci. Technol.* 1 (1970): 313-326.

50. Moses, E., Yarin, A.L., and Yoseph, P.B., "On Knocking Prediction in Spark Ignition Engines," *Combust Flame* 101, no. 3 (1995): 239-261.

51. Rothe, M., Heidenreich, T., Spicher, U., and Schubert, A., "Knock Behavior of SI-Engines: Thermodynamic Analysis of Knock Onset Locations and Knock Intensities," SAE Technical Paper 2006-01-0225, 2006, doi:10.4271/2006-01-0225.

52. Cowart, J.S., Haghgooie, M., Newman, C.E., Davis, G.C. et al., "The Intensity of Knock in an Internal Combustion Engine: An Experimental and Modeling Study," SAE Technical Paper 922327, 1992, doi:10.4271/922327.

53. Douaud, A. and Eyzat, P., "Four-Octane-Number Method for Predicting the Anti-Knock Behavior of Fuels and Engines," SAE Technical Paper 780080, 1978, doi:10.4271/780080.

54. Livengood, J. and Wu, P., "Correlation of Autoignition Phenomenon in Internal Combustion Engines and Rapid Compression Machines," *Fifth Symposium (International) on Combustion* 5, no. 1 (1955): 347-356, doi:10.1016/S0082-0784(55)80047-1.

55. Yates, A. and Viljoen, C., "An Improved Empirical Model for Describing Auto-Ignition," SAE Technical Paper 2008-01-1629, 2008, doi:10.4271/2008-01-1629

56. Vandersickel, A., Hartmann, M., Vogel, K., Wright, Y.M. et al., "The Autoignition of Practical Fuels at HCCI Conditions: High-Pressure Shock Tube Experiments and Phenomenological Modeling," *Fuel* 93 (2012): 492-501.

57. Worret, R., Bernhardt, S., Schwarz, F., and Spicher, U., "Application of Different Cylinder Pressure Based Knock Detection Methods in Spark Ignition Engines," SAE Technical Paper 2002-01-1668, 2002, doi:10.4271/2002-01-1668.

58. Tanaka, S., Ayala, F., and Keck, J., "A Reduced Chemical Kinetic Model for HCCI Combustion of Primary Reference Fuels," *Combustion & Flame* 132 (2003): 219-239.

59. Andrae, J., "Comprehensive Chemical Kinetic Modeling of Toluene Reference Fuels Oxidation," *Fuel* (2013) 107: 740-748, doi:10.1016/j.fuel.2013.01.070.

60. Bozza, F., De Bellis, V., Minarelli, F., and Cacciatore, D., "Knock and Cycle by Cycle Analysis of a High Performance V12 Spark Ignition Engine. Part 2: 1D Combustion and Knock Modeling," *SAE Int. J. Engines* 8, no. 5 (2015): 2002-2011, doi:10.4271/2015-24-2393.

61. Ferrari, G., *Motori a Combustione Interna* (Il capitello, 2005).

62. Watson, N., "A Combustion Correlation for Diesel Engine Simulation," SAE Technical Paper 800029, 1980, doi:10.4271/800029.

63. Whitehouse, N.D., "The Effect of Changes in Design and Operating Conditions on Heat Release in Direct-Injection Diesel Engines," SAE Technical Paper 740085, 1974, doi:10.4271/740085.

64. Mauviot, G., Albrecht, A., and Poinsot, T.J., "A New 0D Approach for Diesel Combustion Modeling Coupling Probability Density Function with Complex Chemistry," SAE Technical Paper 2006-01-3332, 2003, doi:10.4271/2006-01-3332.

65. Dulbecco, A., Lafossas, F.-A., and Poinsot, T.J., "A 0D Phenomenological Approach to Model Diesel HCCI Combustion with Multi-Injection Strategies Using Probability Density Functions and Detailed Tabulated Chemistry," SAE Technical Paper 2009-01-0678, 2009, doi:10.4271/2009-01-0678.

66. Bruneaux, G., "Mixing Process in High Pressure Diesel Jets by Normalized Laser Induced Exciplex Fluorescence Part II: Wall Impinging Versus Free Jet," SAE Technical Paper 2005-01-2097, 2005, doi:10.4271/2005-01-2097.

67. Chmela, F., Pirker, G., Losonczi, B., and Wimmer, A., "A New Burn Rate Simulation Model for Improved Prediction of Multiple Injection Effects on Large Diesel Engines," Thiesel, 2010.

68. Pirker, G., Chmela, F., and Wimmer, A., "ROHR Simulation for Diesel Engines Based on Sequential Combustion Mechanisms," SAE Technical Paper 2006-01-0654, 2006, doi:10.4271/2006-01-0654.

69. Barba, C., Burkhardt, C., Boulouchos, K., and Bargende, M., "A Phenomenological Combustion Model for Heat Release Rate Prediction in High-Speed DI Diesel Engines with Common Rail Injection," SAE Technical Paper 2000-01-2933, 2000, doi:10.4271/2000-01-2933.

70. Dulbecco, A., Lafossas, F.-A., Mauviot, G., and Poinsot, T.J., "A New 0D Diesel HCCI Combustion Model Derived from a 3D CFD Approach with Detailed Tabulated Chemistry," *OGST* 64, no. 3 (2009): 259-284, doi:10.2516/ogst/2008051.

71. Rudloff, J., Dulbecco, A., and Font, G., "The Dual Flame Model (DFM): A Phenomenological 0D Diesel Combustion Model to Predict Pollutant Emissions," SAE Technical Paper 2015-24-2388, 2015, doi:10.4271/2015-24-2388.

72. Maiboom, A., Tauzia, X., Shah, S.R., and Hetet, J.F., "New Phenomenological Six-Zone Combustion Model for DI Diesel Engines," *Energy and Fuels* 23, no. 2 (2009): 690-703.

73. Finesso, R., Spessa, E., Mancaruso, E., Sequino, L. et al., "Spray and Soot Formation Analysis by Means of a Quasi-Dimensional Multizone Model in a Single Cylinder Diesel Engine under Euro 4 Operating Conditions," SAE Technical Paper 2015-24-2416, 2015, doi:10.4271/2015-24-2416.

74. Arsie, I., Di Genova, F., Pianese, C., Sorrentino, M. et al., "Development and Identification of Phenomenological Models for Combustion and Emissions of Common-Rail Multi-Jet Diesel Engines," SAE Technical Paper 2004-01-1877, 2004, doi:10.4271/2004-01-1877.

75. Rether, D., Schmid, A, Grill, M., and Bargende, M., "Quasi-Dimensional Modeling of CI-Combustion with Multiple Pilot and Post Injections," SAE Technical Paper 2010-01-0150, 2010, doi:10.4271/2010-01-0150.

76. Kuleshov, A., "Model for Predicting Air-Fuel Mixing, Combustion and Emissions in DI Diesel Engines over Whole Operating Range," SAE Technical Paper 2005-01-2119, 2005, doi:10.4271/2005-01-2119.

77. Musculus, M. and Kattke, K., "Entrainment Waves in Diesel Jets," SAE Technical Paper 2009-01-1355, 2009, doi:10.4271/2009-01-1355.

78. Desantes, J.M., Pastor, J.V., García-Oliver, J.M., and Pastor, J.M., "A 1D Model for the Description of Mixing-Controlled Reacting Diesel Sprays," *Combustion and Flame* 156 (2009): 234-249.

79. Nishida, K. and Hiroyasu, H., "Simplified Three-Dimensional Modelling of Mixture Formation and Combustion in a D.I. Diesel Engine," SAE Technical Paper 890269, 1989, doi:10.4271/890269.

80. Jung, D. and Assanis, D.N., "Multi-Zone DI Diesel Spray Combustion Model for Cycle Simulation Studies of Engine Performance and Emissions," SAE Technical Paper 2001-01-1246, 2001, doi:10.4271/2001-01-1246.

81. Varde, K.S., Popa, D.M., and Varde, L.K., "Spray Angle and Atomization in Diesel Sprays," SAE Technical Paper 841055, 1984, doi:10.4271/841055.

82. Lefebvre, A.H., *Atomization and Sprays* (Taylor & Francis, 1989).

83. Siebers, D.L., "Scaling Liquid-Phase Fuel Penetration in Diesel Sprays Based on Mixing-Limited Vaporization," SAE Technical Paper 1999-01-0528, 1999, doi:10.4271/1999-01-0528.

84. Chmela, F. and Orthaber, G., "Rate of Heat Release Prediction for Direct Injection Diesel Engines Based on Purely Mixing Controlled Combustion," SAE Technical Paper 1999-01-0186, 1999, doi:10.4271/1999-01-0186.

85. Dulbecco, A., "Modeling of Diesel HCCI Combustion and Its Impact on Pollutant Emissions," Ph.D. thesis, Institut National Polytechnique de Toulouse, 2010.

86. Naber, J. and Siebers, D., "Effects of Gas Density and Vaporization on Penetration and Dispersion of Diesel Sprays," SAE Technical Paper 960034, 1996, doi:10.4271/960034.

87. Siebers, D., Higgins, B., and Pickett, L., "Flame Lift-Off on Direct-Injection Diesel Fuel Jets: Oxygen Concentration Effects", SAE Technical Paper 2002-01-0890, 2002, doi:10.4271/2002-01-0890.

88. Inagaki, K., Ueda, M., Mizuta, J., and Nakakita, K., "Universal Diesel Engine Simulator (UniDES): 1st Report: Phenomenological Multi-Zone PDF Model for Predicting the Transient Behavior of Diesel Engine Combustion," SAE Technical Paper 2008-01-0843, 2008, doi:10.4271/2008-01-0843.

89. Yang, Q. and Grill, M., "A Quasi-Dimensional Charge Motion and Turbulence Model for Diesel Engines with a Fully Variable Valve Train," SAE Technical Paper 2018-01-0165, 2018, doi:10.4271/2018-01-0165.

90. Dec, J., "A Conceptual Model of DI Diesel Combustion Based on Laser Sheet Imaging," SAE Technical Paper 970873, 1997, doi:10.4271/970873.

91. Anderlohr, J.M., Bounaceur, R., Pires Da Cruz, A., and Battin-Leclerc, F., "Modelling of Auto-Ignition and NO Sensitization for the Oxidation of IC-Engine Surrogate Fuels," *Combustion and Flame* 156, no. 2 (2009): 505-521.

92. Magnussen, B.F. and Hjertager, B.H., "On Mathematical Modeling of Turbulent Combustion with Special Emphasis on Soot Formation and Combustion," *16th International Symposium on Combustion*, Cambridge, MA, 1976.

93. Neely, G.D., Sasaki, S., Huang, Y., Leet, J.A. et al., "New Diesel Emission Control Strategy to Meet US Tier 2 Emissions Regulations", SAE Technical Paper 2005-01-1091, 2005, doi:10.4271/2005-01-1091.

94. Guardiola, C., Lopez, J.J., Martin, J., and Garcia-Sarmiento, D., "Semiempirical In-Cylinder Pressure Based Model for NOX Prediction Oriented to Control Applications," *Applied Thermal Engineering* 31 (2011): 3275-3286.

95. Koci, C., Svensson, K., and Gehrke, C., "Investigating Limitations of a Two-Zone NOx Model Applied to DI Diesel Combustion Using 3-D Modeling," SAE Technical Paper 2016-01-0576, 2016, doi:10.4271/2016-01-0576.

96. Bayer, J. and Foster, D., "Zero-Dimensional Soot Modeling," SAE Technical Paper 2003-01-1070, 2003, doi:10.4271/2003-01-1070.

97. Dulbecco, A. and Font, G., "Development of a Spray-Based Phenomenological Soot Model for Diesel Engine Applications," SAE Technical Paper 2017-24-0022, 2017, doi:10.4271/2017-24-0022.

CHAPTER 4

Compressor and Turbine as Boundary Conditions for 1D Simulations

Oldřich Vítek, Jan Macek, and Zdeněk Žák
Czech Technical University in Prague

This chapter describes the main principles of the physics-based modeling of turbomachinery and positive displacement compressors. When focusing on radial/axial turbines, the models are related to mass flow rate and power/torque at defined pressure ratio and speed while considering single-, twin- and parallel-scrolls of different sizes coupled with control devices (e.g., waste-gate, variable geometry nozzle). Although the standard approach uses machine maps, the deeper insight calls for the models of different accuracy, which are discussed below.

The turbine-related topics start with a standard 0D map-based model. The semi-predictive 0D central streamline (map-less) physical model is shown, which creates a base for 1D detailed, fully unsteady physical model of a centripetal turbine. It is extended to describe choking of a radial turbine and specific issues of twin- or parallel-scrolls. Moreover, the calibration of models, based on experiments, is mentioned.

Turbocharger mechanical layout is described briefly after turbine features. This includes 1D rotor dynamics and friction/mechanical losses, needed to amend the differential equation of a rotor motion to determine a turbocharger shaft speed.

When dealing with compressors, main focus is put on centrifugal ones. The models are (again) related to mass flow rate, and shaft input power at defined pressure ratio and speed. The compressor topics are discussed start with 0D map-based model, again. Similarly to a turbine, a 0D central streamline (map-less) quasi steady physical model is created, which is able to predict choke limit. The surge limit and surge phenomena can be simulated to a certain extent but the model is not predictive enough for determining them.

Finally, positive displacement compressors are presented in a similar way, the last on being unsteady physics-based model with complicated rotor geometry, obtained from 3D CAD model.

Keywords

Radial centripetal turbine, Axial turbine, Positive displacement compressor, Roots blower, Lysholm compressor, Radial centrifugal compressor, One-dimensional model, CFD model, Unsteady flow, Simulation, Turbocharging, Map-less turbocharger representation, Transonic flow, Impeller vortex, Incidence loss, Windage loss, Reynolds number, Leakage flow, Model calibration, Experiments, Speed non-uniformity, Friction losses

4.1 Requirements for 1D Simulation Boundary Conditions

The one-dimensional (1D) model of an IC engine with cylinders, piping systems, heat exchangers, and turbomachines can be built using modules, solving basic partial differential equations with initial and boundary conditions, using module interfaces, as described, e.g., in Ciesla [18] and Morel [28] or [51].

There are many possibilities how a turbomachine or positive displacement compressor model can be linked to such a virtual engine. In any case, it has to link total pressure and temperature at the end of a 1D module to mass flow rate (MFR) through a machine and to static or total pressure downstream of a machine—Jenny [33], as shown in Figure 4.1 for the example of a turbine, similar to Zinner [83]. The similar situation occurs in the case of a compressor.

The machine can link in general more interfaces with different MFRs, as well known, e.g., from twin-scroll turbines (Figure 4.2). Some model parameters (brake mean effective pressure (bmep) or relative air-to-fuel (A/F) ratio in Figure 4.1) are set during simulation as controlled model outputs, but the other additional parameters of a model (e.g., shaft speed or a rack position, yielding reference area of a turbine nozzle) have to be found from adjacent models coupled to the thermo-aerodynamic machine model (Figure 4.3).

In any case, a compressor or a turbine creates a component of complicated thermal, aerodynamic, and mechanical system of the whole engine. Nevertheless, any turbomachine inside is a complicated system, as well, unsteady in time and spread in space, with not negligible dimensions. The example of a radial turbine, shown in Figure 4.2 and

FIGURE 4.1 Standard use of a turbine mass flow rate (MFR) model.

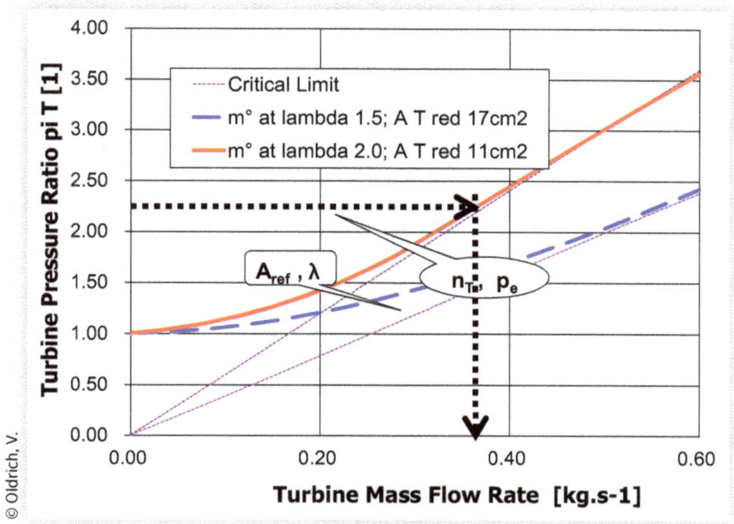

FIGURE 4.2 The simplification of a turbine to a 1D model: scroll(s) 1′ and 1″, nozzle(s) 2′ and 2″, common nozzle exit after mixing 2N, virtual module for total state transformation 2N-rel2N, incidence loss rel2N-rel2I, impeller inlet (relative to rotating channel) rel2I, rotating channel with impeller exit rel3I, virtual module for total state transformation rel3-3, outlet diffuser 3-4—if used.

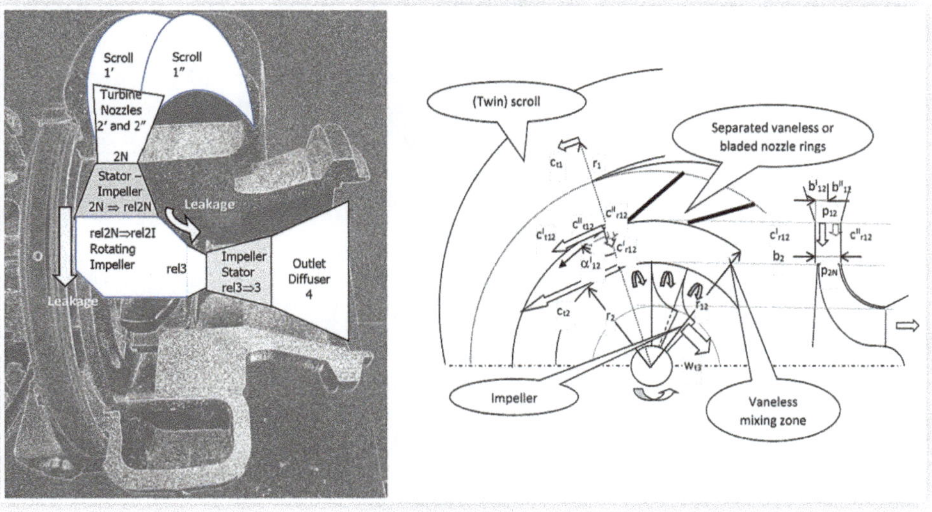

with length/time scales mentioned below in Table 4.1, presents those facts transparently. On one hand, the question is how far the lumped parameter, map-based models, ignoring three-dimensional (3D) unsteady flow inside those machines, can be representative for boundary conditions. On the other hand, the already developed computational fluid dynamic (CFD) codes offer options for the description of inside thermo-/aerodynamics today. Even 1D models themselves can be applied for the deepening of internal unsteady aerodynamic simulations of turbomachines, as described below. The outputs of those virtual machines are MFR, power and outlet temperature at the currently known inlet state of gas and shaft speed, as presented in the example of Figure 4.1.

FIGURE 4.3 Basic interfaces between TC and ICE models.

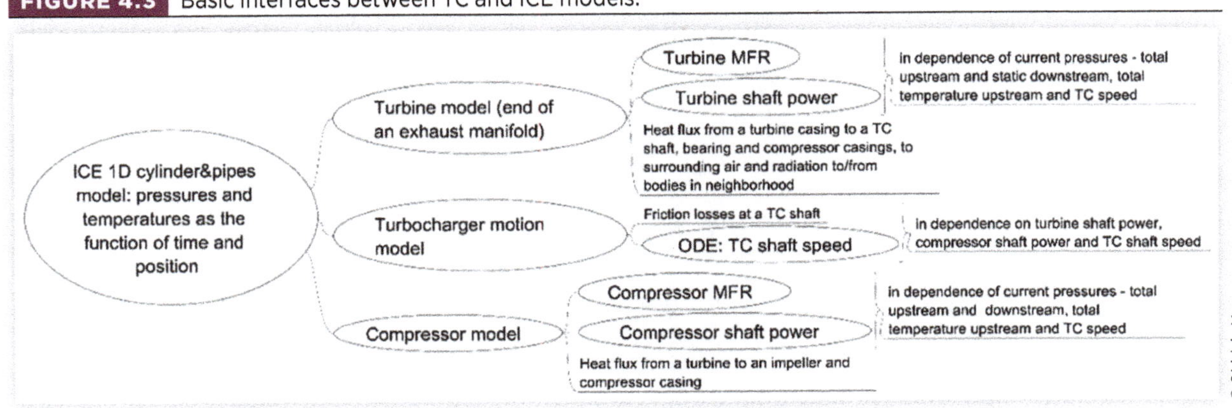

TABLE 4.1 Comparison of length and time scales for two typical DI and SI vehicle engines with radial centripetal turbine

Engine geometry	Bore	mm	120	80	80
	Number of cylinders		6	4	4
	Lenqth of manifold	mm	468	208	208
	Turbine rotor dia	mm	100	44	44
	Stator wave path	mm	188	84	84
	Rotor wave path	mm	95	42	42
Operation	Enqine speed	r.p.m.	2000	4000	6000
	Turbocharger speed	r.p.m.	88265	205108	205108
	Gas temparature	K	800	850	1000
	Sonic speed	m.s-1	551	567	615
	Flow velocity	m.s-1	167	100	150
Duration of impeller revolution		ms	0.7	0.3	0.3
		deq CA	8.2	7.0	10.5
Duration of pressure wave single passage	Manifold	ms	0.85	0.37	0.34
	Turbine stator	ms	0.34	0.15	0.14
	Turbine rotor	ms	0.17	0.07	0.07
	Manifold	deq CA	10.2	8.8	12.2
	Turbine stator	deq CA	4.1	3.5	4.9
	Turbine rotor	deq CA	2.1	1.8	2.5
Duration of pressure single passage	Manifold	ms	2.81	2.08	1.39
	Turbine stator	ms	1.13	0.84	0.56
	Turbine rotor	ms	0.57	0.42	0.28
	Manifold	deq CA	33.7	49.9	49.9
	Turbine stator	deg CA	13.6	20.1	20.1
	Turbine rotor	deg CA	6.8	10.1	10.1

The links between turbomachines inside a turbocharger (TC), including not only obvious friction losses of a shaft but also electric motor drive in the case of e-turbo, are among other features to be described by the current models.

The shaft speed is determined from dynamic equations of motion of a TC rotor, which may be coupled to an engine or electric motor, in some cases via gears with known transmission ratio. In any case, acting torques have to be determined for solving ordinary

FIGURE 4.4 Available types of a TC models.

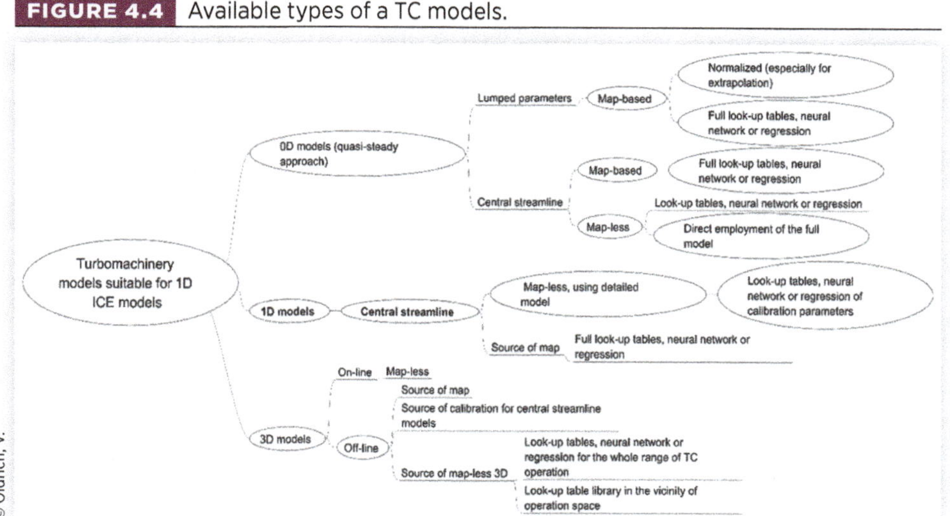

differential equation (ODE) describing system dynamics. If real control system is used, as for a variable nozzle turbine (rack position and turbine flow area), the phase shift of control signals in time due to actuator dynamics and the computational speed of an electronic control unit have to be taken into account. Additional heat fluxes in a machine casing can be taken into account using suitable solid phase models.

The result of model equation integration is important for modelling of power-passive machines, e.g., for a compressor air delivery, described by MFR at instantaneous pressures and upstream inlet temperature, formally in the same manner as for an active turbine. All those links are described in Figure 4.3.

Fulfilling these tasks, the different types of models may be used. The main possibilities are presented in Figure 4.4. The number of coordinates starts with lumped parameter models for zero-dimensional (0D) quasi-steady map-based tools. The disadvantage of the data- and map-based models of this type is caused by poor extrapolation capabilities. Normalized curves of parameters are used for this purpose frequently.

The use of map-based models requires the definition of ideal processes, typical, e.g., for isentropic efficiency, discharge coefficients or for blade-speed ratio (BSR). The definition of ideal case may cause difficulties, e.g., if mixing of flows occurs, as in a twin-scroll turbine, since it is influenced by losses. The better results can be yielded by using map-less data-based models, but the extrapolation in this case is even worse, because there are no generalized forms of relations, as it is for maps or even normalized maps.

Physical models according to Figure 4.4 may feature distributed parameters. If advanced description is applied (i.e., better than quasi-steady algebraic equations), they can be easily adapted for internal unsteadiness (wave phenomena), if length and time scales need such a detail. Length dimensions of turbomachine parts may be comparable to the pipes used in manifolds. The same is valid for time scales. The quasi-steady approach may cause errors under such circumstances. The example of length, time, and crank angle scales for radial turbines is presented in Table 4.1 [41].

1D physical, central streamline models use loss (drag), lift, or discharge coefficients. Therefore they can be calibrated by experiments. It is a disadvantage due to the need to employ the measurements of machine integral parameters and to fit the model to them, but simultaneously it creates an advantage of reliable results based on reality.

The physical models can easily simulate the real processes inside a machine, which is important especially if flow mixing or splitting takes part after flow state change,

FIGURE 4.5 Standard map-based, lumped parameter approach for a twin-scroll turbine requires fictitious connection of manifold branches at pressures different from those governing the mixing zone for using maps in 1D code.

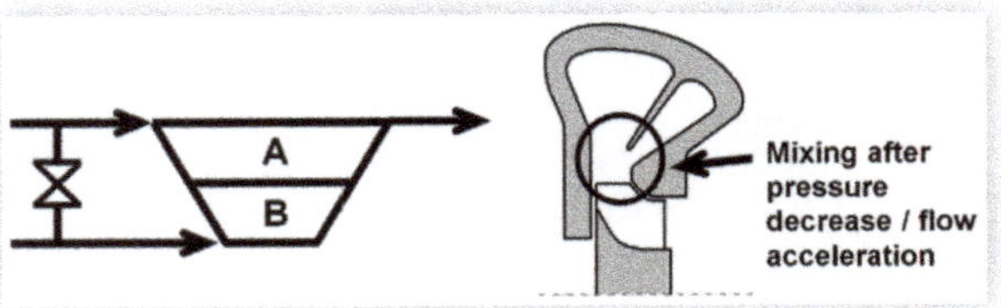

e.g., due to the acceleration inside a machine, as presented in Figure 4.5 from [84]. If standard lumped parameter model with turbine maps is used, it requires adding fictitious pipe connections, which have to be calibrated individually case to case.

As mentioned above, the current 1D unsteady-flow models offer possibilities of using the 1D modules (solvers) as building stones even for turbomachine representation. Such a possibility is represented schematically in Figure 4.6 (compare it to Figure 4.2). Turbine blade cascades are modelled by pipes and connecting orifices with cross sections adjusted to real flow areas including flow separation. The losses are represented by pressure loss coefficients, wall friction, and discharge coefficients, with the most suitable type of loss for every blade section. The mixing of flows from both scroll partitions (if twin scroll is modelled) is realized in flow-split (FS) module, in which different momentum is taken into account. The rotor blade channels are subjected to centrifugal acceleration, if flow velocity features non-zero radial component. The fictitious heat transfer and change of flow area are applied in interfaces between stator and rotor, mimicking transformation of velocities and total states from steady blades to moveable ones and back. Real heat transfer to/from the walls may be modelled additionally. In the real unsteady case, the local rotor torque is integrated from Eulerian theorem. Sealing leakages are considered, as well. Moreover, the amendments of model boundary conditions, e.g., considering heat transfer inside a machine or the detailed tribology models of friction losses in bearings, can be adapted to the model. It will be described, using some simplified examples, below.

The 3D model employment creates the highest level of simulations. Those models are extremely time consuming and a routine use online cannot be recommended yet. The offline application has more forms, either with look-up tables or regression model interfaces, prepared in advance for the whole operation domain mapping (in fact, using map-based approach with 0D or 1D based on 3D numerical experiments) or simulating

FIGURE 4.6 General unsteady model of a single-scroll turbine and a parallel twin-scroll turbine with basic modules transferable to 1D solver.

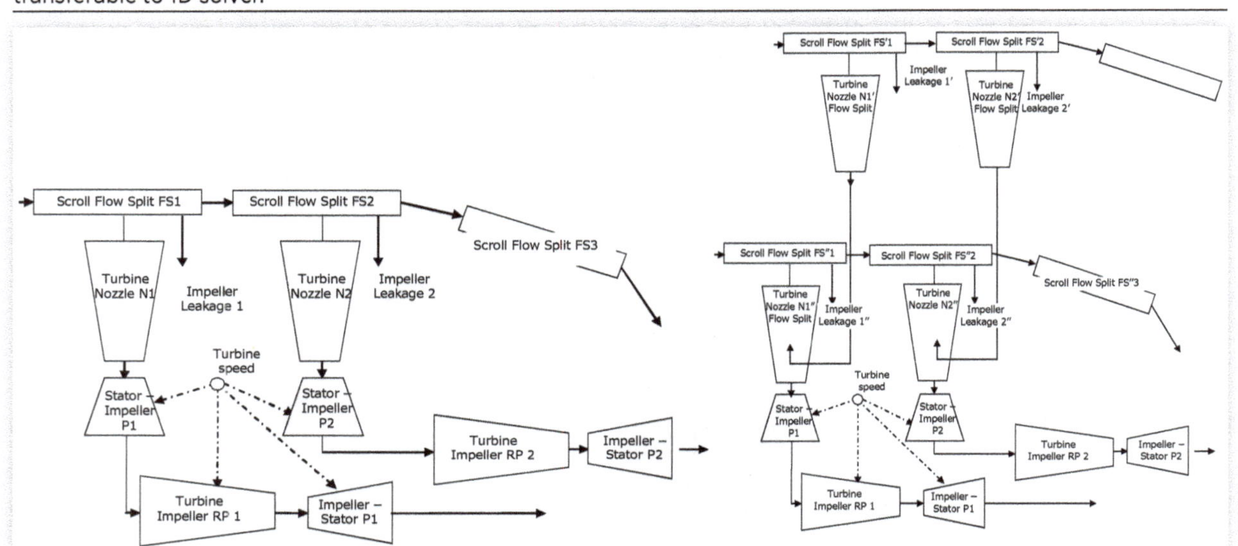

just neighborhood of operation states, similar to approach used in Macek [43]. The important condition for the use of 3D models is the reflection of physical reality at sufficient accuracy level, e.g., due to the use of mesh fine enough, since if linked directly to 1D models, there is no possibility to calibrate these models otherwise than using a very general way, e.g., by choice of turbulence model.

4.2 General Principles of Physical Modelling of Turbomachinery and Positive Displacement Machines

4.2.1 Basic Equations

The conservation laws for an open thermodynamic system in 1D simplification yield for mass, momentum, and energy of the following partial differential equations for density, relative velocity toward walls of rotating channel, and total energy in conservative form, suitable for application of well-known numerical schemes

$$A\frac{\partial\rho}{\partial t}+\frac{\partial(A\rho w)}{\partial x}=0 \tag{4.1}$$

$$\frac{\partial(A\rho w)}{\partial t}+\frac{\partial\left(A\rho w^2\right)}{\partial x}=-A\frac{\partial p}{\partial x}-A\rho\tau+A\rho C_\varepsilon;\quad \tau=\lambda\frac{O}{4A}\frac{|w|w}{2} \tag{4.2}$$

In this equation, C_ε is component of centrifugal acceleration due to rotation of a channel with angular speed ω in the direction of a streamline. λ_f, O, and A are boundary layer (BL) friction coefficient, wetted surface, and cross section of channel, respectively.

Total energy conservation of internal energy u_e, kinetic energy for relative velocity w, and potential energy (in centrifugal force field in dependence on the distance from axis of rotation r) is amended by convective heat flux and gas thermal conductivity, whereas the divergence of channel walls by angle α is taken into account together with angle $\boldsymbol{\varepsilon}$ between a streamline and distance from axis of rotation, represented by the radius r

$$\begin{aligned} c_v\rho\frac{\partial T}{\partial t}+u\frac{\partial\rho}{\partial t}+\rho w\frac{\partial w}{\partial t}-\rho\omega r^2\frac{d\omega}{dt}&=-c_p\rho w\frac{\partial T}{\partial x}+h\frac{\partial\rho}{\partial t}-\rho w^2\frac{\partial w}{\partial x}+ \\ &+\rho\omega^2 r\cos\varepsilon+\frac{O}{A\sin\alpha}\alpha_Q\left(T-T_{wall}\right)-\lambda_Q\frac{\partial^2 T}{\partial x^2}-\lambda_Q\frac{\partial T}{\partial x}\frac{d\ln A}{dx} \end{aligned} \tag{4.3}$$

The last equation is written for general direction of centrifugal acceleration, for a diabatic machine especially due to the heat transfer at walls and for the non-uniform angular speed of TC shaft $\boldsymbol{\omega}$. The time rate of angular speed in Equation (4.3)—angular acceleration of the shaft—might be quite significant due to pulsating driving torque. It may be applied in energy conservation relation as a specific source term, increasing the total energy if angular speed increases. The equations above are used, if unsteady-flow 1D model is created. It is of advantage, if the model uses already available modules with well-elaborated 1D solvers, as [18, 28] or [51], connected by newly developed interfaces respecting 3D flow substance and transformed to the central streamline simplification. Such an approach has been successfully adopted in the past (Pohorelsky [54] or Brynych [10]).

If open thermodynamic system is subjected to quasi-steady conditions, the integration of Equations (4.1)–(4.3) yields well-known conservation theorems of Stodola (energy) or that of Euler (angular momentum). Energy conservation yields the power delivered from a system with heat flux from environment and steady MFR

$$\dot{Q} + \dot{m}\left(h_1 + \frac{c_1^2}{2} - h_2 - \frac{c_2^2}{2}\right) = P \tag{4.4}$$

where c is a velocity relative to the fixed coordinate system. Static enthalpy h increases by kinetic energy to stagnation enthalpy. If coordinate system of channel walls rotates with a circumferential velocity u and w is velocity relative to the channel walls, no power is delivered by the machine. The potential energy of fluid in the field of centrifugal acceleration changes, if flow changes the distance from axis of rotation. Integrating work of inertia forces and assuming adiabatic case, Equation (4.4) can be reduced to the conservation of total enthalpy in flow relative to the coordinate system fixed toward walls of rotating channel

$$h + \frac{w^2}{2} - \frac{u^2}{2} = h_0 - \frac{u^2}{2} = h_t = const. \quad \text{where } u = \omega r \tag{4.5}$$

If speed of sound is reached by flow velocity, critical state occurs. The critical static temperature depends on total inlet temperature and the change of potential energy (if any)

$$T_{out}^* = \frac{1}{\kappa r}\left[\kappa r T_{0in}\frac{2}{\kappa+1} + \frac{\kappa+1}{\kappa-1}\left(u_{out}^2 - u_{in}^2\right)\right] \tag{4.6}$$

The relation is valid for any adiabatic case including irreversible processes, from which pressures can be found, if appropriate relations with loss coefficient for accelerated flow are used—e.g., from Equation (4.17). Equations for quasi-steady flow may be used for algebraic-type central streamline models or used as interfaces between 1D modules.

The specific power of an adiabatic machine in a fixed-in-space coordinate system, where certain work may be performed due to the movement of channel walls, is given by

$$\Delta\left(h + \frac{c^2}{2}\right) = \Delta h_t = \Delta h_0 = \frac{P_T}{\dot{m}} = \Delta\left(uc_t\right) = u_2 c_{2t} - u_3 c_{3t} \tag{4.7}$$

The Euler turbine theorem was applied at the right-hand side of Equation (4.7) using tangential components of absolute velocities according to velocity triangles, plotted, e.g., for a turbine in Figure 4.10 (see coordinate system axes). In the role of a flow rate determining component of flow velocity, the radial component is used at an impeller inlet/outlet for a turbine/compressor impeller, respectively. An axial turbine impeller exducer outlet or compressor inducer inlet is used, respectively, as well. The angles are measured from the flow direction, i.e., from radial or axial axes, respectively.

If mixing of flows takes part and pressures are the same at the both inlet and outlet of mixing region, equation for momentum conservation yields

$$\left(\dot{m}' + \dot{m}''\right)\vec{w}_{mixed} = \dot{m}'\vec{w}' + \dot{m}''\vec{w}'' \tag{4.8}$$

It has to be transformed to velocity components, as shown in Figure 4.10.

The application of the above-mentioned equations to elements of turbine or compressor layouts, e.g., the specific cases of a vaneless nozzle ring, a radial-axial diffuser with rotating flow or case of incidence loss at an impeller inlet, will be mentioned below. The integration of friction losses and other dissipation of kinetic energy would require 3D description with turbulence models. For the goals of the current section it is much simpler to employ empiric knowledge of losses, calibrated by experiments. Transformation of enthalpies to pressures is possible, using properly defined loss or efficiency coefficients, which will be done in the following section.

4.2.2 Mean Value of Rothalpy and Centrifugal Force at Central Streamline

The real channel inside a blade cascade can be represented by a series of finite volumes (FVs) or pipes, constructed along the central streamline as shown in Figure 4.2. It yields the general scheme shown in Figure 4.7, Macek [40], *a* being axial direction and the axis of rotation at the same time. In the most cases, the general splitting of a channel is done using one radial, one axial, and additionally one or two diagonal pipe sections.

If flow in the channel of a rotating impeller is being described, the influence of external - centrifugal and Coriolis - acceleration is considered. The former one acts in a direction of radial component of flow direction and causes the changes of potential energy (rothalpy)—Equation (4.5). The latter one yields tangential pressure field gradient only and transfers velocity changes into the work of an impeller. The position of a central streamline, especially for long blades, should be set according to pressures and potential energy in the field of centrifugal force. The total averaged enthalpy flux in axial blade cascade is

$$\int_{r'}^{r''}\left(h+\frac{w^2}{2}-\frac{\omega^2 r^2}{2}\right)\rho w_r 2\pi r dr = \left(h+\frac{w^2}{2}\right)\rho w_r \pi\left(r''^2 - r'^2\right) + \\ -\frac{\omega^2}{2}\rho w_r \frac{\pi}{2}\left(r''^4 - r'^4\right) = \dot{m}\left(h+\frac{w^2}{2}-\frac{\overline{u}^2}{2}\right) \tag{4.9}$$

since the mean diameter for the same MFR is

$$\pi\left(r''^2 - \overline{r}^2\right) = \pi\left(\overline{r}^2 - r'^2\right) \\ \overline{r}^2 = \frac{r''^2 + r'^2}{2} \approx \frac{r''^2}{2}; \quad \overline{r} \approx \frac{r''}{\sqrt{2}} \approx \frac{r''}{1.4} \tag{4.10}$$

The last formula assumes negligible root (hub) radius of an impeller just to compare specific numbers. The central streamline with equal MFR above and below is representative of rothalpy, calculated from the blade velocity at streamline radius.

FIGURE 4.7 General scheme of a 1D rotating finite volume (radius of rotation *r*, flow coordinate *x*).

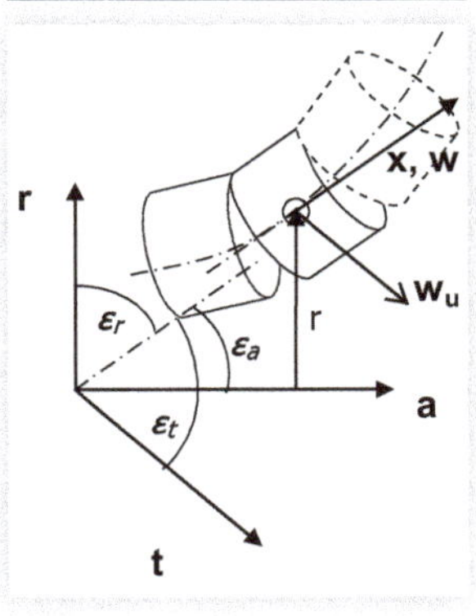

Assuming constant density along a blade, mean centrifugal force is

$$\bar{a}_C = \frac{\int_{r'}^{r''} \omega^2 r \rho 2\pi r b dr}{\int_{r'}^{r''} \rho 2\pi r b dr} = \frac{\rho\omega^2 \frac{2\pi}{3} b\left(r''^3 - r'^3\right)}{\rho b\pi\left(r''^2 - r'^2\right)} = \frac{\rho\omega^2 \frac{2\pi}{3} b\left(r''^2 + r''r' + r'^2\right)}{\rho b\pi\left(r'' + r'\right)}$$

$$= \frac{\omega^2 \frac{2}{3} r''^2 \left(1 + \frac{r'}{r''} + \frac{r'^2}{r''^2}\right)}{r''\left(1 + \frac{r'}{r''}\right)} \approx \frac{2}{3}\omega^2 r'' = \omega^2 \frac{r''}{1.5} \tag{4.11}$$

Both conditions of energy and force averaging feature certain difference. The mean radius for calculation of mean acceleration is close to the center of streamline radius, nevertheless, but centrifugal force from mean acceleration acts from the whole height of an axial channel on the smaller radius of a radial part (Figure 4.7), if it is connected to adjacent radial or diagonal module. The relations above can be corrected for variable density in pressure field of centrifugal force. In any case, the radii for calculation of mean centrifugal force at inlet and outlet of radially positioned pipe modules should be subjected to optimization during calibration against experiments.

The tangential position of a central streamline in a radial part of a channel should be shifted, as well, taken into account at least one half of Coriolis acceleration, caused by magnitude change of circumferential speed if flow proceeds along radius. Its second half, caused by radial velocity direction change, is negligible due to the flow slip toward the solid body rotation of impeller blades (Stodola's relative vortex). The tangential position of central streamline is not important for 1D model, however, and need not to be done.

4.2.3 Transformations between Rotating Impeller and Stator or between Cartesian and Cylindrical Coordinates

4.2.3.1 TRANSFORMATION OF TOTAL STATES BETWEEN IMPELLER AND STATOR

Transferring flow from a stator to a rotor and backward, the changes of velocities and kinetic energy (the stagnation states of fluid) have to be respected.

Defining flow angles by α or β for a stator or an impeller, respectively, the velocity triangles and splitting vectors into tangential or axial/radial components yield, e.g., from Figure 4.10

$$\begin{aligned} w_2 \cos\beta_2 &= w_{2r} = c_{2r} = c_2 \cos\alpha_2 \\ \vec{c}_{2t} &= u_2 + \vec{w}_{2t} = u_2 + w_{2r} \tan\vec{\beta}_2 \end{aligned} \tag{4.12}$$

The symbols with arrows show, by their signs, the real direction (positive in the direction of rotation). The symbols without signs are always positive. The relations have to be modified according to the specific situation in different parts of machines. The mass transporting component of velocity (axial or radial) is kept even for compressible fluids at almost constant and reasonably small value by the blade design. These components are calculated from MFR if static density is known.

If the standard 1D unsteady-flow modules with external acceleration are available in the form of solvers, they can be used for simulation of channel flow. The stagnation temperature inside them has to be decreased or increased at the stator-rotor linking interfaces to keep the same static temperature despite changing stagnation enthalpy.

It can be done by fictitious heat transfer, in the case of a turbine from the fluid at both inlet and outlet interfaces of an impeller. The reason for it, i.e., reduction of downstream velocities at the both interfaces, is clearly visible in h-s diagram in Figure 4.9 and from velocity triangles in Figure 4.10. The sum of both fictitious heat fluxes equals to the power extracted from an impeller. In the case of compressor, it is just reversed, since an impeller receives driving power (see Figure 4.23). The stagnation pressure can be calculated from both static and stagnation temperatures via isentropic change if no incidence loss occurs. If it does, the additional drop of static pressures at interface occurs—see h-s diagrams, e.g., Figure 4.9. The heat sink at impeller inlet has to change the stagnation state by

$$\dot{Q} = \Delta H_0 = \dot{m}_I \left(h_{0rel} - h_0 \right) = \frac{w_2^2 - c_2^2}{2} \tag{4.13}$$

which can be calculated from MFR using angles of flow and radial or axial components of velocities and static density. A similar procedure with reversed processes is used for impeller-to-stator transformation.

4.2.3.2 GEOMETRICAL TRANSFORMATION OF BLADE CASCADES

Useful geometrical transformation occurs, if standard Cartesian coordinates and axial blade cascade geometry is to be transformed into polar or cylindrical coordinates (Sherstjuk [65]) used in radial machines. In this case, conformal (angle keeping) transformation yields, if small corrections of axial or radial dimension of blade height are allowed to compensate for density changes

$$d\phi = \frac{b_{ref} K_2}{br} dx_t; \quad \frac{d(br)}{br} = \frac{b_{ref} K_2}{br} dx_a \tag{4.14}$$

$$k_2 \left(x_{a,out} - x_{a,in} \right) = k_2 c \cos\gamma = \ln\left(\frac{(br)_{out}}{(br)_{in}} \right)$$

$$\Delta\phi = k_2 \left(x_{t,out} - x_{t,in} \right); \quad k_2 = \frac{2\pi}{zs}; \quad z = \frac{2\pi \cos\gamma}{\frac{s}{c} \ln\left(\frac{(br)_{out}}{(br)_{in}} \right)} \tag{4.15}$$

These transformations help in transferring results from axial blade cascades to radial ones [49] (Figure 4.8).

FIGURE 4.8 Conformal transformation of general straight line from polar coordinates.

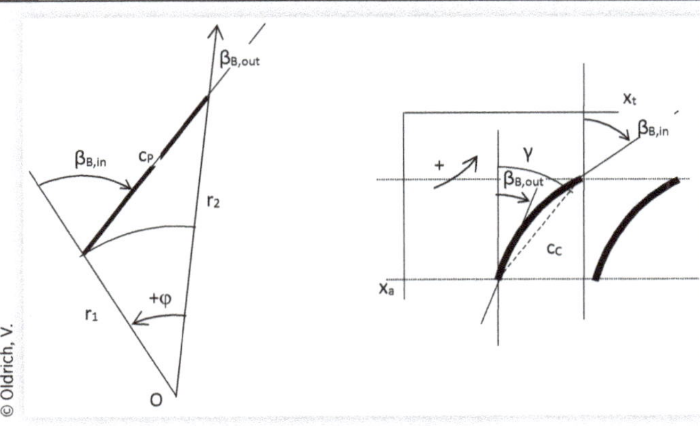

4.2.4 Loss Coefficients in Compressible Fluid Flow

The loss coefficients are defined from unused ("lost") enthalpy head if real adiabatic case is compared to isentropic expansion or compression, as demonstrated in Figure 4.9 or Figure 4.23, for a turbine or a compressor, respectively

$$\frac{w_{2s}^2}{2} = h_{01} - h_{2s} = h_{01} - h_2 + \Delta h_{lost} = \frac{h_{01} - h_2}{\eta_{out}} = \frac{w_2^2}{2\eta_{out}}$$

$$\Delta h_{lost} = c_p T_{01}\left[1 - \left(\frac{p_{02}}{p_{01}}\right)^{\frac{\kappa-1}{\kappa}}\right] = \frac{\zeta_{out}}{2} w_2^2; \quad p_{01} = p_{02}\left[1 - \zeta_{out}\left(1 - \left(\frac{p_2}{p_{02}}\right)^{\frac{\kappa-1}{\kappa}}\right)\right]^{\frac{1-\kappa}{\kappa}} \tag{4.16}$$

$$\eta_{out} = \frac{1}{1+\zeta_{out}}; \quad C_D = \sqrt{\eta_{out}}; \quad C_P = \zeta_{out}$$

Equation (4.16) considers accelerated (nozzle) flow with reference enthalpy head calculated from the achieved outlet speed. It defines isentropic efficiency of expansion, equivalent loss or pressure coefficient, and, for the case of nozzle, MFR discharge coefficient, as well. Those parameters are often used in different solvers instead of isentropic efficiency of expansion.

In the case of diffuser flow, i.e., compression, the relations are slightly modified, using inlet kinetic energy as normalizing factor

$$\eta_{in} = \frac{w_{1s}^2/2}{w_1^2/2} = 1 - \frac{\Delta h_{lost}}{w_1^2/2} = 1 - \zeta_{in} \tag{4.17}$$

The empiric formulas for the estimation of kinetic energy losses valid locally or for the whole blade cascade are used in the following text. The example of local loss is an

FIGURE 4.9 h-s diagram of a centripetal radial turbine with the representation of specific enthalpy heads, kinetic and potential energies together with definition of loss coefficients.

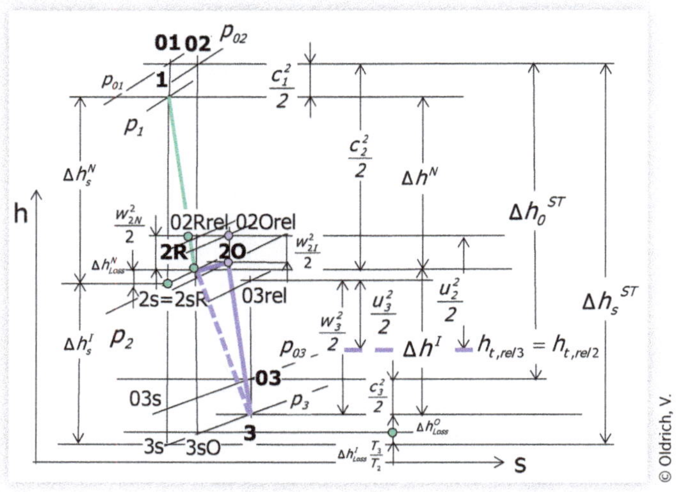

incidence angle loss at inlet to radial channel surrounded by prismatic radial vanes (Benson, see Dixon [25] and Figure 4.9)

$$\zeta_{I,inc} = 2\frac{p_{rel02N} - p_{rel02I}}{\rho_{2I} w_{2I}^2} = \left(\tan\beta_2 + K_{inc}\right)^2 \quad (4.18)$$

The global profile losses of an axial compressor blade cascade are described, e.g., by famous Howell theory [30]. Since the aim of this chapter is to define boundary conditions, it is advisable to calibrate all empiric formulas using the best fit of integral simulated results (MFR, pressure ratios, power under defined values of independent variables, as speed, inlet temperature and pressure, etc.) according to experiments.

4.3 Radial-Axial Centripetal Turbine

The different types of turbine boundary condition are presented in Figure 4.4. The overall representation of processes in a radial centripetal turbine, which should be simulated by models at different levels, is shown Figure 4.9 in h-s diagram for a single-scroll turbine.

4.3.1 0D Map-Based Models

The 0D map-based model is widely spread and used in any 1D code. It uses mostly turbine discharge coefficient or reduced MFR in dependence on pressure ratio and speed for MFR turbine mapping and isentropic efficiency in dependence on PR and BSR for power mapping.

The isentropic efficiency, PR, and BSR is defined for a single-scroll turbine as

$$\pi_T = \frac{p_{01}}{p_3}; \quad c_s = \sqrt{2c_p T_{01}\left(1-\pi_T^{\frac{1-\kappa}{\kappa}}\right)}; \quad c_T = \sqrt{2\Delta h_T} = \sqrt{2c_p\left(T_{01}-T_{03}\right)}$$
$$\eta_{Ts} = \frac{c_T^2}{c_s^2}; \quad BSR = x = \frac{u_2}{c_s} \quad (4.19)$$

In the case of a twin-scroll turbine, the parameters of an ideal turbine have to be defined rather artificially. The problem is not caused by split flow in nozzles but by mixing both of them upstream of or inside an impeller. Isentropic enthalpy head can be averaged using weighting by MFR in a natural way

$$\pi_T^i = \frac{p_{01}^i}{p_3}; \quad c_s^i = \sqrt{2c_p T_{01}^i\left(1-\pi_T^{i\frac{1-\kappa}{\kappa}}\right)}; \quad x^i = \frac{w_{u2}}{c_s^i}; \quad c_s = \sqrt{\frac{c_s^{I2}\dot{m}^I + c_s^{II2}\dot{m}^{II}}{\dot{m}^I + \dot{m}^{II}}} \quad (4.20)$$

There is no simple procedure for averaging pressure ratio to define MFR dependence or discharge coefficient, however. Considering MFR, turbine nozzles could be simulated separately, but both turbine MFR and power, created by an impeller with flows from both nozzles, cannot be physically split.

Many attempts were done to find the suitable form of maps from the measurements at a turbine steady-flow test bed for turbine matching by simulation. Recently, specific test facilities were built for this purpose, e.g., [13] or [84]. Test results are available at turbine manufacturers, mostly as confidential data. Published results can be found, e.g., in Aymanns [2], Lückmann [38], Uhlmann [69], Brinkert [9], or Winkler [78], including asymmetric turbine scrolls and influence of waste-gate (WG) valve flow. Using a turbine model, composed of two (nearly) independent turbines with the same speed, has to be amended by a virtual cross-link between scroll partitions to take the other partition pressure waves into account. Even in this case, the pressures governing a mixing process are not respected in an appropriate way and the orifice area has to be calibrated [28] (Figure 4.5).

In the case of pulse operation, which is typical for any exhaust system and especially increased if twin-scroll is used, the isentropic process relations need averaging, if the results are to be compared with steady test bed results. It is an old issue, analyzed by Zinner in the 1960s [83] and tested, simulated, and reanalyzed many times after him—new experiments by Capobianco [13, 14, 15, 16], simulation and analysis Bulaty [11], De Bellis [23], Lujan [36], Westin [77], Winterbone [80] and Zehnder [82] or Macek [39], [47] and Mikula [50]. The physical modelling using experiments at a specific test bed with pulsating flow is useful for qualitative understanding, e.g., [13, 14, 15, 16]. Since the gas at a test bed is often cold, and the pulses are not similar to working engine conditions, this method cannot yield final results for other engine sizes, especially if not associated with analysis of in-turbine phenomena. The combination of different depths of physics with experiments is needed, and dedicated experiments, as in Serrano [61, 62] or Zak [84], are worthwhile.

The described situation calls for a change from lumped parameter, map-based approach to more detailed models.

4.3.2 Central Streamline Models of a Radial Turbine

The nature of in-turbine processes at possibly separated BL and blade tip leakages (Dixon [25], Sherstjuk [65]), Coriolis acceleration field (Dixon [25] or Watson [75]), and transonic conditions (Shapiro [64]) is rather complicated for direct and sufficiently fast 3D simulation, since the needs for mesh density are extraordinary high. Therefore, simulation approaches offer a wide range of models, based on 1D or 3D, e.g., Canova [12], Frederickson [27], Gurney [29], Hu [32], Winkler [78], or, in 3D, Osako [53]. 3D seems to be a solution for the future, but it still needs the same calibration as 1D does. Moreover, the detailed geometry data are not usually released by TC manufacturers. Time requirements of 3D simulations are still rather far from optimization feasibility, and the calibration possibilities of 3D models are limited.

The current state of the art offers well-proven and fast 1D solvers, sufficiently opened for the purposes intended by this task—e.g., Morel [51]. There is no need to redevelop them, especially if they are integrated to the whole engine and powertrain models. The procedure based on experience with similar approach to pressure-wave supercharger or Roots blower modelling applied to larger systems (Pohorelsky [54] or Brynych [10]) will be described below.

The simultaneous impact of requirements and possibilities have motivated the current research, amending the results described in Macek et al. [40], [41], [44, 45, 46],

Vitek [73], and Zak [84]. The use of 1D solver modules is easier, if the solver contains boundary conditions with momentum mixing, as in FS—e.g., [28].

First, the quasi-steady algebraic model is described. Then, the full unsteady model using modules of 1D solver is mentioned.

4.3.3 Central Streamline Model using Quasi-Steady Impeller Flow

Central streamline model uses the partial processes, which have been already mentioned, combining them into a series of changes, as plotted in entropy-enthalpy (temperature) diagram in Figure 4.9. After an experience with convergence of models finding MFR to known upstream and downstream pressure and upstream temperature in derated points of operation (Macek [40] and [41]), new model has been developed, iterating pressure differences at assumed MFR and impeller speed. Its direct use, as a turbine boundary condition, needs additional iteration from known pressures and temperatures to MFR, but it can be performed fast.

Basic blade cross sections can be found using blade flow angles, which represent mean values of the total flow (Figure 4.10). They are subjected to calibration. All capacity losses, related to MFR, are involved in discharge and kinetic energy loss coefficients. Pressures, temperatures, and turbine power (as a side effect, isentropic efficiency, as well) are found as a function of MFR at fixed speed and (initially) fixed outlet temperature.

The exducer state creates starting point for the numerical procedure, going backwards, i.e., upstream. It is not possible without iterations, of course, since at least inlet stagnation temperatures have to be kept at fixed values. Outlet temperature has to be estimated to start upstream calculations. At the end of the single run of a procedure, it is corrected to inlet temperatures. Inlet stagnation pressures are the direct output of the numerical procedure. If fixed inlet pressures have to be kept at boundary condition level, another iteration of MFR is needed.

The backward going basic iteration is based on impeller total enthalpy conservation (including potential energy, i.e., rothalpy). Constant pressure thermal capacity is assumed to be independent of small temperature differences; therefore reference temperature for zero enthalpy does not occur in the energy conservation equation. No leakage along impeller blades is assumed (all leakage flow leaves the impeller blades upstream of an interblade channel).

FIGURE 4.10 Velocity triangles and components of velocities for the impeller inlet and outlet of a centripetal radial turbine.

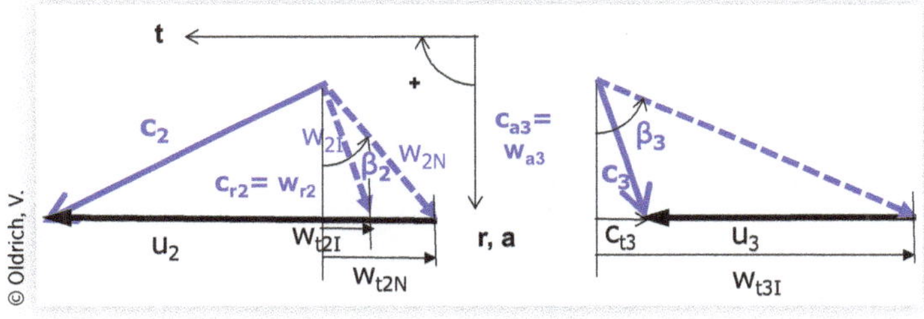

Using initially assumed MFR and the gas state downstream of an impeller p_3, T_3 and ρ_3, the following relations are valid for subsonic conditions, if isentropic change assumed for the first approximation is corrected by iteration coefficient k_p

$$c_pT_{2I}+\frac{w_2^2}{2}-\frac{u_2^2}{2}=c_pT_3+\frac{w_3^2}{2}-\frac{u_3^2}{2}$$

$$c_pT_3k_T\left(\frac{p_{2I}}{p_3}\right)^{\frac{\kappa-1}{\kappa}}+\frac{1}{2}\left(\frac{\dot{m}_I}{A_{2I}\rho_3}\right)^2k_\rho\left(\frac{p_3}{p_{2I}}\right)^{\frac{2}{\kappa}}=c_pT_3+\frac{1}{2}\left(\frac{\dot{m}_I}{A_{3I}\rho_3}\right)^2-\frac{u_3^2-u_2^2}{2}=P \tag{4.21}$$

$$c_pT_3k_Tx^{\frac{\kappa-1}{\kappa}}+\frac{1}{2}\left(\frac{\dot{m}_I}{A_{2I}\rho_3}\right)^2k_\rho x^{-\frac{2}{\kappa}}-P=0$$

The last equation may be solved using Newton-Raphson method for static pressure ratio x.

If critical pressure ratio in impeller exducer relative flow is reached, the outlet blade pressure is no more equal to outlet exhaust pressure, but the inlet density has to be found, which delivers the MFR through a nozzle with sonic speed at the outlet. It is triggered by critical temperature from Equation (4.6). The pressure-temperature dependence similar to Equation (4.34) can be found combining Equations (4.6) and (4.16). The relation between inlet stagnation temperature, pressure, and critical MFR can be derived then. If initially assumed MFR through a turbine should be reached, the necessary inlet stagnation pressure can be found iteratively, as mentioned above. If critical MFR occurs, the reduced MFR is fixed, and to reach assumed MFR, the inlet pressure has to be changed accordingly. Then, the outlet blade pressure is no more equal to the pressure in a following exhaust pipe. The details can be found in Macek [46].

Using the backward going procedure, all velocities and relative stagnation state variables can be simply found at impeller inlet (i.e., after incidence loss took part). Nozzle choking check and correction, if necessary, can be performed using procedure similar to an impeller. After it, the partition outlet velocities and both tangential and radial speed components, using known exit angles, can be found. Immediate mixing can be assumed at radius of scroll partitions outlet with no pressure change. It can be corrected by calibration of a loss coefficient. The last linking condition considers static pressures at a nozzle outlet $2N$ and an impeller inlet $2I$ using Equation (4.18), an empirical equation for a pressure loss due to the off-design incidence angle at impeller (radial) blades.

The model may be amended by the more detailed description of vaneless nozzle, as described in Macek [41]. Some principles of it are mentioned in the section on compressor modelling (vaneless diffuser, Equations (4.47) and (4.48)). Unlike bladed nozzles, the vaneless nozzle is seldom choked, since the sonic line has to be circular due to rotational symmetry. The nozzle is choked only if the radial component of velocity achieves sonic limit. This feature can be modelled in quasi-steady description easily, but it is not easily transferable to 1D pipe elements.

Knowing the state parameters in the gap between a nozzle ring and an impeller, leakage at both shroud side and the impeller disk (hub side) may be found.

Before the turbine power is calculated, windage losses of an impeller should be derived from the total head obtained in expansion. The power lost by windage is predicted simply by general similarity laws of rotating disc and a turbulent BL friction. The simplification of this model is compensated by the calibration coefficient K_{wind}:

$$\Delta h_{wind}=\frac{K_{wind}\rho_{2N}D_{2I}^2w_{u2}^3}{\dot{m}_T\,\mathrm{Re}^{0.2}},\quad \mathrm{Re}=\frac{w_{u2}D_{2I}}{\nu_{2N}} \tag{4.22}$$

FIGURE 4.11 SAE map—reduced mass flow rate (solid lines) and isentropic efficiency (dashed lines) for a turbine at equal distribution of flow between entries and for a turbine at fixed flow of 0.01 kg/s in one entry, variable flow from 0.05 till 0.5 kg/s in the other one.

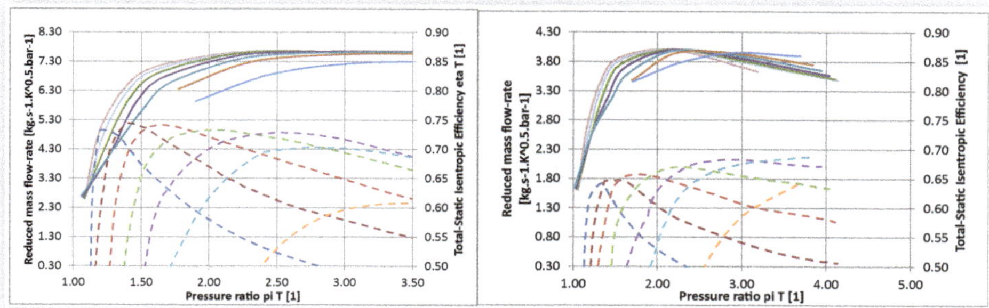

Turbine power can be found from MFR and specific power, usually from both Stodola (difference of total enthalpies) (Equation (4.4)) or Euler (angular momentum conservation) (Equation (4.7)) using parameters at impeller inlet and exit in absolute (i.e., fixed in space) coordinate system. The differences between them check the accuracy of numerical procedure. According to experience, they are less than 1%.

Averaged values of turbine discharge coefficient and isentropic efficiency may be determined as a side effect, because they are not directly needed for simulation, if map-based approach is abandoned. It has been clarified in the introductory part of this contribution: the simulation needs turbine immediate permeability (i.e., MFR for all branches of exhaust manifold as a boundary condition for pressure wave reflections) and turbine immediate power for integrating dynamic equation of a turbine shaft, loaded by a compressor and friction resistances, nothing else. The artificial, physically compromising methods for reduction and averaging of turbine dimensionless parameters are not needed directly for simulations, which increase model accuracy.

Altogether, the model has seven calibration coefficients for a single-scroll layout (two expansion efficiencies, three separation coefficients, incidence loss correction, and windage loss correction); another three have to be added for the second partition of a nozzle.

The examples of results are in the form of SAE maps presented in Figure 4.11.

The model is important for understanding interactions of loss sources and for rough optimization of turbine main dimensions together with details of a trim. It is very important for optimum matching of a compressor, which determines turbine BSR. 3D CFD models are not yet able to help in understanding of these interactions.

This quasi-steady model may substitute standard maps in 1D codes, in which case they are extrapolated by in-code procedures, sometimes inaccurately. The results of the code may be transferred in the form of look-up tables (data libraries) found for a specific case of simulated operation, i.e., very close to real operation modes. The further analysis of results and full description of the code can be found in Vitek [73] or Macek [46].

4.3.4 Unsteady-Flow 1D Turbine Model

The way of enhanced 1D turbomachinery model had been introduced by E. Jenny (Brown Boveri AG) [30] even in the early 1950s of the last century, using a classic method of characteristic lines for hyperbolic partial differential equation system, extended by many authors for the domain in consideration, e.g., Seifert [59] or [60] and Zehnder [81]. The isolated attempts to do so by numerical solution of schematized radial rotating channel are available, as in Hu [32]. The reality of impeller design cannot be schematized into purely radial or purely axial channel for the modern turbine impellers of rather diagonal design, which is fully justified by the absolute optimum efficiency for today's turbine specific speeds (Vitek [73]).

The physical to 1D reduced model of a radial turbine should consist of a set of gas channels featuring total pressure and/or temperature changes and losses. It has to describe an inlet volute with stepwise flow splitting to nozzles, mixing of flows at twin-scroll volute exit, nozzle blade flow, rotating impeller channels of general shapes and radial-to-axial bending, an outlet casing with a diffuser—if used, WG ports and valves, etc. Obviously, these demands cannot be fulfilled by a single-purpose model with sufficient flexibility.

The recent advances in 1D solvers, as [51], with high level of both integration and flexibility—like components described in [26], [51]—offer much more general description, however yet unused. The reason for it consists in the second important problem, which forces engine designers to use the lumped parameter, map-based turbine models: the lack of turbine data suitable for detailed 1D model calibration. They are neither directly measurable nor available from 3D simulations, as mentioned above. Moreover, the details of turbine design are frequently classified know-how of a manufacturer. Nevertheless, the extrapolation of maps delivered by a turbine manufacturer to extreme BSR or (w_u/c_s) using 1D steady model was elaborated and successfully tested, as mentioned already in Macek [40]. The remaining problem consists in a more comfortable way for finding calibration parameters of a model applicable in engine design practice.

4.3.5 Model Structure

The described model, according to the previous sections, can be transferred into modules of a 1D code with description of pressure waves and mass/enthalpy accumulation. It is schematically shown in Figure 4.12 for the simplest representation of in-turbine channels for a single-scroll turbine and in more general form in Figure 4.6. The main elements are pipes P or rotating pipes RP (split for solution of unsteady flow into FVs), boundary conditions: throttling orifices between two pipes with defined discharge coefficients μ and cross-section areas A in the form of flow connections and FS, i.e., pipes with several outlets, and auxiliary "signal" sensors, transformers, and "actuators", transferring, and transforming additional values needed for correction of module features (e.g., TC shaft speed as in Figure 4.12, results of interpolations in look-up tables, etc.). The assignment of model modules to a real single- or twin-scroll

FIGURE 4.12 The simple level of unsteady turbine model using serial connection of general pipes, flow connections, and rotating pipes; the model starts at volute inlet(s). The 1D modules for a twin scroll are plotted at the right.

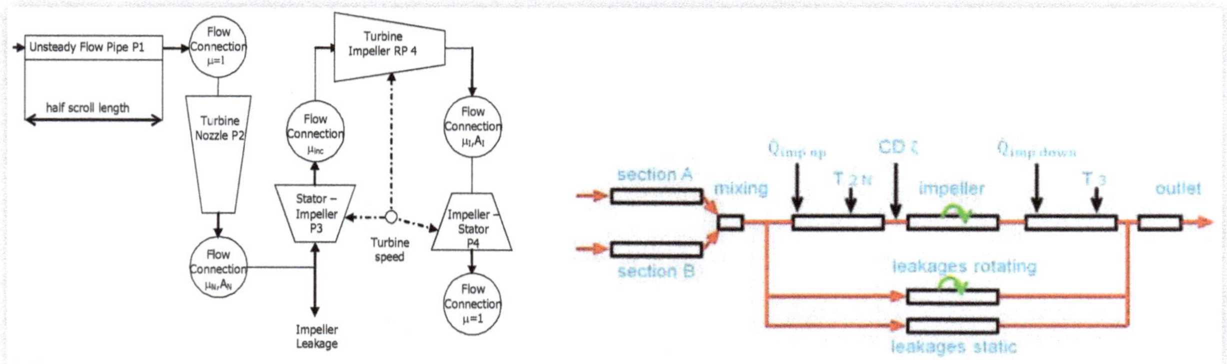

FIGURE 4.13 Layouts of single or twin scrolls and main dimensions of a turbine impeller [84].

turbine is shown in Figure 4.2, the schematic layout of a twin-scroll turbine in Figure 4.12—details in Macek [41, 44, 46], Vitek [73], and Zak [84].

The whole 1D model is described by geometrical, directly measurable parameters from Figure 4.13. The mean diameter of axial blades is calculated according to Equation (4.10). The angles at turbine nozzles are used for flow cross-section area with possible flow separation (including the case of vaneless nozzle); the deviations of angles are applied for torque/power simulation. In the case of vaneless nozzle, velocity of sound can be exceeded (the reason is in tangential component acceleration, as mentioned above; see also more detailed description at compressor vaneless diffuser below); otherwise sonic limit is the highest exit nozzle velocity. In the case of exducer exit, relative angle valid for momentum and separation coefficient are used. The main parameters are as follows:

- Alpha 2—nozzle exit angle: The coefficient directly influences the dimension at the outlet of the vaneless nozzle ring. The pipe area is calculated accordingly.
- Delta Alpha 2—deviation of nozzle exit angle: Coefficient represents deviation between mass and momentum averaged exit angles. Delta alpha does not influence the outlet diameter of nozzle ring but directly influences the velocity c_{2t} in Equation (4.7).
- Beta 3—impeller exit angle: It influences the dimension at the turbine wheel outlet.
- K sep—flow separation coefficient at exducer exit: The effective flow cross section is reduced by the flow separation coefficient.
- R C imp—radius for determination of centrifugal accelerations at the boundaries of the sections (FVs) of an impeller channel, avoiding false mean accelerations from a centerline; see remarks to Equation (4.11).
- p loss coefficient impeller: Pressure loss coefficient in impeller pipe according to C_P in Equation (4.16).
- p loss coefficient nozzle (A or B): Pressure loss coefficient in scroll partition pipe according to C_P in Equation (4.16).
- K inc—correction of impeller incidence loss: This coefficient corrects the additional incidence loss to impeller according to Equation (4.18).
- K wind—coefficient of windage losses: Tuning coefficient stated in Equation (4.22).
- mu leakage static: Discharge coefficient of static leakage orifice.
- mu leakage rotating: Discharge coefficient of leakages under influence of centrifugal force.

4.3.6 Calibration Procedure for a Model

The model parameters can be found from measurements at steady-flow test bed, fitting the values of pressure ratios in individual scroll partitions and turbine power at different MFRs, speeds, inlet temperatures, and levels of partial admission (MFR in a partition/

total MFR in both partitions). The genetic algorithm with initially fixed ranges of parameters can be applied, as published in Macek and Vitek [41, 44] or for partial admission in Zak [84]. Recent examples of calibration coefficients are presented in Figures 4.14-4.17 for full admission with an admission level of 50%, partial admission of the level of 87% and closed inlet of the other partition, i.e., the admission level of 100% in the partition in consideration. The incidence loss coefficient was kept at constant value of 0.1 according

FIGURE 4.14 Exit angles from a nozzle and its momentum deviation in dependence on pressure ratio and partial admission levels of 50% (full), 87%, and 100% (closed) with different BSR values.

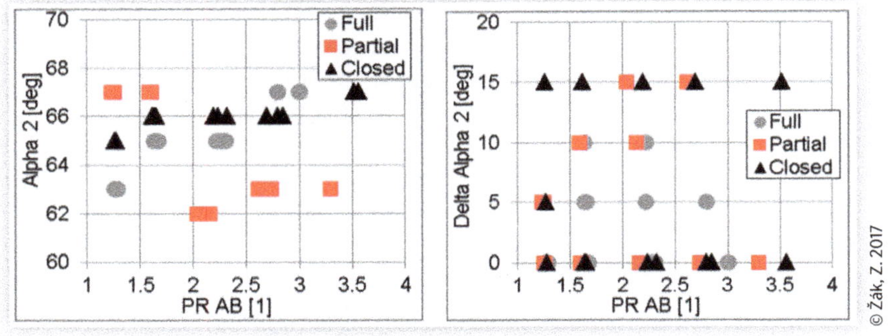

FIGURE 4.15 Exit angle and separation coefficient at exducer outlet in dependence on pressure ratio and partial admission levels of 50% (full), 87%, and 100% (closed) with different BSR values.

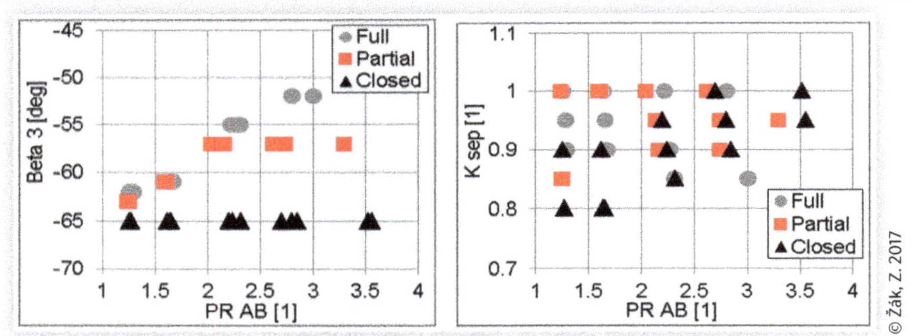

FIGURE 4.16 Pressure loss coefficient and windage coefficient in dependence on pressure ratio and partial admission levels of 50% (full), 87%, and 100% (closed) with different BSR values.

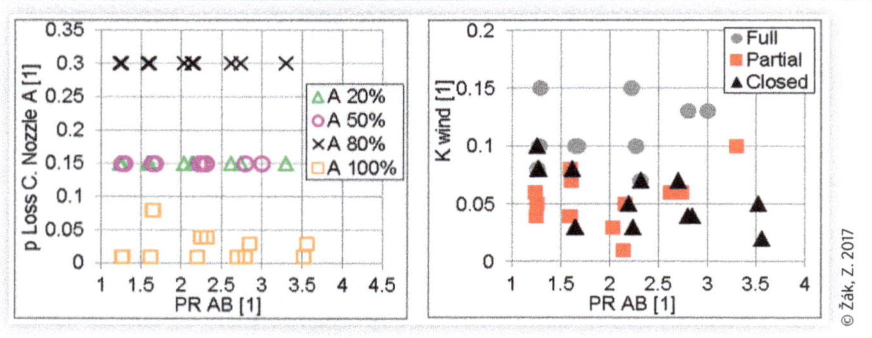

to Equation (4.18), and both discharge coefficients of nozzles are at the constant level of 0.99.

Figures 4.14-4.16 present the results in dependence on pressure ratio PR at variable BSR. The results show reasonably stable calibration coefficients inside expected ranges. The angles of flow were compared to the known blade geometry. The optimization of centrifugal acceleration radii-Equations (4.10) and (4.11)—has not yet been performed. The first attempts to do it prove yet better agreement.

This conclusion is supported by overall averaged relative errors in Figure 4.17. The increase of error in the right part for high BSR is not very important, since those operation modes yield only very low power. Moreover, the sensitivity to centrifugal force is very high in this domain, and the additional optimization of decisive radii would yield better accuracy.

The investigation of turbines is still in progress, but the features of physical model, especially its insensitivity to deviations from rated operation and extrapolation capabilities, are very promising.

FIGURE 4.17 Overall error (average of pressure ratio errors in both partitions and power error) plotted vs. blade-speed ratio BSR AB (approximate pressure ratio level PR AB = 2.2); full admission of an impeller (gray circle), partial admission level 87% (red square), one section closed (black triangle).

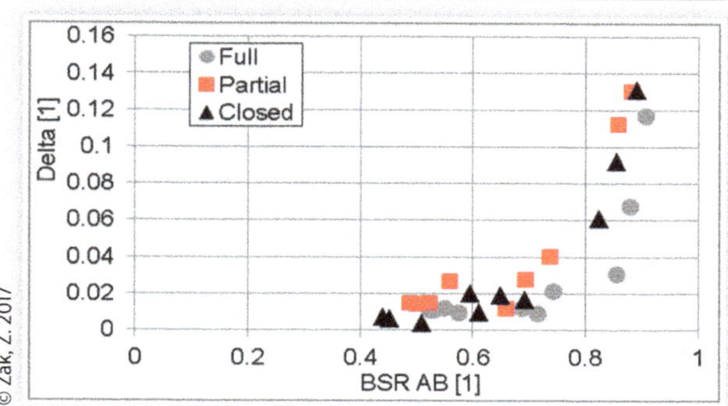

The use of calibration parameters in the 1D unsteady model is straightforward and resembles the central streamline quasi-steady model, already described. However, it does not need any additional iteration during its use.

Transformation of total states and velocities is done in short specific "transformation" 1D pipes stator-impeller or impeller-stator (Figure 4.12) with different and variable area at inlet and outlet. The variable area was already used for nozzle or impeller channel model, as well. Unlike those standard pipes, the transformation pipes feature additional enthalpy source term for changing total state inside. The pipes are often short (in most cases single FV). The determination of a source term is based on Equation (4.13). Similar procedure is applied for impeller exit. In both cases, the heat is extracted from a system. Sum of both enthalpy reductions would be equal to turbine power if turbine operates at steady conditions.

Turbine torque has to be integrated step-by-step at any instant to correct the results for unsteady mass/momentum accumulation. The integration schemes can be based on the contributions of total enthalpy changes (energy conservation) or angular momentum changes (Euler turbine theorem) for every FV with different unsteady MFR. For a single rotating FV (angular speed ω) with inlet (subscript in, positive flow rate) and outlet (out) MFRs at appropriate radii, the outer torque to a blade channel are

$$\Delta M_T = -\int_{in}^{out} \frac{d\left[\dot{m}r\left(\omega r + w\cos\varepsilon_t\right)\right]}{dx} dx +$$

$$-r\left(\omega r + w\cos\varepsilon_t\right)\int_{in}^{out} \frac{d\dot{m}}{dx} dx - mr\frac{d\left[\left(\omega r + w\cos\varepsilon_t\right)\right]}{dt} \quad (4.23)$$

Both time changes of rotation and relative tangential velocity are taken into account. The angular momentum is calculated in cylindrical coordinates, using radius r and

FIGURE 4.18 Pressures at both twin-scroll turbine entries A and B for different models (1D and lumped parameters 0D) and bmep approx. 9.5 bar, speed 2000 min^{-1}, boost pressure approx. 1.65 bar. From [46].

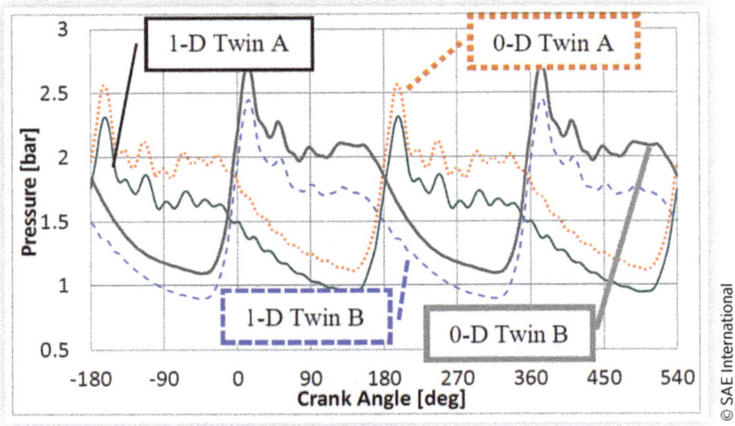

tangential component of relative velocity *w*, given by the projection of relative velocity into tangential direction (cosε_t, Figure 4.7).

The turbine power has to be corrected by windage loss of an impeller disc, as shown in Equation (4.22). Moreover, inertia torque, due to unsteady impeller rotation, is respected in calculating the available turbine power:

$$P_T = P_{Ti} - \Delta P_{wind} - I_T \frac{d\omega}{dt} \tag{4.24}$$

An example of calculated pressures for different exhaust systems is shown in Figure 4.19 for the -twin-cylinder branches of an exhaust manifold with high-pressure pulsations of a four-cylinder engine with a twin-scroll turbine. Obvious shift to higher pressure in the case of 0D model with separated "turbines" is demonstrated in Figure 4.18. The 0D model should be amended by fictitious connection of both exhaust branches as was done in Figure 4.5.

The MFR would be still too high in the case of using 0D realistic turbine flow area found by steady-flow experiments, since the counteracting influence of the pressure pulse from the other manifold does not exist in this model. That is why the turbine test bed results have to be calibrated by experiments for an engine, if 0D model is used.

The comparison of pressures simulated by different models to measured values from [84] is presented in Figure 4.19. The ad hoc fit of a fictitious connecting branch upstream of a turbine helps to 0D being comparable to a physical model, but it has to be found for every case of an exhaust manifold.

FIGURE 4.19 Pressures at inlet of turbine section A, experiment (black solid line), simulation with full 1D unsteady turbine (dashed and dotted line above); a six-cylinder engine, speed 2100 min^{-1}, bmep = 9.8 bar - from [84].

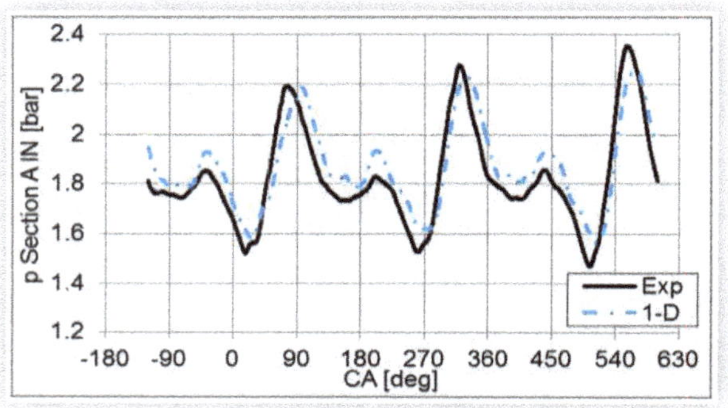

4.3.7 Heat Exchange Parameters of a Physical Turbine Model

Disregarding the assumption of turbine and compressor adiabaticity, the basic energy conservation can take into account different heat transfer modes. The heat transfer to a TC casing is mostly internal one, i.e., it comes from a turbine to a compressor (Aghaali [1], Bohn [5, 6, 7], Casey [17], Cormerais [19] and [20], Rautenberg [57], Serrano [59], or Westin [76]). The heat is removed from exhaust gas mostly before it supplies turbine power to the turbine impeller, in reality, reducing turbine power output and—if temperature difference is used for calculation of compressor specific power as difference of enthalpies—fictitiously increasing compressor power input. The high effort to estimate this heat exchange is important for the assessment of low pressure ratio operation of a TC. In the case of measurements, the false results of compressor power may be easily avoided, if tests are conducted with a turbine driven by a cold air (Macek [46]), if using compressor power calculated from 0D compressor model based on Eulerian theorem; see below. Moreover, for the IC engine pumping loop work, both inaccuracies are mutually compensating each other (Macek [47]).

The above-described physical model is well prepared for simulation of diabatic processes in appropriate modules of pipes, using simplified FEM or thermal resistance method.

4.3.8 Other Layouts of Radial Turbines

The both physical models are well prepared for changes describing additional instruments for turbine control. If WG valve is added to a turbine in the way, which influences aerodynamics of scroll partition(s) or decisive flow parameters downstream of a turbine, the identification of parameters at a test bed should be done including the waste-gating. The asymmetry of both scroll partitions or coupling of WG to singe partition only are among those items. If WG is far enough for avoiding the impact on scroll flow, separate valve can be used in a model. It should be confirmed at test bed using, e.g., different shapes of a WG valve surrogates and calibrating the WG discharge coefficient for the valve alone with plugged inlet to an impeller.

The serial scroll partitions instead of parallel ones, according to the center part of Figure 4.13, may be simulated by the individual scroll partitions coupled directly to an impeller partition with minimized mixing zone, representing narrow transition part from one partition to the other. The impeller partition content has to be mixed along the path in an impeller before reaching exducer nozzle, which mimics the mixing due to impeller rotation. This procedure allows for respecting the losses of momentum by filling and emptying the impeller channels in the case of non-uniform admission. The suitable procedures for specific problems of this partial admission are described for axial turbines in Dibelius [24] and Bulaty [11]. The efficiency of serial scroll partitions may be lower than the efficiency of longitudinally parallel partitions, which create a natural pulse converter (Zehnder [82] or Mikula [50]).

4.3.9 3D CFD Models

The 3D CFD approach is very general and it has high predictive ability, hence it can be applied in all cases of turbomachinery modeling. In this particular case, only turbine data are shown, hence the text is a part of the turbine section. From 3D CFD modeling point of view, the compressor case is more difficult due to possible existence of shock waves, surge phenomena and strong flow separation of decelerated flow.

As it is shown in Figure 4.4, a 3D CFD turbine/compressor model can be applied in different ways. The online application would be the preferred option as these kinds of models have high predictive ability due to their deep physical basis (solving complete 3D Navier-Stokes equation set including turbulence model). However, 3D CFD models are too slow to be applied online for internal combustion engine (ICE) system simulations. Hence, the offline option is widely used at the moment. This concerns mainly the option to provide sound data for a calibration of faster models, as calibration procedure of the abovementioned models (ranging from 0D map-based ones to unsteady 1D CFD ones) is a critical step. It requires reliable data. One possibility is to apply measured data; however it is not always possible to measure all information needed for proper model calibration. The other possibility is to use 3D CFD which can provide detailed information regarding thermodynamic status inside turbine/compressor. Moreover, if no experimental data are available, it is the only source of reasonable turbine/compressor performance data.

Another offline possibility corresponds to "map-less 3D" option in Figure 4.4. This is based on the procedure of 1D and 3D CFD combination to predict engine transient performance as described in Macek [43]. When using this kind of approach, turbine/compressor 3D CFD model can predict complete TC performance under unsteady operation, hence providing MFR and both pressure level and temperature one upstream/downstream of turbine/compressor.

The example of detailed 3D CFD model of radial turbine is presented in Figure 4.20 for the case of twin-scroll radial turbine. The shown data concern both full admission setting and partial one. It seems that 3D CFD can predict both MFR and thermodynamic efficiency very well for low and medium level of BSR. At high BSR levels (more than 0.88), significant difference between simulation and measurement is observed. Similar trend is seen when comparing unsteady 1D CFD model with experimental data (Figure 4.17). Hence, further research is needed to improve simulation performance in high BSR region. Fortunately, the presence of these high BSR levels is usually limited under engine operation (even for a three-cylinder engine in Figure 4.21), and turbine power is very low there.

FIGURE 4.20 Comparison between predicted data by 3D CFD SW tool (continuous lines) and measured data (dashed lines)—left column corresponds to full admission for the case of twin-scroll turbine, while right column represents partial admission case.

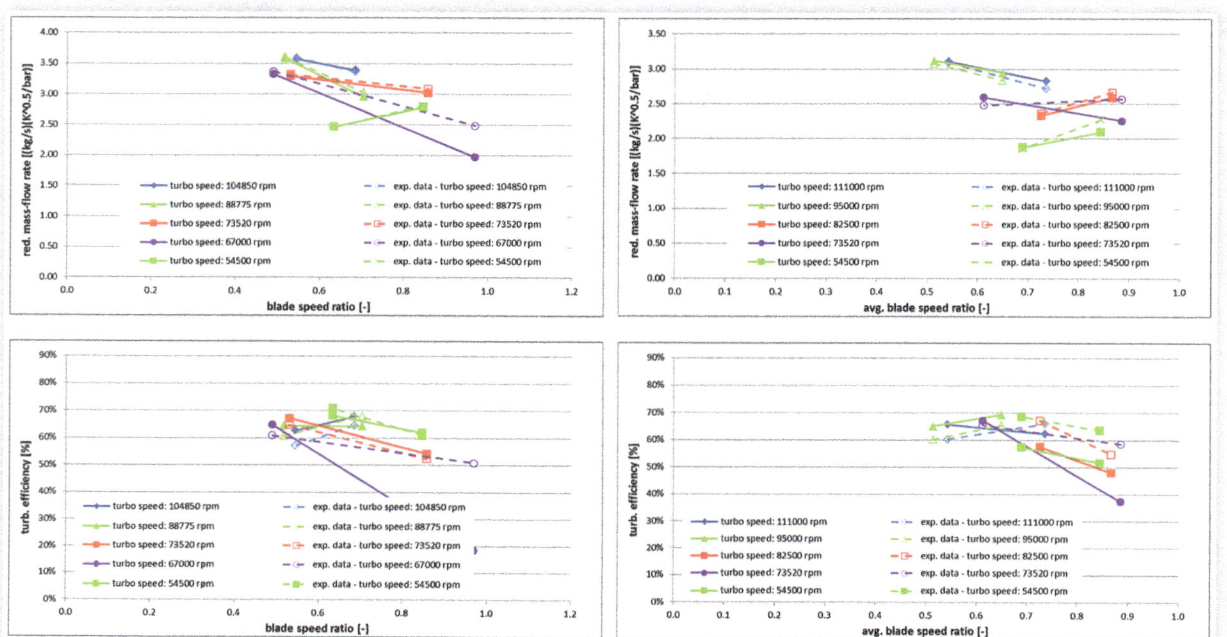

Hence, the influence of poorly predicted turbine MFR and/or thermodynamic efficiency at high BSR on overall TC performance (i.e., values which are averaged over the whole engine cycle) is negligible.

The data presented in Figure 4.20 correspond to thermodynamic performance of a twin-scroll turbine. It means that the predicted efficiency is the thermodynamic one. This includes windage losses caused by braking torque in a very thin gap between turbine (back) disc and turbine housing-this is predicted by empirical model while using experience with similar turbine wheels (Equation (4.22)). This type of a loss is difficult to predict by 3D CFD as the gap is very thin depending on actual rotor position—manufacturing tolerances also have important influence on that phenomenon. Another effect, which needs to be considered, is related to friction losses. This is not easy to estimate—more details can be found below. Both windage losses and friction ones are important when determining overall torque/power balance of a TC (Equation 4.25) which yields TC speed. The TC shaft speed has to be known when using 3D CFD models to take into account all inertia effects correctly.

FIGURE 4.21 Instantaneous turbine BSR and efficiency for the case of highly turbocharged three-cylinder SI engine at 2000 rpm—red curve corresponds to compact exhaust manifold, while blue one represents exhaust system with large volume between ICE and turbine.

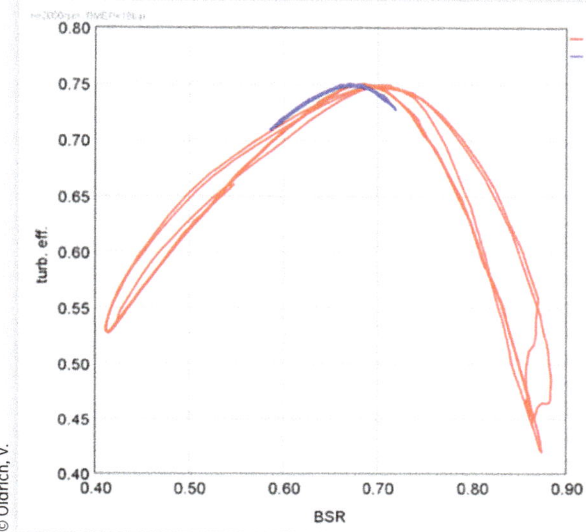

Based on abovementioned facts, the 3D CFD methods can provide reliable TC performance data under assumption that boundary conditions of a 3D CFD model (e.g., rotor speed, inlet/outlet pressure) are set correctly. Empirical submodels are needed (e.g., windage losses or friction ones) to estimate effects which are difficult to be modelled by 3D CFD approach directly. Hence, 3D CFD approach is suitable when boundary conditions (representing a TC, i.e., turbine and/or compressor) are needed for thermodynamic system simulations of ICE.

It is obvious that even if a TC operates at constant rotor speed, the flow inside turbine/compressor is unsteady due to stator-rotor interaction in Figure 4.22. In the case of twin-scroll turbine with partial admission, there is additional effect due

FIGURE 4.22 Example of flow field inside twin-scroll turbine with partial admission—left subfigure corresponds to Mach number, while right subfigure represents turbulence kinetic energy.

to interaction of flows from different scroll channels. This leads to high-frequency oscillations of thermodynamic parameters—this can be usually neglected when dealing with engine system simulations under steady operation. However, it can be important al low TC speeds during engine transient operation—these effects cannot be usually predicted by simpler models. Moreover, when dealing with NVH issues, these high-frequency phenomena have to be taken into account.

4.4 Speed Non-uniformity of an Impeller and Friction Losses

The speed of a TC, which is an important variable in all above relations, can be found in every step of integration of gas dynamic equations by the momentum equation for a TC rotor

$$I_{TC}\frac{d\omega_{TC}}{dt}=\frac{P_T-P_C-P_{bearings}}{\omega_{TC}}+\frac{P_{elM/G}}{\omega_{TC}} \tag{4.25}$$

The relation involves even electric motor power for electrically assisted turbocharging (EAT), which is, however, out of the scope of the current book (Vitek [74]).

The compressor power will be explained in the next section. The current section is focused on adding the bearing friction loss into the turbocharger model. Semi-empiric regression model was developed. It has to respect bearing load of both radial and thrust bearing, oil viscosity, and turbine size.

The power dissipated in bearings can be found from a simple regression relation, taking into account Sommerfeld number of a bearing in hydrodynamic mode and the loss in a thrust bearing, fixed by dependence on all four pressures at a turbocharger. The relation has to be calibrated by the measurement of lubricating oil enthalpy increase in bearings, determined from oil mass flow rate and temperature difference. The cumulative difference of enthalpy, summarizing power dissipation in bearings and enthalpy increase due to turbocharger internal heat transfer can be evaluated and compared to the mentioned regression formula, if the latter one is amended by terms taking into account the temperature difference, causing thermal conduction. The representative temperature for heat transfer from a turbine is averaged between turbine inlet and outlet. The heat conduction depends on difference between turbine and oil temperature. The compressor side temperatures are not significant, being close to oil ones. Then, the measured enthalpy flow difference is

$$\Delta\dot{H}_T=\dot{m}_{oil}c_{p,oil}\left(T_{oil2}-T_{oil1}\right)=P_{bearings}+\Delta\dot{H}_{HT}=-\left(\frac{D_T}{D_{T,ref}}\right)^2\left(\frac{\nu_{oil}}{\nu_{oil,ref}}\right)*$$
$$*\left[\begin{array}{l}\alpha_0+\alpha_1\left(\frac{n_T}{n_{ref}}\right)+\alpha_2\left(\frac{n_T}{n_{ref}}\right)^2+\alpha_3\left(\frac{n_T}{n_{ref}}\right)^3+\alpha_4 p_{T1}+\alpha_5 p_{T2}+\\+\alpha_6 p_{K1}+\alpha_7 p_{K2}+\alpha_8\left(T_{avT1-T2}-0.5(T_{oil1}+T_{oil2})\right)+\alpha_9\left(T_{avT1-T2}-0.5(T_{oil1}+T_{oil2})\right)^2\end{array}\right] \tag{4.26}$$

This calibration of this relation can be done avoiding false results of friction loss power, corrupted by heat transfer. The mitigation of this influence is simple: measurements with different turbine inlet temperature are done (including even cold air turbine operation, if possible) and the averaged temperature terms in Equation (4.26) are calibrated for measured enthalpy flow differences. Then, averaged temperature difference in the last two terms of Equation (4.26) is fixed to zero and the result is very close to accurate value of bearing friction loss, which is used for simulations.

4.5 Centrifugal Compressor

As in the case of a turbine, different models—map-based or map-less—can be applied to find MFR and shaft input power at defined pressure ratio and speed (Figure 4.3 and Figure 4.4), namely, 0D map-based models and quasi-1D central streamline (map-less) physical models.

The general processes in a compressor with total, stagnation, and static states in h-s diagram are shown in Figure 4.23, the layout of an impeller and diffuser(s) in Figure 4.27 (Dixon [25], Jerie [34], Kousal [35], or Macek [49]).

4.5.1 0D Models of a Compressor

The standard map-based lumped-parameter model is well known and used in quasi-steady operation. There are many attempts made to find direct regression representation of it—e.g., Llamas [36]. There are several issues considering such representation:

- Extrapolation potential, which should be based, at least partially, on physics, especially compressor performance at pressure ratio less than one and prediction of choking limit(s).
- Prediction of surge limit.
- Representation of performance after surge line is reached or exceeded.

The surge limit prediction has not been found with sufficient reliability even if 3D modelling is used. The complicated dynamic behavior of BL close to separation is too demanding even for the near future. Other two items can be partially solved even in 0D models with good prospects for 1D models.

It is important to realize that the compressor isentropic efficiency is not directly needed if MFR—pressure ratio—speed curves are determined together with specific power (enthalpy difference in adiabatic case). This map-less approach simplifies the task substantially. The appropriate regression substitution of Eulerian theorem in

FIGURE 4.23 h-s diagram of a centrifugal radial compressor with the representation of specific enthalpy heads, kinetic and potential energies together with definition of loss coefficients.

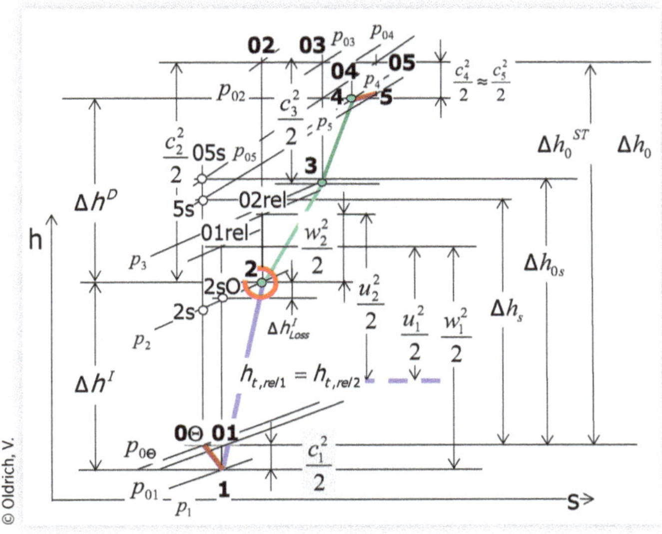

FIGURE 4.24 The speed map of a centrifugal radial compressor measured and represented by regression with surge limit, throttling parabolas at the boundary of static stability of flow and pressure ratio reached with zero MFR.

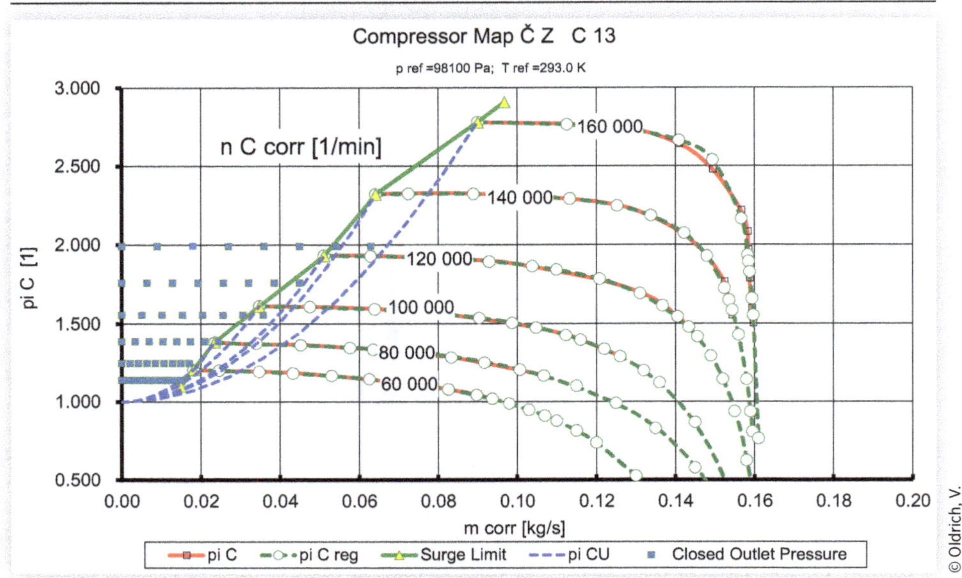

Equation (4.7) is based on impeller outlet velocity triangles (inlet seed is mostly axial) in Figure 4.27, coefficients K_1 till K_4, and empirical knowledge from the best fits of measured and evaluated compressor maps (Vitek [72], Macek [44])

$$\frac{P_C}{\dot{m}_C} = \left(K_1 + K_3 \frac{w_{2r}^4}{u_{C2}^2} + K_4 \frac{u_{C2}^2}{w_{2r}^2} \right) u_{C2}^2 + K_2 w_{2r} u_{C2} = h_{0C2} - h_{0C1}; \quad w_{2r} = \frac{\dot{m}_C}{\pi D_{2C} b_{2C} \rho_2} \tag{4.27}$$

The additional terms with constants K_3 and K_4 are suitable especially for compressors with anti-surge internal recirculation channel (IRC). The coefficients K have to be evaluated from compressor power measurement according to enthalpy (from total temperature) difference under adiabatic condition, i.e., if a turbine is driven by cold air. The regression coefficients for specific power are presented in Macek [44].

The trend of speed lines, being deformed from the specific power (almost straight) lines with slightly negative gradient by changes of internal energy, is presented in Figure 4.24. The regression should take into choking, which limits maximum MFR by reaching sonic velocity in an inducer (the case presented in Figure 4.24) or in bladed or vaneless diffuser. In the latter case, the choking lines are speed dependent, which is the frequent case of larger compressors. The suitable form of pressure ratio dependence on MFR, in which coefficients, as functions of corrected speed, can be found by linear regression after transformation of variables, is

$$\pi_C = \frac{a\left(n_{C,corr}\right)\dot{m}_{C,corr} + b}{-c\left(n_{C,corr}\right)\dot{m}_{C,corr}^{\;2} - d\left(n_{C,corr}\right)\dot{m}_{C,corr} + 1} \tag{4.28}$$

The regression coefficients of speed lines shown in Macek [44] feature reasonable stability. After simple algebraic treatment, the formula from Equation (4.28) can be used for finding asymptotic choking limit and for maximum pressure ratio (if any) inside the

FIGURE 4.25 The efficiency map of a centrifugal radial compressor measured and calculated from regressions for pressure ratio and enthalpy difference.

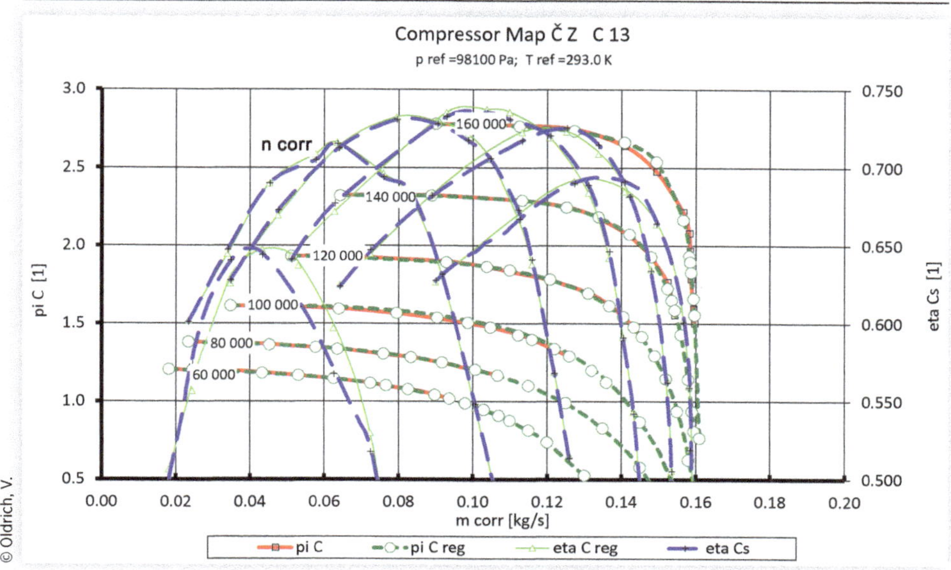

range between choking and surge limits. If maximum exists, the solution of possible MFR for the given pressure ratio (see requirement on boundary condition solution in Figure 4.1) features two roots. The selection of appropriate root depends on previous history of MFR/pressure trace. In most cases, the more stable larger MFR plays this role. The efficiency map, found as a result from specific compressor power and isentropic difference of enthalpies from pressure ratio, is predicted accurately (see Figure 4.25), obviously with good extrapolation capability to pressure ratio less than one.

Surge limit has to be defined separately as the dependence of pressure ratio on corrected MFR; see Figure 4.26. The behavior of a compressor on the left side of surge

FIGURE 4.26 The speed map of a centrifugal radial compressor represented by regression for pressure ratio with both limits of operation.

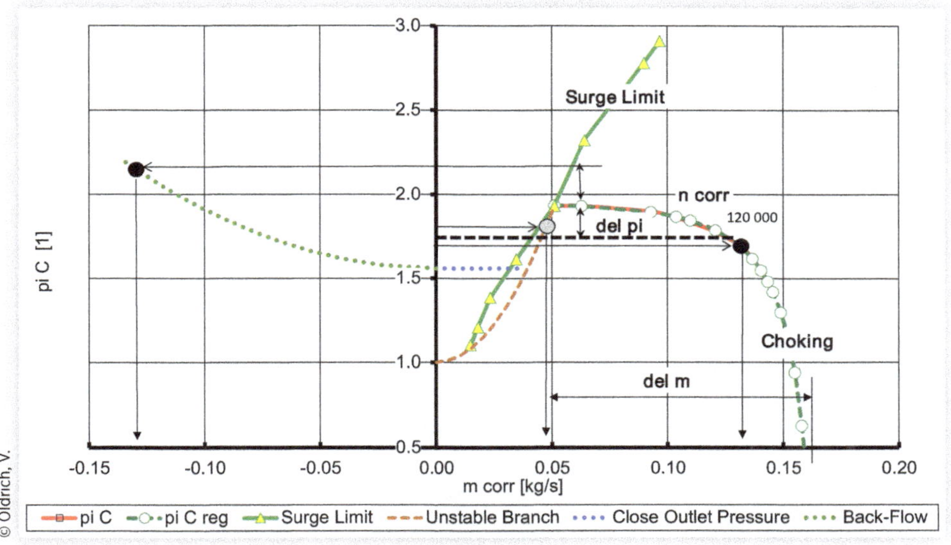

line can be described only very roughly, since the real surge process depends on dynamic separation of BL.

If compressor description is used as a boundary condition in the sense of Figure 4.1 for known pressure ratio and speed, there are two possible cases: the pressure ratio is below the maximum of pressure speed line or it exceeds it. The former case has been already described—the closest point to previous one is selected if two roots exist. Since the danger of surge exists close to maximum point, negative pressure ratio tolerance should be added to maximum pressure ratio of the current speed line (Figure 4.26). The latter case needs to take into account the distance of the current pressure from the speed line (Figure 4.26). If the difference is closer than empiric tolerance ***del pi***, the mild surge occurs between surge limit and stability boundary parabola. Stability boundary parabola—Equation (4.29)—defines the last static stable dependence between pressure ratio and MFR, intersecting the surge point line for the current speed.

$$\pi_{C,u} = 1 + \left(\pi_{C,surge} - 1\right)\left(\frac{\dot{m}_{C,u}}{\dot{m}_{C,surge}}\right)^2 \tag{4.29}$$

The point of temporary surge is $\dot{m}_{C,u}$ for the current pressure ratio instead of solution on the right branch of a speed line. If the problem with the pressure ratio is too close to the speed line maximum keeps during the next integration step, MFR is moving between stability boundary parabola and surge line for the current pressure ratio until the distance to maximum of speed line is sufficient. It mimics mild surge.

If the deviation of the current pressure ratio is greater than ***del pi*** tolerance, deep surge may occur. The point of zero MFR has to be found from the balance between pressure and centrifugal force first. From energy conservation in Equation (5), the pressure ratio at zero MFR is

$$\begin{aligned} u_e + r_C T_{C1} + \frac{\dot{m}_C^2}{2A_r^2\rho^2\cos^2\beta} - \frac{r^2\omega^2}{2} &= const. \\ \pi_{C0} - 1 \approx \frac{1}{r_C T_{C1}}\frac{r_2^2\omega^2}{2}\left(1 - \frac{r_1'^2 + r_1''^2}{2r_2^2}\right) & \quad \text{for } \dot{m}_C = 0 \end{aligned} \tag{4.30}$$

which is plotted in both Figure 4.24 and Figure 4.26 as constant pressure ratio line. To estimate the unstable left branch of a speed line, this constant value is used until intersection with the stability boundary occurs at the MFR

$$\dot{m}_{C0} = \sqrt{\frac{\pi_{C0} - 1}{\pi_{C,surge} - 1}}\,\dot{m}_{C,surge} \tag{4.31}$$

The negative MFR is, very roughly, estimated by the MFR from the standard Saint-Vénant-Wantzel relation governed by the current pressure ratio and the pressure ratio caused by a centrifugal force according to Equation (4.30). The flow area is estimated from choking at the same speed with atmospheric pressure and temperature at compressor inlet, assuming the inducer choking. The result should be reduced by discharge coefficient, estimated by for sharp-edged orifice, i.e., of the value of 0.6. The backflow capacity of a compressor is still huge, as can be seen in Figure 4.26, plotted with appropriate scales. The deep surge starts at negative backflow, and the next development depends on capacity of the inlet manifold system, since the backflow is coupled with significant pressure waves. Then, the positive flow rate can be found at stability parabola or—if pressure ratio falls down below the centrifugal force pressure ratio—at the right branch of a speed line.

The proposed procedure is qualitatively right, but the details should be calibrated by experiments in the future.

4.5.2 Physical 1D Central Streamline Model of a High-Pressure Ratio Centrifugal Compressor

The high-pressure ratios of TC compressors, needed for the current bmep levels, call for better description and understanding of processes inside centrifugal compressors. Even using 1D approach only, suitable for repeated optimization simulations, the model can yield interesting results, if it is based on physical description of processes. There are several attempts to describe the behavior of compressors by 1D, e.g., Bozza [8] or Nakhjiri [52].

The quasi-steady, central streamline model of a centrifugal compressor with axial-radial flow, suitable for 1D engine models with unsteady conditions during both transient load and speed of a car engine, will be described. The developed model of a centrifugal compressor describes the aerodynamics of flow from compressor inlet if MFR and impeller speed are known. The current model tries to treat transonic phenomena and use the available knowledge from axial compressor cascades taking the real asymmetric incidence angle influence into account, better than old NACA shock loss theory (Dixon [25]) for today's shapes of inlet blade profile. The main goal is a development of the compressor physical model for compressible fluid flow at the reasonable level of simplification for compressor performance description. The developed model will be validated by fitting to known compressor maps, aiming at extrapolation of maps and better prediction of surge and choke lines. The philosophy of the model is very similar to the one described in the section on the central streamline turbine model.

4.5.3 Geometry of Flow in Radial Compressor

The compressor performance will be described at fixed MFR and impeller speed with known geometry—at any location defined by the radius r, axial width b, step of blades in cascade s, blade chord c, and blade angle measured from radius α_B or β_B for a stator or impeller, respectively. The positive direction of angle is measured in sense of impeller rotation. General scheme of radially axial centrifugal compressor is plotted in Figure 4.27. Defining flow angles by α or β for a stator or impeller, respectively, Equation (4.12) yields the velocity triangles and splitting vectors into tangential or axial/radial components.

A radial blade cascade may be converted into the axial one using conformal (angle conserving) transformation from the radial-tangential cylindrical coordinates into Cartesian axial-tangential coordinates, used usually for profile cascade, as described by Equation (4.14) and Figure 4.8.

In the case of compressor impeller, the angles from radial direction are small, sometimes even zero (radial vanes), but mostly backswept (Figure 4.27). Even in the case of radial vanes, the flow inside impeller channels is not equivalent to purely straight diffuser channel. Lift force from flow direction change in an axial cascade is replaced, in the case of radial flow, by the lift force created by Coriolis inertia force. Instead of centrifugal force

FIGURE 4.27 A compressor impeller (1-2) and diffuser (2-3 vaneless, 3-4 bladed) with positive sense of flow and blade angles. In reality, outlet from impeller features negative vane angle with backswept blades below. "Jet and wake" velocity profile at outlet of impeller vanes.

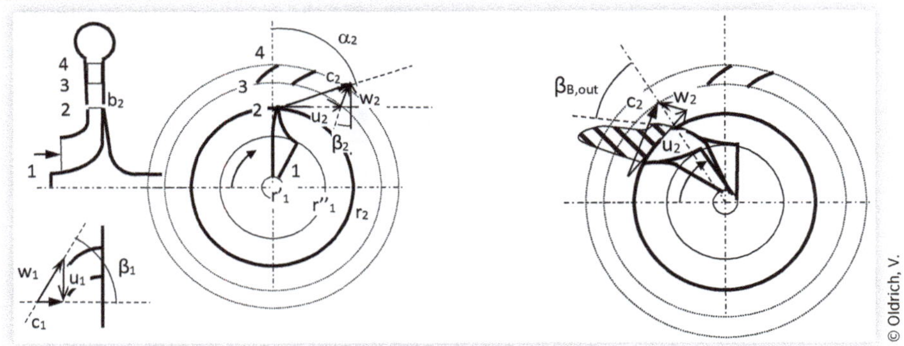

due to channel curvature in an axial cascade, Coriolis force acts on the flow, if radial velocity component exists. Coriolis force causes pressure distribution in tangential direction with increase of pressure in counterrotation direction (counterclockwise in Figure 4.27). This pressure distribution is followed by the flow separation at the suction side of a vane being ahead in the sense of rotation. It is reflected by the outlet velocity profile, called "jet and wake" with flow separation bubble behind the leading vane and with jet part of flow close to the pressure side of the following vane. The equivalence of the centrifugal force in a curved axial channel and Coriolis force in an impeller channel can be used for the estimate of deviation angle and losses in an impeller, as described in Macek [49]. Only one half of Coriolis acceleration has to be used since the flow features (almost) zero angular speed due to the slip of flow relative to vanes. Using the same flow-bearing velocities for both equivalent cascades in radial (subscript r) and axial direction (a), respectively, the force acting on the element of flow with the mass of Δm is

$$w_r = \frac{\dot{m}}{\left(2\pi r_r - z b_{B,r}\right) b_r \rho} = w_a = \frac{\dot{m}}{\pi\left(r_a''^2 - r_a'^2\right)\rho} = w \cos \beta_{B,a}$$
$$F_{Co} = \Delta m \omega w_r \equiv F_{cent} = \Delta m \frac{w^2}{R} \Rightarrow \frac{\omega}{w_r} = \frac{1}{\cos^2 \beta_{B,a}} \frac{y''}{\sqrt{\left(1+y'^2\right)^3}} \tag{4.32}$$

where y is an axial blade centerline shape, described by a function of axial coordinate and derived for finding local curvature. Then, the differential equation for transformed axial blade cascade centerline can be found from its curvature using obvious

$$\frac{1}{\cos^2 \beta_{B,a}} = \tan^2 \beta_{B,a} + 1 = y'^2 + 1$$
$$y'' = \frac{\omega}{w_r} \sqrt{1+y'^2} \tag{4.33}$$

which can be integrated for $\tan \beta_{B,a}$ using Euler's substitution in the integral. The described transformations can help to some extent for the estimation of losses, based on old, but yet worthwhile, generalizations of axial diffuser blade cascades—e.g., Howell [30, 31] or Kousal [35]. Calibration based on measured compressor maps is necessary, nevertheless.

4.5.4 Total and Static States with Transonic Limits and Kinetic Energy Losses

Compressor modelling features high flow velocities from the inlet to an inducer. Static state has to be found from known total state upstream and MFR. The procedure resembles to that described in the case of a radial turbine. Details are described in [49].

The isentropic efficiency and loss coefficient of diffuser flow are defined according to Equation (4.17). Should resulting temperature or density be determined, the procedure described above has to be changed, taking losses of kinetic energy and potential energy centrifugal force field into account. Then, it yields for compressor flow through a generally rotating diffuser cascade with known state at blade inlet *in* the static state at outlet *out* the basic relation for Newtonian iteration, similar to

$$y = T_{0in} - \frac{u_{in}^2 - u_{out}^2}{2c_p} - T_{out} - \frac{1}{2c_p}\left(\frac{\dot{m}}{A\rho_{0in}}\right)^2 \left(\frac{T_{out}/T_{0in}}{\left[\frac{T_{out}}{T_{0in}} - \zeta_{in}\left(1 - T_{in}/T_{0in}\right)\right]^{\frac{\kappa}{\kappa-1}}}\right)^2 \quad (4.34)$$

The derivative of y can be easily found in analytical way and applied to Newton-Raphson method.

4.5.5 Generalization of Results for Axial Blade Cascade

4.5.5.1 GEOMETRY OF AXIAL PROFILE CASCADE

The angles of flow are measured from axial direction. The blade angles are β_B, the angles of flow in coordinate system of blade cascade (relative flow coordinates) are β, and the angles of flow in steady (absolute) coordinate system are $\propto$. The flow turn angle $\boldsymbol{\varepsilon}$, incidence angle $\boldsymbol{\iota}$, and outlet deviation angle δ are defined in a standard way—see, e.g., Dixon [25] or Macek [49].

Moreover, the airfoils are described by length of chord c, cascade step s, maximum distance of airfoil centerline from chord p, and angles of tangent to centerline measured from axial direction β_B.

4.5.5.2 HOWELL THEORY OF COMPRESSOR BLADE CASCADES

The total loss consists of profile loss, secondary loss, and blade tip loss. The profile loss of planar airfoil cascade is caused by surface friction and wake pressure difference. The secondary losses are caused by induced vortices and the blade tip losses depend on leakages and induced vortices, as well.

Profile loss coefficient is calculated from drag coefficient at central streamline. Profile cascade features can be found using Howell's approach generalizing angle of flow turn and drag coefficient for profile cascades.

The normalized angle of flow turn $\varepsilon/\varepsilon^*$can be found from the empirical dependence on normalized angle of incidence $(\iota - \iota^*)/\varepsilon^*$ in Figure 4.28. Drag coefficient c_X depends on normalized step of cascade. All curves can be substituted by polynomial regressions.

FIGURE 4.28 Normalized angle of flow turn and profile drag coefficient for compressor blade cascade. Drag coefficient for different s/c 0.5, 1.0, and 1.5. Comparison of published data and regression model.

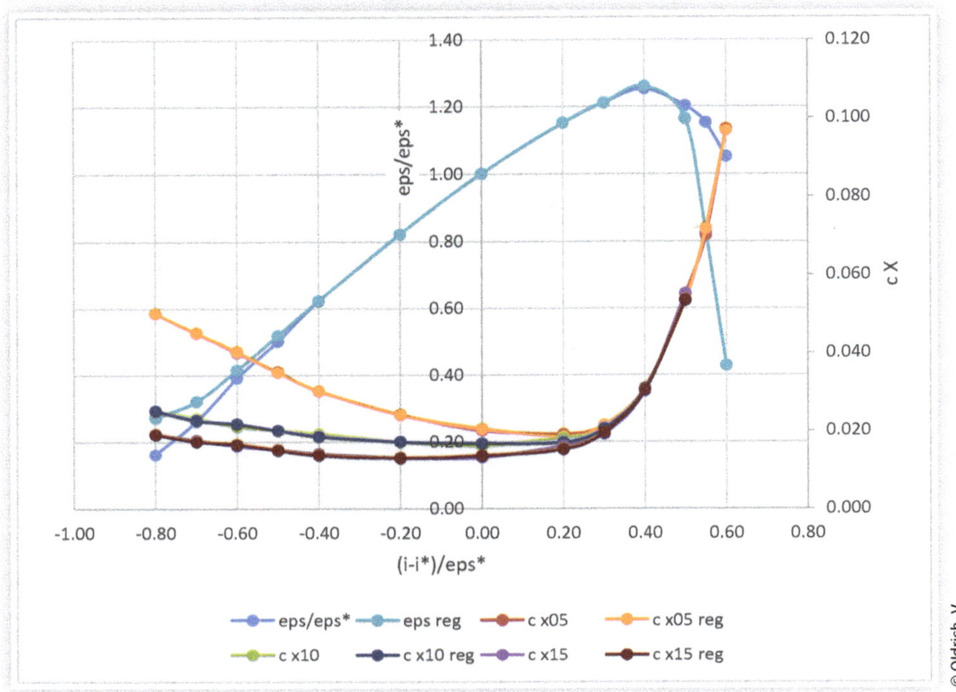

In the case of flow turn angle, it is amended additionally by exponential correction to separation of BL

$$\frac{\varepsilon}{\varepsilon^*} = \left(a_0 + \sum_1^6 a_i \left(\frac{\iota - \iota^*}{\varepsilon^*} \right)^i + a_{13} \left(\frac{\iota - \iota^*}{\varepsilon^*} \right)^{13} \right) \left(1 - e^{\left(\frac{\frac{\iota-\iota^*}{\varepsilon^*} - 0.95}{0.4} \right)^7} \right) \tag{4.35}$$

Reference values can be found from reference deviation angle (Constant rule, NACA (Dixon [25]))

$$\left|\beta_{out}^*\right| = \left|\beta_{B,out}\right| + \delta^*; \quad \delta^* = \frac{0.23\left(\frac{2p}{c}\right)^2 + \frac{\left|\beta_{B,out}\right|}{500}}{\frac{1}{\left|\theta\right|\sqrt{\frac{s}{c}}} - \frac{1}{500}}; \quad \left|\theta\right| = \left|\beta_{B,in}\right| - \left|\beta_{B,out}\right| \tag{4.36}$$

and Howell's relation

$$\tan\left|\beta_{in}^*\right| - \tan\left|\beta_{out}^*\right| = \frac{1.55}{1 + 1.5\frac{s}{c}} = S \tag{4.37}$$

Angles are bound by definitions, which for reference flow turn angle yields

$$\left|\varepsilon^*\right| = \arctan \frac{-\frac{1}{\left(\cos\beta_{out}^*\right)^2} + S\tan\left|\beta_{out}^*\right| + \sqrt{\left(\frac{1}{\left(\cos\beta_{out}^*\right)^2} - S\tan\left|\beta_{out}^*\right|\right)^2 + 8S\tan\left|\beta_{out}^*\right|}}{4\tan\left|\beta_{out}^*\right|} \tag{4.38}$$

Reference values are calculated once for the whole compressor map prediction. Radial cascades in a vaned diffuser are transformed to axial ones using Equation (4.14). Then, using previous relations, drag coefficient can be found.

4.5.5.3 FORCES IN A BLADE CASCADE

Tangential T and axial A forces in an axial compressor blade cascade, acting from fluid to airfoils, can be found from lift and drag forces using mean angle of flow (Dixon [25]) with positive directions defined in Figure 4.8. Drag force acts on fluid against mean velocity; lift force is perpendicular to it according to the sign of velocity circulation. If results of momentum conservation are combined with energy conservation, tangential and axial forces are determined

$$\vec{T} = \overline{\rho b}\, s w_a^2\left(\tan\vec{\beta}_{in} - \tan\vec{\beta}_{out}\right); \quad \vec{\beta}_{out} = \vec{\beta}_{in} - \varepsilon \tag{4.39}$$

$$\vec{A} = -F_Y \sin\left|\vec{\beta}_m\right| + F_X \cos\beta_m = \frac{s\,\overline{b\rho}}{2} w_a^2\left[-\frac{1-\zeta_{in}}{\cos^2\beta_{in}} + \frac{1}{\cos^2\beta_{out}}\right] \tag{4.40}$$

The procedure is prepared for Howell cascade results, in which flow angle change ε and drag coefficient c_x are generalized from experiments. Combining relations above, it yields for loss coefficient

$$\zeta_{in,p} = 1 - \tan\vec{\beta}_m \cos^2\beta_{in}\left[2\tan\vec{\beta}_{in} - 2\tan\left(\vec{\beta}_{in} - \varepsilon\right) - \frac{c}{s}\frac{c_X}{\cos\alpha_m}\tan\vec{\beta}_m\right] + $$
$$+ \frac{c}{s}\frac{c_X\cos^2\beta_{in}}{\cos\beta_m} - \frac{\cos^2\beta_{in}}{\cos^2\beta_{out}} \tag{4.41}$$

and lift coefficient can be found from

$$c_y = \frac{2\cos\beta_m}{sign\beta_m}\frac{s}{c}\left(\tan\vec{\beta}_{in} - \tan\left(\vec{\beta}_{in} - \varepsilon\right)\right) - c_X\tan\left|\beta_m\right| \tag{4.42}$$

4.5.6 Application of Profile Blade Cascade Theory to Compressor Components

Outlet angle from an inducer axial blade cascade can be used for estimation of local flow separation at the start of radial impeller part, using empirical chord length of separated bubble with additional calibration coefficient

$$\Delta s = K_{sep}\left(\sin\left|\vec{\beta}_{out}\right| - \sin\left|\vec{\beta}_{B,out}\right|\right) \tag{4.43}$$

The velocities are found from continuity equation then. The profile drag coefficient is found from regression, described in Equation (4.35), and recalculated to loss coefficient according to Equation (4.41).

Secondary losses depend on lift coefficient square, using classic Glauert results. Secondary loss coefficient has to be added to the profile loss—see Howell [25], Kousal [35], and Vavra [70]—for blade length b and radial shroud clearance k including tip losses according to Vavra [70]

$$\zeta_{in,s} = \left[\frac{0.04}{\frac{b}{c}} + \frac{0.025\frac{k}{b}}{\frac{s}{c}\cos\beta_{out}} \right] \frac{c_Y^2}{\frac{s}{c}} \frac{\cos^2\beta_{in}}{\cos^3\beta_m} \tag{4.44}$$

If $\mathrm{Re}_{c,w\,in} < 200{,}000$ (it may occur at high-pressure compressor stages stages), correction to Re should be done before loss coefficients are summarized (Dunham [26])

$$\frac{\zeta_{2p} + \zeta_{2s}}{\left(\zeta_{2p} + \zeta_{2s}\right)_{\mathrm{Re}=2.10^5}} = \left(\frac{2.10^5}{\mathrm{Re}_{c,w_2}} \right)^{0.2} \tag{4.45}$$

otherwise no correction is applied. Then

$$\zeta_{in} = \zeta_{in,p} + \zeta_{in,s} + \zeta_{in,l} \tag{4.46}$$

All estimations have to be corrected by mentioned calibration coefficients.

4.5.6.1 IMPELLER INDUCER

The relations for flow turn angle and loss coefficient can be directly applied to quasi-axial inducer blades with correction coefficients taking into account the influence of Stodola vortex and centrifugal force stabilization of BL in radial part of blades.

4.5.6.2 BLADED DIFFUSER

Howell theory [30] or [31] can be used after transformation from polar coordinates to Cartesian ones. The procedure is described by Equations (4.14) and (4.35).

4.5.6.3 VANELESS DIFFUSER

The classic vaneless diffuser theory assumes free vortex (i.e., angular momentum conservation) for tangential velocity component and mass conservation with constant density for radial velocity component. If constant axial width b of vaneless diffuser (as plotted between positions 2 and 3 in Figure 4.27) is assumed, well-known logarithmic spiral streamline is achieved.

Both assumptions are too much idealized, since recent compressors achieve transonic flow at an impeller outlet, and the compressibility of fluid and friction loss at side walls of a vaneless diffuser should be taken into account. Velocity components in absolute space of inlet to a vaneless diffuser are

$$\begin{aligned} c_{2t,in} &= u_2 - K_S \frac{\pi u_{2out}}{z_{2out}} \cos\beta_{B2out} + w_{2r} \tan\beta_{B2out} \\ c_{2r,in} &= \frac{\dot{m}}{K_{sep,r} 2\pi R_{2,in} b_{2,in} \rho_{2,out}} \end{aligned} \tag{4.47}$$

Angular momentum conservation for radius greater than the inlet radius of a vaneless diffuser, if turbulent friction at side walls is assumed, yields

$$c_t = \frac{R_2 c_{2t,in}}{r} + \frac{\frac{K_f}{\mathrm{Re}_{2b,c}^{0.2}} \frac{\pi}{b} \frac{\bar{\rho}\left(c_{2t,in}^2 + c_{2r,in}^2\right) R_2^2}{2}}{\dot{m}} \left(\frac{R_2}{r} - 1\right) \tag{4.48}$$

Simplified assumptions have been used for friction torque estimate, considering constant channel with b, angular momentum, and mean constant density.

Centrifugal force, inertia force from change of radial velocity, and pressure equilibrium yield the relation in cylindrical coordinates, which can be amended by mass and energy conservation—Macek [49]—and solved for radial component numerically, if tangential component derivative is expressed by means of Equation (4.48). The vaneless diffuser has to be divided into several radial sectors for at least approximate integration of those differential equations. Narrow circular strips should be used for higher Mach numbers. If Mach number is less than 0.5, the sensitivity to density change is small. This procedure is useful if radial component may exceed sonic velocity, causing choking of a vaneless diffuser. However, this phenomenon is not probable.

4.5.6.4 TRANSONIC PERFORMANCE

The above deduced procedure can be applied for all blade cascades in a compressor if Mach number is less than approximately 0.7. Even before inducer inlet choking is reached, the local relative velocity Mach number, namely, the blade tip Mach number, can exceed transonic limit.

In the dependence of MFRs for Mach number greater than 1, the loss coefficient of inducer axial blades should be reduced before critical MFR for the whole blade height is reached (Kousal [35]).

In the case of a vaneless diffuser, the subsonic assumption might not be the case of current high-pressure compressors. The step-by-step integration of density history can be simultaneously used with assessment of transonic flow issues downstream of an impeller, as mentioned in Macek [49].

4.5.7 Compressor Performance Parameters

The overall picture of processes inside a compressor is plotted in h-s diagram in Figure 4.23, using energy conservation for rotating channel and definition of total and stagnation states and velocity triangles. The pressure losses in yet not described parts (an inlet casing or outlet scroll) may be estimated using empirical loss coefficient for friction loss at walls and local losses with approximately constant velocity and density, which yields input for the following part of a compressor.

Flow angles are calculated from Howell theory with correction to relative vortex in an impeller (Equation (4.41)). Flow area and radial velocity is corrected to local BL separation (Equation (4.43)). Loss coefficients are found according to regression similar to Equation (4.35) after recalculation from drag coefficient to profile loss coefficient (Equation (4.41)) adding all partial losses to it in Equation (4.46). If inlet and outlet velocities at an impeller are known, the power can be calculated from Eulerian theorem and checked by Stodola for the adiabatic case. Windage loss of an impeller can be estimated from windage power in similar way as in the case of a turbine. Windage power is subtracted from the internal power. Details are published in Macek [49]. Reduced MFR, reduced speed, and isentropic efficiency are calculated according to standard definitions.

The whole model of a compressor has to be calibrated in a similar way as in the case of a turbine against measured data. The transformation to modules of a 1D code is not straightforward as in the case of a turbine, and it is currently under development.

4.6 Positive Displacement Compressor

The Roots supercharger is the most famous representative of positive displacement, two-rotor blower with outer compression. The gas is delivered to inter-teeth space at pressure in charger inlet, and its compression occurs after the inter-teeth volume (positive displacement chamber) is connected to outlet piping. Smoother compression can be realized if helical gears and greater number of teeth are used instead of spur gears. The helical gears appeared first with Lysholm type of screw compressors. The advantage of continuous compression is compensated by the fixed built-in compression ratio, which cannot be used for lossless decrease of boost pressure until complicated variable port geometry is used. If helical gears are used, movement of teeth contact in axial direction can change the volume of inter-teeth gap compressing the gas, while both inlet and outlet ports are still closed. Therefore, large amount of development work was devoted in the past to screw supercharger (Lysholm compressor) with continuous compression (Takabe [67] and Miyagi [68]). New concept of helical gear Roots compressor combines features of both designs.

Map-based models are used as a standard. Speed is, in most cases, directly dependent on transmission ratio from an engine crankshaft. The only important feature is valid for any positive displacement machine: there is no need to respect Mach number in a machine for velocity transformation. However, the inlet density determines the mass of air per single delivery cycle. For this reason, speed is not corrected, but MFR has to be corrected using reference density at inlet, i.e.,

$$\dot{m}_{corr} = \dot{m}\frac{p_{ref}}{p_{0C1}}\frac{T_{0C1}}{T_{ref}} \tag{4.49}$$

All complicated processes in a real compressor may be modelled and optimized using the experience with implementation of advanced solver modules developed in 1D codes. The intentionally changed volumes of virtual cylinders, describing inter-teeth volumes and variable connecting orifices, simulating not only ports but also clearance leakages, may be applied if appropriate geometric dimensions are investigated in suitable 3D CAD software.

Based on data from 3D CAD model, 1D model was built, assuming that each rotor tooth is a piston and each gap is a cylinder. The details are in Brynych [10]. The modelled Roots supercharger can be substituted by a piston compressor with many cylinders controlled by ports and other orifices simulating leakages. Chamber and port pressures

FIGURE 4.29 Cylinder and port pressures with inlet and outlet valve lifts at compressor speed of 16 000 min^{-1}.

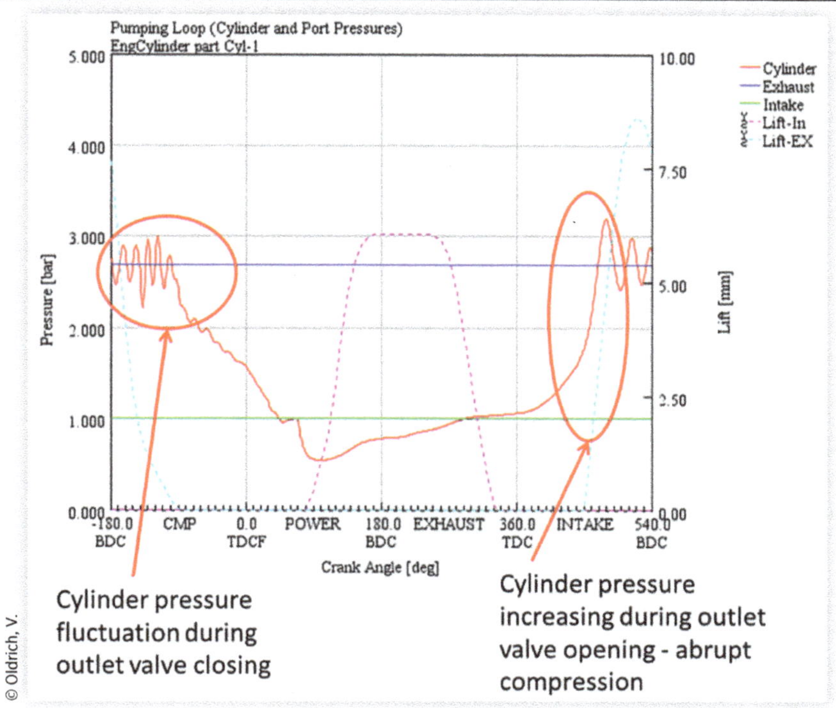

with inlet and outlet valve lifts, simulating the free areas of inlet or outlet ports, respectively, are shown in Figure 4.29. During outlet port opening, the rapid increase of cylinder pressure is obvious due to backflow of compressed air into the chamber as described above. The pressure oscillations caused by it in the dynamic system of volumes and connecting pipes (or FS boundary conditions) are clearly visible. In consequence, should the Roots-type compressor be used, large mufflers at outlet and reduction of NVH are needed, which complicates the compressor application. The simulation offers a tool for reduction of experiment time during the phase of development. The coupling of internal pressure oscillations with pipe pressure waves is weak because of large difference in eigen frequencies.

4.7 Conclusions

The model of a single- or twin-scroll turbine, using central streamline 1D approach, has been developed for both steady and unsteady flows, improving the standard map-based models. They need to use the fictitious parts of an exhaust system, connecting the branches of a pulse manifold, if applied to twin-scroll solutions. Due to issues in defining the isentropic reference processes, the new models are map-less, being fit directly to the MFR, pressure ratios, and power of the turbine in consideration. The calibration procedure, based on genetic algorithms, was developed and tested with the models.

The quasi-steady model can be aligned with any 1D solver, since it is based on algebraic equations only. The unsteady model is based directly on 1D solvers from available 1D simulation software.

The physically based turbine model has been tested by experiments at a turbine steady-flow test bed for several types of turbines, in one case for twin-scroll partial admission, as well. The prediction of integral parameters and pressure wave traces were performed on a six-cylinder engine. The reliability and extrapolation capacity of the model were tested by simulations of many different engines. The numerical features and qualitative predictive capabilities are very satisfactory. The model can be improved and expanded by some additional modifications, e.g., taking reversed flows at very low admission ratios into account. Nevertheless, even the current state of the model is sufficient for its practical application up to high, supercritical pressure ratios, causing choking phenomena at turbines.

The model is available for parallel scroll partitions, but its layout makes it possible to extend it to serial scrolls just by changing mixing location of both scroll flows.

The model can be expanded to more detailed description of turbine components, splitting the scroll and impeller into smaller sections.

The application into 1D commercial codes can be done for both models without significant problems, however, except for transonic flow through vaneless nozzle, which is not very frequent. It can create an efficient tool for optimization of downsized highly dynamic engines, based on physics of unsteady flow and thus insensitive to changes of a current engine layout.

In addition, both models can be used as the sources of virtual sensor signals from phenomena inside a turbine. They can contribute to better understanding and design optimization of current radial-axial and diagonal turbines.

The still remaining items of the further development will cover:

- Adding vaneless nozzles to quasi-steady code and finding the efficient way of transferring it into 1D model
- Improving the FS mixing modules in 1D code
- Implementing additional nozzle downstream of mixing area to mimic the rest of flow path before entering an impeller and combining it with the module for total enthalpy change
- Expanding 1D model to greater details using more sections of a scroll and internal heat exchange in a TC
- Feedbacking of mechanical systems to variable pressure loads (e.g., flutter of blades or thrust bearing performance)
- Extending 3D models for calibration of 1D simulation tools

The presented physical model of a centrifugal compressor is suitable for compressor simulation in 1D codes, if calibrated according to measured compressor maps. On one hand, it uses the basic generalization of experiments valid for axial blade cascades, although amended by certain adoption of radial cascades features, which has to be corrected by available experiments. On the other hand, it treats the transonic flow in the compressors of high-pressure ratio, which yields an opportunity to extrapolate the maps with certain reliability. It is important for choke limit especially. The extrapolation or prediction of surge limits, especially under influence of engine pulsating inlet flow, has not been tested yet. The dynamic surge limit is still certainly a big issue. The procedures for transonic flow prediction are stable and controllable from the numerical point of view, which ensures reliable behavior during calibration done by optimization.

In any case, the further development and validation of the model is inevitable. The still remaining items of the further development will cover in a similar way to a turbine model:

- Compressor inlet duct loss including the optional use of pre-swirl blades
- Leakages at shroud and hub sides influencing backflow to inducer blades and windage loss at hub side of an impeller
- Inducer flow inlet angle corrected to the backflow through a shroud clearance adding angular momentum to inlet flow
- IRC for surge limit modification
- Influence of relative Stodola vortex to secondary vortices in inducer blades (amplification of asymmetry of counterrotating secondary vortex couple)
- Scroll friction and flow separation losses
- The impact of an outlet diffuser located downstream of a scroll
- Heat transfer in a compressor casing
- Transfer of a model from algebraic quasi-steady appearance to 1D solvers
- Extending 3D models for calibration of 1D simulation tools

The model of Roots- or Lysholm-type superchargers was developed, using the principles of 1D module application for unconventional purposes, as used for turbines and compressors. The geometry was generated by 3D CAD model, including time history of volumes, port timing and dimensions, leakage, and clearance analysis. The obtained data have been used for 1D model, using virtual cylinders and valves. The deeper analysis of power consumption and efficiency of a supercharger is under preparation.

The model development will focus on more detailed analysis of phenomena, including wave propagation inside a supercharger and influence of heat transfer. Moreover, there is a good perspective in NVH behavior analysis and optimization. In further work, the 1D model will be connected to an engine model and used for studying the supercharger-TC-engine interactions at steady and unsteady engine operation.

Acknowledgments

This work has been realized using the supports of Technological Agency, Czech Republic, program Centre of Competence, project #TE01020020 Josef Božek Competence Centre for Automotive Industry, EU Regional Development Fund in OP R&D for Innovations (OP VaVpI), and Ministry of Education, Czech Republic, project #CZ.1.05/2.1.00/03.0125 Acquisition of Technology for Vehicle Center of Sustainable Mobility and using the support of The Ministry of Education, Youth and Sports program NPU I (LO), project # LO1311 Development of Vehicle Centre of Sustainable Mobility. In addition, the support was realized by financial donations of Dr. Tom Morel to AFEUS, Czech Republic, and to Zvoníček's Foundation, Czech Republic, project "Development of a 1D Model of a Radial Turbocharger Turbine." The authors would like to express their gratuity to Antonin Mikulec, who corrected language, and to all colleagues from the Josef Bozek Research Centre for their help during experiments and testing.

These supports are gratefully acknowledged.

Definitions, Acronyms, and Abbreviations

Symbols

A - Flow area [m^2]; axial force component {N]

a - Velocity of sound [$m \cdot s^{-1}$]; axial distance or axial coordinate [m]; acceleration [$m \cdot s^{-2}$]

b - Vane axial width or radial height [m]

C - Correction to compressibility [1]; pressure, drag, or lift coefficients related to dynamic pressure [1]; discharge coefficient related to isentropic process MFR [1]

c - Absolute velocity in spatially fixed coordinate system [$m \cdot s^{-1}$]; blade chord length [m]

c_p - Isobaric specific heat capacity [$J \cdot kg^{-1} \cdot K^{-1}$]

c_s - Velocity after isentropic expansion of a total enthalpy head [$m \cdot s^{-1}$]

D - Diameter [m]

E - Total energy [J]

e - Specific total energy [$J \cdot kg^{-1}$]

H - Enthalpy [J]; enthalpy flow (with dot) [W]

h - Specific enthalpy [$J \cdot kg^{-1}$]

I - Moment of inertia [$kg \cdot m^2$]

K - Tuning coefficient [1]

k - Heat transmittance coefficient [$W \cdot m^{-2} \cdot K^{-1}$]

M - Torque [$N \cdot m$]; Mach number [1]

m - Mass, mass flow rate MFR (with dot) [kg, $kg \cdot s^{-1}$]

n - Angular speed [min^{-1}]

O - Wetted perimeter for friction loss [m]

P - Power [W]

p - Pressure [Pa]; position of maximum distance between airfoil centerline and chord [m]

pe - Brake mean effective pressure (bmep)

Re - Reynolds number [1]

r_X - Specific gas constant of specie X [$J \cdot kg^{-1} \cdot K^{-1}$]

R, ***r*** - Radius [m]

s - Specific entropy [$J \cdot kg^{-1} \cdot K^{-1}$]; distance of flow path or blade cascade step [m];

T - Temperature [K]

t - Time [s] or distance in tangential coordinate direction [m]; blade airfoil thickness [m]

u_e - Specific internal energy [$J \cdot kg^{-1}$]

w - Relative velocity [$m \cdot s^{-1}$]

w_u.***u*** - Tangential velocity of rotating solid body [$m \cdot s^{-1}$]

x - Blade tip velocity ratio w_{u2}/c_s [1]; distance along central streamline or in coordinate directions [m]; Cartesian coordinate [m]

y - Iteration variable

z - Number of blades [1]

α - Angle of absolute velocity (measured from radial or axial direction) [deg]; heat transfer coefficient [$J \cdot m^{-2} \cdot K^{-1}$]

α - Regression coefficient

β - Angle of relative velocity (measured from radial or axial direction) [deg]; turbine power correction to pressure pulsations [1]

β_B - Angle of tangent to blade centerline

γ - Turbine efficiency correction to pressure pulsations [1]; angle of blade chord from axial or radial direction [deg]

Δ - Difference

δ - Flow deviation outlet angle [deg]
ε - Pressure ratio <1 [1]; angle from coordinate axis [deg]; flow turn angle [deg]
η - Isentropic efficiency or compression/expansion efficiency related to kinetic energy [1]
ζ - Kinetic energy loss coefficient [1]
ι - Flow incidence angle [deg]
κ - Isentropic exponent [1]
λ - Power input coefficient [1]; thermal conductivity [$W \cdot m^{-1} \cdot K^{-1}$]; friction coefficient [1]; coefficient of secondary losses [1]
μ - Discharge coefficient
ν - Kinematic viscosity [$m^2 \cdot s^{-1}$]
π - Pressure ratio >1 [1]
ρ - Density [$kg \cdot m^{-3}$]
σ - Angle of oblique shock wave measured from flow velocity direction [deg]
θ - Profile centerline turn angle; flow deviation angle in oblique shock [deg]
τ - Shear stress [Pa]
φ - Polar or cylindrical coordinate angle [rad]
ψ - Flow function [1]
ω - Angular velocity [$rad \cdot s^{-1}$]

Subscripts and Superscripts

a - Axial
app - Approximation
ARD - Axially radial diffuser
avT1-2 - Average temperature between T1 (inlet) and T2 (outlet)
B - Blade
b - Brake
bearings - Related to turbocharger bearings
C - Compressor; Cartesian
comp - Compressible
D - Diffuser, discharge
el M/G - Electric motor/generator
F,f - Friction
HT - Heat transfer form a turbine to rotor bearings
hub - Hub (blade root) side
I - Impeller
in - Inlet
inc - Incidence, angle of attack
incomp - Incompressible
K - Compressor
leak - Leakage
loss - Lost, dissipated
m - Mean
max - Maximum
min - Minimum
N - Nozzle ring, vaneless turbine scroll and nozzle
nom - Nominal, at maximum efficiency
oil - Lubricating oil
out - Outlet
P - Polar
p - Profile loss; at constant pressure

Q - Thermal, heat
r - Radial
red - Reduced
ref - Reference value
reg - Regression model
rel - Relative flow
s - Isentropic
sep - Flow boundary layer separation
shroud - Shroud (blade tip) side
vent - Windage
T - Turbine
t - Tangential; total = stagnation state + potential energy
TC - Turbocharger
u - Rotating disc tangential (circumferential) velocity
v - At constant volume
vl, VL - Vaneless
w - Wall
wind - Windage of rotating disc
X - Drag
Y - Lift
0 - Stagnation state
1 - Inlet
2 - Outlet or nozzle ring outlet or compressor impeller outlet
3 - Turbine impeller outlet or compressor bladed diffusor inlet
' - Scroll partition 1; blade root
" - Scroll partition 2; blade tip
* - Critical (sonic); reference (nominal) for airfoils
— - Averaged
→ - Vector (if value is used, it has to feature appropriate sign acc. to axis direction)

Acronyms

A/F - Air-to-fuel ratio
BL - Boundary layer
bmep - Brake mean effective pressure (p_e)
bsfc - Brake specific fuel consumption
BSR - Blade-speed ratio w_u/c_s
CFD - Computational fluid dynamics
CR - Centripetal radial
CRT - Centripetal radial turbine
EAT - Electrically assisted turbocharging
ECR - External recirculation channel
EGR - (external) exhaust gas recirculation
FS - Flow split (or flow junction) element
FV - Finite volume
GT - Gamma Technologies Inc.
HP - High-pressure stage
ICE - Internal combustion engine
IRC - Internal recirculation channel
IGR - Internal exhaust gas recirculation

IMEP - Indicated mean effective pressure
LP - Low-pressure stage
MFR - Mass flow rate
ORC - Organic Rankine cycle
PIMEP - Pumping mean effective pressure
piT,P R - Pressure ratio
Re - Reynolds number
SQT - Sequential turbocharging (twin turbo)
STC - Super/turbocharging (combined turbocharging)
TC - Turbocharger
TST - Two-stage turbocharging
VG - Variable geometry
VGT - Variable geometry turbine
VL - Vaneless
WG - Waste-gate
WHR - Waste-heat recovery
WOT - Wide open throttle
0D - Zero dimensional (depending on time only)
1D - One dimensional
3D - Three dimensional

References

1. Aghaali, H. and Angstrom, H.-E., "Improving Turbocharged Engine Simulation by Including Heat Transfer in the Turbocharger," SAE Technical Paper 2012-01-0703, 2012, 10.4271/2012-01-0703.
2. Aymanns, R., Scharf, J., Uhlmann, T., and Pischinger, S., "Turbocharger Efficiencies in Pulsating Exhaust Gas Flow," *MTZ* (2012): 07-08.
3. Baines, N., Wygant, K.D., and Dris, A., "The Analysis of Heat Transfer in Automotive Turbochargers," *Journal of Engineering for Gas Turbines and Power* 132 (2010): 0402301.
4. Bogomolov, S., Macek, J., and Mikulec, A., "Development of Design Assistance System and Its Application for Engine Concept Modeling," SAE Technical Paper 2011-37-0030, 2011, 10.4271/2011-37-0030.
5. Bohn, D., "Conjugate Flow and Heat Transfer Investigation of a Turbocharger. Part II: Experimental Results," ASME Conference Paper GT2003-38449, 2003.
6. Bohn, D., "Conjugate Calculation of Flow Field and Heat Transfer in Compressor, Turbine and Casing of a Gas Turbine," *VGB Powertech* 83, no. 11 (2003): 54-59.
7. Bohn, D., Heuer, T., and Kusterer, K., "Conjugate Flow and Heat Transfer Investigation of a Turbocharger," *ASME Journal of Engineering for Gas Turbines and Power* 127 (2005): 663-669.
8. Bozza, F., De Bellis, V., Marelli, S., and Capobianco, M., "1D Simulation and Experimental Analysis of a Turbocharger Compressor for Automotive Engines under Unsteady Flow Conditions," *SAE Int. J. Engines* 4, no. 1 (2011): 1365-1384, https://doi.org/10.4271/2011-01-1147.
9. Brinkert, N., Sumser, S., Schulz, A., Weber, S. et al., "Understanding the Twin-Scroll Turbine-Flow Similarity," *ASME Turbo Expo* 49 (2011): 2207-2218.

CHAPTER 4

10. Brynych, P., Macek, J., Vitek, O., and Cervenka, L., "1 D Model of Roots Type Supercharger," SAE Technical Paper 2013-01-0927, 2013, 10.4271/2013-01-0927.

11. Bulaty, T., "Specific Problems of 1-D Models of Gas Exchange at Turbocharged Engines (in German)," *Motortechnische Zeitschrift* 35 (1976): 4.

12. Canova, M., Naddeo, M., Liu, Y., Zhou, J. et al., "A Scalable Modelling Approach for the Simulation and Design Optimization of Automotive Turbochargers," SAE Technical Paper 2015-01-1288, 2015, 10.4271/2015-01-1288.

13. Capobianco, M. and Gambarotta, A., "Variable Geometry and Waste-Gated Automotive Turbochargers: Measurements and Comparison of Turbine Performance," *Journal of Engineering for Gas Turbines and Power* 114 (1992): 553-560.

14. Capobianco, M. and Marelli, S., "Transient Performance of Automotive Turbochargers: Test Facility and Preliminary Experimental Analysis," SAE Technical Paper 2005-24-66, 2005, 10.4271/2005-24-66.

15. Capobianco, M. and Marelli, S., "Turbocharger Turbine Performance under Steady and Unsteady Flow: Test Bed Analysis and Correlation Criteria," *8th International Conference on Turbochargers and Turbocharging, IMechE*, London, 2006.

16. Capobianco, M. and Marelli, S., "Unsteady Flow Behaviour of the Turbocharging Circuit in Downsized Automotive Engines," *FISITA Congress 2006*, paper F2006P119, Yokohama, 2006, 12.

17. Casey, M.V. and Fesich, T.M., "The Efficiency of Turbocharger Compressors with Diabatic Flows," *Journal of Engineering for Gas Turbines and Power* 132 (2010): 072302.

18. Ciesla, C., Keribar, R., and Morel, T., "Engine/Powertrain/Vehicle Modeling Tool Applicable to All Stages of the Design Process," SAE Technical Paper 2000-01-0934, 2000, 10.4271/2000-01-0934.

19. Cormerais, M., Hetet, J., Chesse, P., and Maiboom, A., "Heat Transfer Analysis in a Turbocharger Compressor: Modeling and Experiments," SAE Technical Paper 2006-01-0023, 2006, 10.4271/2006-01-0023.

20. Cormerais, M., "Caractérisation expérimentale et modélisation des transferts thermiques au sein d'un turbocompresseur automobile. Application à la simulation du comportement transitoire d'un moteur Diesel à forte puissance spécifique," Ph.D. thesis, l'Université de Nantes, 2007.

21. Costall, A., Szymko, S., Martinez-Botas, R.F., Filsinger, D. et al., "Assessment of Unsteady Behaviour in Turbocharger Turbines," *Proceedings of ASME TurboExpo 2006*, paper GT2006-90348, Barcelona, 2006.

22. Curtil, R. and Magnet, J.L., "Suralimentation a convertisseur d`impulsions modulaire et temperature de porte de soupapes d`echappement," CIMAC 1979 Wien, D38, 1979.

23. De Bellis, V., Bozza, F., Schernus, C., and Uhlmann, T., "Advanced Numerical and Experimental Techniques for the Extension of a Turbine Mapping," *SAE Int. J. Engines* 6, no. 3 (2013): 1771-1785, 10.4271/2013-24-0119.

24. Dibelius, G., "Partial Admission of Turbocharger Turbines (in German)," *Brown Boveri Mitteilungen* 52 (1965): 3.

25. Dixon, S.L., *Fluid Mechanics, Thermodynamics of Turbomachinery* (Oxford: Pergamon Press, 1975).

26. Dunham, J. and Came, P., "Improvements to the Ainley-Mathieson Method of Turbine Performance Prediction," *Transaction of the ASME* 92 (1970).

27. Fredriksson, C.F., Qiu, X., Baines, N.C., Müller, M. et al., "Meanline Modeling of Radial Inflow Turbine With Twin-Entry Scroll," *ASME Turbo Expo* (2012), doi:10.1115/GT2012-69018.

28. GT-Power v 7.4, *User's Manual and Tutorial GT-Suite™ version 7.4* (Westmont IL: Private Publication Gamma Technologies Inc, 2013).

29. Gurney, D., "The Design of Turbocharged Engines Using 1-D Simulation," SAE Technical Paper 2001-01-0576, 2001, 10.4271/2001-01-0576.

30. Howell, A.R., "The Present Basis of Axial Flow Compressor Design: Part 1 - Cascade Theory and Performance," ARC R&M 2095, 1942.

31. Howell, A.R., "Fluid Dynamics of Axial Compressors," *Proceedings of the Institutional of Mechanical Engineers* 153 (1945): 441-452.

32. Hu, X. and Lawless, P.B., "Predictions of On-Engine Efficiency for the Radial Turbine of a Pulse Turbocharged Engine," SAE Technical Paper 2001-01-1238, 2001, 10.4271/2001-01-1238.

33. Jenny, E., "Berechnungen und Modellversuche über Druckwellen grosser Amplituden in Auspuff-Leitungen," Dissertation ETH Zürich, Ameba Druck Basel, 1949.

34. Jerie, J., "Theory of Aircraft Engines (in Czech)," CTU in Prague, 1985.

35. Kousal, M., "Stationary Gas Turbines (in Czech)," SNTL Prague, 1965.

36. Llamas, X. and Eriksson, L., "Control-Oriented Compressor Model with Adiabatic Efficiency Extrapolation," SAE Technical Paper 2017-01-1032, 2017, 10.4271/2017-01-1032.

37. Lujan, J.M., Galindo, J., and Serrano, J.R., "Efficiency Characterization of Centripetal Turbines under Pulsating Flow Conditions," SAE Technical Paper 2001-01-0272, 2001, 10.4271/2001-01-0272.

38. Lückmann, D., Uhlmann, T., Kindl, H., and Pischinger, S., "Separation in Double Entry Turbine Housings at Boosted Gasoline Engines," *MTZ* 10 (2013).

39. Macek, J. and Mikula, M., "Optimization of Exhaust Systems for Turbocharged Diesel Engines," (in Czech) *Technical Contributions of CKD PRAHA Heavy Industries* 18 (1985): 11–28, ISSN:0322-8523.

40. Macek, J., Vávra, J., and Vítek, O., "1-D Model of Radial Turbocharger Calibrated by Experiments," SAE Technical Paper 2002-01-0377, 2002, 10.4271/2002 -01-0377.

41. Macek, J., and Vítek, O., "Simulation of Pulsating Flow Unsteady Operation of a Turbocharger Radial Turbine," SAE Technical Paper 2008-01-0295, 2008, 10.4271/2008-01-0295.

42. Macek, J., Vítek, O., Burič, J., and Doleček, V., "Comparison of Lumped and Unsteady 1-D Models for Simulation of a Radial Turbine," *SAE Int. J. Engines* 2, no. 1 (2009): 173-188, 10.4271/2009-01-0303.

43. Macek, J., Vítek, O., Doleček, V., Srinivasan, S. et al., "Improved Simulation of Transient Engine Operations at Unsteady Speed Combining 1-D and 3-D Modeling," SAE Technical Paper 2009-01-1109, 2009, 10.4271/2009-01-1109.

44. Macek, J., Vítek, O., Žák, Z., and Vávra, J. "Calibration and Results of a Radial Turbine 1-D Model with Distributed Parameters," SAE Technical Paper 2011-01-1146, 2011, 10.4271/2011-01-1146.

45. Macek, J., "1-D Model of Radial Centripetal Turbine with Variable Exducer By-Pass for the Use of Exhaust Gas Energy 1-D Simulation," Report Z 13 – 11, Czech Technical

University in Prague, Faculty of Mechanical Engineering, Josef Božek Center of Vehicles For Sustainable Mobility, 2013.

46. Macek, J., Zak, Z., and Vitek, O., "Physical Model of a Twin-scroll Turbine with Unsteady Flow," SAE Technical Paper 2015-01-1718, 2015, 10.4271/2015-01-1718.
47. Macek, J. and Vítek, O., "Determination and Representation of Turbocharger Thermodynamic Efficiencies," SAE Technical Paper 2016-01-1042, 2016, 10.4271/2016-01-1042.
48. Macek, J., Vítek, O., and Doleček, V., "Basic Issues of High-Efficiency Turbochargers with Radial Turbines for Diesel Engines," *Paper A54, THIESEL 2016 Conference on Thermo- and Fluid Dynamic Processes in Direct Injection Engines, UPV CMT Valencia*, 2016, ISBN:978-84-9048-535-4.
49. Macek, J., "Physical 1D Model of a High Pressure Ratio Centrifugal Compressor for Turbochargers," *Journal of Middle European Construction and Design of Cars (MECCA)* 15, no. 3 (2017).
50. Mikula, M, "Modelling and Optimization of Pulse Converters for Highly Turbocharged Medium Speed Engines (in Czech)," Ph.D. thesis, Czech Technical University, 1985.
51. Morel, T., Keribar, R., and Leonard, A., "Virtual Engine/Powertrain/Vehicle Simulation Tool Solves Complex Interacting System Issues," SAE Technical Paper 2003-01-0372, 2003, 10.4271/2003-01-0372.
52. Nakhjiri, M., Pelz, P., Matyschok, B., Däubler, L. et al., "Physical Modeling of Automotive Turbocharger Compressor: Analytical Approach and Validation," SAE Technical Paper 2011-01-2214, 2011, 10.4271/2011-01-2214.
53. Osako, K., Higashimori, H., and Mikogami, T., "Study on the Internal Flow of Radial Turbine Rotating Blades for Automotive Turbochargers," SAE Paper 2002-01-0856, 2002, 10.4271/2002-01-0856.
54. Pohořelský, L., Žák, Z., Macek, J., and Vítek, O., "Study of Pressure Wave Supercharger Potential using a 1-D and 0-D Approach," SAE Paper 2011-01-1143, 2011, 10.4271/2011-01-1143.
55. Pohořelský, L., Brynych, P., Macek, J., Vallaude, P.-Y. et al., "Air System Conception for a Downsized Two-Stroke Diesel Engine," SAE Technical Paper 2012-01-0831, 2012, 10.4271/2012-01-0831.
56. Polášek, M. and Macek, J., "On the Possibilities of Turbocharging of a Small Automotive Diesel," *Conference KOKA 2000*, Žilina: Technical University of Žilina, 2000, 177–182, ISBN:80-7100-736-6.
57. Rautenberg, M., Mobarak, A., and Molababic, M., "Influence of Heat Transfer between Turbine and Compressor on the Performance of Small Turbochargers," JSME Paper 83-Tokyo-IGTC-73, International Gas Turbine Congress, 1986.
58. Salkin, A., "Utilization of mode Frontier as a Calibration Tool for Mechanical Models," CVUT Prague/ECN Nantes. Internship Report, unpublished, 2006.
59. Seifert, H., *Instationäre Strömungsvorgänge in Rohrleitungen an Vebrennungskraftmaschinen* (Berlin: Springer, 1962).
60. Seifert, H. et al., "Die Berechnung instationärer Strömungsvorgänge in den Rohrleitungssystemen von Mehrzylindermotoren," *MTZ* 33, no. 11 (1972): 421–428.
61. Serrano, J.R. et al., "An Experimental Procedure to Determine Heat Transfer Properties of Turbochargers," *Measurement Science and Technology* 21 2010: 035109.

62. Serrano, J., Arnau, F., Novella, R., and Reyes-Belmonte, M., "A Procedure to Achieve 1D Predictive Modeling of Turbochargers under Hot and Pulsating Flow Conditions at the Turbine Inlet," SAE Technical Paper 2014-01-1080, 2014, 10.4271/2014-01-1080.

63. Shaaban, S. and Seume, J., "Analysis of Turbocharger Non-Adiabatic Performance," *8th International Conference on Turbocharger and Turbocharging*, 2006, 119-130.

64. Shapiro, A.H., *The Dynamics and Thermodynamics of Compressible Fluid Flow* (New York: The Ronald Press Comp, 1953).

65. Sherstjuk, A.N. and Zarjankin, A.E., *Radial-Axial Turbines of a Small Power* (Moscow: Mashinostroenije, 1976).

66. Sirakov, B. and Casey, M., "Evaluation of Heat Transfer Effects on Turbocharger Performance," *Journal of Turbomachinery* 135 (2013): 021011.

67. Takabe, S., Hatamura, K., Kanesaka, H., Kurata, H. et al., "Development of the high Performance Lysholm Compressor for Automotive Use," SAE Technical Paper 940843, 1994, 10.4271/940843.

68. Miyagi, Y., Takabe, S., Miyashita, K., and Ikaya, N., "Experimental Study of New Lysholm Supercharger with a Simple Unloading System," SAE Technical Paper 960952, 1996, 10.4271/960952.

69. Uhlmann, T. et al., "Development and Matching of Double Entry Turbines for the Next Generation of Highly Boosted Gasoline Engines," *34th International Vienna Motor Symposium*, 2013.

70. Vavra, M.H., *Aero-Thermodynamics and Flow in Turbomachines* (New York: Wiley & Sons, 1960).

71. Vávra, J., Macek, J., Vítek, O., and Takáts, M., "Investigation of Radial Turbocharger Turbine Characteristics under Real Conditions," SAE Technical Paper 2009-01-0311, 2009, 10.4271/2009-01-0311.

72. Vítek, O., Macek, J., and Polášek, M., "New Approach to Turbocharger Optimization using 1-D Simulation Tools," SAE Technical Paper 2006-01-0438, 2006, 10.4271/2006-01-0438.

73. Vítek, O., Macek, J., and Žák, Z., The Physical Model of a Radial Turbine with Unsteady Flow Used for the Optimization of Turbine Matching, Giakoumis, E.G. ed., *Turbocharging and Turbochargers* (Nova Science Publishers, 2017), ISBN 9781536122398.

74. Vitek, O. and Macek, J., "Thermodynamic Potential of Electrical Turbocharging for the Case of Small Passenger Car ICE under Steady Operation," SAE Technical Paper 2017-01-0526, 2017, doi:10.4271/2017-01-0526.

75. Watson, N. and Janota, M.S., *Turbocharging the Internal Combustion Engine* (London: MacMillan Publishers, 1982), ISBN: 0333242904.

76. Westin, F., Rosenqvist, J., and Angström, H-E., "Heat Losses from Turbine of a Turbocharged SI-Engine- Measurements and Simulation," SAE Technical Paper 2004-01-0996, 2004, 10.4271/2004-01-0996.

77. Westin, F. and Angstrom, H.E., "Calculation Accuracy of Pulsating Flow through the Turbine of SI-Engine Turbochargers," SAE Technical Paper 2005-01-0222, 2005, 10.4271/2005-01-0222.

78. Winkler, N., Angstrom, H.E., and Olofsson, U., "Instantaneous On-Engine Twin-Entry Turbine Efficiency Calculations on a Diesel Engine," SAE Technical Paper 2005-01-3887, 2005, 10.4271/2005-01-3887.

79. Winkler, N., Angstrom, H.E., and Olofsson, U., "Instantaneous On-Engine Turbine Efficiency for an SI Engine in the Closed Waste Gate Region for 2 Different Turbochargers," SAE Technical Paper 2006-01-3389, 2006, 10.4271/2006-01-3389.

80. Winterbone, D.E. et al., "A Contribution to the Understanding of Turbocharger Turbine Performance in Pulsating Flow," *International Conference on Internal Engine Research*, Paper C433/011, Instn. Mech. Engrs. London, 1991.

81. Zehnder, G., "Berechnung von Druckwellen in der Aufladetechnik," *Brown Boveri Mitteilungen* 58, no. 4/5 (1971): 172-176.

82. Zehnder, G. and Meier, E., "Abgasturbolader und Aufladesysteme für Hochaufladung," *Brown Boveri Mitteilungen* 64, no. 4 (1977): 185-198.

83. Zinner, K., *Supercharging of Internal Combustion Engines* (Heidelberg: Springer, 1978).

84. Žák, Z., "1-D Unsteady Model of a Radial Centripetal Turbine for Turbocharging Optimization," PhD thesis, Czech Technical University, 2017.

CHAPTER 5

3D-CFD Combustion Models for SI and CI Engines

Tommaso Lucchini
Politecnico di Milano

Yuri Wright
ETH

5.1 Introduction to CFD Simulation of In-Cylinder Flows

Although flows in internal combustion (IC) engines have been studied for over a century, many of the physical processes involved are still poorly understood [1]. The ability to predict the thermophysical processes in IC engines has therefore attracted the attention of engine developers for decades, since spatially and temporally resolved information of in-cylinder flow fields and chemical composition can aid tremendously the interpretation of thermodynamic data, in most cases the only quantity available from engine test rigs. While optical engines enable quantitative diagnostics of flow quantities, fuel-air mixing and flame topologies, and select species concentrations, the information is often only available in individual planes or small sections of the domain at high resolution, although significant recent advances are reported, e.g., in [2, 3, 4, 5, 6, 7, 8, 9] and references therein. 3D-CFD on the other hand provides data for the entire computational domain at the discretized resolution, including areas which are difficult to access (e.g., within the valve curtain or inside bowls). In addition, numerical tools can be readily used to rapidly assess changes in geometrical configurations (runners, bowl shapes, etc.), which are time-consuming experimentally compared to changes in operating conditions.

Early efforts toward modeling flow, heat transfer, sprays, and combustion in engine-like configurations are reported in [10, 11, 12, 13, 14, 15] and references therein. While computational resources available at the time restricted many of the studies to simplified configurations, the considerable speedup of computer hardware in conjunction with advances in parallel programming and numerical schemes enable computations of full engine cycles nowadays. Nonetheless, numerical methods suffer from similar challenges as experiments, namely, time-varying grids with moving pistons and protruding valves, necessitating advanced mesh motion treatment as will be discussed. The highly turbulent nature of the flow undergoing compression leading to high pressures place considerable requirements for resolving all scales of the flow and directly solving the governing Navier-Stokes (NS) equations; scaling laws for engine speed and size can be found, e.g., in [16]. Aside from turbulence, combustion events lead to further pressure increase: flame structures are of the order of tens of micrometers making fully resolved calculations of entire engine domains prohibitively expensive at typical engine conditions. In agreement with projections of [17] in the context of aerodynamics, first fully resolved Direct Numerical Simulations (DNS) are however emerging for engine-like configurations, cf. [16, 18, 19] and references therein. Such numerical studies complement high-fidelity optical experiments and provide vast amounts of data for model development.

For engineering and design purposes, where rapid turnaround times allowing for parametric studies are a prerequisite, models are needed for turbulence, two-phase flow, and combustion. In use for many decades, Reynolds-averaged Navier-Stokes (RANS) has been the backbone of engine simulation, which most often in conjunction with two-equation models for turbulence [20] will also be discussed. Since unsteady RANS of engine cycles provides ensemble-averaged quantities, the results are ideally suited for direct comparison with phase-averaged experimental data. The ensemble-averaged nature of the equations however precludes prediction of cycle-individual events such as knock or misfire. Large Eddy Simulation (LES) has the potential to address the main RANS shortcomings. It solves the large structures on the grid (and hence a part of the turbulence spectrum) and models only the subgrid scales. However, to get converged statistics, it is necessary to compute many cycles for a consistent comparison with experiments. Early work using LES in the IC engine context is reported in [21, 22], while reviews on LES of turbulent combustion [23, 24] and specifically for IC engines can be found in [25], respectively. As a consequence of their specific merits, both RANS and LES methods will continue to play a role in engine development depending on the nature of the question to be addressed, while DNS can be increasingly applied to study relevant phenomena to gain insights with respect to physical processes and to improve modeling.

Aside from turbulence, many other phenomena in IC engines require considerable abstraction of the process to make it tractable for simulation and hence necessitate significant modeling efforts. These include among others two-phase flows, high-pressure gaseous injections, and, most notably, turbulent combustion. Although an exhaustive discussion of all these is impossible within this chapter, the challenges and a selection of modeling approaches are summarized, along with some examples and suggestions for further reading.

5.2 The Finite Volume Method Applied to Simulation of IC Engines

Flows in IC engines are compressible and turbulent. Liquid fuel injection requires consideration of two different phases and during the combustion process chemical reactions need to be taken into account. For a correct description of the in-cylinder flow field, fuel-air mixing, and combustion processes, transport equations are solved for mass (continuity), momentum, energy, and composition-related variables, like mixture fraction, progress variable, or chemical species mass fractions. The reader is referred to [26, 27, 28, 29] for the derivation and the final expression of such equations for compressible, reactive flows with sprays. For a generic variable ϕ representing a scalar, vector, or tensor property of the fluid, the transport equation has the form:

$$\frac{\partial \rho \phi}{\partial t} + \nabla \cdot (\rho \mathbf{U} \phi) - \nabla \cdot (D \nabla \phi) = \dot{\phi}_s + \dot{\phi}_{chem} \tag{5.1}$$

The left-hand side (LHS) of Equation (5.1) includes time-derivative, convection, and diffusion terms. ρ is the fluid density, D the diffusion field (scalar or tensor), and $\mathbf{U}$ the fluid velocity. On the right-hand side (RHS) of Equation (5.1), $\dot{\phi}_s$ and $\dot{\phi}_{chem}$ represent contributions from gas-spray interaction and combustion, respectively.

The purpose of the Finite Volume Method (FVM) is to transform Partial Differential Equations (PDE) into a corresponding system of algebraic equations. The solution of this system produces a set of values corresponding to the solution of the original equations at some predetermined locations in space and time. The FVM involves the following steps: mesh generation, discretization of the PDE system, and solution of the system of algebraic equations [30].

5.2.1 Mesh Generation and Management

The problem domain is decomposed into a finite number of control volumes, called *mesh* (or *grid*). Each control volume is called *cell* and, in general, it can be composed by an arbitrary number of faces (*polyhedral* mesh). IC engine geometries are complex and for this reason unstructured grids are employed to properly account for the design of piston bowl, cylinder head, and valves. To accommodate the piston and valve motion, the grid is deformed during the simulation and, sometimes, its topology is changed by adding or removing cells, faces, and points [31, 32]. In case deforming meshes are used, multiple meshes are required to cover the full-cycle simulation. Each one is valid within a certain crank angle interval, and, at the end of it, the computed solution is interpolated on a new mesh [33]. The use of topological changes makes possible to use, in principle, only one mesh which is continuously modified to accommodate the piston and valve motion.

5.2.2 Discretization

For any cell with volume V_p, Equation (5.2) is expressed in its *second-order, integral, semi-discretized* form [34]:

$$\int_t^{t+\Delta t} \left[\left(\frac{\partial \rho \phi}{\partial t} \right)_P V_P + \sum_f F \phi_f - \sum_f D_f \mathbf{S} \cdot (\nabla \phi_f) \right] dt = \int_t^{t+\Delta t} \left(\dot{\phi}_s V_P + \dot{\phi}_{chem} V_P \right) dt \tag{5.2}$$

In Equation (5.2), volume integrals of convection and diffusion terms are converted into surface integrals by the Gauss theorem. P is the center of the computational cell, f refers to any face of the cell, and $\mathbf{S}$ is the face area vector pointing outward of the cell faces.

F is the mass flow through the face, computed as $F = \mathbf{S} \cdot (\rho \mathbf{U})_f$; in case of moving grids [26], the relative mass flux has to be used instead $F = \mathbf{S} \cdot [\rho(\mathbf{U} - \mathbf{V}_{mesh})]_f$ where $\mathbf{V}_{mesh}$ is the face displacement velocity.

Numerical integration and interpolation techniques approximate volume integrals, derivatives, and field values at face centers. For stability reasons, implicit discretization is commonly used where the face values of ϕ and $\nabla\phi$ are determined from the new time values.

5.2.3 Solution of the System of Algebraic Equations

Since ϕ_f and $(\nabla\phi_f)$ also depend on values of ϕ in the surrounding cells, Equation (5.3) produces an algebraic equation which is used to determine the new value of ϕ_P:

$$a_P\phi_P^n + \sum_N a_N\phi_N^n = R_P \tag{5.3}$$

For every cell, one equation of this form is assembled. The value of ϕ_P depends on the values in the neighboring cells, thus creating a system of algebraic equations:

$$[A][\phi]=[R] \tag{5.4}$$

where

[A] is a sparse matrix with coefficients
a_P on the diagonal and a_N off the diagonal
[ϕ] is the vector of ϕ-s for all control volumes
[R] is the source term vector

The system of equations is generally solved with *iterative methods* starting with an initial guess and then continuing to improve the current approximation of the solution until some *solution tolerance* is met. Iterative methods are more economical, but they usually pose some requirements on the matrix, which must be diagonally dominant. In order to improve the solver convergence, it is desirable to increase the diagonal dominance of the system by means of preconditioning techniques.

5.2.4 Influence of Discretization on the Computed Results

Irrespective of the physical models used, the quality of the simulation results is strictly related to mesh size, mesh quality, and the used numerical schemes. Increasing the number of cells is expected to produce a solution, which approaches the exact one as the grid spacing goes to zero. Mesh quality parameters are skewness and non-orthogonality. Skewness reduces the accuracy of surface integrals used to approximate the convection terms, while non-orthogonality potentially introduces unboundedness in the computation of the diffusion term. The presence of moving grids with topological changes normally suggests to use first-order discretization in time. Limited, second-order schemes are used for convection discretization because they represent a good compromise between accuracy and stability. The diffusion term is discretized by decomposing the $\nabla\phi_f$ in two parts, orthogonal and non-orthogonal, which are treated implicitly and explicitly, respectively.

To increase the simulation stability, generally the non-orthogonal part is limited or neglected. The reader is referred to [34], where the effects of mesh quality, size, and numerical schemes is evaluated on different test cases.

5.2.5 Segregated Approach for Equation Solution

In Figure 5.1, a flowchart summarizes the main operations performed by a typical computational fluid dynamics (CFD) solver for the simulation of IC engines at any crank-angle θ, the mesh is moved and, if necessary, equations for the liquid phase are solved with the Lagrangian approach. Afterward, conservation equations for mass, momentum, energy, and composition are sequentially solved using the so-called segregated approach. Three different loops, one inside the other, are used to achieve a smooth convergence of the flow solution at each time-step and ensure consistency between the different state variables of the system. The first loop is called *outer loop* and includes solution of mass conservation equation and creation of the matrix of coefficients from the momentum equation which will be used in the pressure-velocity coupling stage. In the *inner loop* energy and composition equations are solved, and local cell temperature is then estimated accordingly. The pressure-velocity coupling stage is then performed to ensure that both pressure and velocity fields satisfy continuity [26]. Since a diffusion term appears in the pressure equation, the suggested number of iterations (n_{NO}) required for the pressure-velocity coupling stage increases with the mesh non-orthogonality. To ensure convergence and stability, it is suggested to use more than one iteration for the outer (n_{out}) and inner (n_{inn}) loops. Moreover, under-relaxation is generally adopted in equation solution for all the outer iterations except the last one.

FIGURE 5.1 IC engine flow and combustion solver flowchart.

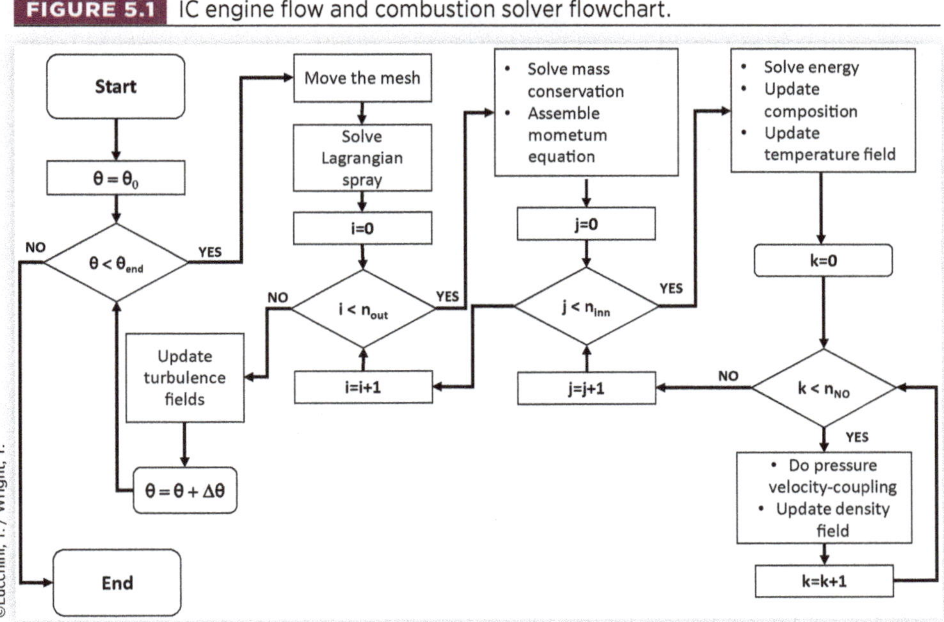

CHAPTER 5

FIGURE 5.2 Variation of a generic quantity f with crank angle at a fixed location inside the cylinder. Continuous line, instantaneous value; dashed line, ensemble averaged value.

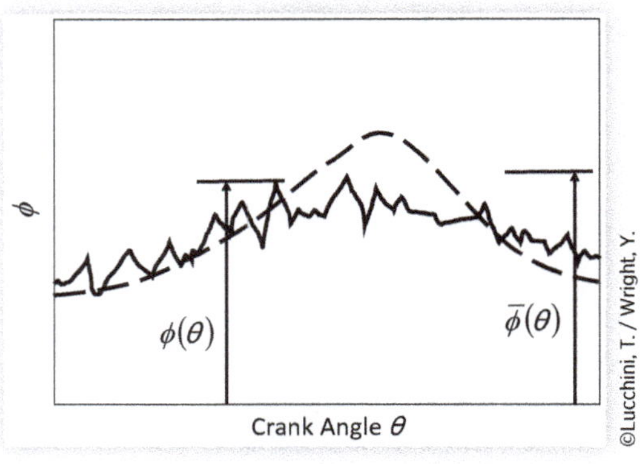

5.3 Turbulence Modeling

In-cylinder flows are turbulent and it is widely recognized that turbulence itself is probably the most complex phenomenon in non-reacting fluid mechanics. Turbulent flows are highly unsteady, random, and three-dimensional characterized by large fluctuations both in length and time scales. The literature about turbulence is enormous and probably proportional to the difficulty of the task. The reader is referred to [35, 36] for in-depth descriptions of turbulent flows and the different modeling approaches.

Figure 5.2 reports the evolution during one engine cycle of a generic quantity ϕ at a specified location inside the cylinder. The continuous line represents the measured instantaneous value of ϕ, while the dashed one is the ensemble average over a large number of experiments for a specific crank angle.

$$\bar{\phi}(\theta)=\frac{1}{N}\sum_{i=1}^{N}\phi_i(\theta) \tag{5.5}$$

The intensity of the fluctuation ϕ' around the mean value $\bar{\varphi}$ is computed for any crank angle as:

$$\phi'(\theta)=\sqrt{\frac{1}{N}\sum_{i=1}^{N}\left[\phi(\theta,i)-\bar{\phi}(\theta)\right]^2} \tag{5.6}$$

Despite NS equations are generally valid for both laminar and turbulent flows, solving them on a grid which is fine enough to resolve the smallest length scales of the flow problem (DNS, which can be considered a "numerical experiment") [37] is still prohibitively expensive for practical engine simulations. Nonetheless, DNS application to simplified but increasingly engine-relevant configurations is starting to emerge [16, 18, 19, 38, 39, 40, 41, 42, 43, 44, 45].

For this reason, in engine CFD simulations, turbulence is usually modeled using two different approaches, namely, the *RANS* method which is widely adopted for industrial purposes and *LES*, although scale-resolving approaches (sometimes also denoted second-generation URANS) are also being increasingly adopted, see, e.g., [46].

5.3.1 RANS

In the RANS approach, the balance equations for mass-weighted averaged quantities are obtained by averaging the instantaneous NS equations. The averaged equations require closure rules: a turbulence model to deal with the flow dynamics and a combustion model to describe chemical species conversion. Solving these equations provide averaged quantities over different cycles for periodic flows like those found in IC engines. A detailed description of the RANS averaging procedure is provided in [29], and here only the main results will be presented.

Any quantity ϕ can be split into mass-averaged (or Favre-averaged) and fluctuating components $\tilde{\phi}$ and ϕ'', respectively:

$$\phi = \tilde{\phi} + \phi'' \quad \text{with} \quad \overline{\phi''} = 0 \quad \text{and} \quad \tilde{\phi} = \frac{\overline{\rho\phi}}{\overline{\rho}} \tag{5.7}$$

By performing the average of the momentum, energy, and species equation, unclosed terms appear. In particular, the average of the fluctuating velocity components in the momentum equation produces the so-called Reynolds stress term $\left(\widetilde{u_i'' u_j''}\right)$, which is modeled using the Boussinesq approach as follows:

$$\overline{\rho}\widetilde{u_i'' u_j''} = -\mu_t\left(\frac{\partial \tilde{u}_i}{\partial x_j} + \frac{\partial \tilde{u}_j}{\partial x_i}\right) + \frac{2}{3}\overline{\rho}\delta_{ij}k \tag{5.8}$$

where

μ_t is the turbulent viscosity

k is the turbulent kinetic energy defined as:

$$k = \sum_{i=1}^{3}\frac{1}{2}\widetilde{u_i'' u_i''} \tag{5.9}$$

The turbulence model provides suitable expressions for k and μ_t, which are then used for the solution of the momentum equation.

Turbulent fluxes $\widetilde{u_j'' e''}$ and $\widetilde{u_j'' Y_i''}$, resulting from averaging energy and species equation, respectively, are closed using a classical gradient assumption:

$$\widetilde{u_j'' e''} = -\frac{\mu_t}{\mathrm{Pr}_t}\frac{\partial e}{\partial x_j} \quad \text{and} \quad \widetilde{u_j'' Y_i''} = -\frac{\mu_t}{Sc_t}\frac{\partial Y_i}{\partial x_j} \tag{5.10}$$

where Pr_t (range: 0.85–1.0) and Sc_t (range: 0.7–0.9) denote the turbulent Prandtl and Schmidt numbers, respectively. In engine CFD simulations, two-equation turbulence models are generally used: one equation is solved for the turbulent kinetic energy k while the other one is solved for either the turbulent kinetic energy dissipation rate ε or the specific rate of dissipation ω.

The k-ε model, originally proposed in [20], is the most widely used in IC engine simulations because it is simple, reliable, and flexible in describing the different processes like generation of charge motions, fuel-air mixing, and combustion. In the original version, equations for k and ε are:

$$\frac{\partial \overline{\rho}k}{\partial t} + \frac{\partial \overline{\rho}\tilde{u}_i k}{\partial x_i} = \frac{\partial}{\partial x_i}\left[\left(\mu + \frac{\mu_t}{\sigma_k}\right)\frac{\partial k}{\partial x_i}\right] + P_k - \overline{\rho}\varepsilon \tag{5.11}$$

$$\frac{\partial \overline{\rho}\varepsilon}{\partial t} + \frac{\partial \overline{\rho}\tilde{u}_i\varepsilon}{\partial x_i} = \frac{\partial}{\partial x_i}\left[\left(\mu + \frac{\mu_t}{\sigma_\varepsilon}\right)\frac{\partial \varepsilon}{\partial x_i}\right] + C_{\varepsilon 1}\frac{\varepsilon}{k}P_k - C_{\varepsilon 2}\overline{\rho}\frac{\varepsilon^2}{k}P_k \tag{5.12}$$

where the source term P_k is given by $-\bar{\rho}\widetilde{u_i'' u_j''}\frac{\partial \tilde{u}_i}{\partial x_j}$ and the Boussinesq expression is used to compute the Reynolds stress. The turbulent viscosity is computed as:

$$\mu_t = \bar{\rho} C_\mu \frac{k^2}{\varepsilon} \tag{5.13}$$

The model constants are usually $C_\mu = 0.09$; $\sigma_k = 1.0$; $\sigma_\varepsilon = 1.3$; $C_{\varepsilon 1} = 1.44$ and $C_{\varepsilon 1} = 1.92$.

The $C_{\varepsilon 1}$ constant is suggested to be increased up to 1.6 for a better prediction of the momentum diffusion along the jet axis during the fuel-air mixing process.

Figure 5.3 and Figure 5.4 report RANS simulation results of the in-cylinder cold flow in two optical engines. In both the cases, the full cycle was simulated using deforming grids with a dominant hexahedral structure. Polyhedral cells are introduced to accommodate the complex combustion chamber shapes at mesh boundary. Figure 5.3 compares experimental and computed velocity fields halfway during the intake stroke (90 CAD after Top Dead Center (TDC)). Measurements were carried out using PIV and micro-PIV techniques [4] for a two-valve, pancake-shaped, transparent combustion chamber (TCC) engine running at 800 RPM. The measurement plane is located on the cylinder

FIGURE 5.3 Comparison between computed and experimental in-cylinder velocity magnitude distributions for a transparent combustion chamber engine halfway during the intake stroke.

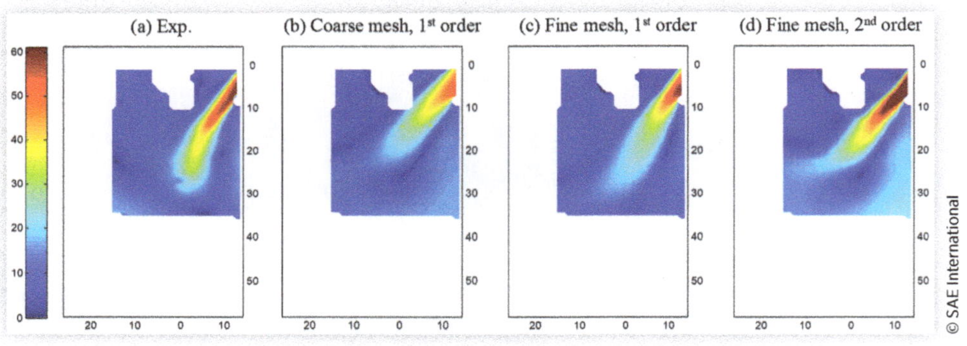

FIGURE 5.4 Comparison between computed and experimental in-cylinder velocity magnitude distributions for a transparent combustion chamber pent-roof engine halfway during the compression stroke.

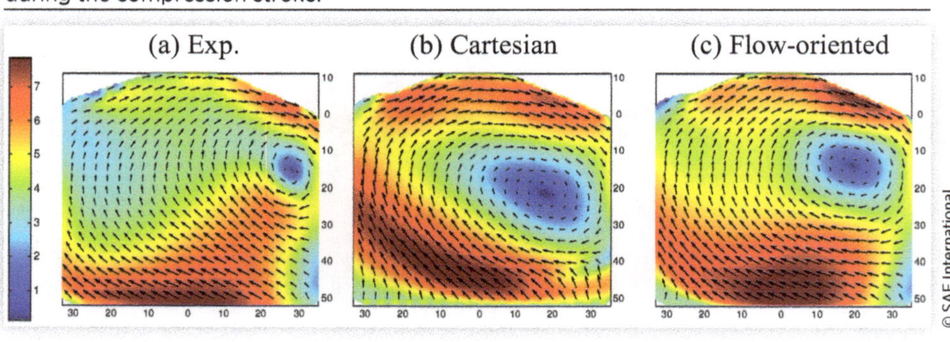

FIGURE 5.5 Comparison between computed and experimental in-cylinder velocity profiles (x and y components) along four horizontal lines located at different distances from the firedeck in the cylinder symmetry plane.

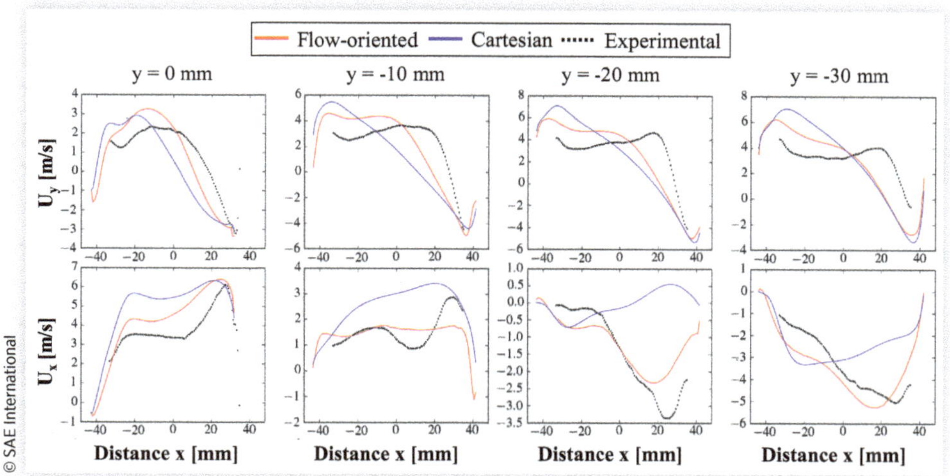

symmetry plane and passes through the intake valve axis. Effects of mesh size and numerical methods were investigated in [47]. In particular, it is possible to see that increasing mesh resolution and discretization accuracy improves the agreement between computed and experimental data.

Effects of mesh structure are reported in Figure 5.4, where a full-cycle under non-reacting conditions was simulated for a four-valve, pent-roof combustion chamber engine running at 850 RPM. The combustion chamber was designed to allow large optical access for a full characterization of in-cylinder flow and turbulence [2].

Experimental in-cylinder flow field on the cylinder symmetry plane is compared halfway during the compression stroke (90 CAD before TDC) with the computed one for a pure Cartesian and a hexahedral flow-oriented grid [48]. The flow-oriented grid better reproduces the location of the tumble vortex as well as its intensity.

The better performance of the flow-oriented grid is further confirmed by Figure 5.5, which shows a comparison between the computed and experimental velocity components along four horizontal lines located at different distances from the firedeck (0, −10, −20 and −30 mm).

5.3.2 LES

While unsteady RANS solves the ensemble-averaged variants of the NS equations, in lieu of averaging, LES applies spatial filtering, thereby the large-scale structures of the flow are resolved and modeling is only required for the unresolved part of the turbulent kinetic energy spectrum below the filter width (the so-called sub-grid scale; cf. Figure 5.6).

Several approaches are in use today to model the SGS, ranging from implicit LES (essentially no model), the broadly established Smagorinsky model,

FIGURE 5.6 Turbulent energy spectrum as a function of the wavenumber k with modeled and resolved scales for LES depending on filter size (mesh resolution) indicated.

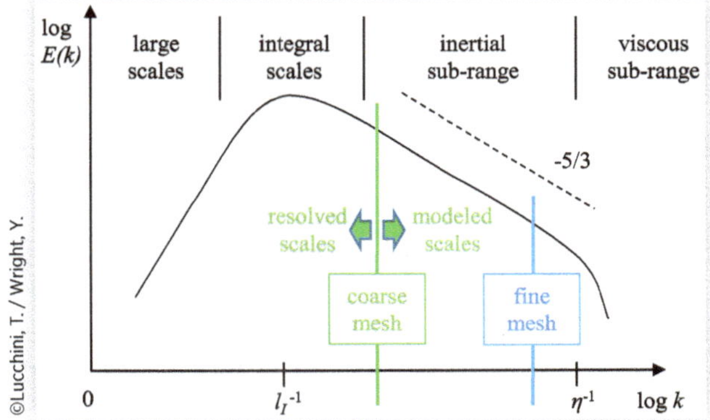

Smagorinsky with dynamic constant determination, models which solve a transport equation for the SGS turbulent kinetic energy [49], the sigma model [50], the WALE model [51], to the recently proposed dynamic structure model; for references and discussion, see also [25].

For FVM, the filter size is normally taken as the cubed root of the cell volume, in principle requiring reasonable cell aspect ratios. LES are typically performed on finer grids than RANS, since according to [35], at least 70% of the kinetic energy should be resolved when using LES; for alternative metrics, see [52].

Conceptually, LES has the following advantages over RANS:

- Part of the turbulence spectrum is resolved.
- The assumption of isotropy applies only to the sub-filter scales.
- Increasing the resolution will increase the amount of resolved kinetic energy and, at the limit of approaching resolution of the finest scales, LES will become a fully resolved calculation (corresponding to a DNS).

The main drawback over RANS is that a sufficient number of realizations need to be computed (e.g., flow-through times or engine cycles) to obtain converged statistics rendering the method computationally expensive. As discussed, e.g., in [54, 55], a large number of cycles may be necessary for converged first- and second-order moments, depending on the problem and quantity of interest. Aside from increased cell counts, which can be addressed by parallelization, fine grids required for LES also lead to a reduction of the integration time-step to satisfy the CFL criterion (this consideration also applies for locally refined grids), further increasing the computational cost of LES. Nonetheless, LES for IC engines is increasingly employed in industry, while it is well established in academic research.

Similar to RANS, special treatment is needed near walls, since the steep velocity and temperature gradients in the evolving boundary layers under IC engine conditions [3, 4, 5, 41, 45, 56] are difficult to resolve also for LES [57]. A range of approaches are in use, including "wall-resolved" methods with near-wall damping, wall functions specifically adapted to LES (e.g., [58]) as well as hybrid LES/RANS approaches [59, 60] (and references therein).

5.4 Modeling Spray Evolution

Spray processes play an important role in many technical systems and industrial applications. In port-fuel gasoline engines, the spray evolution governs the fuel-air mixture formation inside the intake manifolds and the cylinder. Spray pattern and timing allow to achieve either homogeneous or stratified combustion in GDI engines. In Diesel engines, fuel injection and spray evolution mainly controls the combustion process.

To model the spray to model the spray penetrating the combustion chamber, two phases need to be considered: the dispersed liquid and the continuous gas phase. The most straightforward approach would be using a very fine grid to describe the size distribution of the different droplets and properly track the liquid-gas interface. However, such approach can be used only in combination with LES or DNS turbulence modeling and is only feasible for the study of the liquid-jet breakup in simplified configurations due to the large amount of required computational resources [61, 62].

5.4.1 The Eulerian-Lagrangian Approach

To model the fuel-air mixing process in real engine geometries, the so-called Eulerian-Lagrangian approach is used. The main flow-field variables (density, temperature, velocity, etc.) of the gas phase are computed by the NS equations in conjunction with a turbulence model. The liquid phase is described in a Lagrangian way: the spray is assumed to be composed by a set of computational parcels, each of them contains a certain number of droplets with the same properties and evolves in the computational domain exchanging mass, momentum, and energy with the (Eulerian) gas phase [28]:

- **Mass conservation equation**

$$\frac{dm_d}{dt} = \dot{m}_{evap} \tag{5.14}$$

  m_d is the mass parcel mass and $\dot{m}_{evap}$ is the evaporation rate depending on a heat transfer coefficient and the fuel concentration at saturation conditions.

- **Momentum conservation equation**

$$m_d \frac{d\boldsymbol{U}_d}{dt} = \boldsymbol{F}_d + m_d \boldsymbol{g} \tag{5.15}$$

  $\boldsymbol{U}_d$ is the parcel velocity, $\boldsymbol{F}_d$, is the drag force acting on the parcel (which is function of the ambient density, drag coefficient, and droplet velocity) and $\boldsymbol{g}$ is the gravity acceleration.

- **Energy conservation equation**

$$\dot{m}_d h_d + m_d c_L \frac{dT_d}{dt} = \dot{m}_d h_v\left(T_d\right) + \dot{Q}_d \tag{5.16}$$

  h_d is the enthalpy of the parcel, c_L is the specific heat, T_d is the parcel temperature, h_v is liquid heat of evaporation, and $\dot{Q}_d$ is the heat exchanged between liquid and gas phase.

Source terms in the conservation equations of the gas phase enable the increase or decrease of mass, momentum and energy in each grid cell. The effect of the gas phase on the dispersed liquid is accounted for by using the actual data (temperature, gas velocity, etc.) of the grid cell the droplet is crossing at the current instant as a boundary condition [28]. With this approach, to correctly describe the spray evolution, additional sub-models are needed for:

1. *Injection*: specification of the initial size of the parcel, velocity, and direction. Generally two solutions are used: parcels are injected with constant diameter equal to the nozzle diameter (blob injection) or a droplet size distribution resulting either from experimental data or a nozzle-flow model [63] is considered. Velocity is derived from the injected mass flow rate profile, while the cone angle derives from semiempirical correlations or measurement.
2. *Breakup*: droplet diameter changes due to the surface instabilities related to turbulence and aerodynamic forces. Breakup models can be classified into primary and secondary breakup. Purpose of the first ones is the description of the liquid jet atomization with the formation of droplets on which secondary

breakup models act. Different breakup models were proposed over the years to consider the different nozzle geometries and breakup regimes [64, 65].
To account for the different forms of surface instability, breakup models are generally used in combination with the most common which are based on droplet instability theories such as Kelvin-Helmholtz and Rayleigh-Taylor [64], or Liquid Sheet Instability and Taylor Analogy [66] etc.

3. *Drag*: computation of the drag coefficient C_D. For a more realistic description, droplet surface deviation from spherical size can be included [28].
4. *Heat transfer*: correlations to compute the heat transfer coefficient affecting the heat transfer between the liquid and the gas phase.
5. *Evaporation*: computation of mass conservation equation under standard conditions 'with the fuel vapor pressure lower than the ambient pressure and under boiling conditions [67]. Suitable models were recently developed to realistically describe the evaporation of multicomponent fuels [68, 69].
6. *Collision*: droplets might interact among each other mainly in presence of sprays emerging from multi-hole nozzles under weak evaporating conditions. Different collision regimes can be identified: separation, coalescence, and shattering. Collision models are generally grid-dependent and many efforts were dedicated in the past to overcome such limitation [70, 71].
7. *Wall:* droplet impingement on the combustion chamber surfaces creating a liquid film. Impinging regimes are stick, rebound, spread, and splash. Correlations [72, 73] are provided to compute the droplet velocity after impingement as well as the diameter of the droplets emerging from the film after the splashing process.
8. *Turbulent dispersion*: the drag force is corrected with a turbulent velocity contribution to reproduce the effects of turbulent 'vortices on spray dispersion.

The Eulerian-Lagrangian approach is flexible and reliable and allows modeling of the spray evolution with affordable computational costs. However, Lagrangian spray simulations are grid-dependent since the quality of the computed results is related to the so-called void fraction parameter which is the percentage of cell volume occupied by the Lagrangian phase. An in-depth discussion of grid dependency (reported, among others, in [74]) is outside the scope of this chapter, and the reader is referred to [27, 28, 75, 76], where possible solutions to such problems are proposed.

This framework can also be combined with models describing formation and evolution of wall films [77], which is particularly important for impinging sprays during cold start leading to soot formation.

5.4.2 Spray Simulations for GDI Engines

In [78], CFD simulations were performed of a spray emerging from a six-hole nozzle under ambient conditions. The Huh-Gosman injection model was used to specify the spray cone angle, while the droplet velocity was corrected using the Nurick correlation [79] to account for the presence of cavitation inside the nozzle. The KHRT model is used to predict primary and secondary breakup, while spray-wall interaction was described by using the approach from Stanton and Rutland [73]. Different injection pressures were considered: Figure 5.7a illustrates the model capability to predict the spray penetration as function of the injection pressure. The coefficients of the breakup model were tuned for the 10 MPa injection pressure conditions. To validate the spray-wall interaction model, optical data of the spray impinging process were used. Figure 5.7b

FIGURE 5.7 (a) Comparison between computed and experimental spray penetrations as a function of the injection pressure. (b) Comparison between computed and experimental spray evolution at impinging conditions.

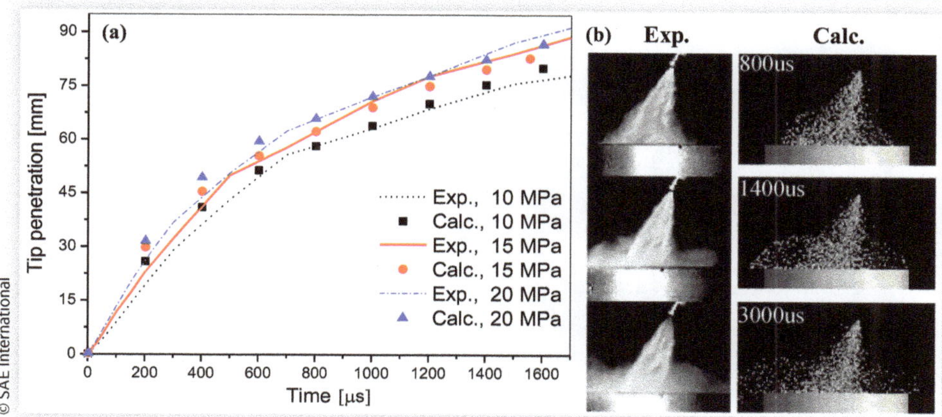

shows that the model properly reproduces the spray morphology, in particular the spray thickness over the liquid film.

5.4.3 Diesel Sprays

The "Spray A" experiment from the engine combustion network (ECN) [80] was simulated. *n*-dodecane is injected by a 90 μm k-nozzle at constant volume conditions under pressure and temperature values which are similar to those encountered in Diesel engines at start of injection (SOI) time [81]. Simulations were carried out on a two-dimensional, axisymmetric mesh, and the KHRT model was used for spray breakup. The spray and turbulence model constants were tuned to match liquid and vapor penetration for the baseline condition (T = 900 K, ρ = 22.8 kg/m^3), and Figure 5.8a shows the model assessment.

FIGURE 5.8 (a) Comparison between computed and experimental data of liquid and vapor penetration for the Spray A "baseline" condition (T = 900 K, ρ = 22.8 kg/m^3). (b) Comparison between computed and experimental radial distributions of mixture fraction at two different distances from the nozzle: 22.5 and 45 mm.

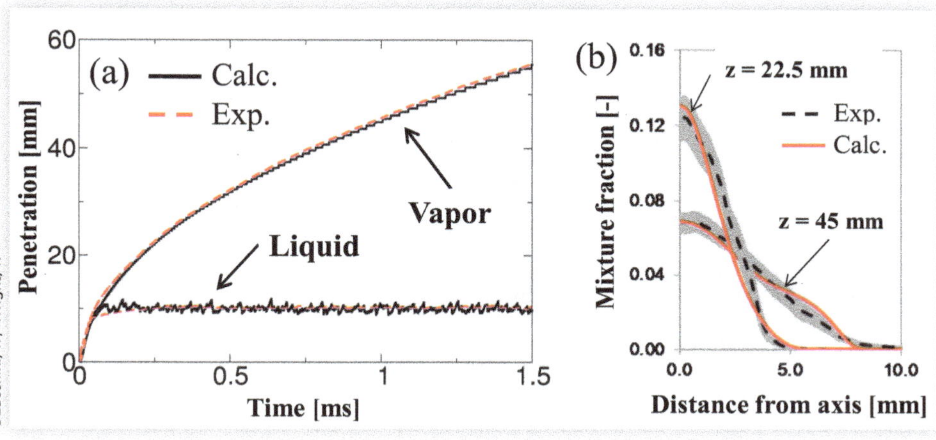

FIGURE 5.9 (a) Combined effects of ambient density and temperature variation on liquid spray and vapor penetration. Dashed line, experimental; continuous line, calculated; (b) effect of ambient temperature on liquid penetration.

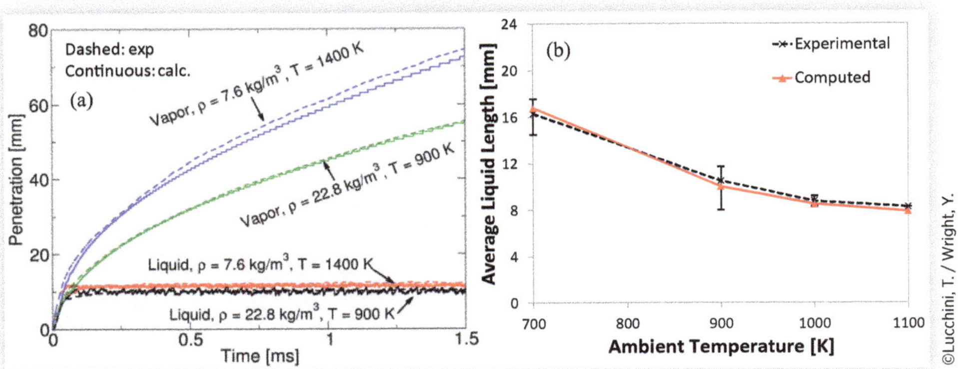

The capability of the proposed methodology to predict fuel-air mixing process in evaporating Diesel sprays was assessed by comparing the computed and experimental radial distributions of mixture fraction in Figure 5.8b at two different distances from the injector and a rather good agreement was achieved. The predictive capability of the proposed methodology was verified in Figure 5.9a-b with satisfactory results. Combined effects of ambient temperature and density variation were analyzed in Figure 5.9(a), while the dependency of the steady-state liquid length on the ambient temperature is illustrated in Figure 5.9(b).

In [82] attention was focused on the simulation of sprays for heavy-duty engines. The fuel is injected in a constant-volume vessel by a k-nozzle with a 205 μm diameter. Three different operating conditions were simulated at 900 K ambient temperature, including variation of ambient density and injection pressure (1: ρ = 22.8 kg/m^3, p_{inj} = 150 MPa; 2: ρ = 40 kg/m^3, p_{inj} = 80 MPa; 3: ρ = 40 kg/m^3, p_{inj} = 150 MPa). The Huh-Gosman model [63] was used to model the spray atomization, while secondary droplet breakup is described by means of the approach proposed by Pilch and Erdman [65]. A consistent comparison between computed and DBI (diffused back illumination) experimental data of liquid penetration was also developed. In particular, a light

FIGURE 5.10 (a) Left panel: 2-D optical thickness maps obtained with DBI experiments. Right panel: numerical reproduction of the optical thickness maps using simulated liquid spray data. (b) Comparison between experimental and calculated liquid penetration values.

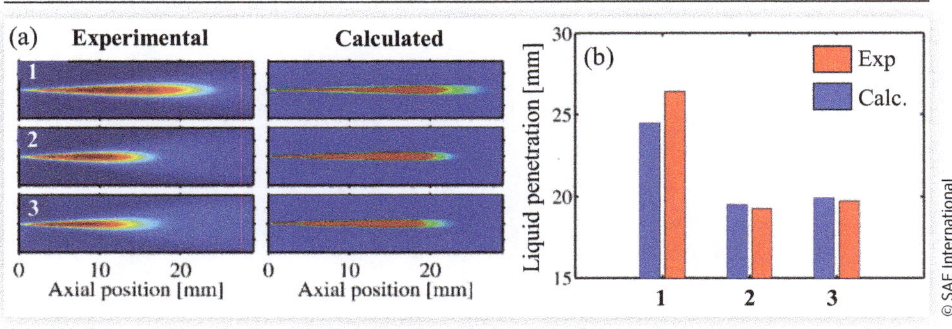

scattering model was implemented by the authors, and the steady-state liquid length was computed by processing experimental and computed optical thickness profiles in the same way. Figure 5.10a compares computed and experimental optical thickness maps for the three operating conditions. Steady-state spray penetration is correctly estimated by the simulations, despite its angle looks smaller compared to experiments. For the sake of completeness, Figure 5.10b compares computed and experimental data of liquid spray penetration directly obtained from DBI maps. Combined effects of ambient density and injection pressure are correctly predicted.

5.5 Turbulent Reacting Flows: An Overview

Compared to non-reacting single and two-phase flows, where conservation equations for momentum, continuity, and enthalpy need to be solved as outlined above, in the case of reactive flows, tens or hundreds of chemical species may be involved, participating in many chemical reactions (often hundreds or thousands, cf., e.g., [83, 84]). For each species, the corresponding transport equation can written as:

$$\frac{\partial(\rho Y_i)}{\partial t} + \nabla\cdot(\rho\mathbf{u}Y_i) - \nabla\cdot\left[\frac{\mu}{\mathrm{Sc}_i}\nabla Y_i\right] = \dot{\omega}_i \tag{5.17}$$

where Y_i denotes the species mass fraction and the source term on the RHS accounts for the production/consumption of the individual species due to chemical reactions.

For the oxidation of hydrocarbons—common to most combustion processes involving fuel and oxidizer—the highly complex chemical system is generally described by means of a mechanism with a number of chemical reactions. For a reaction k, the rate of reaction can be calculated as:

$$\dot{\omega}_k = k_{fk}\prod_{i=1}^{n}\left(\frac{\rho Y_i}{W_i}\right)^{\nu'_{ik}} - k_{bk}\prod_{i=1}^{n}\left(\frac{\rho Y_i}{W_i}\right)^{\nu''_{ik}} \tag{5.18}$$

where the exponent ν_{jk} denotes stoichiometric coefficients in forward (′) and backward (″) directions for elementary reactions involving a number n of chemical species each one of them having molecular mass W Likewise, k_{fk} and k_{bk} denote the rate coefficients in forward and backward directions, which in general, are described by Arrhenius expressions. In the modified form shown here, the forward rate coefficient is expressed as a function of its pre-exponential factor A_{fk}, a pre-exponential temperature dependence and an exponential dependence on the activation temperature T_{Afk}:

$$k_{fk} = A_{fk}T^{b_{fk}}\exp\left[-\frac{T_{Afk}}{T}\right] \tag{5.19}$$

CHAPTER 5

For species i, the rate of change is obtained from summation over the total number of reactions (r) as:

$$\dot{\omega}_i = W_i \sum_{k=1}^{r} \dot{\omega}_k \left(\nu''_{ik} - \nu'_{ik} \right) \tag{5.20}$$

From Equations (5.18) and (5.19), it is evident that the individual species consumption and production rates are nonlinear functions of species concentrations and, in particular, of temperature. The chemical source terms appearing in the species transport equation (5.17) can therefore not simply be evaluated using the Favre-averaged (in the case of RANS) or filtered (for LES) temperature and species mass fraction due to the strong nonlinearity of the rate expressions [53, 85, 86]. Following [87], conceptually for the under-resolved calculation of turbulent reacting flows, models are needed for the turbulence, the chemistry, the effect of turbulence on chemistry, and vice versa. Even if for LES a higher resolution will increase the level of the resolved scales of the flow as discussed above, modeling is nonetheless required since combustion occurs at the molecular level, i.e., at the modeled scales.

In the following, a selection of combustion models which are well established for different combustion modes will be presented, which address these difficulties. These approaches cover the range of models derived from conceptual considerations to statistical methods.

5.6 Combustion in Diesel Engines

Diesel combustion is a complex process to model, since it is determined by physical and chemical phenomena related to mixture formation, autoignition, and mixing-controlled combustion. Conceptual models describing the main features of the flame structure have been derived from detailed optical investigations in engines and constant-volume vessels [88,89]. From them, different approaches were proposed over the years to model combustion in Diesel engines. They can be mainly classified depending on the way fuel oxidation and mixing effects are described:

- Simplified chemistry vs detailed chemistry.
- Turbulence-chemistry interaction vs well-stirred reactor models

The next sub-paragraphs present three different approaches: they all include turbulence-chemistry interaction. Simplified chemical kinetics is included only by the first one.

5.6.1 The Characteristic Time-Scale Combustion (CTC) Model

The Characteristic Time-Scale Combustion (CTC) model considers a limited number of chemical species (fuel, O_2, N_2, CO_2, CO, H_2O, H_2) and provides suitable expressions for chemical species reaction rates under autoignition and mixing-controlled combustion. Auto-ignition (AI) reaction rate $\dot{\omega}_{AI}$ is computed by means of the Shell model, which is an eight-step reduced mechanism which is able to describe the main features of hydrocarbon autoignition [90]. A characteristic time scale τ_c governs mixing-controlled combustion reaction rate for any chemical species $\dot{\omega}_{MIX,i}$:

$$\dot{\omega}_{MIX,i} = -\frac{\left(Y_i - Y^*\right)}{\tau_c} \tag{5.21}$$

where

Y_i is the actual species mass fraction
Y_i^* is the corresponding one at equilibrium conditions
τ_c includes laminar chemistry and turbulent time scales as follows:

$$\tau_c = \tau_l + f\tau_t \tag{5.22}$$

where τ_l is derived by a single-step reaction expression for hydrocarbon fuel oxidation while τ_t is proportional to the turbulent time scale k/ε. f represents the progress of combustion being equal to one when all the fuel is converted into products. A temperature threshold value, normally set to 1100 K, is used to switch from autoignition to mixing-controlled combustion. To correctly match the autoignition time and heat release rate during the mixing-controlled combustion phase, the CTC model requires to perform some preliminary tuning both on the shell model constants and the parameter τ_t. Detailed description of the CTC is provided in [91, 92], where it was applied for both CI and spark ignition (SI) combustion. The CTC model is reliable and fast. However, its predictive capabilities are drastically limited by the use of reduced chemistry to describe autoignition and flame structure: this represents an important limitation in case a wide range of operating conditions has to be simulated with different exhaust gas recirculation (EGR) rates and complex injection strategies. This is the reason why currently the CTC model was almost completely replaced by approaches based on detailed chemistry.

5.6.2 Multiple Representative Interactive Flamelet Model (mRIF)

The Multiple Representative Interactive Flamelet (*mRIF*) model is based on the laminar flamelet concept: the turbulent time and length scales are much larger than the chemical ones and reactions occur in a locally undisturbed sheet which can be treated as an ensemble of stretched counterflow diffusion flames called *flamelets* [93, 94]. All the reacting scalars only depend on the mixture fraction variable, Z, which is related to the local fuel-to-air ratio for non-premixed combustion. Hence, local chemical composition can be estimated from the Z field in the CFD domain, assuming that its sub-grid distribution can be represented by a β-PDF. To this end, transport equations for both Z and its variance Z''^2 need to be solved [29, 53]. Mixing is enhanced by the scalar dissipation rate χ which is a function of the turbulent time scale and mixture fraction variance:

$$\chi = C_\chi \frac{\varepsilon}{k} \widetilde{Z''^2} \tag{5.23}$$

where the constant C_χ is normally set to 2. The local flame structure is defined by the flamelet equations that are solved assuming unity Lewis number [95] in mixture fraction domain:

$$\rho\frac{\partial Y_i}{\partial t}=\rho\frac{\chi}{2}\frac{\partial^2 Y_i}{\partial Z^2}+\dot{\omega}_i \tag{5.24}$$

$$\rho\frac{\partial h}{\partial t}=\rho\frac{\chi}{2}\frac{\partial^2 h}{\partial Z^2}+\dot{\omega}_h+\frac{dP}{dt} \tag{5.25}$$

where

Y_i denote the mass fraction of species i
ρ is the density
h is the enthalpy

The RHS of Equation (5.24), $\dot{\omega}_i$, is the chemical source term of species i due to chemical reactions. In addition to the enthalpy source term, $\dot{\omega}_h$, the second term on the RHS of Equation (5.25) is due to the rate of change of pressure in the closed combustion chamber.

Equations (5.24) and (5.25) are solved on a one-dimensional (1D) mesh, and effects of mixing related to turbulence and flow field are grouped into the scalar dissipation rate term χ having an erfc function profile in the mixture fraction space (AMC model [96]) and proportional to the average stoichiometric scalar dissipation rate value for any flamelet. The use of multiple flamelets [97] makes possible to better account for the effects of the local flow and turbulence on the flame structure and to predict flame stabilization. To this end, the distribution of the so-called flamelet marker fields M_j has to be computed by solving specific transport equations [97, 98].

Figure 5.11 summarizes the operation of the *mRIF* combustion model, illustrating the mutual interactions between the CFD and the flamelets domains. At each time-step, average stoichiometric scalar dissipation rate values are passed to each flamelet that solves Equations (5.24) and (5.25) accordingly. The chemical composition in the CFD domain is computed from the mixture fraction, its variance, and the flamelet marker distribution. Temperature is updated from new chemical composition and total enthalpy, whose variation is only due to flow and spray evaporation. For further information, the reader is referred to [95].

FIGURE 5.11 mRIF combustion model schematic and interfacing to CFD solver.

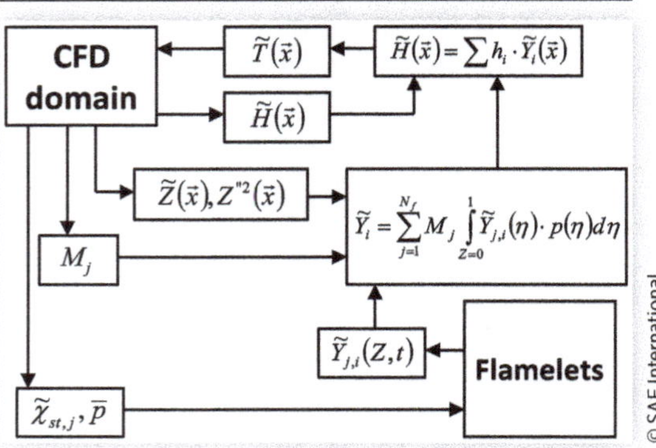

To predict pollutant emissions, specific sub-models are incorporated in mRIF, as discussed in [99, 100]. Concerning prediction NO_x, they are assumed to be only NO and generally only the thermal formation is considered. Two different approaches are used. In the first one, the concentration of NO is directly extracted from the flamelet domain as it happens for the other chemical species. In the second one, a transport equation for NO mass fraction is solved in the CFD domain, and its reaction rate is directly taken from the flamelet domain presuming a β-PDF. Semiempirical models are employed for soot: in the CFD domain, transport equations are solved for soot volume fraction and particle number density considering inception, surface growth, coagulation, and oxidation steps. Normally acetylene (C_2H_2) is assumed to be the soot precursor, and its concentration is taken from the flamelet domain [101, 102].

5.6.3 Light-Duty Engines

In [103], the CTC model was applied to optimize the injection strategy and the EGR rate for a passenger-car, common-rail, Diesel engine. Simulations were carried out in a one-eighth sector of the combustion chamber with a spray-oriented grid to minimize the numerical diffusion. The grid size ranges from 58,000 cells at IVC to 8,000 cells at TDC. Concerning pollutant emissions, the extended Zeldovich mechanism was used to predict NO, assumed to be dominant NOx species, while soot was estimated by the five-step reduced model suggested by Fusco [104] including formation of soot precursor (acetylene) from fuel, soot inception, surface growth, and coagulation. The purpose of the investigation was the reduction of pollutant emissions. A postinjection was used at full-power conditions to enhance the oxidation of soot, while at mid-high load different strategies were considered including variation of the EGR rate and advanced injection profile. The CTC model constants were tuned to match the in-cylinder pressure and heat release rate profiles at full-power condition, and Figure 5.12a reports the model assessment. Effects of post-injected Diesel fuel mass on soot and NOx emissions are shown in Figure 5.12b: it is possible to see that simulation agrees with experimental data in terms of the amount of fuel to be delivered in the postinjection event to minimize the soot emissions.

The same model setup was then applied to evaluate the combined effects of EGR rate and injection strategy at medium-high load conditions, normally occurring during acceleration in the NEDC cycle. Figure 5.13c displays that both experiments and CFD show that increase of EGR corresponds to higher soot and lower NOx emissions when using the conventional injection strategy consisting in pilot, main, and post-injection (see Figure 5.13a). At the highest EGR rate, a further reduction of soot emissions is possible by changing the injection strategy with the so-called *boot* injection profile which is shown in Figure 5.13b. The influence of the injection strategy is correctly predicted by the CTC model: for the same amount of EGR rate, soot is reduced.

FIGURE 5.12 (a) Application of the CTC model for prediction of Diesel combustion: comparison between computed and experimental heat release rate profiles; (b) effects of post-injected amount of fuel mass on soot and NO_x emissions prediction for a passenger-car Diesel engine using the CTC model.

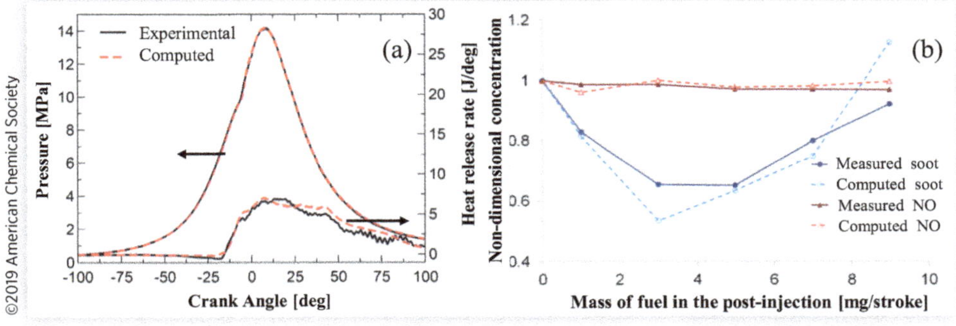

FIGURE 5.13 Effects of multiple injection configuration on emissions applying the characteristic time-scale combustion model, Fusco model for soot, and Zeldovich mechanism for NO_x.

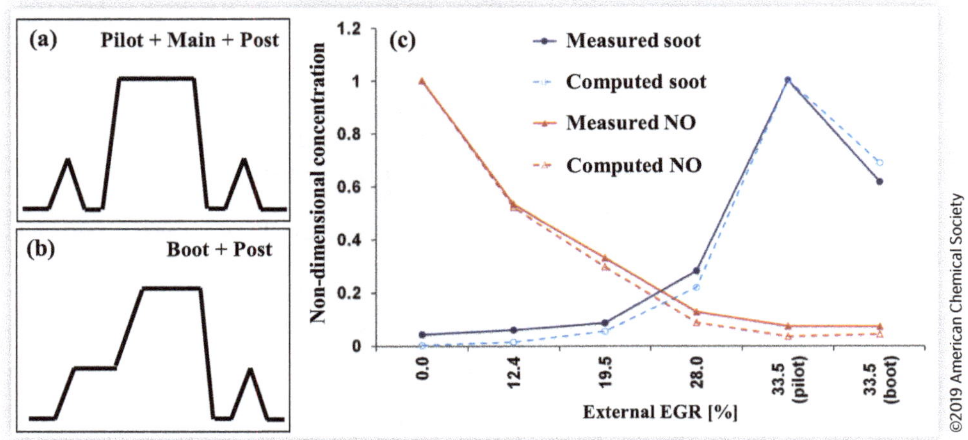

5.6.4 Sandia Optical Engine for Heavy-Duty Applications

Although restricted in terms of engine speeds and cylinder peak pressure levels, optical engines have proven tremendously useful to study combustion. Especially in combination with high-speed imaging, temporally and spatially resolved measurements of a variety of properties of interest are possible, providing insights into spray plume evolution, mixing processes, the low- and high-temperature ignition event and subsequent high-temperature fuel conversion producing soot and the interdependency of all these processes. Compared to spray combustion chambers, the time-varying geometry introduces a p/T evolution, allowing for investigations into the effects of phasing of the injection/combustion event. As discussed, e.g., in [105] studying four prototypical operating condition of the Sandia heavy-duty engine, an accurate description of the entire process chain is paramount for accurate prediction of pollutants. Tracer-based LIF diagnostics was used to validate the mixing field evolution and the combustion event for both "classical" Diesel combustion as well as low-temperature combustion (Figure 5.14 and Figure 5.15).

A CMC [106] combustion model enabled excellent prediction of ignition time/location and heat release rate, forming the basis for good predictions concerning also total in-cylinder soot volume evolutions as well as soot spatial distributions, as shown in the right panels of Figure 5.14 and Figure 5.15.

In [107, 108] the study was extended to sweeps in TDC temperature, SOI, and EGR, reporting excellent predictions for a wide range of conditions. Developments reported in [109] enabled the extension to multiple injections to study in particular the effect of postinjections on soot oxidation as well as the impact of EGR under such conditions [110, 111].

Transported PDF methods [112, 113] are also being used to simulate spray combustion, see, e.g., [114, 115].

FIGURE 5.14 "Classical" Diesel combustion with injection shortly before TDC. From [105].

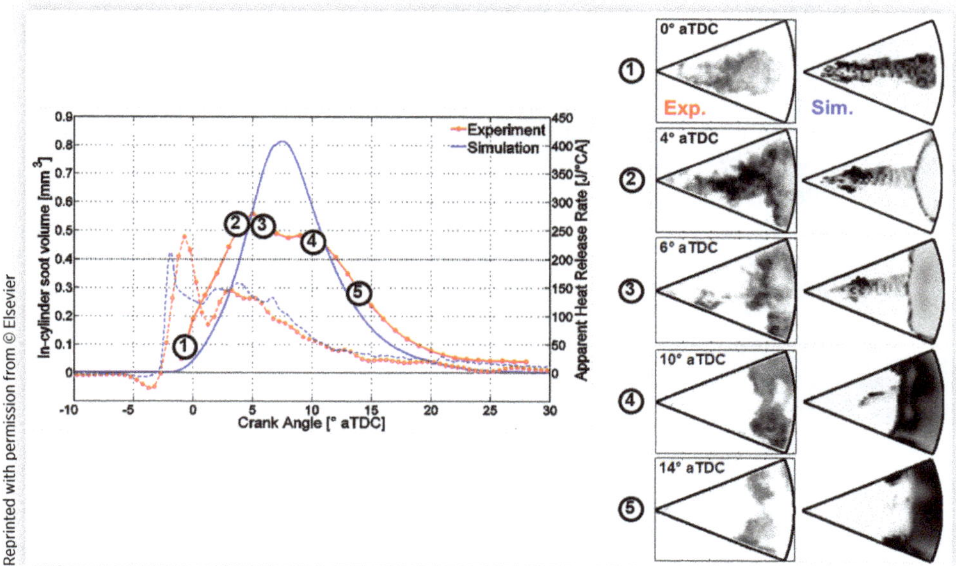

FIGURE 5.15 Low-temperature combustion early injection combustion mode. From [105].

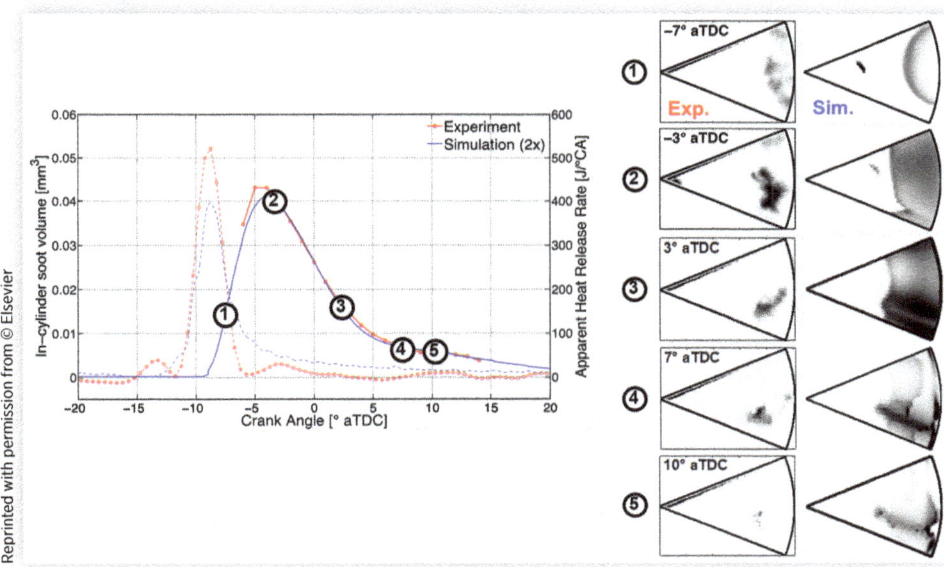

5.6.5 Heavy-Duty Engines

The validation of the *mRIF* model applied to a small-bore, heavy-duty engine was performed considering 14 operating points at different load and speeds as illustrated on the engine map reported in Figure 5.16a. The engine computational mesh is reported in Figure 5.16b: it is possible to see that the grid is spray-oriented and also progressively refined to ensure sufficient mesh resolution to capture the flame structure even far from the nozzle. The KHRT model was used to describe the spray evolution and, in all the considered operating points, fuel is delivered by means of a single injection event.

The *mRIF* capabilities to reproduce in-cylinder pressure evolution are reported in Figure 5.17. For any operating point, Figure 5.17a compares the computed maximum cylinder pressure with the corresponding experimental one. The agreement is rather good for the

FIGURE 5.16 Map of operating points used for the validation of the mRIF combustion model on a heavy-duty Diesel engine (a); (b) computational spray-oriented mesh used for the simulations.

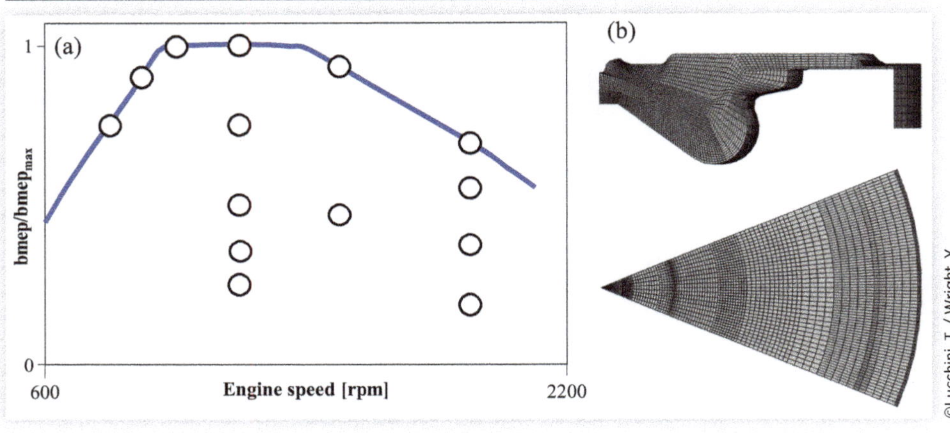

FIGURE 5.17 Validation of the RIF model applied to heavy-duty Diesel engine combustion (a) computed vs experimental cylinder peak pressure; (b) computed vs experimental gross indicated work (normalized with respect to the maximum experimental one).

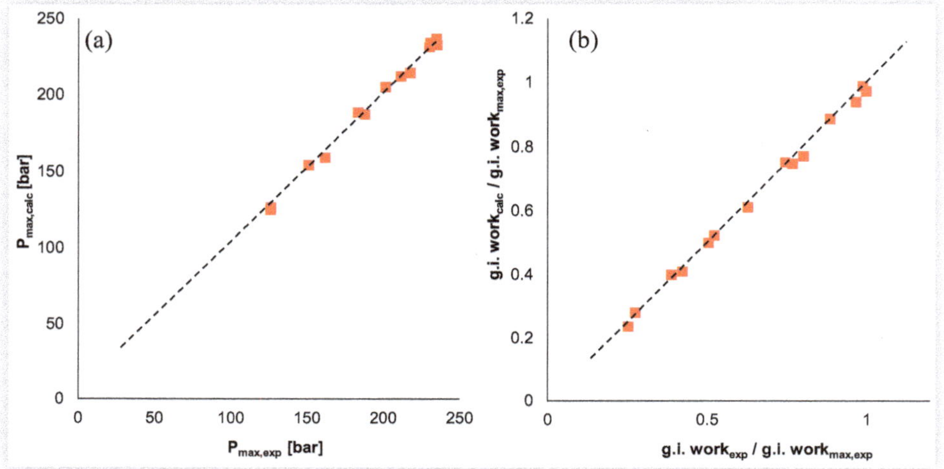

FIGURE 5.18 Validation of the RIF model applied to heavy-duty Diesel engine combustion (a) computed vs experimental NOx emissions using two different models; (b) computed vs experimental CO emissions.

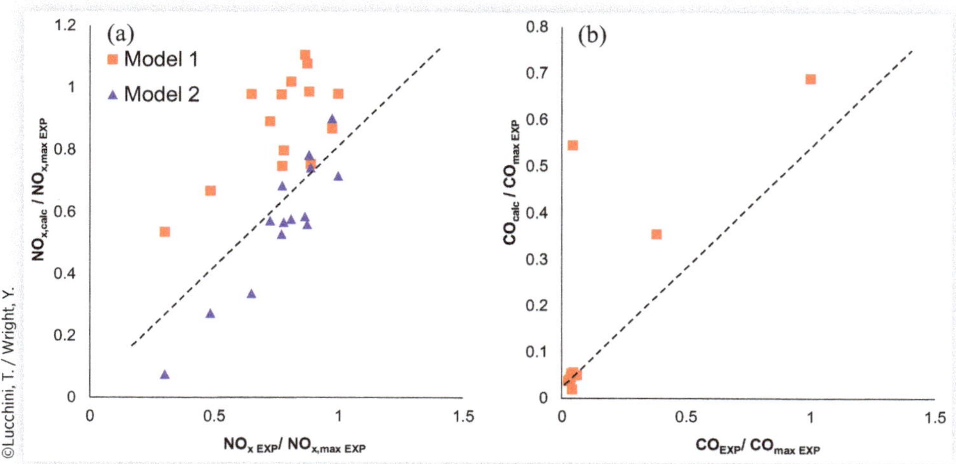

different tested speeds and load. A similar chart is reported in Figure 5.17b for what concerns the gross indicated work. The rather good agreement between computed and experimental data shows that the *mRIF* model can be applied as a predictive tool to estimate the engine performance. Prediction of pollutant emissions is illustrated in Figure 5.18. In particular, Figure 5.18a shows that both the models used are able to reproduce the experimental trend of NO. When NO concentration is directly taken from the flamelet domain (Model 1), its evolution depends on flamelet temperature history and mixture fraction distribution. For this reason, the time scales that are typical of NO formation are neglected, and the predicted concentrations are higher than the experimental ones. When a transport equation is solved to computed NO concentration (Model 2), the reaction rate is consistent with the mRIF model, and the effects of flow are correctly taken into account. However, this last model cannot completely consider the effects of local NO concentration and temperature distribution [99]. Consequently, predicted concentrations are lower than the ones from Model 1 but, in this specific case, closer to experimental data. CO concentration is directly estimated from the flamelet domain, and Figure 5.18b illustrates that for most of the operating points, the agreement between computed and experimental data is rather good.

5.6.5.1 MARINE ENGINES

While the four-stroke marine engines are similar to their heavy-duty counterparts discussed above, two-stroke engines used for the largest vessels differ considerably:

I. Ratios between stroke and cylinder bore are typically around 4, necessitating a crosshead linking the connecting rod to the crank shaft, enabling engine speeds down to 60 RPM allowing for direct drive of very large screws.

II. Contemporary two-stroke engines employ uniflow scavenging, with one centrally located exhaust valve in the cylinder head and ports at the bottom of the liner for the incoming scavenge air. Fuel injectors hence need to be placed at the circumference at two or three locations depending on engine size. Due to the strong swirling air motion present in two-stroke engines, this results in a jet-in-cross-flow configuration and strong interaction between the reactive jets of the individual injectors [116].

III. Very large injector orifice diameters of the order of 1 mm are required to admit fuel quantities of the order of tens of grams per stroke and cylinder leading to very large spray penetration length [117].

CHAPTER 5

FIGURE 5.19 Lift-off length versus liquid length for automotive vs heavy-duty injectors. From [118].

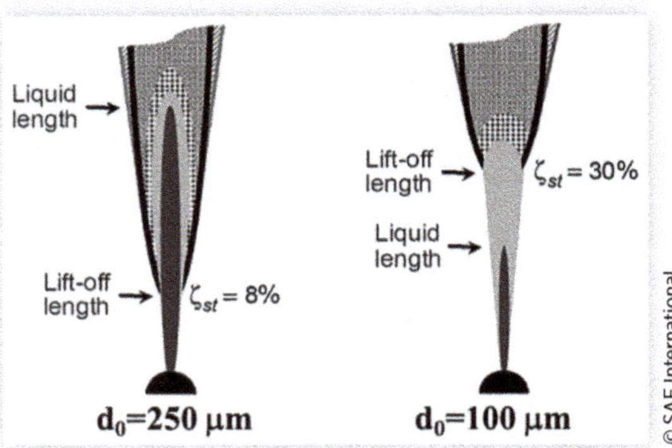

As discussed in [118], cf. Figure 5.19, while the liquid length is roughly proportional to the nozzle diameter, the flame lift-off length does not scale accordingly, leading to flame/spray overlap for larger injector diameters necessitating also special combustion modeling considerations [119, 120] (Figure 5.20 (right)).

In contrast to smaller heavy-duty configurations, spray ignition typically does not occur near the spray tip but at the leeward side of the spray as shown in [121, 122] (cf. Figure 5.20 (left)) and is only weakly sensitive to the orifice diameter.

Fuel used in marine engines typically contain considerably longer-chain hydrocarbons and up to several percent sulfur. The strongly varying composition of fuels from different bunkers in addition complicates modeling thermophysical properties of the liquid. The low volatility and considerable flame/spray overlap results in substantial fuel pyrolysis with poorly understood chemical kinetics and leads to considerable soot formation and radiation, the latter were estimated to account for up to 50% of the entire wall heat transfer in large engines [123]. Recent investigations of soot radiation at ECN "Spray A" conditions are reported in [124] and for heavy-duty engines in [125], while extensions thereof to two-stroke marine dimensions are underway.

5.7 Combustion in Spark Ignition (SI) Engines

As for Diesel engine combustion, modeling is required also in the case of premixed combustion; flame thicknesses computed for freely propagating laminar flames at engine-relevant conditions amount to the order of tens of micrometers, evidently well below any resolvable scale in engineering-type full-cycle engine simulations. In the following, the challenges pertaining to premixed combustion are illustrated, highlighting the need for sophisticated ignition treatment as well as the models typically employed for turbulent flame propagation.

FIGURE 5.20 Comparison of simulated liquid (blue) and flame (red) spray regions for a sweep in injector diameters from 0.2 to 1.2 mm at the time of ignition ~1.2 ms after SOI (left) and 6ms after SOI during the quasi-steady period (right). The grey isoline denotes the stoichiometric mixture fraction. From [119].

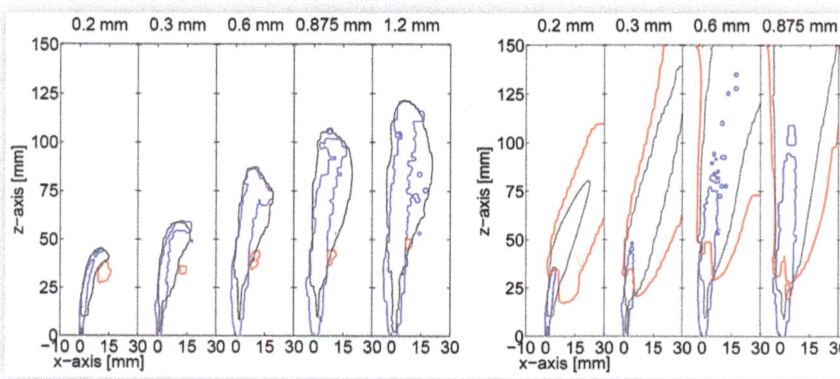

FIGURE 5.21 Peak pressure versus the duration between ignition and 2% converted fuel for all calculated cycles and configurations. From [55].

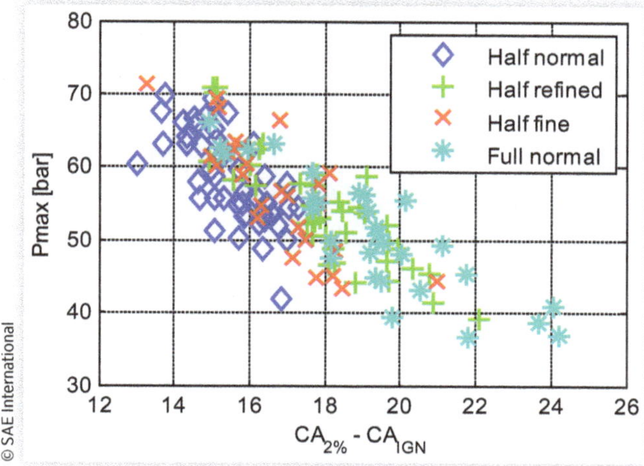

Combustion in spark-ignited engines is characterized by significant cycle-to-cycle variation, as documented, for example, in the early experimental work of Hill [126]. As shown also in [127], this early phase is particularly susceptible to the mixture reactivity, e.g., changes to equivalence ratio. Similar observations have also been made for large EGR rates [128]. Numerical studies [55, 129] confirm that the peak pressure strongly correlates with the first few percent of total energy released (e.g. CA02 or CA05), corresponding to the early flame kernel development phase, as illustrated in Figure 5.21.

The accurate modeling of ignition hence plays an important role as will be discussed next.

5.7.1 Ignition Modeling

Conventional spark plug ignition is characterized by a large number of complex processes, including breakdown of a high voltage across the spark plug electrodes leading to a pressure wave and the subsequent formation of a plasma channel which is partially ionized [130, 131]. Thermal expansion leads to significant growth of the established kernel, which is subject to interaction with the surrounding flow field. Aside from the mixture reactivity discussed above, the impact of large-scale bulk motion as well as turbulence length and velocity scales varying between individual engine cycles plays a decisive role with respect to the transition toward successful establishment of a propagating turbulent flame.

In the early work of Herweg et al. [132], a 1D model was proposed, which has been widely used in the engine community and seen successful application also to spherically expanding turbulent flames in combustion vessels and lean-burn gas engines, see, e.g., [133]. In the DPIK model proposed in [134], Lagrangian markers were introduced to track the surface of the expanding flame kernel which is initialized in the spark plug gap. Once a critical radius is reached (which is related to the integral length scale), a CTC model is initialized. The model was reported to improve combustion predictions of the studied SI engines considerably. The AKTIM model [135] developed in the context of a coherent flame model also uses Lagrangian markers to track the flame kernel; however the particles interact with the flow field. Eulerian concepts have also been used, cf. [136]. For the level-set combustion modeling framework, corresponding developments are reported in [137] and the SparkCIMM model presented in [138, 139]. Recent LES/DNS of ignition processes are reported in [130].

As discussed in [140, 141], for accurate predictions of the early flame phase, to accurately represent the ignition process, the following phenomena require accurate modeling:

1. Energy transfer to the gas phase from the electric circuit
2. Thermal expansion of the ignition kernel after the spark discharge
3. The subsequent transition to flame propagation
4. Effects of local flow conditions leading to kernel displacement and possible restrike

5.7.2 Flame Propagation Modeling

Variants of the CTC model—similar to the one described in Section 5.6.1, "The Characteristic Time-Scale Combustion (CTC) Model"—have also been used for premixed combustion in early applications. Later developments for propagating turbulent premixed flames in IC engines were derived using topological considerations since combustion in open chamber spark plug-ignited IC engines occurs predominantly in the thin flames or corrugated flames regime (cf. Figure 5.22) where the axis correspond to the length scale ratio and velocity scale ratio of the respective chemical (s_L^0 and δ_{th}, the laminar flame speed, and flame thickness) and flow quantities (u' and l_I, the turbulence intensity, and integral length scale):

Common to most approaches is the solution of a transport equation of the reaction progress variable (c), which contains a source term due to combustion:

$$\frac{\partial \bar{\rho}\tilde{c}}{\partial t} + \nabla \cdot \bar{\rho}\tilde{\mathbf{u}}\tilde{c} - \nabla \cdot \left(\overline{\rho D_c \nabla \tilde{c}} - \overline{\rho \mathbf{u}'' c''} \right) = \overline{\dot{\omega}} \tag{5.26}$$

FIGURE 5.22 Regime diagram with typical domains for premixed combustion in gas turbines (GT) and IC engines. Color trajectories illustrate length- and velocity-scale ratio evolutions throughout the combustion event, averaged at the flame front in two different SI engines. Adapted from [165].

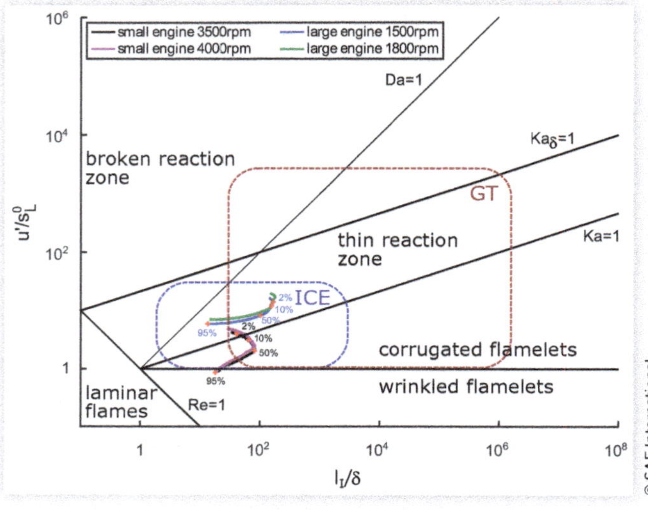

Solutions to provide a closure for the strongly nonlinear chemical source term (RHS of Equation (5.26)) include the Weller flame area model [142], which employs a so-called wrinkling factor Ξ:

$$\overline{\dot{\omega}_{Weller}} = -\rho_u S_u \Xi |\nabla b| \tag{5.27}$$

S_u and ρ_u denote the laminar flame speed and unburnt density, respectively, in conjunction with $|\nabla b|$ representing the absolute of the gradient of the regress variable (1-c). The wrinkling factor can be obtained either by simple algebraic correlations or by solving an additional transport equation.

Flame surface density models [143] calculate the source terms as:

$$\overline{\dot{\omega}_{FSD}} = -\rho_u \langle S_c \rangle_s \Sigma \tag{5.28}$$

where the turbulent consumption speed $\langle s_c \rangle$ is calculated from the laminar flame speed s_L^0 and I_0 models the effects of stretch:

$$\langle S_c \rangle_s = I_0 s_L^0 \tag{5.29}$$

Σ denotes the flame surface density (a measure of the active flame area per cell volume) for which a transport equation with terms for production and destruction is solved. The approach is attractive and hence widely adopted because it allows to decouple chemistry (contained in $\langle s_c \rangle$) from turbulence-chemistry interaction which is all contained in Σ.

Level-set methods view the flame as an infinitesimally thin interface separating the burnt from the unburnt side, where the flame front is represented by an isosurface of a field scalar G, normally defined by $G = 0$. In the LES context, following [144], this can be written as:

$$\frac{\partial}{\partial t}\left(\bar{\rho}\tilde{G}\right)+\frac{\partial}{\partial x_i}\left(\bar{\rho}\tilde{u}_i\tilde{G}\right)=\bar{\rho}s_T\left|\nabla\tilde{G}\right|-\bar{\rho}D_t\tilde{\kappa}\left|\nabla\tilde{G}\right| \tag{5.30}$$

Often a variance equation is solved in conjunction to reconstruct the turbulent flame brush thickness and the reaction progress variable (cf., e.g., [53]). The entire turbulence-chemistry interaction is contained in the formulation of the turbulent flame speed s_T, for which a variety of closures have been proposed; see, e.g., [145, 146, 147, 148, 149] although many more exist.

In particular in conjunction with LES, flame-thickening approaches originally proposed in [150] have also been used to model turbulent combustion, starting from gas turbine combustors [151] and later on adapted by [152] to cater for changing pressures in engines. In this TFMLES approach, the flame is artificially thickened in order to be resolvable on the computational mesh. This is motivated by the fact that the flame speed can be expressed as:

$$\delta_L^0 \propto \sqrt{D_{th}A} \tag{5.31}$$

where A denotes the pre-exponential term of an Arrhenius-type expression of the reaction rate; see, e.g., Equation (5.19) and D_{th} the thermal diffusivity.

The flame thickness can be written as:

$$\delta_L^0 \propto \frac{D_{th}}{\delta_L^0}=\sqrt{\frac{D_{th}}{A}} \tag{5.32}$$

By increasing the thermal diffusivity and accordingly reducing the pre-exponential in the flame speed calculation, the overall propagation speed can be maintained, and the flame can be resolved on an appropriately fine LES grid (doing so however changes the Damkohler number). Further details on the dynamic determination of the thickening can be found in [152].

Flamelet methods involving pre-calculation of flamelet tables (parameterized by reaction progress) as functions of pressure and unburnt temperature have also been employed for premixed combustion in IC engines [153, 154]. During the calculations, the averaged chemical source term in the reaction progress variable transport equation as well as species mass fractions is then retrieved from the table. For reactive scalars, an appropriate treatment of the reaction progress scalar dissipation rate is particularly important, since the relaxation hypothesis (widely adopted for non-premixed combustion) cannot be employed [155] and more sophisticated expressions are needed [156].

CMC models for premixed combustion initially validated by means of bluff body flames [157, 158] have recently also been applied to engine-like configurations [159]. Multiple mapping conditioning has also found its application to premixed problems (cf. e.g. [160]).

5.7.3 Examples of Applications

A vast body of literature exists also for non-premixed combustion, and both RANS and LES turbulence models are used in different combination of the models presented above.

FSD models were used for combustion following the comprehensive ignition treatment studies presented in [141, 161] in a RANS context. LES in conjunction with flame surface density models was used to study cyclic variability in research and full metal engines, e.g., in [162, 163, 164].

Level-set methods have also been used extensively, for RANS applications; the reader is referred, e.g., to [133, 165, 166] and references therein. As an example, Figure 5.23 shows the response to changes in engine speed in terms of pressure traces and heat release rates in two distinct engine configurations for two different s_T closures.

Efforts using LES are documented in [55, 137, 138, 159, 167, 168] among others looking into cyclic variability, correlations between early fuel conversion, and peak pressure as well as flame topology influence. The study of [169] investigates natural gas combustion in a direct injection SI engine, showing good agreement in terms of the spread in pressure evolutions compared to the experiment (Figure 5.24 (left)) and number of cycles required to achieve converged results (right). The study further elucidates the impact of the high-pressure gas injection event on the turbulence at spark timing, suggesting that while the turbulence produced by the injection itself is rapidly dissipated, the momentum of the injection is stored in the tumble and becomes available late during the cycle when this breaks down.

FIGURE 5.23 Comparison of pressure traces and heat release rates in two different engines (small engine upper, large engine lower). The impact of a change in engine speed (reference left two columns, increased speed right two columns) is assessed for two different turbulent flame speed closures. From [165].

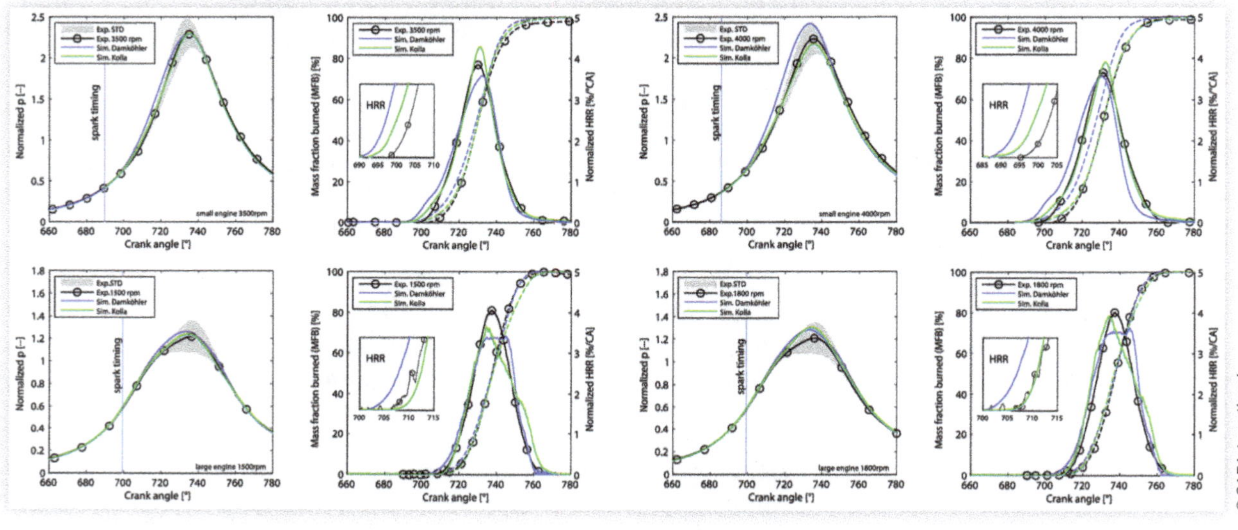

FIGURE 5.24 Comparison of pressure traces from LES with envelope of experimental data (left) and evolution of COV of P_{max} (right). From [169].

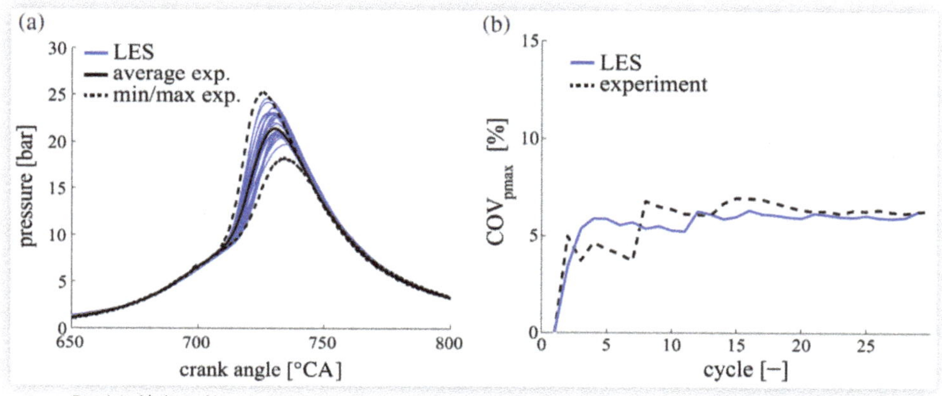

5.7.3.1 CMC PREMIXED

CMC models using the reaction progress as a conditioning variable have also been used to study combustion in IC engine like geometries [159]. Initial flow fields at different turbulence levels from a DNS have been filtered at the LES resolution to ensure well-defined initial conditions. Figure 5.25 compares the results from CMC and G-equation combustion models to the DNS at the same overall reaction progress (x_b) for the three different turbulence levels. While G-equation is strictly only valid in the thin and extends to the corrugated flame regime, CMC makes no assumptions on the regime and shows considerably better agreement with the fully resolved calculation at the highest turbulence level, approaching the broken reaction zones regime where the notion of an interface separating the burnt from the unburnt is strictly no longer valid.

FIGURE 5.25 Progress variable c on a section through the center plane. DNS (left), LES-CMC (middle), LES G-equation (right) for low (upper), reference (middle), and high (bottom) turbulence levels for comparable levels of combustion progress x_b. From [159].

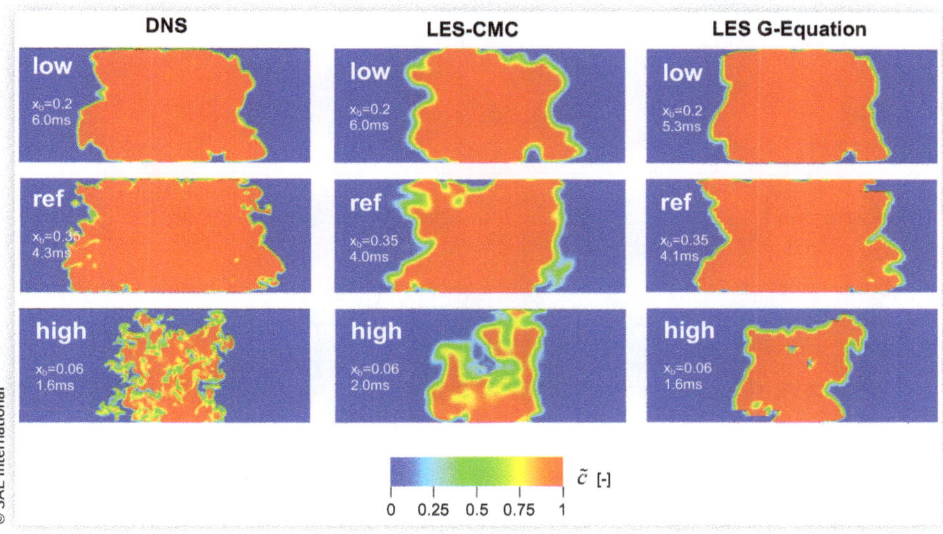

CHAPTER 5

Abbreviations

0D, 1D, 2D, 3D - Zero-, one-, two-, three-dimensional
AKTIM - Arc and kernel tracking ignition model
AMC - Amplitude mapping closure
BDC - Bottom dead center
CAD - Crank angle degree
CI - Compression ignition
CFD - Computational fluid mechanics
CMC - Conditional moment closure
CTC - Characteristic Time-Scale Combustion
DBI - Diffused back illumination
DI - Direct injection
DPIK - Discrete particle ignition kernel
DNS - Direct Numerical Simulation
ECN - Engine combustion network
EGR - Exhaust gas recirculation
FGM - Flamelet generated manifolds
FSD - Flame surface density
FVM - Finite Volume Method
GT - Gas turbine
IC - Internal combustion
LES - Large-Eddy Simulation
LHS - Left-hand side
NEDC - New European Driving Cycle
PDE - Partial Differential Equation
PDF - Probability density function
RANS - Reynolds-averaged Navier-Stokes
RHS - Right-hand side
RIF/mRIF - (multiple) Representative Interactive Flamelets
RPM - Revolutions per minute
SI - Spark ignited
SPARKCIMM - SPARK channel ignition monitoring model
TDC - Top Dead Center
TFM - Thickened flame model

References

1. Borée, J. and Miles, P.C., In-Cylinder Flow, in *Encyclopedia of Automotive Engineering* (John Wiley & Sons, Ltd, 2014).
2. Baum, E. et al., "On the Validation of LES Applied to Internal Combustion Engine Flows: Part 1: Comprehensive Experimental Database," *Flow Turbulence and Combustion* 92, no. 1 (2014): 269-297.
3. Renaud, A. et al., "Experimental Characterization of the Velocity Boundary Layer in a Motored IC Engine," *International Journal of Heat and Fluid Flow* 71 (2018): 366-377.

4. Alharbi, A.Y. and Sick, V., "Investigation of Boundary Layers in Internal Combustion Engines Using a Hybrid Algorithm of High Speed Micro-PIV and PTV," *Experiments in Fluids* 49, no. 4 (2010): 949-959.

5. Jainski, C. et al., "High-Speed Micro Particle Image Velocimetry Studies of Boundary-Layer Flows in a Direct-Injection Engine," *International Journal of Engine Research* 14, no. 3 (2013): 247-259.

6. Peterson, B. et al., "High-Speed PIV and LIF Imaging of Temperature Stratification in an Internal Combustion Engine," *Proceedings of the Combustion Institute* 34, no. 2 (2013): 3653-3660.

7. Peterson, B. et al., "Spray-Induced Temperature Stratification Dynamics in a Gasoline Direct-Injection Engine," *Proceedings of the Combustion Institute* 35, no. 3 (2015): 2923-2931.

8. Sick, V., Drake, M.C., and Fansler, T.D., "High-Speed Imaging for Direct-Injection Gasoline Engine Research and Development," *Experiments in Fluids* 49, no. 4 (2010): 937-947.

9. Ding, C.P. et al., "Flame/Flow Dynamics at the Piston Surface of an IC Engine Measured by High-Speed PLIF and PTV," *Proceedings of the Combustion Institute* (2018).

10. Spalding, D.B., Predicting the Performance of Diesel Engine Combustion Chambers," *Proceedings of the Institution of Mechanical Engineers* (1970).

11. Gosman, A.D. and Johns, R.J.R., "Development of a Predictive Tool for In-Cylinder Gas Motion in Engines," SAE Technical Paper 780315, 1978, 10.4271/780315.

12. Gosman, A.D. and Johns, R.J.R., "Computer Simulations of In-Cylinder Flow, Heat Transfer and Combustion: A Progress Report," in *13th CIMAC Congress*, Vienna, Austria, 1979.

13. Cartellieri, W. and Johns, R.J.R., "Multidimensional Modelling of Engine Processes: Progress and Prospects," in *Proceedings of the 15th CIMAC Congress*, 1983.

14. Gosman, A.D., "Multidimensional Modelling of Cold Flows and Turbulence in Reciprocating Engines," SAE Technical Paper 850344, 1985, 10.4271/850344.

15. Bracco, F.V., "Modeling of Engine Sprays," SAE Technical Paper 850394, 1985, 10.4271/850394.

16. Frouzakis, C.E. et al., "Direct Numerical Simulations for Internal Combustion Premixed Gas Engines: First Steps, Challenges and Prospects," in *13th International Congress on Engine Combustion Processes (ENCOM)*, Ludwigsburg, Germany, 2017.

17. Spalart, P.R., "Strategies for Turbulence Modelling and Simulations," *International Journal of Heat and Fluid Flow* 21, no. 3 (2000): 252-263.

18. Giannakopoulos, G.K. et al., "Direct Numerical Simulation of the Flow in the Intake Pipe of an Internal Combustion Engine," *International Journal of Heat and Fluid Flow* (2017).

19. Schmitt, M. et al., "Direct Numerical Simulation of the Compression Stroke under Engine Relevant Conditions: Local Wall Heat Flux Distribution," *International Journal of Heat and Mass Transfer* 92 (2016): 718-731.

20. Jones, W.P. and Launder, B.E., "The Calculation of Low-Reynolds-Number Phenomena with a Two-Equation Model of Turbulence," *International Journal of Heat and Mass Transfer* 16, no. 6 (1973): 1119-1130.

21. Haworth, D.C. and Jansen, K., "Large-Eddy Simulation on Unstructured Deforming Meshes: Towards Reciprocating IC Engines," *Computers & Fluids* 29, no. 5 (2000): 493-524.
22. Celik, I., Yavuz, I., and Smirnov, A., "Large Eddy Simulations of In-Cylinder Turbulence for Internal Combustion Engines: A Review," *International Journal of Engine Research* 2, no. 2 (2001): 119-148.
23. Janicka, J. and Sadiki, A., "Large Eddy Simulation of Turbulent Combustion Systems," *Proceedings of the Combustion Institute* 30, no. 1 (2005): 537-547.
24. Pitsch, H., "Large-Eddy Simulation of Turbulent Combustion," *Annual Review of Fluid Mechanics* 38 (2006): 453-482.
25. Rutland, C.J., "Large-Eddy Simulations for Internal Combustion Engines - A Review," *International Journal of Engine Research* 12, no. 5 (2011): 421-451.
26. Ferziger, J.H. and Peric, M., *Computational Methods for Fluid Dynamics* (Springer, 2002).
27. Stiesch, G., *Modeling Engine Spray and Combustion Processes* (Springer Science & Business Media, 2013).
28. Baumgarten, C., *Mixture Formation in Internal Combustion Engines* (Springer Science & Business Media, 2006).
29. Poinsot, T. and Veynante, D., *Theoretical and Numerical Combustion* (RT Edwards, Inc, 2005).
30. Schäfer, M., *Computational Engineering: Introduction to Numerical Methods* (Springer, 2006).
31. Lucchini, T. et al., "Automatic Mesh Motion with Topological Changes for Engine Simulation," SAE Technical Paper 2007-01-0170, 2007, 10.4271/2007-01-0170.
32. Senecal, P.K. et al., "A New Parallel Cut-Cell Cartesian CFD Code for Rapid Grid Generation Applied to In-Cylinder Diesel Engine Simulations," SAE Technical Paper 2007-01-0159, 2007, 10.4271/2007-01-0159.
33. Lucchini, T. et al., "Development of a CFD Approach to Model Fuel-Air Mixing in Gasoline Direct-Injection Engines," SAE Technical Paper 2012-01-0146, 2012, 10.4271/2012-01-0146.
34. Jasak, H., *Error Analysis and Estimation for the Finite Volume Method with Applications to Fluid Flows* (London: Imperial College of Science, Tecnology and Medicine, 1996).
35. Pope, S.B., *Turbulent Flows* (Cambridge, UK: Cambridge Univesity Press, 2000).
36. Tennekes, H., Lumley, J.L., and Lumley, J., *A First Course in Turbulence* (MIT Press, 1972).
37. Moin, P. and Mahesh, K., "Direct Numerical Simulation: A Tool in Turbulence Research," *Annual Review of Fluid Mechanics* 30, no. 1 (1998): 539-578.
38. Schmitt, M. and Boulouchos, K., "Role of the Intake Generated Thermal Stratification on the Temperature Distribution at Top Dead Center of the Compression Stroke," *International Journal of Engine Research* 17, no. 8 (2016): 836-845.
39. Schmitt, M. et al., "Direct Numerical Simulation of the Effect of Compression on the Flow, Temperature and Composition under Engine-like Conditions," *Proceedings of the Combustion Institute.*
40. Schmitt, M. et al., "Direct Numerical Simulation of Multiple Cycles in a Valve/Piston Assembly," *Physics of Fluids* 26, no. 3 (2014): 035105.

41. Schmitt, M. et al., "Direct Numerical Simulation of the Effect of Compression on the Flow, Temperature and Composition under Engine-Like Conditions," *Proceedings of the Combustion Institute* 35, no. 3 (2015): 3069-3077.
42. Schmitt, M. et al., "Investigation of unsteady wall heat flux under engine relevant conditions using Direct Numerical Simulation," in *4th International Conference on LES for Internal Combustion Engine Flows (LES4ICE)*, IFPen: Rueil-Malmaison, Paris, France, 2014.
43. Schmitt, M. et al., "Investigation of Cycle-to-Cycle Variations in an Engine-Like Geometry," *Physics of Fluids* 26, no. 12 (2014): 125104.
44. Schmitt, M. et al., "Investigation of Wall Heat Transfer and Thermal Stratification under Engine-Relevant Conditions Using DNS," *International Journal of Engine Research* 17 (2015): 63-75.
45. Schmitt, M. et al., "Direct Numerical Simulation of the Compression Stroke under Engine-Relevant Conditions: Evolution of the Velocity and Thermal Boundary Layers," *International Journal of Heat and Mass Transfer* 91 (2015): 948-960.
46. Buhl, S. et al., "Comparative Study of Turbulence Models for Scale-Resolving Simulations of Internal Combustion Engine Flows," *Computers & Fluids* 156 (2017): 66-80.
47. Lucchini, T. et al., "Automatic Mech Generation for Full-Cycle CFD Modeling of IC Engines: Application to the TCC Test Case," SAE Technical Paper 2014-01-1131, 2014, 10.4271/2014-01-1131.
48. Lucchini, T. et al., *Automatic Mesh Generation for CFD Simulations of Direct-Injection Engines* (SAE International, 2015).
49. Sone, K. and Menon, S., "Effect of Subgrid Modeling on the In-Cylinder Unsteady Mixing Process in a Direct Injection Engine," *Journal of Engineering for Gas Turbines and Power* 125, no. 2 (2003): 435-443.
50. Nicoud, F. et al., "Using Singular Values to Build a Subgrid-Scale Model for Large Eddy Simulations," 23, no. 8 (2011): 085106.
51. Nicoud, F. and Ducros, F.J.F., "Turbulence, and Combustion, Subgrid-Scale Stress Modelling Based on the Square of the Velocity Gradient Tensor 62, no. 3 (1999): 183-200.
52. di Mare, F., Knappstein, R., and Baumann, M., "Application of LES-Quality Criteria to Internal Combustion Engine Flows," *Computers & Fluids* 89 (2014): 200-213.
53. Peters, N., *Turbulent Combustion* (Cambridge University Press, 2000).
54. Goryntsev, D., Sadiki, A., and Janicka, J., "Cycle-to-Cycle Variations Based Unsteady Effects on Spray Combustion in Internal Combustion Engines by Using LES," SAE Technical Paper 2012-01-0399, 2012, 10.4271/2012-01-0399.
55. Koch, J. et al., "LES Multi-Cycle Analysis of the Combustion Process in a Small SI Engine," *SAE Int. J. Engines* 7, no. 1 (2014): 269-285.
56. Ma, P.C. et al., "Development and Analysis of Wall Models for Internal Combustion Engine Simulations Using High-speed Micro-PIV Measurements," *Flow, Turbulence and Combustion* 98, no. 1 (2017): 283-309.
57. Mandanis, C. et al., "Wall Heat Flux and Thermal Stratification Investigations during the Compression Stroke of an engine-like Geometry: A comparison between LES and DNS," *Flow, Turbulence and Combustion* 100, no. 3 (2017): 769-795.
58. Plengsaard, C. and Rutland, C., "Improved Engine Wall Models for Large Eddy Simulation (LES)," SAE Technical Paper 2013-01-1097, 2013, 10.4271/2013-01-1097.

59. Keskinen, K. et al., "Numerical Assessment of Wall Modelling Approaches in Scale-Resolving In-Cylinder Simulations," *International Journal of Heat and Fluid Flow* 74 (2018): 154-172.
60. Keskinen, K. et al., "Hybrid LES/RANS with Wall Treatment in Tangential and Impinging Flow Configurations," *International Journal of Heat and Fluid Flow* 65 (2017): 141-158.
61. De Villiers, E., Gosman, D., and Weller, H., "Detailed Investigation of Diesel Spray Atomization Using Quasi-Direct CFD Simulation," in *The Sixth International Symposium COMODIA*, 2004.
62. Lebas, R. et al., "Numerical Simulation of Primary Break-Up and Atomization: DNS and Modelling Study," *International Journal of Multiphase Flow* 35, no. 3 (2009): 247-260.
63. Huh, K.Y. and Gosman, A.D., "A Phenomenological Model of Diesel Spray Atomization," *Proc. Int. Conf. on Multiphase Flows*, Tsukuba, Japan, 1991.
64. Reitz, R.D., "Modeling Atomization Processes in High-Pressure Vaporizing Sprays," *Atomisation and Spray Technology* 3, no. 4 (1987): 309-337.
65. Pilch, M. and Erdman, C.A., "Use of Breakup Time Data and Velocity History Data to Predict the Maximum Size of Stable Fragments for Acceleration-Induced Breakup of a Liquid Drop," *International Journal of Multiphase Flow* 13, no. 6 (1987): 741-757.
66. Schmidt, D.P. et al., "Pressure-Swirl Atomization in the Near Field," SAE Technical Paper 1999-01-0496, 1999, 10.4271/1999-01-0496.
67. Zuo, B., Gomes, A.M., and Rutland, C.J., "Modelling Superheated Fuel Sprays andVaproization," *International Journal of Engine Research* 1, no. 4 (2000): 321-336.
68. Camm, J. et al., "The Effect of Non-Ideal Vapour-Liquid Equilibrium and Non-Ideal Liquid Diffusion on Multi-Component Droplet Evaporation for Gasoline Direct Injection Engines," SAE Technical Paper 2015-01-0924, 2015, 10.4271/2015-01-0924.
69. Keller, P. et al., "The Influence of Differential Evaporation on the Structure of a Three-Component Biofuel Spray," *International Journal of Engine Research* 16, no. 5 (2015): 610-626.
70. Schmidt, D.P. and Rutland, C.J., "A New Droplet Collision Algorithm," *Journal of Computational Physics* 164, no. 1 (2000): 62-80.
71. Nordin, N., Complex Chemistry Modeling of Diesel Spray Combustion, in *Department of Thermo Fluid Dynamics* (Chalmers University of Technology, 2001).
72. Bai, C.X., Rusche, H., and Gosman, A.D., "Modeling of Gasoline Spray Impingement," *Atomization and Sprays* 12, no. 1–3 (2002): 1-27.
73. Stanton, D.W. and Rutland, C.J., *Modeling Fuel Film Formation and Wall Interaction in Diesel Engines* (SAE International, 1996).
74. Pizza, G. et al., "Evaporating and Non-Evaporating Diesel Spray Simulation: Comparison between the ETAB and Wave Breakup Model," *International Journal of Vehicle Design* 45 (2007): 80-99.
75. Lippert, A.M. et al., "Mesh Independence and Adaptive Mesh Refinement For Advanced Engine Spray Simulations," SAE Technical Paper 2005-01-0207, 2005, 10.4271/2005-01-0207.

76. Kösters, A. and Karlsson, A., "A Comprehensive Numerical Study of Diesel Fuel Spray Formation with OpenFOAM," SAE Technical Paper 2011-01-0842, 2011, 10.4271/2011-01-0842.

77. Bai, C. and Gosman, D., "Mathematical Modelling of Wall Films Formed by Impinging Sprays," SAE Technical Paper 960626, 1996, 10.4271/960626.

78. Montanaro, A. et al., "Experimental Characterization of High-Pressure Impinging Sprays for CFD Modeling of GDI Engines," *SAE Int. J. Engines* 4, no. 1 (2011): 747-763.

79. Nurick, W.H., "Orifice Cavitation and Its Effect on Spray Mixing," *Journal of Fluids Engineering* 98, no. 4 (1976): 681-687.

80. Pickett, L.M., *Engine Combustion Network*, 2012. http://www.sandia.gov/ecn/dieselSprayCombustion.php.

81. Meijer, M. et al., "Engine Combustion Network (ECN): Characterization and Comparison of Boundary Conditions for Different Combustion Vessels," *Atomization and Sprays* 22, no. 9 (2012).

82. Maes, N. et al., "Experimental and Numerical Analyses of Liquid and Spray Penetration under Heavy-Duty Diesel Engine Conditions," *SAE Int. J. Fuels Lubr.* 9, no. 1 (2016): 108-124.

83. Lu, T.F. and Law, C.K., "Toward Accommodating Realistic Fuel Chemistry in Large-Scale Computations," *Progress in Energy and Combustion Science* 35, no. 2 (2009): 192-215.

84. Pitz, W.J. and Mueller, C.J., "Recent Progress in the Development of Diesel Surrogate Fuels," *Progress in Energy and Combustion Science* 37, no. 3 (2011): 330-350.

85. Bolla, M. et al., "Influence of Turbulence–Chemistry Interaction for n-Heptane Spray Combustion under Diesel Engine Conditions with Emphasis on Soot Formation and Oxidation," *Combustion Theory and Modelling* 18, no. 2 (2014): 330-360.

86. D'Errico, G. et al., "Comparison of Well-Mixed and Multiple Representative Interactive Flamelet Approaches for Diesel Spray Combustion Modelling," *Combustion Theory and Modelling* 18, no. 1 (2014): 65-88.

87. Bray, K.N.C., "The Challenge of Turbulent Combustion," *Symposium (International) on Combustion* 26, no. 1 (1996): 1-26.

88. Dec, J.E., "A Conceptual Model of DI Diesel Combustion Based on Laser-Sheet Imaging*," SAE Technical Paper 970873, 1997, 10.4271/970873.

89. Musculus, M.P.B., Miles, P.C., and Pickett, L.M., "Conceptual Models for Partially Premixed Low-Temperature Diesel Combustion," *Progress in Energy and Combustion Science* 39, no. 2–3 (2013): 246-283.

90. Halstead, M.P. et al., "A Mathematical Model for Hydrocarbon Autoignition at High Pressures," *Proceedings of the Royal Society of London (Series A, Math. and Phys. Sciences)* 346, no. 1647 (1975): 515-538.

91. Kong, S.-C., Han, Z., and Reitz, R.D., "The Development and Application of a Diesel Ignition and Combustion Model for Multidimensional Engine Simulation," SAE Technical Paper 950278, 1995, 10.4271/950278.

92. Kuo, T.-w. and R.D. Reitz, Computation of Premixed-Charge Combustion in Pancake and Pent-Roof Engines. *SAE Technical Paper* No. 890670, 1989.

93. Peters, N., "Laminar Diffusion Flamelet Models in Non-Premixed Turbulent Combustion," *Progress in Energy and Combustion Science* 10, no. 3 (1984): 319-339.

94. Peters, N., "Laminar Flamelet Concepts in Turbulent Combustion," *Symposium (International) on Combustion* 21, no. 1 (1988): 1231-1250.

95. Pitsch, H., Barths, H., and Peters, N., "Three-Dimensional Modeling of NOx and Soot Formation in DI-Diesel Engines Using Detailed Chemistry Based on the Interactive Flamelet Approach," SAE Technical Paper 962057, 1996, 10.4271/962057.

96. O'Brien, E.E. and Jiang, T.L., "The Conditional Dissipation Rate of an Initially Binary Scalar in Homogeneous Turbulence," *Physics of Fluids A* 3, no. 12 (1991): 3121-3123.

97. Lehtiniemi, H. et al., "Efficient 3-D CFD Combustion Modeling with Transient Flamelet Models," SAE Technical Paper 2008-01-0957, 2008, 10.4271/2008-01-0957.

98. Barths, H., Antoni, C., and Peters, N., "Three-Dimensional Simulation of Pollutant Formation in a DI Diesel Engine Using Multiple Interactive Flamelets," SAE Technical Paper 982459, 1998, 10.4271/982459.

99. Lucchini, T. et al., "A Comprehensive Methodology for Computational Fluid Dynamics Combustion Modeling of Industrial Diesel Engines," *International Journal of Engine Research* 18, no. 1–2 (2017): 26-38.

100. Felsch, C. et al., "Evaluation of Modeling Approaches for NOx Formation in a Common-Rail DI Diesel Engine within the Framework of Representative Interactive Flamelets (RIF)," SAE Technical Paper 2008-01-0971, 2008, 10.4271/2008-01-0971.

101. Leung, K.M., Lindstedt, R.P., and Jones, W.P., "A Simplified Reaction-Mechanism for Soot Formation in Nonpremixed Flames," *Combustion and Flame* 87, no. 3–4 (1991): 289-305.

102. Bolla, M. et al., "Soot Formation Modeling of n-Heptane Sprays Under Diesel Engine Conditions Using the Conditional Moment Closure Approach," *Combustion Science and Technology* 185, no. 5 (2013): 766-793.

103. D'Errico, G. et al., "Computational Fluid Dynamics Simulation of Diesel Engines with Sophisticated Injection Strategies for In-Cylinder Pollutant Controls," *Energy and Fuels* 26, no. 7 (2012): 4212-4223.

104. Fusco, A., "Application of a Phenomenological Soot Model for Diesel Engine Combustion," *COMODIA* (1994): 571-576.

105. Bolla, M. et al., "Modelling of Soot Formation in a Heavy-Duty Diesel Engine with Conditional Moment Closure," *Fuel* 117 (2014): 309-325.

106. Klimenko, A.Y. and Bilger, R.W., "Conditional Moment Closure for Turbulent Combustion," *Progress in Energy and Combustion Science* 25, no. 6 (1999): 595-687.

107. Farrace, D. et al., "Predicting In-Cylinder Soot in a Heavy-Duty Diesel Engine for Variations in SOI and TDC Temperature Using the Conditional Moment Closure Model," *SAE Int. Journal of Engines*, 2013. 6(3): p. 1580-1593, 10.4271/2013-24-0016.

108. Farrace, D. et al., "Numerical Study of the Influence of EGR on In-Cylinder Soot Characteristics in a Heavy-Duty Diesel Engine Using CMC," *SAE Int. J. Engines* 7, no. 1 (2014): 256-268, 10.4271/2014-01-1134.

109. Bolla, M. et al., "An Extended CMC Model for the Simulation of Diesel Engines with Multiple Injections," in *International Multidimensional Engine Modeling Users' Group Meeting*, Detroit, Michigan, 2014.

110. Frapolli, N. et al., "Simulations of In-Cylinder Processes in a Diesel Engine Operated with Post-Injections using an Extended CMC Model," SAE Technical Paper 2014-01-2571, 2014, 10.4271/2014-01-2571.

111. Pandurangi, S.S. et al., "Influence of EGR on Post-Injection Effectiveness in a Heavy-Duty Diesel Engine Fuelled with n-Heptane," *SAE Int. J. Engines* 7, no. 4 (2014): 1851-1862, 10.4271/2014-01-2633.

112. Haworth, D.C., "Progress in Probability Density Function Methods for Turbulent Reacting Flows," *Progress in Energy and Combustion Science* 36, no. 2 (2010): 168-259.

113. Haworth, D.C. and Pope, S.B., Transported Probability Density Function Methods for Reynolds-Averaged and Large-Eddy Simulations, in *Turbulent Combustion Modeling: Advances, New Trends and Perspectives*, T. Echekki and E. Mastorakos eds. (Dordrecht, the Netherlands: Springer, 2011), 119-142.

114. Chishty, M.A. et al., "Soot Formation Modelling of Spray-A Using a Transported PDFApproach," SAE Technical Paper 2015-01-1849, 2015, 10.4271/2015-01-1849.

115. Pei, Y., Hawkes, E.R., and Kook, S., "A Comprehensive Study of Effects of Mixing and Chemical Kinetic Models on Predictions of n-heptane Jet Ignitions with the PDF Method," *Flow, Turbulence and Combustion* (2013): 1-32.

116. Weisser, G. et al., "Integrating CRFD Simulations into the Development Process of Large Diesel Engine: A Status Report," in *CIMAC World Congress*, 1998.

117. Bolla, M. et al., "3D-CFD Lagrangian Spray Simulations for Large Two Stroke Marine Diesel Engines Compared with Experimental Data of a Combustion Chamber," ASME ICES2012-81016, 2012.

118. Siebers, D.L. and Higgins, B., "Flame Lift-Off on Direct-Injection Diesel Sprays Under Quiscent Conditions," SAE Technical Paper 2001-01-0530, 2001, 10.4271/2001-01-0530.

119. Bolla, M. et al., "Influence of Injector Diameter (0.2-1.2 mm range) on Diesel Spray Combustion: Measurements and CFD Simulations," SAE Technical Paper 2014-01-1419, 2014, 10.4271/2014-01-1419.

120. Srna, A. et al., "CMC Model Applied to Marine Diesel Spray Combustion: Influence of Fuel Evaporation Terms," SAE Technical Paper 2014-01-2738, 2014, 10.4271/2014-01-2738.

121. Bolla, M., Wright, Y.M., and Boulouchos, K., *Application of a Conditional Moment Closure Combustion Model to a Large Two-Stroke Marine Diesel Engine Reference Experiment* (Fukuoka, Japan: COMODIA, 2012).

122. Schmid, A. et al., *Investigation of Marine Diesel Ignition and Combustion at Engine-Like Conditions by means of OH* Chemiluminescence and Soot Incandescence* (Fukuoka, Japan: COMODIA, 2012).

123. Borman, G. and Nishiwaki, K., "Internal-Combustion Engine Heat-Transfer," *Progress in Energy and Combustion Science* 13, no. 1 1987: 1-46.

124. Bolla, M. et al., "Influence of Turbulent Fluctuations on Radiation Heat Transfer, NO and Soot Formation under ECN Spray A Conditions," *Proceedings of the Combustion Institute* 36, no. 3 (2017): 3551-3558.

125. Paul, C., Haworth, D.C., and Modest, M.F., "A Simplified CFD Model for Spectral Radiative Heat Transfer in High-Pressure Hydrocarbon–Air Combustion Systems," *Proceedings of the Combustion Institute* (2018).

126. Hill, P.G., "Cyclic Variations and Turbulence Structure in Spark-Ignition Engines," *Combustion and Flame* 72, no. 1 (1988): 73-89.

127. Aleiferis, P.G. et al., "The Nature of Early Flame Development in a Lean-Burn Stratified-Charge Spark-Ignition Engine," *Combustion and Flame* 136, no. 3 (2004): 283-302.

128. Granet, V. et al., "Large-Eddy Simulation and Experimental Study of Cycle-to-Cycle Variations of Stable and Unstable Operating Points in a Spark Ignition Engine," *Combustion and Flame* 159, no. 4 (2012): 1562-1575.

129. Enaux, B. et al., "LES Study of Cycle-to-Cycle Variations in a Spark Ignition Engine," *Proceedings of the Combustion Institute* 33 (2011): 3115-3122.

130. Colin, O. et al., "DNS and LES of Spark Ignition with an Automotive Coil," *Proceedings of the Combustion Institute* 37 (2018).

131. Thiele, M. et al., "Numerical Simulation of Spark Ignition Including Ionization," *Proceedings of the Combustion Institute* 28, no. 1 (2000): 1177-1185.

132. Herweg, R. and Maly, R.R., "A Fundamental Model for Flame Kernel Formation in S. I. Engines," SAE Technical Paper 922243, 1992, 10.4271/922243.

133. Xu, G. et al., "CFD-Simulation of Ignition and Combustion in Lean Burn Gas Engines," SAE Technical Paper 2016-01-0800, 2016, 10.4271/2016-01-0800.

134. Fan, L. and Reitz, R.D., "Development of an Ignition and Combustion Model for Spark-Ignition Engines, SAE Technical Paper 2000-01-2809, 2000, 10.4271/2000-01-2809.

135. Duclos, J.M. and Colin, O., "Arc and Kernel Tracking Ignition Model for 3D Spark-Ignition Engine Calculations," *COMODIA*, Nagoya, Japan, 2001.

136. Colin, O. and Truffin, K., "A Spark Ignition Model for Large Eddy Simulation Based on an FSD Transport Equation (ISSIM-LES)," *Proceedings of the Combustion Institute* 33, no. 2 (2011): 3097-3104.

137. Tan, Z. and Reitz, R.D., "Modeling Ignition and Combustion in Spark-Ignition Engines Using a Level Set Method," SAE Technical Paper 2003-01-0722, 2003, 10.4271/2003-01-0722.

138. Dahms, R.N. et al., "Understanding Ignition Processes in Spray-Guided Gasoline Engines Using High-Speed Imaging and the Extended Spark-Ignition Model SparkCIMM. Part A: Spark Channel Processes and the Turbulent Flame Front Propagation," *Combustion and Flame* 158, no. 11 (2011): 2229-2244.

139. Dahms, R.N. et al., "Understanding Ignition Processes in Spray-Guided Gasoline Engines Using High-Speed Imaging and the Extended Spark-Ignition Model SparkCIMM Part B: Importance of Molecular Fuel Properties in Early Flame Front Propagation," *Combustion and Flame* 158, no. 11 (2011): 2245-2260.

140. Lucchini, T. et al., "A Comprehensive Model to Predict the Initial Stage of Combustion in SI Engines," SAE Technical Paper 2013-01-1087, 2013, 10.4271/2013-01-1087.

141. Cornolti, L. et al., "A Comprehensive Lagrangian Flame–Kernel Model to Predict Ignition in SI Engines," *International Journal of Computer Mathematics* 91, no. 1 (2014): 157-174.

142. Weller, H.G. et al., "Prediction of Combustion in Homogeneous-Charge Spark-Ignition Engines," *International Symposium COMODIA* 94 (1994): 163-169.

143. Colin, O. and Benkenida, A., "The 3-Zones Extended Coherent Flame Model (Ecfm3z) for Computing Premixed/Diffusion Combustion," *Oil & Gas Science and Technology* 59, no. 6 (2004): 593-609.

144. Pitsch, H. and Duchamp de Lageneste, L., “Large-Eddy Simulation of Premixed Turbulent Combustion Using a Level-Set Approach,” *Proceedings of the Combustion Institute* 29, no. 2 (2002): 2001-2008.

145. Peters, N., “The Turbulent Burning Velocity for Large-Scale and Small-Scale Turbulence,” *Journal of Fluid mechanics* 384 (1999): 107-132.

146. Gülder, Ö.L., “Twenty-Third Symposium (International) on CombustionTurbulent Premixed Flame Propagation Models for Different Combustion Regimes,” *Symposium (International) on Combustion* 23, no. 1 (1991): 743-750.

147. Kolla, H., Rogerson, J.W., and Swaminathan, N., “Validation of a Turbulent Flame Speed Model across Combustion Regimes,” *Combustion Science and Technology* 182, no. 3 (2010): 284-308.

148. Kobayashi, H. et al., “Burning Velocity Correlation of Methane/Air Turbulent Premixed Flames at High Pressure and High Temperature,” *Proceedings of the Combustion Institute* 30, no. 1 (2005): 827-834.

149. Zimont, V. et al., “An Efficient Computational Model for Premixed Turbulent Combustion at High Reynolds Numbers Based on a Turbulent Flame Speed Closure,” *Journal of Engineering for Gas Turbines and Power* 120, no. 3 (1998): 526-532.

150. Butler, T.D. and O'Rourke, P.J., “A Numerical Method for Two Dimensional Unsteady Reacting Flows,” *Symposium (International) on Combustion* 16, no. 1 (1977): 1503-1515.

151. Thibaut, D. and Candel, S., “Numerical Study of Unsteady Turbulent Premixed Combustion: Application to Flashback Simulation,” *Combustion and Flame* 113, no. 1 (1998): 53-65.

152. Thobois, L., Lauvergne, R., and Poinsot, T., “Using LES to Investigate Reacting Flow Physics in Engine Design Process,” SAE Technical Paper 2007-01-0166, 2007, 10.4271/2007-01-0166.

153. Ahmed, I. et al., “Spark Ignition Engine Simulation Using a Flamelet Based Combustion Model,” SAE Technical Paper 2015-24-2402, 2015, 10.4271/2015-24-2402.

154. Ghiasi, G. et al., “Sensitivity of Flamelet Combustion Model to Flame Curvature for IC Engine Application,” SAE Technical Paper 2017-24-0038, 2017, 10.4271/2017-24-0038.

155. Langella, I. and Swaminathan, N., “Unstrained and Strained Flamelets for LES of Premixed Combustion,” *Combustion Theory and Modelling* 20, no. 3 (2016): 410-440.

156. Dunstan, T.D. et al., “Scalar Dissipation Rate Modelling for Large Eddy Simulation of Turbulent Premixed Flames,” *Proceedings of the Combustion Institute* 34, no. 1 (2013): 1193-1201.

157. Farrace, D. et al., “A LES-CMC Formulation for Premixed Flames Including Differential Diffusion,” *Combustion Theory and Modelling* (2018): 1-21.

158. Farrace, D. et al., “Unstructured LES-CMC Modelling of Turbulent Premixed Bluff Body Flames Close to Blow-Off,” *Proceedings of the Combustion Institute* 36, no. 2 (2017): 1977-1985.

159. Koch, J. et al., “Assessment of Two Premixed LES Combustion Models in an Engine-Like Geometry,” SAE Technical Paper 2018-01-0176, 2018, 10.4271/2018-01-0176.

160. Straub, C. et al., "Multiple Mapping Conditioning Coupled with an Artificially Thickened Flame Model for Turbulent Premixed Combustion," *Combustion and Flame* 196 (2018): 325-336.

161. Sforza, L. et al., "Modeling Ignition and Premixed Combustion Including Flame Stretch Effects," SAE Technical Paper 2017-01-0553, 2017, 10.4271/2017-01-0553.

162. Fontanesi, S., A. d'Adamo, and C.J. Rutland, "Large-Eddy Simulation Analysis of Spark Configuration Effect on Cycle-to-Cycle Variability of Combustion and Knock," *International Journal of Engine Research* 16, no. 3 (2015): 403-418.

163. Richard, S. et al., "Towards Large Eddy Simulation of Combustion in Spark Ignition Engines," *Proceedings of the Combustion Institute* 31 (2007): 3059-3066.

164. Vermorel, O. et al., "Towards the Understanding of Cyclic Variability in a Spark Ignited Engine Using Multi-Cycle LES," *Combustion and Flame* 156, no. 8 (2009): 1525-1541.

165. Koch, J. et al., "Comparison and Sensitivity Analysis of Turbulent Flame Speed Closures in the RANS G-Equation Context for Two Distinct Engines," *SAE Int. J. Engines* 9, no. 4 (2016): 2091–2106, 10.4271/2016-01-2236.

166. Xu, G. et al., "Characterization of Combustion in a Gas Engine Ignited Using a Small Un-Scavenged Pre-Chamber," accepted for publication in the *International Journal of Engine Research* (2019).

167. Ewald, J. and Peters, N., "On Unsteady Premixed Turbulent Burning Velocity Prediction in Internal Combustion Engines," *Proceedings of the Combustion Institute* 31, no. 2 (2007): 3051-3058.

168. Zhao, L. et al., "Examining the Role of Flame Topologies and In-Cylinder Flow Fields on Cyclic Variability in Spark-Ignited Engines Using Large-Eddy Simulation," *International Journal of Engine Research* 19, no. 8 (2017): 886-904.

169. Schmitt, M. et al., "Multiple Cycle LES Simulations of a Direct Injection Natural Gas Engine," *Flow, Turbulence and Combustion* 95, no. 4 (2015): 645-668.

CHAPTER 6

Control-Oriented Gas Dynamic Simulation via Model Order Reduction

Stephanie Stockar and Marcello Canova
The Ohio State University

6.1 Introduction and Motivations

Meeting the current emission regulations, fuel economy standards, and customer demands requires the application of advanced control and estimation algorithms to optimize the engine air-path system and the combustion process in heavily transient operating conditions. These needs have recently become more critical with the advancements in engine design dictated by the industry-wide need to reduce CO_2 emissions. The adoption of downsizing, boosting, flexible valve actuation technology, and variable displacement technology has resulted in a significant increase in the air-path system complexity, with several degrees of freedom now available to control and optimize the pressures and flow rates throughout the system [1, 2, 3, 4, 5, 6, 7, 8, 9, 10]. Hence, sophisticated, multivariable control techniques are nowadays required to fully exploit the potentials of the current technology.

While model-based control is today an established practice in the automotive industry, the state of the art for control development and verification is based on lumped-parameter, cycle-averaged models (or "mean-value" models) [11, 12, 13, 14]. In these models, the air-path system of the engine is represented as a network of control volumes and restrictions, and the model variables are assumed spatially and cycle-averaged [15, 16, 17, 18, 19, 20]. Due to the approximations introduced, the predictive ability of such models is notoriously limited, as evidenced by the presence of many parameters that typically require extensive calibration. A well-known example is the manifold filling dynamics (MFD) equation, which is the common control-oriented model utilized to control the engine air flow rate [16, 19, 20]. Due to the spatially and cycle-averaged approximation, all high-frequency phenomena associated with pressure wave propagation are inevitably neglected, and the unmodeled effects are lumped into a single empirical parameter, the volumetric efficiency. This parameter is heavily calibrated from experimental data, typically in steady-state conditions only [15, 21, 22, 23, 24, 25].

During the past few years, the rapid improvement in accuracy and computation time of numerical solvers for engine system simulation has allowed powertrain control engineers to exploit accurate, first-principle models as "virtual engines." To this extent, full engine performance simulation is conducted almost exclusively using 1D wave action models, whose ability to predict the effects of pressure wave propagation on the intake and exhaust flows proves very valuable for an accurate prediction of the cylinder charge and its composition [26, 27, 28, 29]. More recently, commercial 1D wave action simulation software products have been extended to incorporate math-based environments and tools that can assist the engine control system engineering process. For example, software-in-the-loop (SIL) and hardware-in-the-loop (HIL) simulation methods offer the opportunity to significantly reduce the development time and costs associated with engine control system design, calibration, and verification processes and ultimately shorten the time to deploy advanced powertrain technologies into production. In particular, SIL simulation offers the opportunity to conduct the calibration process in a virtual environment before transitioning to an engine test bench. In this case, the compiled engine controller software is incorporated into a simulation tool that contains a physical model of the complete engine system. The ability to achieve the co-design of engine and powertrain hardware and controls systems has had considerable impact in terms of quality, cost, improvements, and development time [15, 21, 30].

Despite the commercial availability of high-fidelity engine performance simulation software, there are fundamental limitations in their current computational capabilities, thereby limiting their application to SIL/HIL environments and ultimately preventing these tools to be fully integrated into control algorithm prototyping and verification tasks. While some of the existing commercial products, such as GT-Power, do allow for co-simulation of model and controller in different environments (for instance, MATLAB/Simulink [31]), these simulations are computationally expensive to execute and cannot be performed in real time.

Such limitations have motivated investigations into numerical and computational methods that could solve the fundamental equations of 1D wave action models in faster than real-time conditions. The objective is not only to provide high-fidelity plant simulators for SIL/HIL environments but also to facilitate the direct use of wave action models into engine air-path system control algorithms. This poses a significant challenge, since the nonlinear partial differential equations (PDEs) that constitute the backbone of 1D gas dynamic models must be converted into low-order linear or nonlinear ordinary differential equations (ODEs) to allow for the application of feedback control design methods. This conversion should preserve the same fidelity of 1D gas dynamic models (including the prediction of pressure wave propagation) however at a significant fraction of the mathematical complexity [32, 35].

In [32, 33, 34], for example, the authors present a quasi-static approach for treating the governing equations for unsteady compressible flows. This results into an extension of the MVM with additional calibration parameters that accounts for one-dimensional effects, such as wave propagations and change in the cross-sectional area. In [36], the method of characteristics (MOC) is applied for the modeling of the wave propagations in the air-path system of a single-cylinder engine. A similar approach is presented in [37], where a finite volume method (FVM) is developed as a solution method for the governing equations and implemented in a time-marching algorithm for the simulation of a single-cylinder engine. Finally, [38] presents a more fundamental approach for the extension of mean-value models to turbocharged engines. The authors identify the different time scales of the air-path system dynamics model in order to separate low-frequency and high-frequency modes. With this procedure, the high-frequency dynamics of the system, mainly induced by the nonlinearities in the governing equations, are neglected, resulting into a lower-order, computationally faster model.

In summary, the traditional methods to develop control-oriented models for engine air-path systems hinge upon the lumped-parameter approximation, leading to a low-frequency (mean-value) representation of the pressures and flow rates throughout the system. This approach significantly approximates the gas dynamic effects due to wave propagation in the intake and exhaust systems, forcing model developers to adopt expensive and time-consuming calibration processes that require large sets of engine test bench data. Additionally, due to the lack of physical basis, the heuristic model development processes lead to models that only perform well in limited set of conditions and are not portable across different engine designs.

6.2 Model Order Reduction

As the automotive industry relies heavily on high-fidelity engine simulation software for component design and performance evaluation, researchers have recently explored the opportunity to leverage the equations of simulation data of physics-based model to aid the synthesis of low-order, control-oriented models. In comparison to the conventional, "bottom-up" model development methods, which attempt at capturing the system dynamics from heuristic and data-driven methods, the idea of a "top-down" approach is appealing for many reasons. First, starting from the conservation laws used in high-fidelity models provides a mathematically rigorous framework to capture both the low-frequency and high-frequency phenomena that affect the flow and pressures throughout the engine breathing system. Furthermore, the physics-based nature of high-fidelity simulation models guarantees much more generality and portability across multiple engine platforms, as such models can accurately simulate different design variations with minimal changes in the parameters. Finally, high-fidelity models typically require a smaller set of engine test data to conduct the parameter identification.

Many engineering models commonly used today in engine simulation (e.g., unsteady gas dynamics, heat and mass transfer) are based on PDEs. However, solving models based on linear or nonlinear PDEs usually requires computing the numerical solution of large system of equations, which is generally time-consuming. Model Order Reduction (MOR) is a discipline that stems from the fields of applied mathematics and system dynamics and aims at approximating models based upon large-scale systems of equations with models of lower order in ODE form while preserving certain aspects of the original system. In contrast with the conventional control-oriented modeling approach, MOR is not based on introducing simplifying assumptions on the physics of the system but rather operates directly on the constitutive equations of high-fidelity models. In other words, MOR allows one to mathematically reduce the complexity of a physics-based, high-fidelity model by directly manipulating the model equations. As a result, models obtained with these techniques are generally calibration-free. The main advantage of MOR is that it allows one to retain the physical parameters setting of the original high-fidelity model; hence if a high-fidelity model is already calibrated on engine test data, the synthesis of the low-order model is calibration-free. Another significant advantage of MOR is that the reduced-order model is typically in the form or linear or nonlinear ODEs, allowing one to apply the resulting equations for control synthesis using a variety of established design methods.

In literature, specific MOR techniques have been proposed to treat large-scale dynamic systems constituted by linear or nonlinear hyperbolic PDEs, the most general mathematical formulation describing the flow of compressible fluids. Such equations are generally expressed as follows:

$$\frac{\partial \boldsymbol{u}}{\partial t}+\mathcal{L}_r(\boldsymbol{u})+S(\boldsymbol{u})=0 \tag{6.1}$$

where

$\boldsymbol{u}(\boldsymbol{r}, t)$ is the vector whose elements are a set of properties of the flow (such as the local value of the density or the velocity vector)
$\mathcal{L}_r$ is a spatial differential operator in $\mathbb{R}^n$ (where n is the number of spatial dimensions)
S represents a source term

Most MOR methods are *global* or *spectral* methods, meaning that they use information from the entire domain to determine the evolution of the system at a point. For example, if an approximate solution is a linear combination of functions, the value of the solution at each point in the field is affected by the value of the solution in the entire domain.

Conversely, *local* methods only use information from neighboring points to determine the evolution of the system at a specific point, which is the common assumption of all numerical methods that discretize the governing PDEs.

The general process to conduct MOR of Equation (6.1) using spectral methods is illustrated in Figure 6.1. The objective of MOR is to find an approximated solution $\hat{\boldsymbol{u}}(\boldsymbol{r},t)$ by minimizing the error $e=\|\hat{\boldsymbol{u}}-\boldsymbol{u}\|$, where $\boldsymbol{u}$ is the true solution of the original PDE model and $\|\cdot\|$ is a predefined measure. Since the exact solution to Equation (6.1) is almost never known, a residual is defined directly from the PDE definition:

$$R(\hat{\boldsymbol{u}})=\left[\frac{\partial \hat{\boldsymbol{u}}}{\partial t}+\mathcal{L}(\hat{\boldsymbol{u}})+S(\hat{\boldsymbol{u}})\right]-\left[\frac{\partial \boldsymbol{u}}{\partial t}+\mathcal{L}_r(\boldsymbol{u})+S(\boldsymbol{u})\right]=\frac{\partial \hat{\boldsymbol{u}}}{\partial t}+\mathcal{L}(\hat{\boldsymbol{u}})+S(\hat{\boldsymbol{u}}) \tag{6.2}$$

FIGURE 6.1 Process for model order reduction of a general PDE using spectral methods.

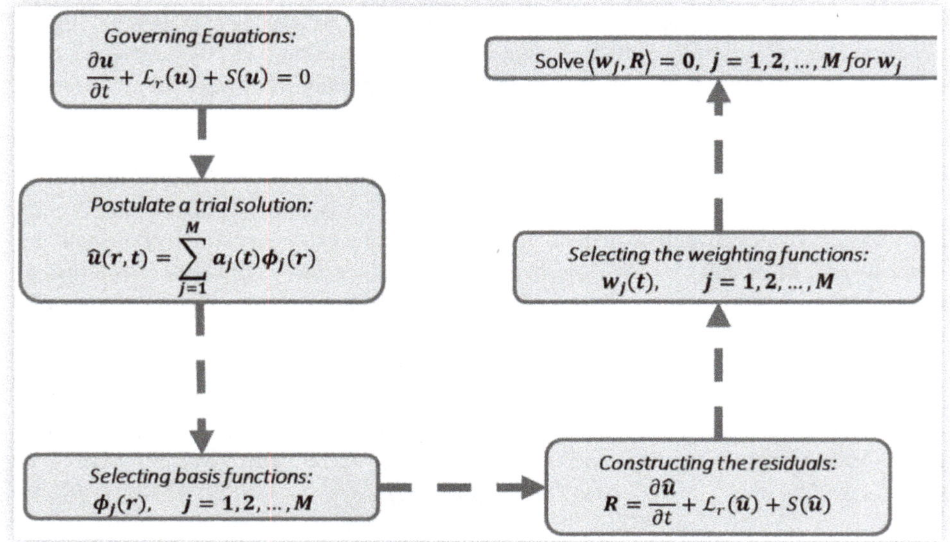

Note that the residual is zero if $\hat{\boldsymbol{u}} = \boldsymbol{u}$ (true solution). The residual is therefore an indirect measure of the approximation error.

The approximate solution to Equation (6.1) is defined as the finite series:

$$\hat{u}(r,t) = \sum_{j=1}^{M} w_j(t)\phi_j(r) \tag{6.3}$$

where

$\phi_j(\boldsymbol{r})$, $\boldsymbol{r} \in \mathbb{R}^n$ is a spatial basis function (SBF)
$w_j(t)$ is a time-varying weighting coefficient

Note that the above representation is not unique, and the expressions for ϕ_j and w_j are unknown. Therefore, the MOR problem requires one to first select a specific basis function ϕ_j that is able to describe the spatial distribution of the PDE solution and then determine the time-varying weighting coefficients by minimizing the residual $R(\hat{\boldsymbol{u}})$ in Equation (6.2). The minimization is typically achieved by enforcing the residual to be orthogonal to a *test space*, which is typically defined as the subspace spanned by the trial solution ϕ_j. The orthogonality criterion requires the definition of an *inner product*, hence a projection operator that minimizes the L_2 norm of the approximation error.

Various techniques have been proposed to define the basis functions and the projection criteria. In the field of nonlinear PDEs describing compressible fluid flows, proper orthogonal decomposition (POD) is a method that aims at obtaining a reduced-order representation of a large-scale set of time and spatially dependent data, by projecting high-dimensional data into a lower-dimensional space [39]. With POD, it is possible to find the unique sequence of orthogonal basis functions $\phi_j(\boldsymbol{r})$, such that the least-square error between the approximate representation and the data is minimized at each truncation level [39, 40, 41].

POD was introduced in fluid dynamics by Lumley et al., to model coherent structures in turbulent flows [42, 43], and subsequently extended in [44, 45], where a series of snapshots obtained from experimental measurements and simulations at a different instants of time are examined. These solution snapshots are used to form an eigenvalue problem that is solved to determine a set of optimal basis functions representing the flow field. Reduced-order models are then generated by projecting the governing PDEs onto the subspace spanned by a specific number of POD basis functions. A complete review of the use of POD in the analysis and modeling of turbulent flow can be found in [45], where it is suggested that the use of POD along with the Galerkin projection method leads to a systematic process to reduce nonlinear hyperbolic PDEs.

Other examples of reducing the Navier-Stokes equations on the basis obtained from POD can be found in [46, 47]. A low-order ODE model describing the dynamics of the flow past a rectangular cavity is derived in [48]. Starting from the CFD solution of the fully compressible, unsteady Navier-Stokes equations, the 2D isentropic and incompressible Euler equations are projected onto the POD basis using the Galerkin method. The same approach has been applied in [49, 50] to fully compressible flows.

Finally, MOR via POD with Galerkin projection has been demonstrated to yield control-oriented models. For instance, [51, 52, 53] present this methodology in the context of model-based feedback control of subsonic shallow cavity flows, to reduce the resonance phenomena. In this work, the reduced-order model is designed based on the Galerkin projection of the Navier-Stokes equations onto POD modes obtained from simulation data.

In general, POD is an established methodology for the solution of nonlinear hyperbolic PDE problems. The basis functions (modes) obtained via POD can be proven to be optimal in approximating the energy content of the data at each truncation level, which is particularly appealing for the synthesis of low-order ODE models for control applications. However, a shortcoming of POD is that the method is data-driven. In fact, the subspace obtained by applying POD is specific to the data (experimental or from simulation) used to generate the solution, which does not guarantee the ability to generalize the results to different operating conditions, particularly in case of nonlinear systems.

For this reason, alternative methods to POD have been proposed to generate the SBFs. In some cases, analytical functions could be defined *a priori*, that is, without knowing the particular nature of the solution from experiments or simulations. Common functional representations proposed in literature are Fourier series (localized in frequency), wavelet (localized in both time and frequency), Legendre polynomials, and Chebyshev polynomials [56, 57, 58, 59, 60].

The use of polynomial basis functions as a possible way to characterize nonlinear wave action phenomena in one-dimensional ducts has been recently shown as an effective method to reduce nonlinear hyperbolic PDEs. The method, described in the following section, was originally proposed in [54] and subsequently applied to a single-cylinder engine simulation with experimental verification [16, 55] and to the simulation of exhaust systems of turbocharged SI engines [61, 62, 63].

6.3 Model Order Reduction of 1D Gas Dynamic Equations via Polynomial Approximation

The general form of the governing equations for a 1D control volume with compressible fluid and constant cross-sectional area was presented in Chapter 1 and is summarized in matrix form as follows:

$$\frac{\partial \boldsymbol{W}}{\partial t}+\frac{\partial}{\partial x}F(\boldsymbol{W})+\boldsymbol{B}(\boldsymbol{W})=0 \tag{6.4}$$

$$\boldsymbol{W}(x,t)=\begin{bmatrix}\rho\\ \rho u\\ \rho e_0\end{bmatrix};\quad F(\boldsymbol{W})=\begin{bmatrix}\rho u\\ p+\rho u^2\\ \rho u h_0\end{bmatrix};\quad B(\boldsymbol{W})=\begin{bmatrix}0\\ \rho G\\ -\rho\dot{q}\end{bmatrix} \tag{6.5}$$

where $u = u(x, t)$, $\rho = \rho(x, t)$, $p = p(x, t)$, $e_0 = e_0(x, t)$, $h_0 = h_0(x, t)$ are the local flow velocity, density, static pressure, stagnation internal energy, and stagnation enthalpy, respectively. The friction force per unit area G is expressed through the Darcy-Weisbach equation, with a correction to account for possible flow reversals [64]:

$$G=\frac{1}{2}f\frac{4}{D}|u|u \tag{6.6}$$

where

f is the Moody friction factor

D the hydraulic diameter relative to the cross section

The term $\dot{q}$ represents the rate of heat exchanged per unit mass, which in most practical cases can be assimilated to convective heat transfer:

$$\dot{q} = -\frac{4h_c}{\rho D}(T - T_w) \tag{6.7}$$

where
h_c is the heat transfer coefficient
T_w is the wall temperature, generally assumed constant

The closure of the problem is finally obtained by specifying a set of initial and boundary conditions (BCs), which are specific to the system considered, for instance, the inlet and outlet sections of a pipe or a restriction element.

Using the ideal gas law, it is possible to redefine Equations (6.4)-(6.5) in terms of density, mass flow rate, and internal stagnation energy:

$$A\frac{\partial \mathbf{W}}{\partial t} + \frac{\partial}{\partial x}F(\mathbf{W}) + \mathbf{B}(\mathbf{W}) = 0 \tag{6.8}$$

$$W(x,t) = \begin{bmatrix} \rho \\ \dot{m} \\ \rho e_0 \end{bmatrix}; \; F(W) = \begin{bmatrix} \dot{m} \\ (\gamma-1)A^2\rho e_0 - \frac{\gamma-3}{2}\frac{\dot{m}^2}{\rho} \\ \gamma \dot{m} e_0 + \frac{\gamma-1}{2A^2}\frac{\dot{m}^3}{\rho^2} \end{bmatrix};$$

$$B(W) = \begin{bmatrix} 0 \\ \frac{|\dot{m}|\dot{m}}{\rho} f \sqrt{\frac{\pi}{A}} \\ -2h_c\sqrt{\pi A}\left[T_w - \frac{\gamma-1}{R}\left(e_0 - \frac{\dot{m}^2}{2\rho^2 A^2}\right)\right] \end{bmatrix} \tag{6.9}$$

The MOR method proposed in [65, 66, 67] consists of applying Equations (6.8)-(6.9) in conservative (integral) form to the staggered grid scheme shown in Figure 6.2. The adoption of a staggered grid presents the computational advantage of not requiring the implementation of a flux vector splitting method (FVSM) or a Riemann approach

FIGURE 6.2 Schematic representation of a staggered grid.

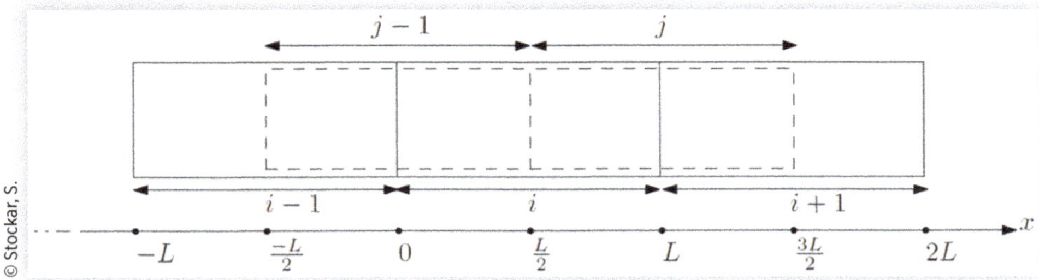

CHAPTER 6

for identifying the upwind direction (the direction in which the information is propagating) [68].

The staggered grid defines cells that are partially overlapping. Here, the mass and energy equations will be integrated over the ith cell, while the momentum equation will be integrated over the jth cell, which is staggered by half a cell length. Integration of Equations (6.8)-(6.9) on the staggered scheme leads to:

$$\int_{\bar{V}} A\frac{\partial \mathbf{W}}{\partial t}d\bar{V} + \int_{\bar{V}} \frac{\partial}{\partial x}F(\mathbf{W})d\bar{V} + \int_{\bar{V}} \mathbf{B}(\mathbf{W})d\bar{V} = 0 \tag{6.10}$$

where $\bar{V}$ is the total volume of the considered control volume. In this case, since the domain is 1D and the cross-sectional area is constant, the integration can be simplified as follows:

$$\int_{cell} A\frac{\partial \mathbf{W}}{\partial t}dx + \int_{cell} \frac{\partial}{\partial x}F(\mathbf{W})dx + \int_{cell} \mathbf{B}(\mathbf{W})dx = 0 \tag{6.11a}$$

Under the assumption of a 1D system, the divergence theorem can be used to equate the second term on the left-hand side of Equation (6.11a) to the value of the flux vector calculated at the boundaries, hence:

$$\int_{cell} A\frac{\partial \mathbf{W}}{\partial t}dx + F(\mathbf{W})\Big|_{bounds} + \int_{cell} \mathbf{B}(\mathbf{W})dx = 0 \tag{6.11b}$$

In the form described in Equation (6.11b), the model is still one-dimensional, where the conservation laws are spatially and temporally dependent. However, the above system could be reduced to a system of nonlinear ODEs by introducing a projection method to the above equations and postulating that the three state variables can be approximated by the finite series:

$$\hat{\mathbf{W}}(x,t) = \begin{bmatrix} \sum_k \alpha_k(t)\phi_k(x) \\ \sum_k \beta_k(t)\chi_k(x) \\ \sum_k \delta_k(t)\psi_k(x) \end{bmatrix} \tag{6.12}$$

Specifically, the above approximant can be defined in a more compact way as:

$$\hat{\mathbf{W}}(x,t) = \begin{bmatrix} \sum_k \alpha_k(t)\phi_k(x) \\ \sum_k \beta_k(t)\chi_k(x) \\ \sum_k \delta_k(t)\psi_k(x) \end{bmatrix} = \begin{bmatrix} \boldsymbol{\phi}^T(x) & 0 & 0 \\ 0 & \chi^T(x) & 0 \\ 0 & 0 & \psi^T(x) \end{bmatrix} \cdot \begin{bmatrix} \alpha(t) \\ \beta(t) \\ \delta(t) \end{bmatrix} = \hat{\mathbf{W}}^x(x) \cdot \hat{\mathbf{W}}^t(t) \tag{6.13}$$

where the terms $\phi_k(x)$, $\chi_k(x)$, $\psi_k(x)$ are the SBFs and $\alpha_k(t)$, $\beta_k(t)$, $\delta_k(t)$ are the time-varying coefficients

The first term of Equation (6.11b), containing the time derivative of the state vector, can be now approximated by defining its volume average over the cell domain:

$$\hat{W}^*(t) = \frac{A\int_{cell} \hat{W}^x(x)\cdot\hat{W}^t(t)dx}{\bar{V}} = \frac{A\int_{cell} \hat{W}^x(x)dx}{\bar{V}}\cdot\hat{W}^t(t) = \frac{\bar{W}^x\cdot\hat{W}^t(t)}{\bar{V}} \tag{6.14}$$

where $\bar{\boldsymbol{W}}^x$ is defined as:

$$\bar{\boldsymbol{W}}^x = \int_{cell} A\begin{bmatrix} \phi^T(x) & 0 & 0 \\ 0 & \chi^T(x) & 0 \\ 0 & 0 & \psi^T(x) \end{bmatrix} dx = \begin{bmatrix} \boldsymbol{\Phi}^T & 0 & 0 \\ 0 & \mathbf{X}^T & 0 \\ 0 & 0 & \boldsymbol{\Psi}^T \end{bmatrix} \tag{6.15}$$

Hence, the volume averages of the flow variables over one cell can be expressed as:

$$\bar{V}\hat{\boldsymbol{W}}^*(t) = \begin{bmatrix} \boldsymbol{\Phi}^T\alpha(t) \\ \mathbf{X}^T\beta(t) \\ \boldsymbol{\Psi}^T\delta(t) \end{bmatrix} = \begin{bmatrix} \sum_k \alpha_k(t)\boldsymbol{\Phi}_k \\ \sum_k \beta_k(t)\mathbf{X}_k \\ \sum_k \delta_k(t)\boldsymbol{\Psi}_k \end{bmatrix} = \begin{bmatrix} \rho^*(t) \\ \dot{m}^*(t) \\ (\rho e_0)^*(t) \end{bmatrix} \tag{6.16}$$

The time derivative and integral operator can be permuted, resulting into:

$$\int_{cell}\left[A\frac{\partial \boldsymbol{W}(x,t)}{\partial t}\right]dx = \bar{\boldsymbol{W}}^x\frac{d}{dt}\hat{\boldsymbol{W}}^t(t) = \bar{V}\frac{d\hat{\boldsymbol{W}}^*(t)}{dt} \tag{6.17}$$

Substituting to Equation (6.11a), the following form is obtained:

$$\bar{V}\frac{d\hat{\boldsymbol{W}}^*(t)}{dt} + F(\boldsymbol{W})\Big|_{bounds} + \int_{cell}\boldsymbol{B}(\boldsymbol{W})dx = 0 \tag{6.18}$$

which, expanded in its three equations, results:

$$\bar{V}\frac{d\rho^*}{dt} + [\dot{m}]_0^L = 0;\quad \bar{V}\frac{d\dot{m}^*}{dt} + \left[(\gamma-1)A^2\rho e_0 - \frac{\gamma-3}{2}\frac{\dot{m}^2}{\rho}\right]_{0.5L}^{1.5L} + \int_{cell}\frac{|\dot{m}|\dot{m}}{\rho}f\sqrt{\frac{\pi}{A}}dx = 0;$$

$$\bar{V}\frac{d(\rho e_0)^*}{dt} + \left[\gamma\dot{m}e_0 + \frac{\gamma-1}{2A^2}\frac{\dot{m}^3}{\rho^2}\right]_0^L - \int_{cell}2h_c\sqrt{\pi A}\left[T_w - \frac{\gamma-1}{R}\left(e_0 - \frac{\dot{m}^2}{2\rho^2A^2}\right)\right]dx = 0 \tag{6.19}$$

where L is the length of the control volume.

If an analytical expression of the SBF is given, a relation could be found to link the spatially averaged state variables to the flux variables, and the assumption illustrated above allows one to achieve a formal model reduction, namely, by projecting Equation (6.19) to the basis defined by the SBF for the given system geometry and solving the resulting nonlinear system of ODEs with the time-varying coefficients as state variables. In principle, this approach is similar to the POD, where a set of eigenfunctions (proper orthogonal modes, obtained from measurements of the flow field) is used as the optimal basis for the decomposition of a set of PDEs. The basis obtained from the POD is optimal in the sense of capturing most of the energy with the least number of terms for any order of truncation.

However, unlike the POD approach, where the determination of the eigenfunctions is often restricted to very simple geometry cases and still requires experimental or simulation results, this approach mimics the POD formalism by assuming that the SBFs are functions that can be chosen to best represent the underlying physics of the problem. In order to close the problem and ensure that Equation (6.19) can be solved, the following conditions for the SBF are introduced:

1. Each SBF has compact support that is coincident with the domain of each cell.
2. $\phi_k(x)$, $\chi_k(x)$, $\psi_k(x)$: $\mathbb{R} \rightarrow \mathbb{R}$ are continuous in each cell.

As a consequence, $\phi_k(x)$, $\chi_k(x)$, $\psi_k(x)$ are Riemann integrable in each cell. Furthermore, it is assumed that $\alpha_k(t)$, $\beta_k(t)$, $\delta_k(t)$: $\mathbb{R} \rightarrow \mathbb{R}$ are continuous and differentiable in each cell. Since it is assumed that both the cross-sectional area A and the SBF are known functions, the integral in Equation (6.19) can be explicitly calculated. Therefore the terms Φ_k, $X_k(x)$, $\Psi_k(x)$, which represent the volume-averaged SBFs over a given cell, simply become numerical coefficients. Note that $X_k(x)$ must be computed on the staggered grid, hence in the domain $[0.5L, 1.5L]$ as shown in Figure 6.2. Similarly, the expression for $F(\boldsymbol{W})$ evaluated at the cell boundaries is a function of $\alpha_k(t)$, $\beta_k(t)$, $\delta_k(t)$. Hence, $F(\boldsymbol{W})$ is a function of the vector $\hat{\boldsymbol{W}}^t(t)$ only, and the final form of the projected conservation laws is given by:

$$\sum_k \frac{d\alpha_k(t)}{dt}\Phi_k + \sum_k \beta_k(t)\left[\chi_k(x)\right]_0^L = 0$$

$$\sum_k \frac{d\beta_k(t)}{dt}X_k + \left[(\gamma-1)A^2\sum_k \delta_k(t)\psi_k(x) - \frac{\gamma-3}{2}\frac{\left(\sum_k \beta_k(t)\chi_k(x)\right)^2}{\sum_k \alpha_k(t)\phi_k}\right]_{0.5L}^{1.5L} + S_1 = 0$$

$$\sum_k \frac{d\delta_k(t)}{dt}\Psi_k + \left[\gamma\frac{\sum_k \beta_k(t)\chi_k(x)}{\sum_k \alpha_k(t)\phi_k}\sum_k \delta_k(t)\psi_k(x) + \frac{\gamma-1}{2A^2}\frac{\left(\sum_k \beta_k(t)\chi_k(x)\right)^3}{\left(\sum_k \alpha_k(t)\phi_k\right)^2}\right]_0^L + S_2 = 0 \tag{6.20}$$

The above equations constitute a system of coupled nonlinear ODEs, which can be expressed in compact form as:

$$\bar{\boldsymbol{W}}^x \frac{d}{dt}\hat{\boldsymbol{W}}^t(t) + \boldsymbol{F}\left(\hat{\boldsymbol{W}}^t(t)\right) + \boldsymbol{S}(t) = 0 \tag{6.21}$$

where the vector $\boldsymbol{S}(t)$ contains all the source terms defined over the computation cell and the only independent variables of the problem are contained in the state vector $\hat{\boldsymbol{W}}^t(t)$. This represents a system of coupled nonlinear ODEs that must be solved for each cell on the defined computation domain. The number of cells over which the equations are solved is determined by the order of the SBF, namely, the number of elements in the vector $\hat{\boldsymbol{W}}^t(t)$.

Despite the apparent complexity, the application of the aforementioned methodology to the solution of an unsteady flow problem in a 1D system is relatively straightforward and is based on the following procedure:

1. Define a set for the SBFs consistent with Equation (6.15) and the conditions listed before. Note that the definition of a SBF is needed in order to be able to express the flux terms (properties at the boundaries) as function of the spatial averages explicitly.
2. Given the geometry of the system, define a staggered grid according to the scheme shown in Figure 6.2. The level of coupling between cells is proportional to the order of the SBF.
3. Given the pipe area and length of each cell, the SBF can be computed off-line. The corresponding numerical values are substituted to Equation (6.24).
4. Solve the system of nonlinear ODEs resulting from applying Equation (6.21) to each cell, predicting the state dynamics in the entire geometry.
5. From Equation (6.13), the dimensional states (density, mass flow rate, stagnation internal energy) can be directly calculated for each cell. All the other flow properties are determined in post processing through algebraic relations.

Ultimately, this solution approach still provides a time-resolved description of the thermo-fluid properties at various locations of the system, in analogy with numerical methods.

6.3.1 Polynomial Spatial Basis Functions

An interesting form of the ODE system described by Equation (6.21) can be derived for a particular class of SBF, namely, polynomial functions. The use of polynomials for the approximation of PDE problems presents practical and computational advantages and it is commonly adopted in many MOR techniques.

Considering SBFs as $\phi_k(x)$, $\chi_k(x)$, $\psi_k(x) = x^{k-1}$, the governing PDEs can be projected onto the following basis:

$$
\begin{aligned}
\hat{\rho}(x,t) &= \sum_{k=1}^{n} \alpha_k(t)\, x^{k-1} \\
\widehat{\dot{m}}(x,t) &= \sum_{k=1}^{n} \beta_k(t)\, x^{k-1} \\
\widehat{\rho e_0}(x,t) &= \sum_{k=1}^{n} \delta_k(t)\, x^{k-1}
\end{aligned}
\tag{6.22}
$$

By applying the above SBF, the projected form of the governing equations can be considerably simplified, which eases the implementation in software. Examples are here shown for the case of constant and quadratic SBF.

6.3.1.1 PIECEWISE-CONSTANT SBF

In this case, a constant SBF is used to model the distribution of the state variables within each cell. Note that this assumption is consistent with the well-mixed approach, wherein each thermodynamic property is assumed constant within the control volume. This approach leads to a solution that is in principle similar to a FVM applied on a staggered grid. Moreover, in case of a constant SBF, the methodology provides the same scheme obtained by applying the method of lines [69]. The presented formalism, however, allows for a generalization of such method and its extension to different distribution of the properties within each computation cell. Starting from Equation (6.22), the assumption $n = 1$ is introduced, leading to $\phi_k(x) = \chi_k(x) = \psi_k(x) = 1$. Hence, the state variables for the ith cell can be expressed as follows:

$$
\begin{aligned}
\rho_i^*(t) &= \alpha(t)\frac{1}{\overline{V}}\int_0^L \left[A\phi_k(x)\right]dx = \alpha_i(t) \\
\dot{m}_j^*(t) &= \beta(t)\frac{1}{\overline{V}}\int_{0.5L}^{1.5L} \left[A\chi_k(x)\right]dx = \beta_j(t) \\
(\rho e_0)_i^*(t) &= \delta(t)\frac{1}{\overline{V}}\int_0^L \left[A\psi_k(x)\right]dx = \delta_i(t)
\end{aligned}
\tag{6.23}
$$

where in this case the index j denotes that the mass flow rate must be evaluated over the staggered grid. In case of a constant distribution of the properties within each control volume, the relation between spatial averages and values at the cell boundaries is straightforward, and the flux terms in Equation (6.19) can be computed as:

$$
\begin{aligned}
&\left[\dot{m}\right]_0^L = \beta_j(t) - \beta_{j-1}(t) \\
&\left[(\gamma-1)A^2\rho e_0 - \frac{\gamma-3}{2}\frac{\dot{m}^2}{\rho}\right]_{0.5L}^{1.5L} = (\gamma-1)A^2\left(\delta_{i+1}(t)-\delta_i(t)\right) - \frac{\gamma-3}{2}\left(\frac{\beta_j^2(t)}{\alpha_{i+1}(t)} - \frac{\beta_{j-1}^2(t)}{\alpha_i(t)}\right) \\
&\left[\gamma\dot{m}e_0 + \frac{\gamma-1}{2A^2}\frac{\dot{m}^3}{\rho^2}\right]_0^L = \gamma\left(\frac{\beta_j(t)\delta_i(t)}{\alpha_i(t)} - \frac{\beta_{j-1}(t)\delta_{i-1}(t)}{\alpha_{i-1}(t)}\right) + \frac{\gamma-1}{2A^2}\left(\frac{\beta_j^3(t)}{\alpha_i^2(t)} - \frac{\beta_{j-1}^3(t)}{\alpha_{i-1}^2(t)}\right)
\end{aligned}
\tag{6.24}
$$

Assuming $S_1 = S_2 = 0$ (no friction or heat transfer), the conservation laws applied to the ith cell become:

$$
\begin{aligned}
AL\frac{d}{dt}\alpha_i(t) &= \beta_j(t) - \beta_{j-1}(t) \\
AL\frac{d}{dt}\beta_j(t) &= (\gamma-1)A^2\left(\delta_{i+1}(t)-\delta_i(t)\right) - \frac{\gamma-3}{2}\left(\frac{\beta_j^2(t)}{\alpha_{i+1}(t)} - \frac{\beta_{j-1}^2(t)}{\alpha_i(t)}\right) \\
AL\frac{d}{dt}\delta_i(t) &= \gamma\left(\frac{\beta_j(t)\delta_i(t)}{\alpha_i(t)} - \frac{\beta_{j-1}(t)\delta_{i-1}(t)}{\alpha_{i-1}(t)}\right) + \frac{\gamma-1}{2A^2}\left(\frac{\beta_j^3(t)}{\alpha_i^2(t)} - \frac{\beta_{j-1}^3(t)}{\alpha_{i-1}^2(t)}\right)
\end{aligned}
\tag{6.25a}
$$

Note that the average density and specific internal energy within the control volume are related to the mass fluxes at the boundaries. Since the mass flow rate is the state of the momentum equations, the values of $\beta_j(t)$ are expressed with respect to the staggered control volume j.

Finally, using the relations of Equation (6.12), it is possible to formulate Equation (6.25a) such that the physical properties (density, flow rate, internal energy) appear as the states of the ODEs, leading to a solution scheme that results first-order accurate.

$$
\begin{aligned}
AL\frac{d}{dt}\rho_i^*(t) &= \dot{m}_j^* - \dot{m}_{j-1}^* \\
AL\frac{d}{dt}\dot{m}_j^*(t) &= (\gamma-1)A^2\left((\rho e_0)_{i+1}^* - (\rho e_0)_i^*\right) - \frac{\gamma-3}{2}\left(\frac{\dot{m}_j^{*2}}{\rho_{i+1}^*} - \frac{\dot{m}_{j-1}^{*2}}{\rho_i^*}\right) \\
AL\frac{d}{dt}(\rho e_0)_i^*(t) &= \gamma\left(\frac{\dot{m}_j^*(\rho e_0)_i^*}{\rho_i^*} - \frac{\dot{m}_{j-1}^*(\rho e_0)_{i-1}^*}{\rho_{i-1}^*}\right) + \frac{\gamma-1}{2A^2}\left(\frac{\dot{m}_j^{*3}}{\rho_i^{*2}} - \frac{\dot{m}_{j-1}^{*3}}{\rho_{i-1}^{*2}}\right)
\end{aligned}
\tag{6.25b}
$$

6.3.1.2 QUADRATIC SBF

Assuming quadratic polynomial basis functions, the states of the system are given by:

$$
\begin{aligned}
\rho_i^*(t) &= \alpha_{i,1}(t) + \alpha_{i,2}(t)x + \alpha_{i,3}(t)x^2 \\
\dot{m}_j^*(t) &= \beta_{j,1}(t) + \beta_{j,2}(t)x + \beta_{j,3}(t)x^2 \\
(\rho e_0)_i^*(t) &= \delta_{i,1}(t) + \delta_{i,2}(t)x + \delta_{i,3}(t)x^2
\end{aligned}
\tag{6.26}
$$

Similar to the case of piecewise-constant SBF, the above expressions can be substituted into Equations (6.20)-(6.21) to obtain the resulting system of ODEs and compute the time-varying coefficients. From the above formulation, it is clear that the use of quadratic SBFs requires knowledge of the value of the states over three adjacent control volumes.

A more practical implementation can be achieved by directly deriving the governing equations as function of the spatial averages of the flow properties within each volume. Such derivation of the governing equation results in a system of nonlinear ODEs whose states are the physical properties of the system (density, mass flow rate, and specific internal energy). This transformation is particularly convenient for control and estimation applications.

In this case, the first step consists in the evaluation of the spatial averages within each control volumes based upon the expressions in Equation (6.26). The spatial averages for the density and specific internal energy are calculated with respect to three volumes, namely, $i-1$, i, $i+1$:

$$
\begin{aligned}
\rho_{i-1}^*(t) &= A/\bar{V}\int_{-L}^{0}\left[\alpha_{i,1}(t) + \alpha_{i,2}(t)x + \alpha_{i,3}(t)x^2\right]dx = \alpha_{i,1}(t) - \alpha_{i,2}(t)\frac{L^2}{2} + \alpha_{i,3}(t)\frac{L^3}{3} \\
\rho_i^*(t) &= A/\bar{V}\int_{0}^{L}\left[\alpha_{i,1}(t) + \alpha_{i,2}(t)x + \alpha_{i,3}(t)x^2\right]dx = \alpha_{i,1}(t) + \alpha_{i,2}(t)\frac{L^2}{2} + \alpha_{i,3}(t)\frac{L^3}{3} \\
\rho_{i+1}^*(t) &= A/\bar{V}\int_{L}^{2L}\left[\alpha_{i,1}(t) + \alpha_{i,2}(t)x + \alpha_{i,3}(t)x^2\right]dx = \alpha_{i,1}(t) + \alpha_{i,2}(t)\frac{3L^2}{2} + \alpha_{i,3}(t)\frac{7L^3}{3}
\end{aligned}
\tag{6.27}
$$

$$(\rho e_0)^*_{i-1}(t) = A/\bar{V}\int_{-L}^{0}\left[\delta_{i,1}(t)+\delta_{i,2}(t)x+\delta_{i,3}(t)x^2\right]dx = \delta_{i,1}(t)-\delta_{i,2}(t)\frac{L^2}{2}+\delta_{i,3}(t)\frac{L^3}{3}$$
$$(\rho e_0)^*_{i}(t) = A/\bar{V}\int_{0}^{L}\left[\delta_{i,1}(t)+\delta_{i,2}(t)x+\delta_{i,3}(t)x^2\right]dx = \delta_{i,1}(t)+\delta_{i,2}(t)\frac{L^2}{2}+\delta_{i,3}(t)\frac{L^3}{3}$$
$$(\rho e_0)^*_{i+1}(t) = A/\bar{V}\int_{L}^{2L}\left[\delta_{i,1}(t)+\delta_{i,2}(t)x+\delta_{i,3}(t)x^2\right]dx = \delta_{i,1}(t)+\delta_{i,2}(t)\frac{3L^2}{2}+\delta_{i,3}(t)\frac{7L^3}{3} \tag{6.28}$$

while the averages for the mass flow rate are evaluated over the staggered grid for the three control volumes, $j-1, j, j+1$:

$$\dot{m}^*_{j-1}(t) = A/\bar{V}\int_{-L/2}^{L/2}\left[\beta_{j,1}(t)+\beta_{j,2}(t)x+\beta_{j,3}(t)x^2\right]dx = \beta_{j,1}(t)+\beta_{j,3}(t)L^2$$
$$\dot{m}^*_{j}(t) = A/\bar{V}\int_{L/2}^{3L/2}\left[\beta_{j,1}(t)+\beta_{j,2}(t)x+\beta_{j,3}(t)x^2\right]dx = \beta_{j,1}(t)+\beta_{j,2}(t)L+\beta_{j,3}(t)\frac{13L^3}{12}$$
$$\dot{m}^*_{j+1}(t) = A/\bar{V}\int_{3L/2}^{5L/2}\left[\beta_{j,1}(t)+\beta_{j,2}(t)x+\beta_{j,3}(t)x^2\right]dx = \beta_{j,1}(t)+\beta_{j,2}(t)2L+\beta_{j,3}(t)\frac{49L^3}{12} \tag{6.29}$$

Solving each of the three sets of equations above, the unknown time-varying coefficients can be determined:

$$\begin{aligned}\alpha_{i,1}(t) &= \frac{1}{3}\rho^*_{i-1}(t)+\frac{5}{6}\rho^*_{i}(t)-\frac{1}{6}\rho^*_{i+1}(t)\\ \alpha_{i,2}(t) &= \frac{1}{L}\left(-\rho^*_{i-1}(t)+\rho^*_{i}(t)\right)\\ \alpha_{i,3}(t) &= \frac{1}{2L^2}\left(\rho^*_{i-1}(t)-2\rho^*_{i}(t)+\rho^*_{i+1}(t)\right)\end{aligned} \tag{6.30}$$

$$\begin{aligned}\beta_{j,1}(t) &= \frac{23}{24}\dot{m}^*_{j-1}(t)+\frac{1}{12}\dot{m}^*_{j}(t)-\frac{1}{24}\dot{m}^*_{j+1}(t)\\ \beta_{j,2}(t) &= \frac{1}{2L}\left(-3\dot{m}^*_{j-1}(t)+4\dot{m}^*_{j}(t)-\dot{m}_{j+1}(t)\right)\\ \beta_{j,3}(t) &= \frac{1}{2L^2}\left(\dot{m}^*_{j-1}(t)-2\dot{m}^*_{j}(t)+\dot{m}_{j+1}(t)\right)\end{aligned} \tag{6.31}$$

$$\begin{aligned}\delta_{i,1}(t) &= \frac{1}{3}(\rho e_0)^*_{i-1}(t)+\frac{5}{6}(\rho e_0)^*_{i}(t)-\frac{1}{6}(\rho e_0)^*_{i+1}(t)\\ \delta_{i,2}(t) &= \frac{1}{L}\left(-(\rho e_0)^*_{i-1}(t)+(\rho e_0)^*_{i}(t)\right)\\ \delta_{i,3}(t) &= \frac{1}{2L^2}\left((\rho e_0)^*_{i-1}(t)-2(\rho e_0)^*_{i}(t)+(\rho e_0)^*_{i+1}(t)\right)\end{aligned} \tag{6.32}$$

Similar to the case of piecewise-constant SBF, the physical quantities can be expressed as functions of the spatial averages within each cell, and then the fluxes at the boundaries of each computation cell are evaluated and substituted in the governing equation. The result is a set of coupled nonlinear ODEs describing the physical states, leading to:

$$
\begin{aligned}
AL\frac{d}{dt}\rho_i^*(t) &= \left(-\frac{1}{24}\dot{m}_{j-2}^* + \frac{13}{24}\dot{m}_{j-1}^* - \frac{1}{24}\dot{m}_j^*\right) - \left(-\frac{1}{24}\dot{m}_{j-1}^* + \frac{13}{24}\dot{m}_j^* - \frac{1}{24}\dot{m}_{j+1}^*\right) \\
AL\frac{d}{dt}\dot{m}_j^*(t) &= (\gamma-1)A\left[\frac{-\rho_{i-2}^* + 2\rho_{i-1}^* + 23\rho_i^*}{24}\frac{-\dot{m}_{j-2}^* + 5\dot{m}_{j-1}^* + 2\dot{m}_j^*}{6}\right. \\
&\quad - \frac{-\rho_{i-1}^* + 2\rho_i^* + 23\rho_{i+1}^*}{24}\frac{-\dot{m}_{j-1}^* + 5\dot{m}_j^* + 2\dot{m}_{j+1}^*}{6}\Bigg] \\
&\quad + \frac{\gamma-3}{2A}\left[\left(\frac{-\dot{m}_{j-2}^* + 5\dot{m}_{j-1}^* + 2\dot{m}_j^*}{6}\right)^2 \frac{24}{-\rho_{i-2}^* + 2\rho_{i-1}^* + 23\rho_i^*}\right. \\
&\quad - \left(\frac{-\dot{m}_{j-1}^* + 5\dot{m}_j^* + 2\dot{m}_{j+1}^*}{6}\right)^2 \frac{24}{-\rho_{i-1}^* + 2\rho_i^* + 23\rho_{i+1}^*}\Bigg];
\end{aligned}
$$

$$
\begin{aligned}
AL\frac{d}{dt}(\rho e_0)_i^*(t) &= \gamma\left[\left(-\frac{1}{24}\dot{m}_{j-2}^* + \frac{13}{12}\dot{m}_{j-1}^* - \frac{1}{24}\dot{m}_j^*\right)\frac{-(\rho e_0)_{i-1}^* + 5(\rho e_0)_i^* + 2(\rho e_0)_{i+1}^*}{-\rho_{i-1}^* + 5\rho_i^* + 2\rho_{i+1}^*}\right. \\
&\quad - \left(-\frac{1}{24}\dot{m}_{j-1}^* + \frac{13}{12}\dot{m}_j^* - \frac{1}{24}\dot{m}_{j+1}^*\right)\frac{-(\rho e_0)_i^* + 5(\rho e_0)_{i+1}^* + 2(\rho e_0)_{i+2}^*}{-\rho_i^* + 5\rho_{i+1}^* + 2\rho_{i+2}^*}\Bigg] \\
&\quad + \frac{\gamma-1}{2A^2}\left[\left(-\frac{1}{24}\dot{m}_{j-2}^* + \frac{13}{12}\dot{m}_{j-1}^* - \frac{1}{24}\dot{m}_j^*\right)^3\left(\frac{6}{-\rho_{i-1}^* + 5\rho_i^* + 2\rho_{i+1}^*}\right)^2\right. \\
&\quad - \left(-\frac{1}{24}\dot{m}_{j-1}^* + \frac{13}{12}\dot{m}_j^* - \frac{1}{24}\dot{m}_{j+1}^*\right)^3\left(\frac{6}{-\rho_i^* + 5\rho_{i+1}^* + 2\rho_{i+2}^*}\right)^2\Bigg]
\end{aligned}
\qquad (6.33)
$$

The key difference between the generalized approach used for the derivation of the equations for the constant SBF and this is the fact that the second-order polynomial SBF introduces a higher degree of coupling among the states. Since the solution at the ith cell depends on the value of the states in the adjacent cells (up to two cells prior), the quadratic SBF results into a solution scheme that is third-order accurate.

6.4 Modeling of the Boundary Conditions

The MOR method described previously leads to sets of low-order ODEs that can be used to simulate the wave propagation phenomena in pipes. On the other hand, the region of validity of such equations is limited to the interior domain, implying that a set of BCs needs to be specified in order to close the problem.

Similar to the common numerical methods for 1D gas dynamics, BCs can be mathematically formulated for specific elements, such as terminations (open or closed ends), inlet regions, and flow restrictions. Intra-pipe BCs are also necessary when the 1D domain is abruptly interrupted, for instance, in the presence of flow splits, manifolds, or restrictions that separate one or more segments of the pipes. For the description of complex system, such as the internal combustion engine, a combination of pipe ends and intra-pipe BCs is generally defined.

Mathematically, BCs are generally defined as zero-dimensional elements. This implies that the physical size of the boundary can be neglected when compared to the total length of the pipe and that the BCs can be applied locally to an infinitesimal small region [64]. Since the implementation of the BCs requires the coupling between zero-dimensional and one-dimensional elements, the equations describing the thermodynamic properties at a boundary need to be defined consistently with the solution scheme proposed above for the 1D nonlinear Euler equations.

To illustrate the assumptions and derivation process, equations for common BCs used for engine air-path systems modeling will be here illustrated. Equations for more complex BCs, including in-cylinder thermodynamics, turbochargers, and flow splits, can be derived with similar considerations, as shown, for instance, in [61, 62, 63, 65, 67].

6.4.1 Boundary Conditions for Open-Ended Pipes

One of the most common BCs for the simulation of wave propagation in pipes is the open end, representing a connection between the pipe and the surrounding environment. In Figure 6.3, a schematic representation of a pipe inlet is presented, which serves as the case study to derive the equations. The process can be easily adapted to the case of pipe outlet.

From basic gas dynamics [64], the values of the thermodynamic properties at the entrance of the pipe should be computed depending upon the direction of the flow. The inflow can be modeled by assuming an isentropic compression from the ambient conditions (stagnation point), while the outflow can be assumed equivalent to an isobaric transformation, in which the pressure at the exit of the pipe is the same as the outside pressure. Note that the same assumptions can be applied to the case of pipe outlet; however, in this case the inflow should be characterized by a negative mass flow rate.

FIGURE 6.3 Schematic representation of the pipe inlet.

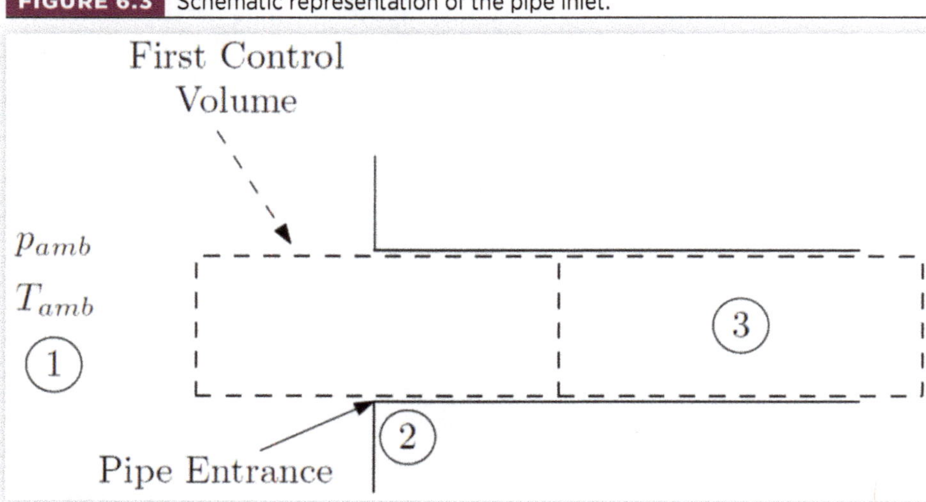

6.4.1.1 MODELING PIPE INFLOW WITH ISENTROPIC ASSUMPTION

The inflow conditions are calculated starting from the stationary mass and energy conservation equations for ideal gases, along with the constitutive equations for an isentropic process. The mass flow rate $\dot{m}_2$ is given by the following expression:

$$\dot{m}_2 = \frac{Ap_{01}}{\sqrt{RT_{01}}}\sqrt{\frac{2}{\gamma-1}}\sqrt{\left(\frac{p_2}{p_{01}}\right)^{\frac{2}{\gamma}}-\left(\frac{p_2}{p_{01}}\right)^{\frac{\gamma+1}{\gamma}}} \tag{6.34}$$

Comparing Equation (6.34) with Equation (6.19), it is evident that the mass flow rate $\dot{m}_2$ corresponds to the state $\dot{m}^*$ of the momentum equation solved on the staggered grid, and the pressure and temperature p_{01}, T_{01} are the known ambient conditions. Therefore, Equation (6.34) can be solved to determine the static pressure p_2 at the entrance of the pipe. For computational ease, the above equation can be implemented off-line to define a map that gives the pressure ratio $\frac{p_2}{p_{01}}$ as a function of the mass flow rate and ambient stagnation conditions.

Once the pressure p_2 is obtained, the density is calculated using the isentropic relation:

$$\rho_2 = \rho_{01}\left(\frac{p_2}{p_{01}}\right)^{\frac{1}{\gamma}} \tag{6.35}$$

And the temperature is obtained from the ideal gas law:

$$T_2 = \frac{p_2}{R\rho_2} \tag{6.36}$$

6.4.1.2 MODELING PIPE OUTFLOW WITH ISOBARIC ASSUMPTION

With reference to the notation in Figure 6.3, the case of outflow corresponds to a negative mass flow rate ($\dot{m}_2 < 0$), and the thermodynamic properties at the inlet of the pipe can be obtained using the following equations:

$$\begin{gathered} \rho_2 u_2 A = \rho_3 u_3 A \\ p_2 = p_1 \\ \rho_2 = \frac{p_2}{RT_2} = \frac{p_2}{RT_1} \\ h_{02} = h_{03} \end{gathered} \tag{6.37}$$

where the specific stagnation enthalpy is defined as follows:

$$h_0 = e_0 + \frac{p}{\rho} = e + \frac{u^2}{2} + \frac{p}{\rho} = c_p T + \frac{u^2}{2} \tag{6.38}$$

The conservation of energy can be therefore written as:

$$c_p T_2 + \frac{u_2^2}{2} = c_p T_3 + \frac{u_3^2}{2} \tag{6.39}$$

Since the velocity can be calculated by combining the conservation of mass with the ideal gas law, Equation (6.39) can be written as:

$$\frac{\dot{m}_2^2 R^2}{2A^2 p_1^2} T_2^2 + c_p T_2 = c_p T_3 + \frac{u_3^2}{2} \tag{6.40}$$

which is a quadratic equation in T_2. Since all the other terms in Equation (6.40) are known, the temperature at the pipe entrance can be solved analytically. Finally, the density at the pipe entrance is obtained using the ideal gas law, as shown in Equation (6.37).

6.4.2 Boundary Conditions for Valves and Restrictions

An important class of BCs for the simulation of IC engine air-path systems includes valves and restrictions, which is generally based upon the model of isentropic compressible flow through a nozzle [19, 70, 71]. This approach is commonly used in 1D gas dynamic models to calculate the flow rate and thermodynamic conditions through the engine intake and exhaust valves.

Similar to the assumptions leading to Equation (6.34), the conservation of mass and energy is applied to a restriction for the cases of subsonic and supersonic flow, leading to:

$$\dot{m}_v = \begin{cases} C_d A \dfrac{p_{ref}}{\sqrt{RT_{ref}}} \sqrt{\dfrac{2}{\gamma - 1}} \sqrt{\left(\dfrac{p_d}{p_{ref}}\right)^{\frac{2}{\gamma}} - \left(\dfrac{p_d}{p_{ref}}\right)^{\frac{\gamma+1}{\gamma}}}, & \text{if } p_d \geq p_{cr} \\ C_d A \dfrac{p_{ref}}{\sqrt{RT_{ref}}} \sqrt{\left(\dfrac{2}{\gamma+1}\right)^{\frac{\gamma+1}{\gamma-1}}}, & \text{if } p_d < p_{cr} \end{cases} \tag{6.41}$$

where

$p_{cr} = p_{ref}\left(\dfrac{2}{\gamma+1}\right)^{\frac{\gamma}{\gamma-1}}$ is the critical pressure

p_d is the static pressure downstream the restriction

Note that the reference conditions p_{ref} and T_{ref} are defined as the stagnation pressure and temperature upstream the restriction based upon the direction of the flow and must be properly implemented to account for forward and reverse flow conditions. This is particularly important in the case of intake and exhaust valves, where backflow can frequently occur due to the pressure fluctuations induced by the wave dynamics in the runners.

The discharge coefficient C_d is introduced to account for losses due to friction and flow contraction and is usually calibrated from experimental data on flow benches or engine test benches.

Finally, the remaining thermodynamic states can be computed by assuming conservation of the stagnation enthalpy across the restriction ($h_{ref} = h_{0d}$), along with the ideal gas law.

6.4.3 Boundary Conditions for Manifolds and Volumes

Engine air-path systems are often characterized by the presence of elements whose dimensions are comparable to its length, for which the 1D assumption is no longer valid. Silencers, air filters, and manifolds are the most relevant examples. These elements are typically modeled using a zero-dimensional approach, where the component is considered as a control volume in which the thermodynamic states are calculated by solving mass and energy conservation only. The volume of the component is assumed sufficiently large to neglect the momentum equation, leading to the simplified expression [19]:

$$\frac{dm}{dt} = \dot{m}_{in} - \dot{m}_{out}$$
$$\frac{dT}{dt} = \frac{1}{mc_v}\left[c_p\dot{m}_{in}T_{in} - c_p\dot{m}_{out}T - c_vT\left(\dot{m}_{in} - \dot{m}_{out}\right)\right] \quad (6.42)$$

Note that the assumption of adiabatic process has been introduced for simplicity. Similar sets of equations can be derived for the case of isothermal control volume or in the presence of heat transfer. The states in Equation (6.42) could be expressed in terms of the state variables of the 1D pipe element, namely, the density ρ and the specific internal energy ρe_0.

In order to couple the BC expressed by Equation (6.42) to the ODEs of the 1D pipe element, specific conditions must be imposed on the mass flow rates entering and exiting the control volume. These can be computed by assuming concentrated losses and the inlet and outlet of the control volume, hence using Equation (6.41). In this case, the orifice equation introduces algebraic relationships between two of the three states of the 1D volume (density and specific internal energy) and the states of Equation (6.42).

6.5 Application to the Shock Tube Problem

The shock tube problem is a well-known benchmark for the evaluation of shock-capturing methods for modeling one-dimensional compressible flows. This initial value problem (IVP) introduces the case of nonlinear wave propagations in the presence of discontinuities and provides an analytical solution under the assumption of nonviscous, isentropic compressible flow [72, 73, 74]. The shock tube consists of a straight pipe in which a gas at low pressure and a gas at high pressure are separated by a membrane. The rupture of the membrane at $t = 0$ generates two waves traveling in opposite directions through the domain. The wave traveling toward the high-pressure region of the tube is a rarefaction

FIGURE 6.4 Schematic of the shock tube problem.

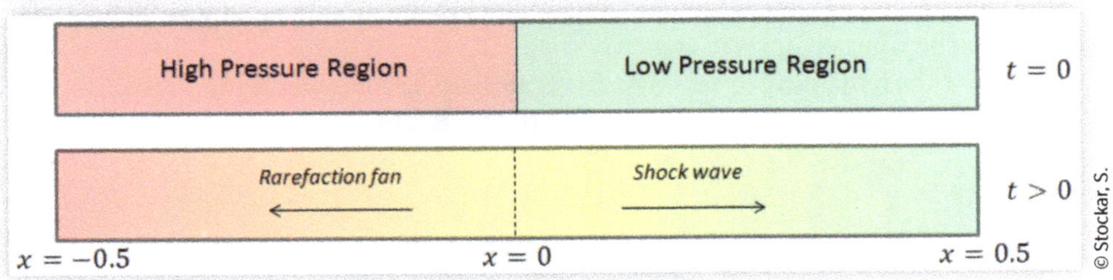

wave, while the wave propagating in the other direction is a shock wave followed by the contact discontinuity.

Figure 6.4 illustrates the case study, which is a common benchmark for verification of numerical methods for nonlinear gas dynamics, particularly to test the ability to predict shock-type discontinuities.

In this study, it is assumed that the straight pipe is long enough to prevent wave reflections at the wall; hence BCs can be neglected. The initial conditions assumed for this problem are described in Table 6.1 and the simulation parameters are summarized in Table 6.2. The initial conditions chosen represent a special case of the one-dimensional Riemann problem, commonly referred to as the *Sod's problem* [73].

The simulation results obtained are shown in Figure 6.5, where flow speed, density, and pressure distribution along the pipe are shown at the end of the simulation.

TABLE 6.1 Initial conditions for the shock tube problem

	$x < 0$	$x > 0$
Density	1.2 kg/m³	1.2 kg/m³
Internal energy	0.8 MJ/kg	0.2 MJ/kg
Flow speed	0 m/s	0 m/s

TABLE 6.2 Summary of model parameters for the shock tube problem

Pipe length	1 m
Courant number	0.3
Simulation duration	0.5 ms
Discretization length	1 mm

FIGURE 6.5 Distribution of the flow variables in the shock tube.

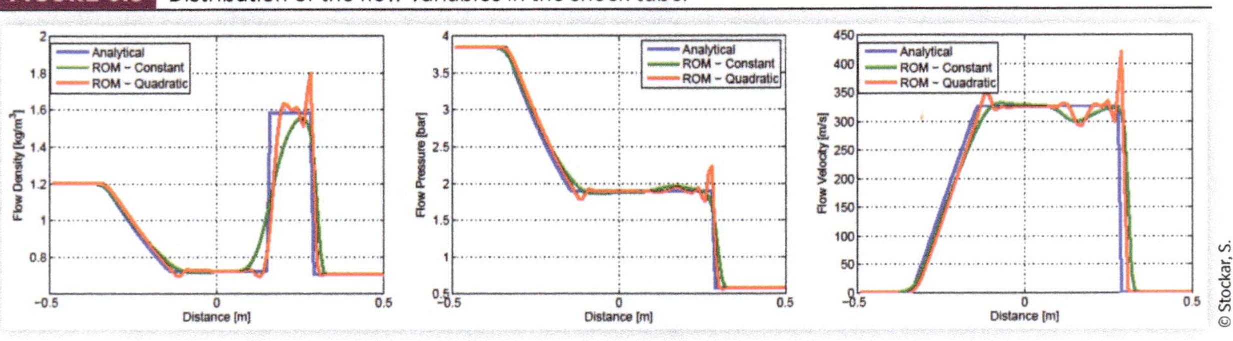

The results shown were obtained using the piecewise-constant and quadratic SBF, with a constant cell length of $L/L_p = 10^{-3}$, where L_p is the total pipe length.

The left plot in Figure 6.5 shows the speed distribution along the pipe predicted by the model with the two SBFs, compared to the analytical solution. In both cases, the proposed modeling approach is able to predict the velocity for the expansion fan and the shock. However, the piecewise-constant SBF, which results into a first-order accurate scheme, introduces artificial dissipation in the solution, while the second-order SBF introduces artificial oscillations, particularly in correspondence of the discontinuity. In general, the velocity profile along the pipe length is predicted more accurately by the quadratic SBF, as this method predicts the location of the changes in the flow speed more accurately than when using the piecewise-constant SBF. A similar behavior can be observed in the central plot, which shows the density profile at the end of the simulation. It is clear that both the piecewise-constant and quadratic SBF present some limitations in proximity of the contact discontinuity and the shock wave. Consistently with the results for the velocity profile, application of the piecewise-constant SBF results in artificial dissipation in the solution, and the resulting solution scheme cannot well predict the peak pressure in the pipe. Better overall prediction is achieved by the higher-order polynomial SBF, although this method introduces spurious oscillations in the solution. Finally, the pressure profile along the pipe is shown in the right plot of Figure 6.5, where the results obtained show good prediction ability of the two different methods.

While the reduced-order model implemented provides accurate predictions, the order of the ODE system requiring solution is very high. For the case $L/L_p = 10^{-3}$, the pipe geometry is discretized using 100 elements, resulting into 300 states. To this extent, the solution is evaluated when increasing the discretization length to $L/L_p = 2 \cdot 10^{-3}$ and $L/L_p = 4 \cdot 10^{-3}$, resulting into 150 and 75 states, respectively.

The comparison is shown in Figure 6.6. The reduction of the number of states introduces an artificial time delay in the rarefaction propagation; however, the amplitude of the oscillations appears to remain almost constant. Table 6.3 shows the computation time normalized with respect to the simulation time obtained when solving the model with the piecewise-constant SBF at the finest discretization length. As expected, the computation time is largely influenced by the spatial discretization, rather than the order or complexity of the SBF (which determines the degree of coupling among states in the model ODEs).

It is worth mentioning that other approaches for the solution of compressible 1D unsteady flow systems based on system dynamics methods have been validated against the shock tube problem; see, for instance, [75, 76]. These approaches are based on two different extensions of the bond graph theory that also account for nonlinearities in the

FIGURE 6.6 Effects of the discretization length on the accuracy of the reduced-order model (quadratic SBF).

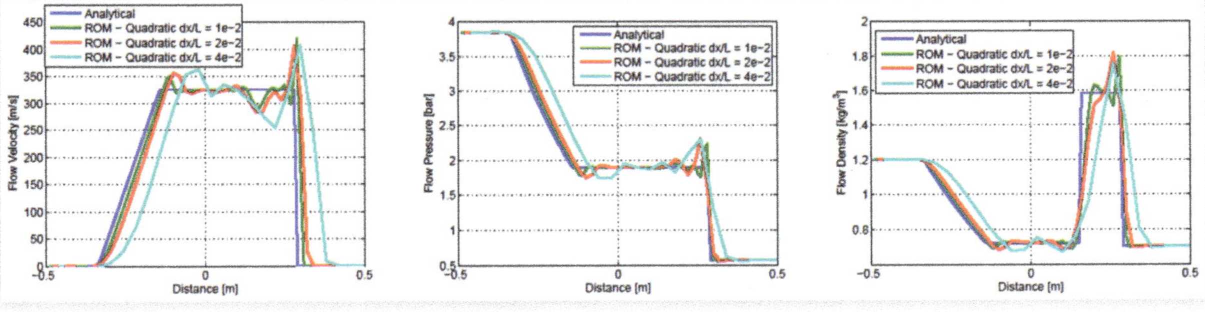

TABLE 6.3 Summary of computation times for the shock tube problem (values normalized)

	$L/L_p = 4 \cdot 10^{-3}$	$L/L_p = 2 \cdot 10^{-3}$	$L/L_p = 10^{-3}$
Constant SBF	0.28	0.28	1
Quadratic SBF	0.32	0.44	1.21

system. Compared to the results shown in the above references, the proposed method presents considerable accuracy improvements for the same number of states.

6.6 Control-Oriented Modeling of Wave Action in Turbocharged SI Engine

The MOR methodology described in this chapter represents a new approach to systematically develop control-oriented models of engine air-path systems that explicitly predict the effects of wave propagation in manifolds and ports. The central idea of this methodology is to develop a *top-down* process to synthesize control-oriented models directly from the constitutive equations (in PDE form) of high-fidelity gas dynamic models. The result is a model in the form of nonlinear ODEs that does not require calibration, as its parameters are directly imported from the high-fidelity model.

The ability to capture the physical phenomena that determine the cylinder charge, thermodynamic conditions, and composition provides the opportunity to design model-based estimators and feedback control algorithms without resorting to lookup tables or extensive parameter calibration.

In this context, the MOR approach described in this chapter was applied to design a control-oriented model of the exhaust manifold of a downsized multicylinder turbocharged SI engine [61, 62]. The objective of the model is to provide an accurate prediction of the exhaust port pressure with crank-angle resolution, with the ultimate goal of developing a model-based estimation of the cylinder charge and residual gas fraction during the valve overlap period in relation with the engine operating conditions, waste-gate position, and variable valve timing (VVT) [63].

A downsized, turbocharged, inline four-cylinder SI engine with VVT and waste-gate valve is considered, and the layout of its exhaust system is illustrated in Figure 6.7.

Characterizing the pressures and flow rates throughout the exhaust system in the crank-angle domain poses several challenges compared to the intake side. For instance, the higher in-cylinder pressure values at exhaust valve opening (EVO) excite the exhaust system causing large fluctuations of the pressure and temperature, which affect the turbine inlet conditions in a highly nonlinear fashion. Furthermore, higher exhaust gas temperatures increase the local value of the speed of sound, which is a known problem in gas dynamic simulation. Finally, the presence of complex BCs due to the turbine and waste-gate valve introduces a challenge in correctly capturing the wave interactions and coupling at that location.

As a starting point and benchmark for comparison, a model of the complete engine was previously developed in GT-Power and validated on experimental data in steady-state and transient conditions.

FIGURE 6.7 Schematic of a L4 turbocharged engine exhaust system.

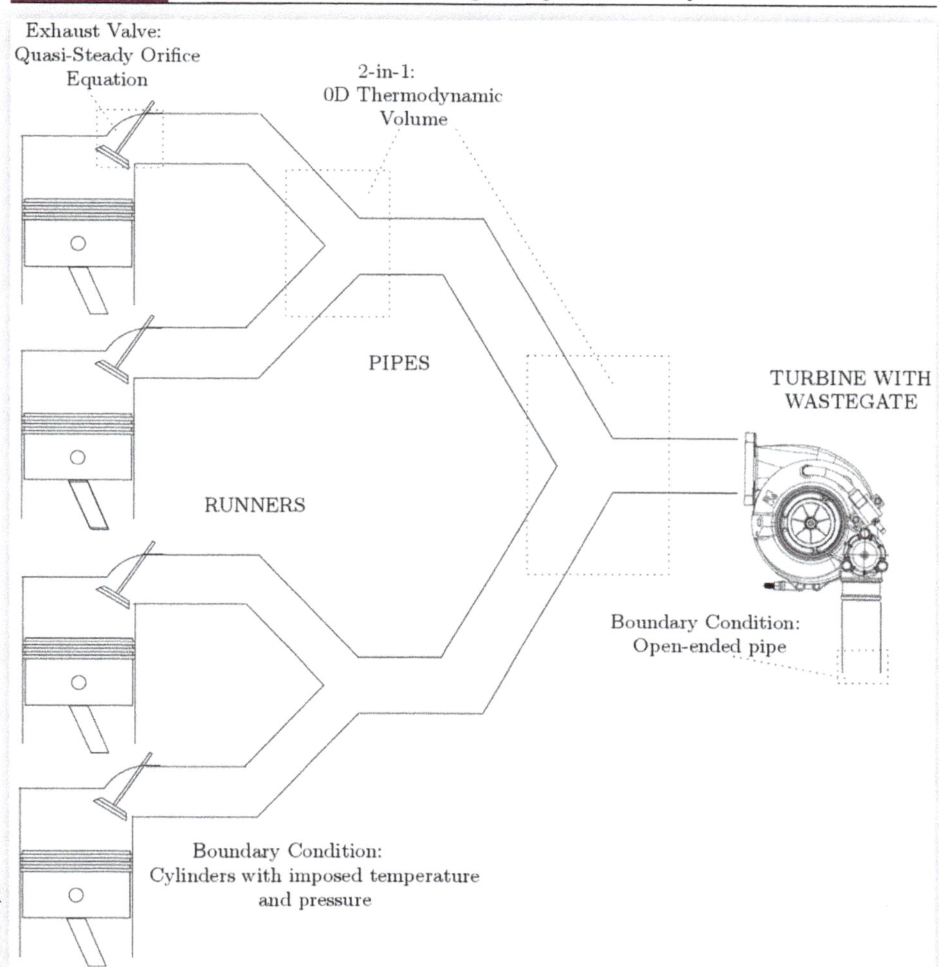

6.6.1 Overview of Model Equations

To simulate the 1D elements of the engine exhaust system, the projection-based MOR with piecewise-constant SBFs is adopted, resulting into the following system of ODEs expressed in terms of the time-varying coefficients:

$$
\begin{aligned}
AL\frac{d}{dt}\alpha_i(t) &= \beta_j(t)-\beta_{j-1}(t) \\
AL\frac{d}{dt}\beta_j(t) &= (\gamma-1)A^2\left(\delta_i(t)-\delta_{i-1}(t)\right)-\frac{\gamma-3}{2}\left(\frac{\beta_j^2(t)}{\alpha_i(t)}-\frac{\beta_{j-1}^2(t)}{\alpha_{i-1}(t)}\right)+S_1(t) \\
AL\frac{d}{dt}\delta_i(t) &= \gamma\left(\beta_j(t)\delta_i(t)-\beta_{j-1}(t)\delta_{i-1}(t)\right)+\frac{\gamma-1}{2A^2}\left(\frac{\beta_j^3(t)}{\alpha_i^2(t)}-\frac{\beta_{j-1}^3(t)}{\alpha_{i-1}^2(t)}\right)+S_2(t)
\end{aligned}
\tag{6.43}
$$

where $S_1(t)$ and $S_2(t)$ represent the friction and heat transfer, respectively, and can be computed starting from Equation (6.9) and applying the MOR process to obtain the corresponding expressions as functions of the time-varying coefficients.

The variables $\alpha_i(t)$, $\beta_j(t)$, $\delta_i(t)$ are the states of the system and can be directly related to the thermodynamic properties averaged over the cell length. In case of uniformly distributed properties, the coefficients coincide to the thermodynamic states:

$$\begin{aligned} \rho_i^*(t) &= \alpha_i(t) \\ \dot{m}_j^*(t) &= \beta_j(t) \\ (\rho e_0)_i^* &= \delta_i(t) \end{aligned} \tag{6.44}$$

The remaining thermodynamic properties are obtained by applying the ideal gas law:

$$\begin{aligned} T_i^* &= \frac{\gamma - 1}{R}\left(\frac{(\rho e_0)_i^*}{\rho_i^*} - \frac{\dot{m}_j^{*2}}{2\rho_i^{*2} A^2}\right) \\ p_i^* &= (\gamma - 1)(\rho e_0)_i^* - \frac{\gamma - 1}{2}\frac{\dot{m}_j^{*2}}{\rho_i^* A^2} \end{aligned} \tag{6.45}$$

where R is the specific gas constant.

For simulating a multicylinder engine exhaust manifold, the reduced-order model derived for the 1D pipe elements must be coupled to BCs. Considering the geometry of the exhaust system in Figure 6.7, BC equations must be specified for the intake and exhaust valves, the flow splits, the cylinder thermodynamics, and the turbine.

For this model, the valves and restrictions are described by Equation (6.41), while the flow splits and junctions are modeled using a combination of 0D thermodynamic volumes (Equation 6.42) and restrictions. The in-cylinder model predicts the charge density, temperature, and composition (air, fuel, residuals) using a simple 0D, single-zone thermodynamic model [67].

The turbine and waste-gate assembly were also modeled by decomposing the system into a network of control volumes and restrictions, as shown in Figure 6.8. The 0D elements introduce accumulation of mass and energy and model the turbine entry scroll, inducer casing, and exducer casing. Each volume is then connected by flow restrictions that represent, respectively, the waste-gate valve and the losses in the stator and in the

FIGURE 6.8 Schematic of the boundary condition model for the turbine.

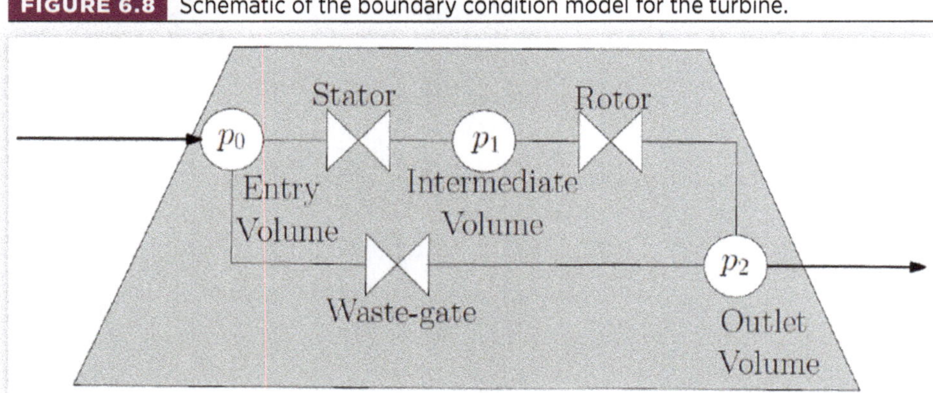

rotor. Since a detailed geometrical characterization of the flow passages through the turbine is typically not available, a set of tunable parameters is introduced and calibrated on the standard turbine characteristic map provided by the turbocharger supplier.

The key assumption introduced to model the turbine as a BC is that the pressure ratio $\epsilon = \frac{p_0}{p_2}$ is evenly split across the stator and the rotor, that is:

$$\epsilon_S = \epsilon_R = \sqrt{\epsilon} \leftrightarrow \frac{p_0}{p_1} = \frac{p_1}{p_2} \tag{6.46}$$

The turbine stator is modeled using a modified orifice equation that differs from the isentropic model presented in Equation (6.41). In this case, the mass and energy conservation equations are applied to a general, polytropic transformation that describes the expansion process of the exhaust gases through the turbine:

$$\dot{m}_{stat} = \begin{cases} (C_d A)_{stat} \dfrac{p_0}{\sqrt{RT_0}} \sqrt{\dfrac{2\gamma}{\gamma-1}} \sqrt{\left(\dfrac{p_1}{p_0}\right)^{\frac{2}{m}} - \left(\dfrac{p_1}{p_0}\right)^{\frac{m+1}{m}}}, & \text{if } \dfrac{p_1}{p_0} \geq \left(\dfrac{2}{m+1}\right)^{\frac{m}{m-1}} \\ (C_d A)_{stat} \dfrac{p_0}{\sqrt{RT_0}} \left(\dfrac{2}{m+1}\right)^{\frac{m}{m-1}} \sqrt{\dfrac{2}{\gamma+1} \dfrac{m-1}{m+1}}, & \text{if } \dfrac{p_1}{p_0} < \left(\dfrac{2}{m+1}\right)^{\frac{m}{m-1}} \end{cases} \tag{6.47}$$

where the upstream stagnation conditions (p_0T_0) are the states of the entry volume in Figure 6.8. The identification of the stator parameters, namely, the equivalent area $(C_dA)_{stat}$ and the polytropic exponent m, is conducted by conducting a least-square minimization of the error between the mass flow rate predicted by Equation (6.47) and the turbine flow map. Figure 6.9 illustrates the agreement between the fitted model based on Equation (6.15) and the GT-Power turbine map.

The effective area of the rotor is then calculated under the assumption that the polytropic exponent m is the same for the stator and the rotor and by using Equation (6.46). In steady-state conditions, the mass flow rate thought the stator must equal to the one through the rotor, which leads to:

$$(C_d A)_{rot} = (C_d A)_{stat} \frac{p_0}{p_1} \tag{6.48}$$

FIGURE 6.9 Results of turbine characteristic map-fitting procedure (units removed).

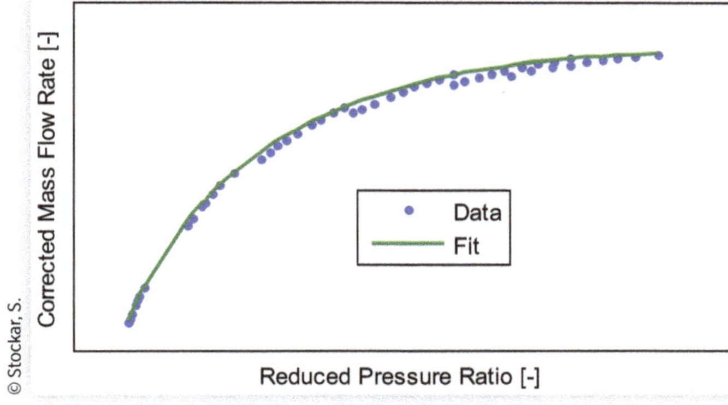

6.6.2 Results and Discussion

The MOR technique described above is compared against a GT-Power model of the engine, previously calibrated on experimental data. The parameters of the different components in the GT-Power model (such as geometric information, friction and heat transfer coefficients, loss coefficients in valves and restrictions) were imported into the reduced-order model. To further evaluate the accuracy and computation time, a low-frequency, mean-value model of the engine exhaust system was developed and calibrated on the GT-Power simulation results for steady-state operating conditions. The simulations of the engine exhaust system are performed by imposing the pressure and temperature profiles for the four cylinders and the waste-gate valve position obtained from GT-Power.

A direct comparison of the model prediction against GT-Power was conducted with respect to a large set of steady-state operating conditions. Table 6.4 lists a few representative points that correspond to high load conditions, where the impact of wave propagation on the pressures and flow rates in the exhaust system is particularly significant.

Case 1 has been selected to validate the turbine main path (closed waste-gate) and analyze the ability of the reduced-order model to capture the interaction of the wave propagation with a closed waste-gate. The normalized mass flow rate though the exhaust valve and the normalized pressure in the exhaust port for cylinder 1 are shown in Figures 6.10 and 6.11. These variables have been selected due to their importance for cylinder charge composition prediction.

The results show that the model is able to capture the pulses in the exhaust port pressure well, accounting for the effects of the blowdown phase and interaction with the wave propagation phenomena. Figure 6.10 shows the agreement between the pressure predicted by the model and GT-Power. In particular, the model predictions agree well with the corresponding results from GT-Power during the exhaust phase, accounting for the interaction among the four cylinders, which is critical for engine charge control.

Since the prediction of the pressure wave propagation is largely affected by the modeling of the BCs, a comparison of the turbine inlet conditions is shown in Figure 6.11.

TABLE 6.4 Steady-state conditions for model comparison

Case	Engine speed [rpm]	Target MAP [bar]	Target TIP [bar]	EVC [deg]	WG Pos. [%]
1	1950	1.7	1.85	368.6	0
2	2200	1.8	1.85	386.6	38
3	2200	1.8	1.85	382.6	37
4	2200	1.8	1.85	278.6	36

FIGURE 6.10 Summary of model verification in steady-state conditions: normalized flow rate and pressure profiles at exhaust port of cylinder 1 (case 1).

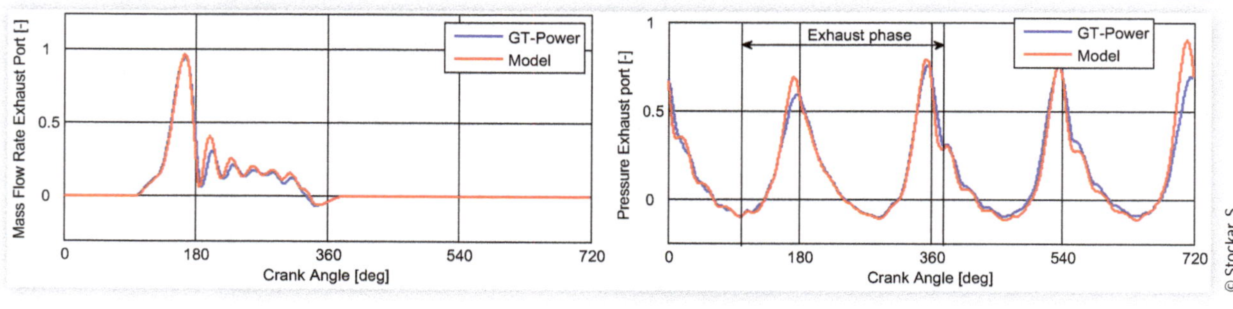

FIGURE 6.11 Summary of model verification in steady-state conditions: normalized flow rate and pressure profiles at turbine inlet (case 1).

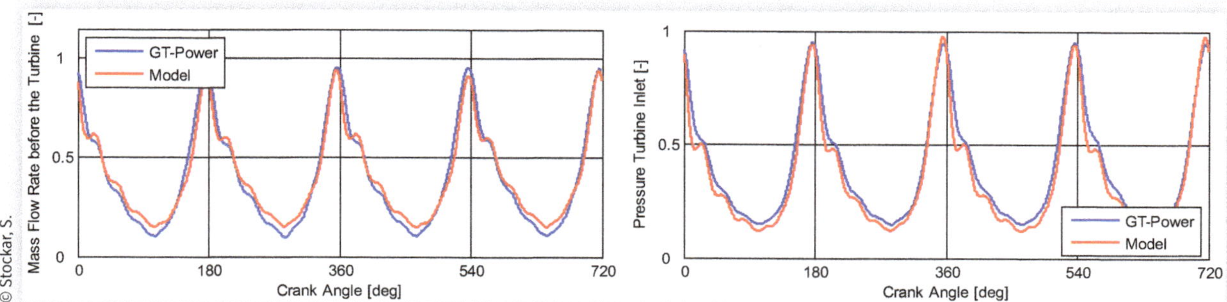

The results show that, at steady-state conditions, the proposed approach for the turbine and waste-gate BC modeling properly captures the wave interaction in the exhaust pipe.

A similar analysis is done here for Case 2, representative of a condition with high throttle opening, increased valve overlap, and partially open waste-gate. The comparison of scaled mass flow rate through the exhaust valve and the scaled pressure in the exhaust port is shown in Figures 6.12 and 6.13. The direct comparison between GT-Power and the reduced-order model shows a good agreement, in particular for the prediction of the mass flow rates at the exhaust port and turbine inlet. In particular, the pressure in the exhaust port shows good agreement for most of the cycle, particularly during the open valve phase, which is critical for charge estimation. There is a slight discrepancy

FIGURE 6.12 Summary of model verification in steady-state conditions: normalized flow rate and pressure profiles at exhaust port of cylinder 1 (case 2).

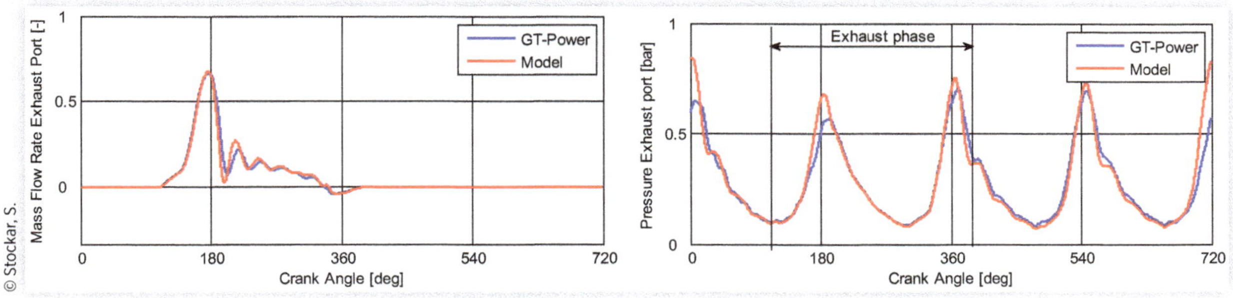

FIGURE 6.13 Summary of model verification in steady-state conditions: normalized flow rate and pressure profiles at turbine inlet (case 2).

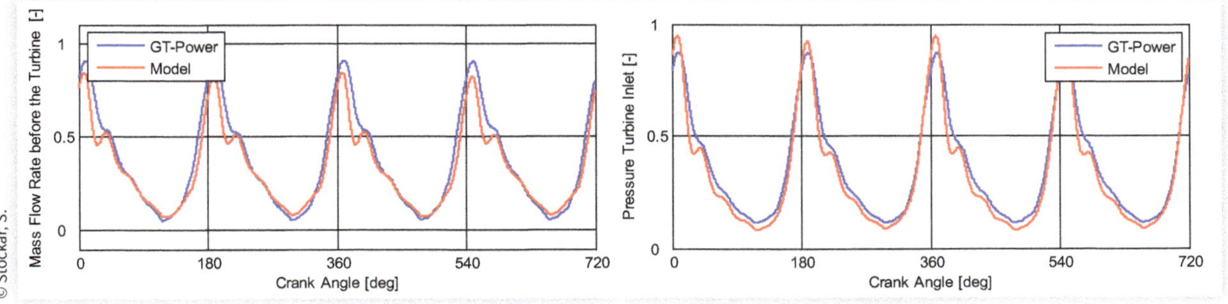

CHAPTER 6

FIGURE 6.14 Input profiles for the transient simulation in GT-Power.

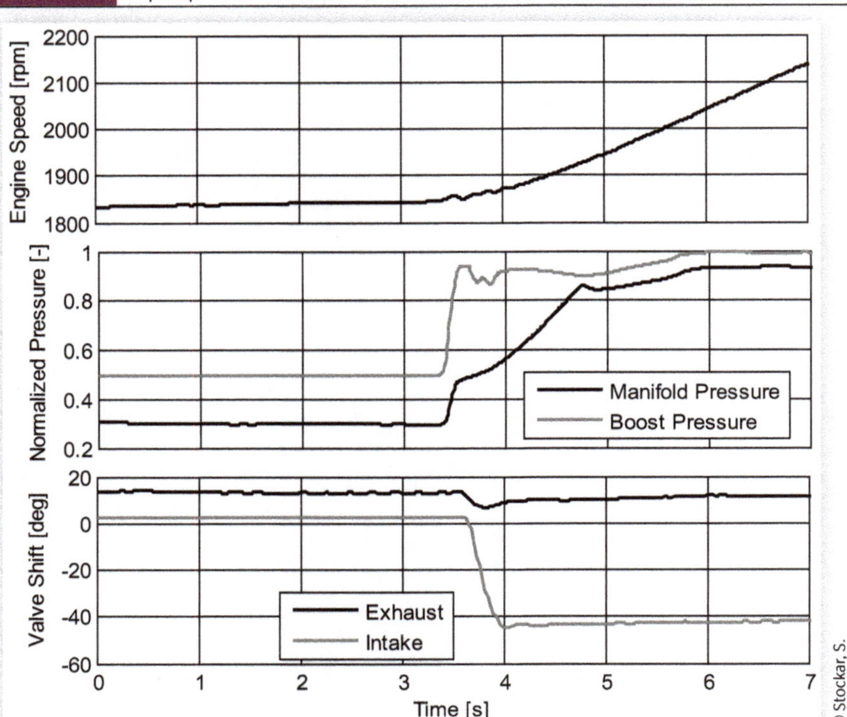

at the end of the cycle that could be due to the cylinder-to-cylinder interaction and the simplifications introduced in modeling the BCs at the flow splits.

To evaluate the accuracy of the reduced-order model in predicting the exhaust system pressures and flow rates during transient conditions, a simple tip-in/tip-out case is considered. For this simulation, the reference profiles for the engine speed, the intake manifold and boost pressure, and the valve timing were used as inputs to the GT-Power model, as shown in Figure 6.14.

In the GT-Power model, two decoupled PI feedback controllers are implemented to modify the throttle and the waste-gate valve positions during the transient in order to track the desired pressure profiles. Since the reduced-order model only describes the exhaust, the waste-gate valve position calculated from GT-Power is used as an additional input to the ROM.

Figure 6.15 compares the predictions of the GT-Power and the ROM in time- and crank-angle domain. During the tip-in/tip-out transient, the results in the time domain for the reduced model are obtained by applying a moving average filter to the crank-angle solution, with a buffer of five engine cycles. The ROM shows very close predictions of the exhaust pressure at the locations compared, namely, exhaust port of Cylinder 1 and turbine inlet. The comparison of the crank-angle resolved variables also shows good agreement between the GT-Power and the ROM prediction. While the ROM slightly underestimates the pressures, the errors in absolute values remain limited and the trend is captured well by the model. During the sharp change in the reference torque profiles between 3 and 5.5 s, the model agrees well with GT-Power. The results are particularly promising given the fact that the ROM is implemented in MATLAB/Simulink in nonlinear state-space form, in which the 72 states correspond to the local values of the thermo-fluid variables throughout the system (density, mass flow rate, stagnation internal energy).

A comparison of computation time between GT-Power and the reduced model was conducted for the simulation of the transient profile described in Figure 6.14. In this

FIGURE 6.15 Transient simulation results and comparison: exhaust port pressure (left) and turbine inlet pressure (right).

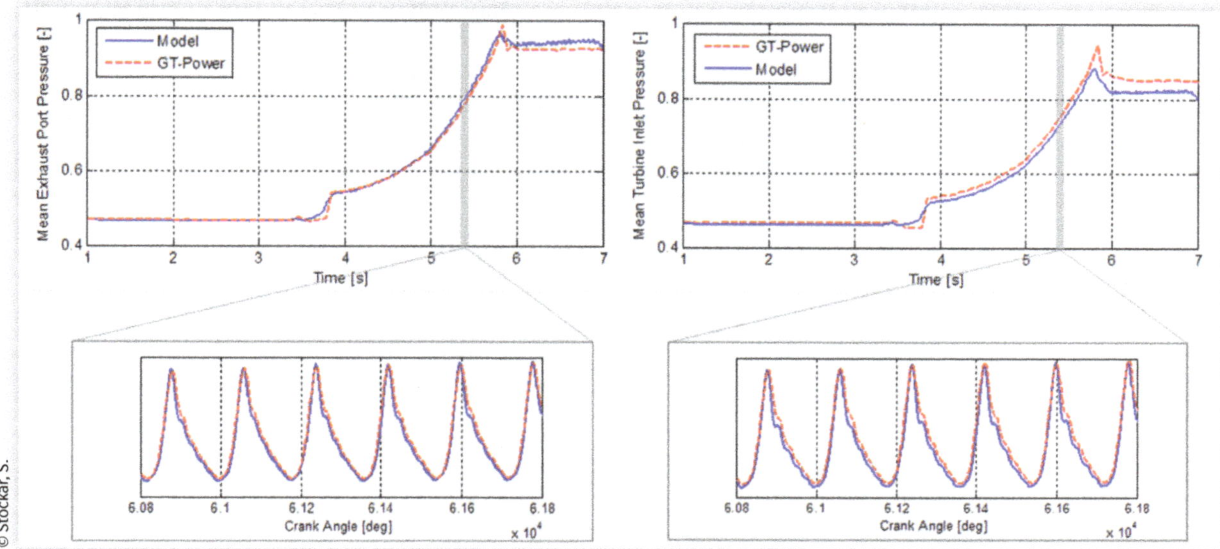

case, both models were executed on an Intel Core i7-4470K CPU at 3.50 GHz with 16 GB of RAM. The real-time factor for GT-Power was 85, while the corresponding value for the reduced model was 8.9, indicating that the reduced model achieves comparable accuracy to GT-Power with computation time almost ten times shorter.

Finally, the ROM was compared against the low-frequency, mean-value model calibrated against GT-Power simulation data for steady-state operating conditions. In particular, at the end of the transient, Figure 6.16 shows that there is a spatially dependent distribution of the average exhaust pressure. This spatial dependence is particularly evident after the waste-gate opens during the load transient (at 6 s). Here, the waste-gate valve opening causes a sudden localized drop of the pressure that, due to the inertance and compliance of the exhaust system, does not fully propagate backward. Because of the geometry of the exhaust and influence of BCs, assuming that a local average pressure in the exhaust plenum can be representative of the entire system (as commonly done in mean-value modeling) is not accurate.

FIGURE 6.16 Comparison between the reduced-order model and the mean-value model of the exhaust system.

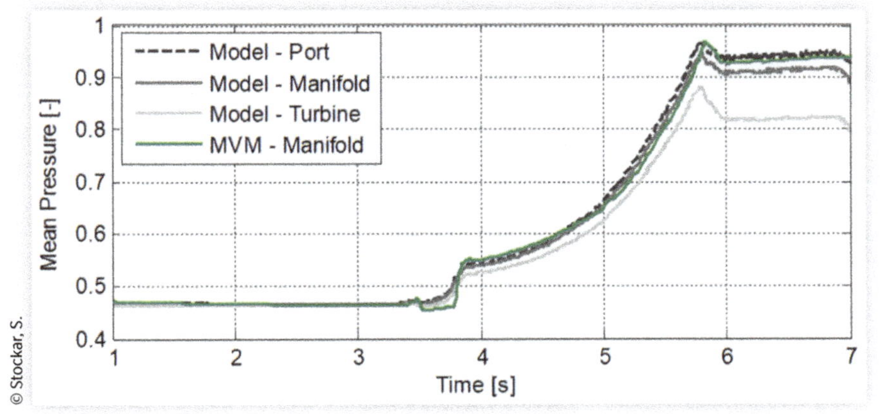

The case study presented shows that using the mean exhaust manifold pressure to calculate the turbine flow rate and power output would lead to significant errors, warranting the need for models that can capture the distributed nature of the physical system and the spatial dependence of the thermo-fluid properties. In this sense, the proposed reduced-order model retains the information about the location and, if necessary, can also provide high-frequency information on the pressure distribution in the exhaust system. On the other hand, a limitation of such approach is the number of states required for a complete characterization of the system, which is compatible for SIL and HIL simulation requirements but relatively large for application to feedback control algorithm design.

6.7 Summary

This chapter introduced the theory of MOR in the context of synthesizing control-oriented models starting from the fundamental equations of 1D gas dynamic models.

Definitions, Acronyms, and Abbreviations

BC - Boundary condition
CFD - Computational fluid dynamics
EVO - Exhaust valve opening
FVM - Finite volume method
FVSM - Flux vector splitting method
HIL - Hardware-in-the-loop
MFD - Manifold filling dynamics
MOC - Method of characteristics
MOR - Model order reduction
MVM - Mean value model
ODE - Ordinary differential equation
PDE - Partial differential equation
POD - Proper orthogonal decomposition
SBF - Spatial basis function
SI - Spark ignited
SIL - Software-in-the-loop
VVT - Variable valve timing

References

1. Moulin, P. and Chauvin, J., "Modeling and Control of the Air System of a Turbocharged Gasoline Engine," *Control Engineering Practice* 19, no. 3 (2011): 287-297.
2. Eriksson, L., Frei, S., Onder, C., and Guzzella, L., "Control and Optimization of Turbo Charged Spark Ignited Engines," *IFAC World Congress*, Barcelona, Spain, Vol. 15, 2002.
3. Pfiffner, R., Weber, F., Amstutz, A., and Guzzella, L., "Modeling and Model-Based Control of Supercharged SI-Engines for Cars with Minimal Fuel Consumption," *Proceedings of the 1997 American Control Conference* (Cat. No. 97CH36041), Piscataway, NJ, Vol. 1, 304-308. IEEE, 1997.

4. Lake, T., Stokes, J., Murphy, R., Osborne, R., and Schamel, A., "Turbocharging Concepts for Downsized DI Gasoline Engines," SAE Technical Paper 2004-01-0036, 2004, doi:10.4271/2004-01-0036.
5. Petitjean, D., Bernardini, L., Middlemass, C., and Shahed, S.M., Advanced Gasoline Engine Turbocharging Technology for Fuel Economy Improvements. SAE Technical Paper 2004-01-0988, 2004, doi:10.4271/2004-01-0988.
6. Weber, F. and Guzzella, L., "Control Oriented Modeling of a Pressure Wave Supercharger," SAE Technical Paper 2000-01-0567, 2000, doi:10.4271/2000-01-0567.
7. Eriksson, L., Nielsen, L., Brugård, J., Bergström, J., Pettersson, F., and Andersson, P., "Modeling of a Turbocharged SI Engine," *Annual Reviews in Control* 26, no. 1 (2002): 129-137.
8. Guzzella, L., Wenger, U., and Martin, R., "IC-Engine Downsizing and Pressure-Wave Supercharging for Fuel Economy," SAE Technical Paper 2000-01-1019, 2000, doi:10.4271/2000-01-1019.
9. Meacham, G.B.K., "Variable Cam Timing as an Emission Control Tool," SAE Technical Paper 700673, 1970, doi:10.4271/700673.
10. Stefanopoulou, A.G., Cook, J.A., Freudenberg, J.S., Grizzle, J.W., Haghgooie, M., and Szpak, P.S., "Modeling and Control of a Spark Ignition Engine with Variable Cam Timing," *American Control Conference, Proceedings of the 1995*, Seattle, WA, Vol. 4, 2576-2581. IEEE, 1995.
11. Doebelin, E., *System Dynamics: Modeling, Analysis, Simulation, Design* (New York: Marcel Dekker, 1998).
12. Karnopp, D., Margolis, D.L., and Rosenberg, R.C., *System Dynamics: A Unified Approach*, 2nd ed., Vol. 514. (New York: Wiley, 1990), xiv.
13. Palm, W.J., *System Dynamics* (New York: Wiley, 2005).
14. Brown, F.T., *Engineering System Dynamics: A Unified Graph-Centered Approach* (Boca Raton, FL: CRC Press, 2006).
15. Dudek, K., Montello, B., Meyer, J., Mohler, S.M., Guezennec, Y., and Yurkovich, S., "Rapid Engine Calibration for Volumetric Efficiency and Residuals by Virtual Engine Mapping," *International Congress on Virtual Power Train Creation*, 2007.
16. Hendricks, E., Chevalier, A., Jensen, M., and Sorensen, S.C., "Modeling of the Intake Manifold Modeling Dynamics," *International Congr. & Expo.*, Detroit, MI, 1996.
17. Fons, M., Muller, M., Chevalier, A., Vigild, C., Hendricks, E., and Sorenson, S.C., "Mean Value Engine Modelling of an SI Engine with EGR," SAE Technical Paper 1999-01-0909, 1999, doi:10.4271/1999-01-0909.
18. Hendricks, E. and Sorenson, S.C., SI Engine Controls and Mean Value Engine Modelling, SAE Technical Paper 910258, 1991, doi:10.4271/910258.
19. Guzzella, L. and Onder, C., *Introduction to Modeling and Control of Internal Combustion Engine Systems* (New York: Springer Science & Business Media, 2009).
20. Eriksson, L. and Nielsen, L., Modeling and Control of Engines and Drivelines (New York: John Wiley & Sons, 2014).
21. Lee, B., "Methodology for Rapid Static and Dynamic Model-Based Engine Calibration and Optimization," Ph.D. thesis, The Ohio State University, 2005.
22. Guerrier, M. and Cawsey, P., "The Development of Model Based Methodologies for Gasoline IC Engine Calibration," SAE Technical Paper 2004-01-1466, 2004, doi:10.4271/2004-01-1466.

23. Rask, E. and Sellnau, M., "Simulation-Based Engine Calibration: Tools, Techniques, and Applications," SAE Technical Paper 2004-01-1264, 2004, doi:10.4271/2004-01-1264.

24. Schlosser, A., Kinoo, B., Salber, W., Werner, S., and Ademes, N., "Accelerated Powertrain Development through Model Based Calibration," SAE Technical Paper 2006-01-0858, 2006, doi:10.4271/2006-01-0858.

25. Follen, K., Canova, M., Midlam-Mohler, S., Rizzoni, G., Lee, B., Matthews, G., and Folkerts, C., "Development of a High-Fidelity Engine Model for Powertrain Control Design and Validation," *Proc. of 9th Stuttgart International Symposium on Automotive and Engine Technology*, Stuttgart, Germany, 2009.

26. Pearson, R.J. and Winterbone, D.E., "A Rapid Wave Action Simulation Technique for Intake Manifold Design," *SAE Transactions* (1990): 1495-1504.

27. Pearson, R.J. and Winterbone, D.E., "The Simulation of Gas Dynamics in Engine Manifolds Using Non-Linear Symmetric Difference Schemes," *Proceedings of the Institution of Mechanical Engineers, Part C: Journal of Mechanical Engineering Science* 211, no. 8 (1997): 601-616.

28. D'Errico, G. and Onorati, A., "Thermo-Fluid Dynamic Modelling of a Six-Cylinder Spark Ignition Engine with a Secondary Air Injection System," *International Journal of Engine Research* 7, no. 1 (2006): 1-16.

29. Morel, T., Silvestri, J., Goerg, K.-A., and Jebasinski, R., "Modeling of Engine Exhaust Acoustics," SAE Technical Paper 1999-01-1665, 1999, doi:10.4271/1999-01-1665.

30. Follen, K., Canova, M., Midlam-Mohler, S., Guezennec, Y., Rizzoni, G., Lee, B., and Matthews, G., "A High Fidelity Lumped-Parameter Engine Model for Powertrain Control Design and Validation," *ASME 2010 Dynamic Systems and Control Conference*, Cambridge, MA, pp. 695-702. *American Society of Mechanical Engineers*, 2010.

31. Benson, R.S., Annand, W.J.D., and Baruah, P.C., "A Simulation Model Including Intake and Exhaust Systems for a Single Cylinder Four-Stroke Cycle Spark Ignition Engine," *International Journal of Mechanical Sciences* 17, no. 2 (1975): 97-124.

32. Cipollone, R. and Sciarretta, A., "The Quasi-Propagatory Model: A New Approach for Describing Transient Phenomena in Engine Manifolds," SAE Technical Paper 2001-01-0579, 2001, doi:10.4271/2001-01-0579.

33. Cipollone, R. and Sciarretta, A., "A New Modelling for the Air and Gas Dynamics in ICE Manifolds Oriented to Air-Fuel Ratio Control," *Spring Technical Conference of the ASME-ICE 1999*, Columbus, IN, 1999, Vol. 32, 103-113.

34. Cipollone, R. and Sciarretta, A., "On the Air Dynamics in ICE Intake Manifolds the Development of a Quasi-Propagatory Model," in A. Tornambe, G. Conte, and A.M. Perdon (eds), *Theory and Practice of Control and Systems* (Singapore: World Scientific, 1998), 189-197.

35. Cavina, N., Borelli, A., Calogero, L., Cevolani, R., and Poggio, L., "Turbocharger Control-Oriented Modeling: Twin-Entry Turbine Issues and Possible Solutions," *SAE International Journal of Engines* 8, no. 5 (2015): 2120-2132.

36. Cavina, N., Migliore, F., Carmignani, L., and Di Palma, S., "Development of a Control-Oriented Engine Model Including Wave Action Effects," SAE Technical Paper 2009-24-0107, 2009, doi:10.4271/2009-24-0107.

37. Zhang, G.Q. and Assanis, D.N., "Manifold Gas Dynamics Modeling and Its Coupling with Single-Cylinder Engine Models Using Simulink," *Journal of Engineering for Gas Turbines and Power* 125, no. 2 (2003): 563-571.

38. Sharma, R., Nesic, D., and Manzie, C., "Model Reduction of Turbocharged (TC) Spark Ignition (SI) Engines," *IEEE Transactions on Control Systems Technology* 19, no. 2 (2011): 297-310.

39. Chatterjee, A., "An Introduction to the Proper Orthogonal Decomposition," *Current Science* (2000): 808-817.

40. Kerschen, G., Golinval, J.-C., Vakakis, A.F., and Bergman, L.A., "The Method of Proper Orthogonal Decomposition for Dynamical Characterization and Order Reduction of Mechanical Systems: An Overview," *Nonlinear Dynamics* 41, no. 1-3 (2005): 147-169.

41. Tropea, C. and Yarin, A.L., *Springer Handbook of Experimental Fluid Mechanics*, Vol. 1 (New York: Springer Science & Business Media, 2007).

42. Lumley, J.L., "Stochastic Tools in Turbulence," Courier Corporation, 2007.

43. Lumley, J.L., "The Structure of Inhomogeneous Turbulent Flows," in *Atmospheric Turbulence and Radio Wave Propagation* (Moscow: Proceedings of International Colloquium, 1967).

44. Holmes, P., Lumley, J.L., Berkooz, G., and Rowley, C.W., *Turbulence, Coherent Structures, Dynamical Systems and Symmetry* (Cambridge, England: Cambridge University Press, 2012).

45. Berkooz, G., Holmes, P., and Lumley, J.L., "The Proper Orthogonal Decomposition in the Analysis of Turbulent Flows," *Annual Review of Fluid Mechanics* 25, no. 1 (1993): 539-575.

46. Ma, X. and Karniadakis, G.E., "A Low-Dimensional Model for Simulating Three-Dimensional Cylinder Flow," *Journal of Fluid Mechanics* 458 (2002): 181-190.

47. Epureanu, B.I., Hall, K.C., and Dowell, E.H., "Reduced-Order Models of Unsteady Viscous Flows in Turbomachinery Using Viscous–Inviscid Coupling," *Journal of Fluids and Structures* 15, no. 2 (2001): 255-273.

48. Rowley, C., Colonius, T., and Murray, R., "POD Based Models of Self-Sustained Oscillations in the Flow Past an Open Cavity," *6th Aeroacoustics Conference and Exhibit*, Lahaina, HI, 2000, 1969.

49. Rowley, Clarence W., Tim Colonius, and Richard M. Murray. "Model Reduction for Compressible Flows Using POD and Galerkin Projection." *Physica D: Nonlinear Phenomena* 189, no. 1-2 (2004): 115-129.

50. Rowley, C., Colonius, T., and Murray, R., "Dynamical Models for Control of Cavity Oscillations," *7th AIAA/CEAS Aeroacoustics Conference and Exhibit*, Maastricht, The Netherlands, 2001, 2126.

51. Caraballo, E., Kasnakoglu, C., Serrani, A., and Samimy, M., "Control Input Separation Methods for Reduced-Order Model-Based Feedback Flow Control," *AIAA Journal* 46, no. 9 (2008): 2306-2322.

52. Samimy, M., Debiasi, M., Caraballo, E., Serrani, A., Yuan, X., Little, J., and Myatt, J.H., "Feedback Control of Subsonic Cavity Flows Using Reduced-Order Models," *Journal of Fluid Mechanics* 579 (2007): 315-346.

53. Yuan, X., Caraballo, E., Little, J., Debiasi, M., Serrani, A., Özbay, H., Myatt, J.H., and Samimy, M., "Feedback Control Design for Subsonic Cavity Flows," *Applied and Computational Mathematics* 8, no. 1 (2009): 70-91.

54. Horlock, J.H. and D.E. Winterbone. *The Thermodynamics and Gas Dynamics of Internal-Combustion Engines*, Vol. II (Oxford, UK: Oxford Science Publications, 1986).

55. Dobner, D.J., "A Mathematical Engine Model for Development of Dynamic Engine Control," *SAE Transactions* (1980): 373-381.
56. Tabib, M.V., Sathe, M.J., Deshpande, S.S., and Joshi, J.B., "A Hybridized Snapshot Proper Orthogonal Decomposition-Discrete Wavelet Transform Technique for the Analysis of Flow Structures and Their Time Evolution," *Chemical Engineering Science* 64, no. 21 (2009): 4319-4340.
57. Robertsson, J.O.A., Blanch, J.O., Symes, W.W., and Burrus, C.S., "Galerkin-Wavelet Modeling of Wave Propagation: Optimal Finite-Difference Stencil Design," *Mathematical and Computer Modelling* 19, no. 1 (1994): 31-38.
58. Rapún, M.-L. and Vega, J.M., "Reduced Order Models Based on Local POD Plus Galerkin Projection," *Journal of Computational Physics* 229, no. 8 (2010): 3046-3063.
59. Mason, J.C. and Handscomb, D.C., *Chebyshev Polynomials* (New York: Chapman & Hall/CRC, 2002).
60. Farge, M., Schneider, K., Pellegrino, G., Wray, A.A., and Rogallo, R.S., "Coherent Vortex Extraction in Three-Dimensional Homogeneous Turbulence: Comparison between CVS-Wavelet and POD-Fourier Decompositions," *Physics of Fluids* 15, no. 10 (2003): 2886-2896.
61. Stockar, S., Canova, M., Xiao, B., Buckland, J., and Dai, W., "Modeling for Estimation of Wave Action in Multi-Cylinder Turbocharged SI Engines," *IFAC-Papers OnLine* 49, no. 11 (2016): 708-713.
62. Stockar, S., Canova, M., Xiao, B., Dai, W., and Buckland, J., "Fast Simulation of Wave Action in Engine Air Path Systems Using Model Order Reduction," *SAE International Journal of Engines* 9, no. 3 (2016): 1398-1408.
63. Stockar, S., Canova, M., Xiao, B., Buckland, J., and Dai, W., "High-Frequency Exhaust Port Pressure Estimation Using a Reduced Order Wave Action Model," *2018 Annual American Control Conference (ACC)*, Milwaukee, WI, 2431-2436. IEEE, 2018.
64. Winterbone, D.E. and Pearson, R.J., *Theory of Engine Manifold Design: Wave Action Methods for IC Engines* (Professional Engineering Pub., 2000).
65. Stockar, S., Canova, M., Guezennec, Y., Della Torre, A., Montenegro, G., and Onorati, A., "Modeling Wave Action Effects in Internal Combustion Engine Air Path Systems: Comparison of Numerical and System Dynamics Approaches," *International Journal of Engine Research* 14, no. 4 (2013): 391-408.
66. Stockar, S., Canova, M., Guezennec, Y., and Rizzoni, G., "A Lumped-Parameter Modeling Methodology for One-Dimensional Hyperbolic Partial Differential Equations Describing Nonlinear Wave Propagation in Fluids," *Journal of Dynamic Systems, Measurement, and Control* 137, no. 1 (2015): 011002.
67. Stockar, S., Canova, M., Guezennec, Y., Della Torre, A., Montenegro, G., and Onorati, A., "Model-Order Reduction for Wave Propagation Dynamics in Internal Combustion Engine Air Path Systems," *International Journal of Engine Research* 16, no. 4 (2015): 547-564.
68. Montenegro, G., Onorati, A., Della Torre, A., and Torregrosa, A.J., "The 3dcell Approach for the Acoustic Modeling of After-Treatment Devices," *SAE International Journal of Engines* 4, no. 2 (2011): 2519-2530.
69. Schiesser, W.E., *The Numerical Method of Lines: Integration of Partial Differential Equations* (Oxford, UK: Elsevier, 2012).
70. Heywood, J.B., *Internal Combustion Engine Fundamentals* (New York: McGraw Hill, 1988).

71. Robert, W.F., Alan, M.T., and Pritchard Philip, J., *Introduction to Fluid Mechanics* (New York: John Wiley & Sons, Inc., 1998), 38.

72. Toro, E.F., *Riemann Solvers and Numerical Methods for Fluid Dynamics: A Practical Introduction* (New York: Springer Science & Business Media, 2013).

73. Sod, G.A., "A Survey of Several Finite Difference Methods for Systems of Nonlinear Hyperbolic Conservation Laws," *Journal of Computational Physics* 27, no. 1 (1978): 1-31.

74. Onorati, A., Perotti, M., and Rebay, S., "Modelling One-Dimensional Unsteady Flows in Ducts: Symmetric Finite Difference Schemes versus Galerkin Discontinuous Finite Element Methods," *International Journal of Mechanical Sciences* 39, no. 11 (1997): 1213-1236.

75. S trand, K. and Engja, H., "Bond Graph Interpretation of One-Dimensional Fluid Flow," *Journal of the Franklin Institute* 328, no. 5-6 (1991): 781-793.

76. Margolis, D., "Bond Graphs for 1-Dimensional Duct Flows Using Nonlinear Finite Lumps," *Simulation Series* 35, no. 2 (2003): 65-71.

CHAPTER 7

Modeling of EGR Systems

José Galindo, Héctor Climent, and Roberto Navarro

7.1 Introduction

Exhaust gas recirculation (EGR) constitutes an effective strategy for NOx reduction and is being applied in internal combustion engines for decades [1]. In compression-ignited (CI) engines, despite the development of complex after-treatment systems that account for NOx abatement, EGR is still attractive to most engine manufacturers and we observe, as emissions legislations evolve, both an increase of the EGR rates and the strategy being active through a widespread zone in the engine operative map. On the other side, in turbocharged SI engines, the EGR strategy arises as a promising technique to reduce fuel consumption and to avoid over-fueling [2], if sufficient energy ignition systems are available.

Recirculation of exhaust gas can be performed internally, with convenient actuation on the intake and exhaust valves, or externally, through elements connecting the exhaust and intake lines. Since internal EGR is not a complex procedure from a modeling perspective, this chapter is devoted to the external EGR operation. EGR system implementation in the engine consists of the well-known high- and low-pressure loops. Both EGR configurations present advantages and drawbacks, which lead to state that the two layouts will be used in future powertrains, even in combined applications. The modeling of these systems is a challenging task due to the events involved, which become relevant depending on the pressure loop configuration. It is obvious that both systems will include the modeling of heat transfer and pressure loss processes inside the EGR line, but more complex phenomena have to be considered.

Fouling phenomena is also a remarkable topic due to its influence on the thermal effectiveness and pressure loss of EGR coolers. Soot deposition and removal mechanisms have to be modeled for the prediction of the cooler performance, which has a strong influence on EGR gas temperature. The chapter includes a review of the modeling techniques in this field.

Optimization of the EGR system components and architecture through all the engine steady state running conditions becomes relevant but we should also remark that the study of engine transient operation is of high interest in homologation and real driving situations. Therefore, the EGR transport in the intake line toward the cylinder is an important matter to ensure the appropriate EGR concentration in the cylinders during transient conditions and, hence, control NOx emissions. These transport delays, which may differ between high-pressure (HP) and low-pressure (LP) configurations, can be analyzed in the frame of a 1D code as shown by Cornolti et al. [3]. In any case, the interaction of the engine transient type with the EGR control valves arises as a big concern.

This chapter will provide an insight into the relevant phenomena mentioned before and its modeling. Hopefully, some hints to interested readers are presented based on real engine applications.

7.2 Modeling of EGR System

In this section, a list of the different EGR system layouts and components is given. Additionally, a discussion on the issues that can appear when modeling such elements.

Generally speaking, EGR system modeling is not essentially different from the exhaust and intake systems as it is explained in other chapters in this book. This modeling is usually carried out by 1D wave action or by 3D CFD, depending on the objectives of the study. However, the nature of the recirculated flow has specific features that will be addressed throughout this chapter.

The first and simplest layout includes only a duct (EGR duct) connecting the exhaust and intake lines and a controlling valve (EGR valve). In this very simple circuit, the flow is essentially 1D within the EGR duct though it can be highly pulsating as it is the exhaust pressure. Wave dynamics is thus an important feature when modeling the EGR flow that can be accurately solved by means of 1D models. The flow in the extraction from the exhaust line, around the valve, and in the EGR introduction in the intake line are highly 3D, but they can be easily characterized with pressure loss coefficients as a function of the valve opening, the junctions' geometry, and air and EGR flow rates [4]. These coefficients can be measured in a steady flow rig and used later within a 1D model in a quasi-steady manner. Yet, there is a phenomenon that can only be accounted by 3D CFD and that can have a significant impact on engine behavior. It is the case of the mixing process between the intake fresh and the exhaust recirculated gas. The mixing is sometimes not sufficiently homogeneous so that the engine cylinders aspirate gas with different EGR rate. This is referred to as EGR dispersion in cylinders and may have a huge impact on engine combustion and pollutant emissions [5]. In order to reduce EGR dispersion, some engines have an EGR mixer. In Section 7.3, a discussion on the modeling issues of EGR mixing is given.

Very soon it was realized that the NOx control was more effective if the recirculated gas was cooled, so that a heat exchanger (EGR cooler) was added to the system [6]. There are two options according to where the EGR valve is located, either upstream the cooler (EGR valve on the hot side) or downstream (EGR valve on the cold side). The effect of cooling the EGR flow has an important side effect, the condensation of part of the components of the gas, mainly hydrocarbons and water vapors [7]. For this reason, cooling down the recirculated gas below the corresponding dew temperature is avoided. Typically, EGR coolers are fed with engine coolant fluid at around 90°C, over dew temperature (Figure 7.1). Nevertheless, during operation at cold ambient temperature or during cold starting condensation may be an issue not only in the cooler but in the entire EGR system.

FIGURE 7.1 Schematic representation of the typical elements found in an EGR line.

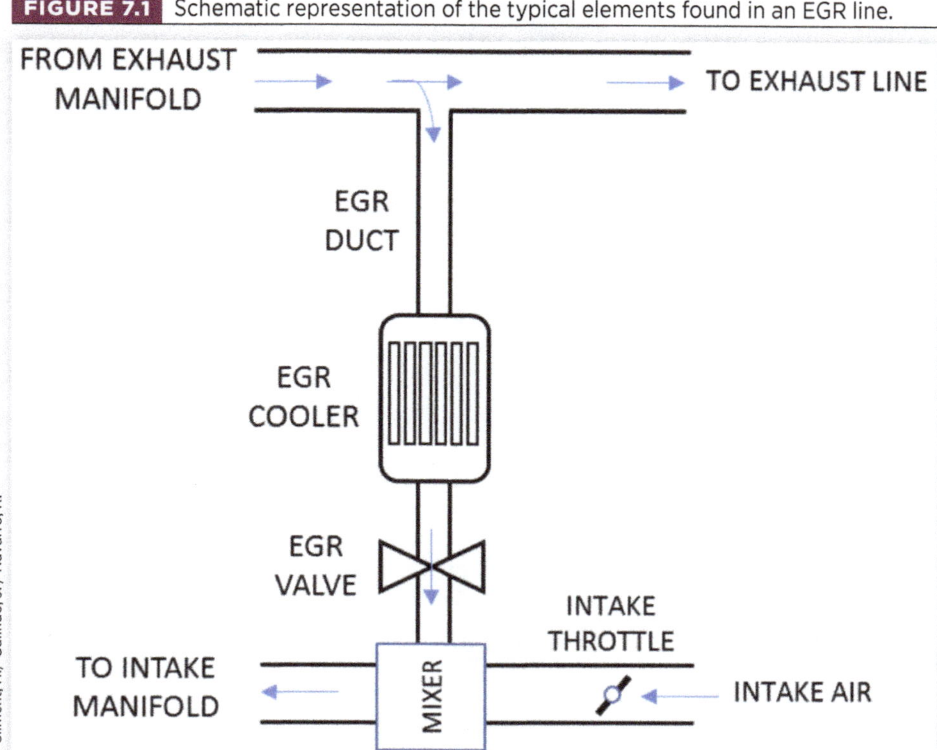

This condensation phenomenon leads to a biphasic flow that is far more complicated to calculate and that may have an impact on the engine behavior. First, the liquid condensates may be harmful for some engine components. In Section 7.4, a discussion of the issues that arise when modeling the condensation phenomenon in the EGR system is given. Secondly, the condensates may also promote the fouling of the engine components, leading in the long run to malfunctions such as the blockage of the EGR valve or cooler clogging. In Section 7.5, a discussion on how to deal with the fouling issue is given.

The recirculated flow is driven by the pressure difference between exhaust and intake circuits, which must be positive to produce the flow in the correct direction. In naturally aspirated engines, exhaust pressure is normally greater than intake pressure so it leads to a positive pressure difference. In turbocharged engines, the exhaust-intake pressure difference depends mainly on exhaust temperature and turbomachine efficiencies. Typically, in smaller displacement engines the pressure difference tends to be positive because small turbomachines with low efficiencies are used. As the engine displacement increases, the exhaust-intake pressure difference is reduced and even sometimes is reversed. In these situations, additional devices are needed to allow exhaust gases recirculation from the exhaust to the intake. In the first case, a valve (throttle valve) can be used either to increase exhaust pressure or to decrease intake pressure leading to a positive pressure difference. Pressure pulses in an exhaust manifold may help promoting at least temporary regular direction flow. The use of variable geometry turbines in turbocharged engines has also been used for this purpose. Closing the turbine mechanism increases upstream exhaust pressure. In petrol engines, the intake throttle used for engine load control can be useful to increase pressure difference. However, throttling the intake or exhaust lines has a penalty on engine fuel consumption. Hence, alternative systems have been developed to increase exhaust to intake pressure difference such as

Venturi pipes, reed valves, rotating disks, and others [8]. Such elements have been implemented in heavy-duty engines, where fuel consumption is a concern.

From the point of view of modeling, the use of components to increase the EGR flow is not a big concern. Intake or exhaust throttles (ETs) can be characterized in steady flow rig by their pressure loss as a function of the throttle position and flow rate. The pressure loss coefficients may be used later in a complete 1D engine model in a quasi-steady manner. Alternatively, 3D CFD modeling may be interesting to optimize the valve design or to quantify the impact of the throttle on the mixing process in the case of intake throttles. Other increasing EGR flows such as variable geometry turbines have modeling issues that are addressed in other chapters in this book. Other components have specific modeling techniques. Reed valves can be calculated as an effective section as a function of the pressure difference if the opening frequency is low, but they need a dynamic modeling of the reed movement, if the opening frequency is of the order of its natural frequency. Rotating disks can be modeled as an effective section as a function of time or crank angle. EGR Venturi pipes can be characterized by steady pressure loss coefficients for simpler 1D modeling or detailed 3D CFD modeling can be used instead if the impact on the mixing process is investigated.

Less common is the use of EGR compressors (EGR pump) to pump the recirculated gas from exhaust to intake lines when the pressure difference is negative. The modeling strategy is usually to employ the steady compressor performance map as explained in Chapter 4.

In some cases, it is preferred to remove particulate matter from the recirculated flow, then a particulate filter is included in the EGR line. More information on the modeling strategies of these elements is given in Chapter 11.

In naturally aspirated engines, the EGR line goes usually from the point with the highest pressure in the exhaust line (exhaust manifold) to the one with the lowest pressure in the intake line (intake manifold) in such a manner that the pressure difference is maximized and so can be the recirculated mass flow. In turbocharged engines different layouts are possible. In the so-called high-pressure EGR or short route EGR system, the exhaust gas is extracted from the manifold upstream the turbine and it is injected downstream the turbocharger compressor. In the so-called low-pressure EGR or long route EGR system, the gas is extracted from the exhaust line downstream the turbine, or even downstream the particulate filter if used. The gas is reintegrated in the intake line upstream the turbocharger compressor (Figure 7.2). The other two possible combinations are hardly ever used. Extracting high-pressure gas upstream the turbine and introducing it at

FIGURE 7.2 Schematic representation of the high- and low-pressure EGR configurations.

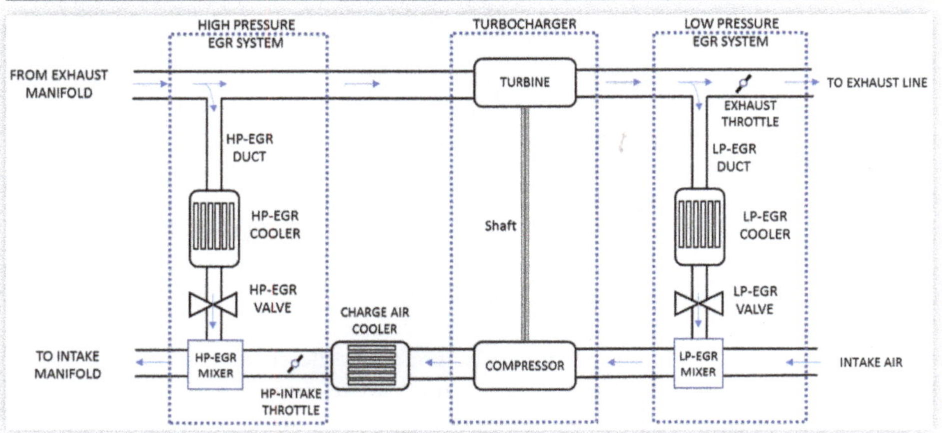

low-pressure compressor inlet has the advantage of a great pressure difference and therefore the possibility of producing high EGR rates. However, it reduces the turbine flow and increases the compressor flow, worsening the turbocharging performance. The opposite possibility of extracting low-pressure gas at turbine outlet and injecting it at compressor outlet would need an EGR pump to overcome the pressure difference. Engines with hybrid EGR systems including short and long route systems are becoming quite usual since they allow to optimize EGR flow in different engine conditions (e.g., speed/load operation, steady/transient operation, and hot/cold operation) [9].

In the case of dual stage boosting systems, the number of possible EGR layouts increases exponentially since there exist three pressure levels in the exhaust and intake systems. However, short route, long route, or hybrid EGR systems are typically used.

The modeling of short route, long route, or hybrid EGR systems does not have particular issues other than those commented above. Yet, it can be remarked that short route EGR normally has an issue related to EGR dispersion since the recirculated gas is usually injected close to or within the intake manifold. This problem will be treated in Section 7.3. Also, long route systems may lead to the deterioration of the turbocharger compressor wheel as the EGR flow may have abrasive particulates or water condensates. The particulates can be removed extracting the exhaust flow downstream the particulate filter or including a filter within the EGR line. As explained above, water condensation could reveal during engine cold starting or even when mixing with cold ambient air just upstream the compressor. The condensation issue in long route EGR systems will be covered in Section 7.4. The introduction of EGR flow upstream the turbocharger compressor may have an impact on its performance because of the flow distortion and the temperature heterogeneity.

The so-called internal EGR is quite a different strategy to the already described EGR systems, where the flow is carried from exhaust to intake lines. In this case, the mixing of the exhaust and fresh gases is done within the cylinder by a proper activation of the intake and exhaust valves. Internal EGR can be performed with fixed valve timing by an activation of the exhaust valve during the intake opening period or by an opening of the intake valve during the exhaust period. Internal EGR is more easily controlled with a variable valve actuation system.

Another strategy that has been used to produce external EGR is to dedicate one of the cylinders of the engine to pump EGR from exhaust to intake [10].

An aspect that is very important in the engine operation is the transient behavior of the EGR flow. Indeed, EGR flow is in nature dynamic since it is driven by the highly pulsating exhaust flow. Moreover, the actuation on the EGR valve leads to the transient evolution of EGR flow. The control of EGR flow within the cylinders can be a challenge if the EGR mixer is far from them, as is the case in long route EGR. 1D modeling is then the right tool to track the EGR movement throughout the intake system and to predict the evolution of recirculated gas arriving to the cylinders. In Section 7.6, some examples of dynamic operation of the LP EGR system are presented.

A final comment considers the difficulty of the assessment of the modeling of EGR systems through experimental tests. First, the nature of EGR flow is hot and pulsating, uneasy to be measured at a running engine. Second, EGR systems are typically very compact and there is little room for instrumentation. The consequence is that EGR flow properties (mass flow species and concentration) are usually assessed by indirect measurements. The situation is even more difficult if a detailed modeling of the more complex phenomena related to the EGR function is wanted. Advanced instrumentation devices in both time and spatial domains are usually needed when modeling phenomena such as air-EGR gas mixing, cylinder-to-cylinder dispersion, vapors condensation, or fouling. Some examples are given in the following sections.

7.3 Modeling EGR Mixing and Cylinder-to-Cylinder Dispersion

As already discussed in Section 7.2, there are two ways of recirculating the exhaust gas, provided that the engine under study is turbocharged. The most popular arrangement in current automotive engines (as of 2019) consists in connecting the high-pressure side of the exhaust line, that is, upstream the turbine, with the intake piping downstream the compressor. This approach is known as HP EGR or short route EGR, due to the geometrical proximity of the piping locations to be connected.

One of the drawbacks of HP EGR is the non-homogeneous EGR distribution among the cylinders, due to the poor mixing between the EGR and the intake of fresh air. This uneven EGR distribution is also an issue for naturally aspirated engines, which can be only overcome by LP EGR applications since the air-EGR mixing is enhanced due to the large distance to the cylinders and the effect of the compressor [11]. Such cylinder-to-cylinder EGR dispersion increases emissions, therefore being a key issue to be considered. Particularly, if a cylinder presents a lower EGR rate than the target one, its NOx generation will be greater than intended. On the other side, when the EGR content of a cylinder is above a certain threshold, particulate matter (PM) formation is boosted. In this section, an approach is sought to model the impact of the relevant parameters (intake manifold and EGR mixer design, amplitude and phase of fresh air and EGR pulses, etc.) on EGR distribution. This methodology should allow engine developers to minimize the NOx penalty due to EGR dispersion.

7.3.1 1D Simulations

To model the cylinder-to-cylinder EGR distribution, one could rely on the usual 0D/1D global engine model. This approach requires no additional computational effort from the baseline engine simulations. However, the manifold is solved in 1D, so the only phenomenon affecting EGR dispersion that can be modeled with this approach is the offset between air and EGR pulses. Indeed, the shift between EGR pulses coming from the exhaust and each cylinder intake event can promote uneven EGR distribution. However, not only the temporal but also the spatial coordinate plays a role in determining the EGR mixing. Hence, the accuracy when predicting dispersion with this simple method is deteriorated as the manifold geometry differs substantially from its simplified 1D representation. Therefore, this approach should be discarded to model cylinder-to-cylinder EGR distribution in manifolds with significant three-dimensional features.

7.3.2 1D-3D (Non-Coupled) Simulations

If a representative prediction of EGR dispersion is sought, the inlet manifold requires a three-dimensional resolution. Depending on how the 1D model is employed for obtaining the boundary conditions for such 3D CFD transient simulation, a non-coupled simulation or a coupled 1D-3D co-simulation could be conducted.

In order to perform a non-coupled 1D-3D simulation, a standard 0D/1D code should be first used to model the whole engine in the working point of interest, with the corresponding 1D depiction of the intake manifold. The objective of such a simulation is to obtain suitable boundary conditions for the subsequent 3D simulation of the manifold. Transient traces of mass flow rate, temperature, pressure, species concentration, etc., located at the manifold main inlet, EGR inlet, and each runner outlet, must be obtained as a result of the periodic solution achieved with the 1D code. With these values, the 3D CFD boundary conditions can be defined to run the simulation until achieving a periodic behavior and then obtaining the EGR distribution. There is no consensus between researchers on the proper selection of boundary condition types. Some works [12] set the velocity/mass flow rate at the inlet and pressure at the outlet boundaries while others [13] employ the opposite scheme, and finally there are researchers [14] that specify the transient mass flow rate traces at all boundaries.

The 3D definition of the inlet manifold allows a proper resolution for convection transport, which covers advection (bulk flow) and diffusion (concentration gradient under turbulent conditions) mechanisms. An accurate resolution of air and EGR transportation into the cylinder has a clear impact on predicting EGR distribution. Even though advection is also resolved in 1D codes, the 3D model of the manifold takes into account geometrical features that affect the bulk flow transportation, which should also improve the prediction of air and EGR pulses phase shift. In fact, Reifhart et al. [15] noticed a significant difference in the attenuation of the EGR pulses between 1D and 3D simulations, since the 1D method does not properly model the air and EGR mixing, resulting in an unrealistic *stiffness* of the pulses.

Sakowitz et al. [13] indicated that the molecular diffusivity presents a negligible impact on EGR mixing in comparison with advection and flow pulsations. Usual automotive flow regimes are turbulent, creating vortices and other secondary flows that increase EGR mixing, which can be considered as a mechanism that boosts the (turbulent) diffusion. However, there is a wide spectrum of turbulence modeling approaches, featuring different solving capabilities and computational efforts. Researchers have employed unsteady RANS ([12, 14]) as well as scale-resolving simulations, such as large eddy simulations (LES) ([13, 15]), to predict cylinder-to-cylinder EGR dispersion. Sakowitz et al. [13] obtained a better agreement with experimental measurements when modeling EGR distribution with LES than with k-ε unsteady RANS, but the strong case dependency does not help to reach a conclusion on the impact of turbulence modeling in EGR mixing. The type of manifold, number of cylinders, and working points are one of the most important variables that will strongly affect the performance of a suitable modeling approach.

Besides, some engines incorporate a mixer in order to reduce the spatial dispersion between the EGR and fresh air streams. In fact, if the EGR mass fraction were to be homogeneous inside the intake manifold, the cylinder-to-cylinder dispersion would be zero. 3D CFD here is mandatory to consider the effect of the EGR mixer, particularly noticing the impact that the mixer design may have on EGR distribution [14].

The 3D consideration of the intake manifold should improve the EGR mixing prediction in comparison with the 1D approach due to the aforementioned reasons. However, the boundary conditions have been obtained by means of an engine 1D model featuring an inlet manifold representation that may be unable to provide an accurate response. One should assess whether the variables at the boundaries of the 3D domain that are not set as boundary conditions (e.g., pressure in the case of an inlet mass flow rate) are coherent with the values predicted by the 1D code or not. If there is a significant discrepancy, it means that the 1D depiction of the intake manifold is behaving in a different way from the 3D simulation. It is worth noting that usually 1D engine models are not fully predictive, so experimental measurements at different locations (from the studied engine or similar ones) are employed to calibrate the 1D model coefficients (friction losses, heat transfer, etc.). In this way, the 1D model calibration coefficients can be fine-tuned to match the response of the 3D manifold. Then, an additional 1D simulation can be performed to obtain new boundary conditions for the 3D domain. Running the 3D code again would hopefully provide a better result provided that the behavior of the manifold in 1D is now closer to its three-dimensional counterpart. This manual feedback loop can be conducted as a sort of soft coupling between the 1D and 3D CFD codes.

7.3.3 1D-3D (Coupled) Co-simulations

When attempting to predict cylinder-to-cylinder EGR distribution, the most ambitious modeling approach is to perform a (hard) coupling between the 1D and 3D codes, that is, the standard procedure of employing a 1D code is followed to model the whole engine save for the intake manifold, which is addressed by a 3D CFD code. In this way, the 1D and 3D codes have a real-time communication (co-simulation) and the boundary conditions received by the 3D model are the most representative ones, which should improve accuracy in the prediction of EGR dispersion.

However, some problems arise when employing the coupled approach. Two CFD codes with different numerical schemes, discretization, etc., are communicating almost at each time step, which sets constraints over the number and type of species to be solved, the time-step size, the location of the boundaries of the domain that act as interfaces [16], etc. Moreover, there is an increase in the computational effort when performing co-simulations due to the additional cycles that are required to achieve a periodic state in comparison with uncoupled 1D-3D simulations [17].

Dimitriou et al. [17] showed that the EGR dispersion with the uncoupled 3D approach is very sensitive to the initial boundary conditions set at the 3D domain, and therefore is strongly dependent on the accuracy of the 1D model. The 1D-3D co-simulation was assessed as more robust regarding the 1D model quality. However, spurious backflows were detected when the intake valves were closed, modifying to the cylinder-to-cylinder EGR dispersion. This fact may affect not only the accuracy but also the stability of the co-simulations, indicating that there is plenty of room for improvement on the coupling of 1D and 3D codes.

7.4 Modeling Generation of Condensates in Low-Pressure Applications

Section 7.2 already introduced the possibility of the condensation of some of the components of the exhaust gas when cooling the EGR flow before reintroducing it in the intake line. If hydrocarbons condense, the formation of deposits can lead to EGR valves sticking [18]. However, in this section only the condensation of water vapor is addressed, since it is the source of potential turbocharger failure when employing LP EGR. Uneven cylinder-to-cylinder distribution due to HP EGR (see Section 7.3) together with the lower engine fuel consumption and higher achievable EGR rates and compressor surge margin when employing LP EGR [3] are tipping the balance to the long route EGR variant. Hence, the generation of condensates is a topic of current interest in the scope of engine modeling.

Psychrometry studies the thermodynamics of gas-vapor mixture, therefore covering humid air (combination of dry air and water vapor) phenomena, which can be represented in a so-called psychrometric diagram, depicted in Figure 7.3. One of the psychrometric properties of such mixture is the dew point, that is, the temperature at which the air is saturated with water vapor for a given specific humidity (w). Assuming thermodynamic equilibrium, one can expect water to be condensed when $T_{air} < T_{dew}$. T_{dew} is increased

FIGURE 7.3 Psychrometric diagram for humid air at p = 1 atm.

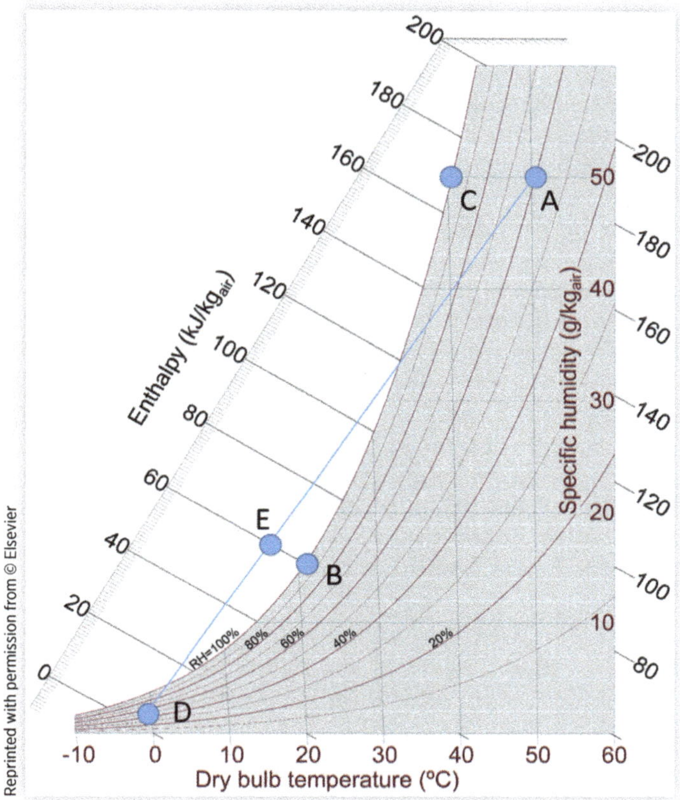

with specific humidity, so the exhaust gas is more prone to present condensation than fresh air, due to its greater amount of water content coming from the combustion process. Serrano et al. [19] show the dependence of specific humidity with fuel-to-air equivalence ratio for a given fuel composition.

Condensation in EGR streams of naturally aspirated engines or turbocharged engines employing HP EGR should not pose a major problem, since the water would be finally digested by the cylinders. When using LP EGR configurations, though, condensation of water at the LP EGR cooler outlet or at the junction located at the compressor inlet is certainly an issue to be addressed. Water droplets could be formed, aggregated, and eventually would impact the compressor wheel, thus damaging and wearing the impeller. The current section is devoted to show how standard 0D/1D engine modeling can be enhanced with psychrometric considerations to predict water condensation due to long route EGR usage. The modeling approach strongly depends on the system at which condensation is produced (LP EGR cooler or LP EGR junction), so separate subsections are devoted to each potential source of condensation.

7.4.1 Condensation in LP EGR Cooler during Engine Warm-Up

Employing a cooler is particularly mandatory for a LP EGR stream. Apart for the increase of EGR flow density that improves the engine volumetric efficiency, standard compressor wheels have a tight overtemperature constraint, which requires the inlet flow to be below a certain temperature threshold. Even though the LP EGR flow can enter the cooler at temperatures greater than 300°C, the high efficiency of such coolers manage the EGR flow to achieve temperatures close to the engine cooling water, which could be about 90°C for a steady-state point. However, if the engine remains stopped for a long period under cold weather conditions, the initial coolant temperature when the engine starts would be that of the ambient (e.g., below 0°C) and water vapor could condensate within the EGR cooler, with the subsequent dragging of the droplets that will arrive to the compressor impeller.

This is issue can be explained by looking at the psychrometric diagram shown in Figure 7.3. For an inlet specific humidity of w_{in} = 50 g/kg, if the EGR stream at the outlet of the cooler is at 50°C (point A), the relative humidity (RH) is 60% and no condensation is expected to happen at the cooler. If the outlet temperature is 42°C instead (point C; dew conditions), the air is saturated with water vapor. Consequently, if the EGR temperature is below the dew temperature, the air cannot withstand the amount of water as vapor, therefore triggering condensation. For example, for an EGR cooler outlet temperature of 22°C (point B), the flow contains a maximum specific humidity of w_{out} = 15 g/kg, so the difference with the incoming specific humidity of w_{in} = 50 g/kg would be transformed into liquid water. Following this theoretical approach, a simple model can be implemented in order to quantify the amount of condensed water m_{cond} produced in the EGR cooler during an engine warm-up time interval (T_{w-up}):

$$m_{cond}\left(T_{w-up}\right) = \int_{0}^{T_{w-up}} \dot{m}_{cond}\left(t\right) \cdot dt \tag{7.1}$$

$$\dot{m}_{cond}\left(t\right) = \left[w_{in}\left(t\right) - w_{out}\left(t\right)\right] \dot{m}_{dry\ air}\left(t\right) \tag{7.2}$$

where $\dot{m}_{cond}\left(t\right)$ and $\dot{m}_{dry\ air}\left(t\right)$ are respectively the condensation and dry air mass flow rate at a certain time.

The proposed approach should be considered as a *postprocessing* tool rather than a *predictive* model by itself. A previous 0D/1D engine is required, being accurate in terms of predicting the flow conditions at the inlet of the EGR cooler, the coolant temperature evolution, and cooler efficiency. With this, an estimation of the water condensed in the EGR cooler can be obtained.

If the results of the suggested model is compared against experimental measurements, an overprediction of 20-40% in water condensates can be expected [20]. The problem is that such a simple approach is neglecting plenty of phenomena that have an impact in the condensates going out of the EGR cooler. To name a few:

- EGR flow temperature is not homogeneous across the cooler. Particularly, the EGR local temperature inside the EGR cooler is below the outlet temperature at the locations close to the coolant passages, which will produce an increase of the condensation rate predicted by Equation (7.2).
- Water droplets need to be dragged by the EGR stream from the place at which they are generated to the EGR cooler and finally to the compressor wheel. This will cause a delay between the generation of condensates and their later impact into the impeller. Moreover, some of the droplets may be trapped inside the passages of the EGR cooler, increasing their residence time and allowing a fraction to be evaporated by the unsaturated EGR stream when the engine warm-up is finished. This effect reduces the condensation predicted by Equation (7.2) and modifies its distribution over time.

- The orientation of the cooler decides whether gravity plays a role evacuating the generated droplets toward the outlet of the EGR cooler or recirculating them to the cooler inlet instead.

The prediction of water condensation in the EGR cooler can be enhanced if a 1D thermal model of the EGR cooler is developed or 3D simulations are considered; however, this is outside the scope of the current chapter.

7.4.2 Condensation in the Fresh Air/LP EGR Junction

The previous section has explained a modeling approach for LP EGR condensation that happens only at engine warm-up, since the coolant temperature is above EGR dew temperature when achieving stationary conditions. However, condensation in the junction between the main air path and the LP EGR branch can happen not only during engine warm-up but also at steady-state points.

Again, Figure 7.3 explains why such a situation may happen. If the EGR flow at the cooler outlet presents the psychrometric conditions represented by point A and meets a cold air stream (point D) in the LP EGR junction, a first approximation of global conservation equations provides that the conditions of the blended stream must be located in the straight line that connects both points, at a particular location dictated by the "lever rule." If the EGR rate is moderate, the psychrometric conditions of this mixture could be represented by point E in Figure 7.3. Since such a point is oversaturated, the actual conditions would be found at the iso-enthalpy line passing through point E but at RH = 100 %, that is, point B. Again, the difference in specific humidity would cause water vapor to be condensed in the junction, just upstream the compressor.

Serrano et al. [19] developed a "perfect mixing" model to predict condensation rate at the junction by considering the solution given by conservation equations to the perfect mix of the fresh air and LP EGR streams. Such a tool can be implemented in standard 0D/1D engine codes to analyze a worst-case scenario in terms of condensation. This model is useful to assess in which working points condensation can pose a threat to the engine.

However, 3D CFD simulations are required in case an accurate prediction of the condensation rate is sought. A condensation submodel to be implemented in 3D codes was developed and verified by Serrano et al. [21] and validated by Galindo et al. [22]. The condensation submodel only increases the computational effort of a standard 3D simulation by 20%, but the 3D calculations are orders of magnitude slower than 0D/1D engine simulations, so it is not feasible to replace the 0D model of a junction by its 3D counterpart. An approach to limit the computational cost could be to perform 3D simulations of the junction alone without the compressor, since its impact on condensation can be neglected [23, 24] only in several representative engine working points. A coefficient of performance of the junction (η_{junction}) could be defined to relate the ability of the geometry to avoid condensation:

$$\eta_{\text{junction}} = \frac{\dot{m}_{\text{PM}} - \dot{m}_{\text{junction}}}{\dot{m}_{\text{PM}}} \cdot 100 \tag{7.3}$$

where $\dot{m}$ is the mass flow rate of condensates at a certain location (e.g., the compressor inducer section) for the actual junction or the perfect mixing (PM) model [19].

An average value of the junction effectiveness could be then obtained from the selected 3D simulations. With this, Equation (7.3) can be employed to obtain the predicted condensation rate at a different working point, considering that the condensates calculated by the perfect mixing model [19] bear the dependence on the psychometric inlet conditions of the junction.

7.5 EGR Cooler Fouling

This section is devoted to the modeling of the fouling phenomena that take place in the gas side of an EGR cooler due to the deposition of substances present in the exhaust gas. Current investigations have already identified the different mechanisms involved in the fouling deposition and growth of the deposit layer. Since the fouling phenomenon leads to both an increase of the EGR cooler pressure loss and a reduction on its thermal efficiency, it is encouraged that these side effects should be taken into consideration when developing the EGR system model.

The following paragraphs provide some insights into the mechanisms that appear in the fouling deposition in the internal walls of the EGR cooler. Some studies [25, 26] have demonstrated that the thermal efficiency stabilizes over time giving the idea that some removal mechanisms could also be happening. However, while fouling deposition descriptions are generally widely accepted, the deep understanding of the removal theories is unknown yet. Finally, this section does not aim to provide a detailed description concerning the fouling phenomenon such as the morphology of the deposition and its chemical or physical properties closely related to the topic. Since it is still nowadays an open issue, it falls beyond the scope of this chapter.

7.5.1 Fouling Deposition

The dominant mechanism in soot particles deposition is the thermophoretic phenomenon. For typical diesel soot particle size, the deposition velocity for different mechanisms respects the following order: thermophoresis, diffusion, gravitation, turbulent impact, and electrostatics, the thermophoretic deposition velocity being more than 100 times higher than any of the other mechanisms [27]. Thermophoresis is a phenomenon that can be observed in, among others, gaseous mixtures where particle motion exhibits a response to the temperature gradient. In the case of EGR coolers, the particles traveling with the flow are carried from the gas (hot) to the inner surface (cold) of the cooler ducts. Afterward, the particles impact the wall and stick to it. The velocity of the particles approaching the wall depend on the temperature gradient, the absolute temperature of the particle, and the kinematic viscosity of the gas, following the expression [28]:

$$V_{\text{th}} = -K_{\text{th}} \frac{\nu}{T} \nabla T \tag{7.4}$$

where K_{th} is the thermophoretic coefficient.

Condensation also plays a relevant role in EGR cooler fouling. Condensation, as commented in Section 6.4, takes place when the inner wall temperature of the cooler ducts falls below the dew point of the species at each partial pressure. Based on the species, condensation types can be classified into three categories: water vapor, unburned hydrocarbon, and acid condensation. Water vapor typically condensates at around 40°C, while sulfuric acid condensation occurs below 105°C [29]. Hydrocarbons from unburned fuel and lube oil could be heavy enough to condensate at temperatures similar to the range in EGR cooler fouling, which may lead to thick wet soot and cooler clogging.

The condensate's mass flux can be obtained using [30]:

$$j_g = K_g \rho_g \ln\left[\frac{1-y_{g,i}}{1-y_{g,o}}\right] \tag{7.5}$$

where $y_{g,i}$ and $y_{g,o}$ are the mol fraction of vapor at the interface and in the bulk mixture, respectively; ρ_g is the gas density; and K_g is a mass transfer coefficient, defined by the binary diffusion coefficient and the length of the diffusion layer:

$$K_g = \frac{D_{AG}}{\delta_m} \tag{7.6}$$

7.5.2 Removal Mechanisms

Figure 7.4 presents the thermal efficiency evolution of three high-pressure EGR coolers with three different cooling technologies when tested for several hours in steady operation on a diesel engine [31]. The thermal efficiency is calculated as the actual gas temperature decreases over the maximum temperature difference between gas and coolant. The conclusion is the same for the three devices: a drop on the thermal efficiency occurs during the first stages and an asymptotic behavior is achieved as time passes. The reduction of the thermal efficiency is linked to the fouling process. Since the thermophoresis should always be taking place (i.e., there is a temperature gradient between the bulk gas and the inner surface), a stabilization in the efficiency implies that removal mechanisms occur and the removal rate equals to the fouling deposition one.

After a closer look at the plots, we can observe some discontinuities in the thermal efficiency evolution that are linked to engine stops when performing the tests. The results are presented in a continuous time evolution but the tests were performed in different

FIGURE 7.4 Thermal efficiency evolution for three different EGR coolers technologies.

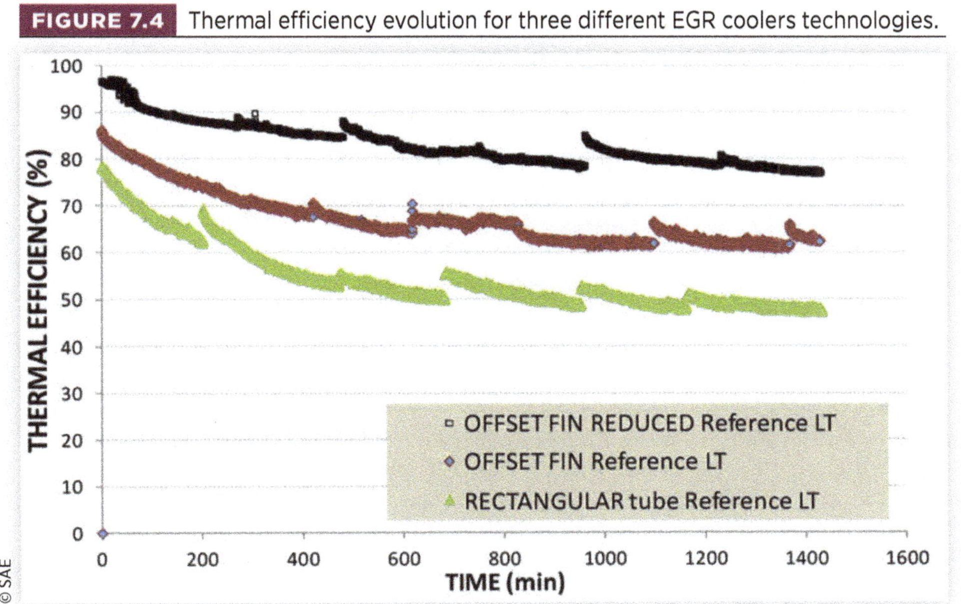

days with engine stops during night hours. The sudden increase in the efficiency could be related to some cleaning process after the engine soaks and starts again. Although the removal mechanisms are still under deep investigation, the following paragraphs describe the factors that are generally mentioned in the literature.

Shear stress created by the gas flow in the inner deposit surface is a source of particle removal. This stress correlates with the EGR mass flow through the channels inside the cooler. The higher the mass flow, the higher the gas velocity and deposits blowing off. Mud cracking, although a different mechanism, is closely related to shear stress since large cracks arise in the inner surface and deposit flakes are taken out. Another cause for deposits removal occurs when a large particle hits the interface. Its landing may throw smaller particles away from the surface and promote erosion in the interface when moving due to the gas flow velocity.

Water vapor condensation when the EGR cooler cools down and reheats reveals as another reason to explain the sudden discontinuities in the thermal efficiency after the engine stops. Water in liquid phase, and even acids from the exhaust gas, can penetrate inside the deposit layer due to its high porosity and detach the particles once it evaporates with temperature when the engine warms up. Finally, as hot exhaust gases flow, unburned hydrocarbons evaporate from the deposit layer, which leads to a shrink of the soot particles and the detachment of small soot particles in the process.

A comprehensive model of the fouling and removal mechanisms in EGR coolers is under heavy development. We have discussed about the complexity and variety of the involved processes and suggest the interested reader to search in the recent literature about the subject in order to be updated. If a detailed model of the fouling deposition and removal mechanisms is not practical in a given situation, a parametric study modifying the EGR pressure loss and thermal efficiency to account for best and worst scenarios would constitute a valuable task to perform. The literature [31, 32] is plenty of results where an interested reader could obtain typical figures for the EGR cooler performance in terms of pressure loss and thermal characteristics, similar to the plots we have presented in the previous section.

7.6 EGR Transport and Control in Transient Operation

This section is focused on the LP EGR configuration. LP EGR, among others, presents a big disadvantage versus HP EGR concerning the exhaust gas transport up to the cylinders: it takes more time to transport the EGR due to the length of the intake line [3]. New homologation cycles, such as the Worldwide harmonized Light vehicles Test Cycles (WLTC) and real driving emissions (RDE) cycles, include more engine transient operation events that are more challenging for the EGR in-cylinder control with LP configuration.

An interesting result of this section would be to quantify the delay of the exhaust gas transport in the intake manifold. In addition, the choice of the most suitable modeling approach to capture the EGR transport phenomenon from the exhaust line to the intake ports is an important outcome of this section too.

To illustrate the issues that arise in transient operation, we will analyze a sudden change in the engine load, which is usually attached also to a sudden change of the EGR rate between initial and final steady situations. Typical EGR rates in diesel engines at partial load are higher than those at full load. The most extreme case would correspond to the situation where the EGR strategy is cancelled at full load and this is the situation

FIGURE 7.5 Transient examples represented in the engine map.

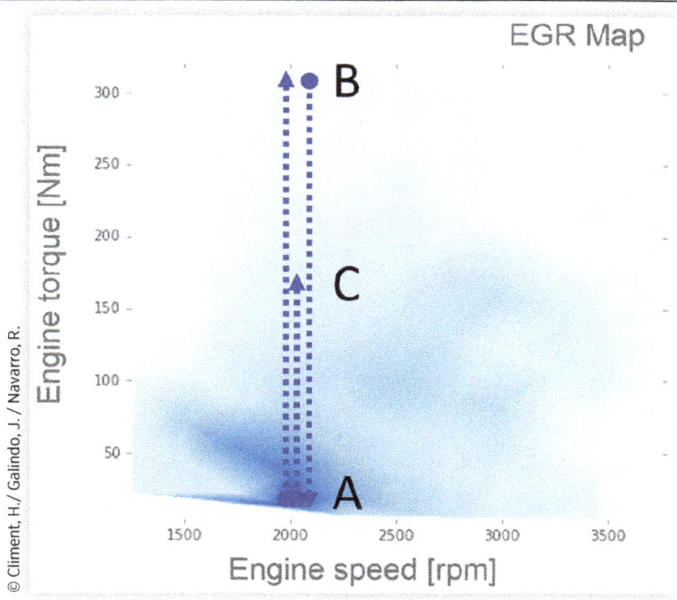

that will be considered. So the initial and final stages of the transient correspond to low and high load engine running conditions and vice versa. The mentioned transient situations are presented in Figure 7.5. In this way, we will analyze both a pedal tip-in (from point A to B) and a tip-out maneuver (B to A). A second transient type to be considered corresponds to the situation where the EGR strategy is not cancelled in the final steady engine conditions (from point A to C).

7.6.1 EGR Control in Steady Operation

Let us consider an engine with the schematic layout represented in Figure 7.6. If the LP EGR strategy is enabled, it is obvious to state that the LP EGR has to be open and, usually, regulating the air mass flow entering into the engine. The ET valve acts as a backpressure valve. It is placed in the exhaust line, downstream of the LP EGR inlet. The objective of this valve is to increase the EGR rate if it is not enough with the LP EGR valve fully open.

FIGURE 7.6 Schematic representation of the EGR system in a turbocharged engine.

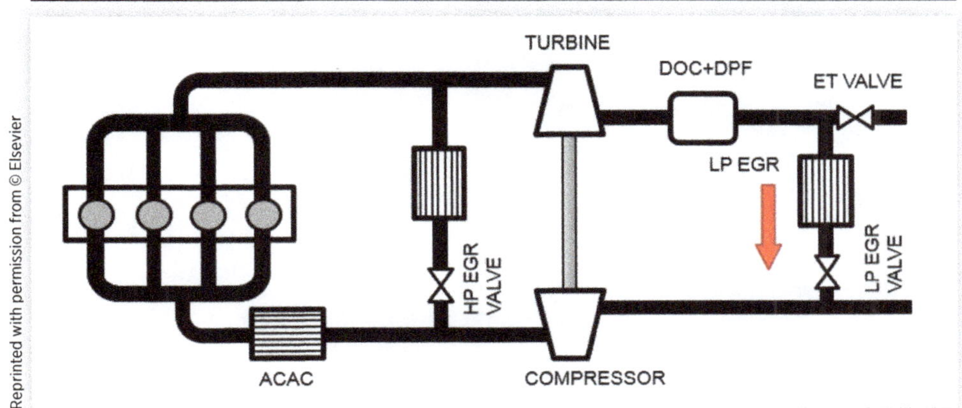

CHAPTER 7

It usually happens at low load and low speed engine conditions. Another possible configuration would be to use an intake throttle valve downstream the air filter instead of the ET. However, this configuration is not considered here.

At full load conditions or during the transient performance, the EGR strategy is usually avoided, since it is desired to allow the maximum air mass flow to enter into the cylinders. Therefore, the LP EGR valve is fully closed and the ET valve is fully open in these conditions.

7.6.2 Tip-In from Low to Full Load

This transient corresponds to the classic engine load response at constant engine speed. The engine transient starts at low load, in an operating condition where the LP EGR strategy is enabled. Suddenly, the pedal is pushed to its maximum position (full load), where the engine does not usually work inside the EGR zone.

Figure 7.7 presents a comparison between measurements and predicted results by the 1D engine model during the tip-in load transient from point A to B. The plots correspond to the EGR valve position on the left and the CO_2 fraction in the intake manifold on the right. Results show a 0.4 s delay between the measurement in the CO_2 at the intake manifold and the EGR valve closing because of the volume of the intake line. This delay highly depends on the intake line dimensions of the engine and the transient type, which impacts the flow velocity (advection speed of concentration wave).

Secondly, it is observed that the calculated results are very close to the experimental results. These results show that a 1D engine model can predict reality and it is concluded that, for this type of EGR transport, the model is valid. The initial increase in the CO_2 fraction measured in the intake line is not realistic so an issue related to the device behavior due to the large intake pressure variations might be happening. Anyway, a good model response to the EGR transport phenomenon is found when leaving the EGR zone.

7.6.3 Tip-Out from Full to Low Load

The opposite situation is the following: the engine starts at full load (outside the EGR zone) and, suddenly, the pedal is released and the engine runs in low load conditions performing LP EGR. Figure 7.8 presents, on the left, the ET valve position and, on the

FIGURE 7.7 EGR valve movement (left) and comparison between measurement and predicted results by 1D model of the CO_2 evolution in the intake manifold (right) in transient operation from point A to B.

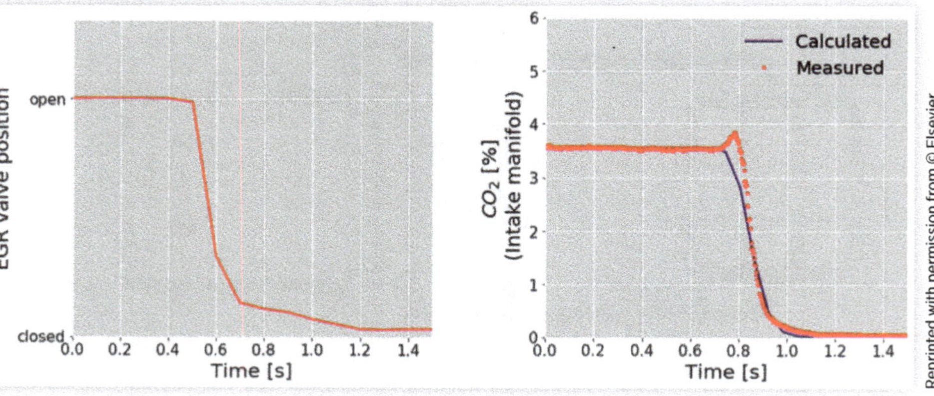

FIGURE 7.8 EGR (top left) and ET (bottom left) valves movement and comparison between measurement and predicted results by 1D model of the CO_2 evolution at the compressor inlet (top right) and in the intake manifold (bottom right) in transient operation from point B to A.

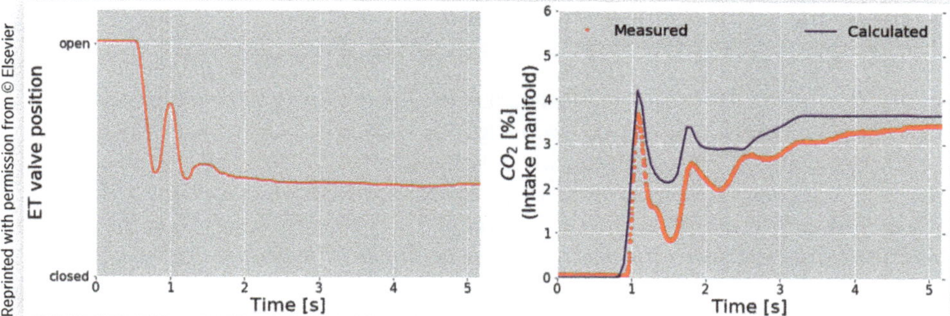

right, the CO_2 concentration in the intake manifold. In this case, the effect of entering into the EGR zone from point B to A is observed. In this tip-out operation, the control strategy makes the EGR open completely and enables the EGR strategy control to the ET valve. This valve, in a first phase, starts to close but an immediate opening peak is observed, followed again by a closing evolution up to the final position. Again, it is observed that the 1D model is able to reproduce the EGR transport in the intake line with a good agreement at least in the signal phasing. The differences in the absolute values during the transient may be attributed to a poor representation of the EGR control valve movements.

7.6.4 Tip-In from Low to Partial Load

The third transient example we want to explore is similar to the first one, but the engine load is increased without leaving the EGR zone. In a fast engine transient, although the final situation is inside the EGR zone, it is very likely that the EGR strategy is switched off in the first stage of the transient and activated in the final part. This leads to a fast operation of both EGR and ET valves in a very short period of time. The synchronization of these valves will affect the EGR transport phenomena from the exhaust to the intake manifold.

Figure 7.9 shows, as previously demonstrated, the EGR (left) and ET (middle) valve positions, and on the right it presents the CO_2 fraction in the intake manifold. In this case, the transient occurs inside the EGR zone from point A to C. As in the transient to

FIGURE 7.9 EGR (left) and ET (middle) valves movement and comparison between measurement and predicted results by 1D model of the CO_2 evolution in the intake manifold (right) in transient operation from point A to C.

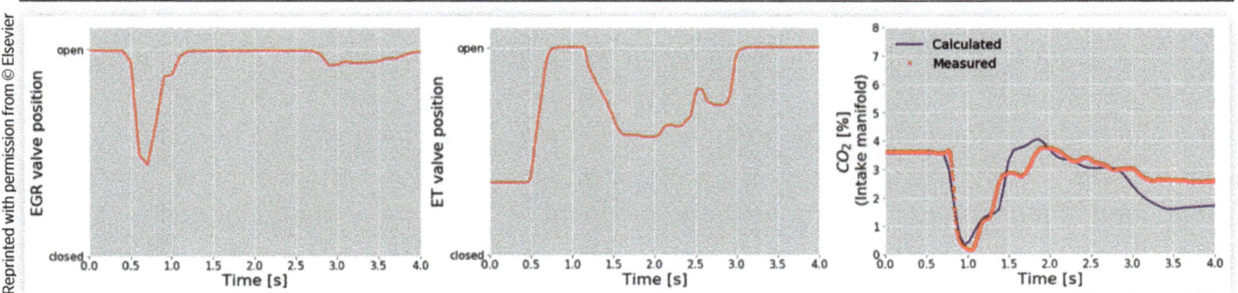

full load, it is also observed that the ECU commands a closing of the EGR valve at the initial part of the transient and an opening of the ET valve. Later, since the engine remains inside the EGR zone, the EGR valve is opened again and the ET performs the control of the air mass flow.

7.6.5 Summary on Transient Modeling

We can state that a 1D model is able to track the species transport during the engine transients when the EGR strategy is cancelled. The predicted delay in the EGR emptying the intake line is properly captured when it is compared with experimental data. In the case of EGR strategy activation or transients within the EGR zone, we remark that it is very important to move all the valves that control the EGR in the same way as the engine does. For instance, the synchronization of the EGR and ET valves is very important to avoid or reduce the overshoot effect. So in this case, the valve motion over time is the challenging factor in properly modeling the EGR transport through the intake and exhaust lines Moreover, in the case of LP EGR, the accuracy of the whole turbocharger simulation may interact with the EGR loop simulation.

7.6.6 Numerical Diffusion

Numerical diffusion is inherent to the computer simulations of fluids since, in an Eulerian approach, time and space are discretized and continuous differential equations turn into finite-difference equations. The difference between the physical and simulated systems depends on the problem conditions and the reader should be aware of this circumstance. If the applied model does not account for the diffusion phenomenon in the real system, numerical diffusion can somehow counteract this situation artificially. From a modeling perspective, the reader should know that large mesh sizes (or, in the limit, filling and emptying models) are prone to having more numerical diffusion than small mesh sizes. The results will largely depend on the situation being modeled, but a mesh size independence study is a good practice when setting up the engine model. As an example, Figure 7.10 shows an engine load transient taking place at 7 s. The pedal is pushed and

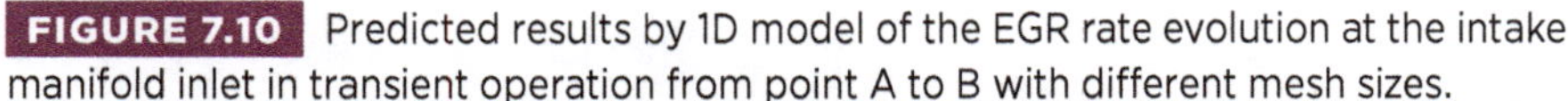

FIGURE 7.10 Predicted results by 1D model of the EGR rate evolution at the intake manifold inlet in transient operation from point A to B with different mesh sizes.

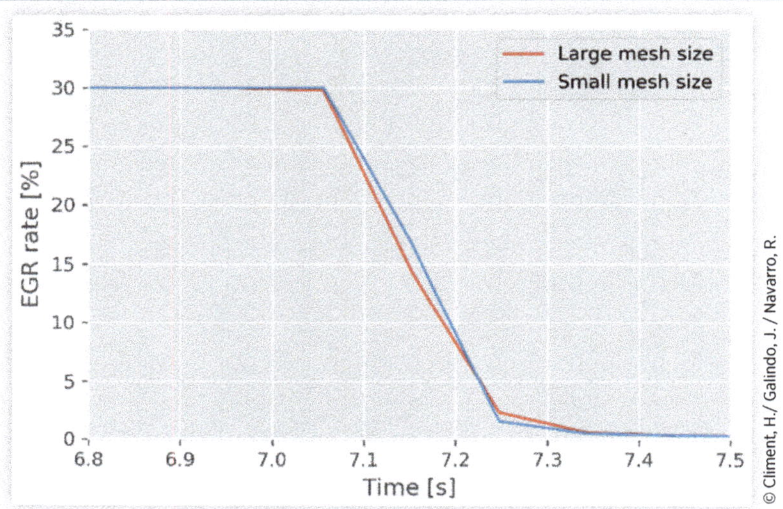

the LP EGR is switched off at the same time. Two simulations are performed with different mesh sizes for the pipes that integrate the intake manifold. The plot shows the EGR rate monitored at the intake manifold inlet. A smoother result is obtained with large mesh size since numerical diffusion is more relevant here, but one should be very careful – the numerical diffusivity is not a physical phenomenon and the mesh-size dependent results may be misleading in general.

7.7 Conclusion

In this chapter, a brief review of the benefits of the use of EGR in compression and spark-ignited internal combustion engines has been given. A description of the different architectures and components that appear in external EGR lines has also been presented.

The modeling issues of EGR lines and components have been mostly explained in other chapters of this book. The issues are related with the features of EGR flow: highly pulsating, high temperature, and multicomponent gas. However, the simulations for engine performance are usually solved with a 1D approach for the EGR ducts and 0D quasi-steady approach for components (valve, coolers, throttle, and boosters). The chapter is focused in four phenomena that cannot be accounted with such simple analysis.

Mixing of EGR gas and intake air can be an issue if it takes place close to the cylinders. A heterogeneous mixing can lead to cylinder-to-cylinder EGR rate dispersion. It has been shown that the proper methodology is a coupled 3D and 1D calculation. In this chapter, a discussion on the setting of this coupled calculation has been done.

Condensation of water, hydrocarbons, or other condensable components is an issue when cooling EGR below dew temperature. A procedure based on psychrometry is given to make estimations of condensation rate in 1D or 3D simulations.

Fouling and clogging is a big issue for EGR components since the sometimes-dirty exhaust gas submitted to condensation can render them useless. Though this phenomenon is very complex to be modeled, the key factors for fouling deposition and removal have been presented in this chapter.

The EGR system simulation in transient operation is a big concern in the engine development process. In this chapter, some transient situations such as tip-ins and tip-outs have been presented. It has been shown that a proper characterization of the valve movements is the key factor for a good modeling. Also, the relevance of the numerical diffusion and mesh size has been discussed.

Definitions, Acronyms and Abbreviations

Acronyms

0D - Zero dimensional
1D - One dimensional
3D - Three dimensional
CFD - Computational fluid dynamics
CI - Compression ignited
EGR - Exhaust gas recirculation
ET - Exhaust throttle
HP - High pressure

LES - Large eddy simulations
LP - Low pressure
NOx - Nitrogen oxides
RANS - Reynolds averaged Navier-Stokes
RH - Relative humidity

Abbreviations

in - Inlet
out - Outlet

References

1. Ladommatos, N., Abdelhalim, S.M., Zhao, H., Hu, Z., "The Dilution, Chemical, and Thermal Effects of Exhaust Gas Recirculation on Diesel Engine Emissions - Part 1: Effect of Reducing Inlet Charge Oxygen," SAE Technical Paper 961165, 1996, doi:10.4271/961165.
2. Wei, H., Zhu, T., Shu, G., Tan L., and Wang Y., "Gasoline Engine Exhaust Gas Recirculation – A Review," *Applied Energy* 99(2012): 534-544.
3. Cornolti, L., Onorati, A., Cerri, T., Montenegro, G., and Piscaglia, F., "1D Simulation of a Turbocharged Diesel Engine with Comparison of Short and Long EGR Route Solutions," *Applied Energy* 111(2013): 1-15.
4. Galindo, J., Climent, H., Guardiola, C., and Doménech, J., "Modeling the Vacuum Circuit of a Pneumatic Valve System," *ASME. J. Dyn. Sys. Meas. Control.* 131, no. 3(2009): 031011-031011-11.
5. Payri, F., Lujan, J., Climent, H., and Pla, B., "Effects of the Intake Charge Distribution in HSDI Engines," SAE Technical Paper 2010-01-1119, 2010, doi:10.4271/2010-01-1119.
6. Lázaro, J., García-Bernad, J., Pérez, C., Galindo, J., Climent, H., and Arnau, F.J., "Cooled EGR Modulation: A Strategy to Meet EURO IV Emission Standards in Automotive DI Diesel Engines," SAE Technical Paper 2002-01-1154, 2002, doi:10.4271/2002-01-1154.
7. Bourgoin, G., Tomas, E., Lujan, J., and Pla, B., "Acidic Condensation in HP EGR Systems Cooled at Low Temperature Using Diesel and Biodiesel Fuels, SAE Technical Paper 2010-01-1530, 2010, doi:10.4271/2010-01-1530.
8. Luján, J.M., Galindo, J., Vera, F., and Climent, H., "Characterization and Dynamic Response of an Exhaust Gas Recirculation Venturi for Internal Combustion Engines," *Proceedings of the Institution of Mechanical Engineers, Part D: Journal of Automobile Engineering* 221, no. 4(2007): 497-509.
9. Lapuerta, M., Ramos, A., Fernandez-Rodriguez, D., and Gonzalez-Garcia, I., "High-Pressure versus Low-Pressure EGR in a Euro 6 Diesel Engine with LNT. Effectiveness to Reduce NOx Emissions," *Thiesel 2018 Conference* (to appear).
10. Alger, T. and Mangold, B., Dedicated EGR: A New Concept in High Efficiency Engines, *SAE International Journal of Engines* 2, no. 1(2009): 620-631.
11. Luján, J.M., Pla, B., Moroz, S., and Bourgoin, G., "Effect of Low Pressure EGR on Gas Exchange Processes and Turbocharging of a HSDI Engine," *Proceedings of the 8th Conference on Thermo- and Fluid Dynamic Processes in Diesel Engines (THIESEL)*, 2008.

12. Page, V., Garner, C., Hargrave, G., and Versteeg, H., "Development of a Validated CFD Process for the Analysis of Inlet Manifold Flows with EGR," SAE Technical Paper 2002-01-0071, 2002, doi:10.4271/2002-01-0071.

13. Sakowitz, A., Reifarth, S., Mihaescu, M., and Fuchs, L., "Modeling of EGR Mixing in an Engine Intake Manifold Using LES," *Oil Gas Sci. Technol.* 69, no. 1(2014): 167-176.

14. Dimitriou, P., Burke, R., Copeland, C., and Akehurst, S., "Study on the Effects of EGR Supply Configuration on Cylinder-to-Cylinder Dispersion and Engine Performance Using 1D-3D Co-Simulation," SAE Technical Paper 2015-32-0816, 2015, doi:10.4271/2015-32-0816.

15. Reifarth, S., Kristensson, E., Borggren, J., Sakowitz, A., and Angstrom, H.E., "Analysis of EGR/Air Mixing by 1-D Simulation, 3-D Simulation and Experiments," SAE Technical Paper 2014-01-2647, 2014, doi:10.4271/2014-01-2647.

16. Riegler, U. and Bargende, M., "Direct Coupled 1D/3D-CFD-Computation (GT-Power/Star-CD) of the Flow in the Switch-Over Intake System of an 8-Cylinder SI Engine with External Exhaust Gas Recirculation," SAE Technical Paper 2002-01-0901, 2002, doi:10.4271/2002-01-0901.

17. Dimitriou, P., Avola, C., Burke, R., Copeland, C., and Turner, N., "A Comparison of 1D-3D Co-Simulation and Transient 3D Simulation for EGR Distribution Studies, *ASME 2016 Internal Combustion Engine Division Fall Technical Conference*, 2016.

18. Furukawa, N., Goto, S., and Sunaoka, M., "On the Mechanism of Exhaust Gas Recirculation Valve Sticking in Diesel Engines," *International Journal of Engine Research* 15, no. 1(2012): 78-86.

19. Serrano, J.R., Piqueras, P., Angiolini, E., Meano, C., and De La Morena, J., "On Cooler and Mixing Condensation Phenomena in the Long-Route Exhaust Gas Recirculation Line," SAE Technical Paper 2015-24-2521, 2015, doi:10.4271/2015-24-2521.

20. Galindo, J., Navarro, R., Tarí, D., and Moya, F., "Development of an Experimental Test Bench and a Psychrometric Model for Assessing Condensation on a LP-EGR Cooler," *International Journal of Multiphase Flow* (under review), 2019.

21. Serrano, J.R., Piqueras, P., Navarro, R., Tarí, D., and Meano, C. M., "Development and Verification of an In-Flow Water Condensation Model for 3D-CFD Simulations of Humid Air Streams Mixing," *Computers and Fluids* 167(2018): 158-165.

22. Galindo, J., Piqueras, P., Navarro, R., Tarí, D., and Meano, C.M., "Validation and Sensitivity Analysis of an In-Flow Water Condensation Model for 3D-CFD Simulations of Humid Air Streams Mixing," *International Journal of Thermal Sciences* 136(2019): 410-419.

23. Galindo, J., Navarro, R., Tarí, D., and García-Olivas, G., "Centrifugal Compressor Influence on Condensation due to Long Route-Exhaust Gas Recirculation Mixing," *Applied Thermal Engineering*, 144(2018): 901-909.

24. Tarí, D., "Effect of Inlet Configuration on the Performance and Durability of an Automotive Turbocharger Compressor, Ph.D. thesis, Universitat Politècnica de València, 2018.

25. Sluder, C., Storey, J., Lance, M., and Barone, T., "Removal of EGR Cooler Deposit Material by Flow-Induced Shear," *SAE Int. J. Engines* 6, no. 2(2013): 999-1008.

26. Warey, A., Bika, A., Vassallo, A., Balestrino, S., and Szymkowicz, P., "Combination of Pre-EGR Cooler Oxidation Catalyst and Water Vapor Condensation to Mitigate Fouling," *SAE Int. J. Engines* 7, no. 1(2014): 21-31.

27. Abarham, M., Hoard, J., Assanis, D., Styles, D., Curtis, E.W., and Ramesh, N., "Review of Soot Deposition and Removal Mechanisms in EGR Coolers," *SAE Int. J. Fuels Lubr.* 3, no. 1(2010): 690-704.

28. Talbot, L., Cheng, R., Schefer, R., and Willis, D., "Thermophoresis of Particles in a Heated Boundary Layer," *Journal of Fluid Mechanics* 101, no. 4(1980): 737-758.

29. McKinley, T.L., "Modeling Sulfuric Acid Condensation in Diesel Engine EGR Coolers," SAE Technical Paper 970636, 1997, doi:10.4271/970636.

30. Collier, J.G. and Thome, J.R., *Convective Boiling and Condensation*, 3rd ed. (New York: Oxford University Press Inc., 1996).

31. Bravo, Y., Arnal, C., Larrosa, C., and Climent, H., "Impact on Fouling of Different Exhaust Gas Conditions with Low Coolant Temperature for a Range of EGR Cooler Technologies," SAE Technical Paper 2018-01-0374, 2018, doi:10.4271/2018-01-0374.

32. Hoard, J., Abarham, M., Styles, D., Giuliano, J., Sluder, C., and Storey, J., "Diesel EGR Cooler Fouling," *SAE Int. J. Engines* 1, no. 1(2009): 1234-1250.

CHAPTER 8

1D Engine Model in XiL Application: A Simulation Environment for the Entire Powertrain Development Process

Feihong Xia and Jakob Andert
RWTH Aachen University

Christof Schernus
FEV Europe GmbH

The development of vehicles and the respective subsystems requires the application of simulation models with different degrees of detail that change throughout the development process. Physical tests are time consuming, costly, and often of limited use in the early stages of product development when hardware components are not yet available.

X-in-the-loop (XiL) applications using a co-simulation methodology have become powerful tools to reduce the overall validation efforts. They enable closed-loop testing of simulated or hardware components in a virtual environment as the joint execution of multiple simulation tools. However, the establishment of a feasible simulation environment faces specific challenges, especially if real-time capability is required.

This chapter provides an overview of the XiL testing approaches and highlights the effects on the virtualized vehicle development process. Exemplary use cases with a real-time capable 1D crank-angle-resolved engine model show the potential to reduce the effort for testing and validation under complex boundary conditions.

8.1 Introduction

Modern vehicles have to meet many different requirements, including high vehicle performance, excellent comfort, low fuel consumption and pollutant emissions, good durability, and low total cost of ownership. Especially new emission legislations, focusing on real-drive conditions, have extended the engine operation to critical map areas like low-end-torque and rated power, even under extreme ambient conditions, like very low or high temperatures or high altitude. To fulfil those requirements, various technologies are being developed for internal combustion engines, such as variable compression ratio, variable valve lift, injection rate shaping, or multi-stage boosting systems. Furthermore, the general trend to powertrain hybridization and electrification opens new dimensions for engine layout and powertrain optimization. The increasing system complexity leads to new challenges in the development process for the powertrain and the entire vehicle. Moreover, the time-to-market is dropping from average values of five years in the 1980s to three years in the twenty-first century [1]. To reduce the development time and costs, most of the engineering tasks are being supported by simulation models, which vary in model depth depending on the application areas and the phase in the development process.

Taking the combustion engine as an example, simple map-based approaches are being used for evaluating different powertrain system layouts based on calculating fuel consumption and vehicle performance [2]. Mean-value (MV) models, which calculate the averaged mass flow, temperature, and pressure based on simplified mass and energy balances over one engine cycle, deliver more details of the gas exchange process and are suitable for function development [3]. More accurate air path modeling can be achieved with zero-dimensional (0D) models based on the "filling and emptying method" and one-dimensional (1D) models based on the numerical solution of the 1D Navier–Stokes flow equations [4]. Those approaches also cover the modeling of processes in the combustion chamber, which can be divided into multiple subzones with ideally mixed gases. Combustion, heat transfer, and emissions are modeled with simple semi-empirical approaches. Subcomponents like valves and charging systems are implemented as characteristic maps [5]. For detailed investigation of the processes in the combustion chamber, three-dimensional (3D) models with suitable physical and chemical methods can be used [6]. Similar variants of models, built with individual tools, exist for nearly every powertrain component throughout the vehicle development process, such as detailed simulations for component design as well as simplified models for system and control applications.

As the modern powertrain is becoming more and more integrated and capable, a single component cannot be optimized without respect for the overall system performance. The system behavior heavily depends on the interactions of the different components. Hence, system integration, optimization, and testing are of essential significance for the development process. They can be supported by the application of virtual components and virtual environment simulations to reduce the high cost of prototype vehicles and to achieve an early assessment of the system behavior. These simulations are referred to as X-in-the-loop (XiL). The term "XiL" covers different in-the-loop applications, including, but not limited to, model-in-the-loop (MiL), software-in-the-loop (SiL), and hardware-in-the-loop (HiL), which are introduced in detail in [7]. To link the virtual components with each other or even with real hardware, high model flexibility regarding modeling depth, computation speed, and system interfaces are mandatory. Here, the co-simulation approach has shown significant advantages. Co-simulation is a joint execution of a set of component simulation models in their individual established tools [8]. On the one hand, it provides the possibility to reuse the component models from different development phases in their individual tools, which avoids the rebuilding and recalibration of models. On the other hand, it creates a modular structure by using standard interfaces and provides high flexibility for individual adaptations on subcomponents.

As XiL simulations for system integration and testing usually require very high computing power, map-based or MV engine models are widely used so far. However, as the computational power of work stations and real-time (RT) simulators is increasing, the use of 1D combustion engine models has become possible for XiL applications. These models have a great potential due to their high accuracy and wide variability in modeling depth. This chapter focuses on the different applications of 1D engine models in XiL simulations, while chapter 6 focuses on the 1D modeling methodologies. The chapter starts with a short introduction on the co-simulation and XiL testing approaches, highlighting the effects on the virtualized vehicle development process. Then the requirements for combustion engine modeling in XiL simulations are discussed. Exemplary use cases show the transformation of the 1D combustion engine model from the component layout phase to XiL simulation in various application areas. Moreover, the potential to reduce the effort for testing and validation under complex boundary conditions is demonstrated.

8.2 XiL Simulations

XiL simulations are widely used in the virtualized vehicle development process. This chapter represents the main idea of XiL simulations and their benefits for the development process. Additionally, two established tools for the building of a XiL platform are introduced.

8.2.1 Advantages for the Development Process

MiL and HiL are well-known terms for testing embedded systems and have proven to significantly improve the quality of the released software [9, 10]. In the last decades, such XiL tests have rapidly evolved from a tool mainly used for algorithm development and hardware validation to a system modeling and testing platform. The testing targets are no longer limited to control units but also include other system components as simulation models or real hardware on the component test bench. Fathy et al. define the HiL simulation, for example, as "a setup that emulates a system by immersing faithful physical replicas of some of its subsystems within a closed-loop virtual simulation of the remaining subsystems" [11]. This definition highlights a main characteristic of the XiL tests, namely, the closed-loop and mutual interactions between its physical or virtual constituents. Such interactions of the target testing component with other submodules are mandatory especially in transient operation conditions and provide the main advantage of XiL compared to conventional stand-alone component testing.

Furthermore, XiL simulations have the following advantages compared to testing on system level with hardware prototypes:

- Early start point: One main advantage of the XiL simulations compared to prototype building is the system integration and testing in early development phases. MiL simulations can be used as soon as the component models are available during system layout to deliver the first estimation of the system behavior. HiL simulations provide the opportunity of concurrent component testing without losing sight of the overall system. Short-term correction and optimization loops lead to early detection of errors and therefore actively reduce the development time and costs [12].

- Low cost and rapid prototyping: XiL simulations provide highly flexible component models at relatively low costs compared to prototype hardware production. By frontloading the system testing and optimization to the XiL platforms, a large number of system variants can be assessed within a much shorter time. As a consequence, the number of necessary prototype parts and the duration of the final system tests can be distinctly reduced. The additional effort for the construction of a XiL platform in the early project phase can always be overcompensated [13].
- Repeatability, safety, and flexibility: As the testing environment of XiL simulations is virtual, it provides high repeatability compared to real prototype testing. Extreme ambient conditions, critical situations, or even destructive events can be easily simulated with safety protection, so that the system and the target component can be tested over a much broader range of operating conditions.
- Transparency: Depending on the simulation method and modeling depth, the virtual components in XiL simulations deliver a certain degree of transparency into the system. This helps testing engineers to understand the system behavior, especially in the case of difficulties with measurements on real prototype hardware.

Finally, it should be noted that different types of XiL simulations have their own advantages and disadvantages. MiL simulations provide very high flexibility in model depth and computation speed. HiL simulations, on the other hand, use real hardware components and often achieve fidelity levels unattainable through purely virtual simulation. The challenges of HiL simulations are the strict RT requirements and the interface definition between virtual components and the real hardware.

8.2.2 Tools

One of the main challenges in XiL simulations is the integration of RT and non-RT systems. Individual solutions for each platform cause high efforts for system vendors and system integrators. Therefore, standardized protocols have been developed for data exchange and the control of heterogeneous systems. Two standards are introduced here: The functional mock-up interface (FMI) for offline simulations and the distributed co-simulation protocol (DCP) for RT applications.

8.2.2.1 THE FUNCTIONAL MOCK-UP INTERFACE (FMI)

The FMI standard version 1.0 [8] and 2.0 [14] were published in 2011 and 2012 as a result of the ITEA2 project MODELISAR and the Modelica Association Project FMI. Today, FMI is particularly used not only in the automotive sector but also in other industrial and scientific projects. It was initiated and coordinated by the Daimler AG. The FMI standard provides a low-level interface to exchange models with the protection of product know-how, which could be recovered from their physical models.

The FMI standard consists of two main parts. FMI for model exchange provides a modeling environment which can generate a dynamic C-Code system model in the form of an input/output block that can be utilized by other modeling and simulation environments. Models are described by differential, algebraic, and discrete equations with time, state, and step events. FMI for co-simulation intends to couple two or more simulation tools in a co-simulation environment. The data exchange between subsystems is restricted to discrete communication points. In the time between two communication points, the subsystems are solved independently from each other by their solver.

Master algorithms control the data exchange between subsystems and the synchronization of all slave simulation solvers (slaves). The interface allows standard as well as advanced master algorithms, for example, the usage of variable communication step sizes, higher order signal extrapolation, and error control [8].

8.2.2.2 DISTRIBUTED CO-SIMULATION PROTOCOL (DCP)

Beyond the well-established FMI standard, the DCP emphasizes the coupling of RT and non-RT systems, commonly found in distributed HiL setups. It has been developed in the context of the ITEA 3 framework project ACOSAR [15], which stands for "Advanced Co-Simulation Open System Architecture," and is subject to standardization as a Modelica Association Project.

DCP enables seamless integration of RT and non-RT systems by providing a tool-independent interface specification. It consists of a data model, a finite state machine, and a communication protocol including a set of protocol data units. It supports a master-slave architecture for simulation setup and control. The specification defines the design of a slave only, while the design of a master is not in the scope of the specification. The master is responsible for the orchestration and configuration of the slaves to fulfil a specific co-simulation scenario. A slave represents a subsystem of the co-simulation scenario. To exchange information between these entities, protocol data units (PDUs) are introduced [16].

8.3 Engine Simulations in the Virtualized Development Process

The engine development process includes the engine hardware and its control unit. It starts with system design, goes through hardware and software implementation, and ends with system integration. The development process is supported by various models and tested by several X-in-the-loop validations, as shown in Figure 1.

During concept design, highly simplified data-based engine models integrated in vehicle models are used to define the engine performance requirements. Thereafter, the engine hardware and the control algorithms are designed separately.

On the engine hardware side, detailed chemical, thermodynamic, and mechanical models are used to support the detailed design, for example, for the layout of the piston module, the optimization of the compression ratio, and the selection of the boosting system. During detailed design, the focus is usually on steady-state engine operating conditions. Since there is often only limited measurement data for model validation available in the design phase, the models used here should have high accuracy and good

FIGURE 1 Model-based development process.

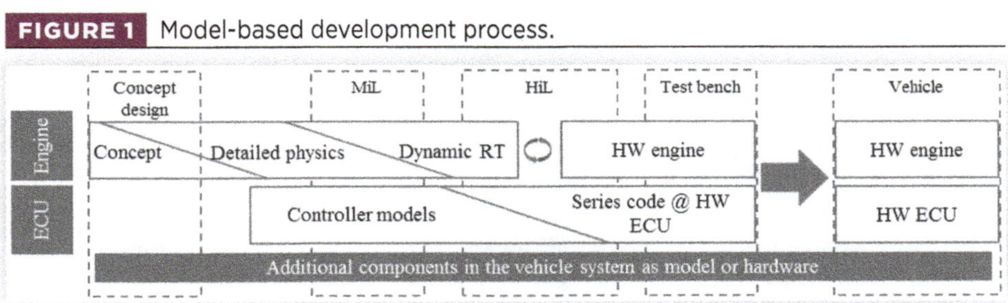

predictability. Typically, there is no stringent requirement on computation speed. Therefore, physical models like computational fluid dynamics combined with phenomenological models are often used. As soon as the first measurement data is available from, for example, single-cylinder engine testing or from first prototype engine tests, the models can be calibrated and used for measurement data analyzation and further system optimization.

On the controller side, the model-based control approach is being widely used due to the increasing complexity of the combustion engine with all its variable systems. Several examples are provided by [17]. Being limited by the computational hardware, the submodels used in the engine control unit (ECU) should be simplified as far as possible. Model inaccuracy can be minimized in the later ECU calibration process, corrected by sensed signals, or be compensated by feedback control logic.

After the detailed design of the engine hardware and its control algorithms, the first system validation can be implemented for MiL tests. At this stage, the control algorithms are coupled with the engine model to close the control loop. The detailed physical engine models from the hardware development can be directly used for the MiL simulation. Alternatively, reduced engine models with less computation effort can be generated to increase the simulation speed. The reduced models usually have less predictability and need inputs from the detailed model simulation results or the test bench measurement data. As the MiL tests are particularly helpful for testing the transient system behavior, inertias and heat capacities in the system, which could be neglected in the steady-state simulations, should be modeled accurately.

After hardware and software development and before the final integration, the combustion engine hardware and the control unit can be tested separately using HiL simulations. The combustion engine is coupled with the rapid prototyping engine control unit for performance and emission calibration. The target ECU, on the other hand, is integrated with a combustion engine model on the ECU HiL test bench. The ECU HiL tests provide not only the possibility for hardware and software validation, but also the opportunity for virtual calibration of the control functions. One of the most important requirements for the engine model in a HiL simulation is the RT capability. The actuators and sensors are also essential parts of the engine model.

The MiL and HiL simulations can be extended by other component models or hardware on test benches to a complete vehicle system simulation. The interaction of the combustion engine with the other vehicle components in transient operation conditions is of particular interest during system integration and testing. Depending on the target of the applications, the model interface for the coupled quantities and the communication frequencies should be carefully defined. While engine brake torque and fuel flow would be sufficient to study the vehicle performance and fuel consumption of a vehicle with a conventional combustion engine, a detailed pressure and temperature modeling is required for the optimization of, for example, an electrified boosting system in interaction with the electric power net components.

8.4 1D Engine Models in XiL Simulations

The various applications with the combustion engine models during different development phases have individual requirements for the modeling depth and the computational effort. Mirfendreski [18] et al. have investigated the computation speed depending on the modeling approaches for the air path and the combustion chamber. Figure 2 shows the

FIGURE 2 Computational speed of different model-concepts in percentage (QDM: quasi-dimensional, MV: mean value, M/S: master/slave cylinder) [18].

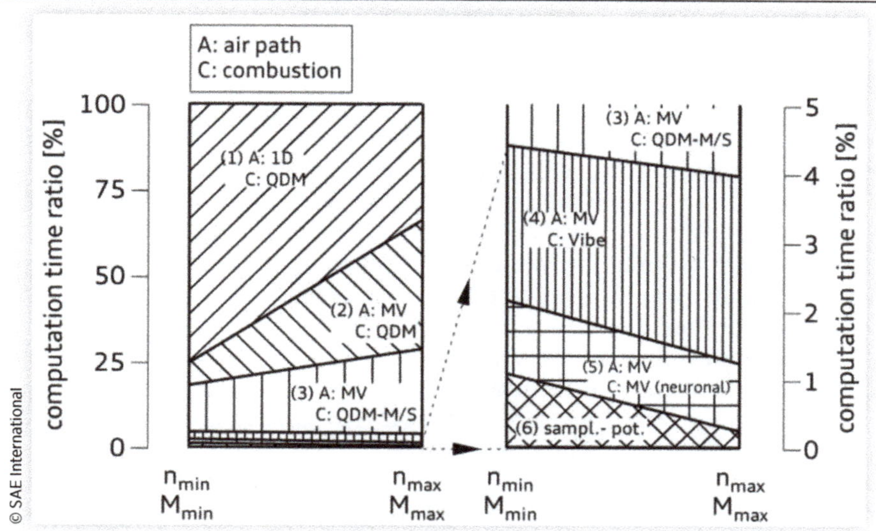

relative difference of computation speed to eliminate the influence of the processing hardware. The results indicate that the 1D air path modeling in combination with quasi-dimensional cylinder modeling could run over 100 times slower than a MV air path model with the combustion and torque production modeled as "black-box" with artificial neural networks [18].

Due to the high computation speed, the MV modeling approach with specific adaptations has been preferred for XiL simulations during the last decades [18, 19, 20, 21]. The main drawback of the MV air path modeling approach is the complete neglect of the gas dynamics. Even if combined with a crank-angle (CA)-resolved cylinder model, the gas exchange process is still simplified with the assumption of constant intake and exhaust manifold pressures. Additionally, in the case of turbocharged engines, the turbine characteristic has a strong nonlinear behavior, which makes the turbine modeling with an average pressure and mass flow inaccurate. These effects are usually compensated with correction factors, which have to be calibrated with the measurement data or simulation results from a more detailed model.

In some application areas without RT requirement, detailed 1D models with slower computation speed have been used instead of the MV models to avoid additional recalibration work [22, 23]. The detailed 1D models also provide higher accuracy in transient operating conditions. However, such simulations usually take several hours or days to calculate a driving cycle.

To find an optimum solution in the trade-off between modeling accuracy and computation speed, several studies have been carried out to simplify the detailed 1D engine model in order to speed up the simulation. One approach is to translate the 1D model with partial differential equations to a lumped parameter model with ordinary differential equations. The wave effects in the air-path are modeled by separating the inertial and capacitive properties of the fluid between finite volumes (capacities) and ducts without modification of the overall air path geometry [24]. This modeling approach has been implemented in the commercial tool WAVE-RT [25]. To further increase the computation speed, the look-up tables (e.g., valve profile and turbocharger maps) are replaced by mathematic functions and polynomials.

In the published example, a faster computation speed than the wall-clock time can be fulfilled on a PC desktop computer with a constant time step corresponding to approximately 1 deg CA at highest engine speed, depending on the CPU frequency. The tool also provides an automatic model generation feature to support the model translation.

Another approach to reach RT without losing CA-level resolution is to reduce the discretization level and increase the simulation time step while keeping the main solver algorithms unchanged. The reduction of the spatial discretization level and the number of flow elements results in a smaller amount of equations being solved within one solver time step. As an explicit flow solver is used, the solver time step is restricted by the dimension of the discrete spatial grid according to the Courant's condition [26]. The increase of the spatial discretization length also enables a larger solver time step, which reduces the times of solver integration within a fixed time window [27].

Compared to the first approach of changing the solver algorithms to ordinary differential equations, the second approach uses the same solver consistently with partial differential equations. Thus, a seamless transfer can be realized from the detailed models with over hundreds of finite volumes to simplified models with several single volumes representing manifolds and pipes. Hence, the model can be adapted freely for different application purposes. However, the balance between spatial discretization, time step size, and computation speed has to be taken into account. Specifically, as a large time step is necessary to reach high computation speed, the resolution of the simulation results is typically lower than that of the first approach. This may lead to model inaccuracies for models that require high sampling rates, for example, for phenomenological emission models that perform accurately only with a high-speed sampling of in-cylinder pressure and temperature.

The second approach with the decreased discretization level is used by the commercial software tool GT-SUITE from Gamma Technologies, LLC. [28], implemented in a separate GT-SUITE-RT solver for XiL applications. This particular solver is speed optimized with limited data logging and can run the model about two times faster than the standard solver and with less computational fluctuations [29]. An additional GT-POWER-xRT solver targeted for RT engine applications further speeds up the simulation by eliminating additional features and tuning all internal iterative algorithms for RT execution. The execution time can be reduced by a factor of ten, compared to the standard solver, allowing for significantly higher sampling rates and thus higher resolution. The tool provides an automated model reduction feature to combine volumes and calibrate heat transfer and pressure loss data.

A more detailed discussion over the model acceleration methodologies can be found in Chapter 6.

8.5 XiL Use Cases for the 1D Engine Models

For both CA-resolved RT modeling approaches, there have been various studies for different application areas of XiL simulations [30, 31, 32, 33, 34]. In the following sections, the focus will be on two use cases to demonstrate the potentials and challenges of using such a model in different stages of the vehicle development process. Both use cases are implemented using the simulation tool GT-Power.

8.5.1 System Layout, Integration, and Testing Using Co-simulation

This use case has been designed to demonstrate the potential of a pure MiL vehicle co-simulation in the system layout and integration phase. The focus is on the prediction of the powertrain performance under real-drive conditions, using component models built with individual simulation tools. A system overview of the co-simulation is shown in Figure 3. The combustion engine has been modeled following a simplified but CA-resolved approach to reach an optimized compromise between simulation accuracy and computation speed. Besides the combustion engine model, the co-simulation also includes a transmission model (built using SimulationX from ESI ITI GmbH) and a vehicle dynamics model with a virtual driver (built using automotive simulation models (ASM) from dSPACE GmbH). The ASM toolchain also provides ambient simulations for roadways and traffic with visualization. The required control functions are implemented as soft control units in SIMULINK from MathWorks, Inc. The submodels are integrated to a co-simulation using the FMI standard with SIMULINK as co-simulation master.

The reference vehicle for the simulation models is a small passenger car. Its performance and emissions have been measured on a chassis dynamometer and on the road with a portable emission measurement system. Also, the reference combustion engine has been thoroughly analyzed on an engine dynamometer test bench. The main characteristics of the reference vehicle are shown in Table 1.

FIGURE 3 System simulation structure: engine, transmission, vehicle dynamics, and control units [35]. (ECU: engine control unit, TCU: transmission control unit)

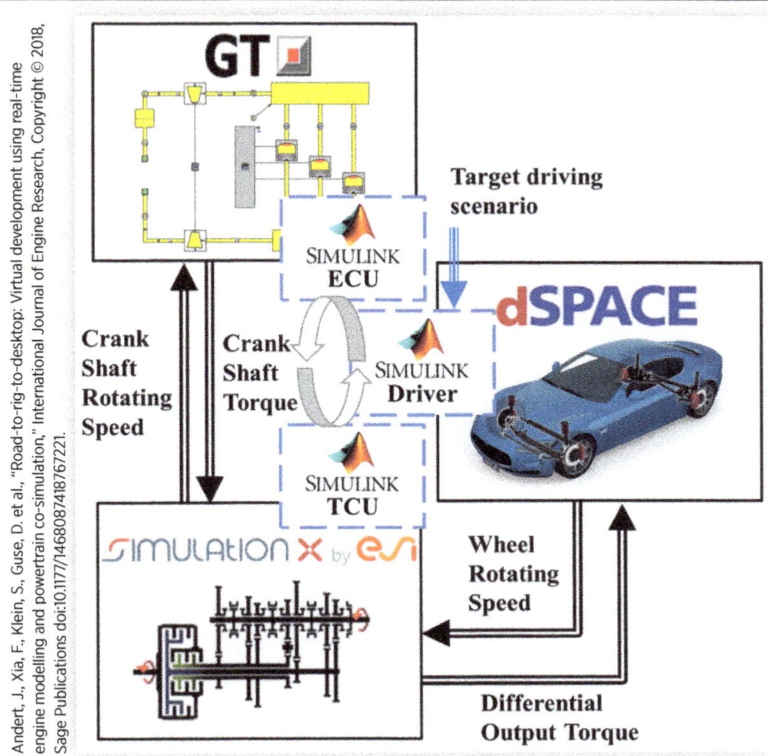

TABLE 1 Main characteristics of the reference vehicle

Vehicle weight	1.2 t
Transmission type	6-gear dual-clutch
Engine displacement	0.9 l
Rated power	66 kW
Number of cylinders	3
Boosting system	Turbocharger with wastegate

8.5.1.1 ENGINE MODEL SETUP AND VALIDATION PROCESS

Due to missing detailed geometric information about the reference combustion engine, the model has been directly generated as a simplified 1D CA-resolved model with 6 subvolumes between the air filter, the compressor, the cylinders, the turbine, and the environment, as shown in Figure 4. The turbocharger is modeled by directly interpolating and extrapolating the characteristic maps over mass flow, pressure ratio, rotating speed, and efficiency. The inertia of the turbocharger is considered in the mechanical shaft. The compressor recirculation valve is modeled using an orifice connection. The wastegate is included in the turbine model. The flow properties of the cylinder head depending on the valve lift are implemented according to detailed measurements taken in a flow laboratory. Wiebe parameters imposed as multidimensional look-up tables based on cylinder pressure measurements on the engine test bench are used for modelling of combustion processes. Since no engine cooling system is modelled yet, the wall temperatures in the system are partly imposed depending on engine power and partly calculated based on the heat capacity of the wall material and the ambient temperature. Engine friction is modeled using the Chen–Flynn [36] model depending on the maximum cylinder pressure and the mean piston velocity.

The model has been validated with measurement data from the engine dynamometer test bench. Three engine operation points are shown here as an example, representing key points in the engine map: low-end torque (n = 2000 1/min, brake mean effective pressure (BMEP) = 18 bar), rated power (n = 5000 1/min, BMEP = 16.5 bar), and a part

FIGURE 4 Overview of the engine model [35].

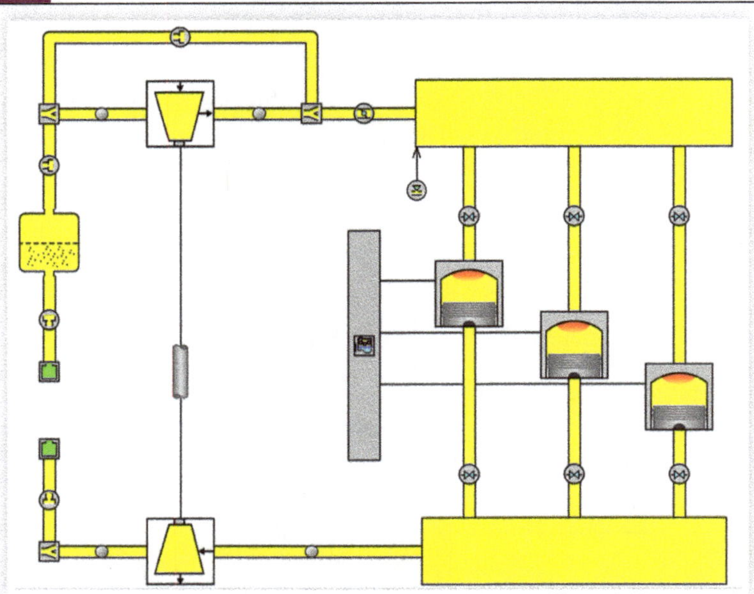

FIGURE 5 Comparison of pressure in the cylinder, intake, and exhaust manifolds between measurement and simulation [35].

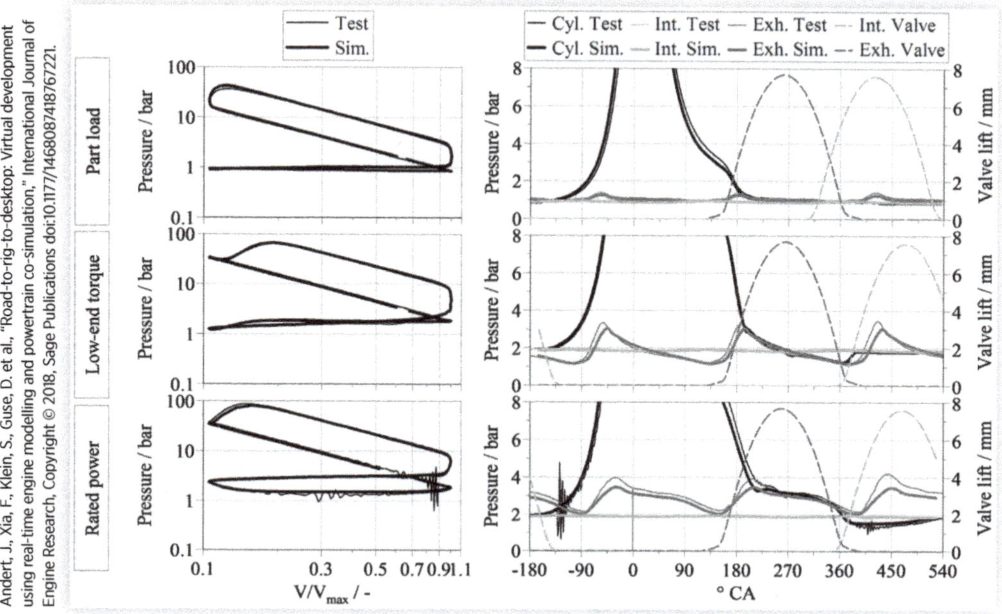

load operating point with moderate throttling and internal exhaust gas recirculation (n = 2000 1/min, BMEP = 6 bar). In Figure 5, the pressure traces in the cylinder and in the intake and exhaust manifolds are compared to the measured data. The displayed measurement data represent the average of 50 engine cycles. In order to achieve a meaningful comparability, the throttle and the wastegate have been controlled to achieve the same indicated engine torque as during the engine measurements. Also, the air/fuel ratio, injection timing, and valve timing were set exactly the same as for the measurements.

Even at the maximum engine speed, no solver instability issue is notable. The cylinder pressure, as well as the pressures in the intake and exhaust manifolds, can be simulated with satisfying accuracy within one engine cycle. This demonstrates good model predictability regarding indicated torque and volumetric efficiency.

In Figure 6, the modeling accuracy of the air path system is represented by the deviations between the modeled temperature, pressure, and brake-specific fuel consumption (BSFC) data and the measurement data. Especially, the pressure data upstream turbine indicate a good modeling accuracy of the turbocharger operation.

The RT capability of the engine model has been verified on a HiL simulator (dSPACE SCALEXIO) [38]. The HiL simulator triggers the simulation with a fixed time step of 0.5 ms, same as the communication frequency in the MiL environment. The time between the task's trigger and the end of task execution is called the turnaround time. An overrun occurs when the turnaround time is longer than the task itself. Which means the scheduler attempts to start the next task when the simulation has not finished its previous execution yet. In this case, the simulation can either be stopped or continued with the last received input values. Overruns can have a strong negative influence on the simulation execution and calculation accuracy and should in principle be avoided in the HiL implementation [37].

With a time step of 0.25 ms and a communication time step of 0.5 ms, the GT-Power solver calculates two time steps within one SCALEXIO task. To evaluate the RT capability of the engine model, different engine operating points are tested with defined engine

FIGURE 6 The absolute or relative deviation (referred to the measurement date) between the engine model and test bench measurement data in steady state engine mapping.

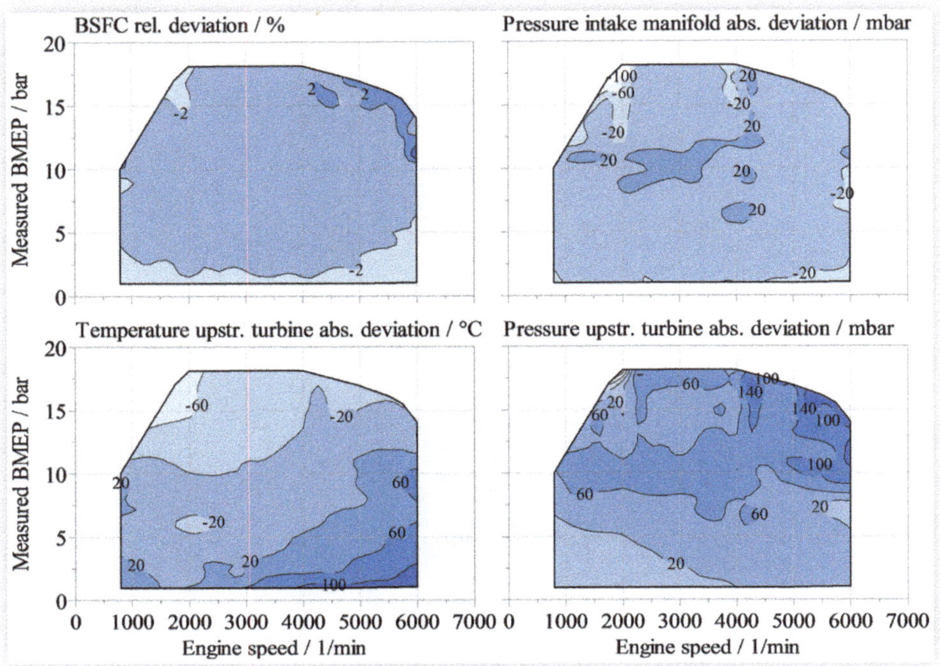

speed, throttle angle, and waste gate position. The turnaround times together with the corresponding engine speed and brake torque values for four representative operating points are shown in Figure 7.

The turnaround time of the engine model shows strong pulsations with slightly different characteristics in the different engine operating points. The amplitude of the turnaround time depends strongly on the engine load. Although a variation of turnaround time, depending on the simulation parameters, is not favorable for a real-time application, the engine model shows that with turnaround times always below 0.5 ms, no overrun in the steady-state tests occurs.

8.5.1.2 SYSTEM INTEGRATION AND SYSTEM VALIDATION

The more detailed the model is, the more control functions are necessary for the transient model operation. One of the main challenges in such a MiL application is the control function generation. As the reference ECU software is not available in this case, neither as a model nor as compiled code, a reconstruction of the absolutely necessary functions is required, which include:

- Throttle angle feed-forward control
- Variable valve timing feed-forward control
- Closed-loop idle control using throttle and ignition timing
- Wastegate feed-forward and feedback control of boost pressure
- Compressor recirculation valve activation along the compressor surge line
- Fuel cut according to torque request and engine speed
- Torque intervention from TCU during shifting

FIGURE 7 Turnaround time of the stand-alone engine model on SCALEXIO for different engine speeds and engine brake torques.

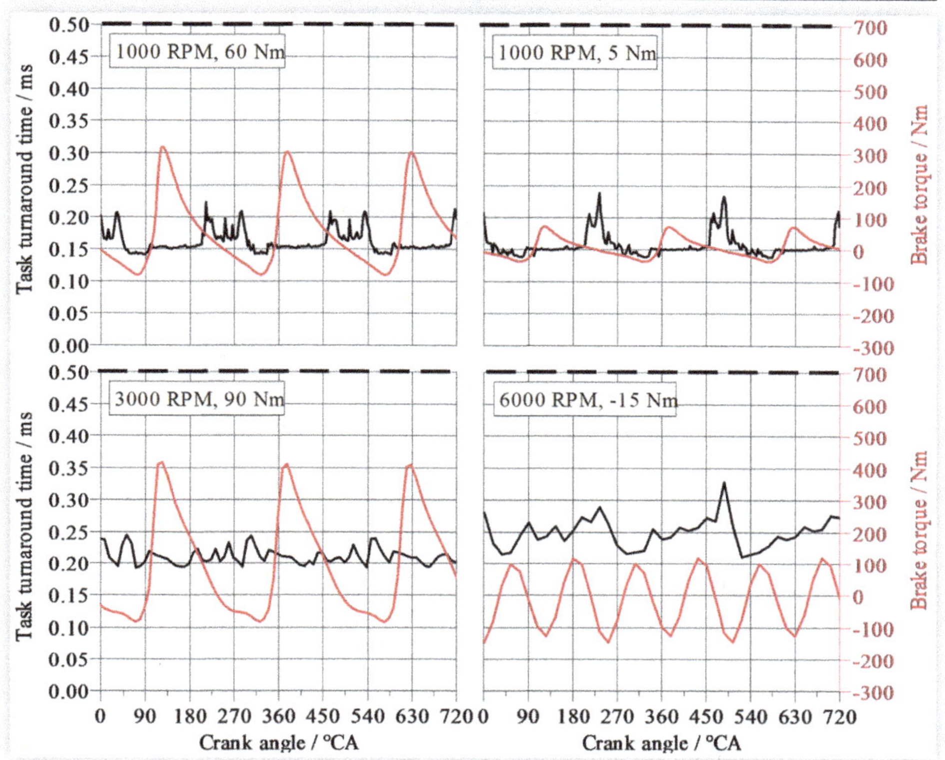

A closed-loop engine operation with the re-constructed ECU software functions has been validated with a load-step measurement on the engine dynamometer test bench. An exemplary comparison at 2000 1/min is shown in Figure 8, demonstrating that the modeled transient engine behavior is close to that of the real engine. Small deviations are caused by inaccuracy in the air path modeling and by the different calibration of the simplified control functions.

FIGURE 8 Comparison between a simulated load step at 2000 1/min and the measured data from an engine dynamometer test bench [35].

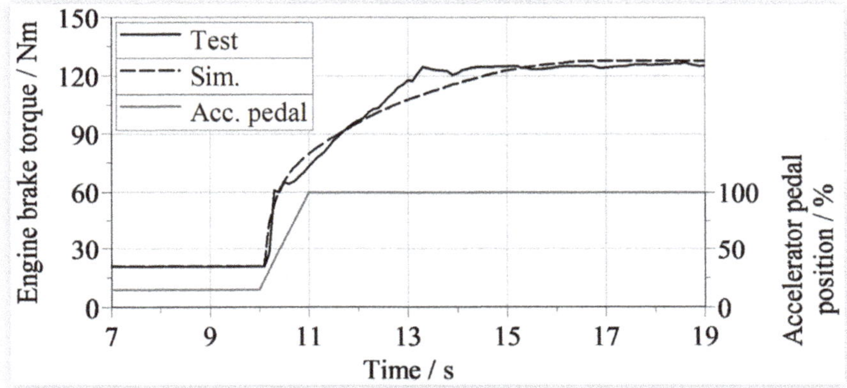

FIGURE 9 Comparison of measured and simulated distributions of the percentage fuel share during the selected 900 s time windows for urban and motorway conditions [35].

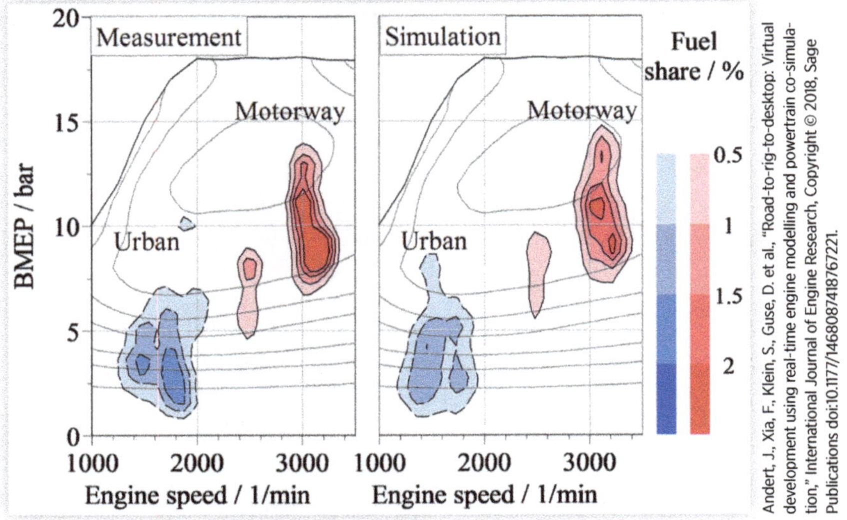

With the reconstructed control functions, the engine model can be operated in transient driving scenarios together with the other vehicle components. Thereto, the system integration with the transmission model, the vehicle dynamics model, the environment simulation, and the driver controller has been implemented in a SIMULINK environment. The vehicle co-simulation has been verified with vehicle measurements under real drive conditions. The whole test drive took 6911 s and covered a distance of 115 km, including urban, rural roads, and motorway operations. To analyze the engine behavior separately for low load and high load conditions, two 900 s time windows are chosen from the vehicle test to represent urban and motorway operation scenarios. The average velocity of the urban part is 37 km/h, including vehicle stops. The average velocity of the motorway part is 116 km/h. Both driving profiles are set up in the model with the corresponding road elevation profiles from the global positioning system (GPS).

The comparison between the simulation and the vehicle measurement shows good agreement. The total fuel consumption deviation caused by the combustion engine modeling is within 5% [35]. The main engine operation area can be predicted with satisfying accuracy in both urban and motorway driving, as shown in Figure 9. In addition, Figure 10 shows a prediction of the simulated turbocharger compressor operation points for the two selected 900 s driving scenarios.

8.5.1.3 APPLICATIONS OF THE CO-SIMULATION PLATFORM

One direct application area of the validated MiL simulation is the virtual vehicle testing under real driving conditions. Various driving scenarios and different ambient conditions with any desired driver behavior can be realized. The MiL simulation predicts not only vehicle performance data like acceleration and fuel consumption but also the thermodynamic system behavior of the combustion engine represented by, for example, turbocharger speed, air/fuel ratio, and exhaust gas temperature.

A further MiL application area is the system design. Based on a current powertrain layout, modifications can be implemented to optimize the system. Modifications regarding engine compression ratio, turbocharger sizing, transmission gear ratio, or

FIGURE 10 Simulated compressor operation points during the selected 900 s time windows for urban and motorway conditions [35].

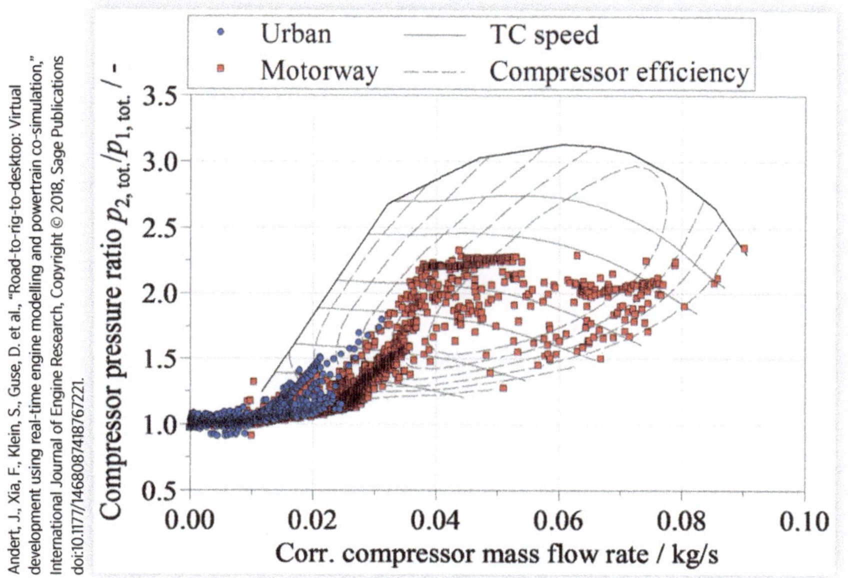

system electrification with additional components can be investigated with relatively small effort. Also, system optimization in transient operation can be done. Additionally, with added hardware components, a function development for the soft control units can be implemented. Especially, the functions with strong connections to multiple hardware components can be tested in the MiL environment.

As an example, a 48 V mild hybridization system has been added to the vehicle co-simulation [39]. Thereto, internal combustion engine was supplemented with an electric compressor and a belt-driven generator. The hardware layout and the software design with a focus on electric energy distribution between the electric compressor and the belt-driven generator can be examined with it.

It is also important that the MiL simulation is RT capable and hence can be integrated to any hardware components. On the one hand, the combustion engine model can be replaced by real hardware for engine-in-the-loop applications [40]. On the other hand, other components like the extended 48 V electric power net with the belt-driven generator and battery can be tested as hardware in closed loop during transient driving cycles [41].

8.5.2 HiL-Based Virtual Calibration

The second use case of the 1D CA-resolved engine model in XiL simulation has been designed to demonstrate its potential for virtual ECU calibration. The focus of the investigation is on the RT capability of the engine model and the closed-loop interaction with the hardware and software of the ECU. The system overview is shown in Figure 11. The model for the compression ignition engine is compiled for FEV's xMOD co-simulation platform, which is running on a workstation with the operating system Windows RTX. The workstation is coupled to an existing ECU HiL simulation platform [42], which includes RT-capable models of powertrain components and a real hardware ECU. It emulates the complete environment for the ECU as if it is operating in a real car. The communication between the co-simulation platform and the HiL platform has been

FIGURE 11 HiL platform layout with ECU and engine model using a co-simulation approach [43].

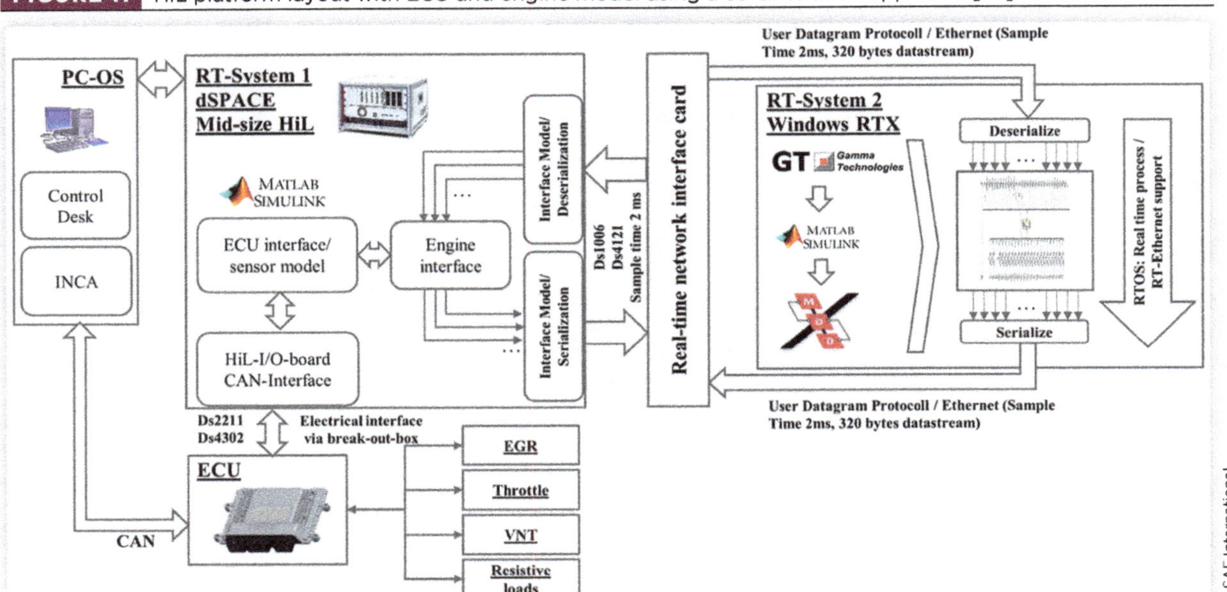

implemented via a 100 Mb/s Ethernet/user datagram protocol (UDP) interface. The distributed hardware for the combustion engine modeling enables flexible system updates without additional high cost to extend the calculation capacity of the HiL simulator.

8.5.2.1 ENGINE MODEL SETUP AND VALIDATION PROCESS

In this use case, the starting point of the engine modeling is a detailed 1D engine model used for turbocharger layout. The reference engine is a 2-L, four-cylinder diesel engine with a single-stage turbocharger with a variable geometry turbine (VGT) and high-pressure/low-pressure exhaust gas recirculation (HP/LP EGR). Extensive engine measurement results provide sufficient data for model parameterization and validation.

The detailed model has been reduced to a run-time optimized model following the modification process described by Cosadia et al. [28]. The discretization level in the air path has been reduced and the solver time step has been increased. Heat transfer and pressure losses have been adjusted to keep the main simulation results unchanged. Compared to the engine model described in the first use case, there are a few differences in the modeling and validation processes due to the engine type and the application purpose.

Firstly, the reference engine is compression ignited with multiple fuel injections and two EGR systems, which means higher system complexity for both combustion and air path modeling. Especially, the EGR system modeling is challenging due to the highly simplified geometry information.

Secondly, the model has been developed for virtual calibration purposes, which means the combination of the control signals (e.g., injection timing, the VGT rack position, the throttle valve position, the EGR valve position, and the air outlet temperature at the charge air cooler) can vary in extremely large ranges outside of the available measurement data for model validation. In this case, the model should be capable of predicting the engine operation precisely enough to avoid huge recalibration effort on the real hardware.

Last but not least, the engine model should work in closed-loop with the hardware ECU with input and output ports. Although the generation and analyzation of the electric signals are taken over by the I/O boards of the HiL simulator, there is a necessary modification of the engine model interface. On the ECU input side, virtual sensors with correct transient behavior are needed in the combustion engine model at the right position to sense the simulated values. On the ECU output side, the most important signals are the injector energizing and the actuation of the EGR valves, which should be transformed into fuel rate and effective flow diameter in the engine model. Only with sufficient sensor and actuator modeling can the closed-loop interaction between ECU and the combustion engine be represented accurately by the HiL simulation.

To fulfil the requirements mentioned above, the RT engine model includes a predictive combustion model reacting on different EGR rates and injection strategies, several thermal couple models for the temperature measurement, an injector model for fuel rate calculation, and calibrated flow characteristics of the EGR valves to reach the exact EGR rates.

The RT engine model is independently verified in simulation on a desktop computer with measurement data for the whole engine map before model transfer to the HiL platform. Ambient temperature, pressure and humidity are adjusted to the test bench conditions. Also, injection events are set according to those used on the engine test bench. PI-controllers are implemented for the turbine rack position and the HP/LP EGR valves. The control targets are intake manifold pressure and EGR rates according to the engine test bench measurements. Representing the modeling accuracy, Figure 12 shows the

FIGURE 12 Offline verification of the RT engine model compared to test bench measurement [43].

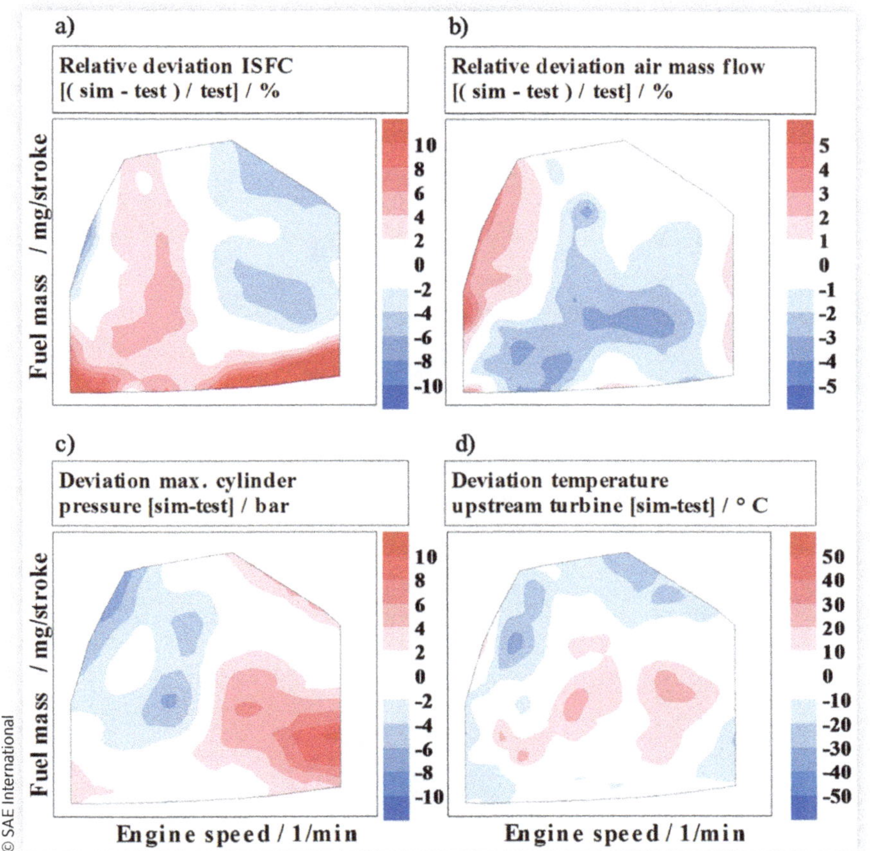

relative deviations between simulation and engine testing for the indicated specific fuel consumption (ISFC), the air mass flow, the maximum cylinder pressure, and the exhaust gas temperature upstream of the turbine in injection quantity vs. engine speed diagrams. The model achieves a similar level of accuracy in the thermodynamic process as the model in the first use case.

8.5.2.2 SYSTEM INTEGRATION AND SYSTEM VALIDATION

After the definition of the signal interface between ECU and the engine model (shown in Figure 13), the engine model has been exported as s-function into the SIMULINK environment. The SIMULINK model is then compiled and uploaded to the RT co-simulation platform xMOD coupled with the dSPACE HiL simulator.

After the verification of the RT capability of the engine model, the integrated system has been tested in steady-state and transient operating conditions. In the steady-state verification with different engine speeds and accelerator pedal positions, the engine model controlled by the ECU shows very similar results to the engine hardware on the dynamometer test bench. The relative deviations between air mass flow and fuel mass flow are smaller than 5% in most operation points, as shown in Figure 14. The actuator positions have relatively higher deviations compared to the measurement data. A deviation of +/- 10% can be observed in Figure 15. The main reason for these deviations is the strong simplification of the air path geometry, which significantly influences the flow characteristics in the EGR systems, which has to be compensated by the actuator positions.

Additionally, the simulated transient system behavior in the accelerator pedal step-in tests shows good accordance with the measurement data. One example is shown in Figure 16 at a constant engine speed of 2000 1/min. The deviations in boost pressure and air mass flow during the load steps are caused by the different injection quantities. In the test bench measurement, the smoke limitation function limited the torque request to around 140 Nm immediately after the increase of the accelerator pedal position. The limited torque request corresponds well to the injection quantity profile in the HiL simulation. However, as the active surge damper function is activated on the engine test bench, the limited torque request is filtered by a PT1 filter, before it is converted to the injection quantity. The deactivated filter function on the HiL platform is the primary reason for the differences in system behavior between HiL simulation and engine test bench.

FIGURE 13 Signal interface between engine model and ECU [43].

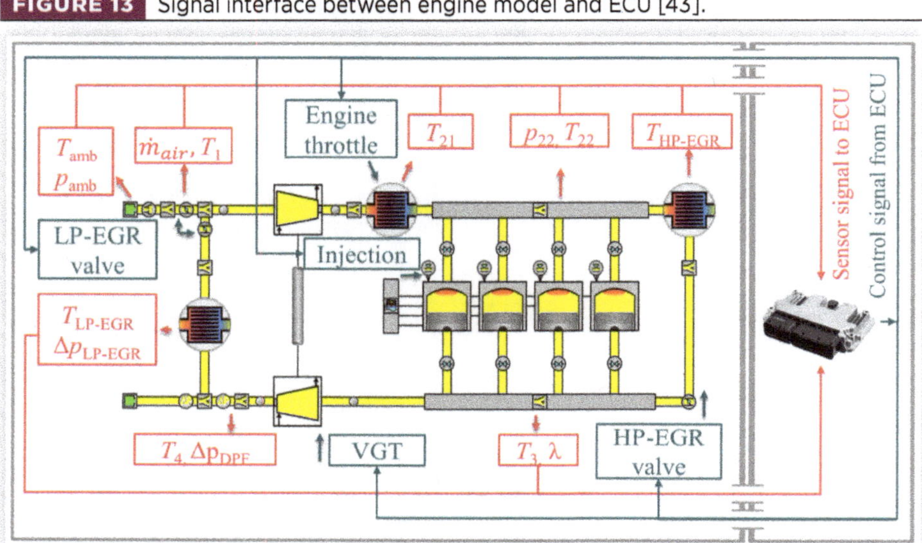

FIGURE 14 Comparison of air and fuel mass flow between the engine test bench and HiL simulation [43].

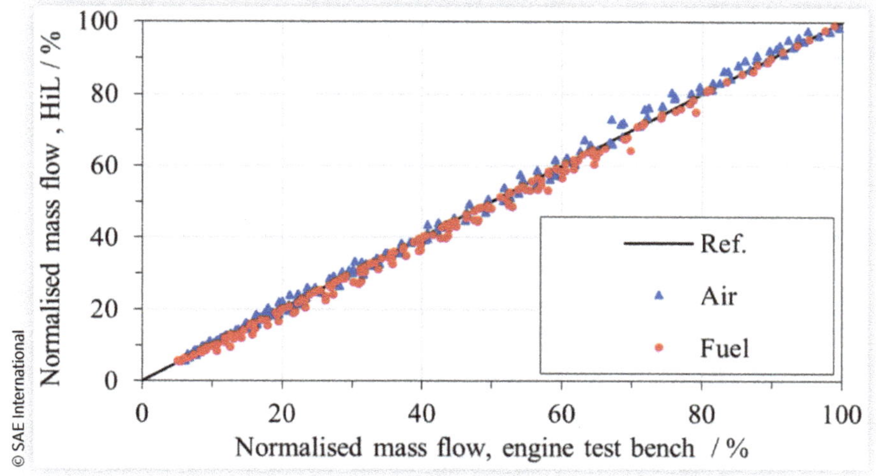

FIGURE 15 Comparison of the actuation signals for LP/HP EGR valves and turbine rack position between engine test bench and HiL simulation [43].

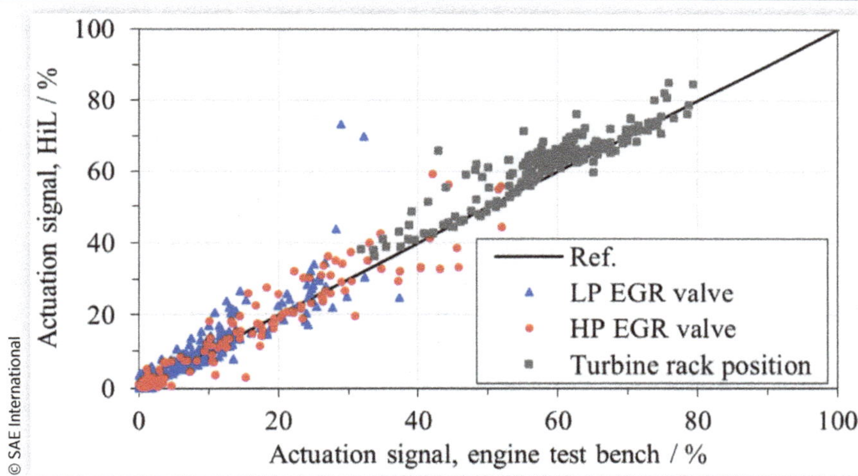

However, from the virtual calibration's point of view, a deactivation of the active surge damper function is beneficial for the fundamental engine calibration tasks. The active surge damper function can be reactivated for the drivability calibration in closed-loop with a vehicle or a vehicle model.

8.5.2.3 APPLICATIONS OF THE HiL SIMULATION PLATFORM

The main advantage of the HiL simulation compared to the MiL simulations, like the one from the first use case, is the availability of the "original" control functions of the ECU. From the system integration point of view, the physical behavior of the engine and its ECU can be reproduced by the modeled combustion engine and the ECU more precisely in the HiL simulation. With the increased system modeling accuracy, other powertrain components can be tested and optimized in a more realistic environment.

From the engine calibration point of view, a certain amount of calibration tasks, which are conventionally implemented on the engine dynamometer test bench or on the

FIGURE 16 Comparison of positive and negative engine load steps at 2000 1/min between engine test bench and HiL simulation [43].

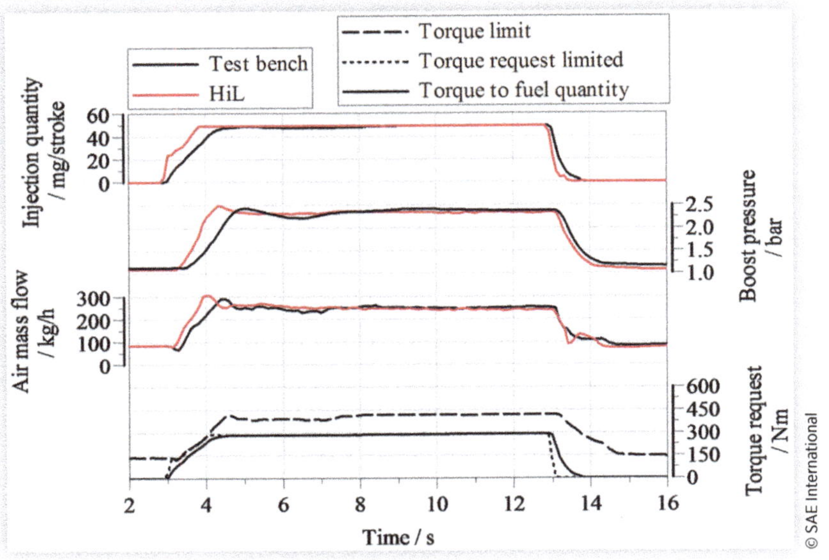

prototype vehicle, can be front-loaded to the virtual environment. One such example refers to the calibration tasks for the air path. Especially, the engine operation under extreme ambient conditions, like high altitude or high temperature operation, can be calibrated with minimal effort. By extending the engine model with thermal management and emissions modeling, more engine control and also on-board diagnostics (OBD) functions can be calibrated. A co-simulation with other powertrain components can provide even more potential for calibration tasks conventionally done on a prototype vehicle.

The calibration quality using the virtual components strongly depends on the modeling accuracy. Therefore, the virtual calibration should be a process with many validation steps and feedback loops. Figure 17 shows the information flow in such a

FIGURE 17 The information flow in a virtual calibration process between XiL environments and vehicle [44].

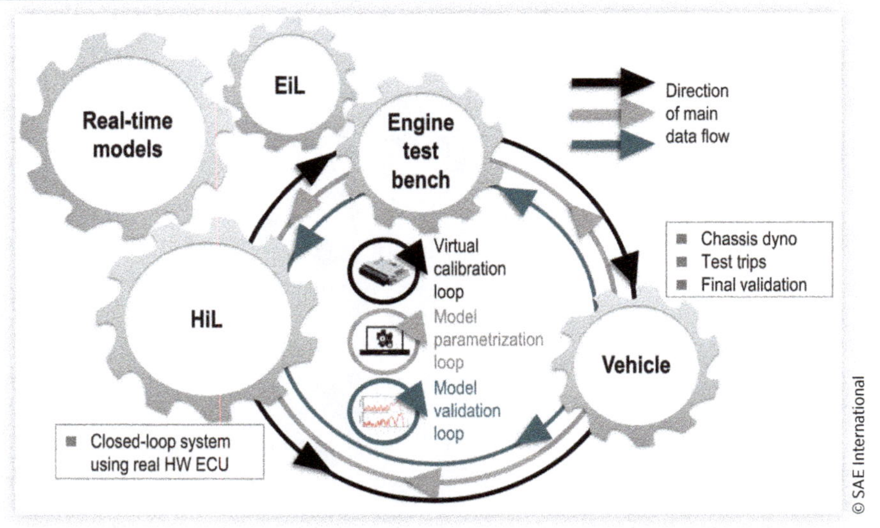

virtual calibration process. The RT models are parameterized and validated with the engine test bench and vehicle measurement data. After integrating the models on the HiL platform with the control units, the control function calibration tasks can be started. The ECU data-set from the virtual calibration is then validated with the real hardware on an engine test bench or in the vehicle. The new measurement data can be used to recalibrate the RT model in the broader operating range [44].

8.6 Summary

XiL simulations are widely used in the modern powertrain development process, not only for control function development but also for system integration and testing. They enable a simulation of the closed-loop interaction between the various submodules and provide clear time/cost advantages compared to the usage of prototype vehicles in later stages of a development program. With the standardization of the implementation process based on FMI and DCP, the effort of building a XiL platform has been significantly reduced.

The combustion engine model can be used for XiL simulations in different vehicle development phases. While detailed 3D/1D models can successfully support the engine hardware layout, the computer run-time optimized 1D/0D or MV models are beneficial for function development/calibration or component integration and testing.

Compared to the MV models, a major advantage of the CA-resolved 1D engine models is the accurate modeling of the wave dynamics. With sufficient model reduction, whether on the solver algorithms or the model geometry with increased simulation time step, a faster simulation speed can be achieved. It enables the model application in areas that require high simulation speed or even RT capability.

The first introduced use case with vehicle co-simulation has shown the application of the CA-resolved 1D engine models in system integration and testing. The second use case presented the application of such models in a HiL simulation for virtual calibration. Both engine models have been validated in comparison to steady-state and transient engine measurement data. An accuracy of over 95% could be demonstrated for the main simulation results (e.g., air/fuel mass flow, boost pressure, and max. cylinder pressure). Also, the integrated systems, whether the vehicle co-simulation or the integrated ECU HiL simulation, have shown good accordance with the measurement data. Moreover, the RT capability of the models has been proven. The use cases have demonstrated the application of the 1D engine modeling approach in the XiL simulations and their potential to effectively support each level of the powertrain development process.

Acknowledgments

The work was partially performed as part of the research project "Advanced Co-Simulation Open System Architecture" (ACOSAR) within the European EUREKA cluster program ITEA3. It also was partially performed within the "Center for Mobile Propulsion" funded by the German Research Foundation (DFG).

Abbreviations

ASM - Automotive simulation models
BMEP - Brake mean effective pressure

BSFC - Brake-specific fuel consumption
CA - Crank angle
DCP - Distributed co-simulation protocol
ECU - Engine control unit
FMI - Functional mock-up interface
GPS - Global positioning system
HiL - Hardware-in-the-loop
HP/LP EGR - High-pressure/low-pressure exhaust gas recirculation
ISFC - Indicated specific fuel consumption
M/S - Master/slave cylinder
MiL - Model-in-the-loop
MV - Mean value
OBD - On-board diagnostics
PDU - Protocol data units
QDM - Quasi-dimensional model
RT - Real-time
SiL - Software-in-the-loop
TCU - transmission control unit
UDP - User datagram protocol
VGT - Variable geometry turbine
XiL - X-in-the-loop

References

1. Prasad, B., "Analysis of Pricing Strategies for New Product Introduction," *Pricing Strategy and Practice* 5, no. 4 (1997): 132–141, doi:10.1108/09684909710184626.

2. Balazs, A., "Optimierte Auslegung von Hybridantriebsträngen unter realen Fahrbedingungen," 2015.

3. Guzzella, L. and Onder, C.H., *Introduction to Modeling and Control of Internal Combustion Engine Systems*, 2nd ed. (Berlin: Springer, 2010), ISBN 978-3-642-10774-0.

4. Richard, A., Xia, F., Franzke, B., Lückmann, D. et al., *Combining Miller Cycle and VTG - A Promising Concept for Future Gasoline Engines?* (Dresden: ATK Aufladetechnische Konferenz, 2016).

5. Merker, G.P., Schwarz, C., and Teichmann, R., *Combustion Engines Development* (2012), doi:10.1007/978-3-642-14094-5.

6. Lehrheuer, B., Morcinkowski, B., Pischinger, S., and Nijs, M., Low Temperature Gasoline Combustion – Potential, Challenges, Process Modeling and Control, in: King, R. (ed.), *Active Flow and Combustion Control 2014, Notes on Numerical Fluid Mechanics and Multidisciplinary Design* (Cham: Springer International Publishing, 2015), 163–179, ISBN 978-3-319-11966-3.

7. Isermann, R., "Engine Modeling and Control: Modeling and Electronic Management of Internal Combustion Engines," 1st ed. (Berlin: Springer, 2016), ISBN 3662506297.

8. Blochwitz, T., Otter, M., Arnold, M., Bausch, C. et al., "The Functional Mockup Interface for Tool Independent Exchange of Simulation Models," *The 8th International Modelica Conference*, Technical Univeristy, Dresden, Germany, *Linköping Electronic Conference Proceedings*, March 20-22, 2011, Linköping University Electronic Press, 2011, 105–114.

9. Isermann, R., Schaffnit, J., and Sinsel, S., "Hardware-in-the-Loop Simulation for the Design and Testing of Engine-Control Systems," *Control Engineering Practice* 7, no. 5 (1999): 643–653, doi:10.1016/S0967-0661(98)00205-6.

10. Ramaswamy, D., McGee, R., Sivashankar, S., Deshpande, A. et al., "A Case Study in Hardware-In-the-Loop Testing: Development of an ECU for a Hybrid Electric Vehicle," SAE Technical Paper 2004-01-0303, 2004, doi:10.4271/2004-01-0303.

11. Fathy, H.K., Filipi, Z.S., Hagena, J., and Stein, J.L., "Review of Hardware-in-the-Loop Simulation and Its Prospects in the Automotive Area," *Modeling and Simulation for Military Applications*, Kissimmee, Florida, USA, *Proceedings of SPIE*, vol. 6228, SPIE, Bellingham, Washington, ISBN 9780819462848, April 18–21, 2006.

12. Pischinger, S., Schernus, C., Lütkemeyer, G., Theuerkauf, H.J. et al., "Investigation of Predictive Models for Application in Engine Cold-Start Behavior," SAE Technical Paper 2004-01-0994, 2004, 10.4271/2004-01-0994.

13. Sternberg, P., Schloßer, A., Richenhagen, J., and Meyer, N., "The Challenge of Early Functional Software Tests," *AUTOREG 2013: Steuerung und Regelung von Fahrzeugen und Motoren; 6. Fachtagung in Baden-Baden*, 05. und 06. Juni 2013, VDI-Berichte, vol. 2196, VDI-Verl., Düsseldorf, ISBN 978-3-18-092196-9, 2013, 399–409.

14. Blochwitz, T., Otter, M., Akesson, J., Arnold, M. et al., "Functional Mockup Interface 2.0: The Standard for Tool independent Exchange of Simulation Models," *9th International MODELICA Conference*, Munich, Germany, *Linköping Electronic Conference Proceedings*, Linköping University Electronic Press, Sept. 3–5, 2012, 173–184.

15. Krammer, M., Marko, N., and Benedikt, M., "Interfacing Real-Time Systems for Advanced Co-simulation - The ACOSAR Approach," *STAF 2016: Software Technologies: Applications and Foundations*, Wien, Austria, July 4–8, 2016.

16. Krammer, M., Benedikt, M., Blochwitz, T., Alekeish, K. et al., "The Distributed Co-simulation Protocol for the Integration of Real-Time Systems and Simulation Environments," *2018 Summer Simulation Multi-Conference*, 2018.

17. Dauron, A., "Model-Based Powertrain Control: Many Uses, No Abuse," *Oil & Gas Science and Technology - Rev. IFP* 62, no. 4 (2007): 427–435, doi:10.2516/ogst:2007054.9

18. Mirfendreski, A., Schmid, A., Grill, M., and Bargende, M., "Presenting a Fourier-Based Air Path Model for Real-Time Capable Engine Simulation Enhanced by a Semi-Physical NO-Emission Model with a High Degree of Predictability," SAE Technical Paper 2016-01-2231, 2016, doi:10.4271/2016-01-2231.

19. Schulten, P.J.M. and Stapersma, D., "Mean Value Modelling of the Gas Exchange of a 4-stroke Diesel Engine for Use in Powertrain Applications," SAE Technical Paper 2003-01-0219, 2003, doi:10.4271/2003-01-0219.

20. Maroteaux, F. and Saad, C., "Combined Mean Value Engine Model and Crank Angle Resolved In-Cylinder Modeling with NOx Emissions Model for Real-Time Diesel Engine Simulations at High Engine Speed," *Energy* 88 (2015): 515–527, doi:10.1016/j.energy.2015.05.072.

21. Wurzenberger, J.C., Heinzle, R., Schuemie, A., and Katrasnik, T., "Crank-Angle Resolved Real-Time Engine Simulation –Integrated Simulation Tool Chain from Office to Testbed," SAE Technical Paper 2009-01-0589, 2009, doi:10.4271/2009-01-0589.

22. Deppenkemper, K., Özyalcin, C., Ehrly, M., Schönen, M. et al., "1D Engine Simulation Approach for Optimizing Engine and Exhaust Aftertreatment Thermal Management

for Passenger Car Diesel Engines by Means of Variable Valve Train (VVT) Applications," SAE Technical Paper 2018-01-0163, 2018.

23. Birckett, A., Tomazic, D., Bowyer, S., Bevan, K. et al., "Transient Drive Cycle Modeling of Supercharged Powertrains for Medium and Heavy Duty On-Highway Diesel Applications," SAE Technical Paper 2012-01-1962, 2012, 10.4271/2012-01-1962.
24. Cipollone, R. and Sciarretta, A., "The Quasi-Propagatory Model: A New Approach for Describing Transient Phenomena in Engine Manifolds," SAE Technical Paper 2001-01-0579, 2001, doi:10.4271/2001-01-0579.
25. Pacitti, G., Amphlett, S., Miller, P., Norris, R. et al., "Real-Time Crank-Resolved Engine Simulation for Testing New Engine Management Systems," *SAE Int. J. Passeng. Cars – Mech. Syst.* 1, no. 1 (2009): 801–809, doi:10.4271/2008-01-1006.
26. Courant, R., Friedrichs, K., and Lewy, H., "On the Partial Difference Equations of Mathematical Physics," *IBM Journal of Research and Development* 11, no. 2 (1967): 215–234, doi:10.1147/rd.112.0215.
27. Gamma Technologies, I., "GT-SUITE Flow Theory Manual: Version 2017," 2017.
28. Cosadia, I., Silvestri, J.J., Papadimitriou, I., Maroteaux, D. et al., "Traversing the V-Cycle with a Single Simulation - Application to the Renault 1.5 dCi Passenger Car Diesel Engine," SAE Technical Paper 2013-01-1120, 2013, doi:10.4271/2013-01-1120.
29. Gamma Technologies, I., "GT-SUITE Controls Coupling and Real Time Manual: Version 2017," 2017.
30. Kouba, A., Hnilicka, B., and Navratil, J., "Downsized Gasoline Engine Cylinder Deactivation MiL Development and Validation Using Real-Time 1-D Gas Code," SAE Technical Paper 2018-01-1244, 2018, doi:10.4271/2018-01-1244.
31. Kouba, A., Navratil, J., and Hnilička, B., "Engine Control Using a Real-Time 1D Engine Model," in: Liebl, J. and Beidl, C. (eds.), VPC – Simulation und Test 2015, *Proceedings* (Wiesbaden: Springer Fachmedien Wiesbaden, 2018), 295–309, ISBN 978-3-658-20735-9.
32. Nanjundaswamy, H., Tatur, M., Tomazic, D., Dahodwala, M. et al., "Development and Calibration of on-Board-Diagnostic Strategies Using a Micro-HiL Approach," SAE Technical Paper 2011-01-0703, 2011, doi:10.4271/2011-01-0703.
33. Nanjundaswamy, H.K., Dahodwala, M., Farsodia, M., Tomazic, D. et al., "Virtual Testing and Simulation Environment [Micro-HiL] for Engine and Aftertreatment Calibration and Development -Part 2," SAE Technical Paper 2012-01-0928, 2012, doi:10.4271/2012-01-0928.
34. Wu, H. and Li, M.-F., "A Hardware-in-the-Loop (HIL) Bench Test of a GT-Power Fast Running Model for Rapid Control Prototyping (RCP) Verification," SAE Technical Paper 2016-01-0549, 2016, doi:10.4271/2016-01-0549.
35. Andert, J., Xia, F., Klein, S., Guse, D. et al., "Road-to-Rig-to-Desktop: Virtual Development Using Real-Time Engine Modelling and Powertrain Co-Simulation," *International Journal of Engine Research*, 2018, doi:10.1177/1468087418767221.
36. Chen, S.K. and Flynn, P.F., "Development of a Single Cylinder Compression Ignition Research Engine," SAE Technical Paper 650733, 1965, 10.4271/650733.
37. dSPACE, "dSPACE HelpDesk Documents: Handling Overrun Situations."
38. Xia, F., Griefnow, P., Klein, S., Tharmakulasingam, R. et al., "Crank Angle Resolved Real-Time Engine Modeling for HiL Based Component Testing," *The 19th Asia Pacific Automotive Engineering Conference & SAE-China Congress 2017*, Shanghai, China, 26.10.2017, 1695–1702, ISBN 978-7-111-58164-2.

39. Griefnow, P., Andert, J., Xia, F., Klein, S. et al., "Real-Time Modeling of a 48V P0 Mild Hybrid Vehicle with Electric Compressor for Model Predictive Control," SAE Technical Paper 2019-01-0350, 2019, doi: 10.4271/2019-01-0350.
40. Klein, S., Griefnow, P., Guse, D., Xia, F. et al., "Virtual 48V Mild Hybridization: Efficient Validation by Engine-in-the-Loop," SAE Technical Paper 2018-01-0410, 2018, doi:10.4271/2018-01-0410.
41. Griefnow, P., Andert, J., Klein, S., Xia, F. et al., "Power-Net-in-the-Loop: Closed Loop Power Net Testing with Real Time Simulation," *VDI Congress ELIV (ELectronics In Vehicles)*, Bonn, 2017.
42. Lee, S.-Y., Andert, J., Quérel, C., Schaub, J. et al., "Virtual Calibration Based on X-in-the-Loop: HiL Simulation of Virtual Diesel Powertrain," *VPC, Simulation und Test, 19. MTZ-Fachtagung*, Hanau, Germany, 26-27.09.2017.
43. Xia, F., Lee, S.-Y., Andert, J., Kampmeier, A. et al., "Crank-Angle Resolved Real-Time Engine Modelling a Seamless Transfer from Concept Design to HiL Testing," SAE Technical Paper 2018-01-1245, 2018, doi:10.4271/2018-01-1245.
44. Lee, S.-Y., Andert, J., Neumann, D., Querel, C. et al., "Hardware-in-the-Loop Based Virtual Calibration Approach to Meet Real Driving Emissions Requirements," SAE Technical Paper 2018-01-0869, 2018, doi:10.4271/2018-01-0869.

CHAPTER 9

Coupling of 1D and 3D Fluid Dynamic Models for Hybrid Simulations

Gianluca Montenegro and Angelo Onorati
Politecnico di Milano

9.1 Introduction

In the last decade, internal combustion engines have experienced a phase of deep and intense investigation finalized mainly to the optimization of performances and fuel consumption. The continuously tightening limitations on pollutant emissions and the restrictions on the emitted noise levels have stimulated the research activity furthermore, making the design phase one of the most critical steps in the engine production process. To accomplish the different requirements imposed by the legislation and by the market, an important role is played by scientists and researchers whose attention is focused on the optimization of strategies to control the engine and the combustion process, on the design of the intake and exhaust systems and on strategies to control the engine (EGR, boosting, and ATS). These last components are becoming of primary importance not only for the role they play on the engine volumetric efficiency, and therefore on the engine performance, but also for their influence on the abatement of the radiated noise and on the aftertreatment performance. On the other side, the optimization of the combustion process, both for gasoline and Diesel engines, and the investigation of alternative combustion modes are becoming key issues, since they can reduce significantly the amount of pollutants in the exhaust gas. In this scenario, fluid dynamic simulation tools have encountered an increased demand to assist the design phase, mainly because of their flexibility and accuracy. As a matter of fact, the prototyping procedure could become sensibly expensive and time-consuming, if many configurations must be tested and the aid of CFD tools can remarkably reduce the duration and the costs of this stage. However, the numerical simulation, if not correctly handled, can produce the opposite effect. For instance, a fully multidimensional analysis of a multi-cylinder engine coupled to its own intake and exhaust system might become challenging even with HPC facilities. This is the main reason why 1D (one-dimensional) models have become a widespread solution for the design of internal combustion engines. They represent a cost-effective solution: low computational burden and extreme flexibility [1, 2, 3, 4]. They can be adopted to model not only the engine itself but also all the other systems that may be included on the vehicle to improve the engine efficiency or to control the engine itself. In this category of 1D tools, we can find other 1D models for the simulation of parallel circuits, such as for the waste heat recovery, based on micro ORC or on similar systems, for the water cooling system of the engine or for other components [5, 6]. In general, the larger is the system, and the more subsystems are integrated in that, the simpler and lighter must be the calculation tool. Large systems, in fact, are often analyzed using 1D simulation tools to save time and simultaneously conserve computational resources. Certain components, however, such as the intake and exhaust manifolds, exhibit a high degree of geometric complexity which cannot be accurately modeled by 1D codes. From a fluid dynamic point of view, 1D models suffer mainly from the limitation of the plane wave assumption, which is not appropriate for the simulation of complex shape components (air-boxes, resonators, asymmetric expansion chambers, multi-pipe junctions, etc.), characterized by a significant non-planar wave motion. This happens whenever there is a variation in the pipe flow section such that the transversal dimension becomes comparable with the length of the pipe. Additionally, symmetry in the geometry is no more present, and this may trigger wave motions transversal to the longitudinal ones, even in geometries with a low degree of complexity. Therefore, in all these cases, the application of multiD codes might be exploited to model a specific component of an intake or exhaust system, where this particular wave motion is triggered, while the 1D approach is applied elsewhere, suitably coupling the two models at their interfaces [7, 8, 9, 10]. This balances the requirements of short runtime and accuracy of the results. The topic of coupling 1D models to 2D or three-dimensional (3D) models is not recent, but it is gaining interest in various application fields due to the improvement of the HPC facilities, which are gaining velocity with an acceptable increase of budget. For this reason, in the automotive field, many 1D commercial software dedicated to the simulation of internal combustion engines offer the possibility to be coupled to commercial and noncommercial 3D codes [4, 9, 11, 12, 13]. In this scenario, it is important to stress out that the way the coupling is realized does not depend only on the type of strategy implemented but also on the possibility that the two codes allow to access the variables involved in the coupling. Usually, the access is possible through the usage of user functions where in most of the cases boundary values can be modified, preventing the user from implementing alternative ways. Conversely, more flexibility can be achieved when coupling proprietary codes with open source software, where any type of variable can be theoretically accessed. From a literature survey, it is possible to find several examples where the 1D/3D coupling has been used to study the fluid dynamics of specific devices such as catalysts, intake systems, mufflers, and similar devices. Simulations have been carried out exploiting 1D-3D coupling to study the flow distribution at the inlet of the catalyst brick for specific configuration of the exhaust manifold [13, 14, 15]. In some applications concerning the study of aftertreatment devices, the coupling between the 3D domain and the 1D code has been realized at channel level. This allows to discretize the reacting zone as a porous zone and to solve the chemistry at 1D level mimicking the existence of a tiny channel in a separate solver [16, 17]. The solution obtained on a single channel is then extended to other channels having similar inlet flow conditions. This allows to speed up the computation runtime since the inclusion of the heterogeneous chemistry in the 3D solver may represent a considerable computation burden. Other applications of 1D coupling strategy are focused on the evaluation of the impact of the EGR on the engine behavior [18, 19]. For this aspect the 1D-3D coupling is very important since the 1D approximation can only determine the average, over all the cylinders, of the EGR for any operating condition. The possibility of modeling in 3D the mixing section and all the path of the gas to the cylinder intake runners allows to account for the efficiency of mixing at the injection section (eventually giving guidelines on the shape of the exhaust gas injector) and of the different distribution of the EGR percentage that is delivered to the cylinders, resulting in different operating conditions from one cylinder to the other. Apart from the adoption of coupling strategies to study or improve the fluid dynamics of internal composition engine systems, another field of application is the optimization of the noise level abatement, since linear and nonlinear 1D approaches are not adequate to capture the non-planar wave motion in complex geometry systems and the related higher order modes, due to the intrinsic limitation of the plane wave assumption. At 3D level, silencers can be studied resorting to linear, such as BEM and FEM tools [20, 21], and nonlinear approaches. However, the adoption of nonlinear models is recommended for the calculation of the silencing system connected to the engine source, to carry out a realistic prediction of the muffler interaction with the engine, when submitted to high amplitude pressure waves. In this case an integrated 1D-multiD fluid dynamic (nonlinear) simulation of the whole system, represented by the engine and the complete intake and exhaust ducting, can be very useful to test and optimize the silencer configuration on a virtual engine test bench and to calculate the tailpipe noise radiated by the inlet and outlet open ends. Coupled simulation have been used to predict the transmission loss (TL) of silencers or the sound pressure level emitted by the intake or exhaust system of internal combustion engines [22, 23, 24, 25, 26, 27] considering also eventual interaction with the deformation of the casing [28]. So far only the application of coupled approach to the study of devices directly connected to the intake or exhaust systems has been considered, but in the scientific scenario there are several types of applications where the combined 1D and 3D simulations have been used, which are related to auxiliary systems for the internal combustion engine or to the analysis of the combustion process. The cooling system of the engine has been studied by means of 1D-3D approaches, where the 1D model was used to simulate the water circuit and the 3D one to model the heat transfer in the cooling jacket on the engine [29, 30]. Similarly, coupled approaches have been used to model the fuel injection system of Diesel engines, where the 1D approximation has been dedicated to the simulation of the pumping unit and of the fuel circuit, whereas the injector sack and nozzle have been modeled relying on the 3D approximation to study the fuel breakup inside the combustion chamber [31, 32, 33]. For the sake of completeness it is worth mentioning applications where the coupling between 3D models and 1D models have been realized to help the simulation of the combustion process or the liquid fuel jet breakup. The 1D model is used to impose time-varying boundary conditions to the 3D simulation where the turbulence generation and decay is simulated during the intake, the compression, the combustion and discharge phases. The information about the turbulence evolution are then sent back to the 1D model to improve the 0D combustion model [34].

In the context of the scenario presented in this section a review of the most common strategies for the coupling will be described, and, at the light of the information about the numerical approaches, some applications of direct 1D-3D coupling will be shown in the field of intake and exhaust system simulation.

9.2 Governing Equations

The coupling between two fluid dynamic codes, 1D and 3D, involves both practical and theoretical aspects. The former depends on the type of the two solvers and how they

FIGURE 9.1 Difference in the definition of the calculation node or computational cell between the finite volume and the finite difference discretization technique.

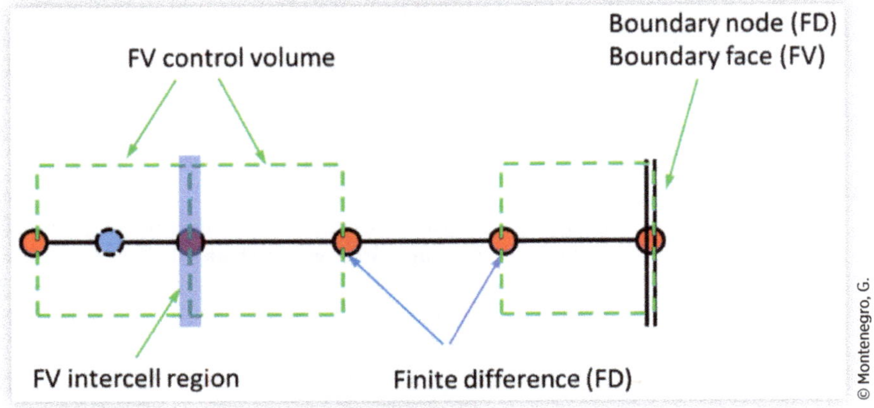

offer access to the calculated fields and variables the latter, instead, depends mainly on numerical considerations. This last factor affects the choice of the strategy that needs to be adopted to couple the two codes and depends on the characteristics of the two solvers, for instance the type of space discretization (finite difference or finite volume, colocated or staggered arrangement, as shown in Figure 9.1) or the time marching strategy (implicit or explicit, second- or first-order accuracy). The scenario of simulation software available on the market, or as research tools in universities, is various and may cover different combinations of numerical solutions. For this reason, the coupling strategies will be discussed referring to a specific case which could be extended, after suitable arrangements, to different combinations of the two solvers. The specific case we shall refer to will be the finite difference approach for the 1D model and the finite volume model for the multidimensional code. For what concerns the 1D model, the finite difference discretization is the one that historically has been adopted by most of the codes. In any case, examples will be also shown for the finite volume approximation. Conversely, in the 3D case, the finite volume approximation is largely used, due to its flexibility in creating the calculation grid and in performing the calculations.

9.2.1 1D Model

The 1D model that we will refer to is based on the finite difference discretization of the nonlinear system of Euler set of equations. The conservation equations of mass, momentum, and energy are expressed in a strong conservative form, with source terms to account for the friction and the heat transfer between the gas and the walls:

$$\frac{\partial W(x,t)}{\partial t}+\frac{\partial F(W)}{\partial x}+B(W)+C(W)=0$$

where

$$W=\begin{bmatrix}\rho F\\ \rho UF\\ \rho e_0 F\end{bmatrix},\quad F(W)=\begin{bmatrix}\rho UF\\ \left(\rho U^2+p\right)F\\ \rho U h_0 F\end{bmatrix},\quad B(W)=\begin{bmatrix}0\\ p\dfrac{\partial F}{\partial x}\\ 0\end{bmatrix},\quad C(W)=\begin{bmatrix}0\\ \dfrac{2}{D}\rho U|U| f_w F\\ -\rho q F\end{bmatrix}$$

FIGURE 9.2 Stencil of the two-step Lax-Wendroff method.

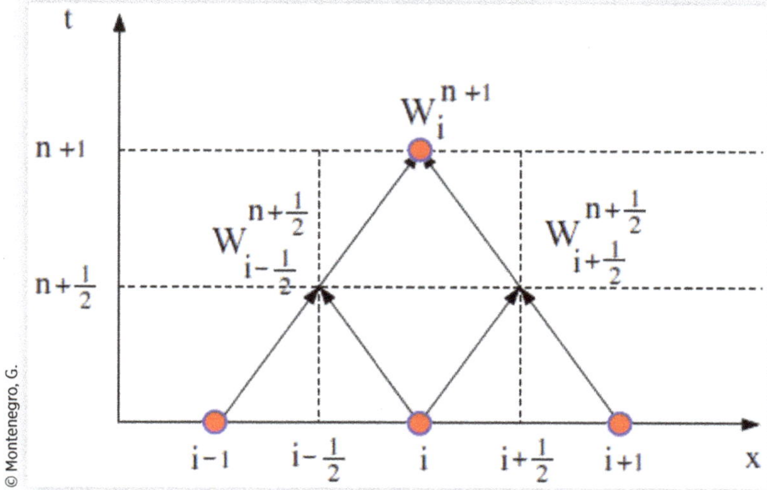

In the above conservation equations, ***W*** is the vector of conserved variables mass, momentum, and energy, while ***F***(***W***) is the flux vector of the conserved variables. The source terms ***B***(***W***) and ***C***(***W***) account, respectively, for the section variation along the duct axis and for the presence of friction and heat flux between the gas and the walls. The system of equations is closed by the perfect gas equation of state. Several numerical schemes can be used for the solution of this set of equations, ranging from the most common Lax-Wendroff with the addition of flux limiters to more advanced ones such as ENO schemes [35, 36]. In the case of the Lax-Wendroff scheme, the computational procedure, whose time-marching stencil is described in Figure 9.2, consists of two steps.

The first step is based on the Lax-Friedrichs method, with space-centered differences in the points (i–1/2, n) and (i+1/2, n):

$$W_{i-1/2}^{n+1/2} = \frac{1}{2}\left(W_i^n + W_{i-1}^n\right) + \frac{\Delta t}{2\Delta x}\left(F_i^n - F_{i-1}^n\right)$$

$$W_{i+1/2}^{n+1/2} = \frac{1}{2}\left(W_{i+1}^n + W_i^n\right) + \frac{\Delta t}{2\Delta x}\left(F_{i+1}^n - F_i^n\right)$$

The second step is a midpoint Leapfrog calculation with the time difference centered at point (i, n+1/2)

$$W_i^{n+1} = W_i^n + \frac{\Delta t}{\Delta x}\left(F_{i+1/2}^{n+1/2} - F_{i-1/2}^{n+1/2}\right)$$

where the flux terms represent the flux of the conserved variables evaluated at time n+1/2. The Lax-Wendroff method achieves second-order accuracy both in space and time, and suffers the presence of spurious oscillations in the solution, according to the Godunov theorem [37, 38], when sharp gradients are present in the solution field. This problem is usually faced by adopting flux-limiting techniques [39], which smooth the solution where numerical overshoots occur. Referring to Figure 9.1, where the computational grid of finite difference space discretization is shown, it can be seen that the computation node (colored in red) are coincident with the boundaries of the 1D domain. This means that, because of its stencil, the method can only be applied between the second and the second last node of the domain. The 1D treatment of the boundary conditions usually relies on a characteristics-based approach, exploiting the mesh method of characteristics (MOCs) [40] only at the boundary cells of the pipes to evaluate the transmission and reflection of waves, in terms of incident and reflected Riemann variables.

Alternatively, in the case of finite volume approximation, the set of equation can be expressed with the more general form:

$$W_i^{n+1} = W_i^n + \frac{\Delta t}{\Delta x}\left(F_{i+1/2}^n - F_{i-1/2}^n\right),$$

derived by the Godunov formulation. The accuracy of this formulation is given by the way in which the intercell flux is determined. This flux can be calculated resorting to exact methods, whose computational burden can become limiting in case of large computational domains, or relying on approximate solutions of the Riemann problem. In this last case, we will describe the HLLC method, since it is one of the most used solvers, also because it has first- and second-order formulation (WAF) [37]. This method is based on the assumption that the conserved variable domain is composed of piecewise constant states, as shown in Figure 9.3. This assumption, which well fits to solvers based on a finite volume discretization, identifies a series of Riemann problems at each interface between two computational cells. The intercell region, therefore, becomes the source of a three-wave centered system, composed of two signals traveling at the speed of sound, which are a compression and a rarefaction wave, and one slow signal, also identified as a contact surface or contact discontinuity, which travels at the fluid velocity. The HLLC Riemann solver was originally proposed by Harten, Lax, and Leer [37], successively modified by Toro [41] to allow the treatment of contact discontinuities. This solver provides an approximate solution of the Riemann problem without recurring to an iterative process to find the exact solution, which, on a large computation domain, might cause a significant burden for the calculation runtime [37]. As shown in Figure 9.3, the intercell-centered wave system defines four different regions: two regions, respectively, on the right and on the left of the S_R and S_L signals and two regions bounded between the contact discontinuity (S^*) and the two fast signals. Crossing the contact surface, the

FIGURE 9.3 Piecewise constant state reconstruction of the solution domain for a Godunov-type method and the centered wave system generated by the Riemann problem located at the intercell region.

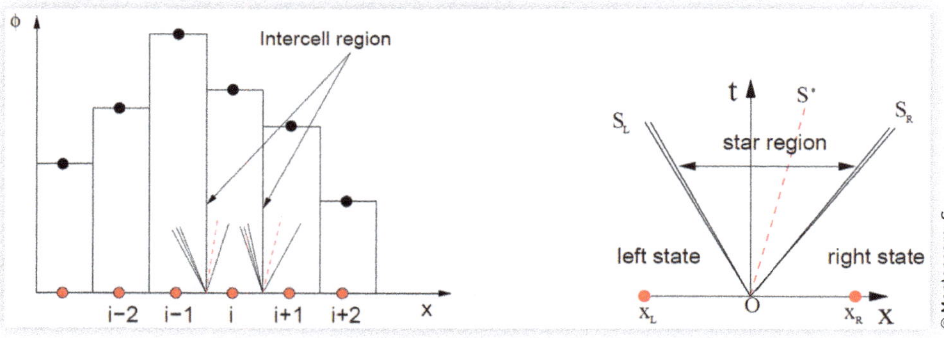

field of temperature and density are discontinuous, while pressure and velocity are constant over the whole *star* region. All the conserved variables are discontinuous across the two signals S_L and S_R. Hence, the HLLC solver, according to the position of each wave with respect to the intercell region, calculates the values of the conserved variables [37]:

$$\tilde{W}(x,t) = \begin{cases} W_L \text{ if } \frac{x}{t} < S_L \\ W_L^* \text{ if } S_L < \frac{x}{t} < S^* \\ W_R^* \text{ if } S^* < \frac{x}{t} < S_R \\ W_L \frac{x}{t} > S_R \end{cases}$$

Hence, assuming that the pressure and velocity fields are constant over the *star* region, the value of the conserved variables in the *star* region is given by the following equations:

$$(\rho)_K^* = \rho_K \frac{S_K - u_K}{S_K - S^*}$$

$$(\rho\vec{U})_K^* = \rho_K \frac{S_K - u_K}{S_K - S^*} \begin{bmatrix} S^* \\ v_K \\ w_K \end{bmatrix}$$

$$(\rho e_0)_K^* = \rho_K \frac{S_K - u_K}{S_K - S^*} \left[e_K + (S^* - u_K) \right] \left(S^* + \frac{p_K}{\rho_k (S_K - u_K)} \right)$$

where the index k can represent the left or the right state.

From the values of the conserved variables at the intercell region, the solver assigns the intercell flux according to the following equations:

$$\mathbf{F}_{i+\frac{1}{2}}^{HLLC} = \begin{cases} \mathbf{F}_L \\ \mathbf{F}*_L = \mathbf{F}_L + S_L (W_L^* - W_L) \\ \mathbf{F}*_R = \mathbf{F}_R + S_R (W_R^* - W_R) \\ \mathbf{F}_R \end{cases}$$

Assuming the same convention on the discontinuity positions adopted for the conserved variable calculation in the star region, the values of the three wave velocities S_L, S_R, and S^* can be evaluated according to the formula proposed by Davis and Batten [42, 43]:

$$S_R = u_R - a_R, \quad S_L = u_L - a_L,$$

$$S^* = \frac{p_R - p_L}{\rho_L (S_L - u_L) - \rho_R (S_R - u_R)} + \frac{\rho_L u_L (S_L - u_L) - \rho_R u_R (S_R - u_R)}{\rho_L (S_L - u_L) - \rho_R (S_R - u_R)}$$

9.2.2 3D Model

On the multidimensional side, the governing equations are the Navier-Stokes equations, where, differently from the 1D assumption, viscous phenomena and thermal conduction effects occurring within the fluid are taken into account. Therefore, the conservation equations solved by the 3D models are the following:

$$\frac{\partial \rho}{\partial t}+\nabla\cdot\left(\rho\vec{U}\right)=0$$
$$\frac{\partial \rho\vec{U}}{\partial t}+\nabla\cdot\left(\rho\vec{U}\otimes\vec{U}\right)=\nabla\cdot\tau.$$
$$\frac{\partial \rho e_0}{\partial t}+\nabla\cdot\left(\rho e_0\vec{U}\right)=-\nabla\cdot\left(p\vec{U}\right)$$

where

$$\tau=-\left(p+\frac{2}{3}\mu\nabla\cdot U\right)\mathbf{I}+2\mu\mathbf{D}.$$

This last term is the shear stress contribution, which is never solved directly but is usually corrected by the Reynolds stress tensor as a result of the turbulence model applied. As in the case of the 1D model, the closure equation is the equation of state of the gas. The adoption of turbulence models introduces additional equations to be solved, usually two, if a two-equation model is adopted. This could be the case of k-epsilon or k-omega turbulence models, which, incidentally, are the most used ones. In some cases, the set of equation is simplified by the neglection of viscous effects and of thermal conduction phenomena. This leads to the adoption of the Euler set of equations and to the neglection of turbulence models, resulting in a large reduction of the computation burden. This simplification is not always acceptable and may be adopted when the viscous phenomena are negligible with respect to the inertial effects, for instance when the Reynolds number is very high or when there is no flow condition (acoustic analysis). The solution approach of this type of system, in the case of a finite volume approach with collocated grid arrangement [44], reduces to the computation of volume and surface integrals [45]. In particular, if we consider the transport equation of a generic intensive quantity ϕ, which can be considered as mass or momentum or energy, the generic conservation equation applied to a control volume V whose surface is S is given by the following:

$$\int_V \frac{\partial \theta}{\partial t}dV+\oint_S \theta\vec{U}\cdot\vec{n}\,dS+\oint_S \gamma\vec{\nabla}\theta\cdot\vec{n}\,dS=Sp,$$

The second and the third terms on the left-hand side of this equation are then transformed into summation over all the faces constituting the computational cell sides, where the face center values are obtained by means of interpolation between the adjacent cell centers with a specific stencil on the basis of the numerical scheme that is used [46].

Due to such interpolation, the value of the variable in the cell center depends on the flux, convective or diffusive, through the surface of the control volume. For this reason, the value of the unknown must be calculated and cannot be overimposed externally by any other calculation. It must be always evaluated as the contribution of fluxes and source terms and not overimposed as cell center value. This means that the coupling should act at boundary condition level or, in case the code allows for that, directly accessing the face center values of the variable we want to assign. The solution procedure of the 3D solver will then calculate the cell center values according to the new fluxes.

9.3 Coupling Strategies

Once the basics of the 1D and 3D CFD have been explained, the focus of this chapter can shift to the different strategies that can be adopted to realize the coupling between these two different approaches and to the solution of the conservation laws. What is clear is that, on the 3D side of a finite volume-based solver, the transfer of information from the 1D side must happen at boundary condition level or adding a correction to the calculation of the flux of the conserved variables at the face centers of the region where the coupling is realized. Having this in mind, the strategy of the coupling can differ for the strategy that is used to overimpose these values or on the way the coupling is realized. For instance, the coupling can be one-way or two-way. In the former case, the 1D code is used to obtain a time-varying trace of a quantity at a certain point which will be assigned as boundary condition to the 3D domain, with no information going back to the 1D domain. The latter is, instead, based on the continuous exchange of information back and forth between the two calculation domains. In any of these two cases, the passage of values can be assessed by assigning uniform values at the boundary where the interface has been placed or adopting specific procedure that account for possible nonuniform distribution of the quantities.

9.3.1 One-Way Coupling

As anticipated, in this type of strategy, the solution is calculated in the two domains separately and not necessarily at the same time. Usually, the solution is calculated in the complete 1D domain, and the 3D domain is a discretization of a portion of the 1D part. The solution at a point corresponding to the position of the interface is then probed and translated into a time-varying boundary condition. In this way the value of the variable assigned is uniformly distributed over the interface region, as shown in Figure 9.4. The 3D simulation runs independently from the 1D simulation and does not provide any information back to the 1D domain. The benefits of this type of approach is that the two simulations are independent, therefore not mutually affected by instabilities that may arise during the calculations, and that the interface region can be placed arbitrarily with less care about the possibility of having unstable behavior of the simulation. The only issue to care about is the compatibility of the boundary condition with the type of flow that is established in the 3D geometry. For example, there may be the case of flow reversals, which in terms of boundary condition assignment means to handle the switch between inflow and outflow conditions. In the case of outflow conditions, if the coupling is realized imposing

FIGURE 9.4 Schematic of the one-way coupling strategy: the two domains are decoupled and there is overlap between the 1D and the 3D.

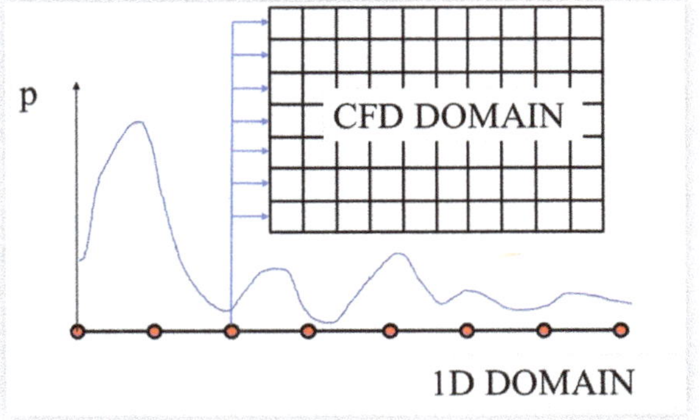

uniform values at the interface and the geometry is such that the flow does not admit a uniform distribution at the interface region, an incompatibility in boundary values and internal fields may arise, generating numerical instabilities. In this type of approach, the 3D analysis allows to achieve a better understanding of the fluid dynamic process happening in a specific device but does not allow, for instance, to evaluate the impact of certain flow conditions on the 3D side on engine parameters such as volumetric efficiency, power output, and so on.

9.3.2 Two-Way Coupling

This strategy is based on the concept that the two coupled codes mutually exchange information at every time step. In this way, what happens in the 3D domain has an impact on the solution of the 1D domain, since it changes its boundary values. Within this strategy there are several possibilities to achieve the coupling based on how the values are assigned from one domain to the other. In this section the main approaches found in literature will be discussed.

9.3.2.1 AVERAGE VALUES

The most straightforward way to couple two codes in the framework of a two-way coupling is the passage of average values referred to the conserved quantity. Referring to Figure 9.5, the passage from 1D to the 3D domain is based on the assignment of the conserved variable in a calculation node of the 1D domain (typically the second node or the second last node) to the boundary faces of the 3D domain.

To do this, a suitable switch between boundary condition types must be accounted for, since not all the variables can be assigned to the face centers (overimposed boundary conditions), but it must be distinguished between inflow and outflow conditions. When inflow for the 3D domain is verified, velocity and temperature are imposed at the face centers of the boundary faces (Dirichlet boundary condition), whereas pressure is determined imposing a null gradient with respect to the internal field (von Neumann boundary condition). In the case of outflow, only pressure may be assigned using the 1D value, and consequently velocity and temperature must be determined imposing null gradient with respect to the internal field. This means that the boundary values advected by the flow are assigned uniformly over the interface region. This strategy may have some problems if the interface region is located in the proximity of corners or regions where the 3D flow cannot be assumed uniform. In the case of flow reversal, when, for example, the 3D domain changes from outflow to inflow, the interface may be crossed by a vortex, which is then forced to vanish by the imposition of a uniform inlet flow. This condition may result in reflections, or generation, of pressure oscillation that may be tracked back in the 1D domain, affecting in this way the prediction of the wave action. For this reason, approaches based on this type of coupling usually suggest placing the interface region where the flow can be considered as 1D as possible, such as straight pipes. On the passage of values from the 3D to the 1D model, the average of quantities is adopted in most of the cases. This approach consists in averaging the values of the conserved variables over a specified region in the 3D domain (usually seen as a cell zone) and then assigned to the first or last node

FIGURE 9.5 Example of two-way coupling based on the passage of average values.

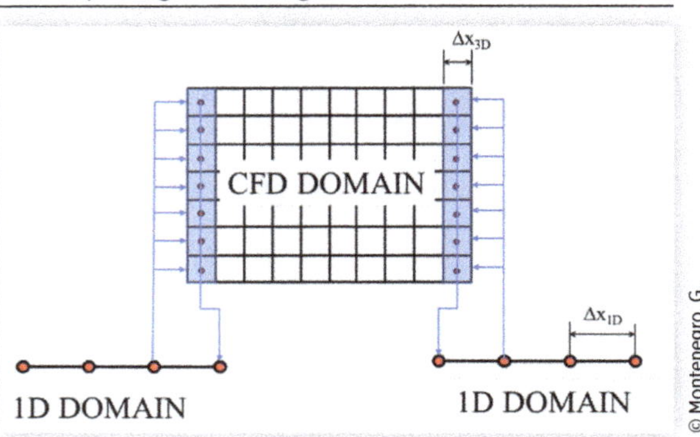

of the 1D domain. The 1D method will then exploit the new thermodynamic states to find the solution at the new time step in the second and second last calculation node of the 1D domain. The extension of the region of cells used to average the conserved variables may vary and can be used to smooth the gradients. The larger is the region, in terms of distance from the coupling interface, the smoother will be the solution. The extension of this region can also be regulated to damp oscillations that can arise at the coupling interface. On the contrary it has the drawback of damping the high frequency component of pressure oscillations that may pass thought the interface, resulting in a strategy not suitable for acoustic simulations.

9.3.2.2 MOC-BASED APPROACH

Another approach to solve the hyperbolic system of equations represented by the conservation of mass, momentum, and energy applied to a compressible and unsteady flow is the MOC. This method has an historical value, since it was applied to solve graphically the system of equations and was abandoned mainly because of its limits in capturing shocks. However, in certain codes it is still used to handle boundary conditions [47]. As a matter of fact, in finite difference-based methods, such as the Lax-Wendroff, the numerical stencil may involve three different nodes, as shown in Figure 9.2, and when the first node or the last node of the 1D domain is involved, the information coming from outside of the domain are missing. Therefore, depending on the particular boundary that has to be modeled, for instance outlet, inlet, or any intra-pipe boundary conditions, the MOC can give the solution at the boundary without requiring the existence of a physical calculation node, relying, instead, on the physical type of boundary that is represented [35, 39]. The MOC reduces a partial differential equation into a set of ordinary differential equations (ODE), which can be integrated and solved. In the case of the Euler system of equations, the main underlying concept is the possibility of finding curves in the space-time domain along which flow properties can be considered constant (in the case of homentropic flow) or can be integrated easily (if the flow is not homentropic). These curves are then linearized according to the solution of the flow field at the previous calculation step. Starting from the set of conservation equations written in non conservative form:

$$\begin{cases} \dfrac{\partial \rho}{\partial t} + \dfrac{\partial (\rho u)}{\partial x} = 0 \\ \dfrac{\partial u}{\partial t} + u\dfrac{\partial u}{\partial x} + \dfrac{1}{\rho}\dfrac{\partial p}{\partial x} = 0 \\ \dfrac{\partial p}{\partial t} + u\dfrac{\partial p}{\partial x} - a^2\left(\dfrac{\partial \rho}{\partial t} + u\dfrac{\partial \rho}{\partial x}\right) = 0 \end{cases}$$

and performing a linear combination of all the equations, the system can be written as:

$$\begin{cases} \left[\dfrac{\partial p}{\partial t} + (u+a)\dfrac{\partial p}{\partial x}\right] + \rho a\left[\dfrac{\partial u}{\partial t} + (u+a)\dfrac{\partial u}{\partial x}\right] = 0 \\ \left[\dfrac{\partial p}{\partial t} + (u-a)\dfrac{\partial p}{\partial x}\right] + \rho a\left[\dfrac{\partial u}{\partial t} + (u-a)\dfrac{\partial u}{\partial x}\right] = 0 \\ \left[\dfrac{\partial p}{\partial t} + u\dfrac{\partial p}{\partial x}\right] - a^2\left[\dfrac{\partial \rho}{\partial t} + u\dfrac{\partial \rho}{\partial x}\right] = 0 \end{cases}$$

The ordinary differentials of the variables p, ρ, and u, represented in square brackets, are calculated along the characteristic lines drawn in the plane (x, t), whose slope depends on the properties of the flow as follows:

$$\begin{cases} \dfrac{dx}{dt} = u + a \\ \dfrac{dx}{dt} = u - a \\ \dfrac{dx}{dt} = u \end{cases}$$

The hyperbolic system can be rewritten in the following form with total derivative, thanks to the equations of directions, evaluated on the lines just described:

$$\begin{cases} \dfrac{dp}{dt} + \rho a \dfrac{du}{dt} = 0 \\ \dfrac{dp}{dt} - \rho a \dfrac{du}{dt} = 0 \\ \dfrac{dp}{dt} - a^2 \dfrac{d\rho}{dt} = 0 \end{cases}$$

The characteristic speeds shown in the previous equations represent the propagation of signals, or disturbances, through the fluid medium. In the first two equations, the disturbance is propagated at the local speed of sound with respect to the fluid. This disturbance is often called a characteristic wave and causes changes in pressure, density, temperature, and velocity of the fluid. The third equation represents a disturbance that propagates at the local velocity of the fluid, u, which carries energy and composition and is called the flow line, or contact surface.

For each one of these curves, there are variables that will remain constant along them. Those variables are known as Riemann invariants and are defined as follows:

$$\lambda = a + \frac{k-1}{2} u \qquad \beta = a - \frac{k-1}{2} u$$

The velocity of the gas, u, and the speed of the sound, a, can be obtained from the *Riemann* invariants, as follows:

$$a = \frac{\lambda + \beta}{2} \qquad u = \frac{\lambda - \beta}{k-1}$$

For what concerns entropy, its value does not change along a line of type $dx/dt = u$. In this case, applying the correlation of an isentropic process along this line, the level of pressure at the new time step can be evaluated:

$$\frac{p}{p_{ref}} = \left(\frac{a}{a_A} \right)^{2k/(k-1)}$$

FIGURE 9.6 Stencil in the space-time plane of the MOC.

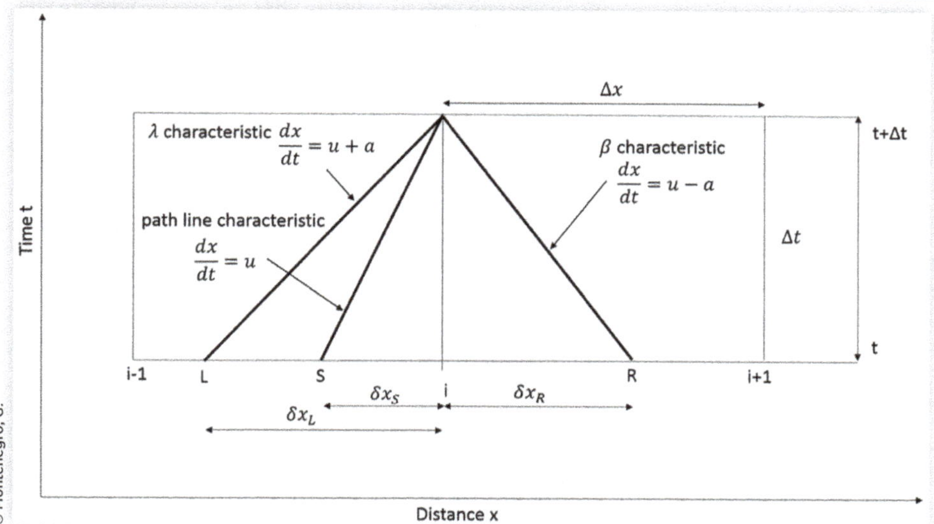

where a_A represents the speed of sound that would be possible to reach, at an arbitrary reference pressure, following an isentropic transformation starting from a pressure p. This theory allows to determine the fluid state, in terms of primitive variables (velocity, temperature, and pressure), knowing the fluid state at the right and at the left side of node i (see Figure 9.6). Some authors have proposed a coupling approach based on the definition of characteristic lines both in the 1D and 3D domain [47], where the 1D code OpenWAM was coupled to a commercial CFD code by means of suitable user functions. To couple the two codes by means of the MOC, it was necessary to know the thermodynamic and fluid dynamic state of the fluid in the 3D solver. To extract initial values at the interface of the two codes, i.e. where the physical boundaries of the two systems are coincident, a process of averaging is required among the several 3D calculation nodes. This can be performed resorting to weighted average processes, keeping the local mass flux as weighting factor. The Riemann invariant can then be calculated based on the averaged thermo-fluid dynamic state. On the 1D side of the coupling, this procedure is straightforward and, once the values of the λ and β invariants are known, the fluid state at the new time step can be calculated and exploited to march forward in time in the two codes. In particular, on the 3D side, the primitive variables at the new time step are used as face centered values for the cells containing the coupling boundary, whereas on the 1D side they are used to determine the new fluid state, mimicking an intra-pipe boundary condition. This procedure is repeated at every time step until the end time of the simulation is reached. As can be seen in Figure 9.7, the time step of the two codes is different due to a different mesh size. To match the correct value of the fluid dynamic quantities at every time step, a linear interpolation must be performed between the actual and the previous time step.

FIGURE 9.7 Time marching procedure of the coupling strategy based on the MOC.

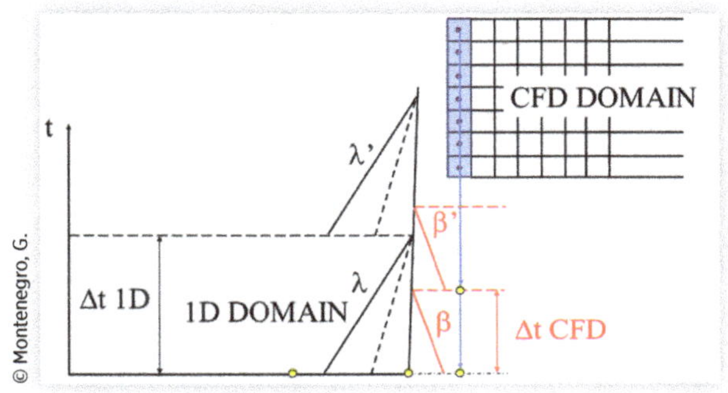

Relying on the MOC means that the value of the primitive variable is the same for both the interface regions of the two domains, unless the characteristic lines are calculated for every cells on the 3D side. This procedure, however, may decouple the two domains, since the primitive variable at the interface between the two domains should be equal. On the basis of this consideration, this approach is suitably applied when the flow is uniformly distributed on the flow area.

9.3.2.3 RIEMANN-BASED APPROACH

The approaches discussed so far are suitably applied where the 1D-3D interface is located far from sharp corners or sudden changes of flow direction. In cases where the geometry is highly complex and there is no location where the flow is nearly 1D, this approach may force the flow at the boundary to values non-compatible with the internal flow. This issue can be overcome relying on a local solution of the coupling, cell by cell, exploiting methods that act on the assignment of the conserved variable flux at the interface boundary. An existing example involves the solution of the Riemann problem for every cell which owns the face constituting the boundary surface [8]. In particular, at the beginning of each time step (as shown in Figure 9.8), the average of the conserved variables in the centroids of the cells is assigned to the last node of the 1D domain, while the conserved variables in the $n-1$ node are assigned to the left state of multiple local Riemann problems. Therefore, all the faces constituting the 1D/multiD boundary identify a Riemann problem, which is solved adopting the HLLC approximated solution. This procedure has the advantage that no field, in particular the velocity, might be forced with non-compatible values. In every flow condition, subsonic or supersonic, the Riemann problem is solved for every cell producing, in certain condition, also nonuniform profiles at the domain interface.

Recalling the previous description of the HLLC Riemann, it can be exploited to calculate the transport term of mass, energy, and momentum conservation equations. Considering the conservation of a generic property ϕ:

$$\frac{\partial(\rho\phi)}{\partial t}+\nabla\cdot\left(\rho\phi\vec{U}\right)=0$$

the transport term is integrated over the cell volume and, by means of the Gauss theorem, reduced to a surface integral:

$$\oint_S \rho\phi\vec{U}\cdot\vec{n}dS=\sum_{f=1}^{n}\rho\phi_f U_f S_f.$$

On the basis of the finite volume discretization, the surface integral becomes a summation of the volumetric fluxes of the conserved variables through the intercell faces, evaluated at the face center. Moreover, the values of the variables at the face center can be evaluated recurring to different interpolation techniques, according to the accuracy desired [45], and the HLLC Riemann solver has been exploited to evaluate directly the volumetric flux at the intercell face:

$$\sum_{f=1}^{n}\rho\phi_f U_f S_f=\sum_{f=1}^{n}F_f^{HLLC}$$

For each boundary face, the Riemann problem has been located at the center of the intercell face (the owner cell centroid

FIGURE 9.8 Procedure adopted for the integration of the two codes. The solution of the local Riemann problem is pointed out.

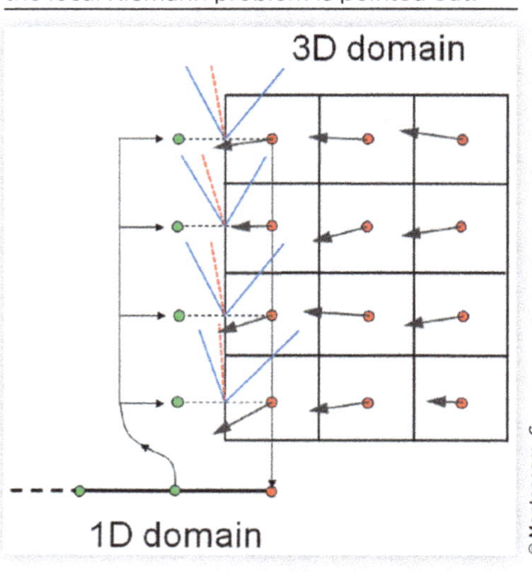

becomes the left state and the neighbor becomes the right one, as shown in Figure 9.9) and the velocity vector U_f has been decomposed into a normal component and a tangential one laying in the face plane:

$$\overrightarrow{U_n} = \vec{n}\left(\overrightarrow{U_f} \cdot \vec{n}\right)$$

$$\overrightarrow{U_t} = \overrightarrow{U_f} - \vec{n}\left(\overrightarrow{U_f} \cdot \vec{n}\right)$$

Since this last component is not responsible of any net flux across the intercell face, it is transported as a fluid property. Consequently, the solution of the centered wave system becomes as follows:

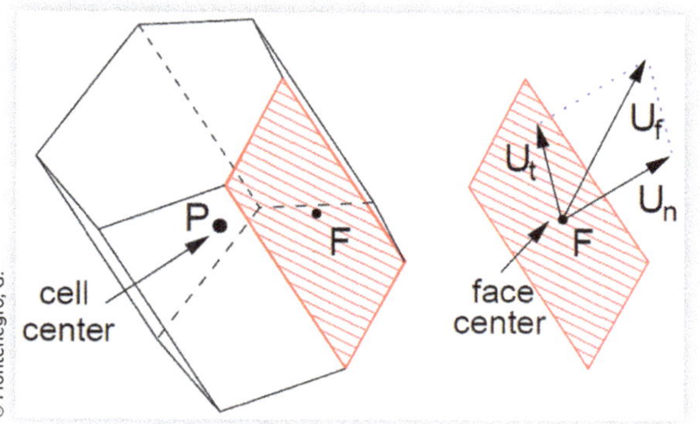

FIGURE 9.9 Generic cell face arbitrarily oriented in the space. The decomposition of the velocity vector is highlighted.

$$(\rho)^*_K = \rho_K \frac{S_K - \overrightarrow{U_K} \cdot \vec{n}}{S_K - S^*}$$

$$\left(\rho\vec{U}\right)^*_K = \rho_K \frac{S_K - \overrightarrow{U_K} \cdot \vec{n}}{S_K - S^*}\left(S^* \cdot \vec{n} + \overrightarrow{U_t}\right)$$

$$(\rho e_0)^*_K = \rho_K \frac{S_K - \overrightarrow{U_K} \cdot \vec{n}}{S_K - S^*}\left[-e_K + \left(S^* - \vec{U} \cdot \vec{n}\right)\left(S^* + \frac{p_k}{\rho_k\left(S_K - \vec{U} \cdot \vec{n}\right)}\right)\right]$$

9.4 Examples of Application

In this section examples of applications of 1D-3D coupling strategies will be discussed exploiting direct experience in the field of internal combustion engine simulations of the authors in their research activity during the last decade. In any case the considerations that will be presented can be applied to similar approaches developed in different codes. The examples will refer to applications in the acoustic field, where the 1D-3D coupling has permitted the study of complex geometries, without resorting to difficult generation of equivalent 1D schematics. In the field of engine performance prediction, it will be shown how the coupling can help in the prediction of global engine parameters such as the volumetric efficiency, aiming at the improvement of engine performance. In addition, it will be shown how the adoption of detailed 3D approaches can bridge the gap the 1D model has when the physics that must be reproduced becomes rather complex, such as intake air systems where the fuel spray is injected (port fuel injection).

9.4.1 Engine Performance

It has been already demonstrated that the integrated 1D-3D approach can give results comparable to 1D calculation if applied to simple geometries [7], justifying the adoption of well-experienced corrective lengths to tune the model, since the computational burden

FIGURE 9.10 Example of a four-cylinder engine schematic with highlighted real components that have been discretized in 1D.

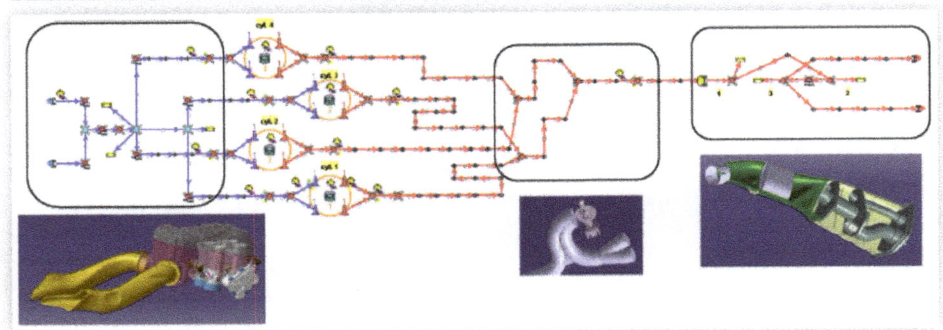

© Montenegro, G.

brought by an eventual 3D domain would not add any additional detail to the final results. Conversely, in the case of complex geometry systems, it can help the understanding of multidimensional effects on the behavior of certain engines.

As an example, we shall refer to a high-performance engine, whose global schematic is depicted in Figure 9.10, where capturing the wave actions in compact and complex systems is a key aspect to better optimize the engine. The air-box of a V4 motorbike engine is shown in Figure 9.11 and it can be seen how complex and compact is the layout. This has been carefully tuned and designed to achieve a specific power output required by the particular application it dedicated to. The full specifications of the engine are summarized in Table 9.1.

The 1D-3D approach was applied to model the air-box coupled to the intake runners (the trumpet shape pipes are modeled as 3D components inside the air-box) and the zip

FIGURE 9.11 Air-box of the V4 engine for motor bike application and the calculation mesh used for the simulations.

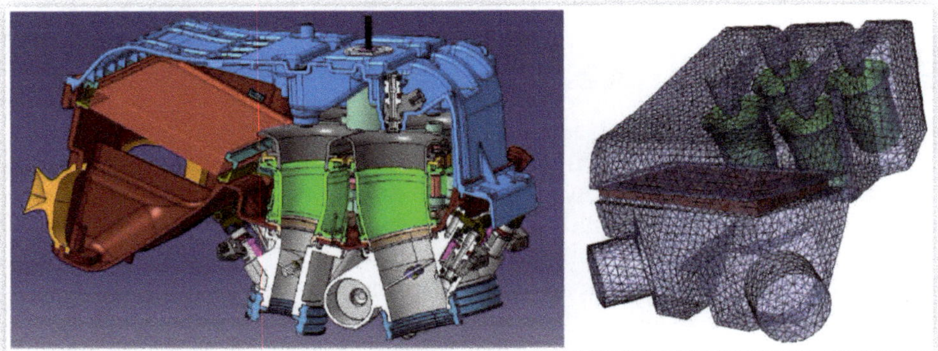

© Montenegro, G.

TABLE 9.1 V4 engine specifications

Engine type	Spark ignition
Number of cylinders	4-V65°
Total displacement	999.6 cc
Compression ratio	13:1
Valves per cylinder	4
Air management	Naturally aspirated
Injection system	PFI (2 injectors per cylinder)

ducts, which represent the two connections between the ambient and the intake system. The mesh of the air-box consists of a fully tetrahedral, whose average edge size is around 5 mm (Figure. 9.11). The mesh reported in this example was generated starting from the triangulated representation of the closed surface. Moreover, in the figure it is possible to see that the filter cartridge is modeled by means of a zone of cells into the air-box, where a source term of flow resistance is introduced to mimic the porous region of the filter cartridge.

In Figure 9.12 two snapshots of the air-box stream lines and pressure fields calculated at 12,500 rpm are shown. As it can be seen, inside the filter cartridge, the pressure waves are propagated with a different velocity, because of the porous material resistance, while the velocity field does not show any evident change in flow direction. This behavior is consistent with the definition of an isotropic tensor to reproduce the flow resistance. The integrated 1D-3D approach was adopted to predict the performances of the engine over a wide range revolution speeds at full load and the approach based on the local solution of the Riemann problem was adopted. An example of the 1D schematic, in which the connection to the 3D model is highlighted, is shown in Figure 9.13.

FIGURE 9.12 Velocity and pressure fields inside the air-box at 12,500 rpm.

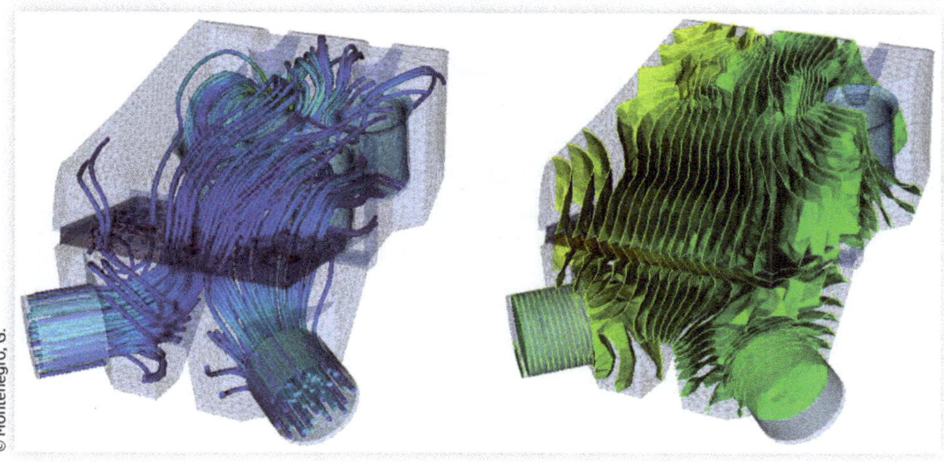

FIGURE 9.13 1D schematic of the intake system.

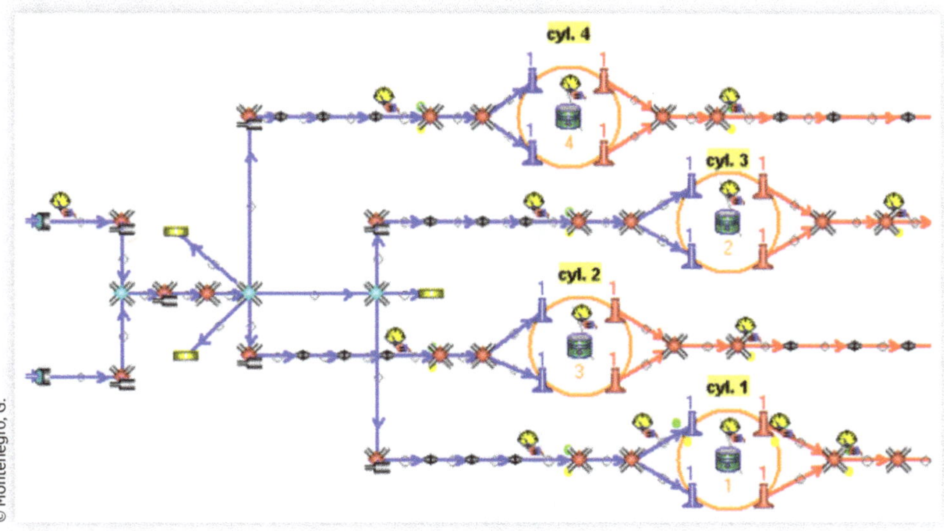

The engine configuration was modeled with the air-box by the 1D-3D hybrid code to point out the effects of flow condition in the zones of the air filter and the trumpets.

These elements are closely coupled to each other inside the air-box due to the engine architecture (4-V65°); for this reason, this zone could present a behavior similar to a four-into-one junction rather than an air-box geometry typical for a four-in-line cylinder engine. Therefore, each cylinder should be penalized or could take advantage of the pressure pulses coming from the other cylinders during the intake stroke, depending on the engine regime. The comparisons between the measured pressure pulses before cylinder 1 and 4 and the calculations carried out adopting the 1D-3D integrated code are shown in Figure 9.14 , and 9.15, respectively. For the sake of brevity, only a selection of the operating points modeled, in particular the most significant, have been chosen: 4500 and 6500 rpm (Figure 9.14a, 9.14b, 9.15a and 9.15b), representative of low engine speed, and 10,500 and 12,500 rpm (Figure 9.14c, 9.14d, 9.15c and 9.15d) the point of maximum volumetric efficiency and maximum power, respectively. At high engine revolution speed, the hybrid model is fairly predictive, exhibiting a good agreement with the measured data both in the absolute value and phase of the wave motion. The other operating conditions, lying in the range between 4500 and 6500 rpm, are affected by a slight delay of the wave motion, which causes discrepancies between the calculated curves and the measured ones. The explanation of this curve shift resides in the absence of a submodel, which accounts for the latent heat of evaporation of the injected fuel droplets.

FIGURE 9.14 Pressure traces upstream of cylinder n. 1: comparison between measured data, fully 1D calculation and 1D-3D calculations. Operating conditions: (a) 4500 rpm, (b) 6500 rpm, (c) 10,500 rpm, (d) 12,500 rpm.

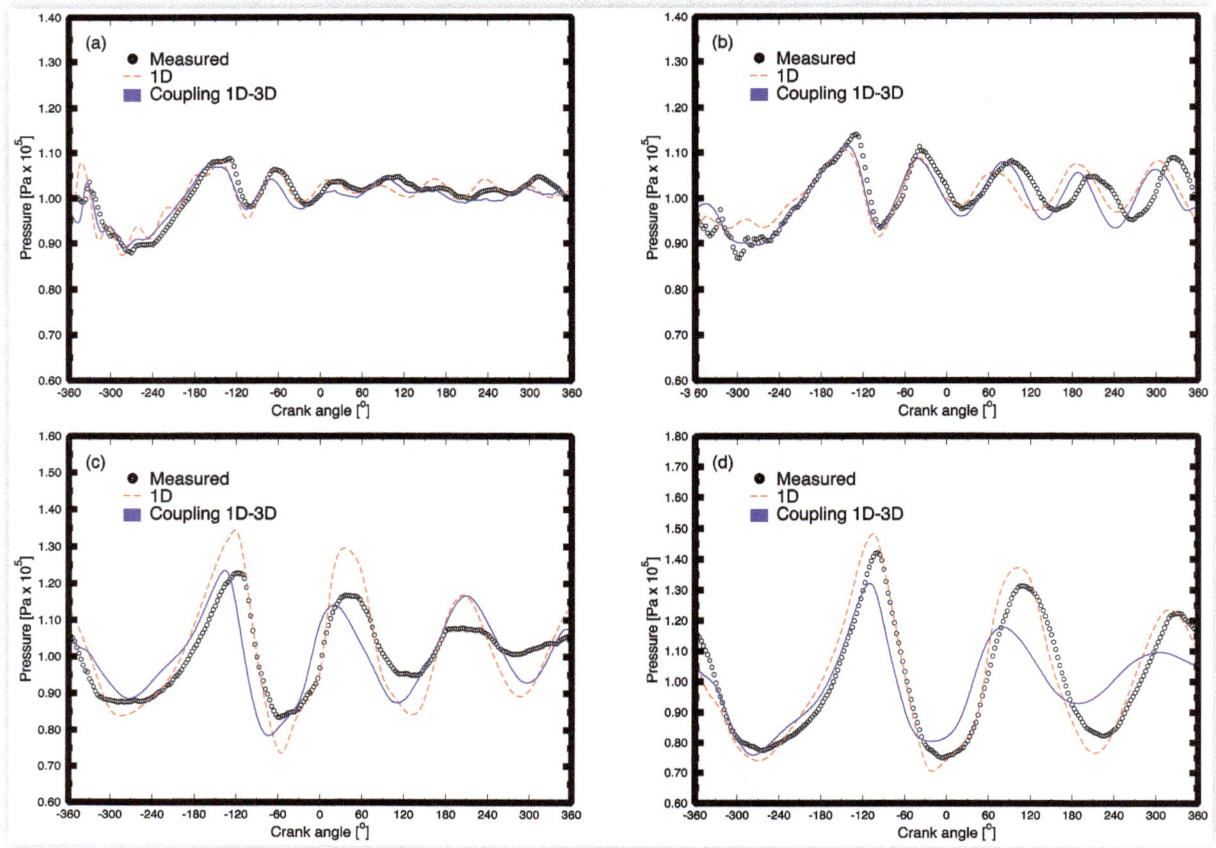

As it will be explained later in this chapter, this phase transition, during which the formation of air-fuel mixture takes place, subtracts the needed energy from the droplet itself, which becomes colder, cooling down the surrounding gas by convection. The reached temperature drop may be around 20°C [48]. Because of this reason, the speed of sound decreases, changing the wave motion into the ducts, resulting in a slower wave propagation. The first steps of this activity required the buildup of the 1D engine schematic on the basis of the known geometrical layout, including complex devices such as the air-box with the filter cartridge, the exhaust junctions, the catalyst, and the silencer. In Figure 9.10 the complete schematic of the engine, as defined by the 1D graphical interface, is shown together with some images of the main components of the intake and exhaust systems. The 1D simulations were performed considering steady-state operating condition, sweeping a range of engine revolution speeds from 4500 to 12,500 rpm, with a step of 1000 rpm at full load. For each one of these operating conditions, the experimental spark advance and air to fuel ratio were used to calculate the engine performances. In Figures 9.8 and 9.9, the pressure pulses calculated by means of the fully 1D approach into the intake runners, respectively, associated to cylinders n. 1 and n. 4, are shown and compared to measured data and calculations carried out by means of the 1D-3D approach. It must be kept in mind that to assess this degree of accuracy with the 1D tool, several configurations were considered, each one with particular features suggested by the analysis of the pressure and velocity fields obtained by the 1D-3D simulations. The analysis suggested that the air-box was resonating along the main longitudinal direction and along the transversal one identified by the width of the device, as can be seen in Figure 9.12. No resonance along the height of the air-box occurred, indicating that the engine breathing is tuned on two main resonances: longitudinal and transversal. As an example of the result of this calibration process, in Figure 9.13 it is shown an equivalent 1D schematic which models the longitudinal dimension of the air-box.

The results obtained were in agreement with the theory of the resonances, neglecting the transversal wave motion. As a proof of this concept, it must be said that the configuration which allowed the best fit of the measured pressure traces was obtained by modeling the plenum volume as a 0D element at boundary condition level [43] and playing with corrective lengths of the adjacent pipes to obtain the desired resonances.

On the exhaust side, a similar study has been carried out to highlight the potential of the 1D-3D simulation. The exhaust system of this engine is characterized by the presence of complex shape junction. This element is a four-into-two-into-one junction composed by two Y-type junctions leading to two main branches (Figure 9.16).

These two branches do not join themselves immediately, remaining separated by a curved baffle for a length of 4 diameters. At the end of the baffle, the two pipes merge in correspondence of a further pipe bend.

In Figure 9.16, it is shown the 1D schematic of the engine in which the 3D model of the junction is visible. The simulations were firstly carried out adopting the integrated approach, and then, on the basis of the analysis of the fluid dynamic behavior, the equivalent 1D schematic was generated. The mesh adopted in the calculations was built once again resorting to a tetrahedral discretization technique with an average cell size of around 5 mm.

In Figure 9.17 two snapshots of the multi-pipe junction velocity and pressure fields calculated at 10,500 rpm are shown. Differently from the air-box, the velocity field points out remarkable directionality effects imposed by the four branches entering the two Y-type junctions and by the curved two-into-one junction. Moreover, the pressure isosurfaces are characterized by evident spherical shape, leading to the consideration that the wave motion, contrary to what can be expected in this part of the system, cannot be reduced to planar, suggesting that a 1D approximation may not consider eventual reflections to lateral walls. These reflections travel up to the exhaust valve

FIGURE 9.15 Pressure traces upstream of cylinder n. 4: comparison between measured data, fully 1D calculation and 1D-3D calculations. Operating conditions: (a) 4500 rpm, (b) 6500 rpm, (c) 10,500 rpm, (d) 12,500 rpm.

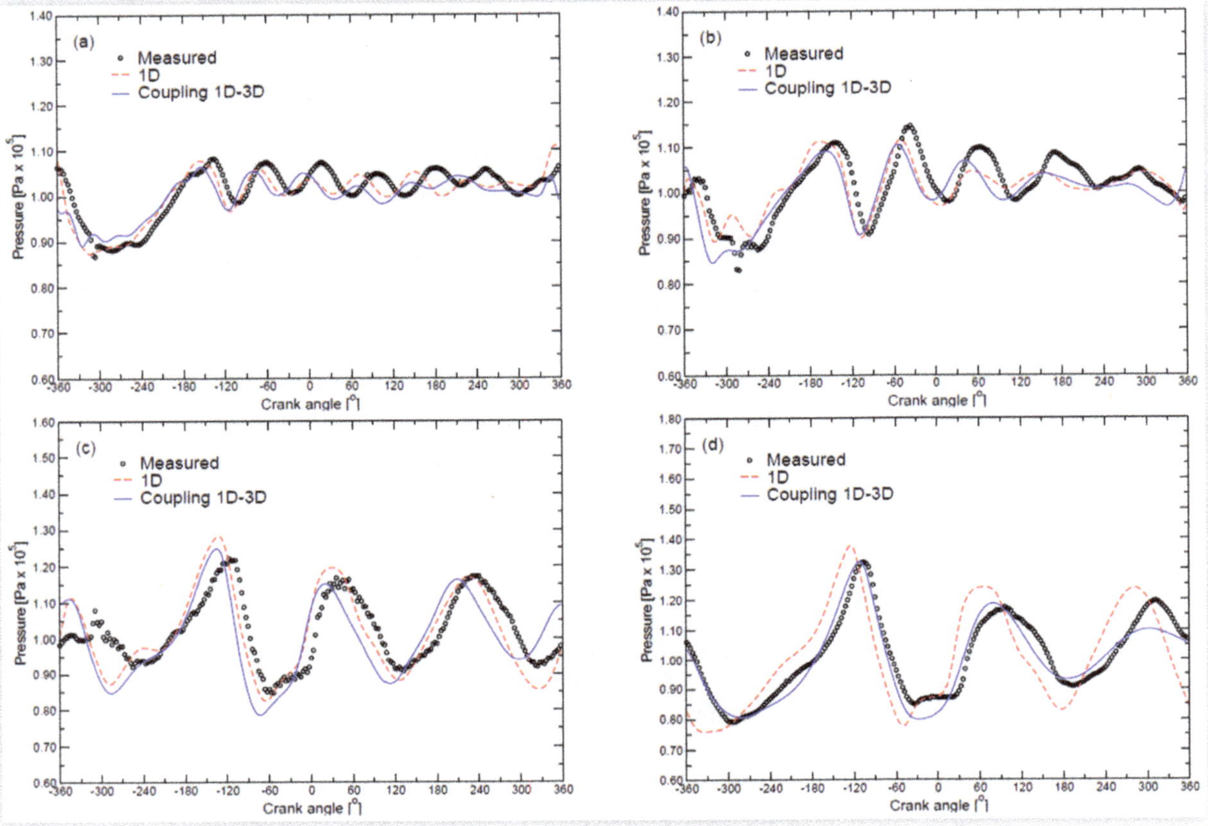

FIGURE 9.16 1D schematic of the V4 exhaust system; the highlighted area shows the multi-pipe junction which was been simulated with 1D code and 1D-3D hybrid code.

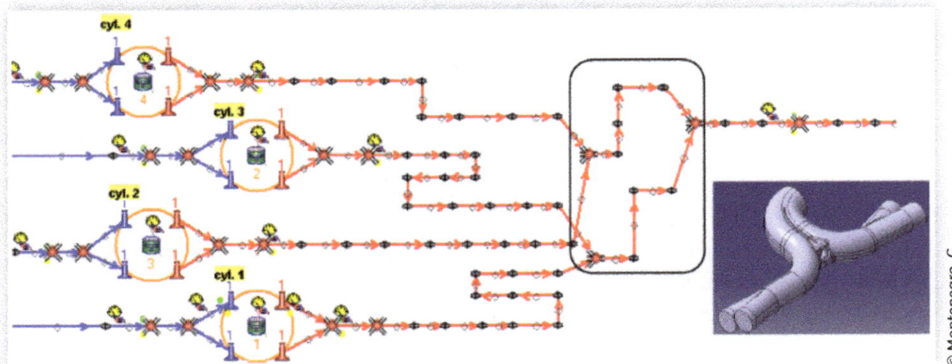

and can affect the burnt gas scavenging process. The visualization of the wave action can indeed help the user to determine the entity of corrective lengths to be used in the 1D calculation to obtain the best calibration of the 1D model. The comparisons between the measured pressure pulses downstream of the cylinder n. 4 and the calculations carried out adopting the two models, namely, the 1D-3D integration and the fully 1D code, are shown in Figure 9.18.

FIGURE 9.17 Snapshot of the solution field at a fixed time step at 10,500 rpm: (a) velocity field, (b) pressure isosurfaces.

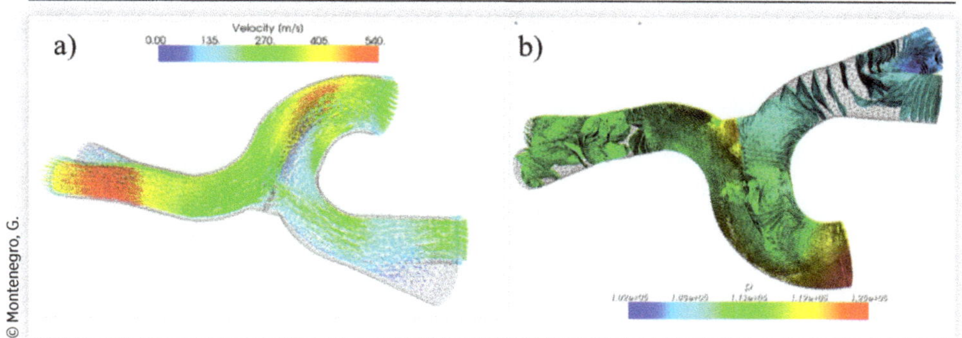

FIGURE 9.18 Pressure traces downstream of cylinder n. 4: comparison between measured data, fully 1D calculation and 1D-3D calculations. Operating conditions (a) 4500 rpm, (b) 6500 rpm, (c) 10,500 rpm, (d) 12,500 rpm.

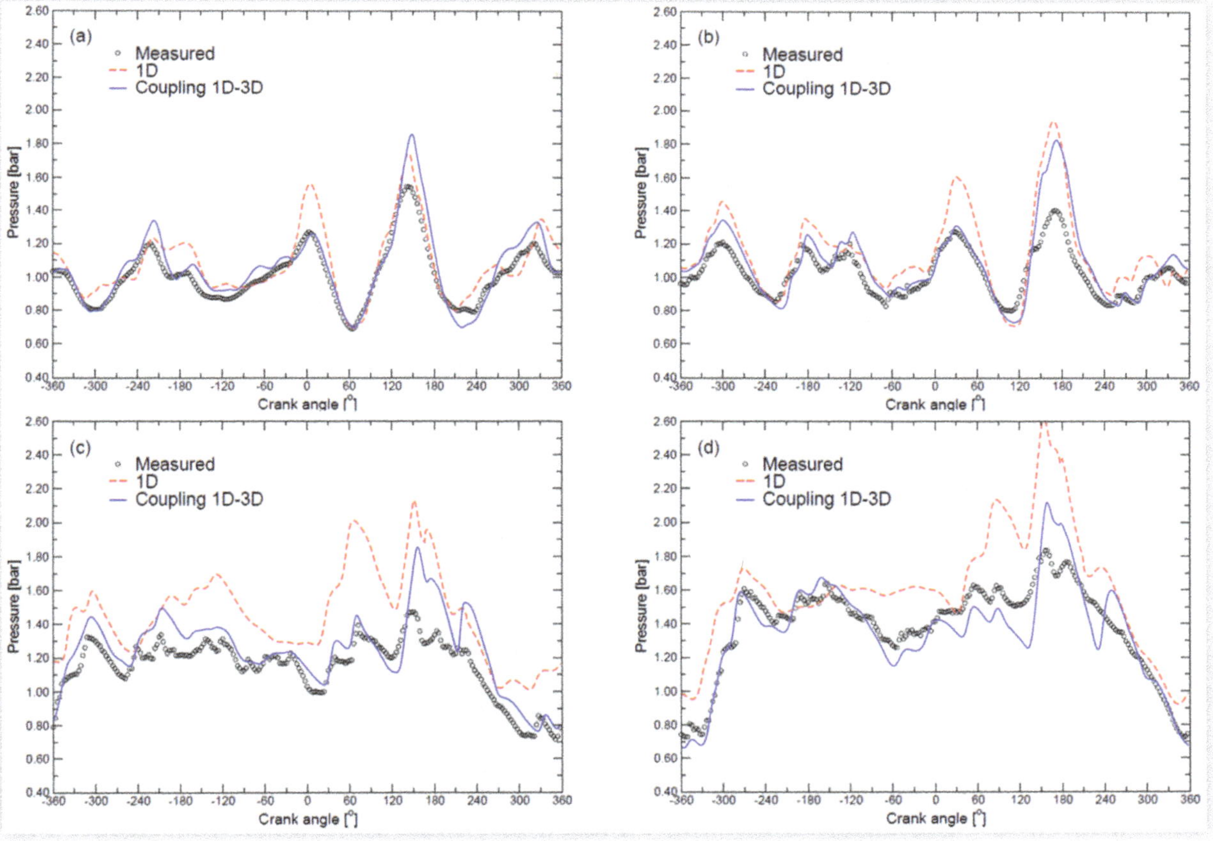

In particular the fully 1D calculation was performed with an advanced 1D model for the multi-pipe junctions [49, 50], which takes into account the pressure loss as a function of the pipe incidence angle on the basis of validated analytical correlations. It can be noticed that the hybrid model is more predictive than the fully 1D approach, exhibiting a fair agreement with the measured data both in the absolute value and phase of the wave motion. It is also interesting to note how the flow uniformities are handled

FIGURE 9.19 Calculated and measured nondimensional volumetric efficiency on the V4 with respect to the maximum volumetric efficiency value. Calculations are referring to fully 1D and 1D-3D.

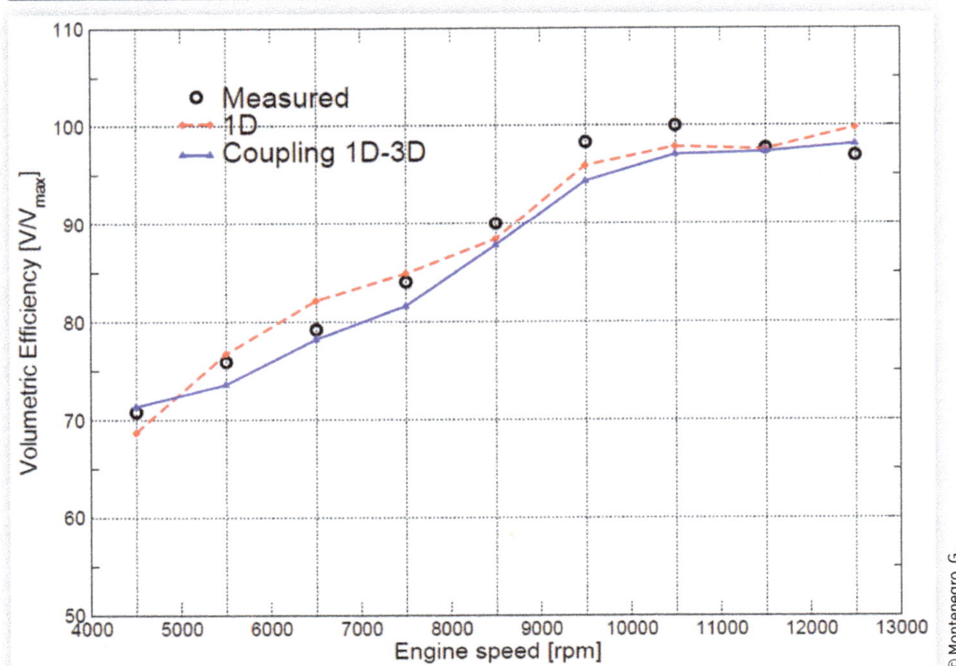

by the numerical procedure at the domain interfaces. In particular, it is shown how the outlet section is dominated by a nonuniform flow, where the gas velocity presents high values and a recirculation region, due to the strong flow directionality imposed by the manifold at the junction inlet. Moreover, an evident backflow is pointed out at the interface section connected to a different cylinder, which is caused by the wave motion during the closed exhaust valve phase.

Figure 9.19 shows the comparison between measured and calculated volumetric efficiency, as a function of the engine speed. The 1D-3D hybrid code shows the best match in the range of the investigated engine speeds. As it can be seen, the prediction is similarly accurate for the two approaches, since both the absolute values and the trend are well captured. At this stage it must also be kept in mind that the volumetric efficiency determined by the fully 1D approach was achieved by a configuration tuned based on 1D-3D simulation results. This means that all the corrective lengths have been set accordingly and that the 1D model cannot be used in a fully predictive way when an optimization on the pipe system is required. In addition, it must be considered that the benefit of the 1D-3D simulations appeared to be in the simulation of the exhaust side, which has a minor impact on the engine volumetric efficiency with respect to the intake. Therefore, considering that the driving phenomena is the resonance of the air-box, the two results are comparable, since both the 1D and the 1D-3D simulations are affected by the neglection of the fuel evaporation effect. In the next section, it will be shown how the inclusion of the fuel spray modeling can improve the prediction of the volumetric efficiency and therefore of the engine overall performance.

9.4.2 Effect of Fuel Injection

The case of the four-cylinder engine has highlighted a great potential of the 1D-3D coupling in capturing the engine performance, with respect to a 1D model carefully tuned to capture the main wave action process that control the engine performance. However, the presence of the fuel injection system in four different cylinders have been highlighted as the main responsible of the incorrect phasing of the wave motion inside intake runner during the valve closure period. For this reason in this section, it will be shown how the inclusion of the fuel spray simulation in the 3D model can help in capturing this key process leading to a better prediction of the intake resonances, and therefore of the engine performance. To this purpose a motorbike single-cylinder engine is considered and modeled resorting to the 1D-3D coupling. The benefits of considering a single-cylinder configuration are several, but the main advantage resides in the fact that there is no effect of cylinder-to-cylinder interaction, allowing to isolate the process and therefore to reduce the uncertainty. The specifications of the engine considered in this case are summarized in Table 9.2.

The engine has been modeled first of all using a fully 1D approach to show the effect of the neglection of the spray modeling and to suggest possible strategies that can be adopted to calibrate the 1D simulation. The schematic of the engine is shown in Figure 9.20.

The simulation of the single-cylinder engine, without performing any calibration on corrective lengths and coefficients, but just relying on a correspondence of geometrical

TABLE 9.2 Single-cylinder engine specifications

Engine type	Spark ignition
Number of cylinders	1
Total displacement	250 cc
Cylinder bore	81 mm
Valves per cylinder	4
Air management	Naturally aspirated
Injection system	PFI (2 injectors)
Injection pressure	0.5 MPa

FIGURE 9.20 1D schematic of the single-cylinder engine model to study the impact of the fuel spray injection.

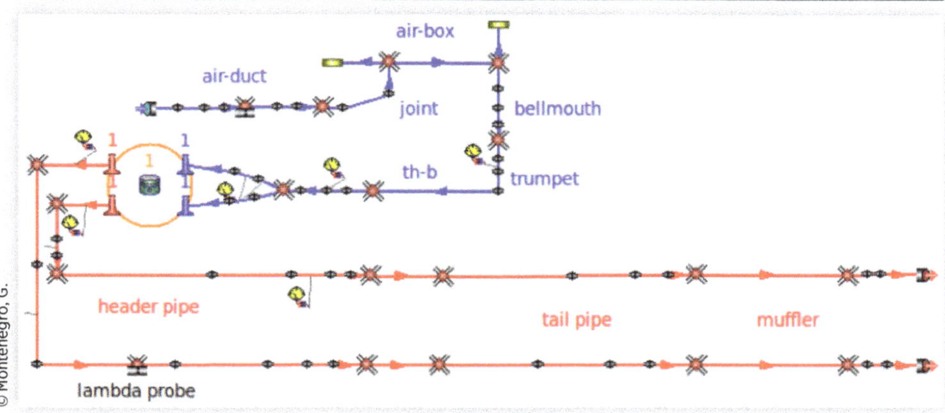

FIGURE 9.21 Comparison between measured and computed normalized volumetric efficiency, as a function of engine revolution speed at WOT operating condition.

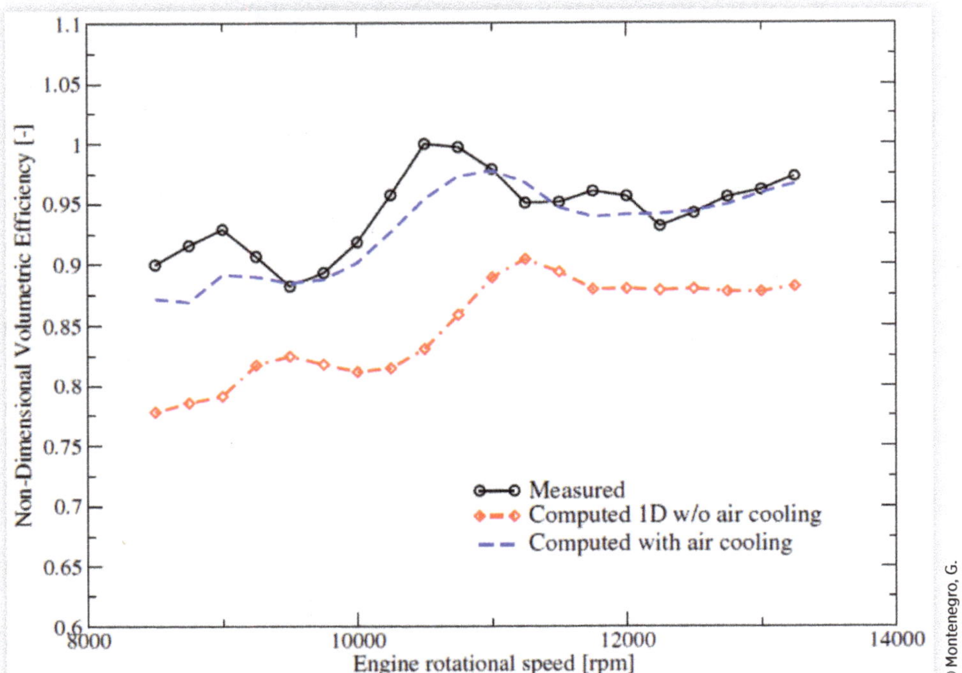

parameters, allows to capture the wave motion associated to the specific geometry and to the air conditions inside the intake system. As can be seen in Figure 9.21, the resulting VE curve shows an overall mismatch between measured and calculated values. The peak of volumetric efficiency appears to be predicted with a shift of around 750 rpm with respect to the rotational speed and underpredicted as absolute value. This, as observed in the previous section, is caused by the wrong capturing of the gas temperature, which, as known, is responsible for the tuning of the intake system. Figure 9.22 shows the comparison between measured and computed intake air temperature downstream of the throttle body, where the gap between the two curves (red and black) is around 25°C.

To consider also this effect in the 1D model, avoiding the inclusion of complex methods for the tracking of an evaporating liquid phase, the reduction of the inlet air temperature has been simulated by enhancing the heat transfer between the gas and the pipe walls of the intake runner. To do so, the pipe wall temperature has been imposed to low values, not corresponding to what happens in the reality, and the heat transfer coefficient has been increased to tailor the desired gas temperature. As a matter of fact, the gas velocity is not so high in the intake runner during the inlet valve closure and therefore this calibration must be done to overcome the limited value of the global heat transfer coefficient. In this way the global acoustic behavior of the intake runner has changed, and the reduction of the inlet air temperature, together with the increase of air density, has boosted the volumetric efficiency curve toward higher values. The result of this particular calibration is shown in Figure 9.22, which points out the impact of the calibration on the air temperature prediction. However, it can be noticed that by increasing the engine speed, the reduction of the time available for the heat transfer prevails against the increase of the convective heat transfer coefficient, resulting in an increasing trend of the gas temperature with the engine speed. In terms of volumetric

FIGURE 9.22 Comparison between measured and computed air intake temperature, as a function of engine revolution speed at WOT operating condition.

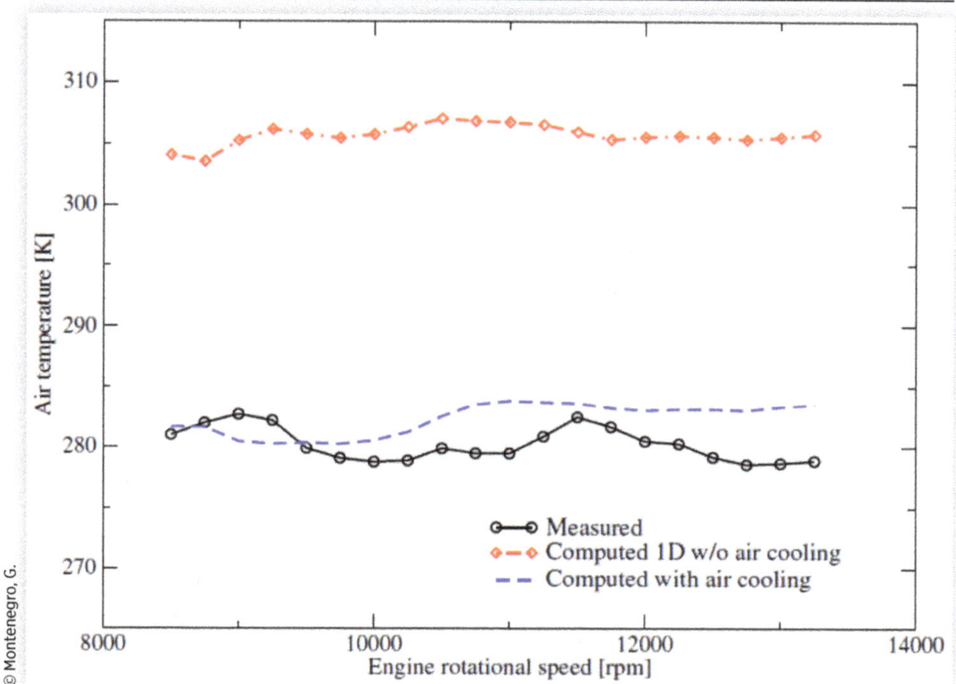

efficiency, the comparison between the measured and computed curves shows an improvement. The charge cooling operated by the pipe walls evidently affects the volumetric efficiency resulting in a curve peak shifted toward lowest revolution speeds and with higher absolute values. By adopting this type of tuning, a better behavior of the engine modeling by means of 1D simulation has been achieved. This behavior could be also highlighted by a wave action analysis, as it can be seen in Figure 9.23.

The strong cooling of the air into the intake runner affects the wave motion: the lower air temperature provides a lower speed of sound. In this way the computed pressure pulses are in better agreement with the measured data, as proved by the improvement of the volumetric efficiency (blue line in Figure 9.21).

FIGURE 9.23 Comparison between measured and computed pressure pulses into the intake runner (downstream the throttle body) for three different rotational speeds: (a) 10,750 rpm, (b) 12,000 rpm, (c) 13,000 rpm.

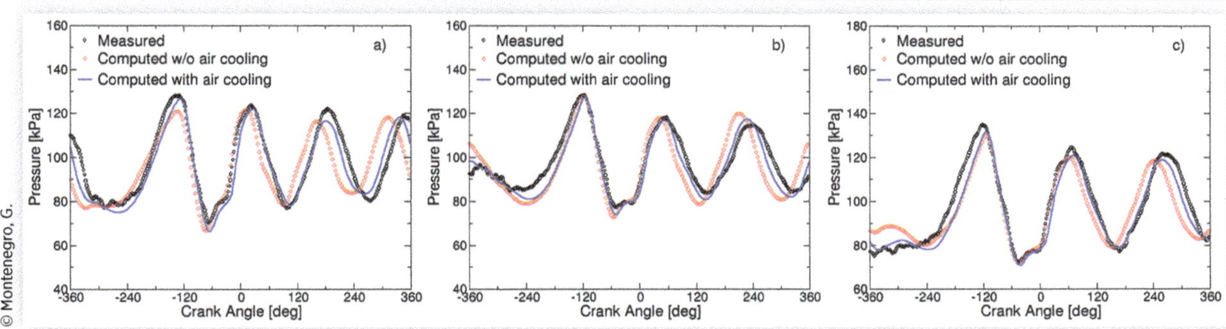

FIGURE 9.24 1D engine schematic of the GP30 engine. The cyan area stands for the 3D domain.

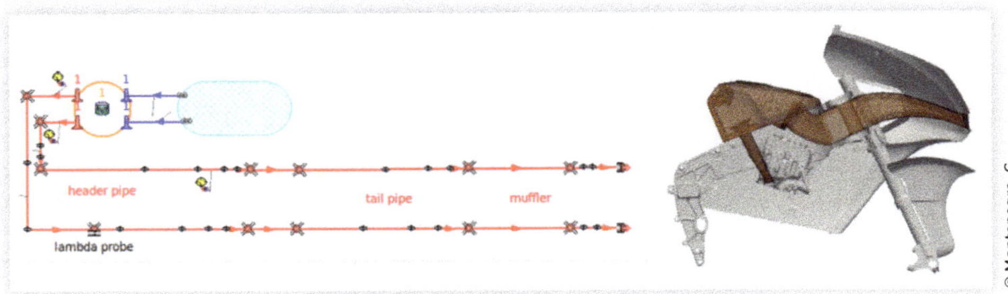

Replacing the intake system with a 3D domain, as shown on Figure 9.24, it is possible to carry out an integrated simulation where the 3D domain is modeled adopting an Eulerian plus Lagrangian approach for the tracking of liquid fuel parcels. The engine is coupled with the 3D domain only at the end of the intake runner, whereas upstream, the 3D domain has a boundary condition of flow stagnation that can mimic the velocity of the motorbike.

The coupling between the 1D and the 3D domain has been realized in a position close to the intake valve, as can be seen in the right side of Figure 9.24, where the 1D head duct before the intake valve is very short. Since the Lagrangian phase is not tracked along the 1D domain, the small 1D pipe permits to have a negligible loss of liquid phase in case of backflow from the 1D to the 3D domain. Additionally, the Riemann-based approach adopted for the coupling allows to place the interface in regions where the flow is not uniformly distributed over the flow area.

The simulation with the coupled approach has already been demonstrated to be an accurate prediction of the wave action inside intake and exhaust systems. In this case, as shown in Figure 9.25, the predicted pressure trace compares well with the measured values, where the measuring point is inside the intake runner not far from the intake valve. The induction phase, occurring at intake valve opened, is well captured at the three engine revolution speeds taken into consideration. Also, the motion during the closed valve period, where the intake runner operates as a quarter wave resonator, is well captured due to the cooling effect operated by the cold droplets subject to evaporation.

When the fuel is injected, the top of the air-box encounters an air column which is subject to a spring-wise oscillating motion inside the intake runner, preventing the fuel spray penetration and allowing the dispersion of the droplet cloud away from the target.

FIGURE 9.25 Comparison between measured and computed pressure pulses into the intake runner (downstream the throttle body) for three different rotational speeds: (a) 10,750 rpm, (b) 12,000 rpm, (c) 13,000 rpm.

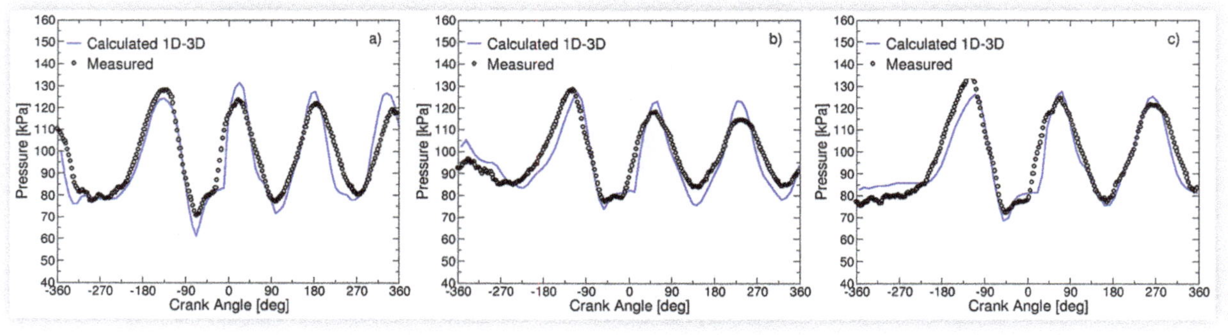

FIGURE 9.26 Example of cloud of droplets at 12,000 rpm (plotted against particle age) with relative fuel vapor mass fraction isosurface. The two snapshots are relative to an instant preceding the fuel injection after 14 thermodynamic cycles.

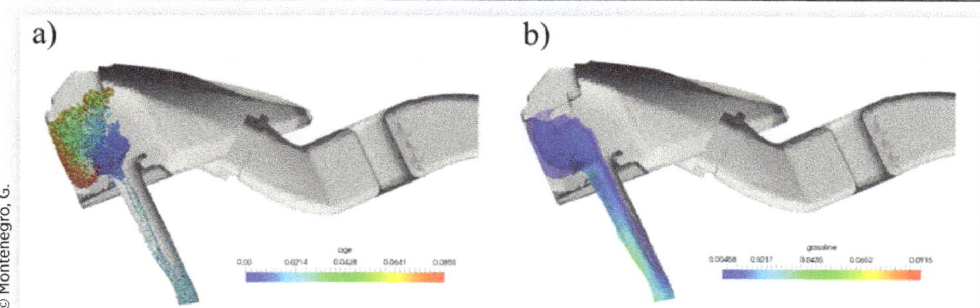

In Figure 9.26 it can be seen the dispersion of the droplet cloud and the consequent fuel vapor formed by the evaporation of the droplets. The two pictures are referred to an instant before the 15th injection, starting from cycle 0. It can be observed that the spray cloud is made by droplets injected during four subsequent injections (particles are scaled with respect to their diameter and colored by their age). The dynamics of the droplets is quite complex and cannot be reduced to simplified conclusions. However, it appears that, regardless of the diameter, as a combination of air motion and particle inertia, the oldest particles are pushed against the back side of the air-box. During the subsequent injections, these particles are subject to evaporation forming sacks of fuel vapor inside the air-box, which represent an undesired effect for drivability during transient conditions (acceleration and deceleration).

The evaporation of the fuels represents the main source of cooling of the fresh charge. In Figure 9.27, it is shown the temperature of the cloud of fuel droplets, where the minimum value reached is around 263°C. This low temperature level, associated with the large heat transfer surface offered by the droplets, causes a vigorous temperature drop of the gas. The resulting comparison between the calculated and measured gas temperature in a position close to the intake valve is shown in Figure 9.28, where the trend is now correctly predicted.

As a consequence of the fact that the cooling effect of the evaporating fuel has been captured, the prediction of the volumetric efficiency is improved. As can be seen from Figure 9.29, both the trend and the absolute value of the volumetric efficiency are well predicted.

FIGURE 9.27 Fuel droplet temperature during the injection.

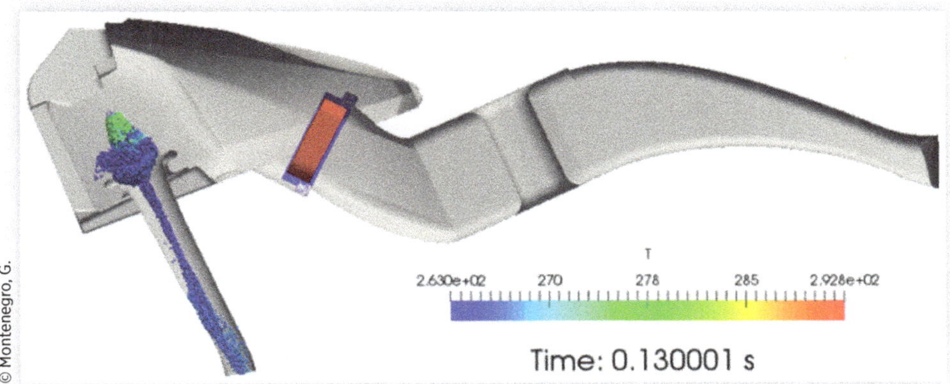

FIGURE 9.28 Comparison between measured and computed air intake temperature by 1D-3D coupling, as a function of engine revolution speed at WOT.

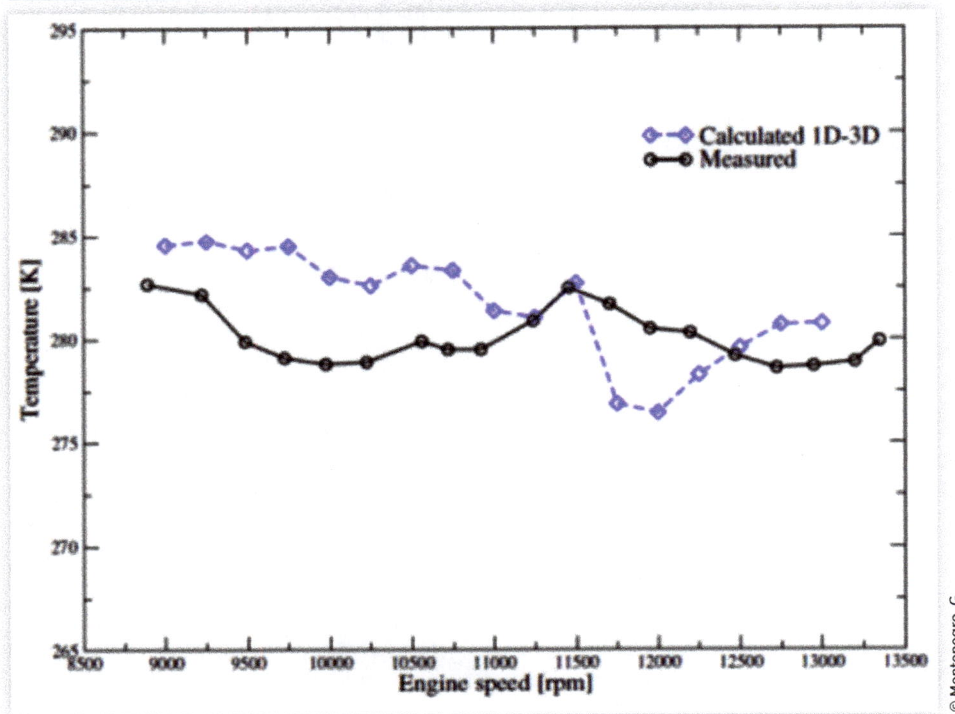

FIGURE 9.29 Comparison between measured and computed volumetric efficiency by 1D-3D coupling with the tracking of lagrangian spray.

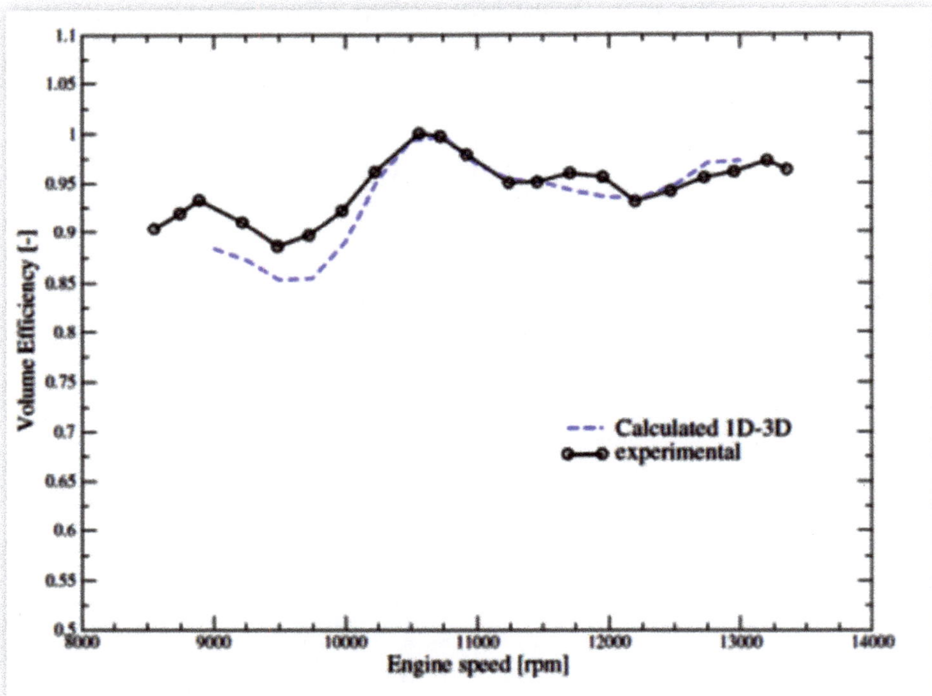

9.4.3 Acoustics of Silencers

Coupled 1D-3D approaches can also be adopted to predict the acoustic emission of a silencer or the sound spectrum emitted by an engine. In this type of application, there is no particular strategy or innovation to emphasize, since the quality of the prediction depends mainly on the accuracy of the numerical methods. In the field of acoustic analysis, there are two main paths that are usually followed: the determination of the TL and the evaluation of the insertion loss. The former is based on the excitation of the silencer, as a standalone device, by means of suitable inlet boundary conditions, and the consequent evaluation of the transmitted signal at the outlet. In this case the evaluation is carried out in the field of the linear acoustic, imposing perturbation with a small amplitude. The latter consist in the evaluation of the acoustic performance of the silencer when it is inserted in the exhaust line of an engine. In this case the silencer is subject to pressure perturbations that have finite amplitude, meaning that the oscillation amplitude is very wide resulting in a process that is typically nonlinear.

9.4.3.1 EVALUATION OF THE TRANSMISSION LOSS

In this case, the 1D-3D coupling is useful to easily impose suitable boundary conditions to the device and allow for a fast evaluation of the acoustic performance. A range of boundary conditions is required for the modeling of silencing duct systems by a fully 1D approach: the upstream excitation source, the tailpipe open termination, abrupt cross-sectional area changes, junctions of pipes, orifices, perforated ducts, axial side branches, etc. These boundary conditions are applied extensively in the case of a fully 1D simulation of the silencer and are based onto the MOC. Conversely, only the upstream excitation source and the tailpipe end (open or anechoic termination) boundary conditions are required to model the system in the case of an integrated 1D-3D simulation, due to the structure of the hybrid calculation. In particular, the use of 1D boundary conditions (coupled to 1D pipes) upstream and downstream of the 3D calculation domain allows a straightforward treatment of nonreflecting boundaries, which conversely could be critical if modeled directly by a CFD approach. In fact, in this case pressure transmissive boundary conditions may result not completely anechoic, reflecting small pressure waves whose magnitude scales with the jump in pressure across the approaching.

The upstream excitation source has been represented by a "white noise" pressure perturbation, covering the whole range of frequencies of interest in the field of IC engines [51]. This represents the most convenient alternative, in terms of computation runtime, to excite a system with several frequencies, instead of repeating the simulation for each harmonic. This technique has proved to be advantageous, for example, in the 1D simulation of perforates, which generally needs a very large computational effort, due to the great number of short ducts required to model the holes, the perforated pipe, and the cavity liner [52]. The white noise perturbation is characterized by a discrete spectrum with constant amplitude components, over a given frequency band. The numerical generation of a white noise periodic pressure perturbation may be carried out as the sum of N sinusoidal pressure oscillations with a fixed amplitude Δp and frequencies which are multiple of the fundamental f_0, with a random phase [53]:

$$p(t) = p_0 + \sum_{1}^{N} \Delta p \sin\left(2\pi f_0 t + \varphi_n\right)$$

where p_0 is a constant value representing the mean ambient pressure upstream of the flow duct system.

Two different muffler configurations can be considered to highlight the benefits of combining 1D codes, with specific boundary conditions. The results of the 1D-3D coupled simulations have been compared to measured TL of data and to calculations carried out by means of the fully 1D approach. To do so, the instantaneous pressure trends at inlet and outlet have been decomposed by the FFT algorithm and the TL calculated, evaluating the incident and transmitted acoustic power [54]. In particular, two expansion chambers with flow reversals have been considered, in which the inlet and outlet pipes are extended inside the muffler, as shown in Figure 9.30. The extensions of inlet and outlet pipes are meant to suppress the passbands of the expansion chamber by means of resonance peaks typical of quarter wave resonators. In general, the expansion chamber with reverse flow is a good test to point out the capability of the simulation technique to capture the attenuation curve in the frequency domain, when the wave motion is significantly multidimensional. Configurations A and B are characterized by different extension lengths, but same dimensions of the main chamber [55]. The multidimensional approach is based on the geometrical reconstruction of the silencer shape, starting from a CAD representation of the wet surface, while 1D approximation is usually tied to the skills of the user in generating equivalent 1D schematic of complex shape silencers. Moreover, in order to capture the effect of non-planar waves, the 1D approximation uses of corrective lengths [56]. The TL calculated by the two approaches, namely, the fully 1D and the 1D-3D is shown in Figure 9.31. In the case of the fully 1D, the agreement is satisfactory up to approximately 700 Hz, due to the good definition of the 1D acoustically equivalent duct systems and to the use of corrective lengths. At frequencies higher than 700 Hz, the predicted TL is poor, due to the excitation of higher-order modes. These are excited at approximately 1000 Hz, in agreement with the nodal lines theory [57]. A good match with experimental data is achieved in the case of the 1D-3D coupling. This has been achieved without the usage of any correction to the length of the pipes.

The importance of multidimensional waves is evident in the 3D calculation reported in Figure 9.32, in which it is possible to notice the non-planar nature of the wave motion.

Similarly to configuration A, the comparison between the measured TL of configuration B and the calculated ones by means of the two different approaches is shown (Figure 9.33). Compared to the previous configuration, this shows the effect of a longer extension of the inlet pipe. Once again it is evident the main limitation of the 1D approach is to capture the correct acoustic behavior at frequencies higher than 800 Hz, similarly to what was highlighted for configuration A. Conversely, for the 1D-3D approach, the agreement can be regarded as good, since the abatement peaks and passbands are

FIGURE 9.30 Layout of the reverse chambers with extended inlet and outlet: configurations A and B.

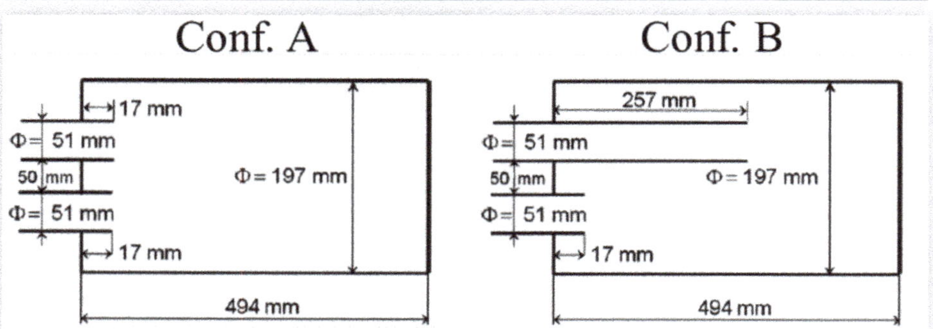

FIGURE 9.31 Transmission loss of the muffler Conf A: (a) fully 1D vs measured, (b) 1D-3D vs measured.

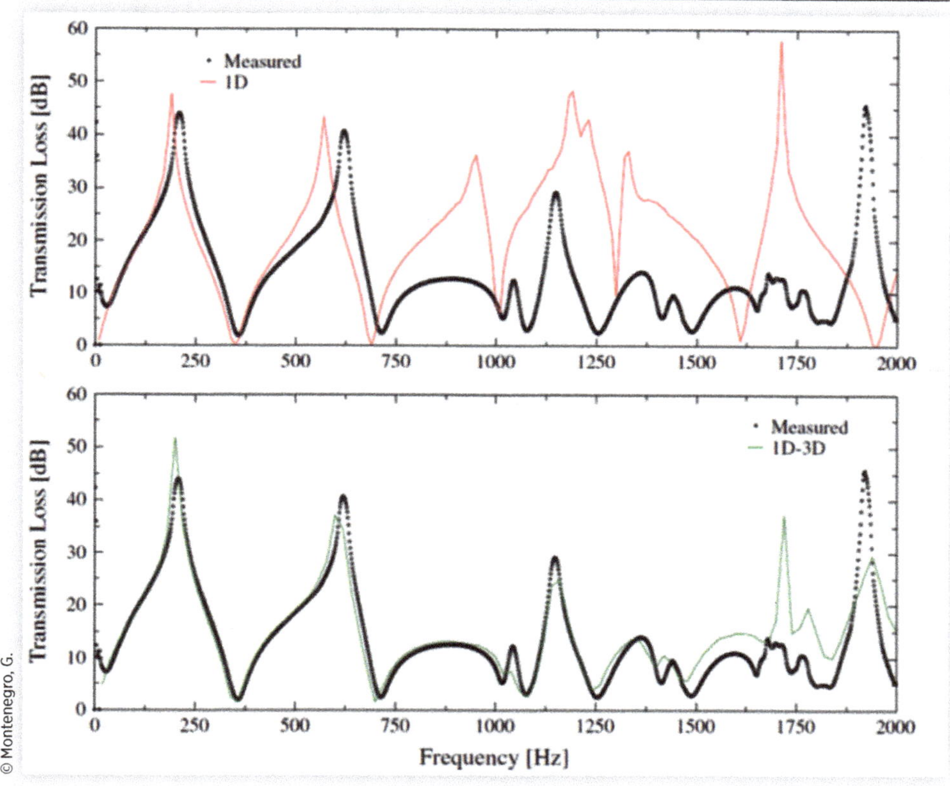

FIGURE 9.32 Non-planar shape of the pressure isosurfaces in the side silencer Conf A.

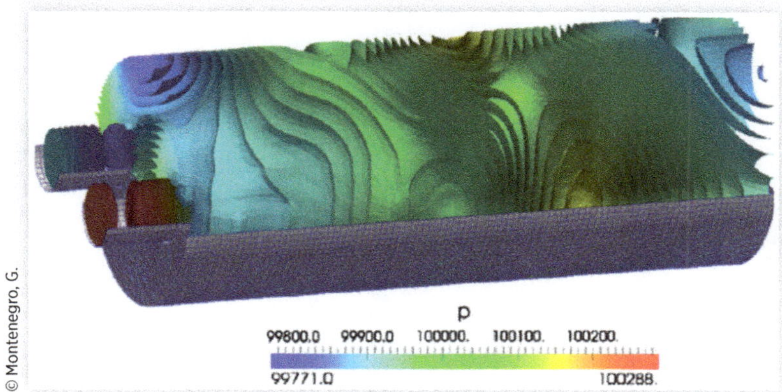

correctly captured over a wide range of frequency, without showing a significant frequency shift. In conclusion, in the case of the evaluation of the TL of a specific silencer, the time required by the 1D-3D coupling is remarkably higher than the computation runtime of the 1D approach; however, the 1D simulation for the two configurations requires a different strategy of generation of the equivalent 1D schematic, which makes this approach not suitable for an automatic process of optimization, since it always requires the contribution of the user.

FIGURE 9.33 Transmission loss of the muffler Conf B: (a) fully 1D vs measured, (b) 1D-3D vs measured.

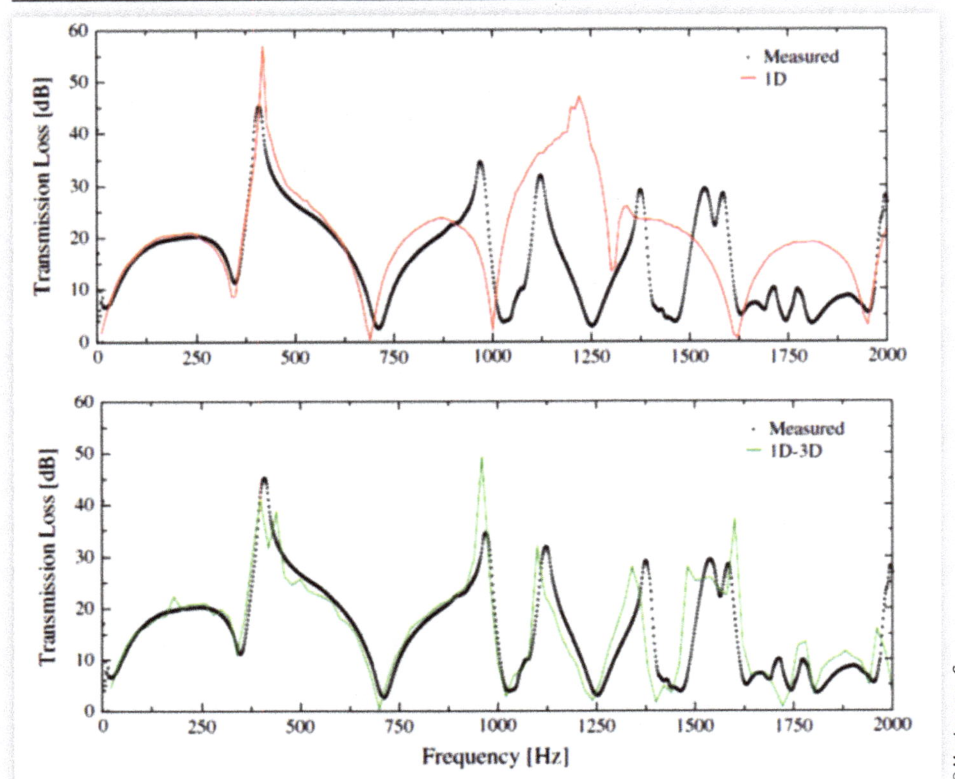

9.4.3.2 EVALUATION OF THE INSERTION LOSS

The evaluation of the TL of a silencer is useful to understand the resonances that can be exploited to damp certain components of the spectrum that are not desired. However, this process is a bit far from what happens in a real exhaust device, where the gas temperature is not uniform, therefore also the speed of sound, and where the input spectrum, i.e. the frequencies of the pressure level, changes when the silencer is mounted on the exhaust system. For this reason, the evaluation of the insertion loss gives a more realistic information about the real behavior of the silencer. An example of application of the 1D-3D methodology for the calculation of the insertion loss is given by [26] where the spectrum of the intake mouth noise has been calculated at different engine operating conditions with and without the presence of a resonator inside the air-box.

The engine taken into consideration is a single-cylinder engine for motorbike application with a swept volume of 250 cc.

The engine was equipped with an air-box, containing also a filter cartridge, without any specific resonator in the intake line. The fluid dynamic analysis, carried out by using both 1D and quasi-3D CFD tools, permitted the optimization of the resonances as a result of the simulation of several configurations.

The most promising has been then implemented on the real configuration respecting the main specifications of the designed Helmholtz resonator, such as the neck diameter, the neck length, and the size of the volume. In Figure 9.34, it is shown a comparison

FIGURE 9.34 View from the top of the motorbike air-box: on the left the original air-box layout, on the right the modified prototype.

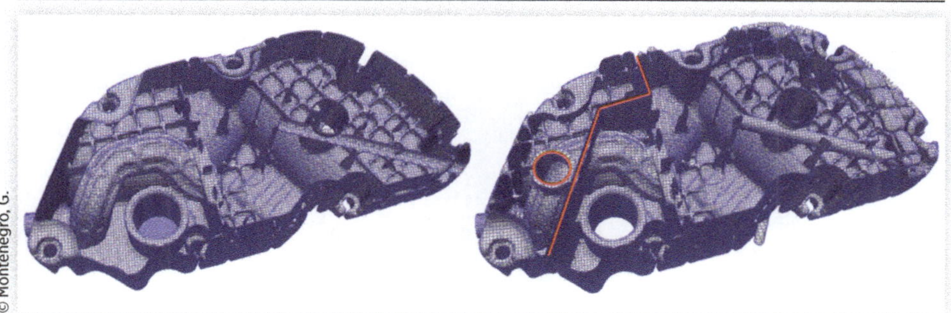

between the air-box in its original configuration and the modified one. The interesting aspect of this analysis consists of the application of the acoustic simulation to a real geometry which includes all the strengthening ribs, the rounded corners, and the seats for screws dedicated to the blockage of the device itself.

The simulation by means of 1D-3D simulation has shown that the resonator has a beneficial effect on the sound pressure level emitted at the intake mouth. As shown in Figure 9.35, where the insertion losses of both the modified and original configuration are presented, the optimized geometry performs better, in the region of low frequencies, then the original one.

In Figure 9.36 it is shown the comparison between the transmission loss measured at the acoustic test bench and the one calculated by means of the 1D-3D coupling for the two configurations, with and without the filter cartridge. This comparison proves that a complex shape of an intake air-box can be modeled accurately by means of coupled simulations. Even though the agreement can be considered satisfactory, there are some frequency ranges where the measurements and the calculation are not matching. This is due to the possibility that the material is subject to deformations when its own resonance frequency is excited by the incident pressure pulsation. As shown in [28], the inclusion of fluid structure interaction submodels in the 3D model can help in capturing these particular effects, which under real engine operating conditions can become so relevant that certain frequencies can be even amplified.

FIGURE 9.35 Insertion loss computed by means of the coupled 1D-CFD approach, partial load: (a) 3000 rpm, (b) 5000 rpm.

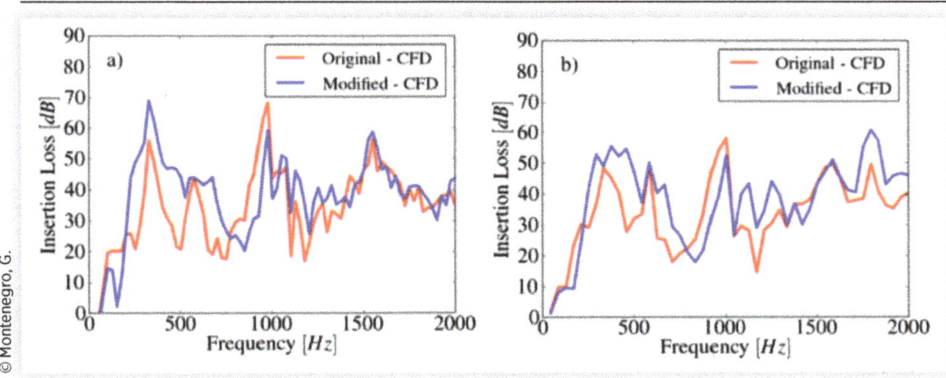

FIGURE 9.36 Comparison between measured and calculated transmission loss: (a) original without filter cartridge, (b) modified without cartridge, (c) original with cartridge, (d) modified with filter cartridge.

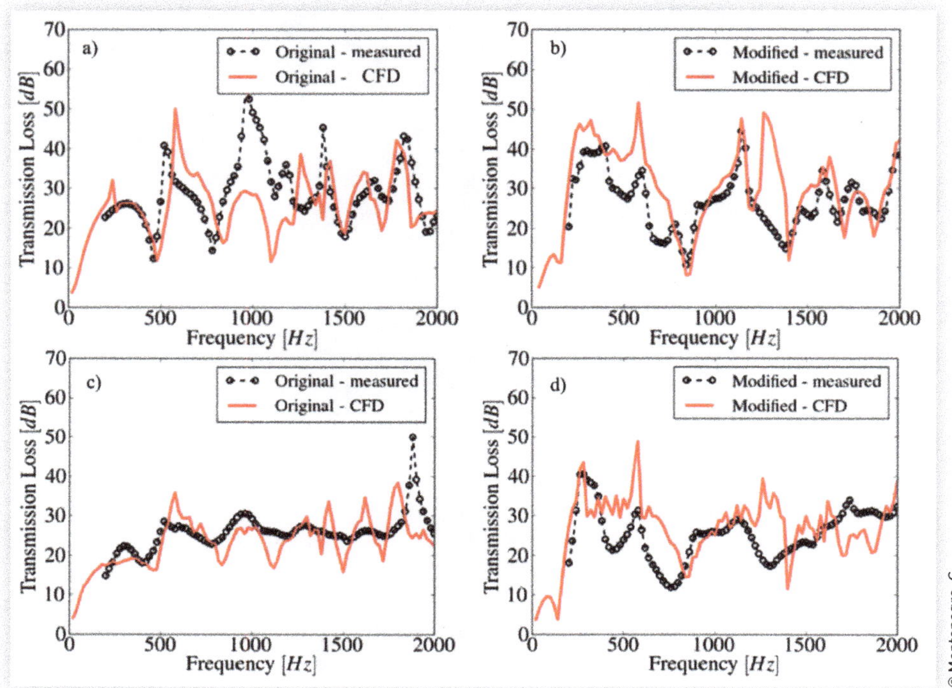

Definitions, Acronyms, and Abbreviations

Symbols

β - Beta characteristic line
λ - Lambda characteristic line
ϕ - Generic intensive quantity
ρ - Density
φ_n - Phase of the nth harmonics
τ - Viscous stress tensor
a - Gas speed of sound
D - Duct diameter
$\mathbf{D}$ - Deviatoric tensor
e_0 - Stagnation specific internal energy
f_0 - Base harmonic frequency
f_w - Friction factor at the duct wall
F - Cross-sectional area of duct
$\mathbf{F}$ - Flux of the conserved variables
k - Ratio between constant pressure and constant volume specific heat capacity
h_0 - Stagnation specific internal enthalpy
p - Pressure
q - Heat transferred per unit mass per unit time

S - Speed of pressure perturbation
Sp - Source term
T - Gas temperature
U - Velocity vector
u - Flow velocity, *x* component
v - Flow velocity, *y* component
w - Flow velocity, *z* component

Abbreviations

0D - Zero-dimensional
1D, 3D - One-dimensional, three-dimensional
BEM - Boundary element method
CFD - Computational fluid dynamics
EGR - Exhaust gas recirculation
ENO - Essentially nonoscillatory
FD - Finite difference
FEM - Finite element method
FV - Finite volume
HLLC - Harten Lax Leer contact surface
HPC - High-performance computing
MOC - Method of characteristics
ODE - Ordinary differential equation
PFI - Port fuel injection
TL - Transmission loss
WAF - Weighed average flux

References

1. Onorati, A., Ferrari, G., D'Errico, G., and Montenegro, G., "The Prediction of 1D Unsteady Flows in the Exhaust System of a S.I. Engine Including Chemical Reactions in the Gas and Solid Phase," SAE Technical Paper 2002-01-0003, 2002, doi:10.4271/2002-01-0003.
2. Gurney, D., "The Design of Turbocharged Engines Using 1D Simulation," SAE Technical Paper 2001-01-0576, 2001, doi:10.4271/2001-01-0576.
3. Claywell, M., Horkheimer, D., and Stockburger, G., "Investigation of Intake Concepts for a Formula SAE Four-Cylinder Engine Using 1D/3D (Ricardo WAVE-VECTIS) Coupled Modeling Techniques," SAE Technical Paper 2006-01-3652, 2006, doi:10.4271/2006-01-3652.
4. Ogink, R. and Golovitchev, V., "Gasoline HCCI Modeling: An Engine Cycle Simulation Code with a Multi-Zone Combustion Model," SAE Technical Paper 2002-01-1745, 2002, doi:10.4271/2002-01-1745.
5. Yang, K., Bargende, M., and Grill, M., "Evaluation of Engine-Related Restrictions for the Global Efficiency by Using a Rankine Cycle-Based Waste Heat Recovery System on Heavy Duty Truck by Means of 1D-Simulation," SAE Technical Paper 2018-01-1451, 2018, doi:10.4271/2018-01-1451.
6. Galindo, J., Dolz, V., Royo-Pascual, L., and Brizard, A., "Dynamic Modeling of an Organic Rankine Cycle to Recover Waste Heat for Transportation Vehicles," *Energy Procedia* 129 (2017): 192-199.

CHAPTER 9

7. Onorati, A., Montenegro, G., and D'Errico, G., "Prediction of the Attenuation Characteristics of I.C. Engine Silencers by 1-D and Multi-D Simulation Models," SAE Technical Paper 2006-01-1541, 2006, doi:10.4271/2006-01-1541.

8. Montenegro, G., Onorati, A., Piscaglia, F., and D'Errico, G., "Integrated 1D-multiD Fluid Dynamic Models for the Simulation of I.C.E. Intake and Exhaust Systems," SAE Technical Paper 2007-01-0495, 2007, doi:10.4271/2007-01-0495.

9. Bella, G., Bozza, F., De Maio, A., Del Citto, F. et al., "An Enhanced Parallel Version of Kiva-3V, Coupled with a 1D CFD Code, and Its Use in General Purpose Engine Applications," Lecture Notes in Computer Science (including subseries Lecture Notes in Artificial Intelligence and Lecture Notes in Bioinformatics), 4208 LNCS, 2006, 11-20.

10. Bozza, F., Gimelli, A., Andreassi, L., Rocco, V. et al., "1D-3D Analysis of the Scavenging and Combustion Process in a Gasoline and Natural-Gas Fuelled Two-Stroke Engine," SAE Technical Paper 2008-01-1087, 2008, doi:10.4271/2008-01-1087.

11. Riegler, U.G. and Bargende, M., "Direct Coupled 1D/3D-CFD-Computation (GT-Power/Star-CD) of the Flow in the Switch-Over Intake System of an 8-Cylinder SI Engine with External Exhaust Gas Recirculation," SAE Technical Paper 2002-01-0901, 2002, doi:10.4271/2002-01-0901.

12. Bohbot, J., Lafossas, F.A., Albrecht, A., Miche, M. et al., A New Coupling Approach Using a 1D System Simulation Software and a 3D Combustion Code Applied to Transient Engine Operation," SAE Technical Paper 2004-01-3002, 2004, doi:10.4271/2004-01-3002.

13. Peters, B.J., Wanker, R.J., Münzer, A., and Wurzenberger, J.C., "Integrated 1D to 3D Simulation Workflow of Exhaust Aftertreatment Devices," SAE Technical Paper 2004-01-1132, 2004, doi:10.4271/2004-01-1132.

14. Liu, Z., Benjamin, S.F., Roberts, C.A., Zhao, H. et al., "A Coupled 1D/3D Simulation for the Flow Behaviour Inside a Close-Coupled Catalytic Converter," SAE Technical Paper 2003-01-1875, 2003, doi:10.4271/2003-01-1875.

15. Štěpánek, J., Kočí, P., Marek, M., and Kubíček, M., "Catalyst Simulations Based on Coupling of 3D CFD Tool with Effective 1D Channel Models," *Catalysis Today* 188, no. 1 (2012): 87-93.

16. Abidin, Z., Das, K., and Roberts, C., "3D-Semi 1D Coupling for a Complete Simulation of an SCR System," SAE Technical Paper 2013-01-1575, 2013, doi:10.4271/2013-01-1575.

17. Wanker, R.R., Wurzenberger, J.C., and Higbie, D.D., "1D and 3D CFD Simulation of Exhaust-Gas Aftertreatment Devices: Parameter Optimization via Genetic Algorithm," *ASME. Pressure Vessels and Piping Conference, Computational Technologies for Fluid/Thermal/Structural/Chemical Systems with Industrial Applications*, Vol. 1, 47-53.

18. Ramanathan, S., Hudson, A., Styron, J., Baldwin, B. et al., "EGR and Swirl Distribution Analysis Using Coupled 1D-3D CFD Simulation for a Turbocharged Heavy Duty Diesel Engine," SAE Technical Paper 2011-01-2222, 2011, doi:10.4271/2011-01-2222.

19. Dimitriou, P., Avola, C., Burke, R., Copeland, C. et al., "A Comparison of 1D-3D Co-Simulation and Transient 3D Simulation for EGR Distribution Studies," *ASME 2016 Internal Combustion Engine Fall Technical Conference, ICEF 2016*, 2016.

20. Shaw, C.E., Moenssen, D., and Kostun, J., "A Correlation Study of Computational Techniques to Model Engine Air Induction System Response Including BEM, FEM and 1D Methods," SAE Technical Paper 2003-01-1644, 2003, doi:10.4271/2003-01-1644.

21. Shaw, C.E., Moenssen, D.J., and Montgomery, W.C., "On the Use of BEA with Engine Simulation as an Input to Predict Air Induction Inlet Noise," SAE Technical Paper 2005-01-2350, 2005, doi:10.4271/2005-01-2350.

22. Bozza, F., Gimelli, A., Piazzesi, R., Fortunato, F. et al., "The Prediction of the Performance and Gasdynamic Noise Emitted by a Medium-Size Spark-Ignition Engine by Means of 1D and 3D Analyses," SAE Technical Paper 2007-01-0380, 2007, doi:10.4271/2007-01-0380.

23. Ghodake, D. and Haque, Z., "The Simulation of Air Induction Noise Using 1D-3D Coupling," SAE Technical Paper 2011-01-0500, 2011, doi:10.4271/2011-01-0500.

24. MacKey, D.O., Blair, G.P., and Fleck, R, "Correlation of Simulated and Measured Noise Emission Using a Combined 1D/3D Computational Technique," SAE Technical Paper 970801, 1997, doi:10.4271/970801.

25. Montenegro, G., Onorati, A., and Della Torre, A., "The Prediction of Silencer Acoustical Performances by 1D, 1D-3D and Quasi-3D Non-Linear Approaches," *Computers and Fluids* 71 (2013): 208-223.

26. Montenegro, G., Della Torre, A., Cerri, T., Lenzi, G. et al., "Fluid Dynamic and Acoustic Optimization Methodology of a Motorbike Intake Airbox Using Multilevel Numerical CFD Models and Experimental Validation Tests," *SAE Int. J. Engines* 6, no. 3 (2013): 1731-1744, doi:10.4271/2013-24-0070.

27. Montenegro, G. and Onorati, A., "Modeling of Silencers for I.C. Engine Intake and Exhaust Systems by Means of an Integrated 1D-multiD Approach," *SAE Int. J. Engines* 1, no. 1 (2009): 466-479, doi:10.4271/2008-01-0677.

28. Della Torre, A., Montenegro, G., and Onorati, A., "CFD Investigation of the Effect of Fluid-Structure Interaction on the Transmission Loss of ICE Silencers," *SAE Int. J. Passeng. Cars – Mech. Syst.* 9, no. 3 (2016): 1063-1070, doi:10.4271/2016-01-1815.

29. Watanabe, N., Kubo, M., and Yomoda, N., "An 1D-3D Integrating Numerical Simulation for Engine Cooling Problem," SAE Technical Paper 2006-01-1603, 2006, doi:10.4271/2006-01-1603.

30. Stroh, C., Reitbauer, R., and Hanner, J., "Increasing the Reliability of Designing a Cooling Package by Applying Joint 1D/3D Simulation," SAE Technical Paper 2006-01-1571, 2006, doi:10.4271/2006-01-1571.

31. Čaika, V. and Sampl, P, "Nozzle Flow and Cavitation Modeling with Coupled 1D-3D AVL Software Tools," SAE Technical Paper 2011-24-0006, 2011, doi:10.4271/2011-24-0006.

32. Čaika, V., Sampl, P., Tatschl, R., Krammer, J. et al., "Coupled 1D-3D Simulation of Common Rail Injector Flow Using AVL HYDSIM and FIRE," SAE Technical Paper 2009-24-0029, 2009, doi:10.4271/2009-24-0029.

33. Marcer, R., Audiffren, C., Viel, A., Bouvier, B. et al., "Coupling 1D System AMESim and 3D CFD EOLE models for Diesel Injection Simulation," *ILASS – Europe 2010, 23rd Annual Conference on Liquid Atomization and Spray Systems*, Brno, Czech Republic, September 2010.

34. De Bellis, V., Bozza, F., Fontanesi, S., Severi, E. et al., "Development of a Phenomenological Turbulence Model through a Hierarchical 1D/3D Approach Applied to a VVA Turbocharged Engine," *SAE Int. J. Engines* 9, no. 1 (2016): 506-519, doi:10.4271/2016-01-0545.
35. Winterbone, D.E. and Pearson, R.J., *Theory of Engine Manifold Design* (London: Professional Engineering Publishing, 2000).
36. Onorati, A. and Ferrari, G., "Modeling of 1-D Unsteady Flows in I.C. Engine Pipe Systems: Numerical Methods and Transport of Chemical Species," SAE Technical Paper 980782, 1998, doi:10.4271/980782.
37. Toro, E.F., *Riemann Solvers and Numerical Methods for Fluid Dynamics* (New York: Springer, 1997).
38. Toro, E.F., Hidalgo, A., and Dumbser, M., "Force Schemes on Unstructured Meshes. I: Conservative Hyperbolic Systems," *Journal of Computational Physics* 228, no. 9 (2009): 3368–3389.
39. Davies, P., "Practical Flow Duct Acoustics," *Journal of Sound and Vibration* (1988): 91–115.
40. Benson, R.S., *The Thermodynamics and Gas Dynamics of Internal Combustion Engines* (Oxford: Clarendon Press, 1982), vol. I.
41. Toro, E.F., Spruce, M., and Speares, W., "Restoration of the Contact Surface in the HLL-Riemann Solver," *Shock Waves* 4, no. 1 (1994): 25-34.
42. Batten, P., Lambert, C., Causon, D., and Clarke, N., "On the Choice of Wave Speeds for HLLC Riemann Solvers," *SIAM Journal on Scientific and Statistical Computing* 9 (1988): 445-473.
43. Davis, S.F., "Simplified Second-Order Godunov-Type Methods," *SIAM Journal on Scientific and Statistical Computing* (1998).
44. Versteeg, H.K. and Malalasekera, W., *An Introduction to Computational Fluid Dynamics: The Finite Volume Method* (Harlow, Essex, UK: Pearson Education Ltd., 2007), Vol. I.
45. Ferziger, J.H. and Perić, M., *Computational Methods for Fluid Dynamics* (Springer, 1997).
46. Hirsch, C., *Numerical Computation of Internal and External Flows: The Fundamentals of Computational Fluid Dynamics* (Butterworth-Heinemann, 2007).
47. Galindo, J., Tiseira, A., Fajardo, P., and Navarro, R., "Coupling Methodology of 1D Finite Difference and 3D Finite Volume CFD Codes Based on the Method of Characteristics," *Mathematical and Computer Modelling* 54, no. 7-8 (2011): 1738-1746.
48. Heywood, J.B., *Internal Combustion Engine Fundamentals* (McGraw-Hill Int. editions, 1988).
49. Basset, M.D., Pearson, R.J., Fleming, N.P., and Winterbone, D.E., "A Multi-Pipe Junction Model for One-Dimensional Gas-Dynamic Simulations," SAE Technical Paper 2003-01-0370, 2003, doi:10.4271/2003-01-0370.
50. Onorati, A., Cerri, T., Ceccarani, M., and Cacciatore, D., "Experimental Analysis and 1D Thermo-Fluid Dynamic Simulation of a High Performance Lamborghini V10 S.I. Engine," SAE Technical Paper 2005-24-081, 2005, doi:10.4271/2005-24-081.

51. Onorati, A., "Prediction of Acoustical Performances of Muffling Pipe-Systems by the Method of Characteristics," *Journal of Sound and Vibration* 171, no. 3 1994: 369–395.

52. Onorati, A., Non Linear Fluid Dynamic Modelling of Reactive Silencers Involving Extended Inlet/Outlet and Perforated Ducts," *Noise Control Engineering Journal* 45, no. 1 (1997).

53. Onorati, A., Numerical Simulation of Unsteady Flows in I.C. Engine Silencers and the Prediction of Tailpipe Noise, in Winterbone, D.E. and Pearson, R.J. eds., *Design Techniques for Engine Manifolds* (London: Professional Engineering Publishing, 1999).

54. Munjal, M.L., *Acoustics of Ducts and Mufflers* (New York: John Wiley & Sons, 1987).

55. Montenegro, G. and Onorati, A., "A Coupled 1D-multiD Nonlinear Simulation of I.C. Engine Silencers with Perforates and Sound-Absorbing Material," *SAE Int. J. Passeng. Cars – Mech. Syst.* 2, no. 1 (2009): 482-494, doi:10.4271/2009-01-0305.

56. Torregrosa, A.J., Broatch, A., Payri, R., and Gonzales, F., "Numerical Estimation of End Corrections in Extended-Duct and Perforated-Duct Mufflers," *ASME Transactions* 121 (1999): 302–308.

57. Eriksson, L.J., "Higher Order Mode Effects in Circular Ducts and Expansion Chambers," *The Journal of the Acoustical Society of America* 68, no. 2 (1980): 545–550.

CHAPTER 10

Extending the 1D Approach to the Simulation of 3D Components: The Quasi-3D Approach

Augusto Della Torre
Politecnico di Milano

Robert Fairbrother
AVL List GmbH

CHAPTER 10

10.1 Introduction

In the last decades, engine-scale modeling has been dominated by 1D simulation tools [1, 2, 3, 4, 5, 6, 7, 8, 9, 10, 11]. Nowadays this approach is certainly considered a mature area, successfully applied for the design and the optimization of the Internal Combustion (I.C.) engine pipe systems. As a matter of fact, the extensive developments of robust numerical methods and specific boundary conditions to simulate all the typical components of the engine gas path (valves, junctions, turbochargers, pressure losses, etc.), together with thermo-dynamic combustion models and emission sub-models for cylinder and after-treatment systems, have allowed the 1D fluid dynamic codes to become consolidated tools for automotive engineers. Commercial codes are daily applied for the optimization of intake and exhaust system geometry, turbocharger matching, valve timings and other applications [12]. However, the main limit of 1D models is clearly represented by the fundamental assumption of one-dimensional flow or planar wave motion in the ducts. This is acceptable in most cases, except for complex shape components (e.g., air-boxes, plenums, multi-pipe junctions, and silencers) in which multi-dimensional effects can be significant and can be captured only by resorting to ad hoc 1D equivalent schematics.

In the last few years, examples of hybrid 1D-multiD fluid dynamic models, aimed at extending the applicability of 1D "codes" to real engine simulations, have been published [12, 13]. In particular, the 3D approach is applied to model the elements in which the non-planar wave motion is predominant, whereas the 1D approach is reserved to simulate simple pipe systems. Hence, the coupled simulation tools can capture the influence on engine performance of both the 3D wave motion in complex devices and the one-dimensional pressure pulsations in the ducts. The drawback of a coupled 1D-3D fluid dynamic code is represented by the significant computational effort required, related to the computational fluid dynamics (CFD) simulations of the 3D domains on the basis of the RANS equations – especially in the case of turbocharged engines, which generally require many thermodynamic cycles to achieve convergence. The use of 3D domains embedded in the 1D schematic of the engine can limit the applicability of the simulation code.

In this scenario, the interest toward quasi-3D models has increased, in order to exploit the advantages of both 1D (simplicity, low computational effort and straightforward boundary conditions) and multi-dimensional models in a convenient way. The main idea behind a quasi-3D approach is to describe the main characteristic lengths of a three-dimensional component keeping low the number of elements that constitute the computational grid. A quasi-3D approach is not designed for capturing the details of the flow field but is able to predict the propagation of non-planar waves as well as the main features of the flow field in complex shape elements. If properly optimized, a quasi-3D tool can provide an accurate description of the pressure wave propagation in the pipe system, allowing the prediction of the overall engine performance and of the acoustic behavior, as well as more time-consuming hybrid 1D-multiD models. A literature survey shows some examples of tools based on a quasi-3D approach [14, 15, 16, 17], either resorting to the generation of network of 1D pipes or clusters of 0D volumes suitably connected between each other.

The quasi-3D approach is not intended as a replacement but as an extension to 1D solutions. Significant portions of the gas path of an internal combustion engine are made up of pipe connections, which can be properly modeled resorting to 1D approximation. The primary application of the quasi-3D solution is the modeling of more complex components and the prediction of acoustics. However, this can also be relevant for performance modeling where determining the pressure drop across components is required. It is also intended that the same models can be used for different applications. That is, the model of a complex component can be developed and validated for acoustics before the exact same model is inserted into a full gas exchange model for performance and acoustic prediction coupled with an engine model. Another advantage of this approach is the elimination of the need to evaluate empirical loss coefficients or end effects [9]. Three-dimensional analysis is also required in order to handle higher frequencies and/or chambers with large transverse dimensions, such that one or more of the higher order modes get cut-on in the chamber, even though the inlet and outlet pipes only permit plane waves [18]. As a further requirement, the approach also has to be able to handle mean flow, perforates, and absorptive material. In this chapter, the general features of the nonlinear quasi-3D approach will be discussed, illustrating at first the details of the numerical model and then providing several examples of application.

10.2 The Quasi-3D Approach

In the framework of 1D simulation codes, the modelling of complex shape devices (e.g., air-boxes, plenums, and silencers) has been traditionally addressed resorting to simplified approaches, namely, 0D approximation and equivalent 1D pipe schemes. The first consists of modeling the component as a simple boundary condition inserted to connect different pipes, without taking into account its effective geometry. The second approach aims to reproduce the effects of the 3D geometry by means of a simplified pipe network and corrective lengths for the ducts. These simplified approaches present two main limitations. First of all, not being geometry-based they need a process of refinement and improvement often supported by experimental results or CFD simulations.

Moreover, even when the model has proved to be reliable, it cannot be adopted for investigating the effects on the overall performance of a change in the geometry of the device. Therefore, it cannot be applied as an optimization tool. Hence, the adoption of a quasi-3D approach represents a convenient way to overcome these limitations. The idea behind a quasi-3D approach is a geometry-based modeling, which is aimed at the reconstruction of the main characteristic lengths of the component in three dimensions. On the other hand, the details of the geometry are neglected, in order to keep low the computational burden. The capability of the approach to provide reliable predictions in terms of acoustic properties and engine performance, requiring a low computational burden, depends primarily on the details of the numerical technique implemented.

FIGURE 10.1 Example of a quasi-3D network: cells and ports.

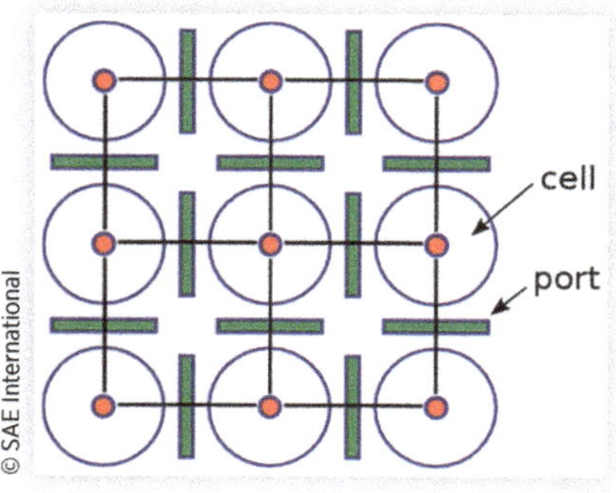

A general quasi-3D approach consists of reproducing a 3D geometry as a network of 0D elements, such as volumes, to which characteristic lengths are assigned. As shown in Figure 10.1, the volume network can be defined by means of two fundamental elements, namely, a cell and a port. The cell element is defined by its volume and by the list of ports attached to the cell. The port element stores information about the connectivity between neighboring cells and geometrical parameters. These include the distance from the adjacent cell centers, the direction with respect to an absolute orientation system and the flow area. The main feature of a quasi-3D approach is that the determination of the thermo-fluid dynamic quantities in the ports is reduced to a 1D problem along the direction of the normal to the port surface.

As shown in Figure 10.2, the quasi-3D geometry reconstruction does not describe the details of the geometry. However, the main characteristic lengths of the component are captured, making it possible to simulate the three-dimensional pressure wave propagation inside the volume with sufficient accuracy.

The approach is based on the Euler formulation of the conservation equations of mass, momentum, and energy for unsteady flows. This is usually an acceptable approximation, since the viscosity in the intake and exhaust gas is very low. Moreover, the velocities inside the typical 3D components included in the engine gas path are usually small compared to the velocity in the pipes. Therefore, the viscous contribution to the overall pressure drop is negligible. However, the viscous phenomena cannot be neglected in the modeling of some specific elements, such as filters, catalytic substrates, perforated ducts and baffles, which are usually included in plenums and after-treatment devices. In these cases, specific sub-models are introduced to take into account their contribution to the pressure drop. The approach can be applied to two main categories of problems, namely, acoustic simulation of stand-alone silencing devices and fluid-dynamic simulation of the entire engine system. For both these applications, the most important physical phenomenon to describe is the unsteady three-dimensional pressure wave propagation in the complex-shape volumes. For acoustic simulations, the accurate description of the pressure waves is necessary for a reliable prediction of the acoustic properties of the device. In a similar way, the accuracy in the prediction of the effective performance of the engine depends on the capability of describing the unsteady pressure wave propagation in the intake and exhaust system, which is strongly influenced by the wave motion in the three-dimensional components.

FIGURE 10.2 Example of quasi-3D reconstruction of an airbox geometry.

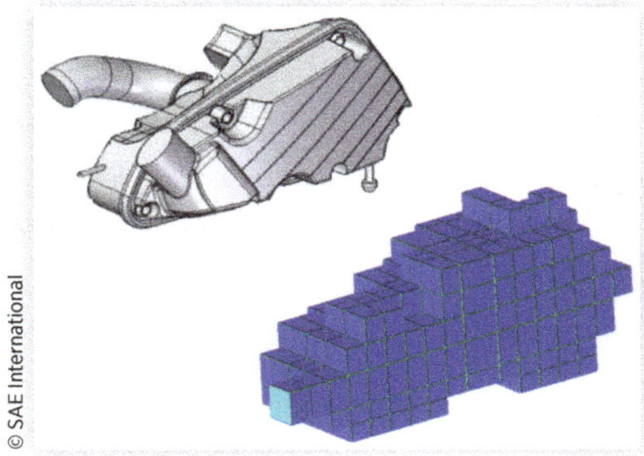

FIGURE 10.3 Example of 3Dcell staggered grid.

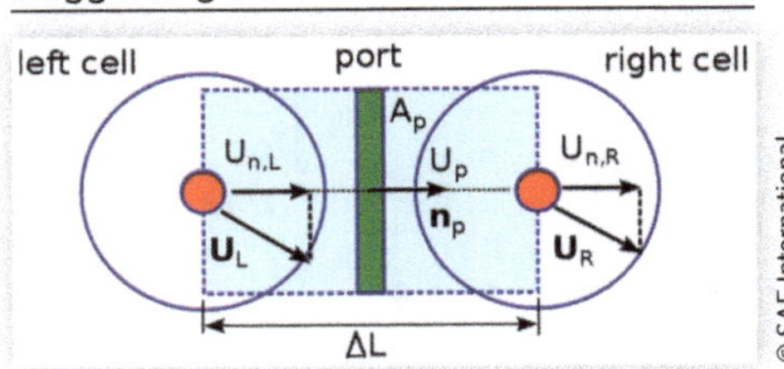

10.3 Numerical Method

The method adopted for the solution of the Euler set of equations operates on a staggered computational grid (Figure 10.3), in which the intensive properties are evaluated in the cell, while velocity and momentum are defined for the ports.

The conservation equations of mass, momentum, and energy are integrated according to a pseudo-staggered arrangement [19, 20]. According to this definition, the fluxes at the port elements are used for the solution of the scalar quantities defined for the cell element. The solution procedure adopted is an explicit time marching staggered leapfrog method in which the momentum equation is used for determining the port flux. This assumption leads to the definition of three different time levels $\left(n,\ n+\frac{1}{2}, n+1\right)$ over which the integration of the three conservation equations is suitably distributed in order to achieve a low diffusive method with second order accuracy in both space and time.

In particular, considering the mass equation, the integration is performed over the cell volume V and the time-step Δt:

$$\int_t^{t+\Delta t}\left[\frac{\partial}{\partial t}\int_V \rho\,\mathrm{d}V+\int_V \nabla\cdot(\rho\mathbf{U})\,\mathrm{d}V\right]\mathrm{d}t=0 \tag{10.1}$$

By means of the Gauss theorem, the transport term is reduced to a surface integral, giving the following:

$$\int_V \nabla\cdot(\rho\mathbf{U})\,\mathrm{d}V=\oint_S \rho\mathbf{U}\cdot\mathbf{n}\,\mathrm{d}S \tag{10.2}$$

which, on the basis of the finite volume discretization, becomes a summation of the mass flows through the ports (N_p). Taking into account the time marching stencil, the mass conservation becomes:

$$\rho^{n+1}=\rho^n+\frac{1}{V}\sum_{p=1}^{N_p}(\rho UA)_p^{n+\frac{1}{2}}\Delta t \tag{10.3}$$

In a similar way, the discretized energy equation has the following expression:

$$(\rho e_0)^{n+1}=(\rho e_0)^n+\frac{1}{V}\sum_{p=1}^{N_p}(\rho UhA)_p^{n+\frac{1}{2}}\Delta t \tag{10.4}$$

where h is the enthalpy evaluated in the port element. The integration of the momentum equation is performed referring to a control volume defined by the port surface A and the distance ΔL between the two neighboring cell centers (Figure 10.3). Since the velocity vector in the port is directed normally to the surface, according to the particular grid adopted, the momentum equation reduces to a 1D problem arbitrarily oriented in space. By naming x_n the coordinate along the normal $\mathbf{n}$ to the port, the integration of the momentum equation over the control volume $A\ \Delta L$ leads to the following:

$$\left(\rho U_n A\right)_p^{n+\frac{1}{2}} = \left(\rho U_n A\right)_p^{n-\frac{1}{2}} + \frac{\Delta t}{\Delta L}\left[\left(\rho U_n^2 + p\right)_{L,n}^{n} - \left(\rho U_n^2 + p\right)_{R,n}^{n}\right] A_p + M_s \tag{10.5}$$

where the subscripts L and R indicate, respectively, the quantities in the left and right cells, whereas M_s is a generic source term that may assume different expressions, depending on the type of the port that has to be modeled. The solution of the momentum equation can be used to determine the port mass flow $\left(\rho U_n A\right)_p^{n+\frac{1}{2}}$, which is then used in the mass and energy conservation equations to march ahead in time. In particular, referring to the solution of the energy equation, the total enthalpy flow is expressed as the product between the mass flow and the enthalpy value $\overline{h}$ at the port location:

$$\left(\rho U_n h A\right)_p^{n+\frac{1}{2}} = \left(\rho U_n A\right)_p^{n+\frac{1}{2}} \cdot \overline{h} \tag{10.6}$$

To achieve second-order accuracy in the description of contact surfaces, the value of $\overline{h}$ can be estimated from linear interpolation of the enthalpy in the neighbouring cells. Otherwise, when the method stability is mandatory and reduced resolution accuracy of contact surfaces does not affect the parameter of interest, the adoption of an upwind approach, by assigning the enthalpy of the left or right cell depending on the flow direction, may be exploited. According to the definition of pseudo-staggered arrangement, in order to construct the momentum equation of the port (Equation 10.5), the velocity must also be defined in the centroid of the cell element. The cell momentum $(\rho \mathbf{U} V)_c$ is a vectorial quantity and its direction is determined by the momentum of all the ports linked to the cell weighted by their distance from the cell center:

$$\left(\rho \mathbf{U} V\right)_c = \sum_{p=1}^{N_p} \left(\rho U A\right)_p \Delta L_{pR/L} \cdot \mathbf{n}_p \tag{10.7}$$

The cell velocity is then determined from the cell momentum as follows:

$$\mathbf{U}_c = \frac{\left(\rho \mathbf{U} V\right)_c}{\left(\rho V\right)_c} \tag{10.8}$$

Figure 10.4 shows how the staggered leapfrog method is applied to a staggered grid arrangement. The center of the computational control volumes for the momentum equation are represented as small squares while the cell control volume centers are represented by circles. They are located at different positions in both space and time.

In this way, the fluxes calculated in the ports at a generic time step $n+\frac{1}{2}$ can be used to update the solution in the cells from time step n to time step $n + 1$. The adoption of time-centered differences for each conservation equation, along with space-centered differences, allows the achievement of second-order accuracy both in space and time.

FIGURE 10.4 Time marching procedure adopted for the solution of the governing equation: (a) represents the cell control volume and (b) represents the port control volume.

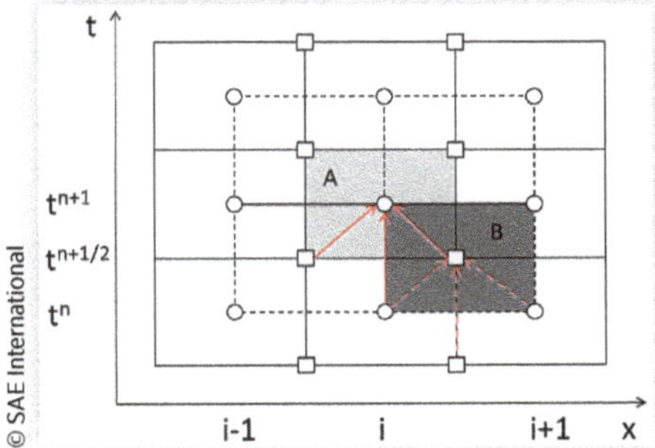

10.3.1 Diffusion Term on the Momentum Equation

The method discussed so far does not satisfy the stability requirement, especially in cases where pressure gradients are significant. For instance, simulations with mean flow or with high amplitude pressure perturbations traveling along the calculation domain (e.g., pulsating flow generated by an engine) will be affected by nonphysical oscillations. In order to apply the method to the solution of unsteady flows in the engine intake and exhaust system, a stabilization technique has been applied to reduce this spurious behavior. Due to the specific structure of the solution procedure, in which the solution of the momentum equation for the port is used to determine the fluxes for mass and energy equations of the cell, it is possible to stabilize the method by acting only on the momentum equation. Therefore, the idea is to add a diffusion source term to the momentum equation (hence the name DTM, diffusion term momentum), which will limit the computed mass flux at the port. In the DTM approach, the momentum equation is modified by adding a source term as follows:

$$\frac{\partial \rho \mathbf{U}}{\partial t} + \nabla \cdot (\rho \mathbf{U}\mathbf{U}) + \nabla p = \nabla \cdot \mathbf{D} \tag{10.9}$$

The tensor **D** is a function of the local velocity fields and of the spatial and temporal discretization and is defined as:

$$\mathbf{D} = \varepsilon \nabla \cdot (\rho \mathbf{U}) \tag{10.10}$$

where ε is a scalar quantity and represents the diffusion coefficient. If the velocity field is smooth, the contribution of the diffusion term $\nabla \cdot \mathbf{D}$ is negligible. However, when large gradients in the solution occur, this term increases in order to damp the numerical spurious oscillations. Projecting Equation (10.9) along the port direction (x_n), it reduces to a 1D problem arbitrarily oriented in space. In this case, only the component of velocity in the port direction x_n is considered, therefore the term $\nabla \cdot \mathbf{D}$ assumes the following expression:

$$\nabla \cdot \mathbf{D} = \frac{\partial D_n}{\partial x_n} = \frac{\partial}{\partial x_n}\left(\varepsilon \frac{\partial (\rho U)_n}{\partial x_n} \right) \tag{10.11}$$

The discretization of Equation (10.9) along the port direction will lead to the following simplified 1D equation:

$$(\rho U_n A)_p^{n+\frac{1}{2}} = (\rho U_n A)_p^{n-\frac{1}{2}} + \frac{\Delta t}{\Delta L}\left[(\rho U_n^2 + p)_{L,n}^n - (\rho U_n^2 + p)_{R,n}^n \right] A_p + \frac{\Delta t}{\Delta L}\left[\tilde{D}_{L,n}^n - \tilde{D}_{R,n}^n \right] \tag{10.12}$$

This definition requires the evaluation in the cells of the diffusivity coefficient ε and of the terms $\tilde{D}_L^n$ and $\tilde{D}_R^n$ (generically referred to as $\tilde{D}_c^n$ hereafter). The coefficient ε is evaluated on the basis of the local velocity, the time-step, and the mesh size:

$$\varepsilon_c = \frac{|\mathbf{U}_c| \Delta x}{2} - \frac{\mathbf{U}_c^2 \Delta t}{2} \tag{10.13}$$

The term $\tilde{D}_c$ is a vector defined in the cell as the product between the diffusion coefficient ε_c and the gradient of the mass flow in each direction:

$$\tilde{D}_c = \varepsilon_c \nabla\left(\rho \mathbf{U} A\right)_c \tag{10.14}$$

The scalar product between $\tilde{D}_c$ and the port direction $\mathbf{n}_p$ gives the term $D_{c,n}$ used in Equation (10.10). The gradient of mass flow in the cell $\nabla(\rho\mathbf{U}A)_c$ is calculated from the mass flow $(\rho\mathbf{U}A)_p$ computed in the attached ports. In particular, it has been computed considering the projections of the port mass flows along each direction x, y, z:

$$\nabla\left(\rho \mathbf{U} A\right)_c = \begin{bmatrix} \dfrac{\Delta\left(\rho \mathbf{U} A\right)_{p,x}}{\Delta x} \\ \dfrac{\Delta\left(\rho \mathbf{U} A\right)_{p,y}}{\Delta y} \\ \dfrac{\Delta\left(\rho \mathbf{U} A\right)_{p,z}}{\Delta z} \end{bmatrix} \tag{10.15}$$

It can be noticed that in the previous definitions the dimensions of the cell Δx, Δy, Δz have been introduced. The cell dimensions have been determined starting from the lengths of the attached ports, since, according to the staggered arrangement, these are properties of the connecting elements and are not initially defined for the cells.

10.3.2 Numerical Tests

In this section, two different test cases are reported to discuss the characteristics of the numerical method in terms of stability and accuracy. The first case is the shock tube test; it sets the initial conditions of a Riemann problem, which will evolve as three traveling waves starting from the center of the system. A 1.0-m long pipe has been discretized with 100 elements setting the Riemann condition at the pipe midpoint imposing two different pressure levels: 5.0 bar and 1 bar. The temperature field is set uniformly along the pipe at 298 K. Starting from time 0.0, a shock wave propagates to the right with a velocity $U + a$ (sum of the flow velocity U and the sound velocity a), a rarefaction wave to the left with a velocity $U - a$, and a contact discontinuity to the right with a velocity U. The capability to reproduce the sharp discontinuities in the thermo-fluid dynamic property profiles (p, ρ, T) provides indications of the stability and accuracy of the method [8, 21]. Figure 10.5a shows that the method without the flux correction distributes the shock wave discontinuity on a small number of nodes but at the same time gives rise to strong numerical oscillations. This behavior is due to the second-order accuracy of the staggered leapfrog method. On the other hand, the method becomes stable when the DTM algorithm is considered (Figure 10.5b), maintaining a good resolution of the sharp pressure discontinuity and damping the numerical oscillations.

Considering the density profile (Figure 10.6), the same considerations about the shock wave resolution can be addressed. Moreover, it can be noticed that the accuracy of both methods reduces to first order when solving for the contact discontinuity, due to the adoption of an upwind approach to determine the enthalpy in the port.

FIGURE 10.5 Calculated pressure profile along the 1D pipe at time 7 ms compared to the exact solution: (a) without the DTM, (b) with the DTM.

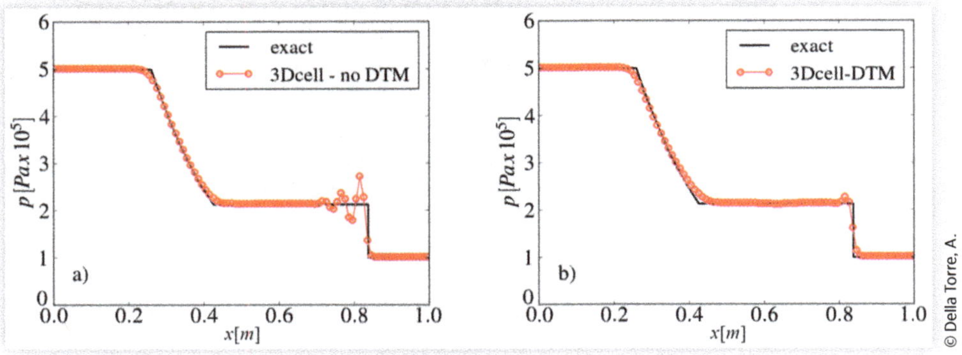

FIGURE 10.6 Calculated density profile along the 1D pipe at time 7 ms compared to the exact solution: (a) without the DTM, (b) with the DTM.

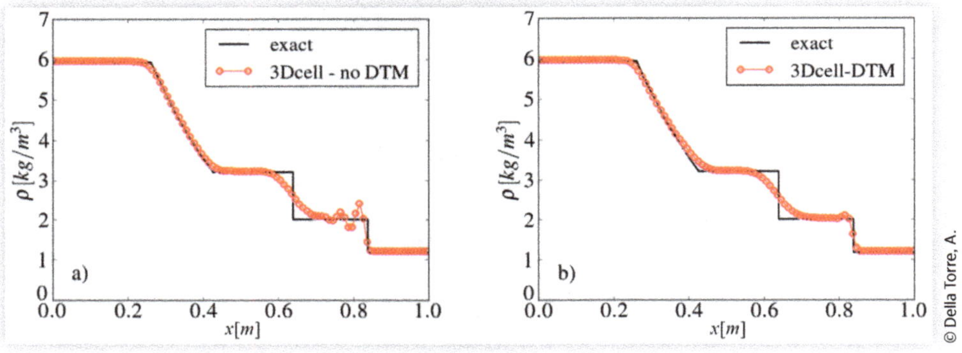

The choice of an upwind differencing for the enthalpy guarantees the stability of the method without applying any flux limiter on the energy equation. The low accuracy in the contact surface resolution is acceptable in most engine applications since it does not have a great effect on the volumetric efficiency prediction. On the other hand, the second-order accuracy in the pressure discontinuity description allows a good prediction of the wave motion in the intake and exhaust systems and of the acoustic performance of these devices.

As a more complex test case, a T junction, in which the secondary branch joins to the principal one with a 90° angle, is considered. Figure 10.7 shows the schematic of the

FIGURE 10.7 Schematic of the test rig with the position of the three pressure probes.

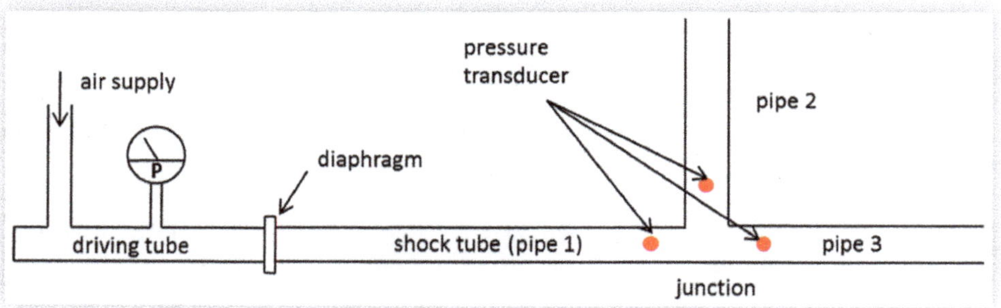

FIGURE 10.8 Pressure wave propagation in the three branches of the junction: (a) after 20 ms, (b) after 28 ms.

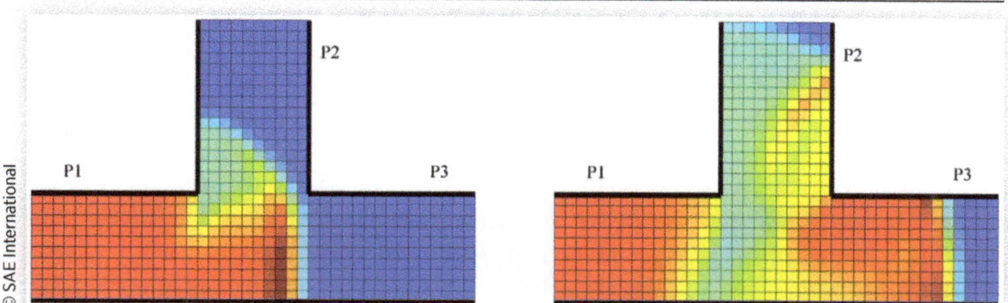

test bench used to probe the wave propagation inside the junction along with the position of the three pressure transducers. When the desired pressure in the driving tube is reached, the diaphragm is burst and a pressure wave propagates in the shock tube, reaches the junction, and is then transmitted into the two exit pipes. The calculated pressure traces, extracted at fixed points, have been compared to experimental measurements available in the literature [8].

In the schematic, the pressure wave reaches the junction traveling from the left side. Since the pipes mounted on the test rig were characterized by a square section, the network of 3D cells used to model the device considered only one cell in the direction normal to the plane containing the junction axis, leading to the construction of a 2D case.

The model of the junction is composed of 550 cells each with a dimension of 5 mm. The three branches of the junction have been connected to three pipes, modeled as 1D. An open-end boundary condition has been applied to the two-outlet pipe, while a Riemann condition has been imposed in the middle of the inlet pipe. An initial pressure of 2 bar at ambient temperature has been imposed in the driving tube, while the initial conditions in the rest of the domain are set to ambient pressure and temperature. The simulation of the shock wave propagation is a strong test for evaluating the effectiveness of the stabilization technique adopted. Figure 10.8 shows an example (two different time steps) of how the pressure wave propagates inside the T junction. The DTM method gives a stable solution and preserves the sharp gradients in the pressure profiles resolving the pressure gradients in two cells.

In Figure 10.9, the instantaneous pressures calculated at the transducer locations are compared with the measured values. It can be seen that the model is able to capture the different pressure levels in the three branches. However, it can be noticed that there is a little overestimation of the pressure wave transmitted in branch 3 and a consequent underestimation of the final pressure level in branch 1. These effects are related to the dissipative phenomena occurring at the corner of the junction, which cannot be predicted by this inviscid model. All in all, it can be seen that the numerical method is able to follow the rapid pressure rise in the branches when the pressure wave arrives, mainly because of its second-order accuracy. Moreover, it is able to describe the high frequency pressure oscillations in branches 2 and 3, which are due to the transversal motion of wave fronts across the pipes.

FIGURE 10.9 Comparison between the computed and measured pressure traces.

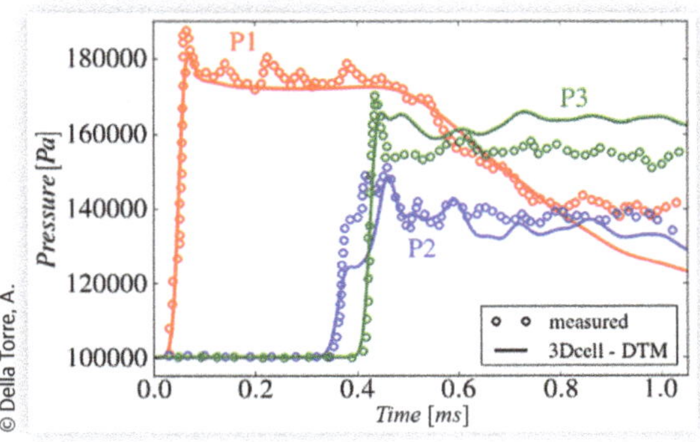

10.4 Sub-models for the Simulation of Intake and Exhaust Systems

Modeling the external shape by means of a suitable arrangement of 0D volumes is fundamental to understand and capture the fluid dynamic response of any device included in the intake and exhaust system. Moreover, in common practice, these devices are usually integrated with elements placed to perform specific operations, such as the removal of pollutants from the exhaust gas and the filtering of the ambient air, namely, catalysts and air filters. This results in a further increase of geometric complexity and in the need for having specific sub-models capable of taking into account their presence. For this purpose, the quasi-3D model has been extended to allow the modeling of typical components existing in intake and exhaust systems such as filtering cartridges, porous layers (for filtering or noise abatement purposes), perforated baffles or pipes, as well as catalyst and particulate filter substrates.

10.4.1 Perforated Surfaces

Perforated elements [22] are modeled by extending the properties of the port element described in Figure 10.3. The perforated port is characterized by the number of holes n_h, hole cross section A_h, hole length L_h, and by the total connector surface A_c. The approach developed for the perforated elements solves the momentum equation for only one single hole, in order to calculate its mass flow. Then the mass flux for the complete connection is determined by multiplying the total number of holes n_h.

As shown in Figure 10.10, the momentum equation for one single hole is solved over a control volume defined by the hole cross section A_h and the length $L_{C,h}$. This length is a function of the hole length L_h and hole diameter d_h according to the following relationship [23]:

$$L_{C,h} = L_h + 0.8\, d_h \tag{10.16}$$

This correction is necessary to take into account the inflow-outflow phenomena and the fact that the flow field is extended outside the hole. The momentum equation (Equation 10.5) applied to the hole control volume then becomes:

$$\left(\rho U_n A\right)_h^{n+\frac{1}{2}} = \left(\rho U_n A\right)_h^{n-\frac{1}{2}} + \frac{\Delta t}{L_{C,h}}\left[\left(\rho_{cv,L} U_{cv,L}^2 + p_{cv,L}\right) - \left(\rho_{cv,R} U_{cv,R}^2 + p_{cv,R}\right)\right] A_h + M_{s,h} \tag{10.17}$$

where ρ_{cv}, p_{cv}, U_{cv} are the density, the pressure, and the normal velocity at the control volume boundaries, respectively. In particular, ρ_{cv} and p_{cv} are assumed equal to the same quantities in the neighboring cell while the velocity U_{cv} is interpolated between the connector and cell centers as follows:

FIGURE 10.10 Schematic of the perforated connector element.

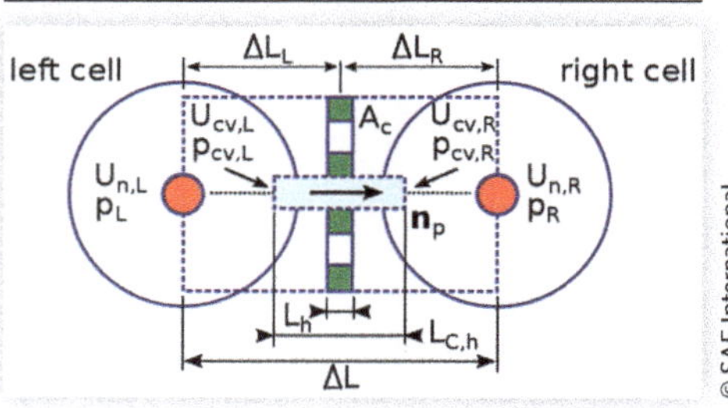

$$U_{cv,L} = \frac{\frac{1}{2} L_{C,h} U_{n,L} + \left(\Delta L_L - \frac{1}{2} L_{C,h}\right) U_h}{\Delta L_L} \tag{10.18}$$

$$U_{cv,R} = \frac{\frac{1}{2} L_{C,h} U_{n,R} + \left(\Delta L_R - \frac{1}{2} L_{C,h}\right) U_h}{\Delta L_R} \tag{10.19}$$

The term $M_{s,h}$ in the momentum conservation equation (Equation 10.17) accounts for the gas-wall friction and is determined as follows [8]:

$$M_{s,h} = (\rho UA)_h |U_h| f_h \frac{2}{d_h} \Delta t \tag{10.20}$$

where the friction factor f_h is defined as:

$$f_h = \frac{\tau_w}{\frac{1}{2}\rho U} \tag{10.21}$$

10.4.2 Sound-Absorptive Material

The quasi-3D model has been extended to account for the effect of sound absorbing material. This material usually fills up cavities and gaps created inside the muffler [24] and, as a consequence, the presence of porous material is considered as a cell property. The flow interacts with the material fiber transferring to it its momentum, and the acoustic energy associated with the pressure waves is dissipated as heat. To take into account this dissipation, the momentum equation has been modified by including an additional source term, which reproduces the effects of the flow resistance caused by the absorptive material. The momentum equation (Equation 10.5) has been modified by adding two source terms accounting for the resistivity properties of the two neighboring cells:

$$M_s = \frac{1}{2}(RU_n A)_L + \frac{1}{2}(RU_n A)_R \tag{10.22}$$

where R expresses the material resistivity. This property is the most important characteristic of a porous material and can be defined by the following equation:

$$R = -\frac{1}{U}\frac{\Delta p}{\Delta x} \tag{10.23}$$

The material resistivity can be calculated by the following correlation [25]:

$$R = R_0 + \gamma R_0^\delta |U|, \quad R_0 = \frac{1}{1000}\frac{\eta}{|r_m^2|}\alpha\mu^\beta \tag{10.24}$$

where

μ is the dynamic viscosity of the gas

$|r_m^2|$ is the mean square average fiber radius

η the ratio between the material packing density and the material density

U is the gas velocity

The α, β, γ, δ coefficients depend on the type of sound absorptive material used. For example, for rock wool the following values are proposed [26]: $\alpha = 4.4$, $\beta = 1.59$, $\gamma = 0.75$, $\delta = 0.57$.

10.4.3 Filter Cartridge

From the fluid dynamic point of view, the filter cartridge creates a flow resistance in the gas stream, leading to the generation of a pressure drop. In order to describe the flow resistance due to the filtration process, the filter can be modeled as a porous layer. In this case, Darcy's law can be conveniently applied, giving an estimation of the pressure drop across the porous wall:

$$\Delta p_{Darcy} = \frac{\mu\, w_{wall}}{K_{wall}} U_p \tag{10.25}$$

where

K_{wall} is the permeability of the media
w_{wall} the thickness of the filtering layer

The modeling of the filter cartridge is realized by extending the properties of the port element. In particular, the source term in Equation (10.5) assumes the following expression:

$$M_s = \Delta p_{Darcy}\, A_p \frac{\Delta t}{w_{wall}} \tag{10.26}$$

As shown in Figure 10.11, the porous layer can have a thickness that is lower than the characteristic length of the connecting port. However, the discretization adopted considers this thickness only for the evaluation of the momentum source term of Equation (10.5), while it integrates the equation over the characteristic length. This avoids the risk of generating excessively small mesh spacing, which usually leads to an undesired increase of computational runtime.

10.4.4 Honeycomb Catalyst Substrate

Monolithic honeycomb catalyst substrates are structures with many small, parallel channels running axially through the core [27]. The catalyst, when subject to a gas flow stream, creates a flow resistance in the gas stream, leading to the generation of a pressure drop. This resistance is mainly due to two different contributions: (i) gas-wall friction inside the tiny channels (distributed pressure loss); and (ii) sudden contraction and sudden expansion at the inlet and outlet of the monolith, respectively (concentrated pressure loss due to vena contracta effects). In the case of catalyst substrates, it is very important to avoid the reconstruction of the channel spacing in order not to increase the computational burden. This means that the mesh size must not resolve the smallest edge length determined by the thickness of the substrate. Therefore, the monolith is discretized as rows of macro-cells having the same mesh size as the fluid cells in the inlet and outlet cones (as shown in Figure 10.12). Each macro-cell models a group of tiny channels, whose total number is given by the cell frontal area and substrate density. The conservation equations are solved only for a single channel and then extended to all the other channels belonging to the same macro-cell. At the interconnection between the external cell

FIGURE 10.11 Schematic of the port element used to model the porous layer. The porous layer is highlighted in yellow.

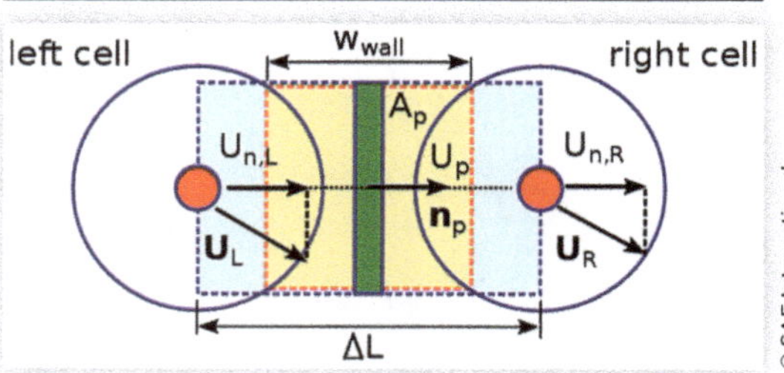

FIGURE 10.12 Example of domain decomposition into macro-cells for the modeling of the catalyst substrate.

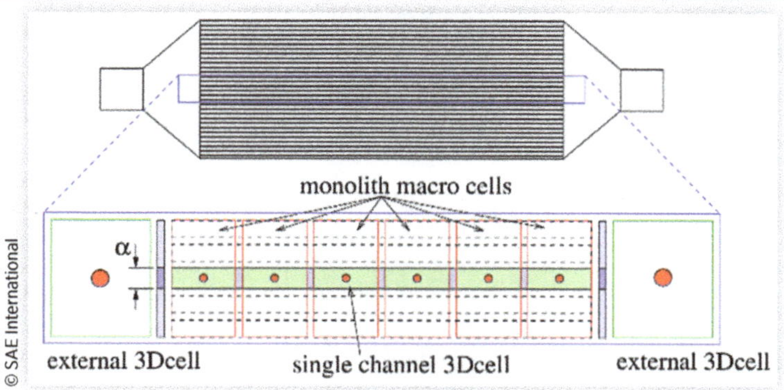

and the catalyst cell, the fluxes of mass and energy are multiplied by the number of the channels in the macro-cell.

The mass and energy conservation equations for the solution of the fluid-dynamic field inside the catalytic converter remain unchanged with respect to Equations (10.3) and (10.4). Instead, the momentum equation solved for the port is modified by the addition of a source term that takes into account the gas-wall friction inside the tiny channels. In particular, the friction term, driven by laminar flow effects, is evaluated resorting to the correlation proposed by Churchill [28]:

$$f_w = 2\left[\left(\frac{8}{\text{Re}}\right)^{12} + \frac{1}{(A+B)^{\frac{3}{2}}}\right]^{\frac{1}{12}} \tag{10.27}$$

where:

$$A = \left[2.21 + 2.46\ln\left(\frac{\varepsilon}{d} + \frac{42.68}{\text{Re}^{0.9}}\right)\right]^{16} \tag{10.28}$$

$$B = \left(\frac{37.53}{\text{Re}^{0.9}}\right)^{16} \tag{10.29}$$

Hence, the source term added to the momentum equation (Equation 10.5) can be written as:

$$M_{s,h} = (\rho UA)|U| f_w \frac{2}{\alpha}\Delta t \tag{10.30}$$

where f_w is the friction coefficient and α is the length of the channel edge. Moreover, at the inlet and outlet of the monolith channel an additional source term is included in the port momentum equation, in order to account for the concentrated pressure losses due to the inflow/outflow phenomena. This can be expressed as:

$$\Delta p_{in-out} = \xi\rho\frac{U^2}{2} \tag{10.31}$$

where $\xi = 0.2$. The corresponding source term is then as follows:

$$M_s = \Delta p_{in-out} A \frac{\Delta t}{\Delta L} \quad (10.32)$$

10.4.5 Particulate Filter Substrate

The particulate filter is a ceramic honeycomb structure whose properties are defined with the same parameters used for catalyst substrates. Its structure is identical to the catalyst one with the only exception that the parallel tiny channels are alternatively closed at one side, as shown in Figure 10.13. In this way, the exhaust gas entering the upstream end is forced through the porous wall separating the channels and exits through the opposite end of an adjacent channel.

The particulate filter is characterized by a higher flow resistance, when compared to the catalyst. The main reason is the lower flow area available for the gas stream, caused by the alternately closed channels. Additionally, the filtration process, although occurring at very low velocity, gives a significant contribution to the flow resistance. The pressure drop introduced by the particulate filter can therefore be split in three different phenomena: (i) gas-wall friction interaction inside the tiny channels, (ii) sudden contraction and sudden expansion at the inlet and outlet of the channels, and (iii) the filtration through the porous wall where the collection of the particulate matter takes place. The last phenomenon gives the most important contribution to the overall pressure drop. In order to determine the pressure drop across the filter wall, Darcy's law for the flow through a porous medium can be conveniently applied [29]. Moreover, Forchheimer's extension can be introduced in order to estimate the inertia effect, as suggested in [30]. This leads to the following estimation of the pressure drop across the porous wall:

$$\Delta p_{por} = \Delta p_{Darcy} + \Delta p_{Forch} = \frac{\mu \, w_{wall}}{K_{wall}} U + \beta \rho w_{wall} U^2 \quad (10.33)$$

where

K_{wall} is the permeability of the media

β is the Forchheimer coefficient

Compared to the case of the catalyst, the particulate filter modeling presents an additional challenge: in this case the problem can no longer be assumed as 1D, since the flow passes through the porous wall from one channel to another. This is an issue, since the solution of the flow through the wall requires the time step to be limited according to the Courant–Friedrichs–Lewy (CFL) criterion on the basis of the small transversal dimension of the tiny channels, resulting in a high computational burden.

Two alternative modeling approaches were investigated, namely, the modeling of the "couple of channels" and the modeling of the monolith as a generic porous media.

FIGURE 10.13 Example of a particulate filter channel: (1) inlet channel, (2) outlet channel.

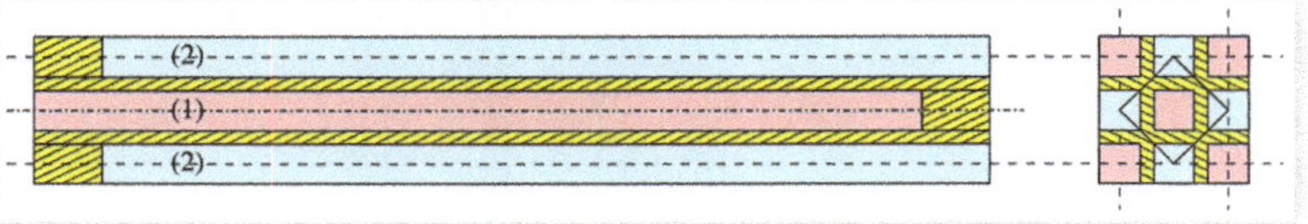

10.4.5.1 MODELING OF COUPLED PARTICULATE FILTER CHANNELS

This discretization approach for the particulate filter monolith is similar to the one adopted for the catalyst substrate. In addition, in this case the particulate filter monolith is discretized in macro-cells, each one representing a segment of the pair of inlet and outlet channels. Within the particulate filter macro-cell, a distinction between inlet channel, outlet channel and porous material is made (Figure 10.14). The two channels are modeled by means of two fluid cells, whereas the porous material is handled with the connector element. In particular, referring to Figure 10.14, the pair of channels can be modeled by means of two cells having a volume $V_{c,\,DPF} = \alpha^2 \Delta x$ linked by a connector of area $A_{c,\,DPF} = 4\alpha\Delta x$.

The momentum equation for the porous connector takes into account (i) the effect of the axial symmetry of the geometry and (ii) the pressure drop caused by the presence of the porous media.

Moreover, the axial symmetry of the channel in the plane orthogonal to the filter axis makes the cell momentum components lying in this plane cancel out, leading to a strong simplification of the connector momentum equation. Therefore, considering the control volume shown in Figure 10.15 as the fluid volume to which the momentum balance is applied, Equation (10.5) becomes as follows:

$$\left(\rho U_n A\right)_{por}^{n+\frac{1}{2}} = \left(\rho U_n A\right)_{por}^{n-\frac{1}{2}} + \left(p_L - p_R\right) A_{por} \frac{\Delta t}{\alpha + w_{wall}} - \Delta p_{por} \frac{\Delta t}{\alpha + w_{wall}} \tag{10.34}$$

where the source term is written according to the definition of filtration pressure drop shown in Equation (10.33). The gas-wall friction is taken into account in the

FIGURE 10.14 Example of domain decomposition into macro-cells for the modeling of the particulate filter.

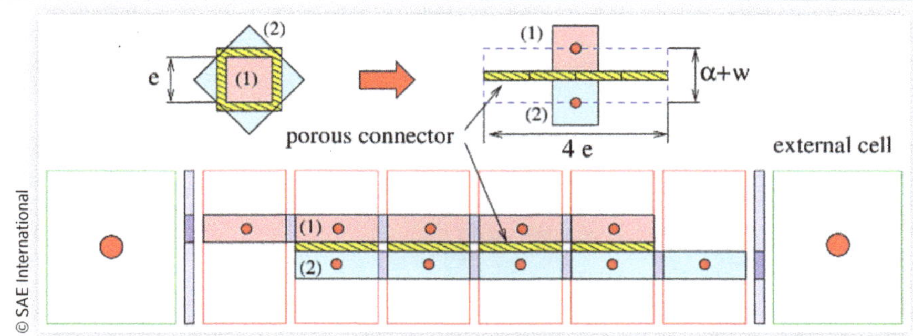

FIGURE 10.15 Schematic of the contribution to the connector momentum balance in a macro-cell.

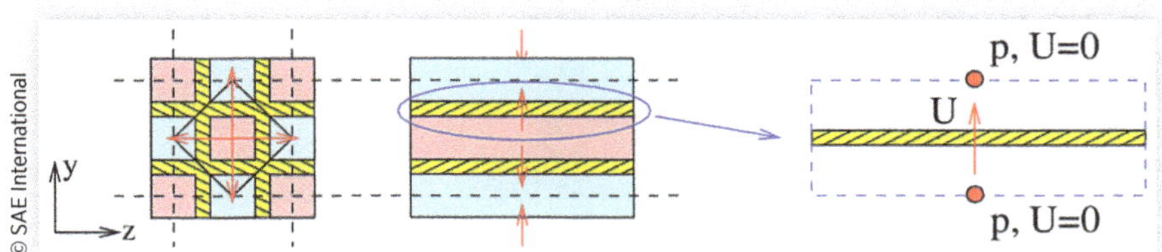

FIGURE 10.16 Modeling of the particulate filter as a generic porous media.

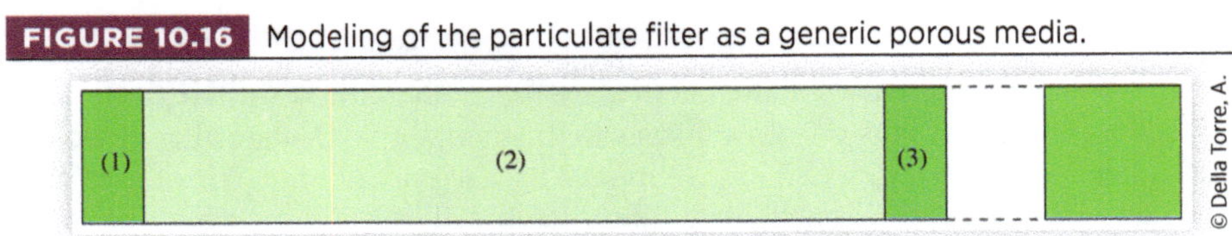

momentum balance of the connector, linking two cells belonging to the same channel. Similarly, the concentrated pressure losses are considered only in the connectors at the beginning and at the end of the substrates. The procedure is identical to the one followed for modeling the catalyst substrate, with the only difference that the pressure drop coefficient is higher than the one used for the catalyst, due to the higher flow section variation.

10.4.5.2 MODELING THE PARTICULATE FILTER MONOLITH AS A GENERIC POROUS MEDIA

This alternative approach consists of modeling the particulate filter monolith as a generic porous media, in which the pressure drop in the axial direction is described by the Darcy's law. The monolith is divided into three different sections in the axial direction (Figure 10.16), in order to take into account that, due to the presence of the plugs, the permeability in the initial and final part of the monolith is considerably lower than that in the middle region. The actual values of the permeability in the three regions can be determined from a detailed CFD simulation of a couple of channels having the same properties, under steady-state flow conditions, or starting from experimental measurements of the pressure drop generated by the device.

The momentum equation (Equation 10.16) is therefore modified adding the following source term:

$$M_s = \left(\frac{\mu}{K_{reg}} U \right) A_p \Delta t \tag{10.35}$$

where K_{reg} is the permeability of the media in a certain region.

10.5 Application to the Acoustic Modeling of Silencers

In this section, the application of the quasi-3D approach to the modeling of silencers will be examined. As previously said, the capability of the quasi-3D numerical method to describe three-dimensional pressure wave propagation, without introducing excessive numerical diffusion, makes it particularly convenient for the calculation of acoustic properties of silencing devices. Moreover, since acoustics in the muffler is essentially influenced by the paramount dimension of the chamber, geometric details can be neglected and a model based on a relatively small number of elements can be set-up, in order to reduce the computational time. In the following sections, the method will be applied and validated on configurations characterized by increasing complexity, from simple empty chambers to after-treatment devices including pollutant removal systems.

10.5.1 Simple Chamber

Trying to prove whether an approach is capable of accurately predicting the wave motion inside muffling devices, one could consider complex geometries, maybe characterized by flow reversals and significant asymmetries, as best candidates for validation test cases. However, the more complex the geometry is, the more difficult it is to understand its acoustic behavior. This is the main reason that leads to the choice of a very simple configuration for initial validation of the proposed approach: a cylindrical expansion chamber. This configuration, if suitably shaped in terms of ratio between its length and diameter, can show significant higher order mode effects. In addition, since the acoustic behavior of a simple expansion chamber is well known from acoustic theory, it is very easy to highlight the excitation frequency of the higher order modes, avoiding the confusion of similar effects generated by the geometry complexity. Moreover, a comprehensive analysis of the simple expansion chamber acoustic behavior was carried out by Eriksson [31, 32], along with a numerical analysis based on the nodal lines theory. As shown in [31], the cut-off frequency of a cylindrical series chamber is mainly driven by the diameter of the component. In particular, Eriksson showed all the modes of transversal resonances for a cylindrical expansion chamber with single inlet and outlet without offset, proposing the following cut-off frequency calculation:

$$f_{\text{asymm}} = 1.84\frac{a}{\pi D} \tag{10.36}$$

At this frequency the first asymmetric high order mode is excited. However, it has been demonstrated that asymmetric wave modes are triggered only in case of eccentricity between the inlet and the outlet pipes. For the case with concentric inlet and outlet pipes only axisymmetric modes can be excited; consequently the first cut-off frequency becomes:

$$f_{\text{axisymm}} = 3.83\frac{a}{\pi D} \tag{10.37}$$

The first cylindrical expansion chamber considered has a length and a diameter of 497 and 297 mm, respectively. It has concentric inlet and outlet pipes with a diameter of 50 mm (Figure 10.17). At first, the quasi-3D model of the chamber was set-up with the

FIGURE 10.17 Schematic of the circular chamber and example of asymmetric and axisymmetric modes.

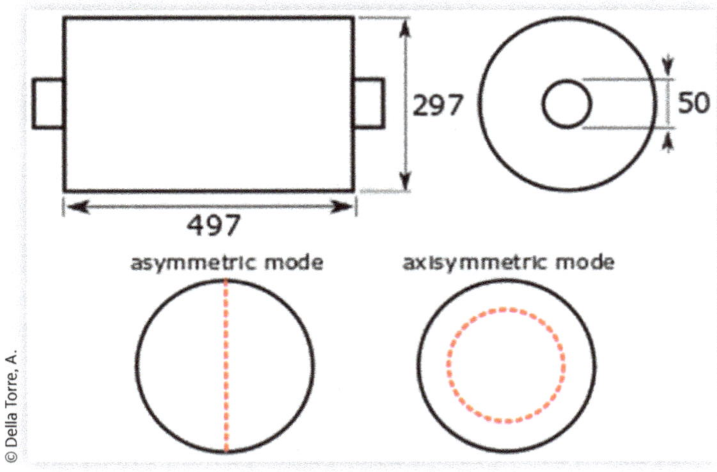

FIGURE 10.18 Quasi-3D reconstruction of a circular chamber and comparison between measured and calculated transmission loss.

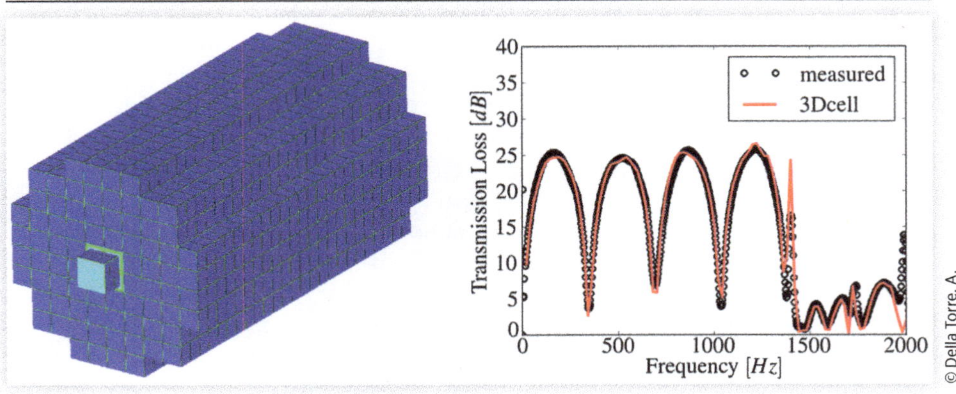

aim of reconstructing the circular shape with a sufficient accuracy in order to capture both the volume and the transversal dimension of the chamber, as shown in Figure 10.18. This is particularly important for an accurate prediction of the acoustic behavior, since the volume of the chamber influences the maximum level of noise abatement while the transversal dimension determines the frequency at which transversal modes are triggered.

The calculations were performed [33, 34] by imposing a white-noise perturbation at the inlet section whereas an anechoic termination was used at the outlet. In Figure 10.18, the computed transmission loss is compared with the measured data, showing a good agreement over the whole frequency range. In particular, the first cut-off frequency, related to a circular symmetric mode, is clearly located around 1416 Hz. The quasi-3D approach is able to capture both the planar wave propagation occurring at low frequencies and the three-dimensional wave propagation when higher order modes are excited at high frequency. To highlight the particular excitation of the symmetric mode, the expansion chamber was also simulated with a pure tone excitation source at the frequency of 1400 Hz. The result of this analysis is shown in Figure 10.19, where the pressure field is sampled along a diameter at different time steps. It can be seen that during an oscillation period the pressure field is constant along the highlighted nodal line.

10.5.2 Reverse Flow Chamber

FIGURE 10.19 Samples of the pressure field along a diameter at different time steps: the nodal line is highlighted in red.

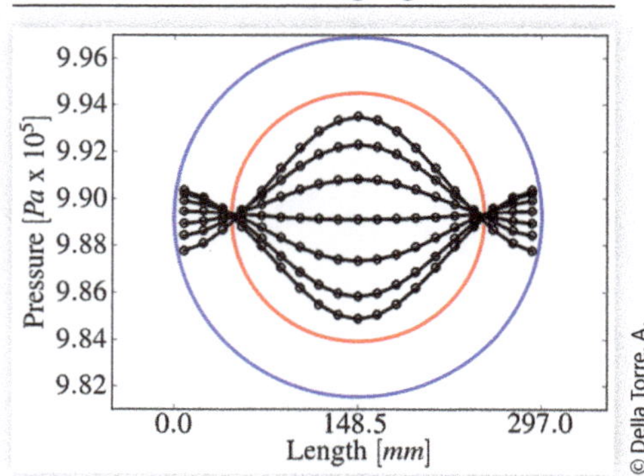

In order to consider a more complex geometry, the example of a reverse flow chamber is reported [35]. The validation case (Figure 10.20) has a length of 494 mm and a diameter of 197 mm; the inlet and outlet pipes are extended into the chamber by 257 and 17 mm, respectively. The quasi-3D model was built using a network of 5782 cells, which reproduce the circular section of the chamber in order to match both volume and transversal dimensions.

Figure 10.20 shows the comparison between the predicted transmission loss and the measured values for the reverse flow chamber. Generally, the agreement is good, since the abatement peaks and pass bands are correctly captured over a wide range of frequency, without showing any frequency shift. In particular, the first resonant peak at 414 Hz corresponds to the quarter wave resonance of the chamber length minus the inlet extension, while the second peak at 974 Hz is related to the inlet extension. The transparency frequencies

FIGURE 10.20 Schematic of the reverse chamber with quasi-3D reconstruction and transmission loss.

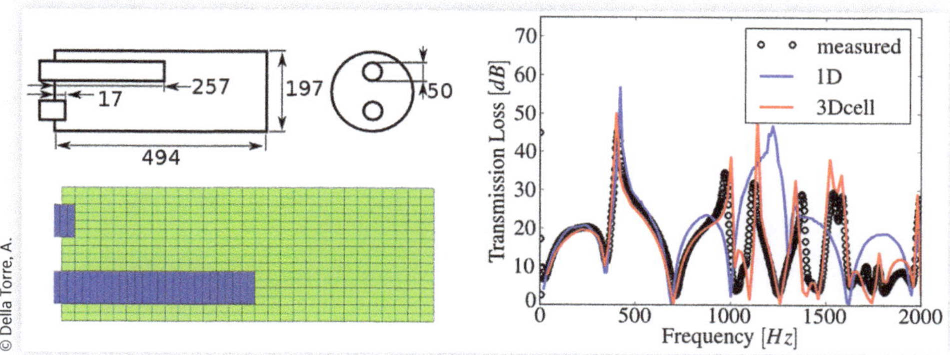

(354 Hz, 720 Hz, 1062 Hz, 1460 Hz) are associated with the main longitudinal resonance. Notice that the quasi-3D method gives a noticeable improvement in the transmission loss prediction at high frequency when compared to the traditional 1D approach.

10.5.3 Perforated Elements and Sound Absorptive Material

Perforated ducts and sound-absorptive materials are usually adopted in the design of silencing devices in order to enhance the noise abatement performance. In the following section, different test cases will be considered, in order to validate the sub-models proposed in Section 10.4. Moreover, the effect of mean flow on the acoustic properties of dissipative silencer will be examined and the accuracy and stability of the numerical approach will be evaluated.

The first configuration considered [17] is a cylindrical chamber having a length of 305 mm and a diameter of 114 mm, with a perforated pipe connecting inlet and outlet (Figure 10.31). This is known as concentric tube resonator [36, 37]. The perforated duct is characterized by a diameter of 50 mm, a thickness of 1.6 mm, and a porosity of 4.6%. The diameter of the holes is 3.2 mm and the corrected length adopted in the simulation, according to Equation (10.16), is 4.8 mm. The quasi-3D model consists of a 7 by 7 by 20 cell matrix; the circular section is approximated by a square shape section. The comparison between predicted transmission loss and measured data is shown in Figure 10.21.

FIGURE 10.21 Schematic of the chamber with perforated pipe and quasi-3D reconstruction as well as transmission loss comparison.

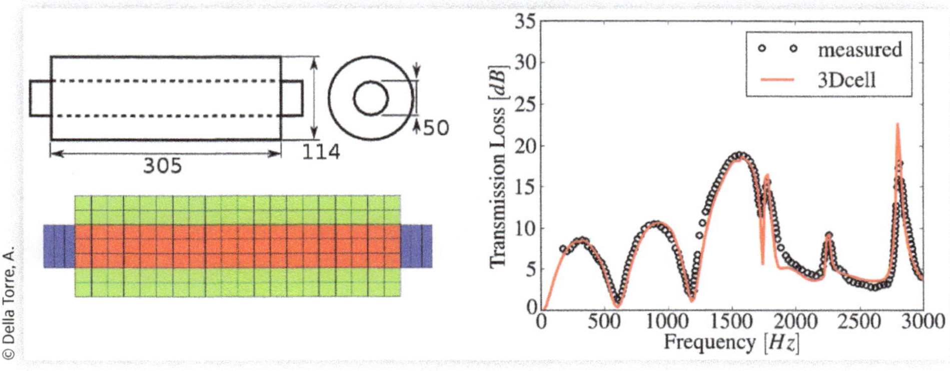

FIGURE 10.22 Schematic of the single plug muffler and quasi-3D reconstruction.

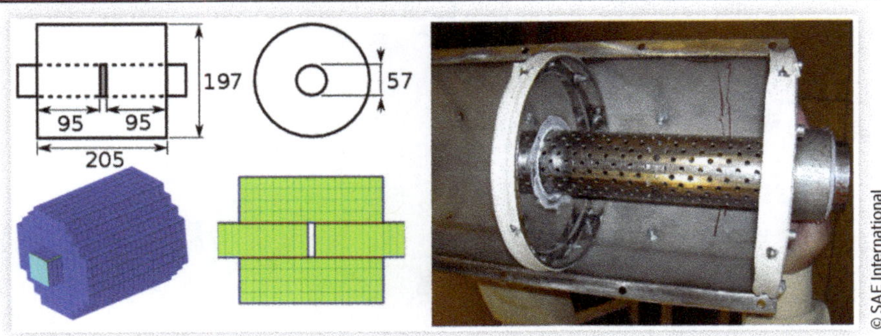

The general agreement is good, with a correct prediction of transparency frequencies associated with the main longitudinal resonance (600 Hz, 1200 Hz) and the resonance peaks associated with the perforated element.

Moreover, acoustic properties in the presence of mean flow have been simulated. For this purpose, two plug flow mufflers have been considered [38]. In these configurations, the presence of a plug element in the inlet perforated duct forces the flow to pass through the holes of the perforated pipe, increasing the dissipation of the acoustic energy and therefore enhancing the acoustic damping of the muffler. This effect becomes more significant in the presence of a considerable flow velocity. Moreover, in this case stability requirements become more stringent for the numerical method and, on the other hand, the numerical dissipation needs to be kept low in order to preserve the accuracy in the description of the pressure wave propagation.

In Figure 10.22, the schematic of a simple plug flow muffler is shown [39]. It consists of a perforated tube running through the center of an expansion chamber. The middle of the perforated tube is blocked so that the flow is forced out through the perforate holes on the inlet half of the pipe and back through the perforate holes on the outlet half of the pipe. Two different porosity levels for perforated sections of the single plug muffler are included in this study: 5% and 12%. The perforation hole diameter for the 5% porosity pipe is 5 mm; the holes are ordered in 14 rows around the pipe surface with a spacing of 28 mm. For the 12% porosity pipe, the diameter is 4 mm; the holes are ordered in 24 rows around the surface with 12 mm spacing. The quasi-3D network is made of 1800 cells and shaped in such a way as to reproduce the circular cross-section of the chamber. The pressure drop across each of the plug mufflers has been measured against the flow velocity. As a first step, before performing the acoustic simulation, the calculation model has been adopted for the evaluation of the pressure drop curve, in order to calibrate the friction coefficient required in the perforated model (Equation 10.20). In Figure 10.23, calculated and measured curves, in case of 5% and 12% porosity, are compared. As can be noted, the increase of porosity causes a decrease of flow resistance, resulting in lower pressure drop.

FIGURE 10.23 Pressure drop of the single plug muffler.

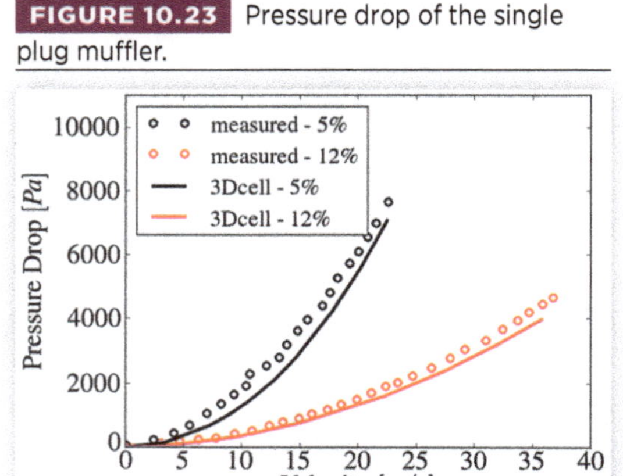

As can be seen in Figure 10.24, in the absence of mean flow, the configuration exhibits the typical behavior of a simple cylindrical chamber, without significant differences when different porosities are adopted. On the other hand, the presence of mean flow causes an increase of the acoustic damping, resulting in a higher value of the abatement peaks. This effect is more significant in the case of a small porosity value.

FIGURE 10.24 Transmission loss of the single plug muffler with 5% and 12% porosity.

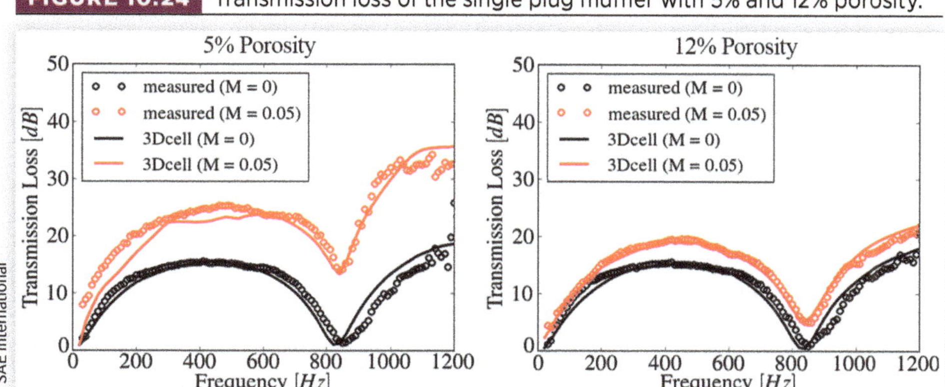

FIGURE 10.25 Schematic of the eccentric plug muffler with quasi-3D reconstruction and pressure loss comparison.

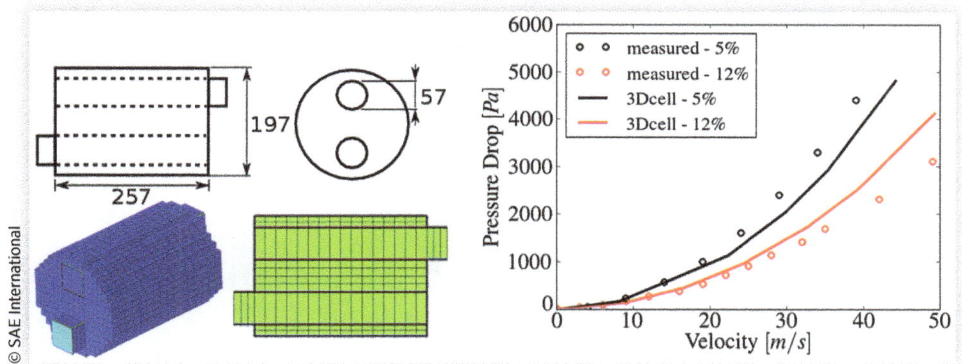

The agreement between the calculations and experiments can been regarded as good, confirming the accuracy of the numerical method even in cases with mean flow.

A second plug muffler is shown in Figure 10.25. This has an eccentric layout, consisting of a chamber with two continuous perforated pipes, which cover the entire length of the chamber. Flow is forced out through the perforate holes of the inlet pipe and in through the perforate holes of the outlet pipe before exiting the muffler itself. Two porosity values of the perforated ducts are considered: 5% and 12%. Also, in this case the pressure drop across the silencer has been computed and compared with the measured values (Figure 10.25).

In this configuration, due to the eccentricity of the geometry, higher order modes related to the transversal plane wave propagation are excited in the frequency range of interest. Transmission losses are plotted in Figure 10.26, showing a good agreement in the prediction of both longitudinal and transversal modes. It can be noticed that the presence of mean flow causes a general smoothing of the abatement function, increasing the damping at the transparency frequencies and smoothing the resonant peaks.

Finally, as a last dissipative silencer, a configuration including sound absorptive material will be examined [40]. The silencer consists of an expansion chamber with an extended outlet pipe. The space between the extended pipe and the muffler shell has been filled with sound absorbing material and (purely for testing purposes) without the use of a perforated plate (Figure 10.27). The quasi-3D matrix is built using a network of 1522 cells with an average mesh spacing of 20 mm. In Figure 10.28, comparisons between

FIGURE 10.26 Transmission loss of the eccentric plug muffler with 5% and 12% porosity.

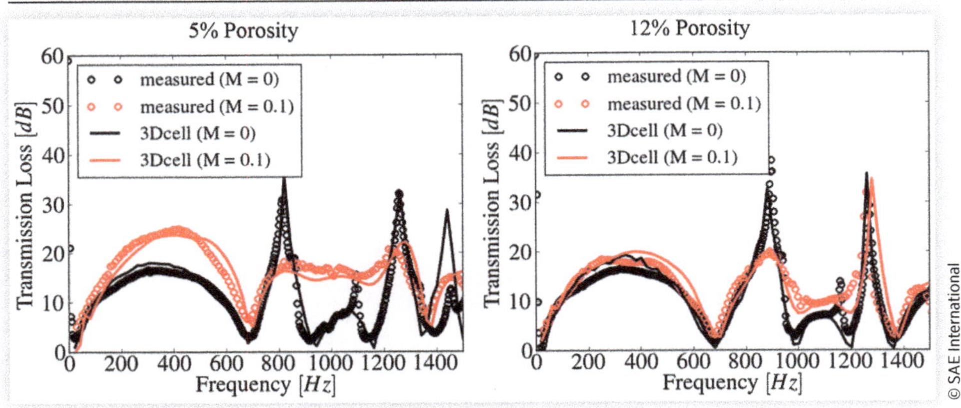

FIGURE 10.27 Schematic of the chamber with sound absorptive material and quasi-3D reconstruction.

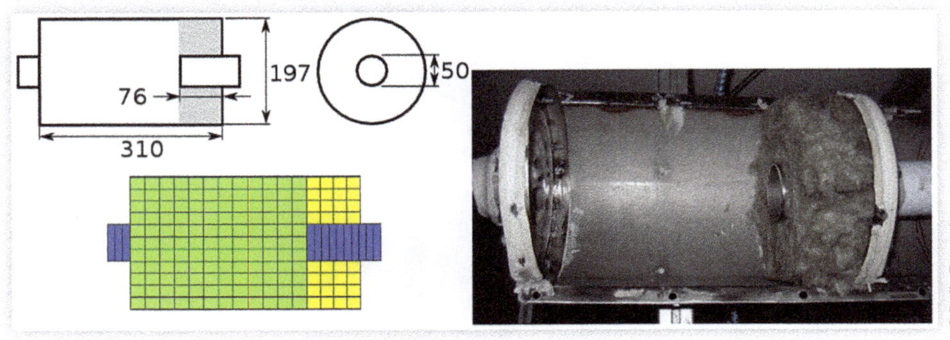

predicted transmission loss and measured data are reported [17]. Moreover, the result for case without the sound absorptive material is added (blue line), in order to better highlight its effect on the acoustic behavior. It can be noticed that the sound absorptive material smooths the acoustic resonances (peaks in the transmission loss) and reduces the transparency of the pass bands.

FIGURE 10.28 Transmission loss of the chamber with sound absorptive material.

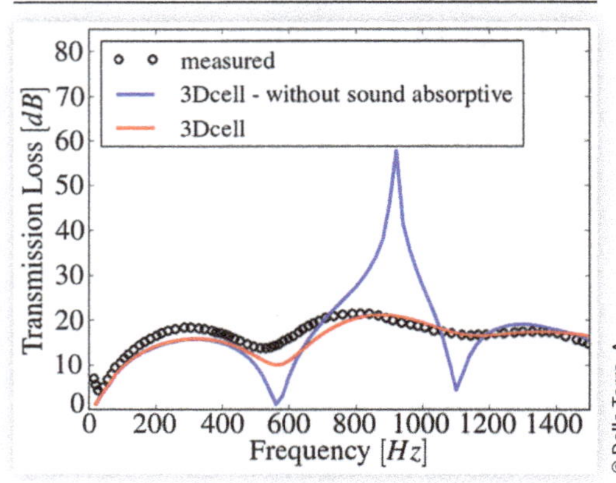

10.5.4 Complex Silencers

Two more complex silencing systems are hereafter considered to reproduce common configurations that are often found in automotive exhaust systems. The first silencer configuration consists of two chambers separated by an unperforated baffle [41]. Both the inlet and outlet pipes go through the baffle and have perforated sections each side of the baffle. The ends of both pipes are closed so the flow is forced through the perforates. The detailed geometry of the muffler is shown in Figure 10.29. The 61 mm length perforated sections consist of 135 holes with a diameter of 5 mm (~24% porosity) and the 70 mm length perforated sections consist of 144 holes with a diameter of 5 mm (~22.5% porosity).

FIGURE 10.29 Schematic and photograph of the two-chamber muffler.

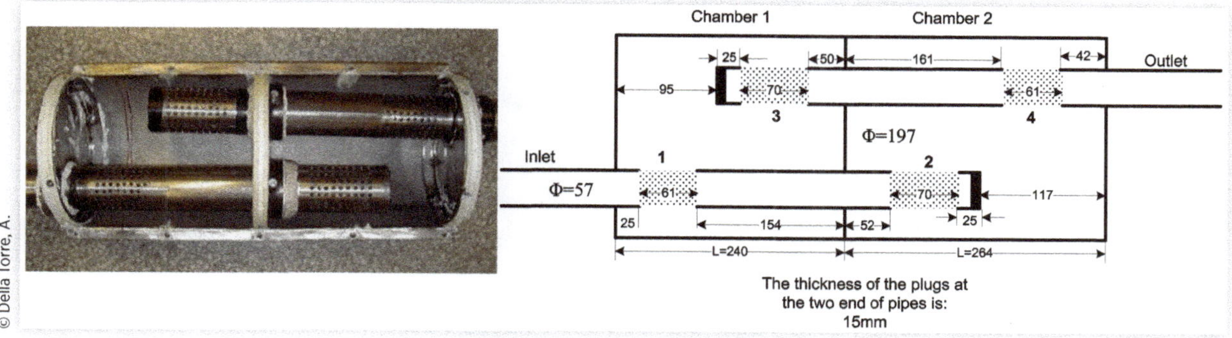

Figure 10.30 shows the 3D model and the structure of mesh. The cells (shown as brown squares) have been reduced in size purely for visualization purposes so that connections (ports) between the cells are visible. There are no connections between cells when they are separated by an unperforated baffle. Internal pipes are typically modeled with standard 1D pipe models. However, if they are perforated then 3Dcells have to be used (at least for the perforated part). This is because the perforations introduce additional dimensions to the flow. The radial (perforate) connections from a perforated pipe can clearly be seen in the end view of the mesh.

The measured and predicted transmission loss for the test case muffler without mean flow is shown in Figure 10.31. This shows a good comparison between the measured

FIGURE 10.30 Two-chamber muffler model and mesh.

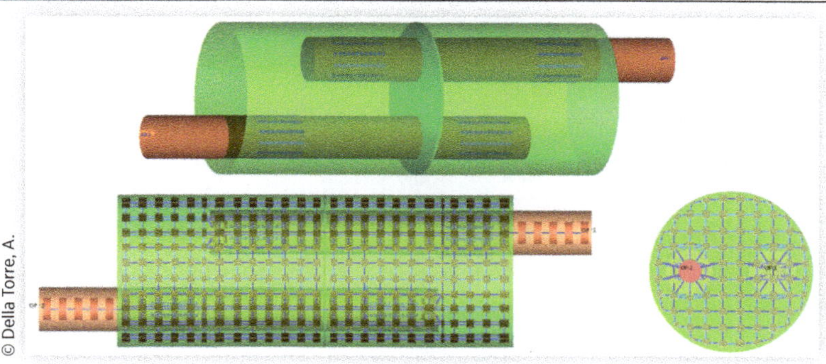

FIGURE 10.31 Two-chamber muffler transmission loss with no mean flow.

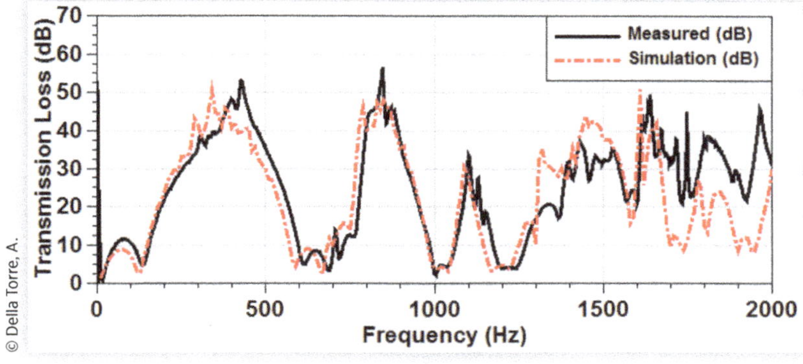

FIGURE 10.32 Two-chamber muffler transmission loss with mean flow.

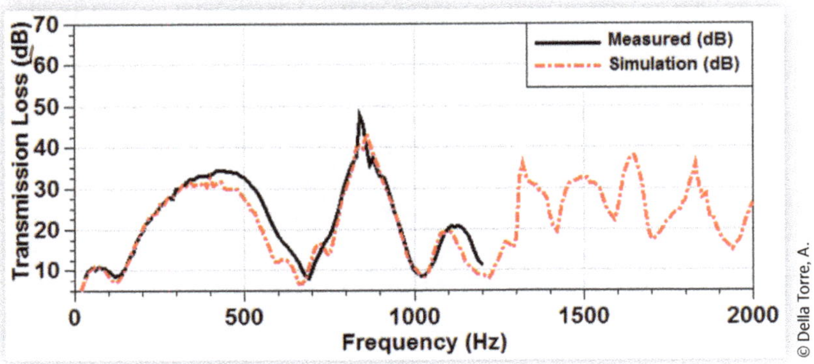

and simulated values. This is especially true for the important resonance ranges at approximately 400, 850, and 1100 Hz, which are very well predicted. The accuracy of the prediction is acceptable up to 1700 Hz, which means that the higher order modes, which are already generated at approximately 1000 Hz, are also predicted well, and from there on it starts to differ slightly.

The same mesh and model also accurately predicts the transmission loss with flow (Mach, M = 0.1) as shown in Figure 10.32. Again, there is good comparison between the measured and simulated transmission loss up to the maximum frequency available for the measured data (1200 Hz). Besides the accurate prediction of the resonance ranges, the decrease of damping due to flow at 400 Hz (from over 50 dB down to 35 dB) and at 1100 Hz (from 30 to 20 dB) is well reproduced with the model.

In addition to the acoustics behavior of the muffler, the pressure drop across the muffler as a function of flow velocity is also important for the accurate prediction of engine performance.

The comparison to the predicted pressure drop from the simulation is shown in Figure 10.33. The pressure drop for the simulation model is taken from measuring points before and after the component. The different flow cases are created by increasing the inlet boundary pressure compared to the outlet boundary. The same 3D mesh network model used for the acoustic simulation is also used for the pressure drop comparison.

As a second complex test case, a muffler with five chambers shown in Figure 10.34 has been considered [42, 43]. This is referred to as a modular muffler since it was designed to test different silencing configurations by changing the pipe and baffles' internal layouts. All the perforated elements have a thickness of 1.5 mm and an orifice diameter of 5 mm.

FIGURE 10.33 Two-chamber muffler pressure loss versus flow.

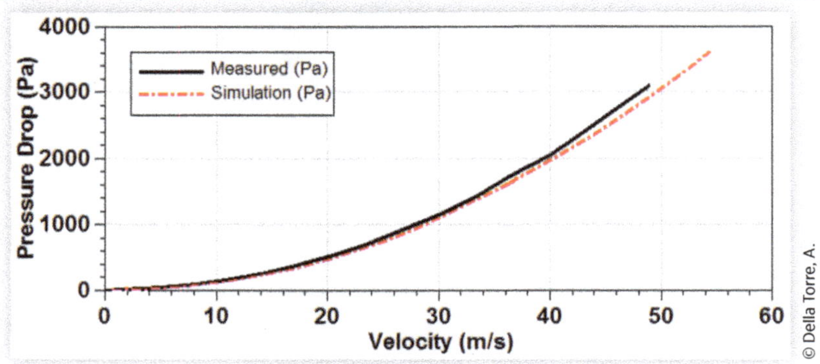

FIGURE 10.34 Five-chamber modular muffler.

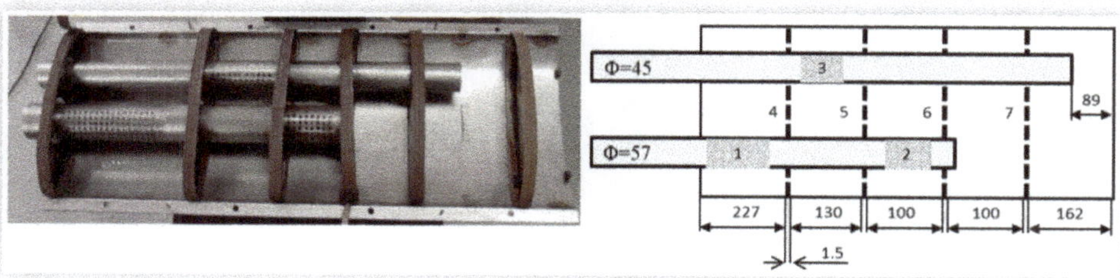

The number of holes for the perforated sections 1–3 are 154, 288, and 144, respectively. The number of holes for the perforated baffle sections 4–7 are 50, 72, 72, and 42, respectively. This test case was specifically designed and manufactured as a complex test example to test out the performance of different simulation methods.

Figure 10.35 shows the 3D model and the structure of mesh. Again, there are no connections between cells either side of an unperforated baffle. If the baffle is perforated, then the connection between 2 cells on either side of the baffle is modeled with a perforated connection. In order to get an accurate prediction of the transmission loss of this relatively complex muffler, it is necessary to model the perforated areas on each baffle in detail. For example, it is not sufficient to simply distribute the perforation evenly across the baffle. The shape and dimensions of the perforated area must be accurately modeled. This puts quite a demand on the pre-processing tool that creates the 3D model. In this case, the perforated areas are modeled by selecting an area (polygon selection) and defining the porosity or the total number of holes as well as the dimensions of the individual holes.

A comparison between measured and simulated transmission loss for the modular muffler with flow (Mach, M = 0.15) and without flow is shown in Figure 10.36. The simulation results are from the same mesh using a 20 mm average size and shows a good agreement with the measurements. The results show that the transmission loss is well predicted up to 500–600 Hz. For the no flow case, each of the three resonant peaks are accurately predicted. The damping of the resonant peaks for the flow case are well predicted by the simulation. Beyond 600 Hz the comparison starts to breakdown but still provides some useful information.

FIGURE 10.35 Modular muffler model and mesh.

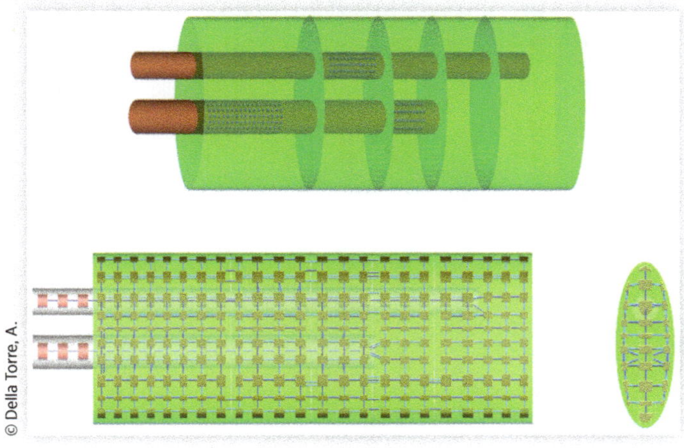

FIGURE 10.36 Modular muffler transmission loss with no mean flow and with a mean flow of M = 0.15.

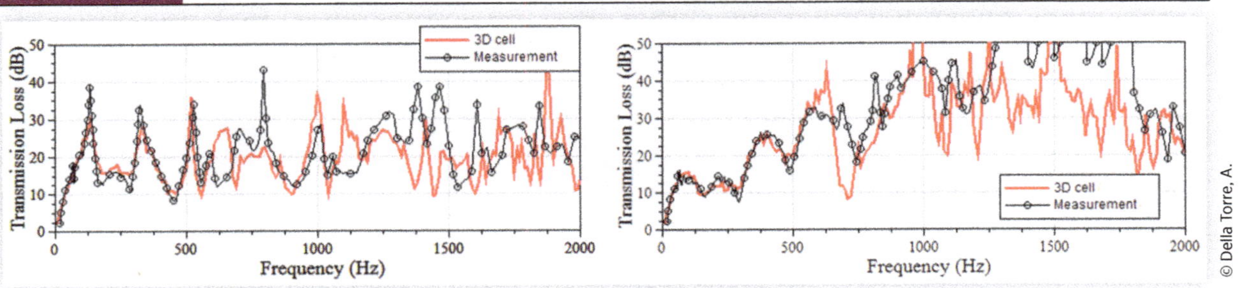

10.5.5 Devices without a Primary Silencing Purpose

Along with silencers, several other devices (e.g., intercoolers and after-treatment systems) are included in the intake and exhaust lines of a modern engine [27, 44, 45]. Although their primary functions are related to pollutant emission abatement or performance enhancement, these devices have also a direct influence on the internal wave dynamics, and thus on intake and exhaust noise. Therefore, the acoustic design of intake and exhaust systems can be affected, since they can have a non-negligible acoustic attenuation potential, which can be exploited and eventually improved. In the following section, the quasi-3D approach will be applied and validated for the acoustic simulation of such devices.

The first configuration considered is a simple cylindrical chamber that includes a catalytic substrate (Figure 10.37). The substrate has a length of 130 mm and a diameter of 121 mm and it is characterized by an open frontal area fraction of 0.612 and a 600 CPSI (cells per square inch) cell density. The inlet and outlet ducts have diameters of 81 and 62 mm, respectively. The lengths of the inlet and outlet cones are 62 and 47 mm. The geometry has been reconstructed with 2790 cells where the circular section has been approximated by a square one [44]. Moreover, the extension of the inlet and outlet ducts, where the microphones are located, have been included in the model. This is done in order to take into account the slight difference in diameter at the connection between the catalyst and the test rig. The comparison between computed and measured transmission loss is shown in Figure 10.37. The acoustic behavior is characterized by dissipative effects due to the gas-wall friction and due to the heat exchanged, resulting in a smooth trend of the TL, where the pass bands are almost cancelled.

FIGURE 10.37 Quasi-3D reconstruction of the cylindrical chamber with the catalytic substrate and transmission loss.

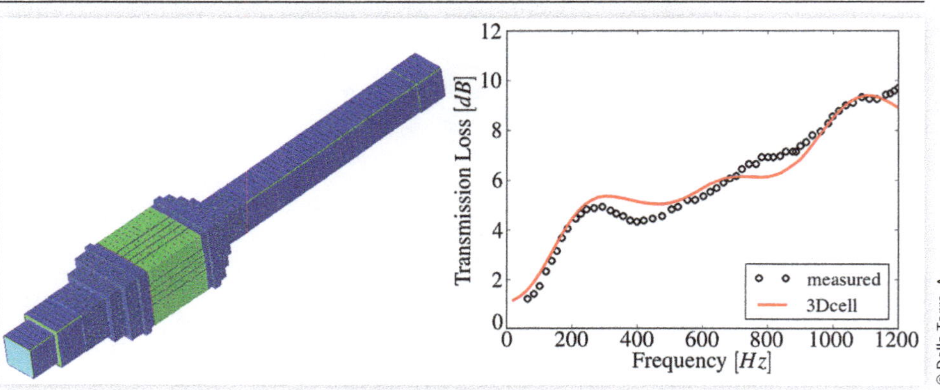

FIGURE 10.38 Quasi-3D reconstruction of the cylindrical chamber with the particulate filter substrate and transmission loss.

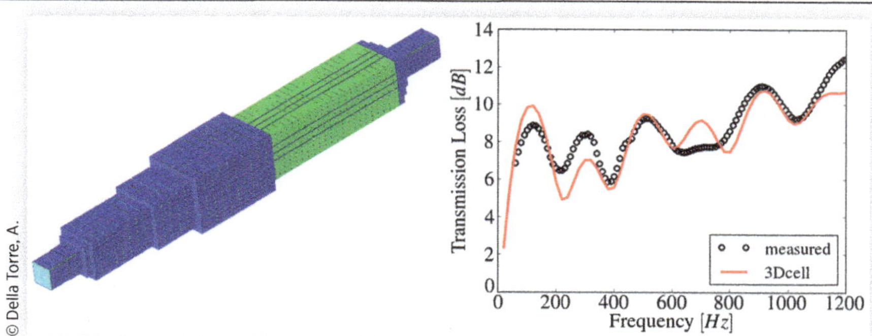

With regards to the particulate filter, the validation of the model was performed with the configuration shown in Figure 10.38. The particulate filter monolith has a length of 311 mm, a diameter of 120 mm, and an open frontal area fraction of 0.610. The quasi-3D model consists of 4076 cells and the circular section has again been approximated by a square one. The substrate has been modeled as a generic porous media, as explained in Section 10.4.5.2. The monolith is divided into three zones with different permeability, higher in the central zone and lower at the extrema where the plugs are located. In Figure 10.38, the comparison between the computed and measured transmission loss is reported, showing a good agreement in terms of abatement level and prediction of the acoustic transparencies.

In order to study the acoustic potential of current after-treatment devices, a relatively complex system, which comprises an oxidation catalyst and a particulate filter, has been considered [45]. The two substrates are included in a cylindrical shell, connected to the inlet and outlet ducts by means of cones. Moreover, the inlet duct is eccentric with respect to the chamber, resulting in the occurrence of transversal resonances in the frequency range of interest. In order to separately analyze the effect of the different components and their eventual mutual interactions, a stepwise approach was adopted. The approach used was based on the removal of the different elements one by one, so that their acoustic characteristics could be studied separately and then their eventual interaction could be addressed.

In Figure 10.39, the schematic of the device with only the catalyst included is shown, along with the comparison between the calculated and measured transmission loss.

FIGURE 10.39 Quasi-3D reconstruction of the after-treatment system without the particulate filter and transmission loss.

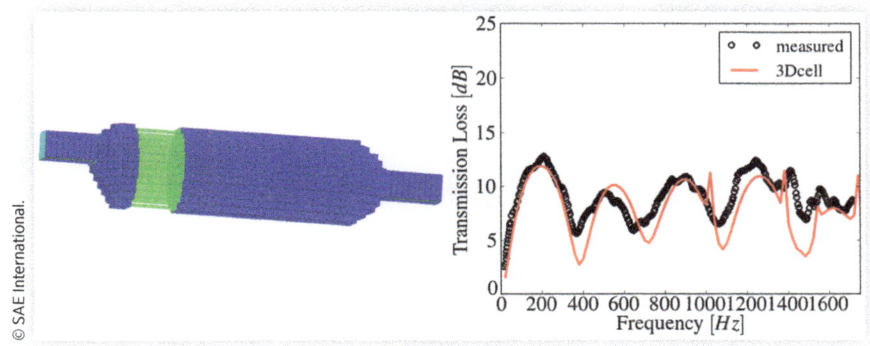

FIGURE 10.40 Quasi-3D reconstruction of the after-treatment system without the catalyst and transmission loss.

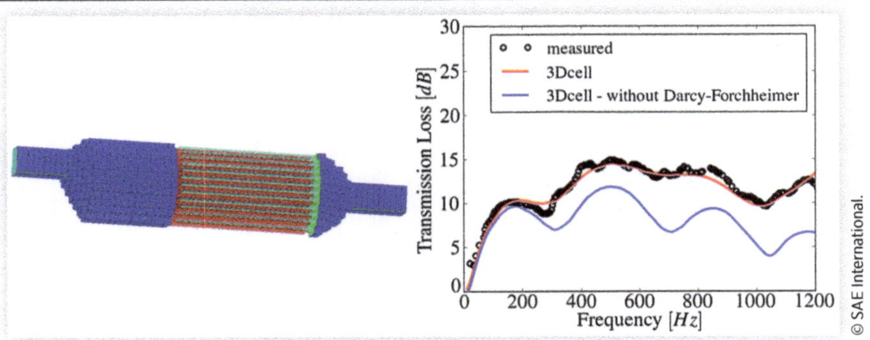

It can be seen that the simulation captures the overall behavior of the device, correctly locating the transparencies and describing the dissipation introduced by the catalytic substrate (resulting in the smoothing of the acoustic function, especially at the pass band frequencies). Moreover, the approach predicts the excitation of higher order modes at 1300 Hz, in agreement with Equation (10.36).

As a second step, the configuration with only the particulate filter substrate included was considered (Figure 10.40). The particulate filter substrate was modeled according to the approach explained in Section 10.4.5.1, where for each macro-cell a couple of inlet and outlet channels is considered. In this way, it was possible to distinguish between the effect of the wave reflection caused by the presence of the plugs and the dissipation introduced across the porous wall. Figure 10.40 shows the comparison between the measured transmission loss and the one calculated by means of the quasi-3D approach, with and without the porous media flow resistance (Darcy and Forchheimer terms). This shows the significant contribution to the noise abatement of the porous material.

Finally, the model for the complete system is considered. The device layout consists of a combination of the two configurations already studied (Figure 10.41). The calculated transmission loss is compared with the measurement results in Figure 10.41, showing a satisfactory agreement in the frequency range of interest.

Moving from the exhaust to the intake system, intercoolers are usually present on modern turbocharged engines and their acoustic behavior needs to be taken into account [44]. The modeling of an air-to-air and an air-to-liquid intercooler will now be considered. The air-to-air intercooler, shown in Figure 10.42, has a geometry in which the inlet and outlet are at an angle of 90 degrees with respect to the tube bundle and are connected

FIGURE 10.41 Quasi-3D reconstruction of the complete after-treatment system and transmission loss.

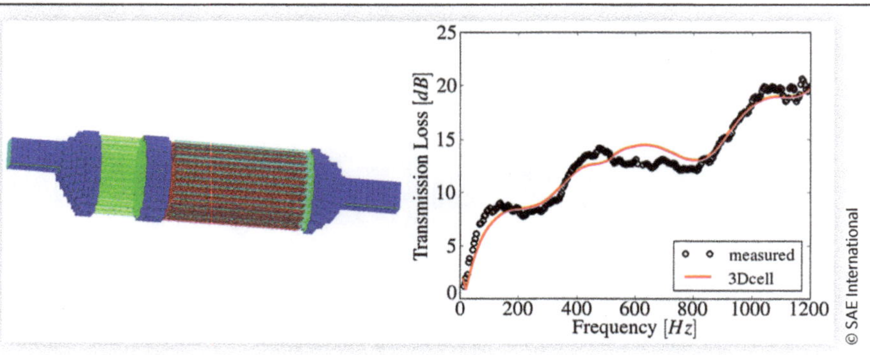

FIGURE 10.42 Quasi-3D reconstruction of the air-to-air inter-cooler and transmission loss.

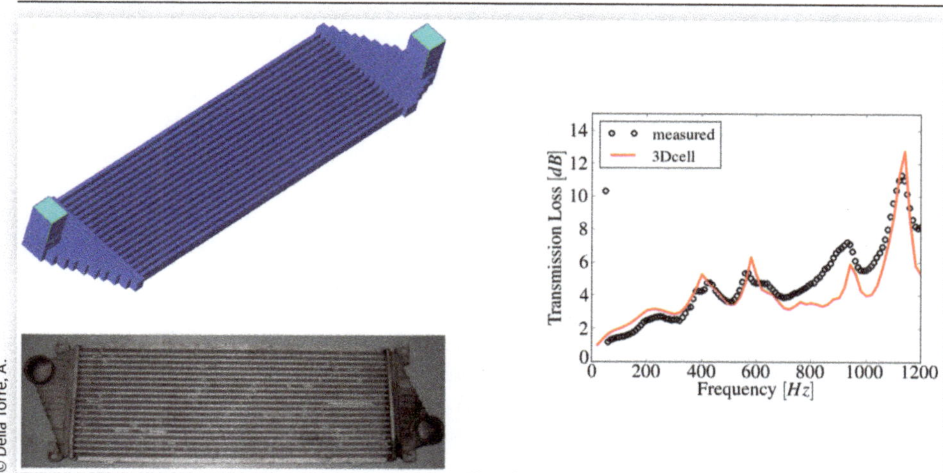

FIGURE 10.43 Quasi-3D reconstruction of the air-to-liquid inter-cooler and transmission loss.

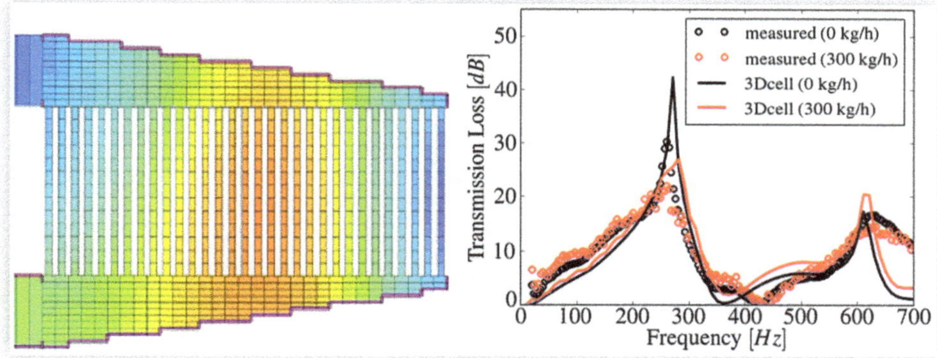

with an offset with respect to each other [46]. The agreement between calculated and measured transmission loss (Figure 10.42) can be regarded as satisfactory.

Finally, the acoustics of an air-to-liquid configuration are simulated considering the presence of mean flow. The configuration (Figure 10.43) is characterized by inlet and outlet ducts perpendicular to the heat exchanger and connected to it by means of two conical collectors [46]. In Figure 10.43, the transmission loss is evaluated in the absence of mean flow and in the case of an air mass flow rate of 300 kg/h. It can be seen that the mean flow increases the dissipation of the acoustic energy, resulting in higher values of the transmission loss with respect to the pass bands.

10.5.6 Computational Runtime

The computational runtime required for the quasi-3D acoustic simulation of common engine components with a complex shape, such as those presented in the previous sections, is usually low when compared to the alternative fully 3D CFD approach. In Table 10.1, the computational runtimes for these approaches are compared for some of the cases previously analyzed.

TABLE 10.1 Comparison of computational runtime [min] for quasi-3D and fully 3D CFD approach.

Case	Figure	Quasi-3D	Fully 3D
Reverse chamber	10.20	4.5	4456
Chamber with perforates	10.21	1.0	5621
Single plug muffler	10.22	2.4	5553

It can be seen that the computational runtime required by the quasi-3D approach is several orders of magnitude lower compared to that required by CFD. This is due to the capability of the quasi-3D to reconstruct the main acoustic lengths of the geometry with a significantly lower number of elements. Moreover, if compared to CFD approaches, it is not required to refine the mesh to model specific parts of the geometry, like perforates, since suitable models are introduced for taking into account the effects of these elements.

10.6 Application to the Engine Modeling

In the previous section, it has been demonstrated that the quasi-3D approach can be an effective tool for the acoustic simulation of the three-dimensional pressure wave propagation inside complex shape silencers. In this section, applications related to the modeling of the entire engine will be considered, in order to describe the phenomena occurring in complex shape devices inserted in the pipe system. For this purpose, the quasi-3D solver has been embedded in a state-of-the-art 1D research code.

10.6.1 1-Cylinder Motorcycle Engine

The first engine system considered [34] is a Piaggio single-cylinder whose specifications are listed in Table 10.2.

The simulation campaign was carried out starting from the generation of the 1D schematic of the whole engine, including the intake and exhaust systems. Figure 10.44 shows a snapshot of the engine schematic along with the two components modeled by means of the 3Dcell approach: the air-box and the rear muffler.

TABLE 10.2 Specifications of the Piaggio single cylinder engine

Engine type	Spark ignition
Number of cylinders	1
Total displacement	124.45 cm^3
Bore	52 mm
Stroke	58.6 mm
Compression ratio	11.2:1
Number of valves per cylinder	3
Air management	Naturally aspirated
Injection system	PFI

FIGURE 10.44 Schematic of the Piaggio motorcycle engine.

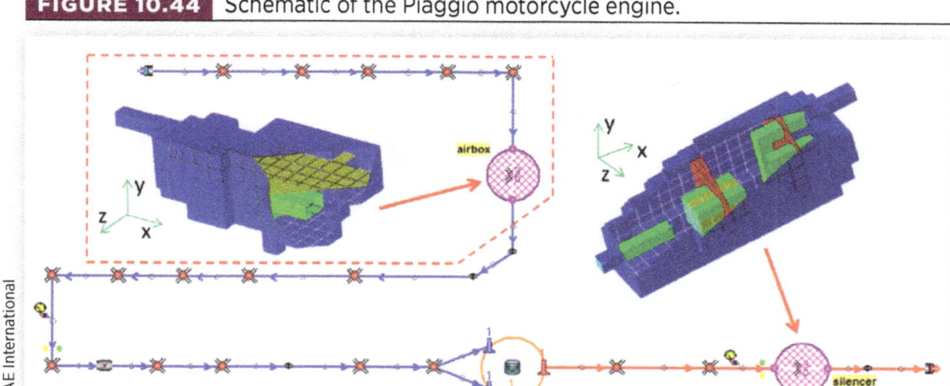

These seemed to be the natural candidates to be modeled by means of the quasi-3D approach. As a matter of fact, the air-box has a 3D shape with the inlet and the outlet made up of curved ducts, located in different planes. Additionally, the filter cartridge is located between the extension of the inlet and outlet pipes, creating a sort of permeable baffle that separates two distinct regions. Figure 10.45 shows the air-box and the equivalent quasi-3D cluster, respectively. The quasi-3D model consists of a network of 314 cells, where the average characteristic length is 20 mm in all coordinate directions. Since it is not possible to exactly follow the surface details with a coarse mesh, the main aim of the network generation was to reproduce the primary acoustic dimensions of the geometry. Moreover, the presence of the filter cartridge has been taken into account by adding a plane of filter-type ports positioned in the middle of the chamber. For the simulation of the porous media, a permeability K_{wall} of 10^{-10} m^2 and a wall thickness w_{wall} of 8 mm have been adopted in Equation (10.25).

Similarly, the silencer, whose geometry and quasi-3D reconstruction are shown Figure 10.46, is characterized by a high degree of complexity: Three resonance chambers have been created inside separated by two rigid baffles. Each chamber is then connected to the next by means of a catalyst and/or resonating elements. The first volume is a series chamber with an extending inlet pipe and is connected to the second chamber by means of a catalyst. In this way, the gas stream is forced to flow through the catalytic converter and is subject to a pressure drop due to the friction of the walls of the tiny channels. Finally, the third chamber is connected to the middle chamber by means of a resonating

FIGURE 10.45 3D view of the Piaggio air-box and network of cells, which reproduce the air-box shape, adopted for the quasi-3D approach.

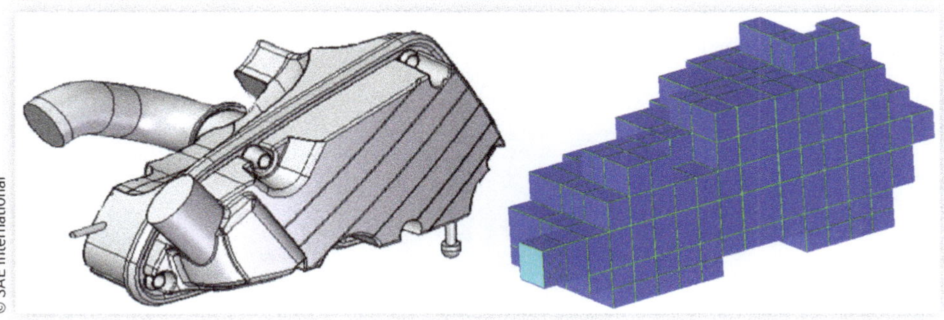

FIGURE 10.46 3D view of the Piaggio silencer and network of cells, which reproduce the muffler shape, adopted for the quasi-3D approach.

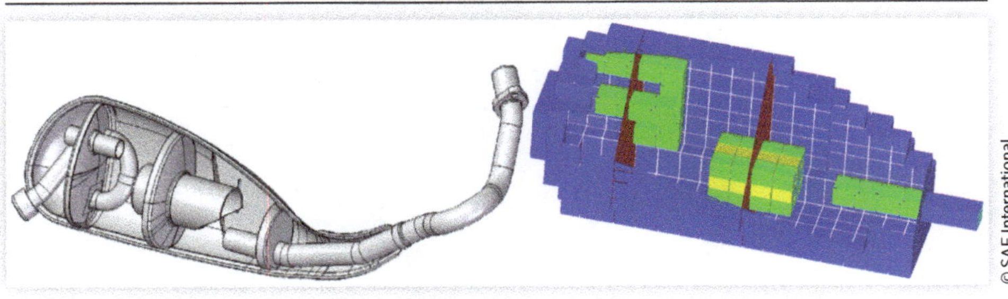

tube that creates a quarter wave resonance effect. The model of the silencer has been built with 780 cells that have an average mesh size of 15 mm. The internal configuration of the silencer depicted in Figure 10.46 consists of different elements: baffles, pipe extensions, and a catalyst. These elements can be easily taken into account using the different types of cells and ports embedded in the model and previously discussed in Section 10.4.

Although the modeling of these components is possible by adopting 1D equivalent schemes or 0D boundary conditions, the application of a quasi-3D technique offers the advantage of a modeling based purely on the geometrical reconstruction of the device's shape and does not need any kind of calibration or corrective length. Moreover, the approach is predictive with respect to the change of any dimension or position of internal elements, for example, baffles, extended pipes, catalysts, or filters.

The numerical simulation has been performed in two steps: (a) engine without the airbox and without the intake line upstream the airbox and (b) engine with the complete intake and exhaust systems. The schematic reported in Figure 10.44 shows in the dashed red line box, the part of the intake removed during step 1. In this case, the trumpet positioned inside the airbox takes the air directly from the ambient. All the simulations have been performed over the entire engine speed range (3000–9250 rpm) in steps of 250 rpm.

Figures 10.47 and 10.48 show an example of the pressure field calculated inside the air-box and the silencer by means of the quasi-3D approach. As previously said, this approach cannot be used to investigate the details of the flow fields; however, it is able to capture the non-uniformities of the thermo-fluid dynamic quantities in the different parts of the component. The pressure field in the vertical plane of the air-box highlights the pressure loss due to the filtration flow through the filter cartridge. The section of the silencer shows the difference in pressure in the three chambers: the pressure loss

FIGURE 10.47 Pressure field inside the air-box along the vertical cutting plane (xy).

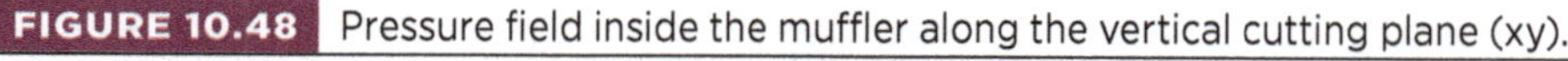

FIGURE 10.48 Pressure field inside the muffler along the vertical cutting plane (xy).

between the first and the second chamber is due to the presence of the catalyst while the one between the second and the third is caused by the restriction represented by the connecting pipe.

In Figure 10.49, the volumetric efficiency curves, calculated by means of the integrated 1D/quasi-3D model, are compared with the experimental data. In each graph, the curves are normalized to the maximum measured value. The agreement between the simulation results and the measured data can be regarded as good, since the trends are correctly described and the absolute values are captured with a maximum difference of 3%. This leads to the conclusion that the adopted models of the intake and exhaust systems are able to correctly describe wave motion in the ducts and correctly predict the pressure losses due to the presence of the airbox and the silencer.

In Figure 10.50, the comparison between the calculated and the measured brake-specific fuel consumption (BSFC) is shown. The capability of describing the BSFC curve depends on the accuracy of the prediction of the air mass trapped inside the cylinder and from the reliability of the combustion and friction models. For the combustion model, the simulations have been conducted by means of an advanced multi-zone model. The engine friction losses (FMEP) used a simplified equation to compute the FMEP as a function of the mean piston speed. Also, for the BSFC curves the agreement between computed and measured data is satisfactory.

The computational runtime required for simulating 25 regimes between 3000 and 9250 rpm is around 15 min when the airbox is removed (case a) and 25 min for the case of the entire engine (case b).

FIGURE 10.49 Comparison between measured and computed volumetric efficiencies with and without airbox; operating point: WOT

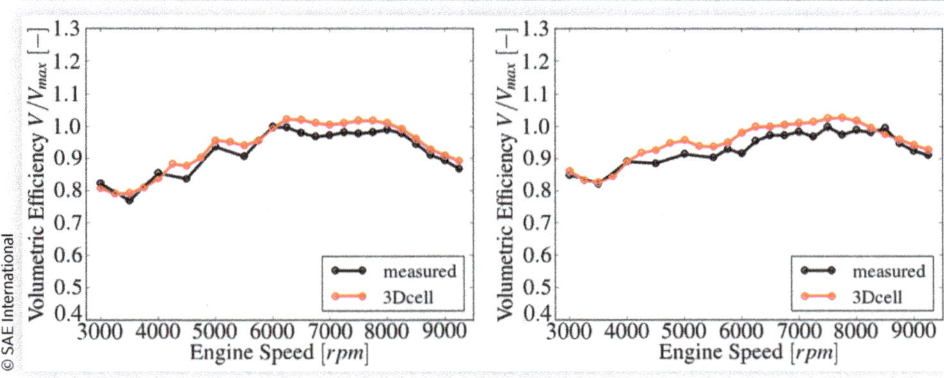

FIGURE 10.50 Comparison between measured and computed BSFC with and without airbox; operating point: WOT

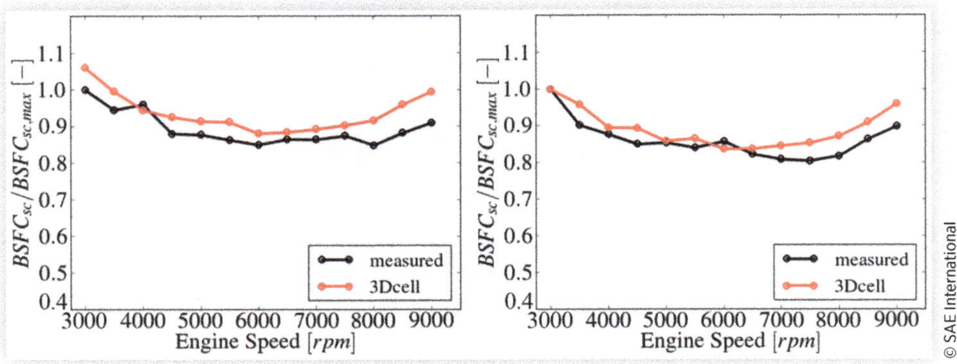

10.6.2 4-Cylinder Motorcycle Engine

This section will describe the modeling of a high-performance Aprilia V4 engine [47]. The engine specifications are summarized in Table 10.3.

The air-box and the silencer of the Aprilia V4 engine have a very complex layout, in order to satisfy all the requirements of the engine manufacturer. Their complex shape, along with the fact that they need, during the design phase, to be optimized in order to meet different targets (performance, noise reduction, and emission control) make these devices good candidates to be modeled by mean of the quasi-3D approach. The complete schematic of the whole engine system, including the components modeled by means of the quasi-3D approach, is shown in Figure 10.51.

The air-box has a 3D shape with two inlets and four outlets that make up the intake runners. The quasi-3D model of the air-box consists of a network of 1634 cells, where the average characteristic length is 20 mm in all directions. During the modeling phase, particular care was taken in order to respect the main dimensions of the internal layout, especially the distances between the axis of the intake runners, which have an influence on the interaction among the cylinders. The filter cartridge is placed between the inlet and outlet pipes and has been taken into account by adding a plane of filter-type ports. The quasi-3D reconstruction of the air-box allowed the main characteristic lengths to be captured, while all the details whose dimensions are smaller than the cell size (or better port characteristic length) are missed.

Similarly, the silencer is characterized by a high degree of complexity: three resonance chambers are created separated by two rigid baffles. Each chamber is then connected to the others by means of ducts and perforations in the baffles. The first volume, which

TABLE 10.3 Specifications of the Aprilia V4 engine

Engine type	Spark ignition
Number of cylinders	4 - V65°
Total displacement	999.6 cm^3
Bore	78 mm
Stroke	52.3 mm
Compression ratio	13:1
Number of valves per cylinder	4
Air management	Naturally aspirated
Injection system	PFI

FIGURE 10.51 Schematic of the Aprilia motorcycle engine, generated for the thermo-fluid dynamic simulations by means of the coupled 1D/quasi-3D model.

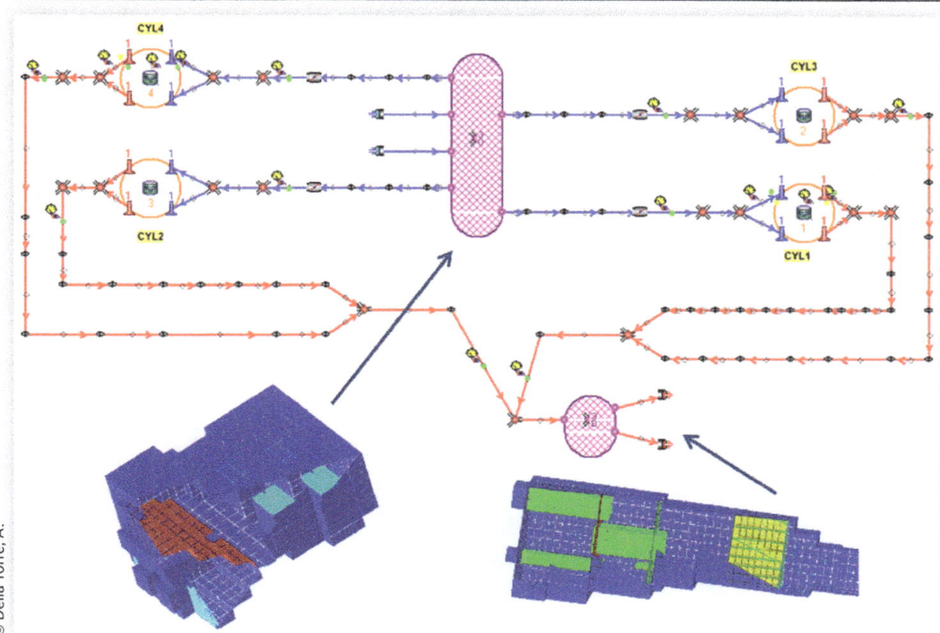

acoustically acts as a series chamber, also contains the catalyst. The gas stream in the first chamber is forced to flow through the catalytic converter, which causes a pressure drop. The model of the silencer was built using 2934 cells having an average mesh size of 1.5 cm.

In order to assess the reliability of the proposed approach for the prediction of unsteady flows in intake and exhaust systems, the computed results were compared to measurements carried out at Aprilia laboratories. The experimental data consist of high-frequency pressure pulses taken from three different points: downstream of the throttle valve in the intake runners of cylinders 1 and 4 and downstream of cylinder 4, in the exhaust duct. Moreover, low-frequency back-pressure measurements have been performed upstream of the silencer along with the volumetric efficiency. This last measurement was obtained by means of lambda sensors placed downstream of each cylinder.

The engine architecture (4-V65°) represents a constraint for the positioning of the intake trumpets inside the air-box, forcing their placement in a close-coupled configuration. For this reason, the intake phase of each cylinder may be strongly affected by pressure perturbations generated by the intake valve opening of the other cylinders. This typically results in effects that can be both beneficial or detrimental for each cylinder.

Several operating conditions were considered; in particular, the range from 3500 to 12,500 rpm was covered at full load with steps of 1000 rpm. However, for the sake of brevity, only two operating points will be discussed. They have been selected as representative of low engine speed operating conditions (4500 rpm) and of maximum power (12,500 rpm). The comparisons between the measured pressure pulses downstream of the throttle valve and the calculations carried out adopting the 1D-quasi-3D integrated code are shown in Figures 10.52 through 10.55.

FIGURE 10.52 Cylinder 1 - intake: 4500 rpm.

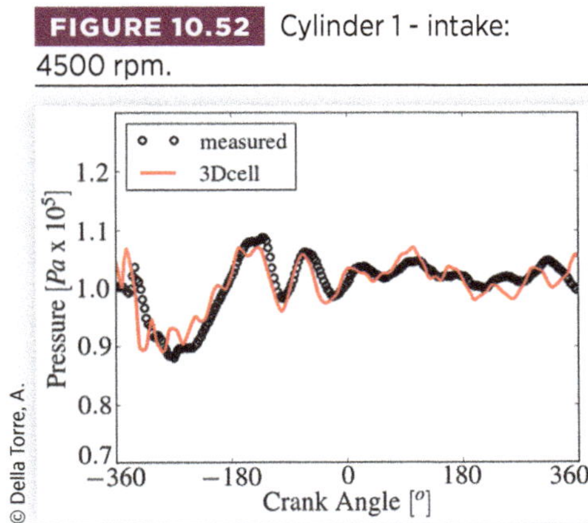

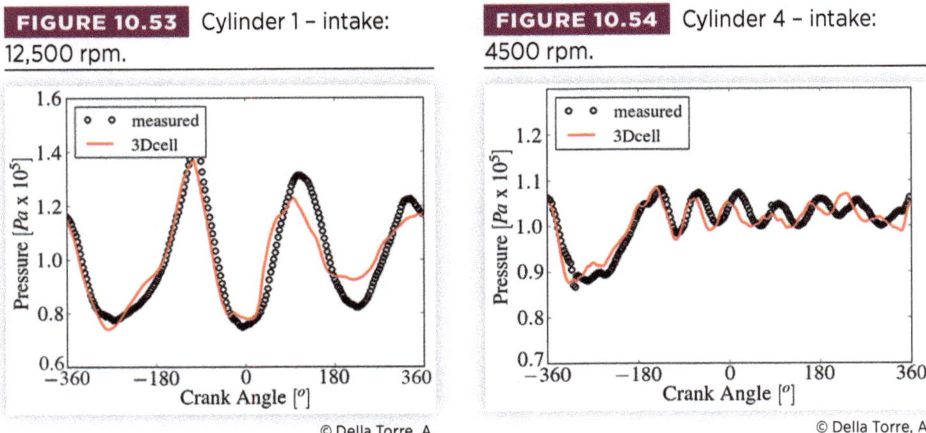

FIGURE 10.53 Cylinder 1 - intake: 12,500 rpm.

© Della Torre, A.

FIGURE 10.54 Cylinder 4 - intake: 4500 rpm.

© Della Torre, A.

It can be seen that at high engine revolution speed, the quasi-3D model is remarkably predictive, exhibiting a good agreement with the measured data both in the absolute value and phase of the wave motion.

At low engine speed operating conditions, 4500 rpm, there is a slight delay of the wave motion, which causes discrepancies between the calculated curves and the measured ones. The explanation of this curve shift resides in the absence of a sub model that accounts for the latent heat of evaporation of the injected fuel droplets. This phase transition, during which the air-fuel mixture formation takes place, removes the energy needed from the droplet itself, which becomes colder, cooling down the surrounding gas by convection.

The comparison between the measured and calculated pressure traces downstream of cylinder 4 are shown in Figures 10.56 and 10.57. The prediction of the wave motion inside the exhaust system is affected by the presence of the muffler and of the 4 into 2 into 1 junction. This last component was modeled resorting to a 1D reconstruction of the pipes in order to capture the overall pressure loss due to the curvature of the pipes, neglecting the influence of small geometric details. It can be seen that the overall prediction is satisfactory, suggesting that the model is able to capture the pressure waves reflected by the silencer.

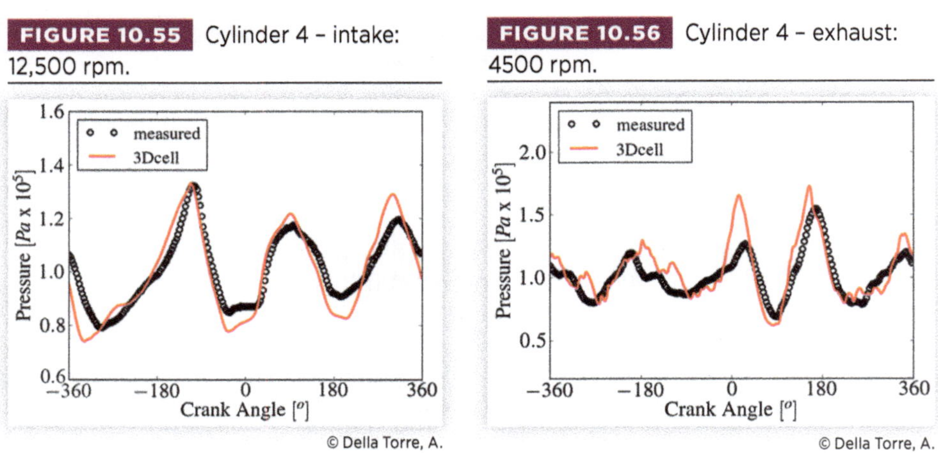

FIGURE 10.55 Cylinder 4 - intake: 12,500 rpm.

© Della Torre, A.

FIGURE 10.56 Cylinder 4 - exhaust: 4500 rpm.

© Della Torre, A.

FIGURE 10.57 Cylinder 4 - exhaust: 12,500 rpm.

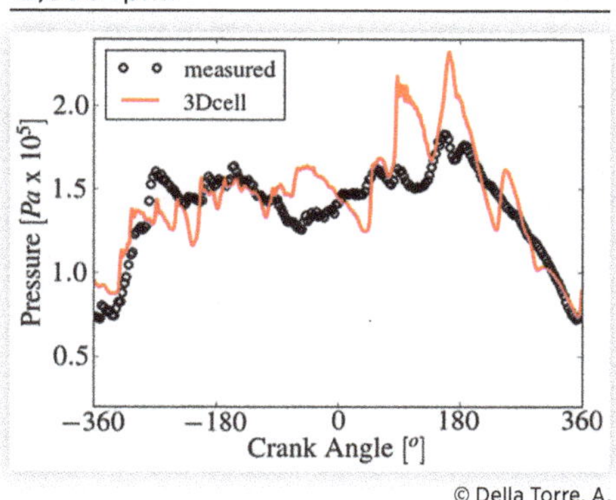

FIGURE 10.58 Back pressure upstream of the silencer.

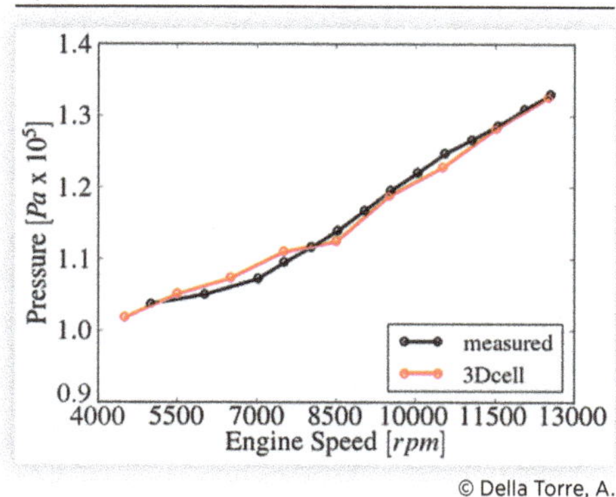

A further validation comes from the analysis of the back pressure (Figure 10.58). The back-pressure trend increases with the engine revolution speed and is well matched by the computed profile.

Due to the quality of the prediction of the wave motion inside the intake and the exhaust systems also, the prediction of the volumetric efficiency is expected to be in agreement with the measured values. Figure 10.59 shows the comparison between the measured and calculated volumetric efficiency curves, in which the volumetric efficiency has been normalized with respect to the maximum experimental value. The agreement between the simulation results and the measured data can be regarded as good, since the trends are correctly described and the absolute values are captured with a maximum difference of 3%.

FIGURE 10.59 Volumetric efficiencies.

Volumetric Efficiency V/V_{max} [−]; measured; 3Dcell; Engine Speed [rpm]

Although the quasi-3D approach can be regarded as a simplification of a 3D CFD solver, it is also possible to reconstruct the shape of pressure iso-surfaces inside the calculation domain. Figure 10.60 shows an example of the pressure field calculated

FIGURE 10.60 Pressure field inside the airbox and silencer.

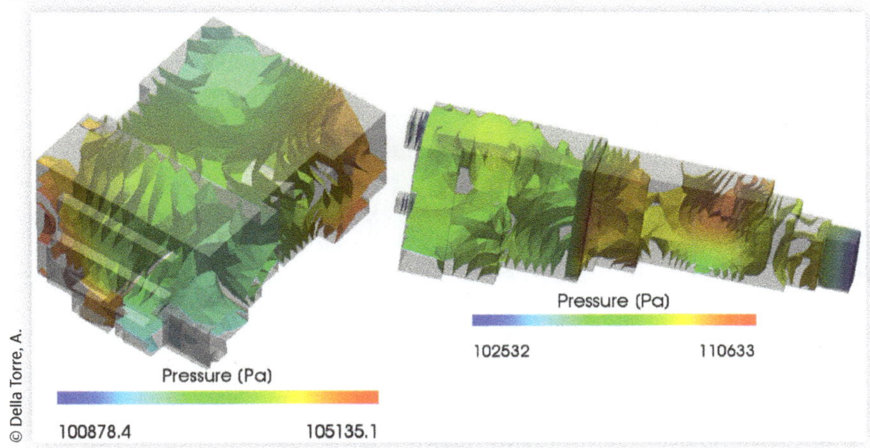

FIGURE 10.61 Sound pressure level calculated at 1 m from the tail pipe at 10,500 rpm.

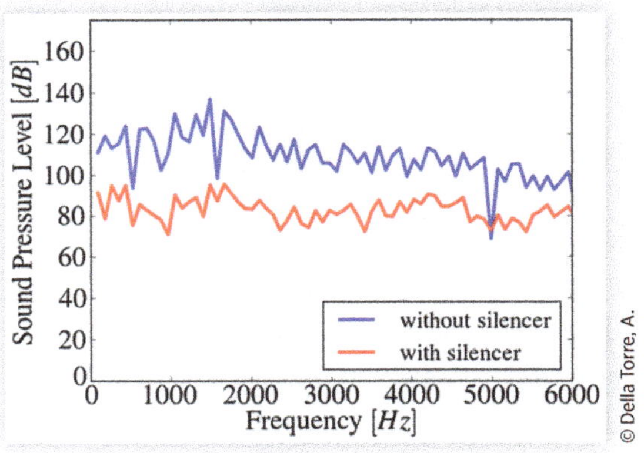

TABLE 10.4 Comparison of computational runtime [min] for quasi-3D and fully 3D CFD approach

Approach	Runtime
1D	15 min
1D/3D CFD	20 h
1D/quasi-3D	45 min

© Della Torre, A.

inside the air-box and the silencer by means of the 3Dcell approach. The flow pattern is clearly 3D and shows nonplanar wave effects. This can be used to estimate which resonance modes are excited and which are not. In particular, looking at the air box, it can be noticed that the main resonances are occurring in two directions, namely, length and width, whereas the wave motion parallel to the direction of the four runners is not excited.

The possibility of capturing the nonlinear propagation of pressure wave inside the pipe systems is fundamental in evaluating the noise abatement performed by a certain silencing device. In Figure 10.61 the sound pressure is shown, calculated 1 m from the tail pipe at 10,500 rpm full load, for two different configurations, without and with the silencer.

The computational runtime required for simulating 16 operating points between 4500 and 12,500 is around 45 min when both the air-box and the silencer are modeled by means of the quasi-3D approach. In addition, in Table 10.4 the computational runtimes related to the 1D, 1D-3D, and 1D-quasi-3D simulations are compared. The 1D-3D simulation has been performed coupling the 1D code GASDYN to the CFD software OpenFOAM, modeling only the airbox by means of 3D CFD.

10.7 Conclusions

In this chapter, the details of the quasi-3D approach have been reviewed and discussed, providing several examples of application. This modeling approach is based on the geometrical reconstruction of the shape of the component, in order to describe its main characteristic lengths in three dimensions. The scope of applicability of this approach consists primarily of the simulation of the unsteady pressure waves propagation and reflection inside complex shape devices. In this framework, the details of the geometry have a minor influence on the overall behavior of the component and, therefore, they can be neglected, making it possible to reconstruct the geometry using a relatively low number of elements.

However, an accurate description of the acoustics and fluid-dynamics phenomena requires the adoption of a numerical technique suitable to operate on the coarse quasi-3D network. The numerical method should be able to closely reproduce the pressure wave propagation in order to provide a sufficient characterization of the acoustic behavior of the I.C. engine intake and exhaust system. At the same time, it should require a low computational burden. It has been shown in this chapter that methods based on a staggered leapfrog technique, eventually coupled with specific approaches for the limitation of the numerical overshoots, represents a good compromise with respect to these requirements.

Moreover, the quasi-3D approach can be easily extended to the modeling of common intake and after-treatment devices, including perforates, sound absorptive material, filtering media, catalysts, and particulate filters. Different applications are presented showing test cases from literature, research projects, and actual automotive applications. Component-scale acoustic simulations demonstrate the capability of the approach to predict the acoustic performance of devices exhibiting a different range of complexity, from simple chambers to after-treatment devices and inter-coolers. Moreover, even for the most complex configurations, the computational runtime is several orders of magnitude lower than traditional fully 3D CFD approaches. Finally, the possibility of adopting the quasi-3D approach for the modeling of complex shape devices inserted in the intake and exhaust engine system in the framework of the entire engine simulation has been investigated. The study has demonstrated that the quasi-3D approach technique can effectively be adopted as an alternative approach and/or extension of the traditional modeling based on 0D approximations or 1D equivalent schemes. Compared to these last approaches, the quasi-3D has the advantage of being based on a pure geometrical reconstruction, which does not need any kind of calibration, making it possible to adopt the approach for a preliminary optimization of the device.

Acknowledgments

Parts of the measurement work were performed by the former Acoustic Competence Center, Graz, meanwhile merged with the Virtual Vehicle Research and Test Center, Graz. Both institutions are funded by the Austrian government, the government of Styria, the Styrian Economy Support (SFG).

Definitions, Acronyms, and Abbreviations

Symbols

A - Surface area [m^2]
a - Sound velocity [m/s]
α - Length of channel edge [m]
β - Forchheimer coefficient [-]
$\mathbf{D}$ - Diffusion tensor [kg/(m s^2)]
D - Diameter [m]
e - Energy [J]
ε - Diffusion coefficient [m^2/s]
f - Friction factor [-]
$\mathbf{h}$ - Specific enthalpy [J/kg]
K - Permeabilty [-]
L - Length [m]
M - Mach number [-]
M_s - Source term on momentum equation
μ - Dynamic viscosity [Pa·s]
p - Pressure [Pa]
ρ - Density [kg/m^3]
R - Absorptive material resistivity [N·s/m^4]
Re - Reynolds number [-]
r - Radius [m]
T - Temperature [K]
t - Time [s]
Δt - Time step [s]
U - Velocity [m/s]
V - Volume [m^3]

CHAPTER 10

w - Thickness [m]
Δx - Cell length [m]
C - Correction
cu - Control volume
h - Perforate hole
L - Cell left of boundary
m - Mean value
n - Time level index
R - Cell right of boundary
reg - Region
x, y, z - Coordinate directions

Acronyms

0D - One dimensional
1D - Two dimensional
3D - Three dimensional
BSFC - Brake-specific fuel consumption
CFD - Computational fluid dynamics
CFL - Courant–Friedrichs–Lewy
DTM - Diffusion term momentum
FMEP - Friction mean effective pressure
I.C. - Internal combustion
PFI - Port fuel injected
RANS - Reynolds-averaged Navier-Stokes
TL - Transmission loss
WOT - Wide open throttle

References

1. http://www.polilink.polimi.it/en/casi-di-successo/gasdyn-software-2/.
2. https://www.avl.com/boost.
3. https://www.cmt.upv.es/OpenWam.aspx.
4. http://www.optimum-power.com/.
5. https://software.ricardo.com/products/wave.
6. https://www.gtisoft.com/.
7. http://www.lotuscars.com/engineering/engineering-software.
8. Winterbone, D.E. and Pearson, R.J., *Theory of Engine Manifolds Design* (Professional Engineering Publishing, 1999).
9. Winterbone, D.E. and Pearson, R.J., *Design Techniques for Engine Manifolds, Wave Action Methods for IC Engines* (Professional Engineering Publishing, 1999).
10. Blair, G.P., *Design and Simulation of Four Stroke Engines* (1999), ISBN:0-7680-0440-3.
11. Alten, H. and Illien, M., "Demands on Formula One Engines and Subsequent Development Strategies," SAE Technical Paper 2002-01-3359, 2002, doi:10.4271/2002-01-3359.
12. Onorati, A., Montenegro, G., D'Errico, G., and Piscaglia, F., "Integrated 1D-3D Fluid Dynamic Simulation of a Turbocharged Diesel Engine with Complete Intake and Exhaust Systems," SAE Technical Paper 2010-01-1194, 2010, doi:10.4271/2010-01-1194.

13. Claywell, M., Horkheimer, D., and Stockburger, G., "Investigation of Intake Concepts for a Formula SAE Four-Cylinder Engine Using 1D/3D (Ricardo WAVE-VECTIS) Coupled Modeling Techniques," SAE Technical Paper 2006-01-3652, 2006, doi:10.4271/2006-01-3652.

14. Morel, T., Morel, J., and Blaser, D., "Fluid Dynamic and Acoustic Modeling of Concentric-Tube Resonators/Silencers," SAE Technical Paper 910072, 1991, doi:10.4271/910072.

15. Morel, T., Silvestri, J., Georg, K.-A., and Jebasinksi, R., "Modelling of Engine Exhaust Acoustics," SAE Technical Paper 1999-01-1665, 1999, doi:10.4271/1999-01-1665.

16. Fairbrother, R., Liu, S., Dolinar, A., Montenegro, G. et al., "Development of a Generic 3D Cell for the Acoustic Modelling of Intake and Exhaust Systems," *ICSV16*, Krakow, Poland, 2009.

17. Montenegro, G., Della Torre, A., and Onorati, A., "A General 3D Cell Method for the Acoustic Modelling of Perforates with Sound Absorbing Material for I.C. Engine Exhaust Systems," *17th International Congress on Sound and Vibration 2010, ICSV 2010.*

18. Munjal, M.L., "*Acoustics of Ducts and Mufflers*, 2nd ed.. (Wiley, 2014), ISBN:978-1-118-44312-5.

19. Versteeg, H.K. and Malalasekera, W., *An Introduction to Computational Fluid Dynamics* (Longman Scientific & Technical, 1995).

20. Chapman, M., Novak, J.M., and Stein, R.A., "Numerical Modeling of Inlet and Exhaust Flows in Multi-Cylinder Internal Combustion Engines," *ASME Winter Annual Meeting*, Phoenix, AZ, November 14-19, 1982.

21. Toro, E.F., *Riemann Solvers and Numerical Methods for Fluid Dynamics* (New York: Springer, 1997).

22. Guo, Y., *Experimental Investigation of Acoustic Properties of Mufflers with Perforated Pipes* (Stockholm, Sweden: KTH, 2007).

23. Torregrosa, A.J., Broatch, A., Payri, R., and Gonzales, F., "Numerical Estimation of End Corrections in Extended-Duct and Perforated-Duct Mufflers," *ASME Transaction* 121 (1999): 302-308.

24. Elnemr, Y., *Investigation of the Acoustic Performance of Dissipative Mufflers: Influence of Different Absorbing Materials and Packing Densities*(Stockholm, Sweden: KTH, 2007).

25. Fairbrother, R. and Jebasinski, R., "Development and Validation of a Computer Model to Simulate Sound Absorptive Materials in Silencing Elements of Exhaust Systems," *IMechE C577/037/2000.*

26. Mechel, F.P. and Ver, I.L., Sound-Absorbing Materials and Sound Absorbers, in *Noise and Vibration Control Engineering: Principles and Applications* (John Wiley & Sons, 1992).

27. Elnemr, Y. et al., "Investigation and Modeling of Catalytic Converter Acoustical Properties in Time Domain," *ICSV 17*, Cairo, Egypt, 2010.

28. Churchill, S.W., "Friction-Factor Equation Spans all Fluid Flow Regimes," *Chemical Engineer* (1977): 91-92.

29. Masoudi, M., Konstandopoulos, A.G., Nikitidis, M., Skaperdas, E. et al., "Validation of a Model and Development of a Simulator for Predicting the Pressure Drop of Diesel Particulate Filters," SAE Technical Paper 2001-01-0911, 2001, doi:10.4271/2001-01-0911

30. Masoudi, M., "Hydrodynamics of Diesel Particulate Filters," SAE Technical Paper 2002-01-1016, 2002, doi:10.4271/2002-01-1016.

31. Eriksson, L.J., "Higher Order Mode Effects in Circular Ducts and Expansion Chambers," *The Journal of the Acoustical Society of America* 68, no. 2 (1980): 545-550.

32. Eriksson, L.J., "Effect of Inlet/Outlet Locations on Higher Order Modes in Silencers," *The Journal of the Acoustical Society of America* 72, no. 4 (1982): 1208-1211.

33. Della Torre, A., Montenegro, G., Cerri, T., and Onorati, A., "A 1D/Quasi-3D Coupled Model for the Simulation of I.C. Engines: Development and Application of an Automatic Cell-Network Generator," *SAE Int. J. Engines* 10, no. 2 (2017): 471-482, doi:10.4271/2017-01-0514.

34. Montenegro, G., Onorati, A., Cerri, T., and Della Torre, A., "A Quasi-3D Model for the Simulation of the Unsteady Flows in I.C. Engine Pipe Systems," SAE Technical Paper 2012-01-0675, 2012, doi:10.4271/2012-01-0675.

35. Veloso, R., Fairbrother, R., and Elnemr, Y., "A 3D Linear Acoustic Network Representation of Mufflers with Perforated Elements and Sound Absorptive Material," SAE Technical Paper 2017-01-1789, 2017, doi:10.4271/2017-01-1789.

36. Guo, Y., *Investigation of Perforated Mufflers and Plates* (Stockholm, Sweden: KTH, 2009).

37. Sullivan, J. and Crocker, M., "Analysis of Concentric-Tube resonators Having Unpartitioned Cavities," *Journal of the Acoustical Society of America* 64, no. 1 (1978): 207-215.

38. Montenegro, G., Della Torre, A., Onorati, A., Fairbrother, R. et al., "Development and Application of 3D Generic Cells to the Acoustic Modelling of Exhaust Systems," SAE Technical Paper 2011-01-1526, 2011, doi:10.4271/2011-01-1526.

39. Dolinar, A. et al., "Linear and No-Linear Acoustic Modelling of Plug Flow Mufflers," *15th ICSV*, Daejeon, Korea, 2008.

40. Fairbrother, R. and Dolinar, A.," Acoustic Simulation and Measurement of a Muffler with a Perforated Baffle and Absorptive Material," *National Symposium on Acoustics*, India, 2007.

41. Fairbrother, R., Tonkovic, D., Putz, N. and Dolinar, A., "Automatic 3D Mesh Generation for more Accurate Prediction of the Acoustics of Intake and Exhaust Components," *Proceedings of the 20th International Congress on Sound and Vibration*, Bangkok, Thailand, July 7-11, 2013.

42. Fairbrother, R., Åbom, M., Elnady, T., and Ollivier, F., "Linear Acoustic Simulation and Experimental Characterisation of a Modular Automotive Muffler," *ICSV13*, Vienna, Austria, 2006.

43. Fairbrother, R. and Varhos, E., "Acoustic Simulation of an Automotive Muffler with Perforated Baffles and Pipes," SAE Technical Paper 2007-01-2206, 2007, doi:10.4271/2007-01-2206.

44. Montenegro, G., Della Torre, A., Onorati, A., Fairbrother, R. et al., "Quasi-3D Acoustic Modelling of Common Intake and Exhaust Components," *19th International Congress on Sound and Vibration 2012, ICSV 2012, 3*, 2012, 2430-2437.

45. Montenegro, G., Onorati, A., Della Torre, A., and Torregrosa, A.J., "The 3Dcell Approach for the Acoustic Modeling of After-Treatment Devices," *SAE Int. J. Engines* 4, no. 2 (2011): 2519-2530, doi:10.4271/2011-24-0215.

46. Veloso, R. and Elnemr, Y., "Linear Acoustic Multiport Modeling of Automotive Intercoolers," *Internoise 2013*.

47. Montenegro, G., Cerri, T., Della Torre, A., and Onorati, A., "Modeling the Unsteady Flows in I.C. Engine Pipe Systems by Means of a Quasi-3D Approach," *Proceedings of the ASME Internal Combustion Engine Division Spring Technical Conference (ICES2012)*, Turin, 6/5/2012-9/5/2012, 2012, 1-13, ISBN:9780791844663

CHAPTER 11

1D Simulation Models for Aftertreatment Components

Federico Millo, Santhosh Gundlapally, Wen Wang, and Syed Wahiduzzaman

CHAPTER 11

11.1 Introduction

There are two broad and interdependent classes of choices available to minimize tailpipe emissions. One of the choices is aimed at the minimization of engine out emissions through optimization of design (e.g., compression ratio, spark/injector location, and in-cylinder geometry) and operating variables (e.g., exhaust gas recirculation [EGR], valve, injection/spark timing, stoichiometry, and injection shaping). However, in order to meet increasingly stringent standards, the second class of actions has to be pursued also. In this class of actions/choices, mitigating devices (i.e., aftertreatment (AT) reactors) are installed downstream of the engine, in the exhaust line.

Catalytic monolith reactors are widely used in the AT applications due to their several advantages over the conventional packed bed reactors [1]. A monolith reactor is a unitary structure usually made of ceramic or metallic material and consists of a large number of parallel flow channels with thin walls (~100 μm) on which a catalyst is deposited as a thin layer (~10 μm). This thin layer of catalyst is usually referred to as washcoat due to the catalyst deposition process. AT components can be classified into two groups based on the nature of the flow inside monolith channels: (i) flow-through monoliths and (ii) wall flow monoliths (WFMs). Flow-through monoliths are used in three-way catalyst (TWC), diesel oxidation catalyst (DOC), selective catalytic reduction (SCR), lean NOx trap (LNT), and ammonia slip catalyst (ASC) applications. WFMs are used in diesel particulate filter (DPF) and gasoline particulate filter (GPF) applications. Discussion of the catalyst technology is outside the scope of this chapter due to the limited space available and interested readers can refer to Heck [1].

An AT system will need a dedicated control system in order to manage various operational requirements (e.g., diesel exhaust fluid (DEF), soot regeneration, and NH_3 storage). Thus, especially in the concept and early design phases, it is imperative that an adequate

FIGURE 11.1 Sample of 1D AT model development and calibration protocol, from synthetic gas bench (SGB) to full-size validation.

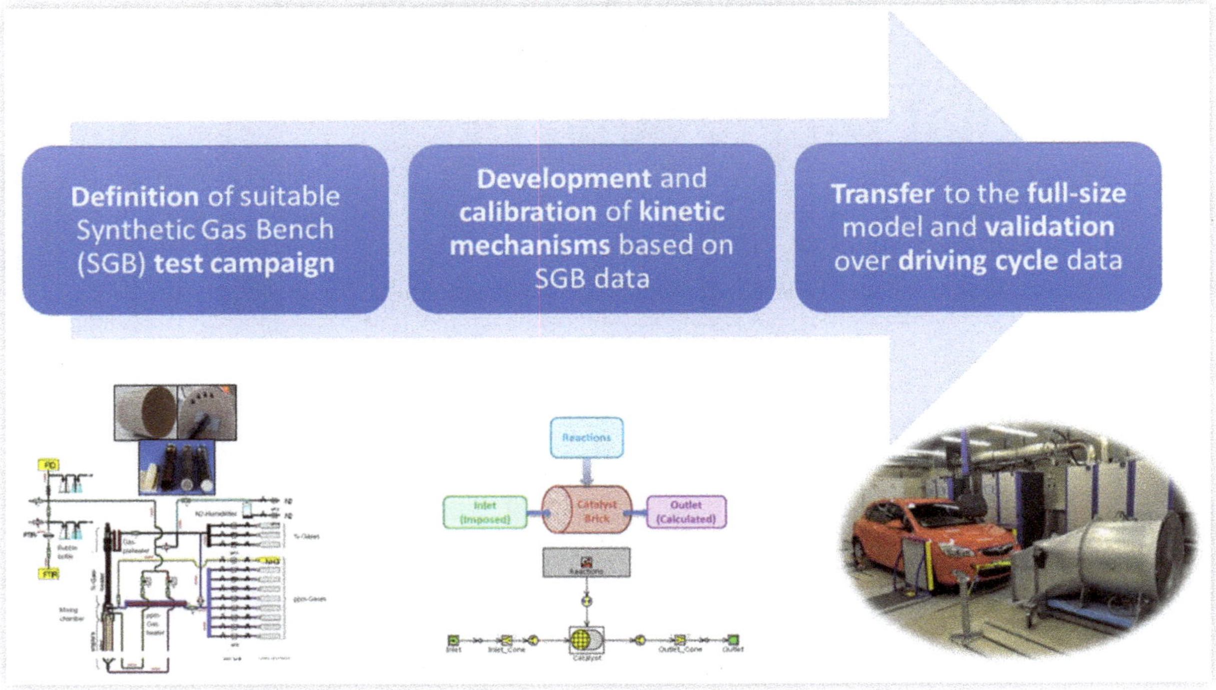

level of design and operating options are evaluated. At this point of the development cycle, much of the hardware may not be available for testing and, even if it were, it would likely to be prohibitively expensive. Consequently, the virtual testing of design options is the only viable option and, in that, 1D simulation tools could be an effective component of the available design tools.

One of the challenges in modeling the engine AT system is to correctly represent the overlapping of several phenomena in a wide range of different operating conditions, including transient operations with complex gas mixtures. To this aim, a dedicated experimental campaign must be designed to characterize the different reaction pathways on catalyst core samples, extracted from the full-size monolith. These data are then used to calibrate kinetic parameters that are finally transferred to the full-size component model for the validation over driving cycle data, thus considering the performance of the AT system with real exhaust gas conditions as the input (Figure 11.1). Such models are essential in industrial applications, especially in the early phase of powertrain development, which could be used for multiple purposes, including sensitivity analysis of different design parameters (platinum group metals [PGM] loading, dimensions etc.), development and optimization of control strategies, preliminary assessment of the capability of a powertrain architecture to fulfill legislation requirements, and virtual assessment of the AT system performance over real driving conditions.

In Section 11.1.1, the literature pertaining to the modeling of AT reactors is briefly reviewed.

11.1.1 Catalyst Technologies

Catalyst formulations used in the monoliths are tailored to a specific application and the exact composition of the catalyst is not usually disclosed by commercial catalyst suppliers. In this section, a brief overview of the commonly used catalyst technologies in the AT applications is given. Interested readers are recommended to consult the book by Heck [1] for further details.

TWC, which oxidizes the unburned hydrocarbons and CO while simultaneously reducing NO, is used to treat the exhaust gas from gasoline engines. TWC is quite effective at stoichiometric conditions and hence gasoline engines are operated at stoichiometric conditions facilitated by a lambda sensor. Current TWC formulations contain Pt/Pd/Rh precious metals and an oxygen storage component (OSC) made of ceria/zirconium. Due to the inherent nature of the feedback controller, the engine cannot be operated exactly with a stoichiometric mixture and therefore operates with an air-to-fuel ratio fluctuating around the stoichiometric value. TWC stores the oxygen during the lean conditions and uses the stored oxygen to oxidize HC and CO during fuel-rich operating conditions. TWC kinetic models can be found in Montenegro et al. [2].

DOC is used to oxidize the unburned hydrocarbons and CO from diesel exhaust. DOC does not reduce NO like TWC but rather oxidizes them to NO_2, which could be beneficial for downstream AT components. DOC formulations commonly have Pt/Pd metals as active components and may also contain hydrocarbon absorbing materials that store unburned hydrocarbons during the cold start and release them once it reaches the light-off temperature. DOC plays an important role in the integrated emission control system, since it could serve multiple purposes. For instance, DOC could be used to produce a specific ratio of NO_2/NO, needed for the efficient operation of downstream SCR and DPF systems. DOC kinetic models can be found in Millo et al. [3].

LNT can be used to reduce the NOx from lean burn gasoline/diesel engines. LNT does not need an external reducing agent unlike SCR but rather depends on the engine control management for its operation. LNT stores NOx during the fuel-lean operations and engine is periodically run fuel-rich for a short time to provide reducing agents that react with the stored NOx. LNT formulations have PGM, OSCs such as ceria, and NOx storage components such as barium. LNT kinetic models can be found in Millo et al. [4].

SCR is used to reduce the NOx from the diesel exhaust gas with the help of a reducing agent. Aqueous urea solution injected at the inlet of the SCR system is commonly used as a reducing agent. Based on the exhaust temperatures encountered, either vanadia- or zeolite-based catalysts are used. The zeolite catalysts are active over a wider operating temperature window than the vanadia-based catalysts but are more prone to sulfur poising. SCR kinetic models can be found in Millo et al. [5].

ASC based on platinum (Pt) is used downstream of a SCR reactor to selectively oxidize NH_3 slipping from a SCR reactor to N_2. ASC also produces some NO, which is undesirable. Dual-layered catalysts, where a SCR layer is coated on top of a Pt layer, are used to reduce the NO formation resulting in higher yield for N_2. Kinetic models for ASC can be found in Colombo et al. [6].

Particulate filters (PFs) are widely used to filter the particles produced from the engine, with efficiencies in the order of 90%. Applications of PF include DPFs and GPFs. PFs are usually made of a ceramic monolith, in which the gas flow is forced through the porous walls where the filtration of particles occurs. PFs are often catalyzed to accelerate the oxidation reactions of the accumulated particles, thus promoting the regeneration of the filter.

Finally, some of the previously mentioned systems can also be combined in a single component, such as TWC on a gasoline particulate filter, DOC on a diesel particulate filter, or SCR on Filter (SCRoF), which combines a SCR with a DPF [5].

11.1.1.1 FLOW THROUGH MONOLITHS

Figure 11.2 shows the important physical and chemical processes taking place in a wash-coated monolithic reactor channel; reactants diffuse from the bulk gas phase to the

FIGURE 11.2 Overview of important physical and chemical processes occurring in a washcoated monolith channel.

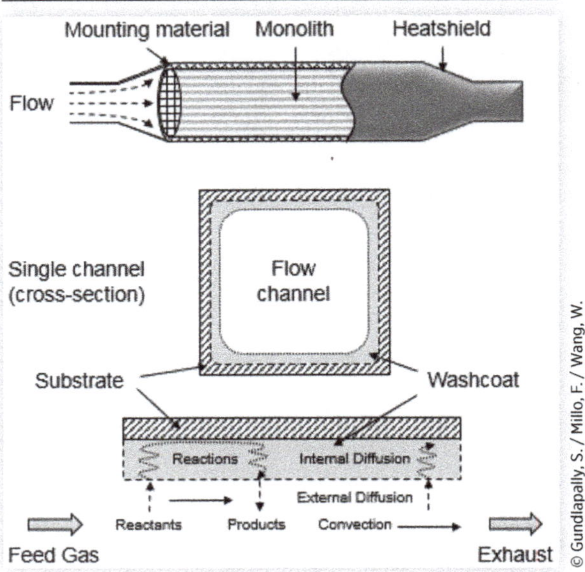

washcoat surface (external mass transfer), reactants diffuse through the washcoat pores (pore diffusion), surface reactions occur at the active sites (adsorption/reaction/desorption), and finally products diffuse back through the washcoat into the bulk gas phase. Depending on the geometric, kinetic, and operating conditions, one or a combination of these steps may control the conversion rate of reactant pollutants. For example, as the inlet gas temperature increases, the controlling step may change from a kinetic-controlled regime at low temperature (before light-off) to an external mass transfer-controlled regime at high temperatures. In between these two extreme regimes a transition regime may exist, where a combination of external mass transfer and pore diffusion steps control the conversion rate of pollutants. There was a substantial number of modeling works published in the early 1970s on the modeling of reactive flows in catalytic monoliths. Kuo et al. [7] were probably the first to model an automotive catalyst as a series of continuously stirred tank reactors (CSTRs), although the concept was fairly common in the chemical engineering community for reactor modeling. Other works (e.g., Young and Finlayson [8]) appeared around the same time, extending the modeling to a two-dimensional domain. Heck et al. [9] published a landmark paper to demonstrate that a one-dimensional mathematical model can be sufficient to model catalytic reactions in a monolithic flow-through catalyst. Since then, an enormous amount of papers have been published in the literature, dealing with the modeling of flow-through catalysts. The majority of them are focused on the development of kinetics for different catalyst formulations, while only a few publications are focused on the physical processes such as heat and mass transfer coefficients and pore diffusion within the washcoat.

11.1.1.2 WALL FLOW MONOLITHS

WFMs are essentially flow-through monoliths with channel ends alternatively plugged, in order to force the gas flow through the porous walls, as shown in Figure 11.3, acting as filters.

This solution is very popular in AT applications to filter engine particle matter (PM), since the filtration efficiency of these devices could easily reach 90%. WFM are usually made by a ceramic substrate, characterized by high temperature resistance and a low thermal expansion coefficient, with pore sizes in the order of micrometers. On the WFM substrate, a catalyzed washcoat could also be included to accelerate oxidation reactions of the accumulated particles. Different filtration mechanisms occur in WFM. Small particles can penetrate inside the monolith pores following the gas flow streamlines. As the particle approaches a wall, it is collected on the surface because of Brownian inertia. This regime of filtration is called deep bed or depth filtration, as shown in Figure 11.4. Once pores are saturated, particles start to accumulate on the external surface (cake filtration regime, Figure 11.4). As this external layer becomes thicker, the pressure drop across the monolith increases, thus the filter must be periodically regenerated.

From a modeling point of view, WFM phenomena involve different length scales. A first class of models describes the

FIGURE 11.3 Flow pattern in a wall flow monolith [10].

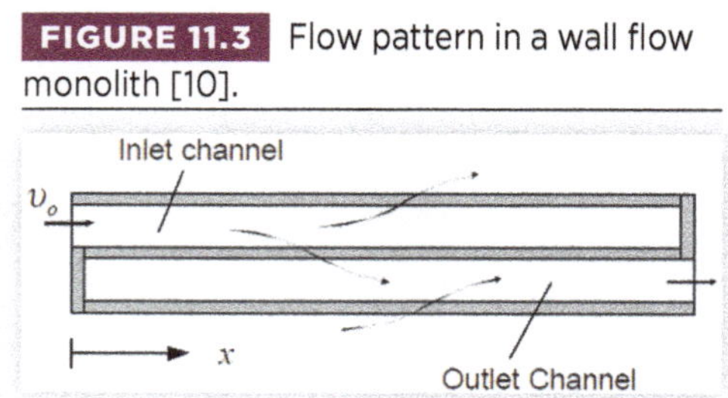

FIGURE 11.4 Filtration mechanisms in a wall flow monolith [11].

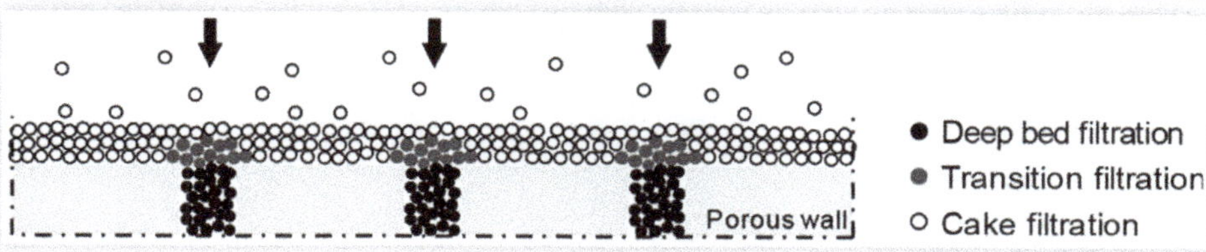

operations occurring at the macroscopic level of the entire filter scale. The second class of models describes the filter channel's length scale while the third one models the phenomena occurring at the microscopic scale of the soot layer and the porous filter walls [12].

Finally, 1D models of single channel WFM have been developed to accurately simulate the monolith filtration capacity, the pressure drop across the monolith, and the PM oxidation kinetics of catalyzed and uncatalyzed PFs [13].

11.2 Mathematical Model for Flow-Through Monoliths

The mathematical model describing the physical and chemical processes shown in Figure 11.2 can be formulated with the following assumptions: (1) conduction and diffusion in the flow direction are negligible in the bulk phase due to high Peclet numbers encountered in these reactors, (2) laminar flow is assumed in the channel due to low Reynolds numbers (<1000), (3) uniform cross-sectional area is considered, (4) reactions occur only in the washcoat (i.e., homogeneous gas phase reactions are negligible), (5) transverse temperature gradients in the solid phase are negligible compared to axial gradients, (6) a uniform radial distribution of the gas at the front face is assumed, while channel-to-channel variations and interactions are ignored, (7) transversal gradients are accounted for using heat and mass transfer coefficients between the fluid and solid.

The last two assumptions allow us to simplify the 2D/3D models into a 1D model where a single representative channel can be used to describe the entire monolith reactor. With the above assumptions, the mathematical model (referred here as 1 + 1D model due to one spatial dimension in the flow direction and second spatial dimension through the washcoat thickness) is described by the following conservation equations and boundary conditions (BCs).

Gas phase species:

$$f_{vd}\rho_g \frac{\partial \omega_{g,i}}{\partial t} = -f_{vd}\rho_g u \frac{\partial \omega_{g,i}}{\partial z} - \rho_g k_i S\left(\omega_{g,i} - \omega_{s,i}\right) \tag{11.1}$$

Gas phase energy:

$$f_{vd}\rho_g C_{pg} \frac{\partial T_g}{\partial t} = -f_{vd}\rho_g C_{pg} u \frac{\partial T_g}{\partial z} + h\,S\left(T_s - T_g\right) \tag{11.2}$$

Solid phase energy:

$$\psi_s \frac{\partial T_s}{\partial t} = \frac{\partial}{\partial z}\left(\Lambda_s \frac{\partial T_s}{\partial z}\right) - h\,S\left(T_s - T_g\right) - h_x S_x\left(T_s - T_x\right) + \sum_{l=1}^{2}\sum_{n=1}^{n_{rxns}} \Delta H_n^{(1)} a_n^{(1)} \overline{r}_n^{(1)} \tag{11.3}$$

where Λ_s and ψ_s are, respectively, effective thermal conductivity and effective heat capacity of solid phase and are given by the following equations:

$$\Lambda_s = f_{sb}\lambda_{sb} + \sum_{l=1}^{2} f^{(l)}\lambda^{(l)} \tag{11.4}$$

$$\psi_s = f_{sb}\rho_{sb}C_{p,sb} + \sum_{l=1}^{2} f^{(l)}\rho^{(l)}C_p^{(l)} \tag{11.5}$$

$\bar{r}_n^{(l)}$ is the average reaction rate across the effective washcoat thickness:

$$\bar{r}_n^{(l)} = \frac{1}{\delta^{(l)}}\int_0^{\delta^{(l)}} r_n^{(l)}\,dx \quad l = 1,2 \tag{11.6}$$

where the effective washcoat thickness is defined as:

$$\delta^{(l)} = \frac{f^{(l)}}{S}, \quad l = 1,2 \tag{11.7}$$

The effective washcoat thickness defined as above accounts for the small additional washcoat volume in the corners of a channel and will be slightly larger than the measured value. Initial and BCs for the above differential equations are trivial and are not given here. The following equations present the species balances and BCs in the washcoat layers.

Washcoat species:

$$f^{(l)}\varepsilon^{(l)}\rho_s\frac{\partial\omega_i}{\partial t} = f^{(l)}\rho_s D_{i,eff}^{(l)}\frac{\partial^2\omega_i}{\partial x^2} + R_i^{(l)}, \quad l = 1,2 \tag{11.8}$$

$R_i^{(l)}$ is the ith species rate in layer l:

$$R_i^{(l)} = MW_i\sum_{n=1}^{n_{rxns}} s_{i,n}a_n^{(l)}r_n^{(l)} \tag{11.9}$$

BC at the surface:

$$k_i\rho_g\left(\omega_{g,i} - \omega_{s,i}\right) = -\rho_s D_{i,eff}^{(l)}\left.\frac{\partial\omega_i}{\partial x}\right|_{x=0} \tag{11.10}$$

Flux continuity at the layer interface:

$$\rho_s D_{i,eff}^{(1)}\left.\frac{\partial\omega_i}{\partial x}\right|_{x=\delta^{(1)}_-} = \rho_s D_{i,eff}^{(2)}\left.\frac{\partial\omega_i}{\partial x}\right|_{x=\delta^{(1)}_+} \tag{11.11}$$

BC at the bottom of the washcoat:

$$\left.\frac{\partial\omega_i}{\partial x}\right|_{x=\delta^{(1)}+\delta^{(2)}} = 0 \tag{11.12}$$

Site balance:

$$A_k^{(l)} \frac{\partial \theta_k^{(l)}}{\partial t} = \sum_{n=1}^{n_{rxns}} \sigma_{k,n} a_n^{(l)} r_n^{(l)} \quad l = 1,2 \tag{11.13}$$

The reaction rates used in the above equations are based on the turnover number, defined as the ratio of moles reacted per moles of active sites per second. Volume rates are obtained by multiplying turnover rates with the active site density. The site density indicates the ratio between moles of active sites (PGM sites, storage sites, etc.) and the catalyst volume, acting as a kinetic scaling factor for different catalyst loadings. It is recommended to use turnover rates as they are independent of catalyst loading and hence are easily portable to different reactors with the same catalyst formulation but different catalyst loadings. Turnover rates are also useful when modeling the aging and poisoning of catalysts, which are generally accounted for by reducing the active site density, allowing the distinction between an ideal case (all the sites are active) and the real case (only some of the sites are active).

11.2.1 Heat and Mass Transfer Coefficients

Heat and mass transfer coefficients determine the rate of heat and mass transfer from fluid to solid walls and are calculated from dimensionless Nusselt and Sherwood numbers, which depend on the channel geometry and on the flow conditions inside the channel. Ideally, their values should be determined from experiments but, due to the uncertainty of experimental data, they are typically derived from the analogy with heat transfer problems. It is important to use appropriate values for these coefficients during the simulations as they determine the conversion efficiency in the external mass transfer–controlled regime. When the boundary layers (flow, thermal, and concentration) are fully developed, Nusselt and Sherwood numbers assume asymptotic/constant values specific to channel geometry and wall BCs, as given in Table 11.1. These asymptotic values are obtained as a solution to the classical Graetz–Nusselt problem with constant flux or constant temperature BCs. Values approach the constant flux limit (Nu_{H2}/Sh_{H2}) when the reactions are slow (before light-off) and approach the constant temperature limit (Nu_T/Sh_T) when the reactions are fast (after light-off) [14].

TABLE 11.1 Asymptotic values of Nusselt/Sherwood numbers and friction factors for commonly used channel geometries [14]

Channel shape	$D_h(4R_\Omega)$	$(fRe)_\infty$	$Nu_{H2\infty}(Sh_{H2\infty})$	$Nu_T(Sh_T)$
Circle (diameter 2a)	2a	16	4.364	3.656
Square (side 2a)	2a	14.23	3.089	2.977
Sinusoidal (base 2a, height 2b)	1.5419a	14.574	0.95	2.496
	1.418a	14.023	1.38	2.536
	1.2103a	13.023	1.55	2.476
	1.0466a	12.234	1.34	2.353
	0.8118a	11.207	0.90	2.110
	0.4673	10.123	0.33	1.719

The gas flow rates encountered in the AT reactors are such that the Reynolds number based on channel diameter is usually in the range 100–1000. Since the dimensionless flow development length (L_h/D_h) for laminar flow in a channel is of the order of 0.05 Re, the entrance length could vary from 5 to 50 times the channel hydraulic diameter. When the boundary layer development lengths are significant, it is essential to use position-dependent heat and mass transfer coefficient values, instead of constant asymptotic values. There is a vast amount of literature, spanning the past four decades and dealing with both theoretical and experimental aspects, on heat and mass transfer coefficients in laminar flows for reacting as well as non-reacting cases. Gundlapally and Balakotaiah [14] compared several different popular correlations and also analyzed the effect of coefficients on the performance of monolith reactors. It was shown that when the transverse Peclet number $\left(Pe = \frac{uD_h^2}{16LD_{m,i}}\right)$ is less than 0.25, flow conditions in the entry region of a monolith channel have negligible effect on the exit conversion. Normally, the transverse Peclet number is less than 0.1 for the monoliths used in the automotive exhaust AT applications and hence the constant values reported in Table 11.1 can be used in practical applications instead of position-dependent correlations.

11.2.2 Bulk and Effective Diffusion Coefficients

There are several correlations available in the literature for calculating molecular diffusion coefficients in binary mixtures. Two correlations that are widely used are the Chapman–Enskog [15] correlation and Fuller correlation [16]. The Chapman–Enskog correlation is:

$$D_{i,j} = \frac{1.858 \times 10^{-7} T^{1.5} \sqrt{\frac{1}{MW_i} + \frac{1}{MW_j}}}{P\sigma_{i,j}^2 \Omega_D} \tag{11.14}$$

where

Ω_D is the collision integral
$\sigma_{i,j}$ is the force constant

The values of Ω_D and $\sigma_{i,j}$ for many species can be found in Bird et al. [17].

The Fuller correlation is based on special atomic diffusion volumes [16] and is given by

$$D_{i,j} = \frac{10^{-7} T^{1.75} \sqrt{\frac{1}{MW_i} + \frac{1}{MW_j}}}{P\left(\left(\sum_i v_k\right)^{\frac{1}{3}} + \left(\sum_j v_k\right)^{\frac{1}{3}}\right)^2} \tag{11.15}$$

where

T is the temperature (K)
P is the pressure (atm)
MW is the molecular weight (g/mol)
v_k is the atomic diffusion volume (cm^3) summed over all the atoms contained in the diffusing species

The mixture-averaged diffusion coefficient $D_{i,m}$ is calculated from the above binary diffusion coefficients $D_{i,j}$

$$D_{i,m} = \frac{1-\omega_i}{\sum_{\substack{j=1 \\ j\neq i}}^{nsp} \frac{X_j}{D_{i,j}}} \tag{11.16}$$

Effective diffusion coefficients within the washcoat can be estimated using either the parallel pore model or random pore model. The parallel pore model is widely used for its simplicity. Using the parallel pore model, the effective diffusion coefficient is estimated from the mixture-averaged diffusion coefficient, Knudsen diffusion coefficient $D_{i,kn}$, and washcoat properties as:

$$D_{i,eff}^{(l)} = \frac{\tau^{(l)}}{\varepsilon^{(l)}}\left(\frac{1}{D_{i,m}} + \frac{1}{D_{i,kn}^{(l)}}\right) \tag{11.17}$$

$$D_{i,kn}^{(l)} = \frac{d_p^{(l)}}{3}\sqrt{\frac{8R_gT}{\pi MW_i}} \tag{11.18}$$

where $\tau^{(l)}$, $\varepsilon^{(l)}$, and $d_p^{(l)}$ are tortuosity, porosity, and pore diameter of washcoat layer l, respectively. The molecular weight of species i, MW_i, is expressed as (kg/mol) in the above equation.

FIGURE 11.5 Bulk and effective diffusion coefficients of selected species as a function of temperature.

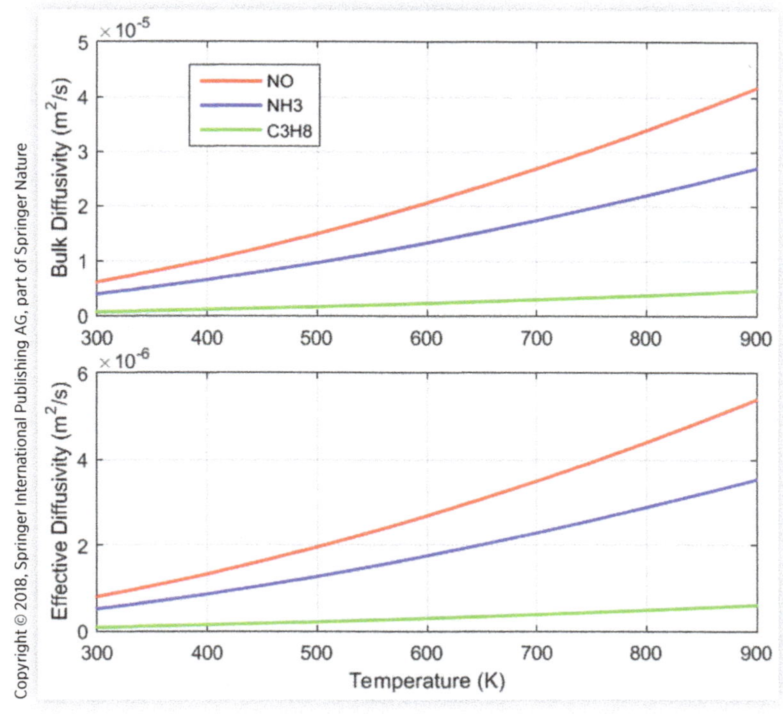

CHAPTER 11

It should be noted that the structure of the washcoat could be highly complex, with many different sizes of pores interconnected with tortuous paths. As such, it would not be realistic to expect a high degree of accuracy with simple models like parallel pore or random pore models while using average washcoat properties (i.e., pore diameter, tortuosity, and porosity). Usually, the average pore diameter reported by catalyst suppliers is in the order of nanometers, which results in small values for effective diffusivities using the above formulas. It is important to check that the estimated effective diffusivities have the same order of magnitude as the measured diffusivity values. Figure 11.5 shows the bulk and effective diffusion coefficients of selected species, in a nitrogen atmosphere, as a function of temperature. Using $\tau^{(l)} = 3$, $\varepsilon^{(l)} = 0.4$, and $d_p^{(l)} = 5\,\mu m$, the calculated effective diffusion coefficients are on the order of $10^{-6} m^2/s$, which are close to the values measured in Zhang et al. [18].

11.2.3 Extending 1D Framework to 2D/3D

As pointed out earlier, assuming uniform gas flow across the front of the reactor and neglecting channel-to-channel interaction allows to use only one channel as representative of the entire monolith [19]. If these conditions are not satisfied, the non-uniform gas distribution at the reactor inlet and channel-to-channel interactions via conduction should be considered. 1D modeling framework can be extended to account for these affects. Instead of solving a single channel, bundles of thermally interacting channels (via conduction), strategically arranged in the 3D space, can be used as an alternative to full 3D CFD simulations [19]. For example, as shown in Figure 11.6, NH_3 concentration near the inlet of a SCR reactor may not be uniform due to the complex processes of injection, decomposition, wall deposition, and evaporation.

FIGURE 11.6 Simulated NH_3 distribution at the front face of an SCR.

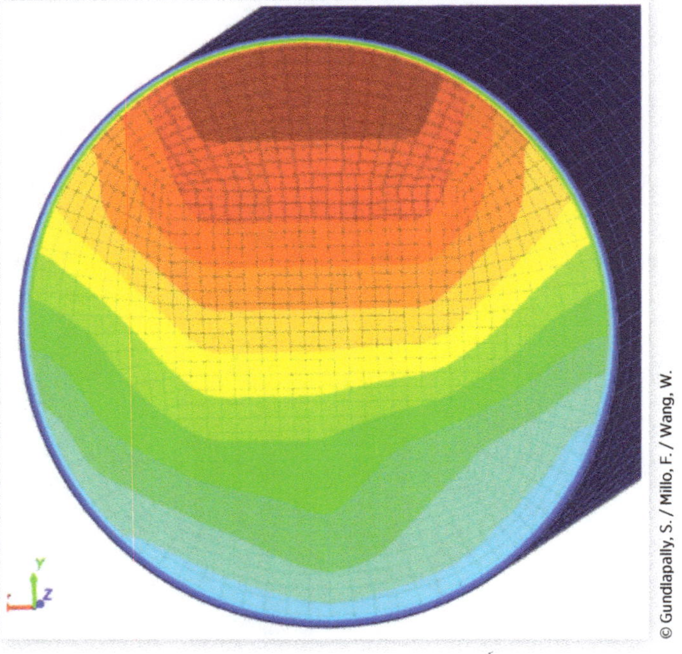

11.2.4 Pore Diffusion

Numerically coupling the washcoat diffusion-reaction equation (Equation 11.8) with 1D equations at each axial location is computationally expensive. The simplest approach to avoid the computational cost of a full numerical solution is to neglect the pore diffusion resistance. In the absence of pore diffusion resistance, species concentration profiles in the washcoat are constant. By integrating the diffusion-reaction equation (Equation 11.8), using BCs as in Equations 11.10 and 11.12, will result in the following equation, expressing the concentration in a single layer washcoat:

$$f^{(1)}\varepsilon^{(1)}\rho_s \frac{d\omega_{s,i}}{dt} = S\rho_g k_i\left(\omega_{g,i} - \omega_{s,i}\right) + R_i^{(1)}\left(\omega_s\right) \quad (11.19)$$

where $\omega_{s,i}$ is the surface concentration of species *i*. Even though this is a reasonable approximation for the AT applications, as the washcoat layers are generally thin and highly porous, the interest in understanding the effect of pore diffusion remains strong, especially for dual layer catalysts, where pore diffusion resistance is intentionally exploited to achieve a higher conversion efficiency [17, 20, 21, 22]. Thus, simplified approaches involving the effectiveness factor [23], the internal mass transfer coefficient [24, 25, 26], and the asymptotic solution [27, 28] are proposed to capture the effect of pore diffusion resistance without the significant computational cost associated with the full numerical solution. For example, Figure 11.7 shows the comparison between asymptotic solution and 1+1D solution for the species concentration profiles within a dual layer washcoat of a SCR reactor [29].

FIGURE 11.7 NH_3 and NO concentrations vs effective washcoat depth, comparison between 1 + 1D and asymptotic approaches [29].

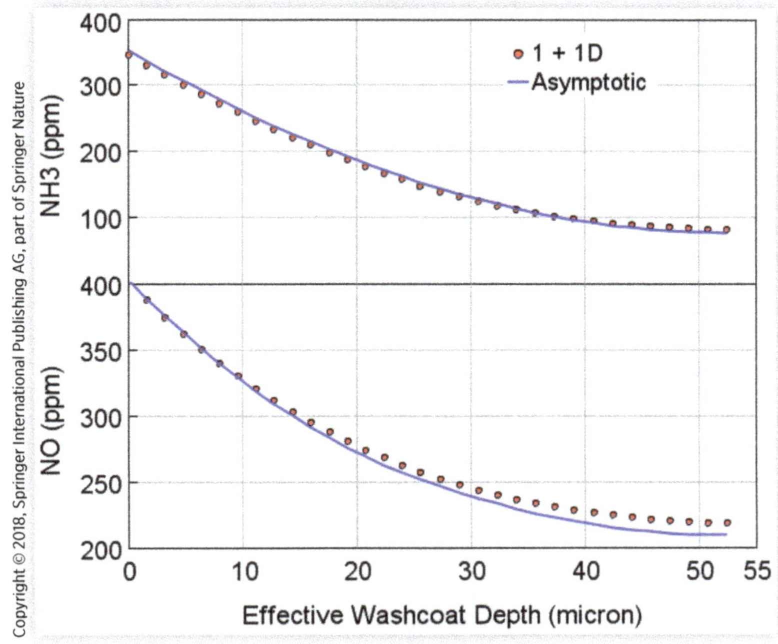

CHAPTER 11

11.2.5 Numerical Solution

The Method Of Lines (MOL) approach is widely used in the literature to discretize the partial differential equations (PDEs) presented in the Section 11.2. After discretizing the spatial dimension, the PDEs become ordinary differential equations (ODEs) that can be solved using any of the ODE solver packages available in the open literature. The most popular ODE solver packages like LSODI, VODE, and DASPK are based on implicit backward difference formulas (BDF), which are best suited for the stiff problems encountered in AT reactors. Note that the diffusion-reaction equation (Equation 11.8) also needs to be discretized along the washcoat thickness for the full numerical solution of the 1 + 1D model. This increases the computational cost tremendously as additional variables (for species concentrations) need to be solved at each axial mesh point, at the transverse mesh points within the washcoat.

It should be noted that there exist some commercial software packages that are customized for modeling AT reactors. These commercial packages automatically formulate and discretize the above differential equations given the user input of reaction mechanism, geometry, and operating conditions. One of the important input parameters for the ODE solvers are the error tolerances, which must be set judiciously for the optimum performance of ODE solvers. Using error tolerances that are too loose may lead to an incorrect solution or to solver failures, whereas the use of too stringent tolerances may degrade the performance of solvers. It is recommended to run the simulations with different tolerances to assess the impact on the solution and to optimize the solver performance. For species fractions, absolute tolerance of 10^{-6} (1 ppm) could be used as a starting point. The absolute tolerance can be set to zero for temperatures as their magnitude will be much larger than zero and hence relative tolerance becomes the controlling parameter. A relative tolerance of 10^{-3} could be used as starting value for species fractions and temperatures.

11.2.6 Quasi-Steady State Approximation

The quasi-steady-state (QSS) approximation is widely used by the reactor modeling community because it reduces the computational time by eliminating shorter time scales associated with the small accumulation terms (time derivatives) in the gas phase species, energy, and washcoat species balance equations. The system of PDEs discussed above becomes a system of differential algebraic equations (DAEs), after invoking the QSS approximation. It is important to note that the QSS approximation introduces additional mathematical complications when the surface concentration changes discontinuously along the reactor length. Why the so-called concentration jumps cause DAE solver failures will be briefly discussed here, while interested readers could refer to Gundlapally [30] for more information on the bifurcation analysis of catalytic reactors. To simplify the discussion, the pore diffusion resistance in the washcoat will be here neglected. For this limiting case, Equation (11.19) becomes the following algebraic equation, after applying the QSS assumption:

$$\mathrm{S}\,\rho_g\,\mathrm{k_i}\left(\omega_{g,i}-\omega_{s,i}\right)+\mathrm{R_i}\left(\omega_s\right)=0 \tag{11.20}$$

Discontinuities in the surface concentration, ω_s, occur when the solution to the above algebraic equation reaches a limit point, where the Jacobian matrix becomes singular. At the limit point, the surface concentration undergoes abrupt changes. Figure 11.8 shows bulk and surface CO concentration profiles along the channel length, where it can be noticed that the discontinuity in the CO surface concentration occurs at the dimensionless channel location of 0.5, where the surface concentration jumps

FIGURE 11.8 Bulk and surface concentration profiles. The surface concentration profile shows an abrupt change in CO concentration at the dimensionless channel location of 0.5 [29].

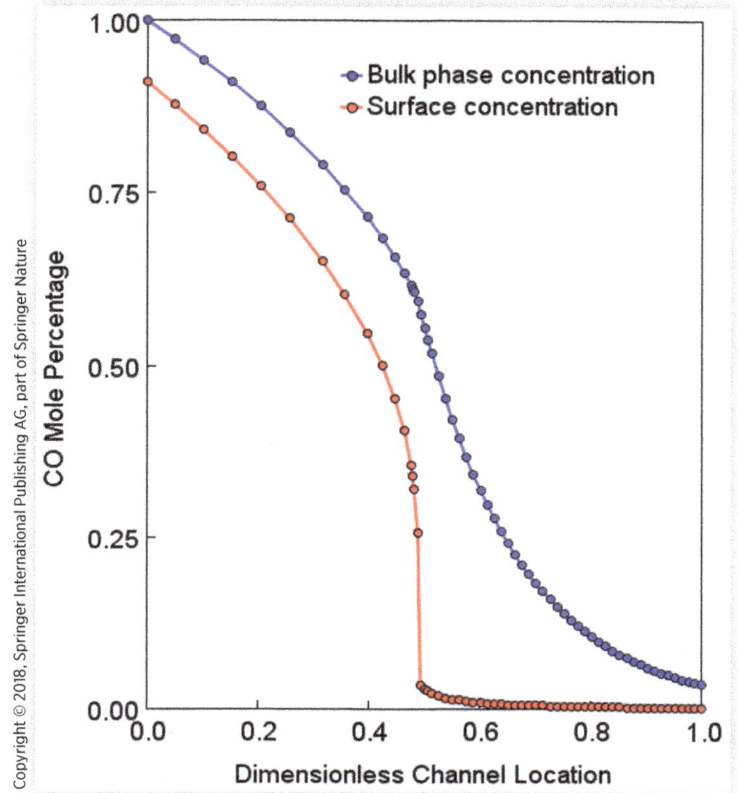

from 0.25% to 0.04%. DAE solvers struggle as the solution approaches discontinuity and often fails as the error test forces very small time-step sizes. Thus, special considerations have to be taken during the simulation to avoid integration failures due to concentration jumps. The Jacobian matrix must be monitored and a special logic should be adopted to help DAE solvers in finding the solution at the limit point. The rest of the variables (ω_g,T_g) do not change discontinuously because axial derivative terms that appear in the governing equations guarantee smoothness.

11.2.7 Adaptive Mesh

Sharp reaction fronts where most of the reactant conversion takes place are routinely present inside the channels of monolith reactors. These reaction fronts dynamically move within the channel in response to the operating conditions and accurately resolving species concentration profiles inside a reaction front is computationally challenging. An adaptive meshing method that dynamically follows the reaction fronts was proposed in Oh et al. [31] but this method has not seen widespread use in the simulation of AT reactors. An adaptive mesh with little computational overhead can be implemented since the solid temperature and surface coverages vary slowly with time due to their large accumulation or capacity terms whereas species concentrations and gas temperature exhibit fast dynamics due to their small accumulation terms. After applying the QSS assumption, solid temperature and coverage equations are the only equations that have explicit time derivatives, while species and gas temperature equations have only axial derivatives. Thus, it is possible to construct the problem as inner and outer integration problems. In the outer problem, the time integration of solid temperature and coverages is carried out on a fixed uniform MOL mesh, whereas in the inner problem the integration of the DAE system consisting of species and temperature equations is carried out. In this dual mesh strategy, the DAE solver automatically creates the spatially non-uniform mesh in accordance with the specified error tolerances. It should be noted that this approach does not decouple or lag the variables in any way but simply takes advantage of the particular structure of the PDE system [31].

The first significant and obvious advantage of the above adaptive mesh technique over the fixed uniform mesh is that the adaptive mesh dynamically adjusts with the moving reaction front inside the channel. The second significant advantage lies in the automatic control of axial integration errors in addition to the time integration errors, whereas with the standard solution methods only time integration errors are controlled. This feature results from the decomposition of the problem into time and axial integration problems. With a non-adaptive mesh, the accuracy of the solution must be accurately checked by varying grid resolution, ensuring that the obtained solution is grid-independent, whereas the adaptive meshing technique discussed above dynamically refines the axial mesh resolution for a grid-independent solution. This is illustrated in Figure 11.9, which shows the comparison of axial CO bulk gas phase concentration profiles obtained using both the adaptive and fixed mesh points. It can be seen that a finer mesh is needed when using a fixed mesh to match the results of the adaptive mesh. In this particular example, the fixed mesh solver needs more than three times axial mesh points than the adaptive mesh to obtain the grid-independent solution.

FIGURE 11.9 CO concentration profiles demonstrating the advantage of automatic control of the axial integration errors [29].

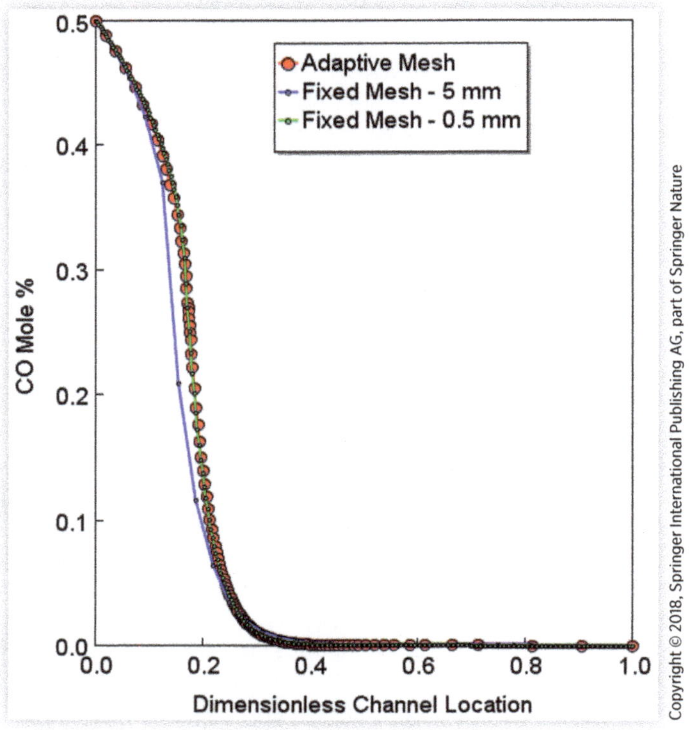

11.3 Mathematical Model for WFMs

Both diesel and direct injection gasoline engine produce particulate matter as a product of the combustion. The amount of tail-pipe emissions depends on the engine operating/design variables and on the mitigating steps taken at the exhaust system. The presence of PF in the exhaust system affects the engine performance in many ways. First, it increases the exhaust back pressure as more soot particles are collected in the filter. Therefore, the filter must be periodically regenerated by burning the stored particles. Thus, additional fuel must be injected to initiate the filter regeneration, increasing the engine fuel consumption. In this context, mathematical models of PFs can be effective in supporting the development of efficient control systems and the whole system design processes.

In order to fully understand design tradeoffs, a physics-based model is needed. The first of such mathematical models was proposed by Bissett [32], which included the conservation equations for mass, energy, and momentum, for both inlet and outlet channels. Channels were coupled through a uniform flow from inlet to outlet, while a constant inlet channel diameter was assumed, regardless of soot cake formation. The deep bed filtration regime was neglected in the model, while the soot cake formation was considered. Nevertheless, the model was remarkably successful and formed the basis of subsequent model improvements.

The successive generation of models added various features such as the deep bed filtration based on the spherical unit collector model [12]. Konstandopoulos and Johnson [33] compared three types of filtration models: the fibrous filter model, the spherical unit collector model, and the periodically constricted tube model. The spherical unit collector model was chosen due to the consideration of computational cost, the estimation of input parameters, and the similarity between the model and the monolith wall structures. Several related works incorporated the diffusion of oxygen [34] as well as the so-called back diffusion of NO_2 [35] and NO_2 assisted soot kinetics [36]. Due to the diffusion in the soot cake and substrate layers, species concentration varies along the channel length. Within the practical application range of a WFM, the soot cake thickness can be neglected compared to the channel width. Then the species equations in the channels can be written as [37]:

$$\rho_1 v_1 \frac{\delta y_{1,j}}{\delta z} = \rho_1 k_{1,j} \frac{4}{D_1}\left(y_{w1,j} - y_{1,j}\right) - \rho_{w1} v_{w1} \frac{4}{D_1}\left(y_{w1,j} - y_{1,j}\right) \tag{11.21}$$

$$\rho_2 v_2 \frac{\delta y_{2,j}}{\delta z} = \rho_2 k_{2,j} \frac{4}{D_2}\left(y_{w2,j} - y_{2,j}\right) - \rho_{w2} v_{w2} \frac{4}{D_2}\left(y_{w2,j} - y_{2,j}\right) \tag{11.22}$$

The species equation in the soot cake and substrate is:

$$\rho_w v_w \frac{\delta y_j}{\delta x} - \rho_w D_{eff,j} \frac{\delta^2 y_j}{\delta x^2} = R_j \tag{11.23}$$

The conservation equations for mass, momentum, and energy can be written as:

$$\frac{\delta\left(\rho_i v_i\right)}{\delta z} = \left(-1\right)^i \frac{4}{D_i} \rho_{wi} v_{wi} \tag{11.24}$$

CHAPTER 11

$$\frac{\delta p_i}{\delta z}+\beta\frac{\delta\left(\rho_i v_i^2\right)}{\delta z}=\frac{2C_f\,Re}{D_i^2}\mu_i v_i \tag{11.25}$$

$$\rho_i v_i C_{p,i}\frac{\delta T_i}{\delta z}=h_i\frac{4}{D_i}\left(T_{w,i}-T_i\right)+\left(-1\right)^{-i}\frac{4}{D_i}\rho_{wi}v_{wi}\left(T_{w,i}-T_i\right) \tag{11.26}$$

where i = 1 (inlet channel), 2 (outlet channel). Based on Bissett [38], the friction factor $C_f Re$, the momentum flux factor β, and the Nusselt/Sherwood numbers are functions of the wall Reynolds number. There is an interesting observation that could be made by looking at the balance equations for the inlet channel. This is most recognizable in the energy equation term when expressed in non-dimensional form:

$$Pe_w\hat{v}_{z,av}\frac{d\hat{T}}{d\hat{z}}=\left(Nu-Pe_w\right)\left(1-\hat{T}_b\right) \tag{11.27}$$

It can be seen that, for a fixed Nusselt number, there exists a wall Peclet number above which the gas temperature in the channel increases with wall cooling, which might result counter intuitive other than unphysical. To address this issue, Koltsakis and Stamatelos [39] wrote the inlet channel gas energy equation by assuming that the gas leaves the channel at the bulk gas temperature T_1, which is different from the approach taken for the outlet channel in which gas enters the control volume at wall temperature T_w. This assumption removed the possibility of unphysical temperature gradients.

$$C_{p,g}\rho_1 v_1\frac{\delta T_1}{\delta z}=h_1\frac{4}{d_1}\left(T_w-T_1\right) \tag{11.28}$$

$$C_{p,g}\rho_2 v_2\frac{\delta T_2}{\delta z}=\left(h_2+C_{p,g}\rho_w v_w\right)\frac{4}{d}\left(T_w-T_2\right) \tag{11.29}$$

A similar assumption was also adopted in the species equations and in the corresponding BCs. However, Bissett et al. [38] took a different approach. They identified that the use of the constant Nusselt number was the cause of the non-physical behavior of the governing equation. In fact, it was common practice to use the Nusselt number of a non-porous wall square channel for the porous WFMs. Bissett et al. [38] found that the Nusselt number and friction coefficients are sensitive functions of the wall Reynolds number and, specifically, that there is a non-monotonic behavior in the inlet channel.

The above work was further extended by Wang and Bissett [40] for other geometries of the porous channels. In order to extend the life span of the monoliths and to maintain low pressure drops with soot and ash loading, other channel shapes other than a square – such as triangle, hexagon, or octa-square – have been studied in the literature. Particularly, it has been shown that the hexagon channel design [41, 42, 43] could increase the regeneration efficiency and the soot oxidation rate compared to the conventional square channel designs. For hexagon channel monoliths, there are two inlet channels for each outlet channel. When there is a negligible soot cake or the soot loading is small, the pressure drop of hexagon channel WFMs is shown to be higher than the square channel monoliths [43] as flow mainly goes through the walls between the inlet channel and outlet channels. However, when the soot cake formation was observed on the walls between two inlet channels, in medium or high soot loading regimes, hexagon channel WFMs showed a lower pressure drop than square channel monoliths.

FIGURE 11.10 Streamlines in hexagon channels with suction ($Re_w = 3$). (Left) symmetric BCs and (right) asymmetric BCs.

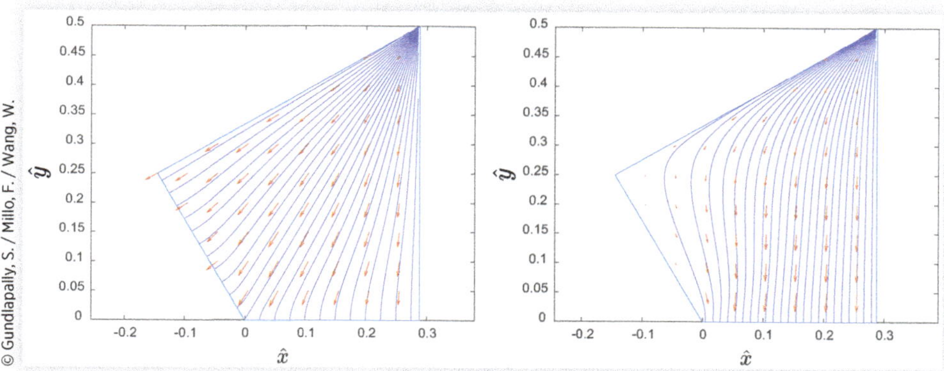

In Wang and Bissett [39], the friction factors and Nusselt number correlation for triangle and hexagon channels with porous walls are derived by solving the three-dimensional (3D) incompressible flow problem. A similar solution has been adopted for the fully developed flow and heat transfer problem, in triangle and hexagon channels with wall suction and injection. Different flow patterns are observed between the symmetric and asymmetric hexagon channels, yet the integrated flow characteristics such as overall friction coefficients and Nusselt numbers are shown to be insensitive to the symmetry of BCs. Figure 11.10 shows the streamlines in one-sixth of a hexagon suction channel with the symmetric and asymmetric BCs. The symmetric BC corresponds to the medium or high loading regime while the asymmetric BC corresponds to clean or low loading regime. Figure 11.11 compares the overall friction factors of the two types of channels. The simulation results for each geometry are fitted into polynomial and limit-point functions. For readers' reference, the fitting functions are listed in Tables 11.2 through 11.6 for the triangle and hexagon channels. The polynomial function fit for the square channel geometry is listed in Table A1 of the reference [38]. The advantages of polynomial function fitting lie in its simplicity and in the fitting accuracy, which can be improved by increasing the order of the functions. On the other hand, the limit-point function fitting shows better accuracy in the proximity of the critical wall Reynolds number, $Re_{w,c}$. In addition, the limit-point function fitting generally shows less error than the polynomial function fittings using the same or fewer coefficients.

With results for several geometries available, it is natural to ask which geometry is better, at least in terms of flow resistance. A comparison of the C_f Re values for all the geometries is depicted in Figure 11.12. However, it should be noted that for this comparison, the wall velocity and the hydraulic diameter are assumed to be constant for all the analyzed geometries. For a more practical comparison, how the channels are combined into the WFM must be considered, since the result might sensitively depend on additional properties of the monolith and on the parameters that are held constant between different geometries. In Wang and Bissett [39], an example of such comparison was shown, by assuming that the temperature, the total mass flow rate, and the

FIGURE 11.11 Friction coefficient comparison in hexagon channels with suction.

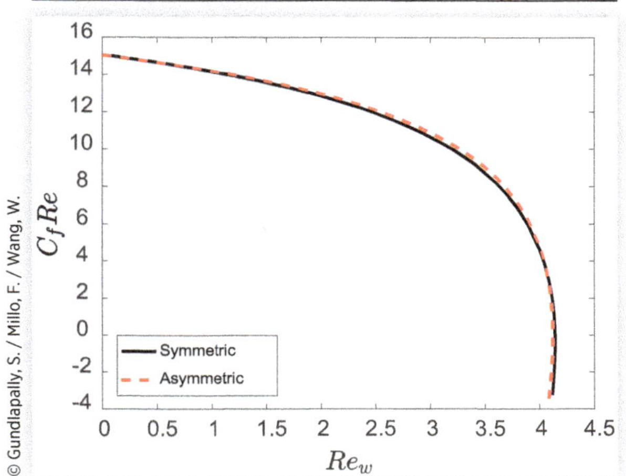

TABLE 11.2 Polynomial fitting functions for the triangle channel [40]

$C_f Re$	Suction ($Re_w \leq 2$)	$C_f Re = 13.34 - 1.50Re_w + 0.220Re_w^2 - 0.387Re_w^3$
	Injection ($Re_w \leq 4$)	$C_f Re = 13.34 + 1.12Re_w - 0.108Re_w^2$
β	Suction ($Re_w \leq 2$)	$\beta = 1.43 + 0.0180Re_w + 0.0497Re_w^2$
	Injection ($Re_w \leq 4$)	$\beta = 1.43 - 0.0253Re_w$
Nu	Suction ($Re_w \leq 2$)	$Nu = 2.49 + 0.591Pe_w - 0.166Re_w$
	Injection ($Re_w \leq 4$)	$Nu = 2.49 - 0.400Pe_w + 0.0189Pe_w^2$

TABLE 11.3 Polynomial fitting functions for the hexagon channel [40]

$C_f Re$	Suction ($Re_w \leq 3.5$)	$C_f Re = 15.06 - 0.965Re_w + 0.166Re_w^2 - 0.114Re_w^3$
	Injection ($Re_w \leq 9$)	$C_f Re = 15.06 + 0.609Re_w - 0.0330Re_w^2$
β	Suction ($Re_w \leq 3.5$)	$\beta = 1.35 + 0.0051Re_w + 0.0135Re_w^2$
	Injection ($Re_w \leq 9$)	$\beta = 1.35 - 0.00923Re_w$
Nu	Suction ($Re_w \leq 3.5$)	$Nu = 3.34 + 0.605Pe_w - 0.126Re_w$
	Injection ($Re_w \leq 9$)	$Nu = 3.34 - 0.437Pe_w + 0.0166Pe_w^2$

TABLE 11.4 Limit point function fittings for the triangle channel [40]

$C_f Re$	$-2.3 \leq Re_w^s \leq 4$	$C_f Re = \left(0.0457 + \frac{0.0519}{\left(Re_w^s + 2.52\right)^{0.621}}\right)^{-1}$
β	$-2.3 \leq Re_w^s \leq 4$	$\beta = 1.04 + \frac{0.524}{\left(Re_w^s + 2.52\right)^{0.295}}$
Nu	$-2.3 \leq Re_w^s \leq 0$	$Nu = 2.49 - \left(0.600Re_w^s - 0.380Pe_w^s\right)\left(1 - \frac{1}{\left(Re_w^s + 2.52\right)^{0.113}}\right) - 0.434Pe_w^s Pr^{0.550}$
	$0 \leq Re_w^s \leq 4$	$Nu = 2.49 - 0.400Pe_w^s + 0.0189\left(Pe_w^s\right)^2$

overall monolith geometry (such as frontal area and length) are held constant for different geometries. In addition, it was assumed that the filtration area was the same to maintain a similar pressure drop across the filter walls. With these assumptions, square and triangle channels resulted in having the same hydraulic diameter and wall Reynolds number, which could be compared directly as depicted in Figure 11.12a. It should be noted that hexagon channels might create additional difficulties for any comparison, since they are configured in a 2:1 inlet to outlet ratio and might also require a choice of flow symmetry, if the flow uses all the inlet channel walls or only uses half of the available channels' areas, or even something intermediate. With these constraints, the hydraulic diameter of hexagon channels could differ from the square and triangle ones, making a comparison using the wall Reynolds number Re_w meaningless. It was shown in Wang and Bissett [39] that the ratio of w and D_h is a function of wS_f, although the function form is slightly different for the hexagon channel monolith compared to the triangle and square monoliths. A typical value of $wS_F = 0.15$ was chosen for the

TABLE 11.5 Limit point function fitting functions for the square channel [40]

C_fRe	$-3.4 \le Re_w^s \le 4$	$C_fRe = \left(0.0447 + \frac{0.0562}{\left(Re_w^s + 3.49\right)^{0.628}}\right)^{-1}$
β	$-3.4 \le Re_w^s \le 4$	$\beta = 0.639 + \frac{0.885}{\left(Re_w^s + 3.49\right)^{0.136}}$
Nu	$-3.4 \le Re_w^s \le 0$	$Nu = 2.98 - \left(0.233Re_w^s - 0.176Pe_w^s\right)\left(1 - \frac{1}{\left(Re_w^s + 3.49\right)^{0.171}}\right) - 0.466Pe_w^s Pr^{0.451}$
	$0 \le Re_w^s \le 4$	$Nu = 2.98 - 0.429Pe_w^s + 0.0165\left(Pe_w^s\right)^2$

TABLE 11.6 Limit point function fittings for the hexagon channel [40]

C_fRe	$-4 \le Re_w^s \le 9$	$C_fRe = \left(0.0449 + \frac{0.0525}{\left(Re_w^s + 4.14\right)^{0.626}}\right)^{-1}$
β	$-4 \le Re_w^s \le 9$	$\beta = 0.984 + \frac{0.504}{\left(Re_w^s + 4.14\right)^{0.215}}$
Nu	$-4 \le Re_w^s \le 0$	$Nu = 3.34 - \left(0.0855Re_w^s - 0.0739Pe_w^s\right)\left(1 - \frac{1}{\left(Re_w^s + 4.14\right)^{0.405}}\right) - 0.483Pe_w^s Pr^{0.403}$
	$0 \le Re_w^s \le 9$	$Nu = 3.34 - 0.437Pe_w^s + 0.0166\left(Pe_w^s\right)^2$

FIGURE 11.12 Friction coefficient comparison for different geometries vs Re number (a) and $wW/(\mu A_F)$ (b).

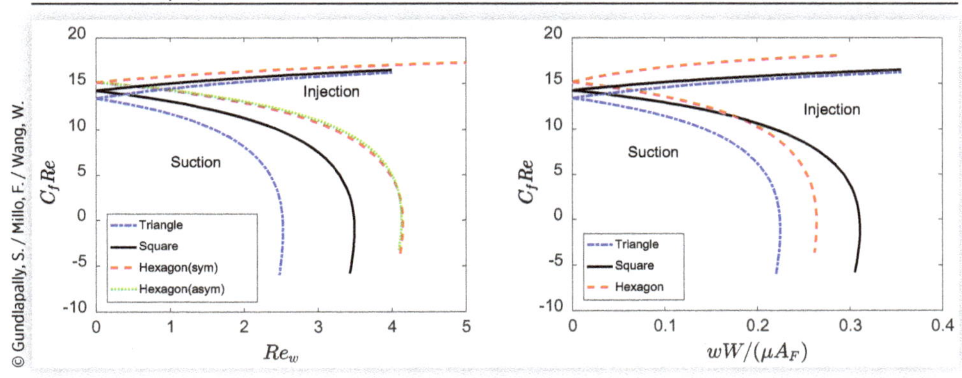

quantitative comparison and the resulting friction factors are plotted in Figure 11.12b against an alternative parameter, the ratio between wW over μA_P, which are the same across the analyzed geometries.

Within the limitations of the comparison defined above, Figure 11.12b ranks the geometries from triangle best (lowest $C_f Re$) to symmetric hexagon worst for lower flow rates. However, for higher flow rates, the symmetric hexagon catches up to and surpasses the square geometry, which is not the case if the comparison were made at the same Re_w as in Figure 11.12a. This demonstrates that "best geometry" questions are seldom easily answered and require the precise definition of other conditions to be maintained during variation of the geometry.

11.4 Measurements and Model Calibration

Reactions play such an important role in the modeling of AT reactors that the majority of the papers published in the literature are focused on the development of kinetic mechanisms for different catalysts. Physical processes in monoliths are reasonably well understood but currently there is no theory available that can reasonably predict the kinetics for a given catalyst. Currently measurements are the only way to reliably measure the kinetics. Kinetic mechanisms are usually divided into micro kinetics and global kinetics. With a micro kinetic approach, all the reaction sub-steps are detailed and modeled, which often requires a significant calibration effort, due to the large number of variables involved, and a deep understanding of the kinetics, which could only be obtained by means of an extensive laboratory-scale characterization. A global kinetic approach instead does not consider all the elementary steps behind each reaction, thus reducing the calibration effort required and the complexity of the problem. For this reason, global kinetics mechanisms are predominantly used in the AT community. In this respect, it is often prudent to take a minimalistic view and, therefore, start with the simplest mechanism that fits a given class of catalyst. Subsequently, experimental results can be used to refine the global kinetic model to fit the observation. Experiments are usually carried out on small monolith samples extracted from larger monoliths provided by the monolith suppliers. Samples can be commercially pre-washcoated or can be washcoated in-house if testing new catalyst formulations. Synthetic gases are used, hence the name SGB test bench, to simulate the exhaust from a real engine. A simple SGB scheme is depicted in Figure 11.13, related to ammonia storage experiments. The gas flow rates in SGBs is selected such that space velocity is the same as a full-scale system (i.e., residence time of the gas inside the channel is the same for both the SGB and the full-scale system). This minimizes the amount of synthetic gases used and hence the cost of the experiments.

It is important to note that these laboratory tests have to be very carefully designed such that some important criteria are met:

The stoichiometry of the mixture should be representative of the application of full-scale reactors

The different batches of mixture for each experiment should be designed (as much as possible) to explore the behavior of the smallest subset of the mechanism

The space velocity and temperature ranges should be chosen in commensurate with the full-scale reactor

Finally, inlet conditions should be also be chosen such that the conversion efficiencies are far away from 0 and 100% such that as much of kinetic information can be extracted as possible.

FIGURE 11.13 Experimental setup: schematic view of laboratory gas bench [4].

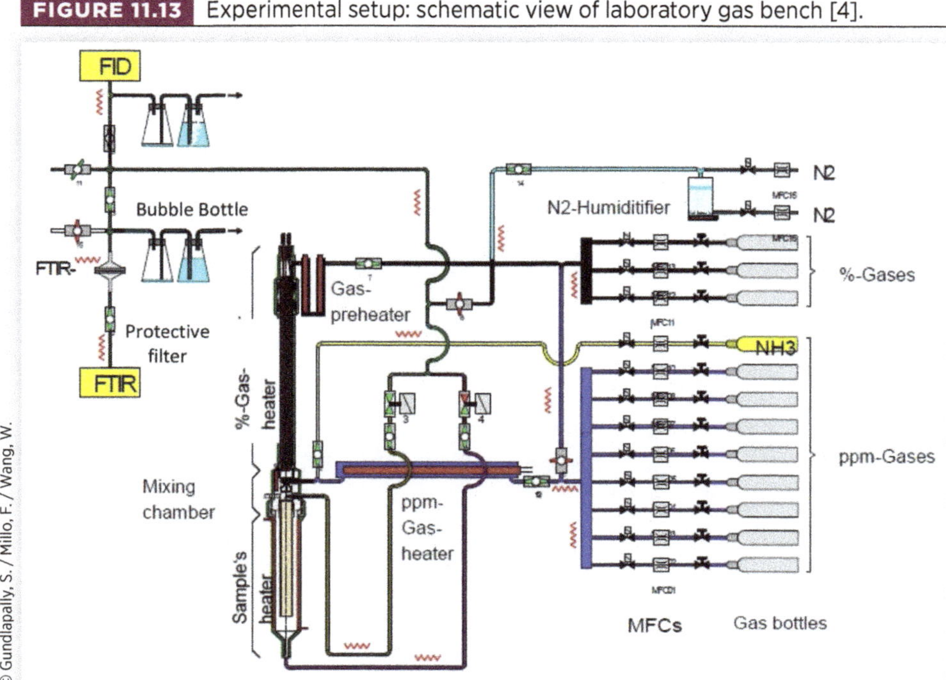

With these criteria in mind, the so-called SGB test protocols are designed and perhaps refined based on the observation. As an example, the LNT test protocol used in Millo et al. [4] is detailed in Table 11.7.

Each experiment must be designed to characterize a particular catalyst operation. As an example, a scheme for light-off, temperature programmed desorption (TPD), and NOx storage and reduction (NSR) experiments is depicted in Figure 11.14.

It should be noted that such protocols may need to be revised based on the formulation of the catalyst and existing knowledge base.

Ideally, kinetics measured in SGB test should be used in full-scale system without any modification of the reaction rate parameters, but it is often found that the model

TABLE 11.7 Test protocol to characterize LNT operations by means of SGB experiments [4]

Characteristic	Test protocol	Inlet reductant feed
Light-off	A temperature ramp, from 100 to 500°C, rate ≤ 5°C/min, is applied to the sample in order to characterize HC, CO, and NO oxidation	H_2 CO HC (propylene, xylene, dodecane)
NOx storage	Temperature programmed desorption: an isothermal NOx loading window is followed by a temperature ramp up to 500°C where stored NOx are released	
NOx reduction	Lean/rich cycling tests are performed in order to observe NOx storage (lean phase) and reduction (rich purge)	H_2 CO HC (propylene)
Oxygen storage capacity	Lean/rich cycling tests are performed in order to characterize oxygen storage during the lean phase and HCs oxidation during the rich purge	H_2 CO HC (propylene)

CHAPTER 11

FIGURE 11.14 Sample of test protocols for light-off, TPD, and NSR experiments.

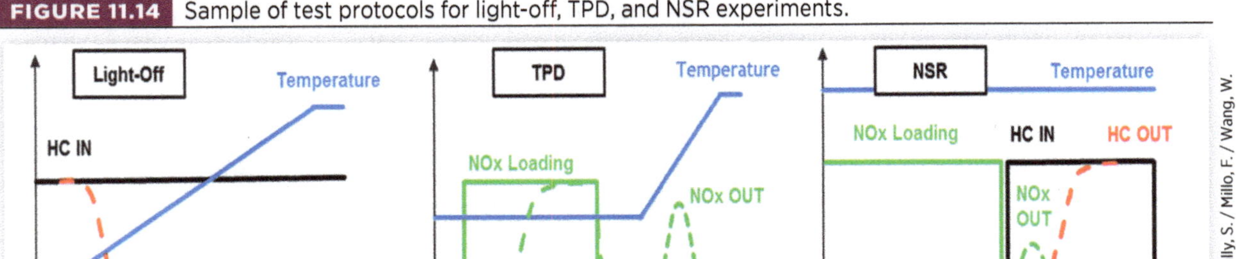

predictions do not agree with full-scale experiments. One of the primary reasons for the departure is due to the small number of HC species used in the SGB tests whereas there is a wide spectrum of HC in the engine out exhaust gas. There are secondary reasons which might compromise similarity in the upscaled system, such as entrance effects, when heat and mass dispersion became significant compared to laboratory-scale conditions [44]. This necessitates the adjustments of kinetics for using in a full-scale system.

11.5 Models for Controls

Map-based models traditionally used in ECUs require great calibration effort to accurately represent the physics over the whole operating range of an engine and hence physics-based models are highly desirable for control purpose. Physics-based models can also estimate difficult/impossible to measure variables such as NH_3 storage [45] but, due to the highly nonlinear reaction source terms, even a 1D single channel simulations would require a lot of computing power (CPU speed and memory), making this approach inappropriate for certain applications [46]. For effective control of AT systems in automobiles and commercial vehicles, one of the main challenges lies in the limited computing power that resides within the engine control unit (ECU). ECU needs a plant model, among other things, for feedback control of DEF dosing, DPF/GPF regeneration, and electrical heating strategies. Traditionally, these models are so-called linearized models that are preferred by control engineers due to the well-developed linear control theories that can be readily adapted for designing optimum control systems [46]. However, linear models are not accurate over the whole operating regions and often not physics-based (e.g., a fit of experimental data). On the other hand, it may be possible to construct simplified physical models of AT components, called grey-box models, able to deliver a high level of accuracy over a wider range of operating conditions, while limiting the computational effort within the constraints of the limited resources in the ECU. One more attractive property of the grey-box models is that the equations are in such a form that they can be linearized without much effort. In this case, the accuracy of linear models might be limited to a small region around the operating point from where the linearized response function is created. However, one possible solution to this problem could be to develop multiple piecewise linear plant models, thus delivering sufficient accuracy for the entire operating range of the SCR. In this case, the controller algorithm should perform the following main steps at each sample: (i) identifies the active linear model based on the values of the exogenous input signals, (ii) measures outputs and estimates the states using a state observer for the selected linear model (iii), and calculates the control law that has been designed for the specific operating area.

There are many control-oriented models presented in the literature and the level of physical details included in those models may vary significantly. Gundlapally et al. [47] recently proposed a systematic procedure for developing grey-box models from detailed models. It is shown that the grey-box model includes all qualitative features of the detailed model and the parameters used in the grey-box model are closely related to the detailed model they are derived from. McMackin et al. [45] presented a SCR grey-box model similar to the one presented by Gundlapally et al. [47] but with an expanded set of reactions. The model was compared with the detailed model over 200 points covering the entire operating map of a diesel engine. Figure 11.15 shows the comparison of NH_3 storage and NOx output predicted from both the grey-box model and the detailed model for 34000 s of simulation.

In order to implement the grey-box model on the ECU, it needs to be compiled into a machine language and flashed (embedded) into the ECU. McMackin et al. [45] imbedded the SCR grey-box model onto a Bosch ECU (ES910) for the purpose of studying various urea control dosing strategies. They reported that grey-box model predictions match, with a limited calibration effort, their existing ECU code performance, which uses map-based models instead. Also, as shown in Figure 11.16, considerable improvements can be achieved with respect to engine out NOx and NH_3 slippage based on strategies that maintain an optimum coverage state of the SCR (θ), as predicted by the embedded grey-box model.

11.6 Concluding Remarks

More stringent emission standards are being adopted worldwide due to a growing concern about pollution effects on the environment and human health. In this context, the AT systems play an essential role in both spark and compression ignition engines.

To comply with the new emission regulations, compression ignition engines require the combination of different AT devices placed upstream in the exhaust line, as close as possible to the engine, to accelerate the warm up and to increase the operating temperature of the system. Therefore, modern exhaust AT systems for compression ignition include a great variety of architectures combining several devices such as a wall-flow PF to collect the particulate matter, often integrated with a SCR coating for NOx abatement,

FIGURE 11.15 SCR grey-box model for control applications. NOx emissions over driving cycle comparison: 1+1D Model (GT model) vs grey-box model (C code) (left-hand side image). Fractional coverage of NH_3 storage sites (theta): 1+1D Model (GT model) vs grey-box model (C code) (right-hand side image).

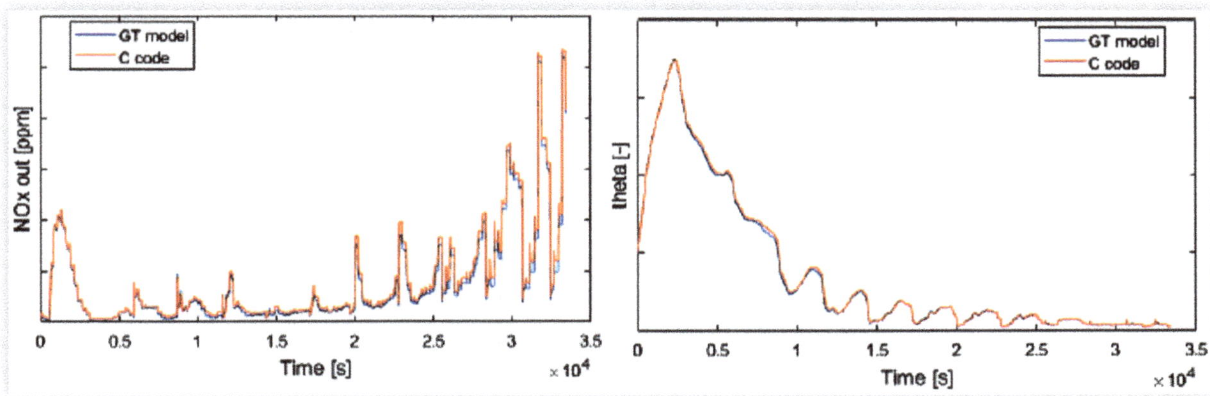

FIGURE 11.16 SCR grey-box model for controls applications. NOx emissions (top image) and ammonia slip (bottom image) comparison between the ECU NO_x-follow control strategy and the grey-box-based NH_3-storage control strategy.

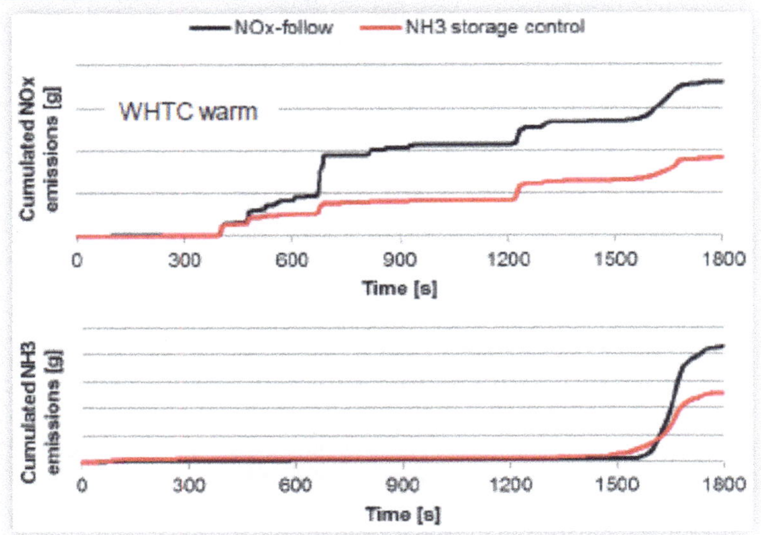

and a series of flow-through monolithic reactors in charge for CO, HC, and NOx low temperature storage and subsequent high temperature abatement.

In spark ignition engines, the introduction of direct injection has allowed an increase in the combustion efficiency thus reducing CO_2 emissions, although requiring the adoption of the wall-flow GPF to fulfill the regulation limits for particulate matter. Due to the stoichiometric operation conditions in spark ignition engines, the TWC can fulfill the simultaneous abatement of CO, HC, and NOx emissions.

On the other hand, upcoming standards are introducing more demanding test procedures, such as real driving emissions, including more stringent durability requirements, in the form of deterioration factors, and severe low temperature tests, widening the BCs under which the engine emissions must be carefully controlled. In this context, the automotive industry demands flexible AT modeling tools that could play a crucial role in both the early engine design phase and in the final vehicle on-board applications.

The aim of the authors has been to support the AT community in such a challenging framework, providing an overview of the state of the art in the AT modeling as well as examples of real-world applications.

In this chapter, after a first description of the principal catalyst technologies used in AT components, a deep insight in modeling techniques has been given, both for flow-through and for WFMs. Main hypotheses have been described and governing equations have been given, covering different AT design options. Different modeling approaches have been analyzed, including detailed physics-based models, which are mainly used in the exhaust AT design and development phase, as well as reduced order models, which combine the physics-based accuracy with the reduced computational effort, making such models suitable for direct integration on the vehicle, for on-board diagnostics, and emission control applications. Practical examples have been included and the approach to experimental measurements for modeling purpose has been analyzed.

Definitions, Acronyms, and Abbreviations

Symbols

$\mathbf{A_k}^{(l)}$ - Active site density for coverage k in layer l (mol-site/m^3)
$\mathbf{a_n}^{(l)}$ - Active site density of reaction n in layer l (mol-site/m^3)
$\mathbf{C_fRe}$ - Friction factor
$\mathbf{Cp_g}$ - Heat capacity of bulk gas (J/(kg K))
$\mathbf{Cp_i}$ - Heat capacity of gas (inlet/outlet channel) (J/(kg K))
$\mathbf{Cp_{sb}}$ - Heat capacity of substrate (J/(kg K))
$\mathbf{D_{eff,j}}$ - Diagonal matrix of effective diffusivities (m^2/s)
$\mathbf{D_h}$ - Hydraulic diameter of the channel (m)
$\mathbf{D_i}$ - Hydraulic diameter of the channel (m)
$\mathbf{D_{i,eff}}^{(l)}$ - Effective diffusivity of species i in washcoat layer l (m^2/s)
$\mathbf{D_{i,j}}$ - Binary diffusion coefficients (m^2/s)
$\mathbf{D_{i,kn}}$ - Knudsen diffusion coefficient (m^2/s)
$\mathbf{D_{i,m}}$ - Mixture-averaged diffusion coefficient (m^2/s)
$\mathbf{D_{m,i}}$ - Mass diffusion coefficient for species i
$\mathbf{dp}^{(l)}$ - Pore diameter of washcoat layer l (m)
$\mathbf{f}^{(l)}$ - Volume fraction of layer l
$\mathbf{f_{sb}}$ - Volume fraction of substrate
$\mathbf{f_{vd}}$ - Void fraction of reactor
h - Heat transfer coefficient (J/(m^2 s K))
$\mathbf{h_i}$ - Heat transfer coefficient (inlet/outlet channel) (J/(m^2 s K))
$\mathbf{H_n}$ - Enthalpy of reaction (J/(mol K))
$\mathbf{h_x}$ - External heat transfer coefficient (J/(m^2 s K))
$\mathbf{k_i}$ - Mass transfer coefficient of species i (m/s)
L - Length of the reactor (m)
MW - Molecular weight (g/mol)
$\mathbf{MW_i}$ - Molecular weight of species i, expressed as (kg/mol)
Nu - Nusselt number
P - Pressure (atm)
Pe - Peclet number
$\mathbf{Pe_w}$ - Wall Peclet number
$\bar{\mathbf{r}}_n^{(l)}$ - Average reaction rate across the washcoat thickness in layer l (mol/(mol-site s))
Re - Reynolds number
$\mathbf{Re_w}$ - Wall Reynolds number
$\mathbf{R_g}$ - Gas constant (J/(mol K))
$\mathbf{R_i}^{(l)}$ - ith species rate in layer l (kg/(m^3 s))
$\mathbf{R_j}$ - Vector of dimensionless species mass rates
S - Specific area per reactor volume (m^{-1})
$\mathbf{s_{i,n}}$ - Stoichiometric coefficient of species i in reaction n
$\mathbf{S_x}$ - External surface area per reactor volume (m^{-1})
T - Temperature (K)
$\hat{\mathbf{T}}_b$ - Dimensionless temperature
$\mathbf{T_g}$ - Gas temperature (K)
$\mathbf{T_s}$ - Solid temperature (K)
$\mathbf{T_x}$ - External temperature (K)

$\mathbf{u}$ - Average gas velocity (m/s)
$\mathbf{v}_k$ - Atomic diffusion volume (cm^3)
$\mathbf{v}_w$ - Flow velocity in the soot cake/substrate (m/s)
$\hat{\mathbf{v}}_{z,av}$ - Dimensionless average axial velocity
$\mathbf{y}_j$ - Mass fraction of species j
$\boldsymbol{\beta}$ - Momentum flux factor
$\boldsymbol{\delta}^{(l)}$ - Effective washcoat thickness in layer l (m)
$\boldsymbol{\varepsilon}^{(l)}$ - Porosity of layer l
$\boldsymbol{\theta}_k^{(l)}$ - Surface coverage of species k in layer l
$\boldsymbol{\Lambda}_s$ - Effective thermal conductivity of reactor (J/(m s K))
$\boldsymbol{\lambda}_{s,b}$ - Thermal conductivity of substrate (J/(m s K))
$\boldsymbol{\mu}_i$ - Viscosity (m^2/s)
$\boldsymbol{\rho}_g$ - Density of bulk gas (kg/m^3)
$\boldsymbol{\rho}_s$ - Gas density within the washcoat (kg/m^3)
$\boldsymbol{\rho}_{s,b}$ - Density of substrate (kg/m^3)
$\boldsymbol{\rho}_w$ - Density of the washcoat (kg/m^3)
$\boldsymbol{\sigma}_{i,j}$ - Force constant
$\boldsymbol{\sigma}_{k,n}$ - Stoichiometric coefficient for coverage k in reaction n
$\boldsymbol{\tau}^{(l)}$ - Tortuosity
$\boldsymbol{\psi}_s$ - Effective heat capacity of reactor (J/(m^3 K))
$\boldsymbol{\Omega}_D$ - Collision integral
$\boldsymbol{\omega}_{g,i}$ - Mass fraction of species i in the bulk gas
$\boldsymbol{\omega}_i$ - Mass fraction of species i in the washcoat
$\boldsymbol{\omega}_s$ - Vector of species fractions at the washcoat surface/channel
$\boldsymbol{\omega}_{s,i}$ - Surface concentration of species i

Acronyms

ASC - Ammonia slip catalyst
AT - Aftertreatment
BC - Boundary condition
BDF - Backward difference formulas
CFD - Computational fluid dynamics
CSTRs - Continuously stirred tank reactors
DAE - Differential algebraic equation
DEF - Diesel exhaust fluid
DOC - Diesel oxidation catalyst
DPF - Diesel particulate filter
ECU - Engine control unit
EGR - Exhaust gas recirculation
GPF - Gasoline particulate filter
LNT - Lean NOx trap
MOL - Method of lines
ODE - Ordinary differential equation
OSC - Oxygen storage capacity
PDE - Partial differential equation
PF - Particulate filter
PGM - Platinum group metals
PM - Particle matter
QSS - Quasi steady state
SCR - Selective catalytic reduction

SCRoF - SCR on filter
SGB - Synthetic gas bench
TPD - Temperature programmed desorption
TWC - Three-way catalyst
WFM - Wall flow monolith

References

1. Heck, R.M., Farrauto, R.J., and Gulati, S.T., *Catalytic Air Pollution Control: Commercial Technology*, 3rd ed. (2016), ISBN 978-0-470-27503-0.
2. Montenegro, G. and Onorati, A., "1D Thermo-Fluid Dynamic Modeling of Reacting Flows inside Three-Way Catalytic Converters," *SAE Int. J. Engines* 2(1) (2009): 1444-1459, doi:10.4271/2009-01-1510.
3. Millo, F., Rafigh, M., Sapio, F., Barrientos, E.J., and Ferreri, P., "Application of Genetic Algorithm for the Calibration of the Kinetic Scheme of a Diesel Oxidation Catalyst Model," 2018, doi:10.4271/2018-01-1762.
4. Millo, F., Rafigh, M., Sapio, F., Wahiduzzaman, S., Dudgeon, R., Ferreri, P., and Barrientos, E., "Modeling NO_x Storage and Reduction for a Diesel Automotive Catalyst Based on Synthetic Gas Bench Experiments," *Industrial Engineering and Chemistry Res.* 57, no. 37 (2018): 12335–12351, doi:10.1021/acs.iecr.8b01813.
5. Millo, F., Rafigh, M., Fino, D., and Miceli, P., "Application of a Global Kinetic Model on an SCR Coated on Filter (SCR-F) Catalyst for Automotive Applications," *Fuel* 198 (2017): 183–192, doi:10.1016/j.fuel.2016.11.082.
6. Colombo, M., Nova, I., Tronconi, E., Schmeißer, V., Bandl-Konrad, B., and Zimmermann, L., "Experimental and Modeling Study of a Dual-Layer (SCR+PGM) NH_3 Slip Monolith Catalyst (ASC) for Automotive SCR Aftertreatment Systems. Part 1. Kinetics for the PGM Component and Analysis of SCR/PGM Interactions," *Applied Catalysis B: Environmental* 142–143 (2013): 861–876, doi:10.1016/j.apcatb.2012.10.031.
7. Kuo, J.C.W. and Lassen, H.G., "Mathematical Modeling of CO and HC Catalytic Converter Systems," SAE Technical Paper 710289, 1971," doi:10.4271/710289.
8. Young, L.C. and Finlayson, B.A., "Mathematical Models of the Monolith Catalytic Converter," *AIChE Journal* 22, no. 2 (1976): 331–344, doi:10.1002/aic.690220217.
9. Heck, R.H., Wei, J., and Katzer, J.R., "Mathematical Modeling of Monolithic Catalysts," *AIChE Journal* 22, no. 3 (1976): 477–484, doi:10.1002/aic.690220310.
10. Liu, Z.G. and Miller, R.K., "Flow Distributions and Pressure Drops of Wall-Flow Diesel Particulate Filters," *SAE 2002 World Congress & Exhibition*, SAE International, 2002, https://doi.org/10.4271/2002-01-1311.
11. Payri, F., Arnau, F.J., Piqueras, P., and Ruiz, M.J., "Lumped Approach for Flow-Through and Wall-Flow Monolithic Reactors Modelling for Real-Time Automotive Applications," *WCX World Congress Experience*, SAE International, 2018, doi:https://doi.org/10.4271/2018-01-0954.
12. Konstandopoulos, A.G., Kostoglou, M., Vlachos, N., and Kladopoulou, E., "Progress in Diesel Particulate Filter Simulation," SAE Technical Paper 2005-01-0946, 2005, doi:10.4271/2005-01-0946.
13. Mohammed, H., Triana, A.P., Yang, S.-L., and Johnson, J.H., "An Advanced 1D 2-Layer Catalyzed Diesel Particulate Filter Model to Simulate: Filtration by the Wall

and Particulate Cake, Oxidation in the Wall and Particulate Cake by NO_2 and O_2, and Regeneration by Heat Addition," SAE Technical Paper 2006-01-0467, 2006, doi:10.4271/2006-01-0467.

14. Gundlapally, S.R. and Balakotaiah, V., "Heat and Mass Transfer Correlations and Bifurcation Analysis of Catalytic Monoliths with Developing Flows," *Chemical Engineering Science* 66, no. 9 (2011): 1879–1892, doi:10.1016/j.ces.2011.01.045.
15. Chapman, S., Cowling, T.G., Burnett, D., and Cercignani, C., *The Mathematical Theory of Non-uniform Gases: An Account of the Kinetic Theory of Viscosity, Thermal Conduction and Diffusion in Gases* (Cambridge University Press, 1990), ISBN 9780521408448.
16. Fuller, E.N., Schettler, P.D., and Giddings, J.C., "New Method for Prediction of Binary Gas-Phase Diffusion Coefficients," *Industrial and Engineering Chemistry* 58, no. 5 (1966): 18–27, doi:10.1021/ie50677a007.
17. Bird, R.B., Stewart, W.E., and Lightfoot, E.N., *Transport Phenomena* (2007), ISBN 0471410772, doi:10.1016/j.ijhydene.2006.08.059.
18. Zhang, F., Hayes, R.E., and Kolaczkowski, S.T., "A New Technique to Measure the Effective Diffusivity in a Catalytic Monolith Washcoat," *Chemical Engineering Research and Design* 82, no. 4 2004: 481–489, doi:10.1205/026387604323050191.
19. Wurzenberger, J.C. and Wanker, R., Multi-Scale SCR Modeling, 1D Kinetic Analysis and 3D System Simulation, SAE Technical Paper 2005-01-0948, 2005, doi:10.4271/2005-01-0948.
20. Metkar, P.S., Harold, M.P., and Balakotaiah, V., "Experimental and Kinetic Modeling Study of NH_3-SCR of NOx on Fe-ZSM-5, Cu-Chabazite and Combined Fe- and Cu-Zeolite Monolithic Catalysts," *Chemical Engineering and Science* 87, no. x (2013): 51–66, doi:10.1016/j.ces.2012.09.008.
21. Václavík, M., Kočí, P., Novák, V., and Thompsett, D., "NOx Conversion and Selectivity in Multi-Layer and Sequential DOC-LNT Automotive Exhaust Catalysts: Influence of Internal Transport," *Chemical Engineering Journal* 329 (2017): 128–134, doi:10.1016/j.cej.2017.05.129.
22. Scheuer, A., Hauptmann, W., Drochner, A., Gieshoff, J., Vogel, H., and Votsmeier, M., "Dual Layer Automotive Ammonia Oxidation Catalysts: Experiments and Computer Simulation," *Applied Catalysis B: Environmental* 111–112 (2012): 445–455, doi:10.1016/j.apcatb.2011.10.032.
23. Aris, R., *The Mathematical Theory of Diffusion and Reaction in Permeable Catalysts: Vol. 1: The Theory of the Steady State* (Oxford Studies in Physics, 1975).
24. Joshi, S.Y., Harold, M.P., and Balakotaiah, V., "Overall Mass Transfer Coefficients and Controlling Regimes in Catalytic Monoliths," *Chemical Engineering and Science* 65, no. 5 (2010): 1729–1747, doi:10.1016/j.ces.2009.11.021.
25. Balakotaiah, V., "On the Relationship between Aris and Sherwood Numbers and Friction and Effectiveness Factors," *Chemical Engineering and Science* 63, no. 24 (2008): 5802–5812, doi:10.1016/j.ces.2008.08.025.
26. Mozaffari, B., Tischer, S., Votsmeier, M., and Deutschmann, O., "A One-Dimensional Modeling Approach for Dual-Layer Monolithic Catalysts," *Chemical Engineering and Science* 139 (2016): 196–210, doi:10.1016/j.ces.2015.09.021.
27. Bissett, E.J., "An Asymptotic Solution for Washcoat Pore Diffusion in Catalytic Monoliths," *Emission Control Science Technology* 1, no. 1 (2015): 3–16, doi:10.1007/s40825-015-0010-2.

28. Bissett, E.J., "Small Washcoat Diffusion Resistance, Further Developments," *CLEERS Workshop*, Dearborn, USA, 2015.

29. Gundlapally, S.R., Dudgeon, R., and Wahiduzzaman, S., "Efficient Solution of Washcoat Diffusion-Reaction Problem for Real-Time Simulations," *Emiss. Control Sci. Technol.* 4, no. 2 (2018): 90–102, doi:10.1007/s40825-018-0083-9.

30. Gundlapally, S.R., *Effect of Non-Uniform Activity and Conductivity on the Steady-State and Transient Performance of Catalytic Reactors* (University of Houston, 2011).

31. Oh, S.H., Bissett, E.J., and Battiston, P.A., "Mathematical Modeling of Electrically Heated Monolith Converters: Model Formulation, Numerical Methods, and Experimental Verification," *Industrial and Engineering Chemistry Research* 32, no. 8 (1993): 1560–1567, doi:10.1021/ie00020a005.

32. Bissett, E.J., "Mathematical Model of the Thermal Regeneration of a Wall-Flow Monolith Diesel Particulate Filter," *Chemical Engineering Science* 39, no. 7–8 (1984): 1233–1244, doi:10.1016/0009-2509(84)85084-8.

33. Konstandopoulos, A.G. and Johnson, J.H., Wall-Flow Diesel Particulate Filters—Their Pressure Drop and Collection Efficiency, 1989, doi:10.4271/890405.

34. Haralampous, O. and Koltsakis, G., "Oxygen Diffusion Modeling in Diesel Particulate Filter Regeneration," *AICHE J.* 50, no. 9 (2004): 2008–2019, doi:10.1002/aic.10181.

35. Haralampous, O.A. and Koltsakis, G.C., "Back-Diffusion Modeling of NO_2 in Catalyzed Diesel Particulate Filters," *Industrial and Engineering Chemistry Research* 43, no. 4 (2004): 875–883, doi:10.1021/ie034187p.

36. Kandylas, I.P. and Koltsakis, G.C., "NO 2 -Assisted Regeneration of Diesel Particulate Filters: A Modeling Study," *Industrial and Engineering Chemistry Research* 41, no. 9 2002: 2115–2123, doi:10.1021/ie010842m.

37. Bissett, E.J. and Wang, W., "On the Implications of Wall Reynolds Number Dependent Nusselt Number and Friction Factor on the Accuracy of Wall - Flow DPF Modeling," ModeGat III, 2013.

38. Bissett, E.J., Kostoglou, M., and Konstandopoulos, A.G., "Frictional and Heat Transfer Characteristics of Flow in Square Porous Tubes of Wall-Flow Monoliths," *Chemical Engineering Science* 84 (2012): 255–265, doi:10.1016/j.ces.2012.08.012.

39. Koltsakis, G.C. and Stamatelos, A.M., "Modes of Catalytic Regeneration in Diesel Particulate Filters," *Industrial and Engineering Chemistry Research* 36, no. 10 1997: 4155–4165, doi:10.1021/ie970095m.

40. Wang, W. and Bissett, E.J., "Frictional and Heat Transfer Characteristics of Flow in Triangle and Hexagon Channels of Wall-Flow Monoliths," *Emiss. Control Sci. Technol.* 4, no. 3 2018: 198–218, doi:10.1007/s40825-018-0093-7.

41. Zhao, C., Zhu, Y., Wang, Y., and Huang, S., "Pressure Drop and Soot Regeneration Characteristics through Hexagonal and Square Cell Diesel Particulate Filters," 2017, ISBN 2017010979, doi:10.4271/2017-01-0979.

42. Iwasaki, K., "Innovative Aluminum Titanate Based - Diesel Particulate Filter Having Asymmetric Hexagonal Cell Geometry," 1–10, 2012, doi:10.4271/2012-01-0838.

43. Tsuneyoshi, K. and Yamamoto, K., "Experimental Study of Hexagonal and Square Diesel Particulate Filters under Controlled and Uncontrolled Catalyzed Regeneration," *Energy* 60 (2013): 325–332, doi:10.1016/j.energy.2013.07.069.

44. Gu, T. and Balakotaiah, V., "Impact of Heat and Mass Dispersion and Thermal Effects on the Scale-up of Monolith Reactors," *Chemical Engineering Journal* 284 (2016): 513–535, doi:10.1016/j.cej.2015.09.005.

45. Mcmackin, M., "SCR Model Reduction Process for Physics-Based ECU Code," *GT European User Conference*, Frankfurt, 2017.

46. Skaf, Z., Aliyev, T., Shead, L., and Steffen, T., "The State of the Art in Selective Catalytic Reduction Control," 2014, doi:10.4271/2014-01-1533.

47. Gundlapally, S.R., Papadimitriou, I., Wahiduzzaman, S., and Gu, T., "Development of ECU Capable Grey-Box Models from Detailed Models---Application to a SCR Reactor," *Emission Control Science and Technology* 2, no. 3 (2016): 124–136, doi:10.1007/s40825-016-0039-x.

CHAPTER 12

3D Simulation Models for After-Treatment Systems

O. Haralampous
University of Thessaly

G. Koltsakis
Aristotle University of Thessaloniki

12.1 Introduction

Exhaust after-treatment modeling has been widely introduced in the research and application of exhaust after-treatment systems in both gasoline and diesel engine applications. The simplified single channel, one-dimensional approach is probably still the preferred approach to simulate in a timely manner complex systems operating at highly transient conditions without significant loss of accuracy. The suitability of the 1D approach is questioned only when the operating conditions (temperature, mass flux, gas composition) is expected to vary substantially among the numerous channels of the catalytic substrate or filter. Exhaust system manufacturers make every effort to ensure the homogeneity of these parameters at the entrance of the device; however this usually requires ample space upfront in order to minimize the effect of flow recirculation that usually appears in such systems. The increasing complexity of modern exhaust systems along with the need to reduce the volume of exhaust after-treatment and minimize backpressure poses problems that can only be solved with certain design compromises resulting to deviations from the ideally uniform inlet flow. In this environment, 3D modeling becomes increasingly important. This chapter examines typical examples of exhaust after-treatment modeling problems addressed via the 3D approach. In the case of gasoline engine applications, multi-dimensional 3-way catalyst modeling has been used to study the effects of the inlet cone shape especially in close-coupled designs where pulsating flow effects are of additional importance. Recent trends in three-way catalyst (TWC) substrate design with nonuniform cell structures are oriented toward homogenizing the inlet flow distribution. We will also discuss the importance of 3D heat transfer modeling in applications with prolonged zero-flow operation, increasingly important in hybrid vehicles. In diesel engines, 3D modeling has been mostly of interest in selective catalytic reduction (SCR) systems modeling targeting at optimum utilization of the injected urea-water solution via maximizing evaporation rates and homogeneity upstream the catalytic substrate. Similar challenges exist in systems where a fuel injector is placed in the exhaust line for thermal management purposes in combination with an oxidation catalyst responsible to oxidize the injected fuel. A significant amount of relevant work has been devoted to multi-dimensional effects in diesel particulate filters during regeneration, since the longevity of such filters is usually determined by their endurance to the thermomechanical stresses developed when thermal gradients develop in the radial direction. In this chapter, the above problems will be presented along with proposed solution methods involving the use of different modeling approaches at reactor and at system level.

12.2 Flow Uniformity

12.2.1 Flow Distribution at 3-Way Catalyst Inlet Face

In order to achieve the legislative tailpipe emissions, the 3-way catalytic converter is expected to attain conversion efficiencies of the order of 95% or above for the main exhaust gas pollutants, namely, CO, hydrocarbons, and NO_x. To achieve this, it is important to make efficient use of the entire catalyst active surface area. From the fluid dynamics point of view, the uniformity of flow distribution at the entrance area of the monolith is of special interest. This subject has attracted much research effort, including both experimental and modeling studies. It is generally claimed that a uniform flow distribution at the inlet monolith face is favorable for both the conversion efficiency, light-off time [1, 2], as well as the durability of the catalytic converter [3]. Therefore, the main problem is to optimize the inlet pipe and diffuser geometry in order to minimize the flow maldistributions at catalyst inlet.

Experimental techniques have been employed to visualize the internal flow structure of a prototype monolith converter [4], concluding that the flow uniformity is a function of the inlet flow Reynolds number. Since then, computational methods relying mostly on CFD software have been widely used to provide more detailed information on the flow field as a function of various design and operating parameters. Three-dimensional flow field simulations at steady state conditions have been presented, including validation of the results with measurements. The effect of inlet geometry on flow distribution has been studied by CFD and a flow uniformity index has been defined as a criterion to quantify the results [5]:

$$\gamma = 1 - \frac{1}{2}\frac{\sum_{i=1}^{n}\left|u_i - \bar{u}\right| A_i}{\bar{u}\,A} \tag{12.1}$$

Several studies have employed a comprehensive CFD-based modeling approach to predict the steady-state performance of the catalytic converter, including the effects of heat and mass transfer in the monolith, surface reactions, heat generation, and ambient heat losses [6, 7].

More recently, transient 3D simulations of complete systems including multiple catalyst monoliths placed along the exhaust line have also been presented [8, 9, 10]. The problem of excessive CPU demands for transient CFD calculations has been addressed using parallel computing techniques, resulting in reasonable calculation times at least for steady-state calculations or simulations of some seconds of real-time operation.

FIGURE 12.1 The computational domain and a detailed view of the grid used inside the channel. Inlet is located at left [6].

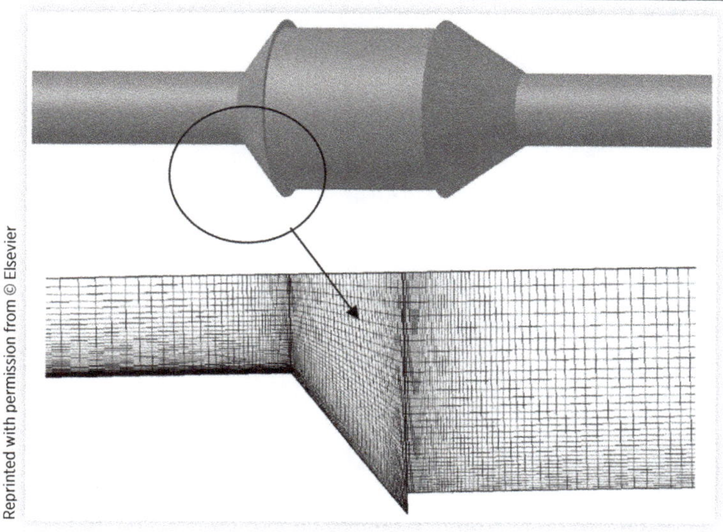

From the engineering point of view, it is of special interest to quantify the effect of flow distribution on the conversion efficiency of the catalytic converter under transient conditions resembling real-world operation. This can be done by coupling the results of the CFD simulations with mathematical models for the transport and chemical phenomena in the monolith channels. Figure 12.1 shows a typical computational mesh used to study the inlet flow distribution in a catalytic converter as presented in [6].

Since the geometry studied was axisymmetric, a 2D approach was employed. The axi-symmetric forms of the turbulent Reynolds-averaged Navier—Stokes equations were discretized by a finite volume technique.

The catalyst monolith effect on the flow is usually simulated by a porous medium with a prescribed pressure drop law. Since the flow through the monolith channels is laminar, the pressure drop per unit length is calculated by using the Hagen—Poiseuille equation:

$$\frac{\Delta p}{L} = \frac{28.5}{d_h^2} \mu u \tag{12.2}$$

where

μ is the dynamic viscosity that is a function of the temperature
u is the velocity through the effective area of the porous medium in the axial direction
dh is the hydraulic diameter of each single passage of the monolith

The pressure drop term is treated as an additional source term in the momentum equation.

The flow upstream and downstream the monolith is treated as turbulent and the low Reynolds version of the k-ε model is used for the turbulence modeling. In addition, the Reynolds stresses are expanded up to the quadratic terms (beyond to the Boussinesq's linear approximation) and thus a nonlinear eddy viscosity model is adopted by modifying accordingly the production term of the turbulence kinetic energy.

The boundary conditions for such a problem include prescribed mass flow rate and temperature at the inlet of the computational domain. For the turbulence quantities, the turbulence kinetic energy is computed by using the equation:

$$k = 1.5\left(Tu \cdot u_{ref}\right)^2 \tag{12.3}$$

where
u_{ref} is the mean velocity corresponding at the selected mass flow rate through the inlet area
Tu is the turbulence level

Usually, the lack of experimental data leads to the use of values close to 5%, which is a typical value for internal channel flows. The value of the turbulence dissipation rate ε is calculated using the semi-empirical equation:

$$\varepsilon = \frac{c_\mu}{50} \rho_{ref} \frac{k^2}{\mu_{ref}} \tag{12.4}$$

where
$c_\mu = 0.09$, ρ_{ref}, and μ_{ref} are the fluid's density and dynamic viscosity at the inlet
k is the turbulence kinetic energy

The pressure at outlet is set equal to the ambient pressure and the mass flow rate through the exit section is corrected at each iteration in relation with the mass flow rate at inlet.

To restrict the study in isothermal conditions, the walls are treated as adiabatic. Special care should be taken to the location of the computational nodes near the wall. Since the low Reynolds variant of the k-ε model is used, the computational mesh is constructed to have the dimensionless local normal distance from the wall, y+, less than 1 for at least five nodes.

A structured grid having 32930 computational nodes was used for this particular simulation of the computational domain. This size was found to give grid-independent solutions for all the test cases examined in this work. Figure 12.1 shows the computational domain and a detailed view of the grid used in the region near the intersections of the inlet pipe with the diffuser and of the diffuser with the porous medium simulating the monolith. The area occupied by the insulating material was accounted for in the formulation of the grid near the diffuser—monolith interface.

The velocity vector field in the axi-symmetric plane of the intersection between the diffuser and the monolith and for a mass flow rate corresponding to 0.021 kg/s is shown in Figure 12.2. There is a large separation region in the diffuser's walls, which strongly affects the radial distribution of the longitudinal velocity at monolith inlet phase.

Previously published experimental results [11] were used to validate the CFD simulation results. Validation is performed by comparing the gas velocity profiles at the monolith outlet, calculated by CFD against the ones obtained experimentally, for the previously described converter set-up. Figure 12.3 presents the respective comparison for three different exhaust gas flow rates, at 300 K. A good agreement is obtained in all three cases, which highlights the accuracy of the employed CFD technique.

The results indicate that for the case of a typical geometry, the flow is expected to be significantly nonuniform mostly at high flow rate conditions. Similar profiles have

FIGURE 12.2 Vector plot in the diffuser walls upstream the monolith's inlet section [6].

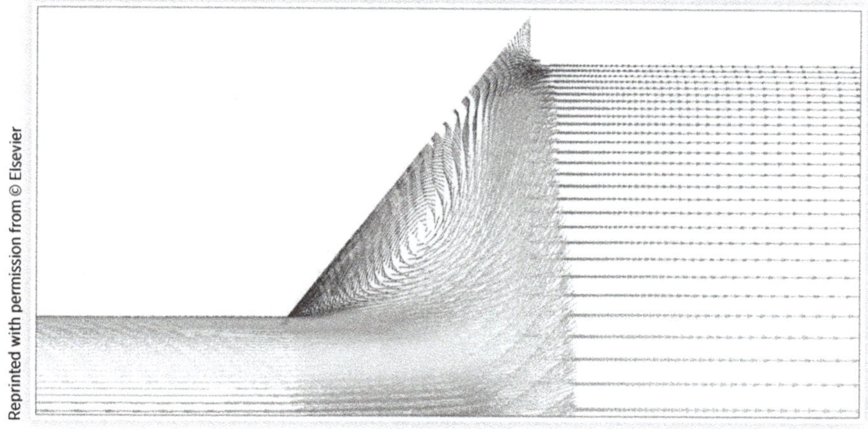

FIGURE 12.3 Comparisons CFD - experimental data. Exhaust gas temperature: 300 K, monolith length: 100 mm [6].

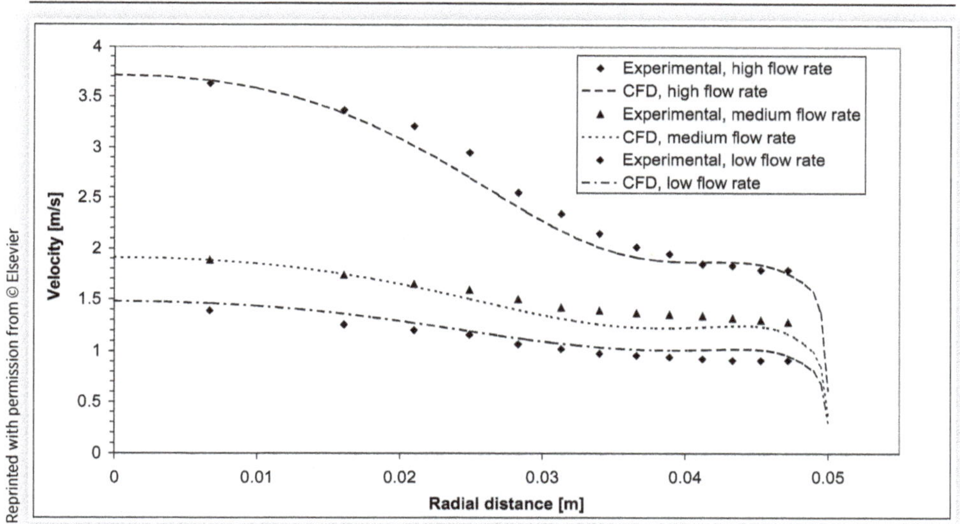

been reported by several researchers. Of note, Hayes et al. [9] showed that the presence of a weld protrusion between the pipe and diffuser has a significant effect on the flow profile. Jeong [10] compared several turbulence models and selected the low Reynolds $k - \omega$ SST based on their agreement with the experimental results of Clarkson [12].

12.2.2 Effect of Substrate Properties

It is important to stress that the flow distribution uniformity is not only a function of the inlet piping and cone geometry, but also depends on the flow resistance induced by the monolith. As described above, the flow resistance is in turn a function of the channel geometry (hydraulic diameter) and the gas viscosity. In the following, we will examine the importance of monolith properties on flow distribution.

During the 1990s, the concept of substrate design with radially nonuniform flow resistance was proposed and studied by Kim and Son [13]. Recently the same concept

FIGURE 12.4 Schematic of a conventional and a radially nonuniform substrate. Red and green represent different cell structures. Blue denotes insulation and canning.

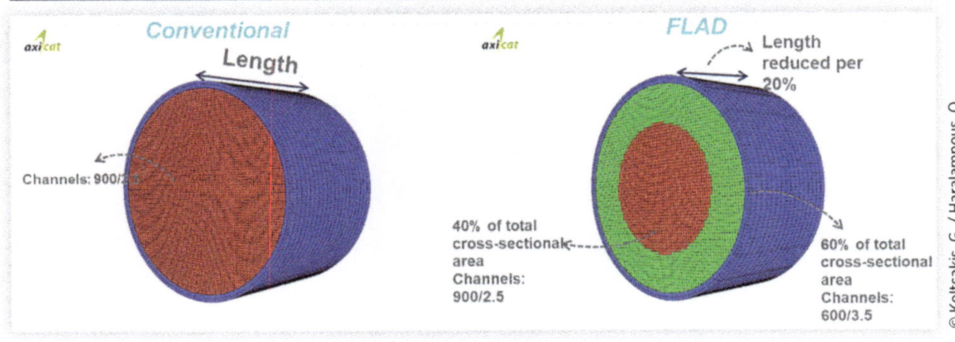

was implemented in an integrally molded substrate [14], named FLAD (flow-adjustable designed cell). FLAD is a registered trademark of Denso Corporation. This newly designed catalyst substrate has a different cell cross-sectional area at the inner portion compared to that at the outer portion.

The target of this concept is to create regions of low flow resistance in the areas where the flow is less favored as a result of inlet cone shape, as shown in Figure 12.4. This is expected to enable a more uniform flow distribution and an improved trade-off between flow uniformity and backpressure. As such, a better use of the precious metal loading can be achieved, resulting in volume and cost reduction with the same aftertreatment performance.

The effect of such a design on the flow distribution can be predicted by solving the flow distribution in a system with an inlet cone and a catalyst, characterized by a porous medium with radially variable flow resistance parameters, corresponding to the local channel shape. This can be conveniently performed by using CFD tools coupled with 3D monolithic reactor tools allowing multiple "radial zones" as shown in Figure 12.4. The results are summarized in Figure 12.5 illustrating that a radially nonuniform channel structure may indeed result in a smoother flow distribution. Such simulations may be extended to benchmark different designs in terms of backpressure and species conversion efficiency in transient cycles.

FIGURE 12.5 Calculated flow distribution for substrates with nonhomogeneous cell structure.

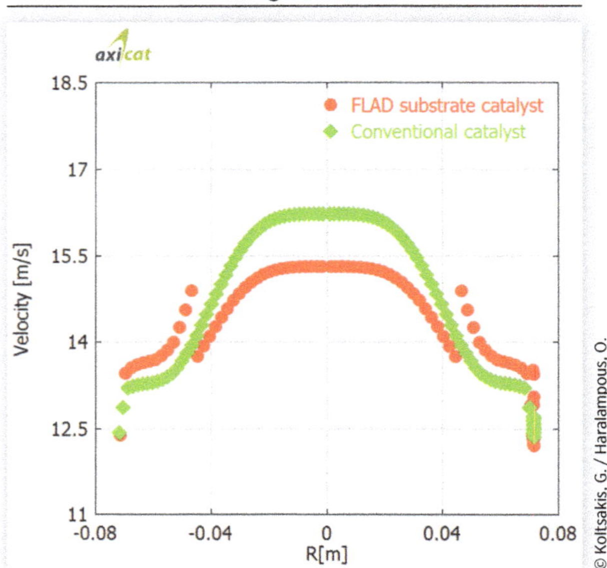

12.2.3 Pulsating Flow Effects

In order to exploit the maximum of the exhaust gas enthalpy, it has become common practice to place the catalyst as close as possible to the exhaust valves in order to reduce light-off times. For such configurations, the flow can be highly pulsating and consequently a fully transient flow analysis becomes necessary as opposed to traditional quasi-steady analysis. In order to develop and optimize such an exhaust system, detailed computational methods need to be employed, taking into account the complex interactions between

FIGURE 12.6 Schematic representation and computational mesh of a close-coupled catalyst system [10].

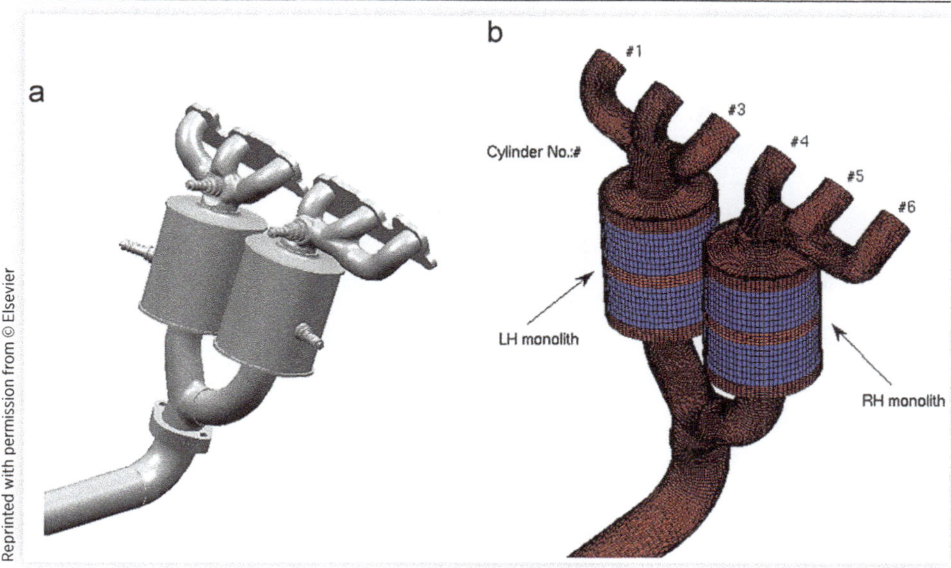

gas pulsations, heat and mass transfer, surface reactions, and heat release, as shown by Missy et al. since 2002 [15, 16].

More recently, Jeong published a detailed computational study of pulsating flow effects from a six-cylinder engine [10]. Figure 12.6 presents the exhaust system geometry with two close-coupled catalysts, as well as the computational mesh. This configuration together with a suitable engine warm-up strategy can be used to achieve early light-off. Figure 12.7 illustrates the gas temperature inside the close-coupled catalyst corresponding to cylinders four to six at the corresponding crank angles of port firing 195°, 435°, and 675°. The applied boundary condition of exhaust gas flow rate and temperature for 1200 rpm WOT is shown in Figure 12.8a. At the start of each pulse, there is a period of about 75° that the exhaust gas remains relatively cool (~700°C) before increasing sharply. Keeping this in mind, the cool areas of Figure 12.7 highlight the locations of fresh exhaust gas flow and offer a qualitative impression of exhaust gas mixing. The same areas can be identified in the concentration results of Jeong for their lower emission conversion.

Furthermore, Figure 12.8b shows the uniformity factor of the same catalyst as a function of crank angle. Over the engine cycle, γ varies from 0.6 to 0.95. Especially at high mass flow rates, the uniformity index reaches high values greater than 0.8. As noted by Missy et al., steady-state simulations typically show higher uniformity indices, even if the higher mass flow rate induced by the pressure wave of the opening valve is applied as a boundary condition.

In a recent work [17], Mat Yamin et al. experimentally investigated the combined effect of flow rate and pulsation frequency on the periodic flow field in a planar diffuser upstream of monoliths of different length. Using particle image velocimetry and hot wire anemometry, they were able to capture these effects in a relatively simple yet representative after-treatment configuration as a function of a non-dimensional parameter J, (reciprocal of the Strouhal number) defined as the ratio of pulse period to residence time within the diffuser. Due to pulsations the flow field varied throughout the cycle. Initially, as the flow accelerated, it remained attached to the diffuser walls for some distance. Separation

FIGURE 12.7 Gas temperature (K) fields in right catalyst after port firing at 1200 rpm. (a) 195°, cylinder 4 (b) 435°, cylinder 5, and (c) 675°, cylinder 6 [10].

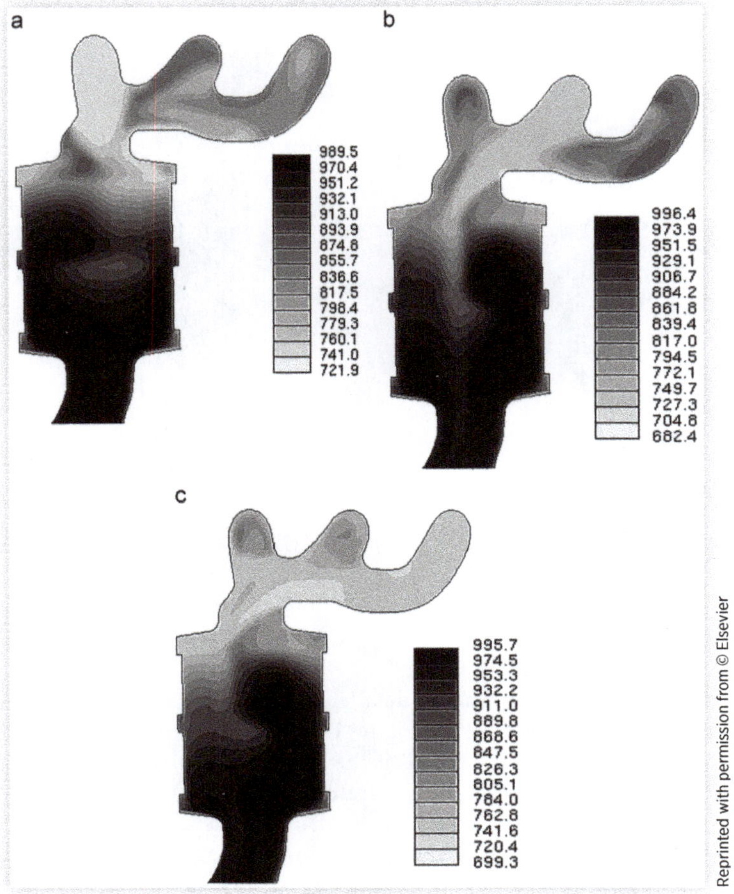

FIGURE 12.8 Transient mass flow rate, temperature, and uniformity index during engine cycle. (a) Mass flow rate and temperature (1200 rpm, WOT). (b) Flow uniformity index [10].

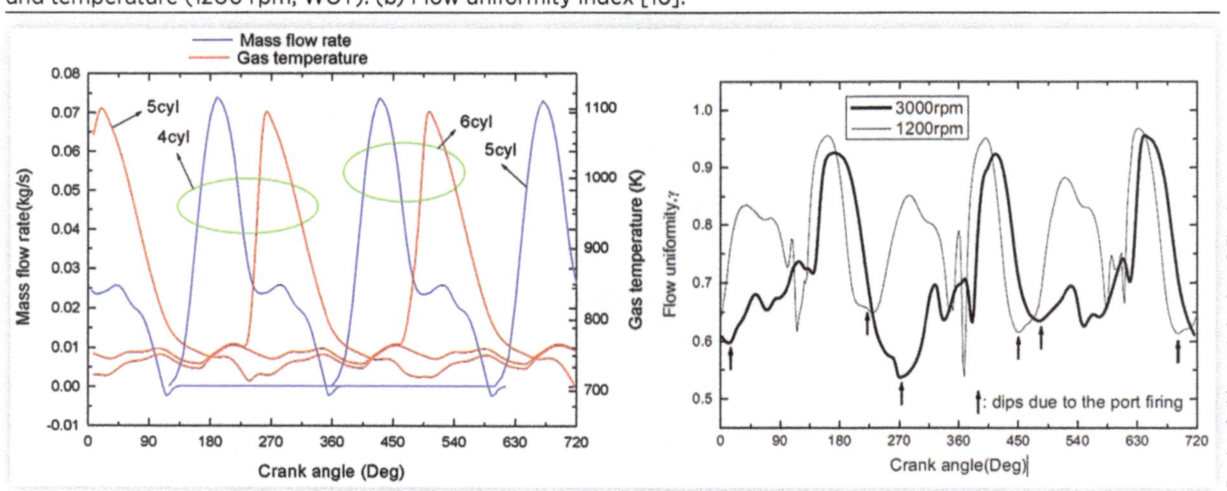

FIGURE 12.9 Steady flow and cycle-averaged velocity profiles at the monolith exit. Velocities are normalized by the cycle-averaged mean velocity downstream of the monolith [17].

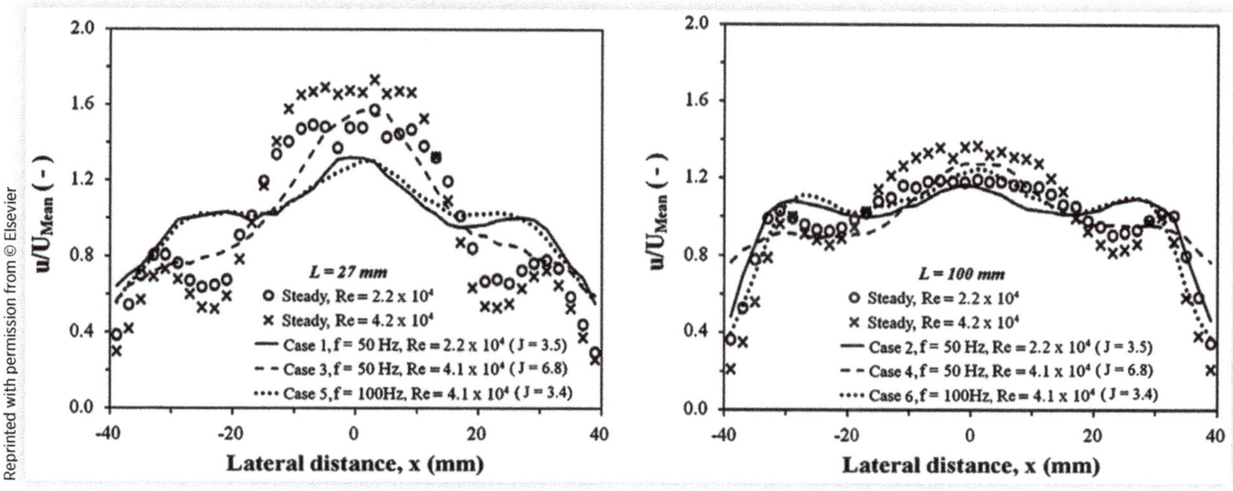

bubbles then formed near the diffuser inlet resulting in the development, later in the cycle, of two large recirculating vortices. These vortices occupied the diffuser volume at the end of the pulse before being transported out during the subsequent cycle.

Figure 12.9 compares the cycle-averaged pulsating flow distributions downstream of a short (27 mm) and a longer (100 mm) monolith with those for steady flow. Such a comparison is useful for assessing the adequacy of deducing flow maldistribution in engine flows based on a steady flow analysis. Maldistribution is lower for J ~ 3.5 (low flow rate or low frequency), increases at J = 6.8, and is highest with steady flow (J infinite). The detailed flow fields show that separation occurs earlier during the cycle for J = 6.8 with larger vortex structures dominating the diffuser. For J ~ 3.5, the flow is able to reattach downstream of the separation bubbles during part of the cycle thus presenting more uniform flow to the monolith. Low Reynolds number (Re) and high-frequency pulses (low J) do not permit the flow to establish sufficient inertia to provide the fully separated flow regimes observed under steady flow conditions and so result in flatter profiles within the monolith. The differences are less pronounced for the longer monolith, as the higher resistance flattens the flow field upstream and within the monolith for both steady and pulsating flows. Mat Yamin et al. conclude that similar relationships would be expected for more practical configurations in agreement to other published studies [18, 19]. Consequently, conditions of high flow, low engine speed, and after-treatment systems with short monoliths (e.g., close-coupled systems) will result in increased flow maldistribution.

12.2.4 Swirl Effects

As a result of the downsizing trend, many engines are equipped with turbocharges and their close-coupled catalysts are exposed to residual swirl. Rusli et al. [20] studied the flow uniformity experimentally and numerically as a function of the swirl number *S*, defined as the ratio of the axial flux of angular momentum to the axial flux of the axial momentum. Figure 12.10 shows that with the introduction of low swirl ($S = 0.23$),

FIGURE 12.10 Velocity distribution 30 mm downstream of the monolith (normalized by the mean velocity in the same cross-section) for $S = 0$ and $S = 0.23$, 63 g/s [20].

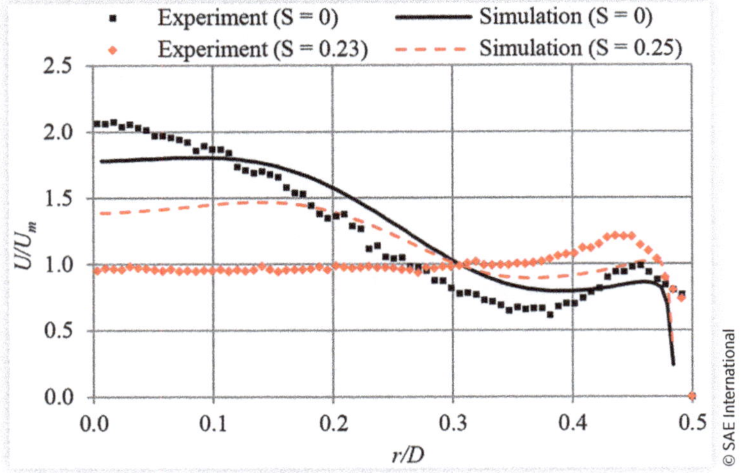

the velocity profile tends to become uniform. As it was explained, the force driving the central jet is balanced by the centrifugal force, forcing the flow to redistribute toward the diffuser wall. Higher swirl numbers have a negative effect on uniformity, as most of the flow concentrates on the periphery. Of importance to modeling, they conclude that although RANS CFD models offer a useful insight into the effect of swirl phenomena in exhaust systems, anisotropic turbulence models, such as RST or LES, are necessary to improve the accuracy of simulations. This is in line with the conclusions of Fisher et al. [21] with respect to swirl mixers, as discussed in more detail in Section 12.4.1.5.

12.2.5 Geometry Optimization Techniques

Optimizing the uniformity of the catalyst and the total backpressure of the exhaust system requires iterative CFD simulations of several geometrical variations. The selection of geometrical parameters to optimize might be straightforward for simple components such as pipes and standard cones but is generally nontrivial for free-form geometries. The continuous adjoint CFD method [22, 23] is an inherently suitable solution for this kind of problem with its automatic adjustment of flow region and mesh, as demonstrated by Hinterberger et al. [24]. Furthermore, the procedure time is in the same order of magnitude as a single flow calculation.

Figure 12.11 presents a typical catalyst with a free-form entrance geometry. The packaging space and the boundary conditions at the inlet and outlet of this geometry are fixed. Within the meshed volume an immersed boundary separates the flow and non-flow (solid) region. This is the optimization element of the problem, while a cost function is used, which typically represents the deviation of velocity profile from uniformity, as well as the total energy loss of the system. The equivalent Lagrange function includes the fluid dynamic equations as constraint and the respective Lagrange multiplier constitutes the so-called adjoint flow field.

A representative example is given Figure 12.12, which depicts the flow velocity, adjoint velocity and sensitivities for backpressure (a and c), and flow uniformity (b and d) at various iteration numbers. After 24 iterations (a), the negative backpressure

FIGURE 12.11 Sketch of package space (a) and of CFD model (b) [24].

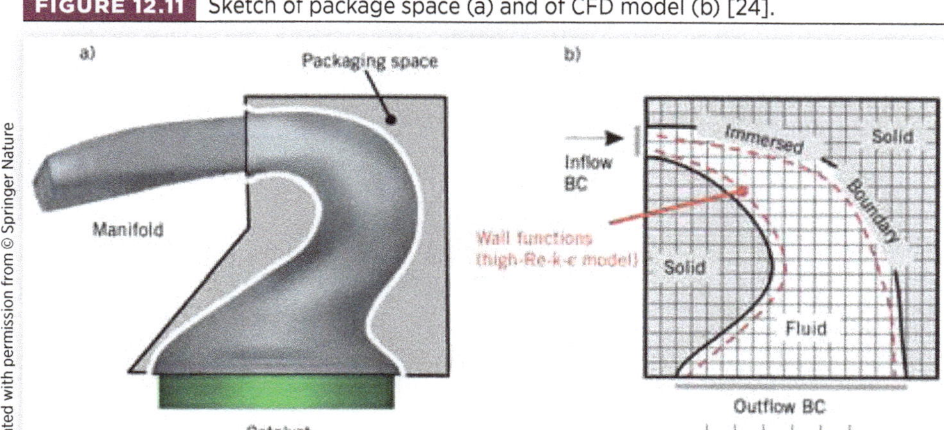

FIGURE 12.12 Sensitivities for backpressure in inlet cone (a, c) and for flow uniformity across the catalyst (b, d) after 24 iterations (a), after 460 iterations (b), and after 2500 iterations (c, d); final cone geometry after 2500 iterations (c, d) [24].

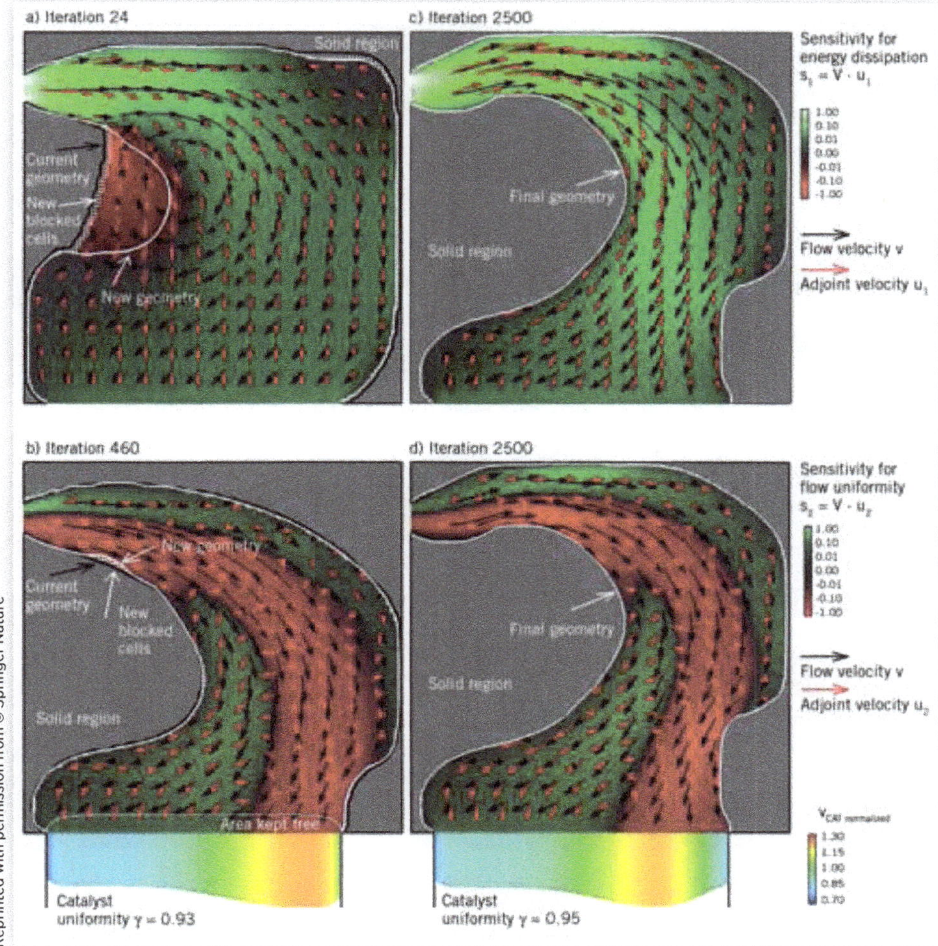

sensitivity zone indicates an area where obstruction of the flow would be beneficial. Iteration 460 shows excessive nonuniformity on the right side of the volume, which can be mitigated by blocking the flow in the lower right corner. The final result (c and d) is a compromise between conflicting targets, as concluded by Hinterberger et al. [24].

12.3 3D Heat Transfer Modeling

12.3.1 Transient Heat Transfer

A light-off scenario or a legislated driving cycle is a typical application for engineers using 1D models, however both these cases are considered challenging in 3D. Transient CFD simulations of reacting flows with conjugate heat transfer are generally very time-consuming and avoided. One of the difficulties in conventional transient CFD approaches is the requirement of time steps in the order of ms to converge. The boundary conditions, as well as the thermal, physical, chemical state of the porous medium, vary with a time scale of 1s, while the gas balances reach steady-state much faster. Thus, it is possible to assume quasi-steady-state conditions for the gas and simulate the transient state of porous medium and pipe walls using large time steps. Such an approach was presented by Missy et al. already in 2002 [15]. Recently, Luo et al. [25] reported a speedup of 15–20 times using a pseudo steady state solver in a transient simulation of an urea/SCR system. As will be discussed later, a similar approach has been recently applied by the authors in full-scale transient drive-cycle simulations.

In the general case, the 3D transient heat transfer equation in the monolith is given by the equation:

$$\frac{\partial}{\partial x}\left(\lambda_{s,x}\frac{\partial T_s}{\partial x}\right)+\frac{\partial}{\partial y}\left(\lambda_{s,y}\frac{\partial T_s}{\partial y}\right)+\frac{\partial}{\partial z}\left(\lambda_{s,z}\frac{\partial T_s}{\partial z}\right)+S=\rho_s C_{p,s}\frac{\partial T_s}{\partial t} \tag{12.5}$$

which includes a source term for the exchange of heat with exhaust gas and reaction exotherms. Furthermore, the honeycomb structure imposes an anisotropy to conductivity, which is taken into account. The control volume can be extended beyond the substrate radius to include the packaging materials, namely, insulation and steel. Thus, the boundary conditions, expressed at the canning periphery, take into account convection (free and forced) as well as radiation to ambient:

$$\lambda_{s,r}\frac{\partial T_s}{\partial r}=h_P\left(T_s-T_\infty\right)+\sigma\,\varepsilon_{rad}\left(T_s^4-T_{sur}{}^4\right) \tag{12.6}$$

Similarly, heat losses by free convection and radiation from the front and rear faces of the device are taken into account as shown below:

$$\lambda_{s,z}\frac{\partial T_s}{\partial z}=h_F\left(T_s-T_\infty\right)+\sigma\left(1-\varepsilon\right)\varepsilon_{rad}\left(T_s^4-T_{sur}{}^4\right) \tag{12.7}$$

The appropriate ambient gas temperature T_∞ and surrounding temperature T_{sur} is used for each boundary ($r = R$, $z = 0$ or $z = L$).

In many cases, besides the substrate, the metal piping energy balance has to be included in full-scale transient models to obtain accurate results. To this end, the modeling approach applied by the authors comprises of OpenFOAM [26] steady-state simulations invoked iteratively by axisuite [27]. At the start of each time step, axisuite calculates the local heat source terms, mass source terms and flow resistance in the porous medium, as well as the time-varying pipe wall temperature profile and imposes these constraints to the CFD simulation. After convergence of the CFD simulation, the results are retrieved by axisuite to continue solving the transient porous medium and pipe balances and advance in time.

Figure 12.13 shows an indicative result of a TWC with bend inlet pipe. Starting from an initial solid temperature of 310 K, the exhaust system undergoes a warmup scenario of constant inlet exhaust gas flow and temperature. The temperature profiles in the inlet cone show a considerable variation in the first 70s due to the gradual heat-up of the pipe walls.

The catalyst inlet temperature nonuniformity and its variation during the first seconds of a driving cycle can be critical for the accurate prediction of conversion performance. In most cases, this phenomenon was roughly approximated with 1D and

FIGURE 12.13 Simulation of a TWC warmup scenario demonstrating the impact of inlet pipe thermal inertia. Constant inlet boundary conditions: 650 K, 0.0152 g/s. Initial solid temperature: 310 K.

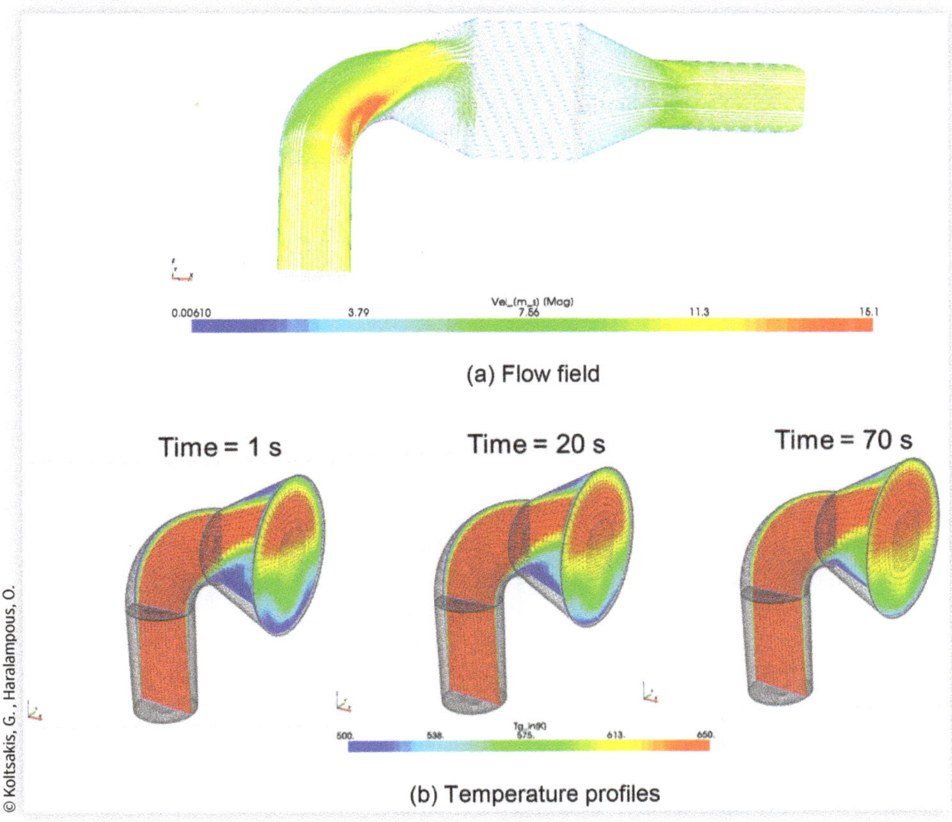

FIGURE 12.14 3D model effect on HC emissions prediction in transient cycle (LA04).

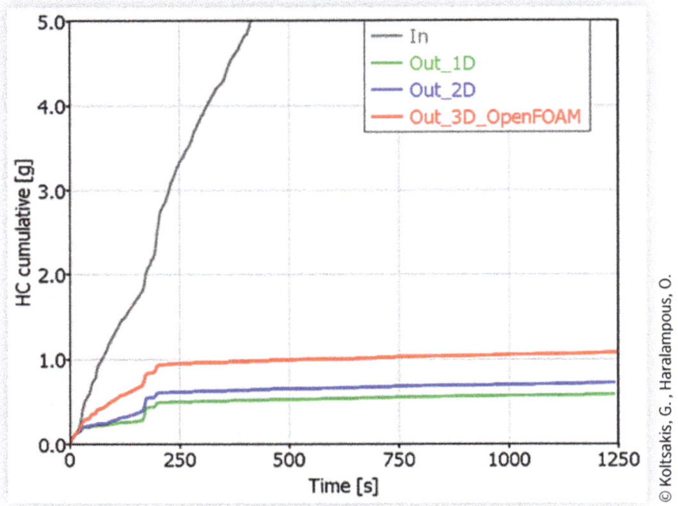

2D models. Using the modeling technique described above and a modern personal computer, the simulation of a full-scale transient 3D cycle of 30 min duration is currently possible in practical CPU times (<50 h CPU). Figure 12.14 compares the cumulative HC emissions calculated with models of various accuracy for the aforementioned system in an LA04 cycle. The 3D model results indicate a more gradual activation of the catalyst taking into account the effect on pipe walls inertia, which leads to considerably higher emissions compared to 1D and 2D.

12.3.2 Heat Losses under Zero Flow Conditions

The pressure to reduce CO_2 emissions by increased powertrain efficiency results in generally lower exhaust gas temperatures. In addition, a hybrid operating system is characterized by frequent engine start-and-stops, thus several cooling-downs of the exhaust system. These cooling periods may result in catalyst temperature decrease below the light-off temperature, which will temporarily deactivate the aftertreatment device. Therefore, it is crucial to study and optimize the thermal behavior of the exhaust system to avoid unwanted light-outs of the catalytic devices after prolonged zero flow operation.

The numerical investigation of the heat transfer under near-zero mass flow is a rather recent technological challenge and only few research works have been published on this subject. Chung et al. [28] performed a numerical simulation of the catalytic converter cool-down process using computational fluid dynamics (CFD) software and validated it with experimental results. Chen et al. [29] provided a 3D model to simulate the thermal and conversion characteristics of monoliths operating under flow maldistribution conditions. Klein and Cheng [30] studied the catalyst temperature behavior in regular engine shut-down and restarting. Braun et al. [31] simulated the cold start-up of an automotive three-way catalytic converter using a CFD-code application. Shu et al. [32] investigated the temperature as a function of time for both steady-state and transient driving operations. Adamczyk et al. [33] studied the device cool-down characteristics for the cold-start converter modeling. Katare and Laing

[34] made a hybrid approach for the exhaust aftertreatment modeling and, more specifically, for a DOC.

Koltsakis et al. [35] used the commercial axisuite® modeling platform to simulate the 3D flow and heat transfer inside the aftertreatment system comprising a DOC and diesel particulate filter (DPF) in-series under transient operating conditions with emphasis on zero flow modes. The numerical method is validated with experimental measurement results from cool-down/light-off experiments, performed on an engine test bench.

In case of near-zero flow conditions, natural convection from the monolith faces becomes important. The convection coefficient from the monolith faces h_F in Equation (12.7) is computed using the equations for natural convection for a vertical plate. For the 3D simulation, the following equations are used:

The local Nusselt number is computed based on the following equation:

$$Nu_x = \left(\frac{Gr_x}{d}\right)^{1/4} g(\mathrm{Pr}) \tag{12.8}$$

with

$$g(\mathrm{Pr}) = \frac{0.75\sqrt{\mathrm{Pr}}}{\left(0.609 + 1.221\sqrt{\mathrm{Pr}} + 1.238\,\mathrm{Pr}\right)^{1/4}} \tag{12.9}$$

and

$$Gr_x = \frac{g\beta\left(T_s - T_\infty\right)x^3}{\nu^2} \tag{12.10}$$

The emissivity ε_{rad} refers to the substrate material. The term (1-ε) refers to the fraction of the face area of a monolith occupied by solid material. These theoretical expressions have been developed for simple geometries, thus corrections in the final heat transfer coefficients might be necessary to obtain agreement with experimental measurements.

The conventional and the hybrid operation mode result in different temperature, mass flow rates, and concentration profiles at the entrance of the DOC, as demonstrated by Koltsakis et al. [35]. The temperatures in the hybrid mode are significantly lower suggesting a high risk of operation below the light-off temperature. It is interesting to note that, even though the engine is switched off for the 42% of the New European Driving Cycle (NEDC) time duration in the hybrid operation mode, the total CO emissions are not lower compared to the conventional cycle. This is due to the high CO emissions observed immediately after restarting the engine. On the other hand, the engine-out HC emissions during hybrid mode are substantially lower compared to those of the conventional mode.

It is important to check whether the model is able to reproduce accurately the temperature field during engine shut-off and the subsequent effect on DOC conversion efficiency performance. Indeed, Figure 12.15 demonstrates that the 3D model, equipped with the free-convection and radiation losses, is able to reproduce the measured temperature field inside the DOC over the complete driving cycle. Additionally, the thermal response during a zero flow rate period is shown in more detail. The predicted temperature agrees with the thermocouple measurement at T10.

In order to understand the thermal behavior of the DOC during engine shut-off, Figure 12.16 shows the calculated temperature field in a cross-section of the DOC, as a function of time. Due to the round shape of the catalyst and the symmetric inlet flow conditions, it is observed that the predicted temperature field is not axi-symmetric.

FIGURE 12.15 Measured vs calculated temperatures in hybrid mode operation, (a) complete NEDC, (b), investigation area [35].

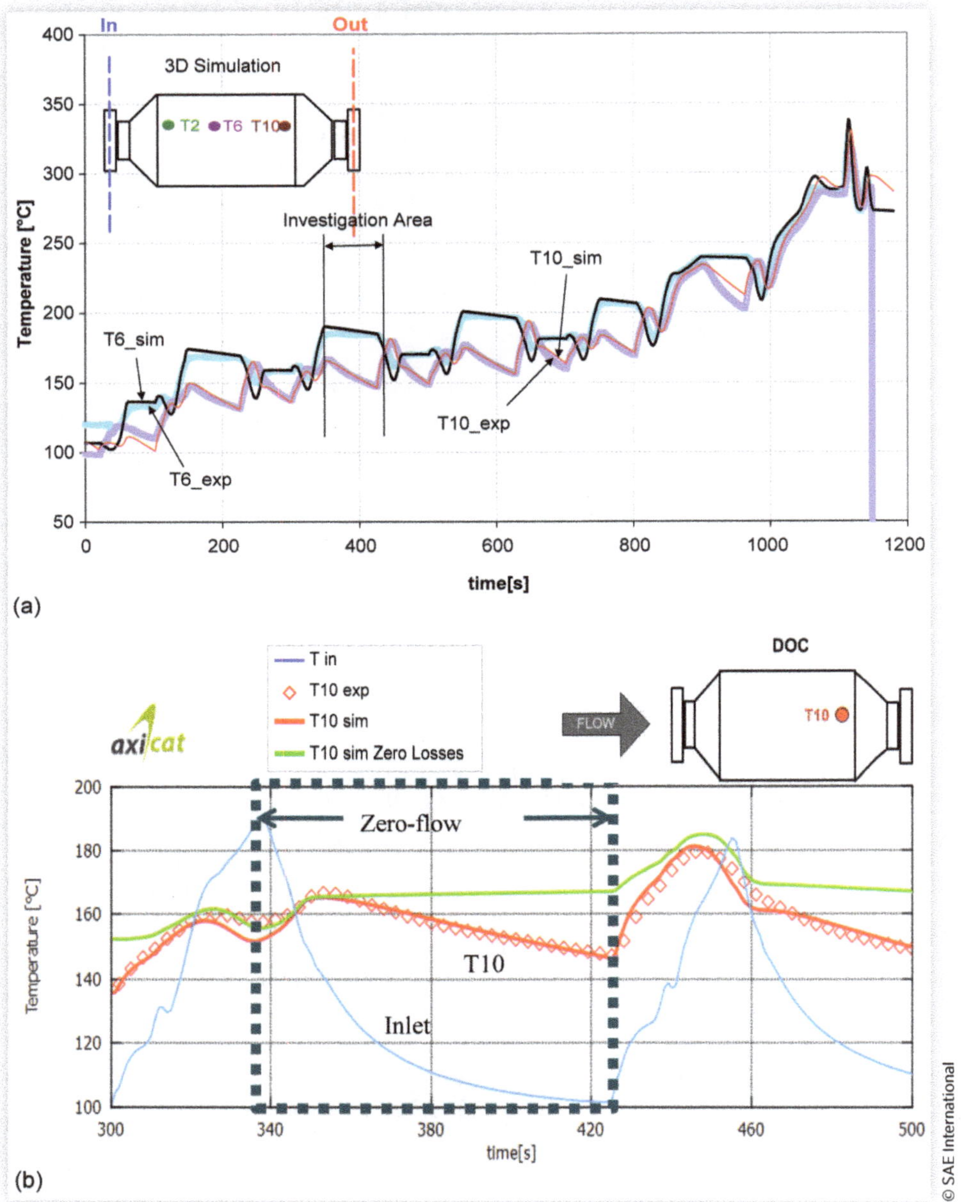

This is due to the gravity-related nature of free-convection losses. In the model, the local heat transfer coefficient depends on the distance from the bottom part of the DOC without any symmetry from the centerline.

Based on the successful calculation of the substrate temperature, the model is able to predict accurately CO and HC emissions during hybrid operation mode, applying the same reaction rate parameters developed for the conventional mode. The cumulative mass error is less than 5%, which is within the repeatability and measurement accuracy margins [35]. These results confirm the applicability of 3D catalyst modeling in hybrid powertrain applications.

FIGURE 12.16 Calculated temperature field in a cross-section of the DOC vs time during zero flow mode [35].

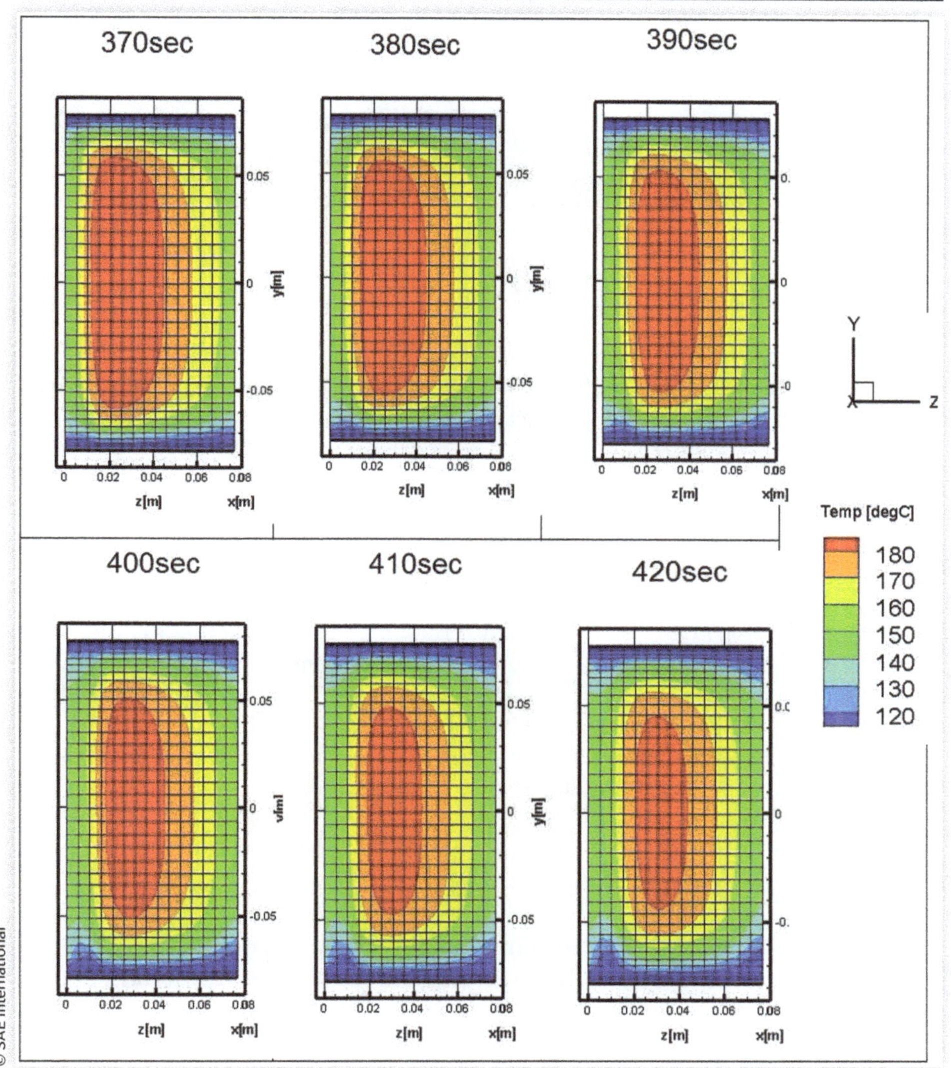

12.4 3D Multiphase Flows in Exhaust Systems

12.4.1 Urea-Water Solution Injection in SCR Systems

The SCR system involves a series of complex phenomena from the injection of urea-water solution (UWS) until the reaction of NH_3 with NO_x in the converter. An overview is given in Figure 12.17. These phenomena interact with the 3D flow and boundary conditions in the exhaust system, as a result 3D modeling is crucial for the design of SCR systems [36] using Eulerian–Lagrangian CFD methods.

FIGURE 12.17 Phenomena under consideration in a 3D SCR model.

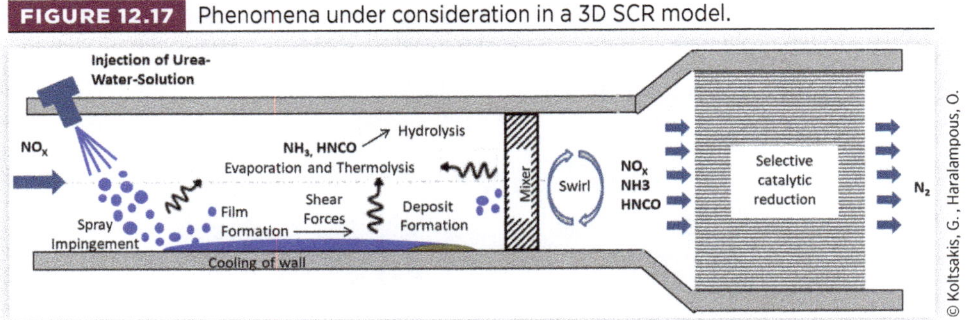

12.4.1.1 SPRAY DYNAMICS

The droplet size distribution in a real-world urea-SCR system depends on the injection system. In passenger cars and trucks, injection velocities typically range from 5 to 25 m/s with a droplet Sauter mean diameter of 20–150 µm and a temperature of 300–350 K [40]. The dosing system has to ensure the proper preparation of the reducing agent at all operating conditions. A Rosin-Rammler distribution is usually employed to model the droplet distribution in conjunction with a uniform injection velocity [37, 38]. A more advanced approach has been recently presented by Edelbauer et al. [39], which involves volume of fluids (VOF) simulation coupled with standard Lagrangian particle tracking (LPT). The VOF method is applied on a high resolution mesh close to the injector to predict the spray formation up to a transition zone, where the droplet information is passed to the LPT for the simulation of the rest of the domain with a coarser mesh.

After injection, the droplets follow a trajectory in the exhaust system, which is determined by the momentum balance:

$$m_d \frac{du_d}{dt} = F_D + F_B \tag{12.11}$$

In Eulerian–Lagrangian modeling of the urea-SCR system, the drag force and the buoyancy force are sufficient to correctly describe droplet motion [37]. The drag force is calculated as a function of the relative velocity U_C:

$$F_D = C_D \frac{\pi d_d^2}{4} \frac{\rho_g |U_C| U_C}{2} \tag{12.12}$$

where the drag coefficient C_D is given for the range $20 < Re_p < 260$ as:

$$C_D = \frac{24}{Re_d}\left(1 + 0.2\, Re_d^{0.63}\right) \tag{12.13}$$

and the buoyancy force is defined as:

$$F_B = m_d \frac{g(\rho_p - \rho_g)}{\rho_p} \tag{12.14}$$

12.4.1.2 WATER EVAPORATION

During their transport, droplets heat up due to heat convection initiating water evaporation. Abramzon and Sirignano [41] proposed an extended particle heat transfer model taking into account the thickening of the thermal and diffusion boundary layers around the droplets due to the Stefan flow induced by evaporation. A mathematical model of a droplet, of uniform temperature and composition, can be expressed as below:

$$\frac{dm_d}{dt} = 2\pi\bar{\rho}_g\bar{D}_g r_g Sh^* \ln\left(1+B_M\right) \tag{12.15}$$

$$\frac{dT_d}{dt} = \frac{\left(\frac{dm_d}{dt}\right)}{m_d \cdot c_{p,d}}\left(\frac{\bar{c}_{p,vap}\left(T_g - T_d\right)}{B_T} - h_{vap}\right) \tag{12.16}$$

where B_M and B_T are the Spalding mass and heat transfer numbers calculated as:

$$B_M = \frac{Y_{vap,s} - Y_{vap,g}}{1 - Y_{vap,g}} \tag{12.17}$$

$$B_T = \left(1+B_m\right)^x - 1, \quad \text{where } x = \frac{\bar{c}_{p,vap}}{\bar{c}_{p,g}}\frac{Sh^*}{Nu^*}\frac{1}{Le} \tag{12.18}$$

If boiling temperature is reached during evaporation, it is assumed that the droplet remains at this temperature, and the following model equations are used:

$$\frac{dm_d}{dt} = 2\pi\lambda_g r_g Nu^* \ln\left(1+B_T\right) \tag{12.19}$$

where B_T is given as:

$$B_T = \frac{\bar{c}_{p,vap}\left(T_g - T_s\right)}{h_{vap}} \tag{12.20}$$

In all the above equations, Sh^* and Nu^* are the modified Sherwood and Nusselt numbers that take into account the thickening of the film layers due to Stefan flow:

$$Sh^* = 2 + \left(Sh_0 - 2\right) / F\left(B_M\right) \tag{12.21}$$

$$Nu^* = 2 + \left(Nu_0 - 2\right) / F\left(B_T\right) \tag{12.22}$$

where $F(B)$ is the correction factor defined as:

$$F\left(B\right) = \left(1+B\right)^{0.7}\frac{\ln\left(1+B\right)}{B} \tag{12.23}$$

As for the values of Sh_0 and Nu_0, they can be obtained using the Frossling correlations, which are:

$$Sh_0 = 2 + 0.522\,\mathrm{Re}^{1/2}\,Sc^{1/3} \tag{12.24}$$

$$Nu_0 = 2 + 0.522\,\mathrm{Re}^{1/2}\,\mathrm{Pr}^{1/3} \tag{12.25}$$

Considering that the vapor is saturated at the droplet surface, $Y_{vap,s}$ is the calculated vapor pressure at the droplet temperature. Wherever mean property values are used for the gas film, they can be calculated using the 1/3 rule [41].

12.4.1.3 UREA DECOMPOSITION

When urea is injected as a solution, it is typically assumed that the liquid portion of the droplet is evaporated first. When all the water has evaporated, urea is treated as a solid particle. The decomposition of the solid particle is modeled based on the same principles as described above for the liquid droplets. The decomposition reaction produces NH_3 and HNCO:

$$(NH_2)_2CO(s) \rightarrow NH_3(g) + HNCO(g)$$

The reaction rate and temperature derivative can be expressed as [40]:

$$\frac{dm_u}{dt} = -2\pi \cdot r_u A e^{(-E/\Re T_u)} \tag{12.26}$$

$$\frac{dT_u}{dt} = \frac{\left(2\pi\lambda_g r_g Nu^*\left(T_g - T_d\right) + \frac{dm_u}{dt} h_{dcmp}\right)}{m_u c_{p,u}} \tag{12.27}$$

Furthermore, a homogeneous gas phase reaction can be assumed for the hydrolysis of isocyanic acid:

$$HNCO(g) + H_2O(g) \rightarrow NH_3(g) + CO_2(g)$$

The rate of hydrolysis is given by Yim et al. [42].

12.4.1.4 SPRAY-WALL INTERACTION AND WALL FILM FORMATION

In actual exhaust configurations, impingement of droplets on the catalyst and the walls cannot be excluded given the slow evaporation and thermolysis of the urea solution [37] and the inertia of the droplets. Under specific conditions, this can lead to the formation of melamine complexes [43], which should be avoided with an appropriate design of the exhaust system. Smaller droplets are generally favorable from this perspective.

The impact of droplet on rigid surfaces may produce a wide variety of consequences, according to the diameter, velocity, temperature of the impacting droplet, and the nature of the surface. Kuhnke [44] has identified four regimes based on two parameters:

$$K = \frac{(\rho D)_d^{3/4} u_d^{5/4}}{\sigma_d^{1/2} \mu_d^{5/4}} \tag{12.28}$$

$$T^* = \frac{T_w}{T_{sat}} \tag{12.29}$$

The critical transition temperature T^* usually takes values in the range of 1.1 although higher values have been used for modeling of urea-water injection experiments [37]. As shown in Figure 12.18, if the wall temperature is sufficiently high, it will remain dry as droplets rebound or break up thermally depending on parameter K. In both cases, the contact of the wall and liquid droplets will lead to heat loss from the former, with an upper limit equal to the evaporation rate of droplets. Given the higher thermal mass of the wall, a transient heat balance for the wall is necessary to account for all the phenomena that might occur in the exhaust system.

It is also possible that the wall temperature might drop below the critical limit under unfavorable conditions. Droplets can adhere to the surface for K values below 150, or even splash at higher K values. The result will be a formation of liquid film on the wall affecting the impinging of further droplets. The phenomena become quite complex in this case. In order to evaluate the heat loss of the wall to the liquid film, the Nukiyama curve has to be employed, taking into account the different boiling modes. The film interacts with the flow, the droplets, and the wall showing a transient and nonuniform behavior. Due to shear stresses, gravity, and pressure gradients, the film develops a velocity profile, which affects the continuity as well as enthalpy and urea

FIGURE 12.18 Regime map for spray/wall interaction [37] based on the original work of Kuhnke.

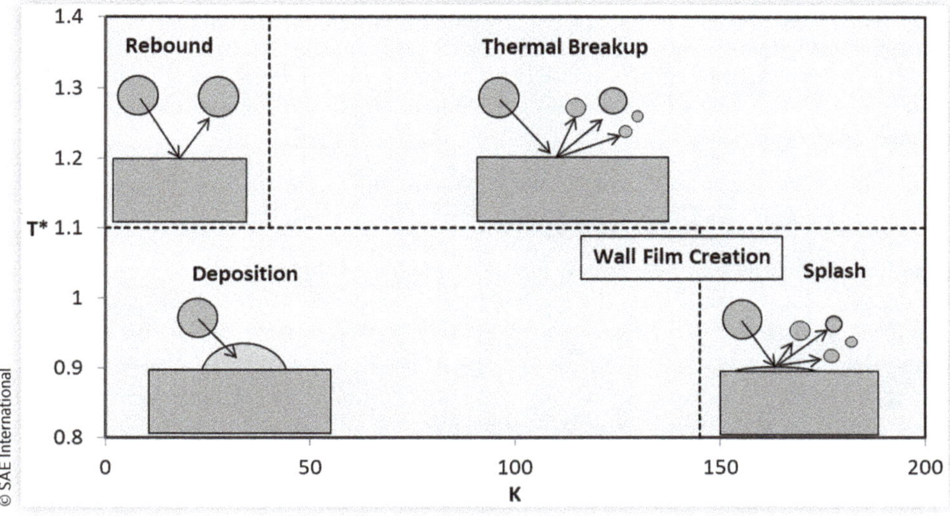

FIGURE 12.19 Simulated and measured stationary temperature field of the pipe wall in the region of the spray impingement [36].

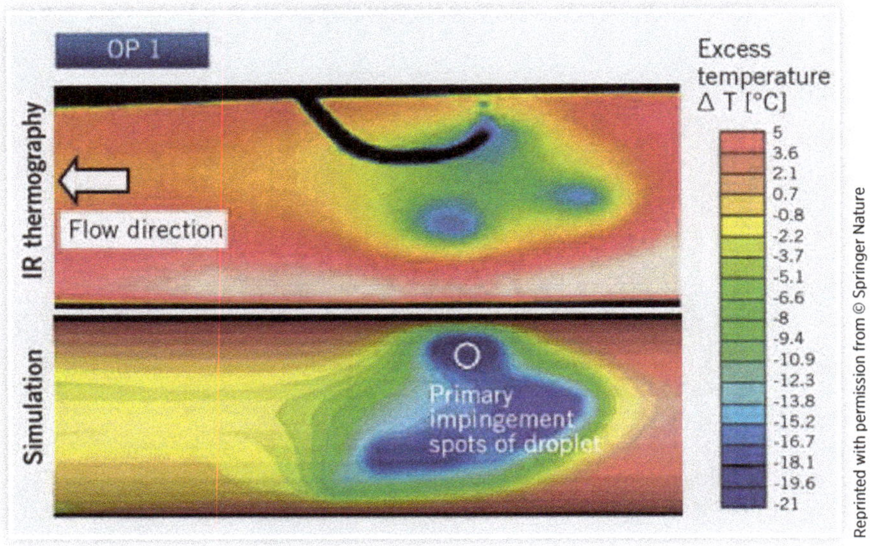

concentration balance. At the same time, droplet deposition, evaporation, and urea thermolysis contribute also in these balances. Additionally, given sufficient residence time and sufficiently low temperature, urea in the liquid film can react with the dissolved isocyanic acid and form solid by-products. The formation of by-products is a problem for the dosing module operation and may also affect the pressure drop or even clog the exhaust pipe. The interested reader is referred to the work of Birkhold et al. [37, 40] for a detailed presentation of the balance equations and to the work of Brack et al. [45, 46] for a kinetic reaction scheme regarding formation and decomposition of urea-derived by-products.

The correct prediction of the wall temperature is critical for the calculation of the well-wetting process. Echtle et al. [36] performed such a calculation and predicted the wall temperature taking into account the heat capacity and thermal conductivity of the pipe walls. The calculated temperature field was experimentally validated against measurements obtained with an infrared camera, as shown in Figure 12.19. For the specific engine operating point, a film is formed immediately due to the deposition of droplets, while free convection or nucleate boiling takes place at the liquid-solid interface. Due to the shear forces of the flow and gravitation, the wall film spreads beyond the primary impingement spot.

12.4.1.5 STATIC MIXER MODELING

A sufficient mixing length is not always available on a vehicle; as a result mixers are commonly placed downstream the injectors to increase turbulence and enhance urea/ammonia mixing with the exhaust gas. Additionally, mixers serve as hot surfaces that enable thermal breakup of droplets that impinge on them. SCR mixers are relatively simple and short, as they must operate at a low pressure drop [47]. The impingement of urea spray against the static mixer structure – while effective in the breakup of urea droplets – comes at a cost of increased pressure drop. As an alternative approach, swirl mixers have

FIGURE 12.20 Impact of two flat-bladed mixers on NH_3 distribution along the centerline of the exhaust pipe [49].

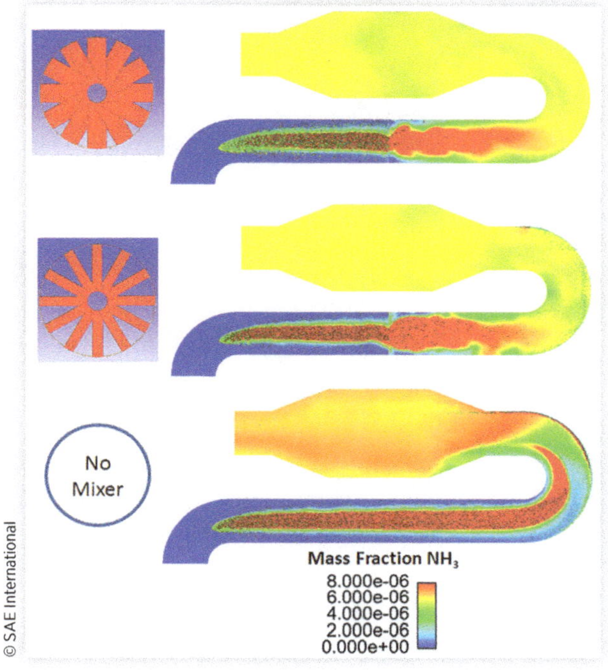

been developed, where the mixing is achieved through a swirl motion of the exhaust gas downstream of the urea injector. Besides typical mixers, geometrical configurations in the pipe and cone can serve as swirl generators. Ottaviani et al. [48] modified the merge pipe of a dual line exhaust system to generate swirl motion and homogenize the ammonia distribution at the entrance of an SCR catalyst just downstream.

The impact of two flat-bladed mixers on NH_3 distribution has been studied by Drennan et al. [49] for a representative exhaust configuration. As illustrated in Figure 12.20, without the mixer, a central core of high ammonia concentration is produced as droplets evaporate downstream the injector. A fraction of the droplets impact on the 180° bend and evaporate, producing regions of high ammonia concentration at the upper edge of the SCR inlet, given a sufficiently high pipe temperature. Much better uniformity is produced using the two mixers. Most of the droplets impact on the mixer blades and evaporate quickly, while a small number pass through the mixer. The impact is more pronounced for the wide blade mixer, although at higher pressure drop.

Figure 12.21 shows an extract of results that were calculated with the CFD model of a close-to-production exhaust system [36]. In the figure, the comparison of the simulation and measurements is shown for an exhaust system with a mixing device. The mixing device creates an airflow with a high vortex density that causes a better convective mixing of the ammonia vapor (left side). A dramatically improved uniform distribution follows from that at the catalyst entry (right side). Besides, the mixing device prevents to a significant extent wall wetting, mitigating wall film formation, and the problems coming along with it.

To validate the simulation results, spatially resolved measurements of the ammonia concentration were carried out. The detection of the reducing agent was

FIGURE 12.21 Simulation of the flow field and the ammonia distribution for a close-to-production exhaust system with mixing device (left), calculated and measured uniformity index along the mixing pipe (right) [36].

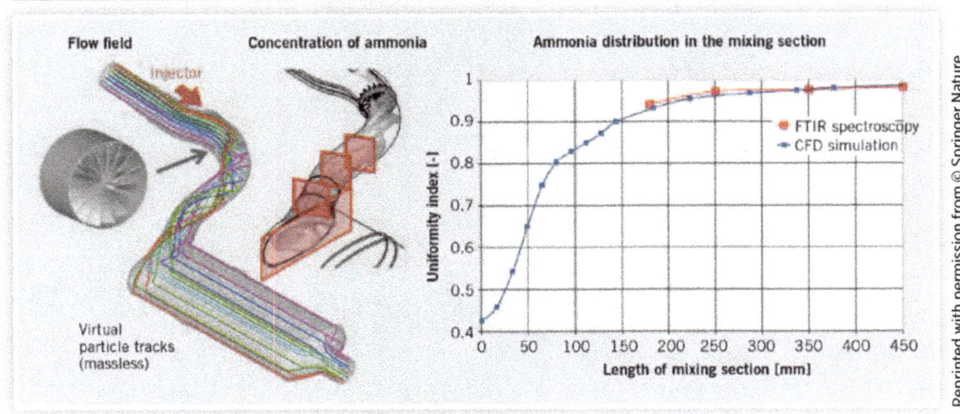

realized with the FTIR methodology at pipe cross sections varying from the beginning to the end of the mixing pipe. The ammonia distribution that was experimentally determined is plotted against the simulation results. An excellent correlation can be observed.

The selection of a suitable turbulence model and numerical special discretization scheme for such an application has been studied by Fisher et al. [21]. Using a representative state-of-the-art exhaust system with SCR and mixer, they compared several $k - \varepsilon$ schemes to the Reynolds stress model (RSM) with second-order discretization and noted considerable differences in terms of swirl number, mean turbulent kinetic energy (TKE), dissipation, and viscosity. These differences were attributed to elevated TKE values across as wide range of pipe cross-section and a qualitatively different energy turbulent dissipation topology, as shown in Figure 12.22. They conclude that the RSM with its anisotropic turbulence description is capable of predicting noticeable coupling between the primary swirl core and outer vortices. Furthermore, the deviation of the predicted uniformity index from the experimental was only 2% for a wide range of conditions. Given that the computational cost of RSM is about 20% higher but far lower than LES, it is considered a practical approach for industrial-type development.

FIGURE 12.22 Turbulent kinetic energy and dissipation on the mid-plane of the exhaust system under variation of the model setup [21].

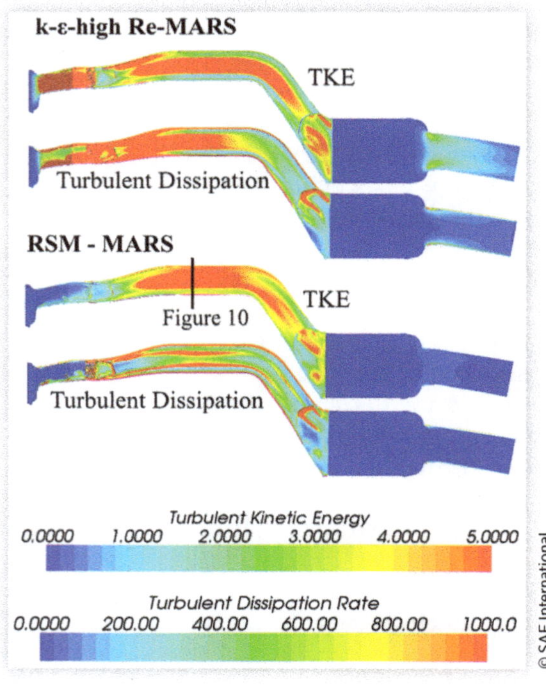

12.4.2 NH_3 Uniformity and Effect on deNOx

As described above, 3D CFD codes are capable of calculating the 2-phase reacting flow phenomena occurring in the gas phase upstream the catalytic converter. This is today a common practice to optimize the geometry of the pipes as well as the mixing regions and improve the injection systems for UWS. The typical target for the system designer is to guarantee a certain minimum value of the uniformity index in terms of NH_3 distribution at catalyst entrance. A nonuniform distribution will result in NH_3 excess in certain areas of the catalyst whereas other parts of the catalyst will receive less NH_3. After prolonged operation, the channels receiving more than the targeted NH_3 amount will be saturated and eventually NH_3 slip will occur. At the same time, the channels that are underfed with NH_3 may show reduced deNOx efficiency. If a certain level of deNOx performance is to be guaranteed, the dosing system will have to inject higher quantities of UWS to ensure that all channels receive the minimum NH_3 necessary for NO_x conversion. This obviously results in higher NH_3 slip from the "overfed" channels. The end result is the need for larger ammonia slip catalysts and the higher UWS consumption, both increasing operational and investment costs.

Even though the use of CFD to calculate the uniformity index of NH_3 at catalyst entrance has become a standard practice, the quantification of its effect on system efficiency is rarely examined. This would require the calculation of the reacting flow in the catalyst region using proper monolithic reactor modeling accounting for the selective catalytic surface reactions as well as the convective mass transfer in the channels and species diffusion within the active material. Such an attempt has been presented in [50] for an exemplary exhaust system geometry consisting of a long pipe providing mixing length for a UWS injector and a catalytic reactor, as shown in Figure 12.23. The simulation was performed in the commercial CFD software ANSYS/Fluent that was coupled with the catalytic reactor modeling software axisuite via a user function approach. In this way, it was possible to obtain nontrivial conclusions with respect to the combined effect of flow and NH_3 nonuniformity on deNOx performance. More specifically, Figure 12.24 shows that the calculated NO_x concentration at the exit of the catalyst is not only affected by the available NH_3 amount at the entrance of the catalyst but also by the local flow velocity. The latter determines the available

FIGURE 12.23 Calculated velocity field in the vicinity of the SCR catalyst [50].

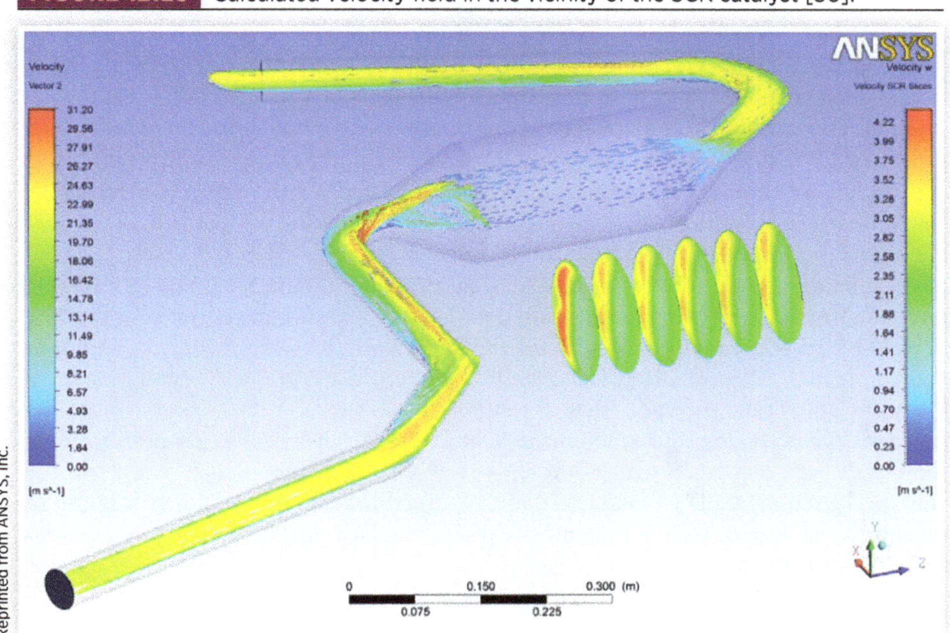

FIGURE 12.24 Calculated NO concentration field in the vicinity of the SCR catalyst [50].

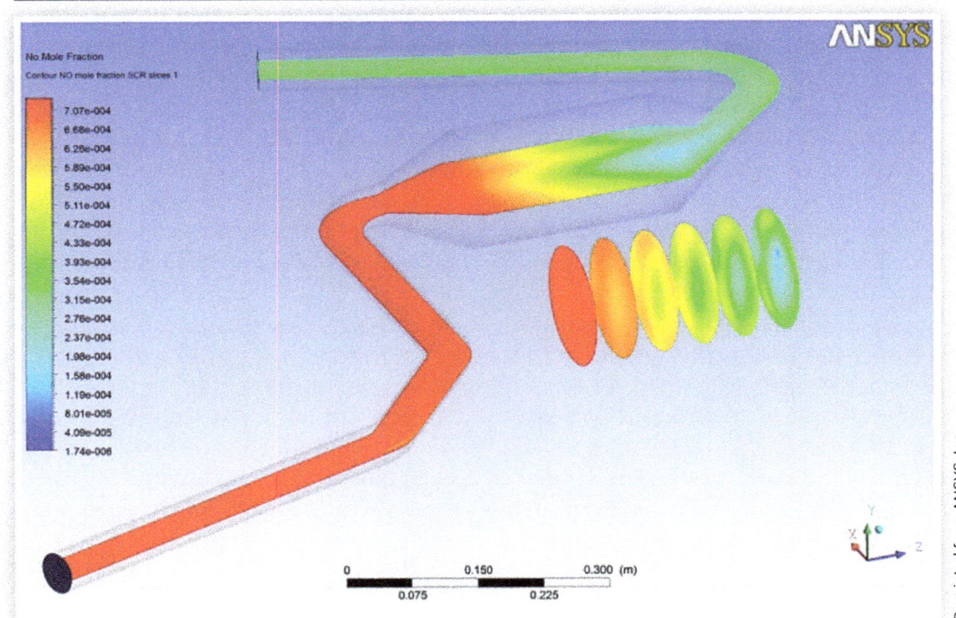

residence time of the reacting species in the active catalyst region that plays an important role in the reactor performance.

CFD analysis, either stand-alone or coupled with catalytic reactor calculations, is applicable at selected steady-state operating points of the engine operating map. The high computational burden involved in these calculations is still prohibitive to allow a fully transient 3D modeling approach for calculations of more than a few seconds of real-world operation. On the other hand, the SCR reactor performance is intrinsically highly affected by the catalyst "memory," that is, the amount of stored NH_3. Since the storage and release of NH_3 is taking place in timescales of minutes, the realistic prediction of deNOx SCR performance remains a challenge.

Recently, a number of alternative, compromised modeling approaches have been proposed to deal with this issue, based on certain simplificating assumptions. Such approaches are expected to play a significant role especially with respect to meeting the challenges of real-driving emissions legislation that requires highest performance under a wide range of transient operating modes. Kalyankar et al. [51] used steady-state 3D CFD simulations with spray analysis to determine the ammonia NO_x ratio (ANR) field and calculate the probability density functions (PDF) of ANR at the entrance of the SCR catalyst for several mass flow rates. Using the PDFs, they estimated the input conditions and conducted 1D simulations for several bins (typically 5) of ANR, representative of the uniformity at the entrance. Thus, it was possible to simulate the SCR system during transient cycles, taking into account the SCR kinetics and the ANR nonuniformity at the entrance in a computationally inexpensive way. On a similar approach, Wurzenberger et al. [52] used 3D CFD to calculate the UWS fractions that ended up evaporating or forming a wall film, as well as the wall thickness, and used this information in fast transient 1D simulations.

12.4.2.1 FUEL INJECTION

Fuel injection upstream of a DOC is one of the thermal management technologies used to raise temperature in the exhaust system. One possible application could be to decouple the DPF regeneration procedure from in-cylinder injection control, which is an advantage for retrofit systems, while it can mitigate oil dilution problems [53].

The Eulerian–Lagrangian approach described for SCR systems can be used for multi-component hydrocarbon injection. The spray dynamics and evaporation rate are similarly calculated from the momentum, mass equation, and energy balance taking into account turbulent dispersion. The design of fuel injection systems is usually aimed at efficient evaporation through finer atomization of the spray and HC impingement on hot exhaust line walls. Under favorable conditions, no wall film is expected and the assumption of immediate evaporation of the droplets at wall-contact can be used.

This approach has been employed by Stepanek et al. [54], who used droplets consisting of five representative hydrocarbon components. Two DOC-DPF exhaust systems with fuel injection were simulated as shown in Figure 12.25. The simulation results shown in Figure 12.26 illustrate the differences in distribution of hydrocarbon concentration at the monolith inlet. For the long pipe model, a high uniformity of the inlet concentration distribution is achieved (HC mole fraction uniformity index 0.995). On the contrary, the short pipe does not provide enough time/space to homogenize the mixture, so that considerable concentration nonuniformity is observed at the catalyst inlet. In the regions of increased inlet HC concentration, higher outlet gas temperature and reduced HC slip is encountered due to the exothermic reactions, as shown in Figure 12.27.

FIGURE 12.25 Indicative system geometries with (a) long pipe system, (b) short pipe system studied by Stepanek et al. [54]. The systems include DOC (multiple color channels), DPF (red), and fuel injector (arrow).

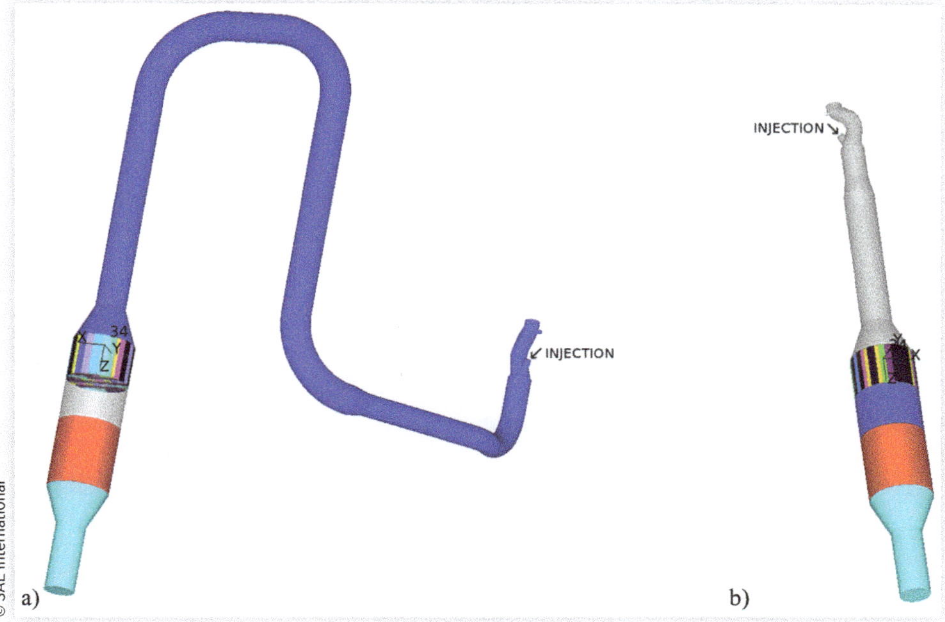

FIGURE 12.26 Hydrocarbon concentration (C16) distribution at the DOC inlet for a (a) long pipe system and (b) short pipe system [54].

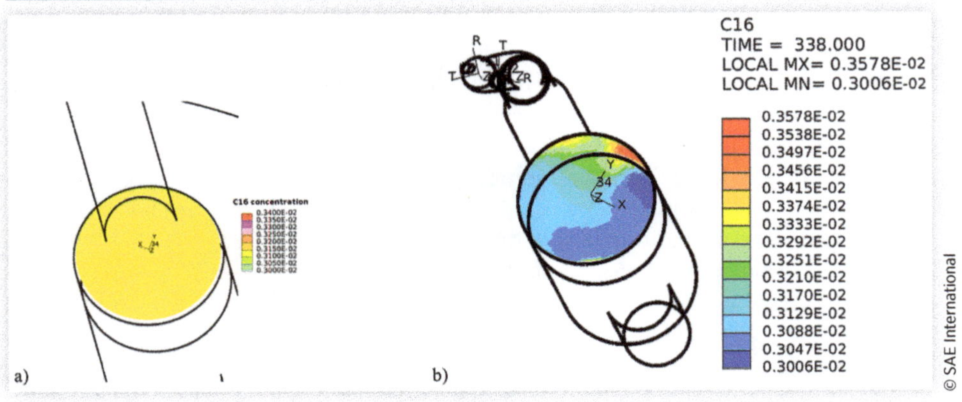

FIGURE 12.27 Hydrocarbon (C16) slip (a) and temperature (b) distributions at the outlet of the DOC in short pipe system [54].

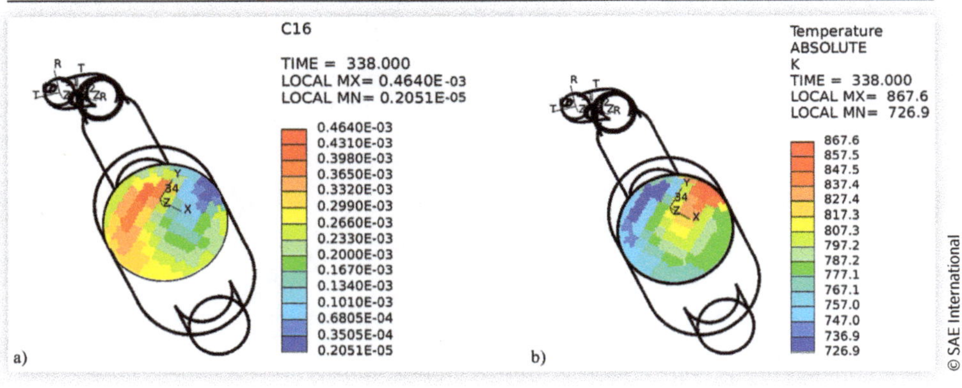

12.5 Coupled 3D Flow, Thermal and Chemical Analysis in DPF Regeneration Events

Although 2D, axi-symmetric modeling of the regeneration process is sufficient in the case of cylindrical nonsegmented filters, the heat transfer problem is more complicated in the case of real-world filters. Most commercialized filters are segmented SiC filters and some of them are oval shaped. The presence of adhesive cement layers inside the filter structure influences local heat transfer and flow resistance in adjacent channels. The cement layer effect on the adjacent channels has been modeled by Koltsakis et al. [55] and has been shown to affect significantly the evolution of filter regeneration.

Moreover, 3D effects may arise from asymmetric flow conditions at the entrance of the filter. Due to the importance of system packaging restrictions, aftertreatment devices require careful mounting and positioning in order to avoid unwanted effects from flow maldistribution [56]. In this case, it is important to calculate the influence of flow and temperature nonuniformities by simulating in a 3D environment the complete exhaust aftertreatment system. For this task, it is possible to couple channel-scale DPF models, and its derivatives, with commercial CFD packages. The porous medium approach has been employed for the simulation of the filter region, with pressure drop resistances, mass, and energy source terms taken from the DPF models [57].

12.5.1 Inlet Flow Distribution

An important aspect of 3D DPF regeneration modeling is the prediction of the transient flow distribution at filter inlet during regeneration. Even if the total gas flow rate at DPF inlet is constant during regeneration, the following parameters influence the flow distribution entering the filter:

- Inlet cone geometry
- Soot distribution in the filter
- Temperature distribution in the filter affecting the local flow resistance

The modeler may be able to neglect the effect of inlet cone geometry if the manufacturer designs the cone shape to eliminate flow maldistribution at DPF entrance due to flow separation. On the other hand, radially nonuniform soot distribution occurs commonly in practice due to partial filter regeneration events. Temperature nonuniformities in the filter also occur, but their effects on inlet flow distribution are comparatively smaller, as shown in [58]. Koltsakis et al. [57] investigated the effect inlet cone geometry on flow and soot distribution during DPF loading, as shown in Figure 12.28. The initially nonuniform flow observed in the case of a clean DPF tends to be homogenized after soot is accumulated in the filter, which acts as a flow-balancer. This seems to be the case in all examined inlet pipe geometries despite the large differences observed in the clean-state calculations.

12.5.2 DPF Regeneration

The importance of 3D modeling in predicting regeneration events and identifying failure-prone conditions has been demonstrated by several researchers [57, 59, 60]. During regeneration, the inlet cone geometry can induce significant nonuniformities in temperature, flow, and soot loading consumption. Lower temperatures are expected near the periphery, due to steady-state heat losses to the ambient and the transient heat-up of the pipes and cone upstream of the DPF. It has been shown that temperature nonuniformity can affect the local regeneration rates inside the filter resulting in nonuniform soot loading, which in turn induces a flow nonuniformity.

As an example, Figure 12.29 shows the regeneration of a filter with bent inlet pipe calculated by Koltsakis et al. [57]. Due to the nonuniform inlet temperature profile, initial soot oxidation focuses on the upper central segments, combined with sharp temperature increase in the same location. Subsequently, the localized soot consumption uncovers regions with lower flow resistance, resulting in higher local velocities and faster cooling. At the same time, neighboring regions start to heat up and regenerate, which results in a slight homogenization of the flow. The final soot loading field remains highly nonuniform and a considerable fraction of the soot fails to regenerate in the periphery. The regeneration implications of several inlet cone geometries are summarized in Figure 12.30, which shows the peak temperatures and final soot fields in correlation with inlet temperature profile.

FIGURE 12.28 Flow and soot distribution vs time during loading for four cone variants [57].

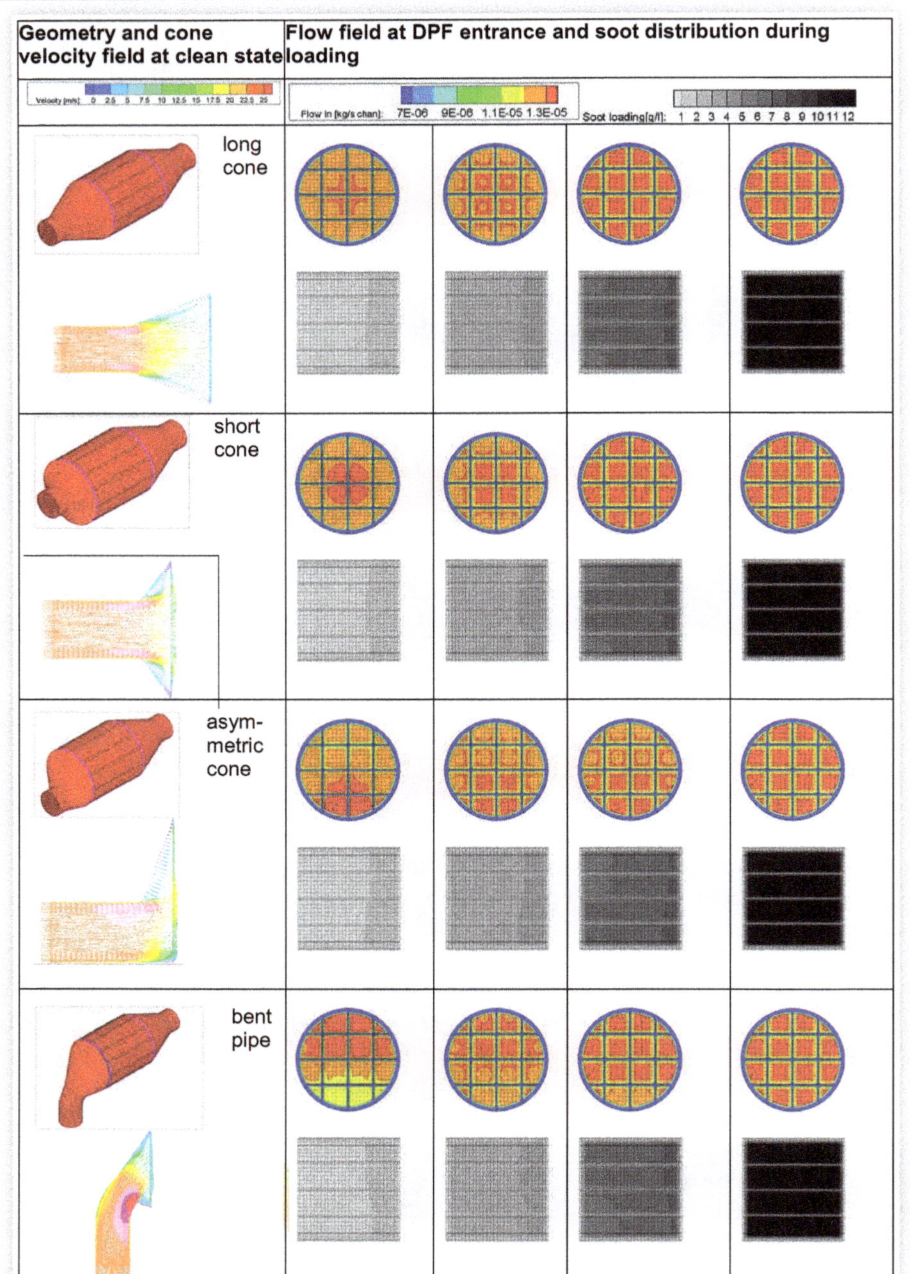

FIGURE 12.29 Computed outlet temperature profiles, soot loading fields, and inlet flow profiles during a transient regeneration of a filter with bent inlet pipe geometry [57].

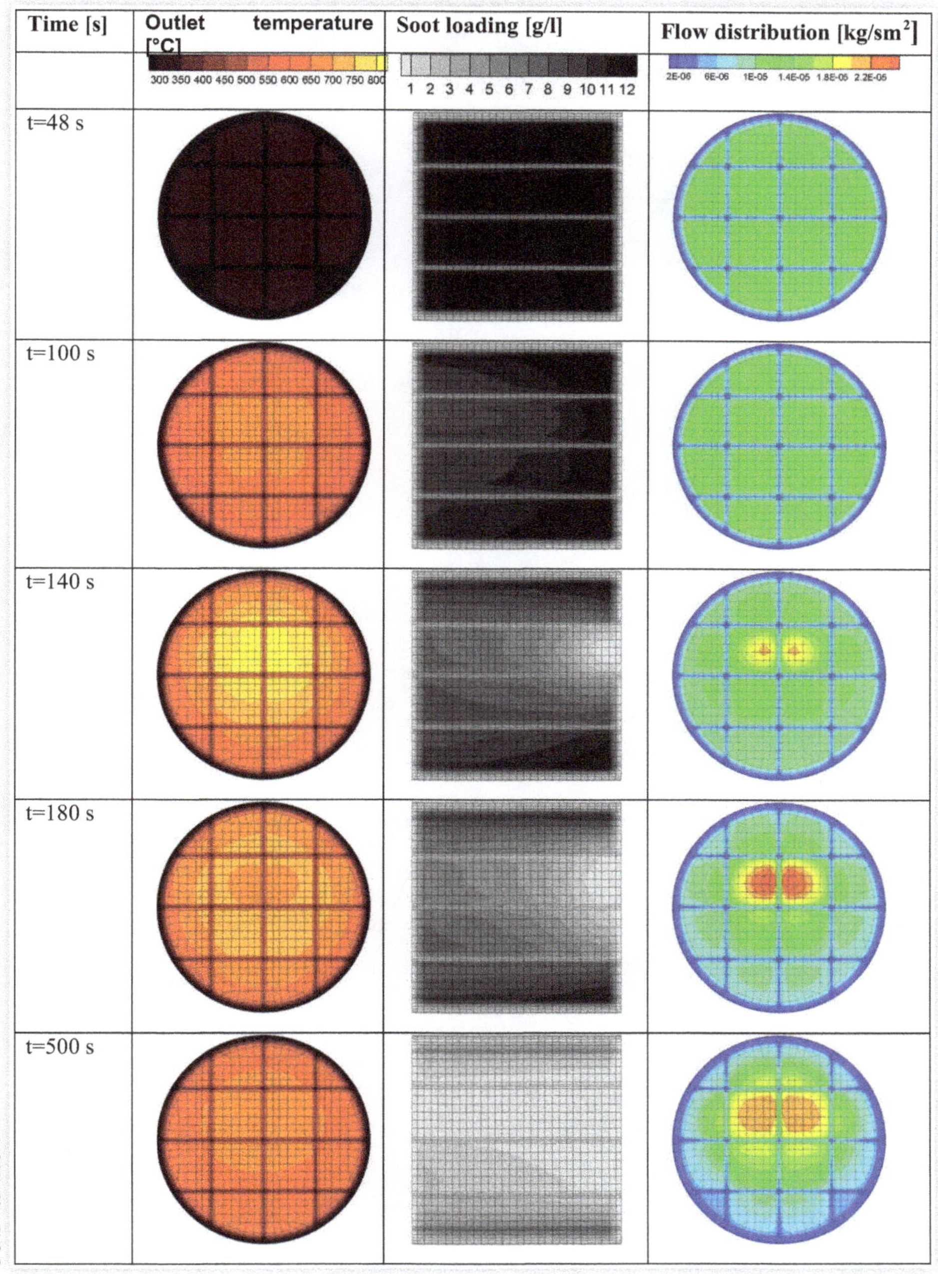

FIGURE 12.30 Effect of diffuser geometry on inlet gas temperature, temperature profile at the peak of the regeneration, and final soot loading [57].

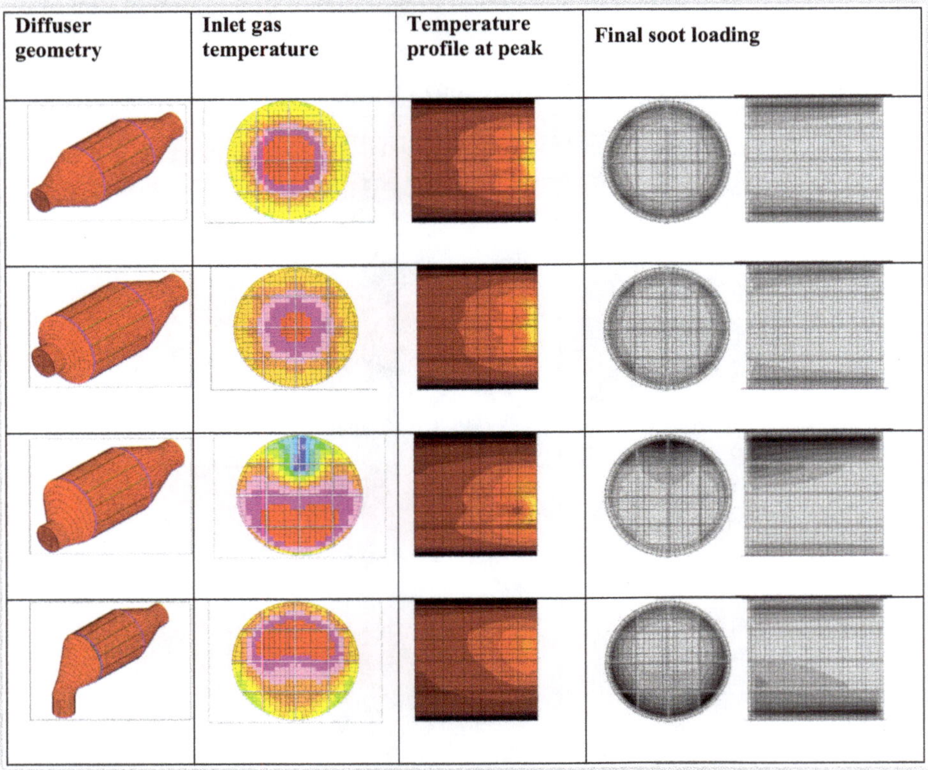

Diffuser geometry	Inlet gas temperature	Temperature profile at peak	Final soot loading

12.5.3 Stress Analysis

Due to the failure problems of DPFs, stress analysis was very early employed in the study of filters. A nonuniform temperature field alone or together with a multi-material structure causes the development of thermally induced stresses, which can be critical in the case of DPF. The mechanical stresses can be easily calculated with stress analysis packages, however it is crucial that the accurate 3D temperature field is defined [59, 61, 62, 63], besides the properties of packaging materials and substrate. Regarding the latter, a common practice is the usage of an equivalent lumped material with anisotropic structural properties as explained by Kim [64], instead of the actual honeycomb geometry. This allows the usage of simplified meshes skipping the channel details. It is also convenient when comparing stresses with failure limits, which are commonly given for equivalent materials due to the measurement techniques applied in practice [65].

The 3D simulation of a DPF regeneration is followed by a stress analysis calculation repeated every time step in order to obtain the full history of stresses and estimate the global critical points. Hajireza et al. [63] calculated the axial and vertical stresses that develop during worst-case regeneration event, as illustrated Figure 12.31. The normal Z stresses are typically responsible for ring-off cracks, while stresses on the XY outlet face can result in crack development in the same area.

FIGURE 12.31 Calculated mechanical stresses during DPF regeneration. Entrance of the flow on the right [63].

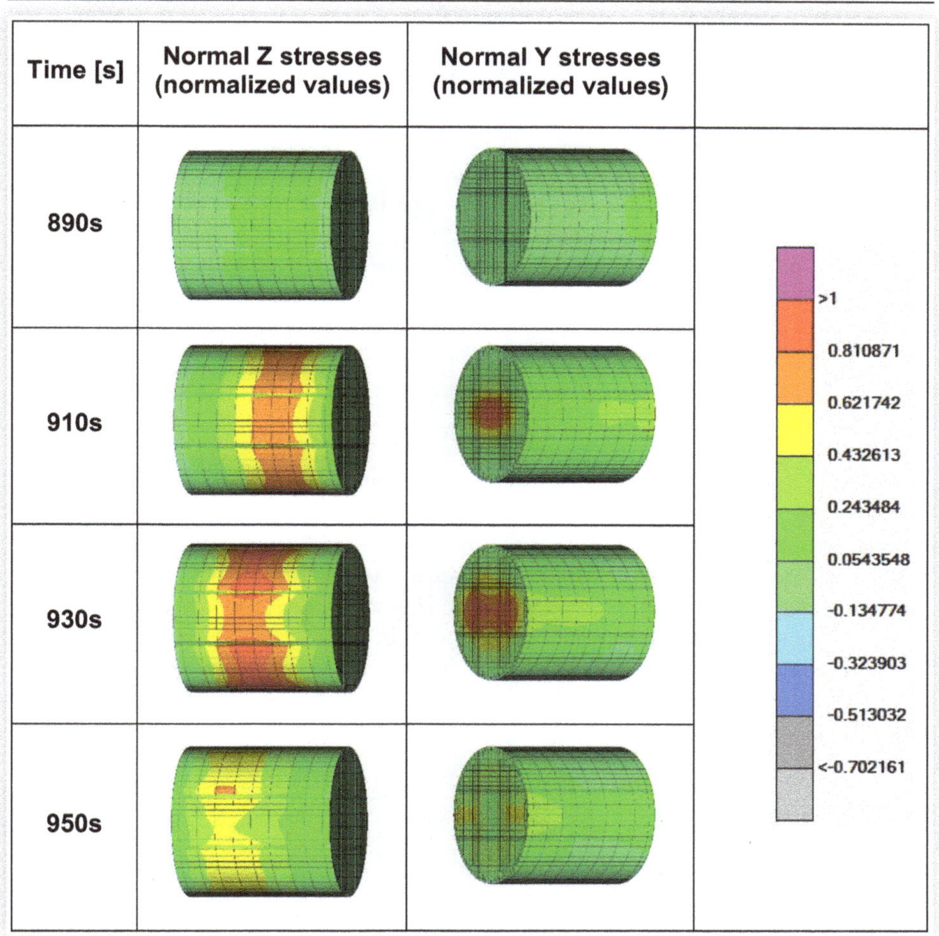

12.6 Concluding Remarks

Over the last years, the emphasis of exhaust after-treatment simulation has shifted to applications related to fully transient driving cycle predictions, usually linked with predicting "real-driving emissions." This task requires models with fast execution times, usually much faster than real time. Despite the fact that the complexity/CPU requirements of 3D models – even with today's computation power – prohibits their use in the above context, their role remains of industrial relevance, especially in the design phase of exhaust after-treatment systems.

The present contribution reviewed some common applications of 3D modeling in both gasoline and diesel engine exhaust systems. The problem of flow uniformity becomes more and more relevant due to the fact that current exhaust systems are placed extremely close to the exhaust valves, to minimize heat losses that become more severe with efficient, low heat rejection engines. The problem of flow pulsations and the inhomogeneous catalyst exposure to flows from multiple cylinder exhausts poses a big modeling challenge that may require further attention in the future.

In addition to the flow velocity patterns, the problem of nonuniform temperature distribution is obviously having an impact on a number of exhaust aftertreatment technologies. Interestingly, this topic has received much less attention, probably due to the fact that the thermal problem is mostly of relevance under transient conditions. 3D thermal problems in the catalyst domain become of increasing importance in systems with long zero flow modes. In this area, a deeper understanding of the heat transfer mechanisms associated with free convection is required, involving both testing work and 3D CFD analysis.

3D simulation in diesel exhaust aftertreatment is expected to continue playing a key role in the design of UWS-based SCR technologies. More emphasis and further technological advances are expected in the area of predictive simulations for wall film and solid deposit formation. The challenge of linking the results of the steady-state analysis to real-life fully transient simulations is also expected to attract the interest of the engineering community.

Even though the majority of the work presented here dealt with flow- and emissions-related predictions, 3D structural analysis in exhaust systems is also well established and will continue to serve the design of thermally stable exhaust systems. In the future, the boundary conditions for such models may increasingly be provided by the respective predictive thermo-fluid models instead of temperature measurements.

In a future of increased virtualization of the development process, the fidelity of 3D modeling approaches is expected to support the time and cost-reduction targets. Building on the experience gained over the three previous decades, the development of customized solutions and workflows is nowadays in the scope of the academic and industrial research.

Definitions, Acronyms, and Abbreviations

Abbreviations

ANR - Ammonia NO_x ratio
DOC - Diesel oxidation catalyst
CPU - Central processing unit
CFD - Computational fluid dynamics
DPF - Diesel particulate filter
FLAD - Flow adjustable designed cell
HC - Hydrocarbon
LES - Large eddy simulation
LPT - Langragian particle tracking
NEDC - New European Driving Cycle
PDF - Probability density function
RANS - Reynolds-averaged Navier-Stokes
RSM - Reynolds stress model
SCR - Selective catalytic reduction
TKE - turbulent kinetic energy
TWC - Three-way catalyst
UWS - Urea-water solution

VOF - Volume of fluids
WOT - Wide open throttle

Variables

A - Frontal area, m^2
or pre-exponential factor, *
B_M - Spalding mass transfer number, -
B_T - Spalding heat transfer number, -
c_f - k-ε turbulence model parameter, -
C_D - Drag coefficient, -
C_p - Thermal capacity, J/(kg*K)
d_h - Channel hydraulic diameter, m
d - diameter, m
D - Diffusivity, m^2/s
E - Activation energy, J/mole
F_B - Buoyancy force, N
F_D - Drag force, N
g - Gravitational acceleration, m/s^2
Gr - Grasshof number, -
h - Heat transfer coefficient, W/m^2K
h_{vap} - Heat of evaporation, J/kg
J - Ratio of pulse period to residence time within the diffuser (reciprocal of Strouhal number)
k - Turbulence kinetic energy, J/kg
K - Spray/wall-interaction model parameter
L - Monolith length, m
Le - Lewis number, -
Nu - Nusselt number, -
m - Mass, kg
p - Pressure, Pa
Pr - Prandtl number, -
r - Radius, m
R - Universal gas constant, J/(mol*K)
Re - Reynolds number, -
S - Heat source term, W/m^3
or swirl number, -
Sc - Schmidt number, -
Sh - Sherwood number, -
T - Temperature, K
T^* - Critical transition temperature, -
Tu - Turbulence intensity, -
t - Time, s
u - Velocity, m/s
U_C - Relative velocity, m/s
x - x coordinate, m
y - y coordinate, m
$y+$ - Dimensionless local normal distance from the wall, -
Y - Mass fraction, -
z - Axial coordinate, m

Greeks

β - Thermal expansion coefficient, 1/K
γ - Flow uniformity index, -
ε - Turbulence dissipation rate, J/(kg*s) or void fraction, -
ε_{rad} - Emissivity, -
λ - Thermal conductivity, W/(m*K)
μ - Dynamic viscosity, kg/m s
ν - kinematic viscosity, m^2/s
ρ - Density, kg/m^3
σ - Stefan-Boltzmann constant, $W/(m^4K)$

Subscripts

* - Characteristic
$_0$ - Initial
∞ - Ambient
d - Droplet
F - Frontal or rear face
g - Gas
i - Cell index
P - Periphery
$_p$ - Particle
$_{ref}$ - Reference
$_s$ - Solid or surface
$_{sat}$ - Saturation
$_{sur}$ - Surroundings
$_u$ - Urea
$_w$ - Wall
$_{vap}$ - Vapor

References

1. Windmann, J., Braun, J., Zacke, P., Tischer, S. et al., "Impact of the Inlet Flow Distribution on the Light-Off Behavior of a 3-Way Catalytic Converter," SAE Technical Paper 2003-01-0937, 2003, doi:10.4271/2003-01-0937.
2. Guojiang, W. and Song, T., "CFD Simulation of the Effect of Upstream Flow Distribution on the Light-Off Performance of a Catalytic Converter," *Energy Conversion and Management* 46 (2005): 2010-2031.
3. Martin, A.P., Will, N.S., Bordet A., Cornet, P. et al., "Effect of Flow Distribution on Emissions Performance of Catalytic Converters," SAE Technical Paper 980936, 1998, doi:10.4271/980936.
4. Wendland, D.W. and Matthes, W.R., "Visualization of Automotive Catalytic Converter Internal Flows," SAE Technical Paper 861554, 1986, doi:10.4271/861554.
5. Weltens, H., Bressler, H., Terres, F., Neumaier, H. et al., "Optimisation of Catalytic Converter Gas Flow Distribution by CFD Prediction," SAE Technical Paper 930780, 1993, doi:10.4271/930780.

6. Tsinoglou, D.N., Koltsakis, G.C., Missirlis, D.K., and Yakinthos, K.J., “Transient Modelling of Flow Distribution in Automotive Catalytic Converters,” *Applied Mathematical Modelling* 28 (2004): 775-794.
7. Tsinoglou, D., Koltsakis, G., Missirlis, D., and Yakinthos, K., “Modelling of Flow Distribution during Catalytic Converter Light-Off,” *International Journal of Vehicle Design* 34 (2004): 231-259.
8. Zhang, X. and Romzek, M., “Computational Fluid Dynamics (CFD) Applications in Vehicle Exhaust System,” SAE Technical Paper 2008-01-0612, 2008, doi:10.4271/2008-01-0612.
9. Hayes, R.E., Fadic, A., Mmbaga, J., and Najafi, A., “CFD Modelling of the Automotive Catalytic Converter,” *Catalysis Today* 188 (2012): 94-105.
10. Jeong, S.-J., “A Full Transient Three-Dimensional Study on the Effect of Pulsating Exhaust Flow under Real Running Condition on the Thermal and Chemical Behavior of Closed-Coupled Catalyst,” *Chemical Engineering Science* 117 (2014): 18-30.
11. Will, N.S. and Bennett, C.J., “Flow Maldistributions in Automotive Converter Canisters and Their Effect on Emission Control,” SAE Technical Paper 922339, 1992, doi:10.4271/922339.
12. Clarkson, R.J., “A Theoretical and Experimental Study of Automotive Catalytic Converters,” Ph.D. thesis, Coventry University, UK, 1995.
13. Kim, J.Y. and Son, S., “Improving Flow Efficiency of a Catalytic Converter Using the Concept of Radially Variable Cell Density—Part I,” SAE Technical Paper 1999-01-0769, 1999, doi:10.4271/1999-01-0769.
14. Toyota Global Newsroom, “Toyota's New Exhaust Purifying Catalyst Reduces Precious Metal Usage by 20%,” February 22, 2017, https://newsroom.toyota.co.jp/en/detail/15121501.
15. Missy, S., Thams, J., Bollig, M., Tatschl, R. et al., “Computer-Aided Optimisation of the Exhaust Gas Aftertreatment System of the New BMW 1.8-Litre Valvetronic Engine,” *MTZ worldwide* 1/2002 63.
16. Benjamin, S.F and Roberts, C.A., “Catalyst Warm-Up to Light-Off by Pulsating Engine Exhaust: Two-Dimensional Studies,” *International Journal of Engine Research* 5, no. 3, JER 03302.
17. Mat Yamin, A.K., Benjamin, S.F., and Roberts, C.A., “Pulsating Flow in a Planar Diffuser Upstream of Automotive Catalyst Monoliths,” *International Journal of Heat and Fluid Flow* 40 (2013): 43-53.
18. Benjamin, S.F., Roberts, C.A., and Wollin, J., “A Study of Pulsating Flow in Automotive Catalyst Systems,” *Experiments in Fluids* 33 (2002): 629-639.
19. Persoons, T., Van den Bulk, E., and Fausto, S., “Study of Pulsating Flow in a Closecoupled Catalyst Manifold Using Phase-Locked Hot-Wire Anemometry,” *Experiments in Fluids* 36, no. 2 (2003): 217-232.
20. Rusli, I.H., Aleksandrova, S., Medina, H., and Benjamin, S.F., “The Effect of Swirl on the Flow Uniformity in Automotive Exhaust Catalysts,” SAE Technical Paper 2017-01-2384, 2017, doi:10.4271/2017-01-2384.
21. Fischer, S., Bitto, R., Lauer, T., Krenn, C. et al., “Impact of the Turbulence Model and Numerical Approach on the Prediction of the Ammonia Homogenization in an Automotive SCR System,” *SAE Int. J. Engines* 5, no. 3 (2012): 1443-1458, doi:10.4271/2012-01-1291.

22. Othmer, C., "A Continuous Adjoint Formulation for the Computation of Topological and Surface Sensitivities of Ducted Flows," *International Journal for Numerical Methods in Fluids* 58 (2008): 861–877.

23. Giannakoglou, K.C. and Papadimitriou, D.I., Chapter 4: Adjoint Methods for Shape Optimization, in *Optimization and Computational Fluid Dynamics*, Thevenin, D. and Janiga, G. eds. (Berlin Heidelberg: Springer-Verlag, 2008), ISBN:978-3-540-72152-9.

24. Hinterberger, C., Kaiser, R., and Olesen, M., "Automatic Shape Optimization of Exhaust Systems," *MTZ* 72 (07-08/2011).

25. Luo, Z., Sukheswalla, P., Drennan, S., Wang, M. et al., "3d Numerical Simulations of Selective Catalytic Reduction of NOx With Detailed Surface Chemistry," *CLEERS Workshop*, October 4, 2017.

26. OpenFOAM, The Open Source CFD Toolbox, Programmer's Guide version 3.0.1, December 13, 2015.

27. axisuite User Guide, version 2017A, December 15, 2017.

28. Chung, C., Geer, L., and Rajadurai, S., "Numerical Simulation and Experimental Validation of the Catalytic Converter Cool Down Process," SAE Technical Paper 2000-01-0204, 2000, doi:10.4271/2000-01-0204.

29. Chen, D.K.S., Oh, S.E., Bissett, E.J., and Ostrom, D.L.V., "A Three-Dimensional Model for the Analysis of Transient Thermal and Conversion Characteristics of Monolithic Catalytic Converters," SAE Technical Paper 880282, 1988, doi:10.4271/880282.

30. Klein, D. and Cheng, W.K., "Spark Ignition Engine Hydrocarbon Emissions Behaviors in Stopping and Restarting," SAE Technical Paper 2002-01-2804, 2002, doi:10.4271/2002-01-2804.

31. Braun, J., Hauber, T., Többen, H., Windmann, J. et al., "Three-Dimensional Simulation of the Transient Behavior of a Three-Way Catalytic Converter," SAE Technical Paper 2002-01-0065, 2002, doi:10.4271/2002-01-0065.

32. Shu, Y., Romzek, M., and Meda, L.G., "Thermal Analysis of Diesel Aftertreatment System," SAE Technical Paper 2010-01-1215, 2010, doi:10.4271/2010-01-1215.

33. Adamczyk, A.A., Hubbard, C.P., Ament, F., Oh, S.H. et al., "Experimental and Modeling Evaluations of a Vacuum-Insulated Catalytic Converter," SAE Technical Paper 1999-01-3678, 1999, doi:10.4271/1999-01-3678.

34. Katare, S. and Laing, P.M., "A Hybrid Framework for Modeling Aftertreatment Systems: A Diesel Oxidation Catalyst Application," SAE Technical Paper 2006-01-0689, 2006, doi:10.4271/2006-01-0689.

35. Koltsakis, G., Samaras, Z., Karvountzis-Kontakiotis, A., Zacharopoulou, T. et al., "Implications of Engine Start-Stop on After-Treatment Operation," *SAE Int. J. Engines* 4, no. 1 (2011): 1571-1585, 10.4271/2011-01-1243.

36. Echtle, H., Schöffel, S., Wenninger, G., Fischer, S. et al., "Optimization of the Selective Catalytic Reduction by Means of Numerical Methods," *MTZ* 71 (10/2010): 54, doi:10.1007/BF03227056.

37. Birkhold, F., Meingast, U., Wassermann, P., and Deutschmann, O., "Analysis of the Injection of Urea-Water-Solution for Automotive SCR DeNOx-Systems: Modeling of Two-Phase Flow and Spray/Wall-Interaction," SAE Technical Paper 2006-01-0643, 2006, doi:10.4271/2006-01-0643.

38. Ström, H., Lundström, A., and Andersson, B., "Choice of Urea-Spray Models in CFD Simulations of Urea-SCR Systems," *Chemical Engineering Journal* 150 (2009): 69-82.

39. Edelbauer, W., Birkhold, F., Rankel, T., Pavlovic, Z. et al., "Simulation of the Liquid Break-Up at an AdBlue Injector with the Volume-of-Fluid Method Followed by Off-Line Coupled Lagrangian Particle Tracking," *Computers and Fluids* 157 (2017): 294-311.

40. Birkhold, F., Meingast, U., Wassermann, P., and Deutschmann, O., "Modeling and Simulation of the Injection of Urea-Water-Solution for Automotive SCR DeNOx-Systems," *Applied Catalysis B: Environmental* 70 (2007): 119-127.

41. Abramzon, B. and Sirignano, W.A., "Droplet Vaporization Model for Spray Combustion Calculations," *International Journal of Heat and Mass Transfer* 32 (1989): 1605-1618.

42. Yim, D.S., Kim, S.J., Baik, J.H., Nam, I. et al., "Decomposition of Urea into NH_3 for the SCR Process," *Industrial and Engineering Chemistry Research* 43, no. 1 (2004): 4856-4863.

43. Fang, H.L. and DaCosta, H.F.M., "Urea Thermolysis and NOx Reduction with and without SCR Catalysts," *Applied Catalysis B: Environmental* 46 (2003): 17-34.

44. Kuhnke, D., "Spray/Wall-Interaction Modelling by Dimensionless Data Analysis," Ph.D. thesis, University of Darmstadt, 2004.

45. Brack, W., Heine, B., Birkhold, F., Kruse, M. et al., "Formation of Urea-Based Deposits in an Exhaust System: Numerical Predictions and Experimental Observations on a Hot Gas Test Bench," *Emission Control Science and Technology* 2 (2016): 115-123.

46. Brack, W., Heine, B., Birkhold, F., Kruse, M. et al., "Kinetic Modeling of Urea Decomposition Based on Systematic Thermogravimetric Analyses of Urea and Its Most Important by-Products," *Chemical Engineering Science* 106 (2014): 1-8.

47. Addy Majewski, W., "Urea Dosing and Injection Systems," DieselNet Technology Guide, retrieved January 2018.

48. Ottaviani, E., Jean, E., Albert, H., Amon, B. et al., "Dual Line Exhaust Design Optimisation to Maximize SCR Catalyst Efficiency thru Improved Ammonia Distribution," SAE Technical Paper 2009-01-0914, 2009, doi:10.4271/2009-01-0914.

49. Drennan, S., Kumar, G., Quan, S., and Wang, M., "Application of Automatic Meshing to Urea-Water Injection Simulation for Engine Aftertreatment," SAE Technical Paper 2015-01-1057, 2015, doi:10.4271/2015-01-1057.

50. Faltsi, R., Mutyal, J., Braun, M., Reinelt, R. et al., "Confidence in Modeling SCR Aftertreatment Systems," *Automotive Simulation World Congress*, 2012.

51. Kalyankar, A., Munnannur, A., and Liu, Z.G., "Implementing the Effect of ANR Non-Uniformity on Transient SCR Performance through Coupling between 3D Spray Analysis and 1D Kinetic Modeling".

52. Wurzenberger, J.C., Nahtigal, A., Mitterfellner, T., Pachler, K. et al., "NOX ABATEMENT—1D/3D Simulation of Urea Dosing and Selective Catalytic Reduction," *CLEERS Conference*, 2017.

53. Fasolo, B., Hardy, J.-P., and Leroy, K., "Exhaust Fuel Injection System for Efficient DPF Regenerations," *MTZ 07-08I2009* 70.

54. Stepanek, J., Koci, P., Weaver, M., Frey, R. et al., "Modeling of Injected Diesel Fuel Conversion and Heat Release in Oxidation Catalyst: 3D-CFD & 1D Channels Approach," SAE Technical Paper 2012-01-1293, 2012, doi:10.4271/2012-01-1293.

55. Koltsakis, G.C., Haralampous, O.A., Margaritis, N.K., Samaras, Z.C. et al., "3-Dimensional Modeling of the Regeneration in SiC Particulate Filters," SAE Technical Paper 2005-01-0953, 2005, doi:10.4271/2005-01-0953.

56. Hinterberger, C., Olesen, M., and Kaiser, R., "3D Simulation of Soot Loading and Regeneration of Diesel Particulate Filter Systems," SAE Technical Paper 2007-01-1143, 2007, doi:10.4271/2007-01-1143.

57. Koltsakis, G.C., Samaras, Z.C., Echtle, H., Chatterjee, D. et al., "Flow Maldistribution Effects on DPF Performance," SAE Technical Paper 2009-01-1280, 2009, doi:10.4271/2009-01-1280.

58. Haralampous, O.A., Koltsakis, G.C., and Samaras, Z.C., "Partial Regenerations in Diesel Particulate Filters," SAE Technical Paper 2003-01-1881, 2003, doi:10.4271/2003-01-1881.

59. Pontikakis, G., Stamatelos, A., Bakasis, K., and Aravas, N., "3-D Catalytic Regeneration and Stress Modeling of Diesel Particulate Filters by ABAQUS FEM Software," SAE Technical Paper 2002-01-1017, 2002, doi:10.4271/2002-01-1017.

60. Konstandopoulos, A.G., Kostoglou, M., Vlachos, N., and Kladopoulou, E., "Progress in Diesel Particulate Filter Simulation," SAE Technical Paper 2005-01-0946, 2005, doi:10.4271/2005-01-0946.

61. Barataud, C., Bardon, S., Bouteiller, B., Gleize, V. et al., "Diesel Particulate Filter Optimization," SAE Technical Paper 2003-01-0376, 2003, doi:10.4271/2003-01-0376.

62. Kuki, T., Miyairi, Y., Kasai, Y., Miyazaki, M. et al., "Study on Reliability of Wall-Flow Type Particulate Filter," SAE Technical Paper 2004-01-0959, 2004, doi:10.4271/2004-01-0959.

63. Hajireza, S., Johannesen, L.T., Wolff, T., Koltsakis, G.C. et al., "A Modeling and Experimental Investigation on an Innovative Substrate for DPF Applications," SAE Technical Paper 2010-01-0891, 2010, doi:10.4271/2010-01-0891.

64. Kim, J.S., "Substrate Failure Prediction During Converter Canning Process," SAE Technical Paper 2004-01-1134, 2004, doi:10.4271/2004-01-1134.

65. Stobbe, P., Petersen, H.G., Høj, J.W., and Sorenson, S.C., "SiC as a Substrate for Diesel Particulate Filters," SAE Technical Paper 932495, 1993, doi:10.4271/932495.

Modeling of IC Engine Silencers and Tailpipe Noise: 1D and 3D Approaches

Mats Åbom
KTH Kungliga Tekniska Högskolan

Francisco D. Denia and Antonio J. Torregrosa
Universitat Politècnica de València

13.1 Introduction

Noise produced by the intake and exhaust processes is one of the main noise sources in automobile vehicles. Traditionally, exhaust noise was the main concern, but intake noise is becoming more and more important due to the widespread use of highly supercharged downsized engines. Apart from the relatively low frequencies associated with engine breathing, which are common to intake and exhaust noise, other frequency bands related to the turbocharger operation are becoming relevant. This, together with the generalized use of plastic parts, increases the importance of vibro-acoustic coupling. In the exhaust side, design criteria are being substantially affected by the mandatory presence of aftertreatment devices.

In the past, intake and exhaust system development was a highly empirical task. However, this situation is no longer acceptable: the objective of reducing development time and costs is quite incompatible with trial-and-error methodologies based on experimental tests. Additionally, the design of the intake and exhaust lines starts in relatively early stages of the vehicle development process, and it is thus possible (and in fact frequent) that the engine is not still fully developed. Therefore, the engine may change during the development of the gas exchange systems, the eventual consequences from the viewpoint of noise not being easy to predict. Moreover, even if this is not the case, engine manufacturers are more and more reluctant to supply engine prototypes, mostly in the case of new or innovative designs. Therefore, on one hand, it is necessary to reduce the required number of prototypes and engine tests, while, on the other hand, it is advisable to avoid the dependence on engine availability, while providing reaction capabilities against eventual changes in the specifications.

Numerical modeling, while not the only possibility (testing in simple and controlled physical models would be another one), is one of the most promising and widespread solutions.

However, in order to evaluate the results in the terms prescribed by the engine manufacturer, it is necessary to obtain results in terms of the noise emitted. In this sense, the only alternative to engine testing is provided by the use of predictive models that should incorporate an adequate description of the effect of the silencers. Those silencer models, aside from permitting its inclusion in exhaust noise prediction tools, may constitute themselves a significant aid to silencer design.

In this chapter, a rather comprehensive view of the different techniques available for silencer modeling and for intake and exhaust noise prediction is provided. Special emphasis is put on frequency domain methods, which are the preferred solution in those cases in which frequency-dependent properties play a major role. In this sense, most of the material presented here complements and extends that provided in Chapter 10.

Then, the contents are structured into two main blocks: the first one devoted to silencer modeling and the second to orifice noise prediction.

In the first block, firstly linear (frequency domain) methods will be considered, comprising 1D techniques (2-port theory and acoustic network analysis) and three-dimensional (3D) methods including matching techniques and full numerical approaches for the solution of the convective wave equation and the linearized Navier-Stokes equations (LNSE). Secondly, nonlinear (time domain) methods will be discussed, comprising classical 1D techniques and full 3D formulations based on different computational fluid dynamics (CFD) approaches. Quasi-3D approaches provide a solution somewhat intermediate between the 1D and 3D techniques described in this chapter and would obviously fit into the category of time domain methods, but all the relevant applications to silencer modeling have been discussed in detail in Chapter 10, to which the interested reader is referred.

In the second block, specific issues related with tailpipe noise prediction are considered, again covering most of the spectrum of available techniques: linear methods, in which the associated acoustic source description is one of the most important issues; nonlinear methods, in which the description of the open end and the associated emission model plays a central role; and hybrid time-frequency methods, including different solutions as those based on the harmonic balance and the use of digital filter models to implement 2-port approaches from the frequency domain into time domain solvers.

13.2 Engine Silencers: Linear (Frequency Domain) Methods

Classical duct acoustics is based on linearizing the fundamental equations of fluid mechanics which, assuming a steady and uniform mean flow with no viscous or thermal losses, leads to the convected wave equation [1, 2, 3]:

$$\left(\frac{1}{c^2}\frac{D^2}{Dt^2} - \nabla^2\right)p'(x,y,z,t) = 0, \tag{13.1}$$

where

t is the time

x, y, and z are the Cartesian coordinates

c is the adiabatic speed of sound

$D/Dt = \partial/\partial t + U \cdot \partial/\partial z$

U is the mean flow velocity assumed parallel to z

$\nabla = (\partial/\partial x, \partial/\partial y, \partial/\partial z)$ is the nabla operator and p' the acoustic pressure

For linear acoustic problems, it is convenient to Fourier transform (Equation (13.1)) which leads to:

$$\left(\left(k - jM\partial/\partial z\right)^2 + \nabla^2\right)p(x,y,z,\omega) = 0, \tag{13.2}$$

where

ω is the angular frequency
$k = \omega/c$ is the wave number
$M = U/c$ is the Mach number
$p(x,y,z,\omega) = \int_{-\infty}^{\infty} p'(x,y,z,t)e^{-j\omega t}dt$ is the complex valued pressure
$j = \sqrt{-1}$

For the case of sound in straight ducts and pipes, the general solution to this equation will be a sum of modes or waves propagating up- or downstream [3]:

$$p(x,y,z,\omega) = \sum_n \left(p_{n+}(\omega)\Psi_n(x,y)e^{-jk_{n+}z} + p_{n-}(\omega)\Psi_n(x,y)e^{jk_{n-}z}\right), \tag{13.3}$$

where

$n = 0, 1, 2, \ldots, \Psi_n(x,y)$ are different mode shapes
$p_{n+/-}$ are modal pressure amplitudes in the positive (+)/negative (−) z-direction and $k_{n+/-}$ are modal wave numbers

For hard walled ducts and pipes, an often used approximation for gas filled systems, the modes are orthogonal and the lowest ($n = 0$ or zero mode) is referred to as the plane wave. The plane wave has a constant pressure over the cross section, i.e., one can put $\Psi_0(x,y) = 1$. The plane wave propagates for all frequencies, while the higher modes $n > 0$ are associated with a cut-on frequency, i.e., they will only start to propagate above a certain frequency. For simple cross-sectional shapes, e.g., rectangular, circular, or elliptical, analytical expressions for the higher-order modes can be found and be used to derive analytical expressions for silencer performance [2].

13.2.1 Multi-port Methods

In general a multi-port is a system where a known relation exists between a set of input **x** and output variables **y** ("state vectors"). Often this type of approach is called a "black-box" model in the literature [4], since it can be applied without a full knowledge of the "inner" properties of the studied system. For the case of acoustics, **x** and **y** are often defined via pressure, force and displacement, and velocity, respectively. The choice of state variables is dictated by both physical (what can be measured!) considerations as well as mathematical, since certain choices will lead to more or less efficient or even noncasual formulations for certain problems. Furthermore, it is normally assumed that **x** and **y** have the same dimensionality (N). The multi-port is then called an N-port where N can be interpreted as the number of degrees of freedom. For the case of linear and time-invariant systems [4], an assumption consistent with classical duct acoustics, the most general multi-port model is given by [5]:

$$\mathbf{y} = \mathbf{Gx} + \mathbf{y}_s, \tag{13.4}$$

FIGURE 13.1 A multi-port of order N (i.e., N-port) is a system where a causal relation exists between an input **x** and an output **y**, where **x** and **y** are state vectors with N-components. A simple example is an ideal spring with one fixed end, which can be seen as a 1-port defined via F = −Kx, where F is the force, x the displacement, and K the spring constant.

©Torregrosa, A.

which is a linear relationship in the frequency domain where **x** and **y** are complex valued state vectors, **G** is a N × N matrix describing the passive part, and the subscript s denotes the source vector related to sound generation, as represented in Figure 13.1. The multi-port model can be applied both in the plane wave or 1D case, which is the main interest for automotive engine applications, and for cases with higher-order modes. For some recent works on multi-port models for duct acoustics including higher-order modes, see, e.g., [6, 7, 8].

A classical choice of state variables for the fluid-borne 1D case is to use acoustic pressure p and volume flow q. Choosing pressure as the output and volume flow as the input variable gives the so-called *impedance matrix* description:

$$\mathbf{p} = \mathbf{Z}\mathbf{q}, \tag{13.5}$$

excluding any internal sources. The frequencies for free oscillations of the system (eigenfrequencies) are obtained by putting ($\mathbf{p} = \mathbf{0}$). The resulting homogeneous equation has nontrivial solutions only when

$$\det(\mathbf{Z}) = 0. \tag{13.6}$$

The resonance frequencies are defined as the real valued frequencies which for a given **p** gives a large $|\mathbf{q}|$. For systems with small losses, the resonance frequencies will be very close to the real part of the eigenfrequencies. An alternative definition is to choose the volume velocity as the output variable and pressure as the input variable. This leads to the so-called *mobility or admittance matrix* definition:

$$\mathbf{q} = \mathbf{Y}\mathbf{p}, \tag{13.7}$$

where $\mathbf{Y} = \mathbf{Z}^{-1}$. It can be noted that the mobility type of description often is preferred in the structure-borne case, see, e.g., [9]. The equation for free oscillations based on this definition is obtained by changing **Z** to **Y** in Equation (13.5). For a given system, this will create two sets of eigenfrequencies, corresponding to two types of boundary conditions (p = 0 or "pressure release" and q = 0 or "blocked/rigid end").

An example of an acoustical multi-port is a volume with N-connected ducts (Figure 13.2). If only plane waves exist in the ducts, this is an N-port. The inlet and outlet states are defined at the duct openings. An example where this approach is used to model a T-junction (a 3-port) can be found in [10]. Assuming an N-port then the multi-port matrix **G** will be N × N. In many cases spatial symmetries exist and the reciprocity principle [11] is valid, which leads to relationships between the matrix elements. This will cut down the number of unknowns and can be used to reduce the work both in numerical or experimental procedures. Concerning acoustical reciprocity it implies that when source and receiver are interchanged, the response (acoustic pressure) is unchanged. This principle is valid for the classical wave equation with *no flow* but effects of yielding boundaries and losses can be included [12]. However, strictly the principle is not valid when a mean flow is present, and care must be taken to apply it for such cases. If reciprocity can be assumed, then it can be shown [11] that the impedance and mobility matrices are symmetric or antisymmetric (depending on the sign convention for q).

FIGURE 13.2 A volume regarded as a multi-port. Note the positive direction of the volume flow is arbitrary. But it will affect the sign of the real part of the impedance/mobility and at no flow the symmetry of the multi-port.

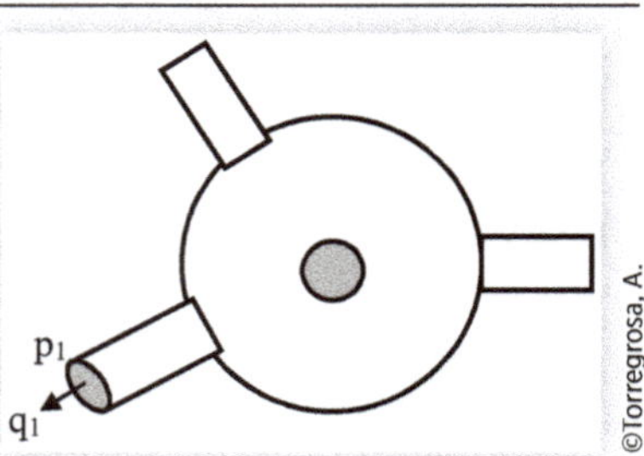

©Torregrosa, A.

13.2.2 1D Methods

The cut-on for the first higher-order mode is given by:

$$f_{cut\text{-}on} = \frac{\alpha c}{d}, \tag{13.8}$$

where

$\alpha = 0.586$ for a circular duct/pipe and 0.5 for a rectangular one

d is the diameter or largest cross-sectional length, only plane waves will propagate in a pipe

Higher-order modes will exist in the form of exponentially decaying non-propagating fields not carrying acoustic power close to sources and discontinuities, e.g., an area change. At a straight pipe cross section, assuming the plane wave field dominates, the sound field can be written based on Equation (13.3) as:

$$p(z,\omega) = p_+(\omega)e^{-jk_+z} + p_-(\omega)e^{jk_-z}, \tag{13.9}$$

where the subscript n = 0 for the plane wave mode has been dropped and the wave numbers are given by $k_+ = k/(1 + M)$ and $k_- = k/(1 - M)$. For sound at audible frequencies in the dimensions typical for intake or exhaust systems, thermal and viscous losses will be only important in the wall boundary layers [2, 11]. The plane wave numbers can be corrected to include this damping assuming that the boundary layers are thin, i.e., much smaller than the pipe diameter. The boundary layer thickness δ is given by, see, e.g., [11]:

$$\delta_v = \sqrt{\frac{2\upsilon}{\omega}} \quad \text{and} \quad \delta_{th} = \sqrt{\frac{2\upsilon}{Pr\,\omega}}, \tag{13.10}$$

where

υ is the kinematic viscosity

Pr is the Prandtl number

subscripts v and th denote "viscous" and "thermal"

In Figure 13.3 the boundary thickness has been plotted for air at 20°C and 400°C and a pressure of 100 kPa.

It is clear from Figure 13.3 that for most cases related to intake or exhaust system acoustics, the boundary layers can be assumed small in the pipes concerned.

FIGURE 13.3 Viscous (solid) and thermal (dashed) boundary layer thickness as a function of frequency for two cases: (a) 20°C and 100 kPa; (b) 400°C and 100 kPa.

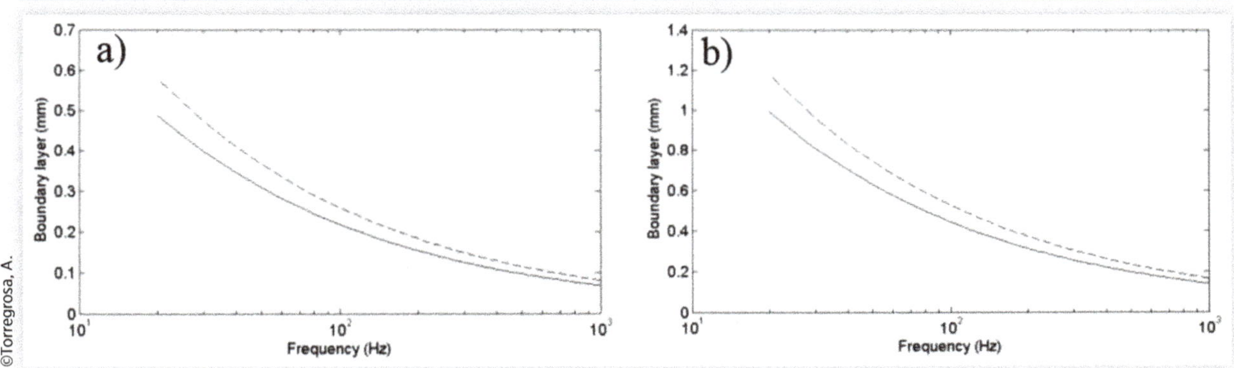

The exception is the small, around 1 mm or less, pipes or channels used for aftertreatment devices. In that case the equations for visco-thermal wave propagation, either based on the so-called Kirchhoff solution or the more accurate approach suggested by Dokumaci [13] must be used. A paper discussing in detail the use of these models for a complete aftertreatment unit is Allam and Åbom [14].

For the thin boundary layer, case formulas for adding the damping to the plane wave model were first proposed ad hoc by Davies [15]. Later Dokumaci [16] presented a derivation assuming small Mach numbers. The problem has recently been analyzed again by Weng and Bake [17], who presented an improved formula not restricted to small Mach numbers:

$$\begin{cases} k_+ = \dfrac{k}{1+M}\left(1+\alpha_0\sqrt{2}e^{-j\pi/4}\left(1+M\right)^{-1/2}\right) \\ k_- = \dfrac{k}{1-M}\left(1+\alpha_0\sqrt{2}e^{-j\pi/4}\left(1-M\right)^{-1/2}\right) \end{cases}, \tag{13.11}$$

where α_0 is the classical Kirchhoff attenuation coefficient for a wide pipe given by:

$$\alpha_0 = 2^{-1/2}\,Sh^{-1}\left(1+(\gamma-1)Pr^{-1/2}\right),$$

where

$Sh = d/2\delta_v$ is the shear wave number

γ is the specific heat ratio

Formally Equation (13.11) is derived for a circular pipe, but for other shapes one can use the formula based on the hydraulic diameter. In addition there is also some damping from turbulence caused by the modulation of the so-called Reynolds stress by the acoustic field. This turbulence damping is only important when the acoustic boundary layers extend beyond the viscous sublayer at the wall, which means that the effects only are important for low frequencies [18, 19]. For intake and exhaust system applications, turbulence damping along a pipe can normally be neglected. The only published example where it has been demonstrated to be of importance is in the modeling of charge air coolers [20].

The acoustic velocity field in the z-direction can be derived from Equation (13.9) by using the linearized equation of motion, see, e.g., [2, 15, 16]:

$$u(z,\omega) = \left(\frac{p_+(\omega)}{Z_+^c}e^{-jk_+z} - \frac{p_-(\omega)}{Z_-^c}e^{jk_-z}\right), \tag{13.12}$$

where $Z_\pm^c = \rho\omega / k_\pm$ is the plane wave (specific) or *characteristic* impedance in the +/– direction. From this result one can obtain the acoustic volume ($q = uS$) or mass ($m = \rho Su$) flow at a cross section by multiplying with the duct cross-sectional area S and density ρ. For computing the efficiency of silencers, it is convenient to have an expression for the time averaged acoustic power W carried by a plane wave [2, 15]:

$$W = \underbrace{\frac{p_{+,rms}^2\left(1+M\right)^2}{Z^o}}_{W_+} - \underbrace{\frac{p_{-,rms}^2\left(1-M\right)^2}{Z^o}}_{W_-}, \tag{13.13}$$

where $Z^o = \rho c/S$ is the plane wave impedance and $p_{rms}^2 = |p|^2/2$ for an harmonic wave. It can be noted that here the small effect of the visco-thermal boundary layers is neglected.

13.2.2.1 2-PORT MODELS: CASCADE SYSTEMS

Most duct elements, e.g., a silencer, are a system with just two openings. If only plane waves exist at these openings, then the system can be described as an acoustical 2-port. In electrical engineering this type of system is often called a four-pole, and this term can also be found in the acoustics literature. In practice this condition is satisfied for sufficiently low frequencies, i.e., below the cut-on frequency for higher modes. This assumes of course that the effect of coupled wall vibrations is negligible, which normally is the case for plane waves in circular gas-filled pipes. For low frequencies strong standing wave effects and coupling between a source and duct system is possible. This makes noise control for low frequencies more difficult and not simply just a question of adding damping to reduce noise levels. Instead a detailed analysis of wave interaction is needed, and this can be achieved by modeling a duct system as a network of two ports.

A commonly used formalism also used here is obtained by choosing acoustic pressure p and volume q to describe the input and output state [11]. Denoting the input side a and the output b, the transfer matrix **T** for a *passive* element (no source term) can be defined via (see Figure 13.4):

$$\begin{pmatrix} p_a \\ q_a \end{pmatrix} = \underbrace{\begin{pmatrix} T_{aa} & T_{ab} \\ T_{ba} & T_{bb} \end{pmatrix}}_{T} \begin{pmatrix} p_b \\ q_b \end{pmatrix}. \tag{13.14}$$

This is called the backward transfer matrix and if a and b are interchanged (= matrix inversion) the so-called forward transfer matrix results. Alternative definitions of the transfer matrix can be found in Munjal [2], where aeroacoustic pressure (= fluctuating enthalpy divided by density) and acoustic mass flow are used as state variables. One can also as suggested by Davies [15] use travelling wave amplitudes (p_+, p_-). As shown below, the transfer matrix formalism is especially useful for treating systems with elements coupled in a chain or cascade.

As long as the interfaces between two 2-ports in a chain are selected in straight duct sections, the continuity of the state variables are satisfied. It then follows from the definition that the total transfer matrix for a cascade coupled system with M passive 2-ports, $\mathbf{T}_1, \mathbf{T}_2, \ldots, \mathbf{T}_M$ is:

$$\mathbf{T} = \prod_{m=1}^{M} \mathbf{T}_m, \tag{13.15}$$

FIGURE 13.4 Acoustical 2-port and circuit representation of a silencer ("pipe element") based on the electric-acoustic analogy [2, 11] (p = "voltage" and q = "current"). Note that a 1D or plane wave state is only assumed at the cross sections a and b, i.e., inside the silencer 3D fields are possible.

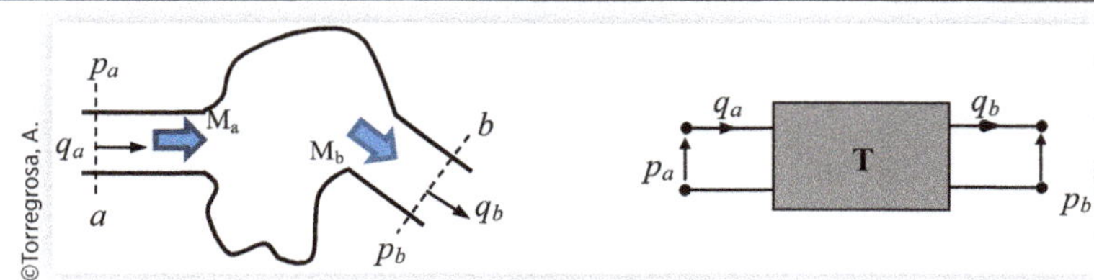

FIGURE 13.5 A cascade with M passive 2-ports excited at the inlet by a 1-port source and terminated by an impedance (a passive 1-port).

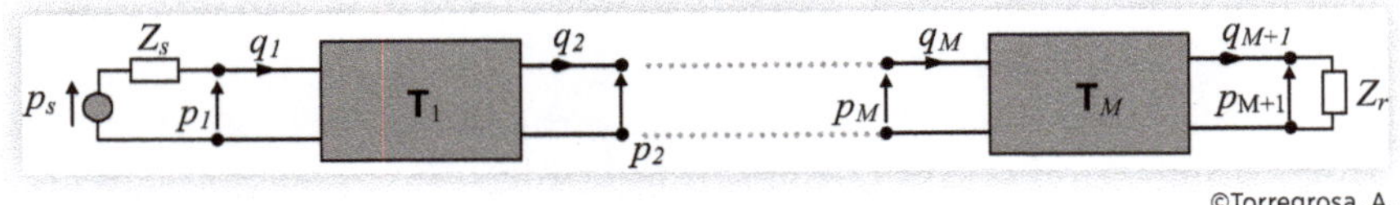

©Torregrosa, A.

assuming the numbering to be from the input to the output of the cascade. In many problems the cascade is excited at the inlet by a source, e.g., an IC engine; see Figure 13.5. The output is normally an open duct or pipe radiating to a surrounding fluid (air). The input can assume a linear and time-invariant source be modeled as an active 1-port. With p and q as state variables, this 1-port can be described via:

$$p_1 = p_s - Z_s q_1, \tag{13.16}$$

where
p_s is the source strength
Z_s is the source input impedance

The minus sign is due to the fact that the volume flow here is defined positive out from the 1-port. Similarly the output can be seen as a passive 1-port described via:

$$p_{M+1} = Z_r q_{M+1}, \tag{13.17}$$

where Z_r is the radiation impedance.

The model in Figure 13.5 can be seen as a representation of a complete intake or exhaust system. The source data can either be obtained by measurements on the IC engine or, as first demonstrated by Fairbrother et al. [21], via simulations using 1D codes for engine gas exchange. The radiation impedance for an open-ended pipe can be estimated from formulas found in the literature; see Section 13.4.1.1. Using the model described above, various measures of the attenuation attained can be calculated. One common is the so-called transmission loss (TL), which normally assumes reflection-free terminations both at the source and termination. This measure is defined as the ratio (in dB) of the incident power W_{in} to the power transmitted W_{tr}. Using this definition and Equations (13.13) and (13.14), it can be shown that [2]:

$$TL = 10\log_{10}\frac{\dfrac{|p_{a+}|^2}{Z_a^o}(1+M_a)^2}{\dfrac{|p_{b+}|^2}{Z_b^o}(1+M_b)^2} = 10\log_{10}\left(\frac{1}{4}\frac{Z_b^o(1+M_a)^2}{Z_a^o(1+M_b)^2}\left|T_{aa}+T_{ab}/Z_b^o+T_{ba}Z_a^o+T_{bb}Z_a^o/Z_b^o\right|^2\right). \tag{13.18}$$

It can be noted here that TL can only be created by two effects—*reflection* of sound and *dissipation* of sound. All silencers are based on one of these principles or a combination of both (hybrid silencers). Another measure, which takes the influence of the source and termination into account, is the insertion loss (IL). This can be defined as the ratio (in dB) between the acoustic power radiated at the outlet of a reference system and the system investigated, with both systems driven by the same source. Either the reference is taken as an existing ("old") silencer, otherwise it is taken as a straight pipe with the same length as the investigated system and the same cross-sectional area as the source outlet. One problem with this choice is that it will introduce the resonances of the straight pipe, which for pipes with a length of a few meters can cause a significant modulation of the IL curve. An alternative to avoid this is to use a short reference pipe instead (say < 0.1 m), with no resonances in the frequency range of interest. Using Equation (13.14), assuming the 2-port is connected to a source and a radiation impedance, see Figure 13.5, it can be shown that:

$$\mathrm{IL} = 10\log_{10}\left|\frac{q_b^{ref}}{q_b}\right|^2 = 10\log_{10}\left|\frac{T_{aa} + T_{ab}/Z_r + T_{ba}Z_s + T_{bb}Z_s/Z_r}{T_{aa}^{ref} + T_{ab}^{ref}/Z_r^{ref} + T_{ba}^{ref}Z_s + T_{bb}^{ref}Z_s/Z_r^{ref}}\right|^2. \quad (13.19)$$

This result assumes that the open end of the intake or exhaust system radiates as a monopole which is correct for low frequencies and Mach numbers < 0.5 [22]. It can also be noted that the radiation impedance for the reference system will be different from the system investigated if the source outlet and the open end have different diameters. Sometimes [15] TL is defined including reflections at the outlet side. This can be seen as special case of Equation (13.19) obtained by assuming a reflection-free source plus a reference system with a reflection-free termination. The two formulas (13.18) and (13.19) will look the same using Munjal's [2] transfer-matrix formalism, as long as the duct acoustic impedance Z^o is modified to ensure that all terms are dimensionless.

13.2.2.2 2-PORT MODELS: GENERAL NETWORKS

For systems with more complex connections than a cascade, the transfer-matrix formalism becomes less useful. As pointed out by Frid [9], a formalism based on mobility matrices (see Equation 13.7), analogous to the one used for structure-borne sound, is then better. Another alternative presented by Glav and Åbom [23] is to use a scattering-matrix formalism, such as that described in Figure 13.6, that also allows for sound generation in all elements.

This formalism is based on travelling pressure waves and is defined by:

$$\begin{pmatrix} p_{a-} \\ p_{b-} \end{pmatrix} = \underbrace{\begin{pmatrix} R_{aa} & T_{ab} \\ T_{ba} & R_{bb} \end{pmatrix}}_{S} \begin{pmatrix} p_{a+} \\ p_{b+} \end{pmatrix} + \underbrace{\begin{pmatrix} p_{a-}^s \\ p_{b-}^s \end{pmatrix}}_{p_-^s}, \quad (13.20)$$

where also an *active* sound-generating part $\mathbf{p}_-^s$ is included that represents the sound generation with reflection-free terminations. The elements in the scattering matrix **S** represent reflection on the a and b side (R_{aa} and R_{bb}) and transmission across the 2-port (T_{ab} and T_{ba}), respectively. Although in principle all 2-port formulations are equivalent and related by linear transformations, the scattering matrix formulation is physically the most correct preserving causality. This is important when coupling time and frequency domain models and for such problems scattering matrix-based formulations are to be preferred. The damping measures discussed previously TL and IL can of course be expressed using the scattering matrix. The TL from a to b is for instance given by:

$$\mathrm{TL}_{ab} = 10\log_{10}\left(\frac{Z_b^o(1+M_a)^2}{Z_a^o(1+M_b)^2}\left|\frac{1}{T_{ba}}\right|\right)^2, \quad (13.21)$$

which is an alternative formulation to Equation (13.18). Interchanging a and b will directly give the formula for transmission from b to a, not requiring a matrix inversion as would be the case for a transfer matrix.

FIGURE 13.6 Definition of positive directions for the scattering-matrix formalism.

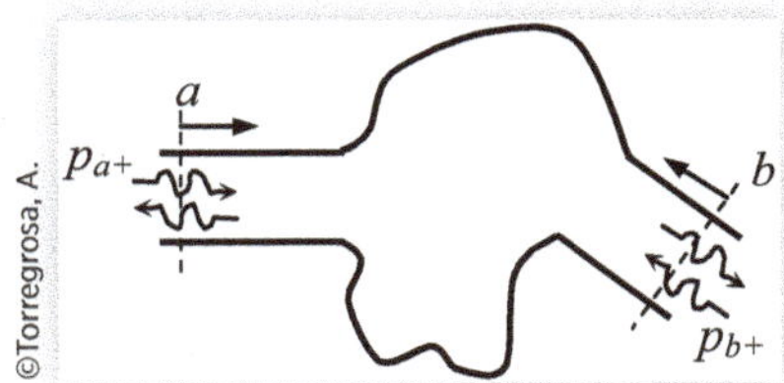

CHAPTER 13

The scattering matrix can also be used for analyzing amplification of sound by vorticity which can lead to whistling [24, 25]. The basis for this method is to transform the travelling wave amplitudes to two variables (X_+, X_-) referred to as acoustic exergy. The exergy is defined so that the power transmitted in the +/– direction is given by $|X_{+/-}|^2$, which based on Equation (13.13) implies that:

$$X_+ = \frac{p_+(1+M)}{\sqrt{Z^o}} \quad \text{and} \quad X_- = \frac{p_-(1-M)}{\sqrt{Z^o}}. \tag{13.22}$$

The scattering matrix formulation using exergy can be written as:

$$\begin{pmatrix} X_{a-} \\ X_{b-} \end{pmatrix} = \underbrace{\begin{pmatrix} R^X_{aa} & T^X_{ab} \\ T^X_{ba} & R^X_{bb} \end{pmatrix}}_{S^X} \begin{pmatrix} X_{a+} \\ X_{b+} \end{pmatrix} + \underbrace{\begin{pmatrix} X^s_{a-} \\ X^s_{b-} \end{pmatrix}}_{X^s_-}. \tag{13.23}$$

The max/min amplification of incident wave power can be analyzed by computing the eigenvalues of the Hermitian matrix: $(S^X)^*S^X$, where * denotes a complex conjugate and matrix transpose [24].

In the exergy formulation, the TL is directly obtained from the transfer elements, e.g., for the direction a to b:

$$TL_{ab} = 10\log_{10}\left(\left|\frac{1}{T^X_{ba}}\right|\right)^2.$$

It is also easy using exergy to, as recently suggested by Kabral and Åbom [26], split the TL in a dissipative ("dis") and reflective ("refl") part:

$$TL^{dis}_{ab} = 10\log_{10}\left(\frac{1-\left|R^X_{aa}\right|^2}{\left|T^X_{ba}\right|^2}\right) \quad \text{and} \quad TL^{refl}_{ab} = -10\log_{10}\left(1-\left|R^X_{aa}\right|^2\right), \tag{13.24}$$

where $TL = TL^{dis} + TL^{refl}$.

Many problems regarding modern intake or exhaust systems involve complex silencers with multiple inlets and outlets. But as demonstrated by Glav and Åbom [23] and Elnady et al. [27, 28], any complex system can be broken down into a network of 2-ports. The concepts of TL and IL can also without problem be generalized to handle such cases; see, e.g., Herrin et al. [29] and Munjal [2]. Concerning the TL, one again prescribes reflection-free terminations at all inlets and outlets. Then it is necessary to introduce the relative amplitude and phase for the incident waves with one inlet assigned a unit (=1) amplitude. For the insertion loss, the main difficulty is to decide on the reference system, e.g., in selecting straight pipes, one must decide to what extent the sources interact. The alternative is to choose a different ("old") silencer with the same number of inlet/outlets which makes it straightforward.

A final note on the strength of 2-port models is that by formulating the basic equations for 1D steady flow as 2-ports, Elnady et al. [30] have demonstrated how a given acoustic 2-port network also can be used to compute the flow distribution in a system.

This is of course useful since the computation of the acoustic 2-port elements then can be made using a correct flow distribution in a system. Efforts to extend the 2-port approach to also handle heat flows and temperature distributions have also been reported [31].

13.2.2.3 2-PORT MODELS: SOUND GENERATION

The complete 2-port model in Equation (13.20) describes both the sound scattering and the sound generation from a component. The scattering or passive part of the properties can today often be computed using the models described in this chapter. Concerning the active or sound-generating part, it is more difficult, in particular for broadband sources, requiring high-fidelity CFD computation; see, e.g., [32]. An alternative is to measure the complete 2-port examples of this for a flow constriction, and a turbo-compressor can be found in Refs. [33, 34].

One can also use the semiempirical formulas for flow noise from a constriction in a duct to derive a complete 2-port model as first suggested by Nygård [35]. This approach is based on the classical work by Nelson and Morfey [36], which identifies a constriction at low Mach numbers as a compact dipole. Based on this, scaling laws for the flow generated noise spectrum (downstream sound power W_+^s in, e.g., one-third-octave bands) can be derived and used to estimate noise from any constriction [37, 38]. For the plane wave range, the dipole can be seen as a point source creating a source strength: $p_+^s = -p_-^s$. This can be related to the downstream sound power from the scaling laws via [37]:

$$W_+^s = \frac{\left|p_+^s\right|^2}{Z^o}, \tag{13.25}$$

where following Nygård [35] the convective effect is neglected considering the uncertainty in the scaling laws. Examples on how this approach is applied to predict flow noise from commercial silencers can be found in Ref. [39].

13.2.2.4 EXAMPLES OF 2-PORT MODELS

In this section a few standard cases will be presented to illustrate how 2-port models can be derived. The derivation will be based on the transfer-matrix (p, q) formalism. More examples in particular for different silencer types, e.g., dissipative with porous lining and perforated, can be found in the book by Munjal [2] and in Section 13.2.3.

13.2.2.4.1 **The Straight Duct.** This is the simplest case and assuming a straight uniform duct of cross-sectional area S containing a homogenous fluid, the sound field can be written as:

$$\begin{cases} p(z) = p_+ e^{-ik_+ z} + p_- e^{ik_- z} \\ q(z) = p_+ e^{-ik_+ z}/Z_+^o - p_- e^{ik_- z}/Z_-^o \end{cases} \Rightarrow \begin{pmatrix} p(z) \\ q(z) \end{pmatrix} = \begin{pmatrix} e^{-ik_+ z} & e^{ik_- z} \\ e^{-ik_+ z}/Z_+^o & -e^{ik_- z}/Z_-^o \end{pmatrix} \begin{pmatrix} p_+ \\ p_- \end{pmatrix}, \tag{13.26}$$

where Equations (13.9) and (13.12) have been used and $Z_\pm^o = Z_\pm^c / S$. For a section of length L with the inlet a at z = 0 and outlet b at z = L, one then obtains:

$$\begin{pmatrix} p_a \\ q_a \end{pmatrix} = \begin{pmatrix} 1 & 1 \\ 1/Z_+^o & -1/Z_-^o \end{pmatrix} \begin{pmatrix} p_+ \\ p_- \end{pmatrix} \quad \text{and} \quad \begin{pmatrix} p_b \\ q_b \end{pmatrix} = \begin{pmatrix} e^{-ik_+ L} & e^{ik_- L} \\ e^{-ik_+ L}/Z_+^o & -e^{ik_- L}/Z_-^o \end{pmatrix} \begin{pmatrix} p_+ \\ p_- \end{pmatrix},$$

CHAPTER 13

from which the backward (or forward) transfer matrix $\mathbf{T}_s$ can be directly obtained:

$$\begin{pmatrix} p_a \\ q_a \end{pmatrix} = \underbrace{\begin{pmatrix} 1 & 1 \\ 1/Z_+^o & -1/Z_-^o \end{pmatrix} \begin{pmatrix} e^{-ik_+L} & e^{ik_-L} \\ e^{-ik_+L}/Z_+^o & -e^{ik_-L}/Z_-^o \end{pmatrix}^{-1}}_{T_s} \begin{pmatrix} p_b \\ q_b \end{pmatrix}. \quad (13.27)$$

Neglecting the visco-thermal losses, this can be reduced to:

$$\mathbf{T}_s = e^{-ikML/(1-M^2)} \begin{bmatrix} \cos\left(kL/(1-M^2)\right) & iZ^o \sin\left(kL/(1-M^2)\right) \\ (i/Z^o)\sin\left(kL/(1-M^2)\right) & \cos\left(kL/(1-M^2)\right) \end{bmatrix}. \quad (13.28)$$

13.2.2.4.2 **An Expansion Chamber.** By putting together a series (cascade) of straight ducts with different cross sections, it is possible to create an "acoustic filter," which reflects sound in certain frequency ranges. The best-known case is the so-called expansion chamber silencer; see Figure 13.7.

Assuming no flow and only considering plane waves, the transfer matrix for this silencer is obtained by multiplying the matrix for three straight duct sections, putting the lengths of the first and last sections to zero. This works since under the 1D and no flow assumptions, the correct coupling conditions at the inlet/outlet area change its continuity of p and q [2, 3, 11]. Based on this result, the TL can be derived [2, 3, 11]:

$$TL = 10\log_{10}\left(1 + \frac{1}{4}(S_1/S_2 - S_2/S_1)^2 \sin^2(kL)\right). \quad (13.29)$$

Extension of the simple 1D analysis for expansion chambers to 3D and non-plane wave effects is possible for circular and rectangular cross sections [2]; see also Section 13.2.3.

For automotive applications the low-frequency damping is often a critical aspect. This can be formulated as achieving a large ratio of $TL_{max}/(c/4L)$, i.e., a steep increase of the damping up to the first TL_{max}. As evident from Equation (13.29), both the expansion factor S_2/S_1 and the chamber length L are then important for a good low-frequency damping.

13.2.2.4.3 **A Conical Pipe Section.** To create a smooth transition reducing the pressure loss, a (straight) conical section can be used. Depending on the flow direction, this either gives an expansion or contraction of the flow. If the flow is not separating but expands or contracts smoothly along the section, one can assume a 1D flow model. For a contracting section, this is no restriction, but for an expansion the top angle of the

FIGURE 13.7 An expansion chamber silencer or muffler and its transmission loss curve exhibiting periodic stopbands (caused by reflection) with maxima (TL_{max}) where the chamber length L equals an odd multiple of a quarter wave-length and minima when it equals a mulitiple of half a wave-length.

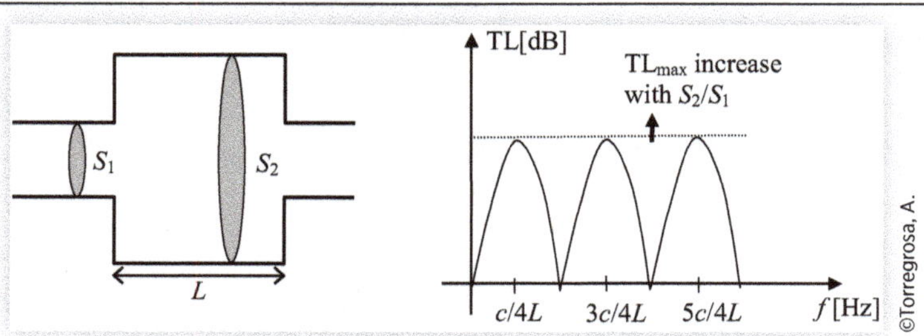

cone should be less than c:a 10 degrees to ensure no separation. There are several papers published on 1D sound propagation in variable area ducts with flow; see, e.g., Miles [40]. Here an alternative method [41] will be described based on approximating the actual duct with piecewise constant area straight ducts as illustrated in Figure 13.8.

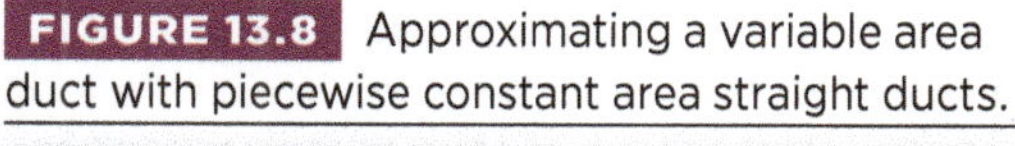

FIGURE 13.8 Approximating a variable area duct with piecewise constant area straight ducts.

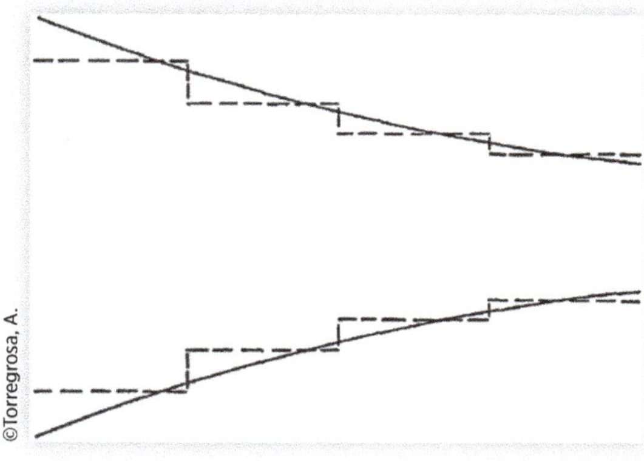

Each straight duct section has a length of $\Delta L = L/N$, where N is the number of straight ducts and L is the length of the variable area duct, e.g., a conical section. Assuming that the duct has an area variation S(z), where z is a symmetry axis with z = 0 at the inlet and z = L at the outlet, the area of the nth straight duct is $S_n = S(z_n)$, where $z_n = (n - 1/2)L/N$ and n = 1, 2, …, N. Since the straight ducts are in cascade, the transfer matrix for the entire section is simply:

$$\mathbf{T}_{cs} = \prod_{n=1}^{N} \mathbf{T}_{s,n}, \qquad (13.30)$$

where $\mathbf{T}_{s,n}$ is obtained from (13.27) or (13.28). The Mach number in each element is assuming incompressible 1D flow given by $M_n = M_0/S_n$, where M_0 is the Mach number at the inlet (z = 0). The assumption of incompressible flow is a good approximation at least up to a Mach number of 0.3. The proposed method will converge when ΔL is much smaller than the wavelength, i.e., at least ten elements per wavelength. Equation (13.30) implicitly assumes continuity of p and q at each intersection, which is not correct with flow at an area jump, but it is correct for the smooth variable area section to which the product should converge.

13.2.2.4.4 **A Side-Branch Resonator.** A side-branch resonator is in general an opening in a duct wall characterized by its impedance Z_{sb}. It is assumed here that the opening is acoustically compact, i.e., that kd << 1 where d is the diameter of the side-branch opening. The principle is based on the reflection created when the impedance or resonance is close to zero, i.e., $Z_{sb} \approx 0$; then the plane wave acoustic pressure at the side branch will also be close to zero. This implies that the incident and reflected waves on the upstream side (a) satisfy (Figure 13.9):

$$p_{+a} + p_{-a} \approx 0 \Rightarrow p_{-a} \approx -p_{+a},$$

FIGURE 13.9 Side branches. (1) general, (2) Helmholtz resonator, (3) quarter-wave resonator (a straight pipe with a rigid termination). The cross sections a and b are assumed to be just up- and downstream of the midpoint of the opening. No flow is assumed to enter into the side branch.

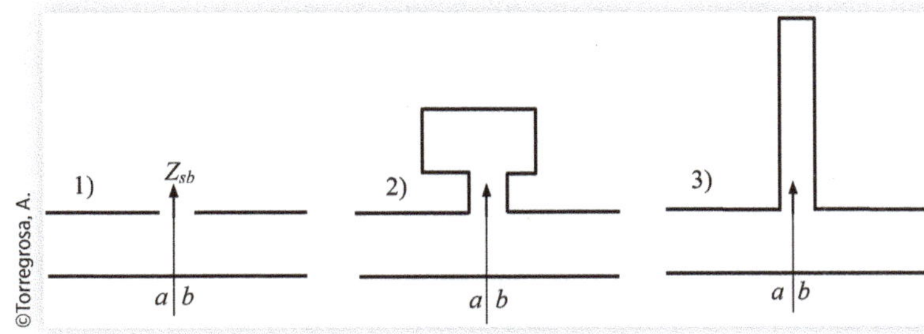

The problem in modeling a side branch with flow is related to that even the concept of impedance is not uniquely defined for this case; see, e.g., Aurégan [42]. This implies that the formulation used should be consistent with the one used for obtaining the impedance most often based on p and q. Using these state variables, the transfer matrix T_{sb} is obtained assuming continuity of pressure across the side branch and conservation of volume flow. This implies:

$$\begin{cases} p_a = p_b \\ q_a = q_{sb} + q_b \end{cases} \Rightarrow \begin{cases} p_a = p_b \\ q_a = p_b / Z_{sb} + q_b \end{cases} \Rightarrow \begin{pmatrix} p_a \\ q_a \end{pmatrix} = \underbrace{\begin{pmatrix} 1 & 0 \\ 1/Z_{sb} & 1 \end{pmatrix}}_{T_{sb}} \begin{pmatrix} p_b \\ q_b \end{pmatrix}.$$

The impedance of the side branch can be split into two parts, an outer (out) and an inner (in), where the inner is defined by the geometry and speed of sound in the branch and the outer will contain the effects of near fields and flow:

$$Z_{sb} = Z_{out} + Z_{in}. \tag{13.31}$$

The near fields correspond to non-propagating higher-order modes excited at the opening in order to adapt the field to the geometry. This effect is often referred to as the "end correction" since it can be shown to be equivalent to adding an extra pipe length equal to 0.85d. The flow effects are mainly related to dissipation created by vortex-sound interaction across the opening. Both effects are included in semiempirical formulas for the impedance of a perforate; see, e.g., Bauer [43] and Elnady et al. [27]. Based on Ref. [27], the outer impedance can for engineering purposes be estimated from:

$$Z_{out}/Z^{\circ} = 0.5M/\sigma + j\cdot\left(0.85kd/2 - 0.3M\right)/\sigma, \tag{13.32}$$

where $\sigma = S_{sb}/S$, i.e., the open area ratio between the side branch and the main duct and d is the side-branch diameter. This result assumes that $kd \ll 1$; neglects the viscous losses compared to the "vortex-sound" effect [3], i.e., losses created by acoustically induced vorticity; and is based on experimental data with $M < 0.3$. As seen from the equation, flow both introduces more losses and reduces the imaginary or inertial part leading to reduced resonance frequencies.

Two common types of side-branch resonators are the Helmholtz (H) and the quarter-wave (Q) resonator. The inner impedance for these two cases are given by [2]:

$$\begin{cases} Z_{in}^{H}/Z^{\circ} = jk_v\left(1 + 0.85d\right)/\sigma + \dfrac{S}{jkV} \\ Z_{in}^{Q}/Z^{\circ} = -j\cot\left(k_v L\right)/\sigma \end{cases}, \tag{13.33}$$

where

k_v is the wave number including viscous damping obtained from Equation (13.11) with $M = 0$

l is the length of the H-resonator neck

V is the H-resonator volume

L is the Q-wave resonator length

The Helmholtz resonator is the equivalent of a mass-spring system with one resonance. The quarter-wave resonator has resonances at odd multiples of the quarter wavelengths corresponding to the pipe length L. Using other shapes such as a conical resonator, it is possible to produce resonances at multiples of half-wavelength harmonics [2]. It can be also be noted that quarter-wave resonators can be parallel and folded to save space [44]. Finally, in automotive applications the opening of a resonator is often covered by a perforate, and the outer impedance should then be computed using the formulas in, e.g., Elnady et al. [27].

13.2.2.4.5 **A Sudden Area Expansion.** The inlet of an expansion chamber silencer creates a sudden area expansion where the flow will separate creating an unstable jet that further downstream expands and breaks down into turbulence. Sufficiently far downstream the flow has expanded into a turbulent pipe flow over the entire cross section. The key to explain the scattering of plane waves at a sudden area expansion is to analyze the acoustic interaction with the unstable jet. This is done by imposing a so-called Kutta condition at the sharp-edged corner where the jet leaves the upstream pipe [45].

The expansion of the jet and turbulence production are neglected in the analysis which is supported by the experimental results. The most complete solution of the problem using these ideas and the Wiener-Hopf method has been presented by Boij and Nilsson [46]. This solution is rather complicated but for low frequencies and Mach numbers, simpler models can instead be applied [15]. For low frequency sound and small Mach numbers (<0.3), one can assume a quasi-stationary and incompressible mean flow model. Applying first conservation of momentum in a control volume between the inlet section "a" and a downstream section "b," where the flow is assumed to be fully expanded, gives (Figure 13.10):

$$(P_a - P_b)\cdot S_b = \rho_a U_b^2 S_b - \rho_b U_a^2 S_a, \tag{13.34}$$

where

P is the fluid pressure
U is the flow speed along the duct axis
ρ is the density

Since the mean flow is assumed to be incompressible, one can put the density equal to ρ_0 (a constant). This will introduce an error $O(M^2)$ which implies that one should only

FIGURE 13.10 A sudden area expansion with three distinctive flow regions. (i) Just downstream of section "a" where the flow leaves the inlet pipe and forms an unstable jet; (ii) the unstable jet starts to expand and breaks down into turbulence; (iii) sufficiently far downstream (section "b") where the flow has expanded and fills the entire cross section. The extended inlet forms a side-branch which assuming a rigid termination can be regarded as a quarter-wave resonator.

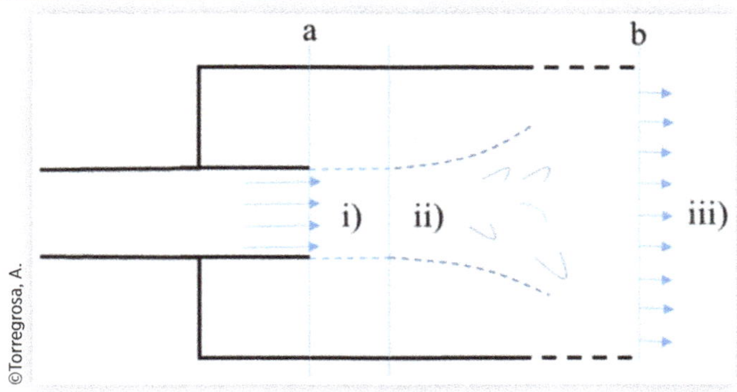

keep linear terms in the Mach number. Writing down this equation, it is assumed that the flow leaving cross section "a" separates and forms a jet with no radial expansion. In principle this assumption is equivalent to the abovementioned Kutta condition and requires that the pressure over the cross section "a" is constant and equal to P_a. The field can now be split into a steady state and a fluctuating (acoustic) part: $P = P_0 + p'$ and $U = U_0 + u'$. Inserting this gives:

$$\left(P_{a,0} + p_a' - \left(P_{b,0} + p_b'\right)\right)\cdot S_b = \rho_0\left(U_{b,0} + u_b'\right)^2 S_b - \rho_0\left(U_{a,0} + u_a'\right)^2 S_a \Rightarrow$$

$$\left(p_a' - p_b'\right)S_b = 2\rho_0 U_{b,0} S_b u_b' - 2\rho_0 U_{a,0} S_a u_a' \tag{13.35}$$

where in the last step the steady-state part cancels and the quadratic terms were omitted since a linear acoustic model is sought. In addition to this, conservation of mass implies:

$$\rho_a S_a U_a = \rho_s S_s U_s + \rho_b S_b U_b, \tag{13.36}$$

where the subscript s denotes the side branch at the expansion and $S_s = S_b - S_a$. Separating again the field into a steady and fluctuating part noting that $U_{s,0} = 0$ gives for the fluctuating part:

$$\rho_0 S_a u_a' + \rho_a' S_a U_a = \rho_0 S_s u_s' + \rho_0 S_b u_b' + \rho_b' S_b U_b \tag{13.37}$$

The above equations can directly be written in the frequency domain simply by dropping the (..)′. The density fluctuations at the cross section "a" is isentropic, and as argued by Cummings [47], the same can be assumed for the downstream cross section "b" ("a fully expanded flow") implying that $\rho' = p'/c^2$. Introducing in addition the volume flow ($q' = Su'$) and the impedance Z_s of the side branch, the two conservation equations above can be written as:

$$\begin{cases} p_a - p_b = 2\rho_0 U_{b,0}/S_b \cdot q_b - 2\rho_0 U_{a,0}/S_b \cdot q_a \\ q_a + p_a U_{a,0} S_a/\rho_0 c^2 = p_s/Z_s + p_b U_{b,0} S_b/\rho_0 c^2 + q_b \end{cases} \Rightarrow \begin{cases} p_a + 2\rho_0 U_{a,0}/S_b \cdot q_a = p_b + 2\rho_0 U_{b,0}/S_b \cdot q_b \\ p_a U_{a,0} S_a/\rho_0 c^2 - p_a/Z_s + q_a = p_b U_{b,0} S_b/\rho_0 c^2 + q_b \end{cases} \tag{13.38}$$

The pressure at the side branch p_s has here been put equal to p_a based on the assumption of no jet expansion at "a." From this result one obtains:

$$\begin{pmatrix} p_a \\ q_a \end{pmatrix} = \frac{1}{\det}\underbrace{\begin{pmatrix} 1 & 2Z_a^o M_a\left(\left(S_a/S_b\right)^2 - S_a/S_b\right) \\ 1/Z_s & 1 + 2Z_a^o M_a\left(S_a/S_b\right)^2/Z_s \end{pmatrix}}_{T_{ae}}\begin{pmatrix} p_b \\ q_b \end{pmatrix}, \tag{13.39}$$

where $\det = 1 + 2Z_a^o M_a\left(S_a/S_b\right)/Z_s$ which is correct to the first order in the Mach number. Concerning the side-branch impedance, it is given by the formula for a quarter-wave resonator; see Equation (13.33).

The result for the area-expansion transfer matrix can be improved by correcting for the 3D acoustic field that is created at the expansion. As proposed by Davies [15], the effect can be included by adding an extra length ΔL or end correction to the side-branch length L. Based on a paper by Karal [48], one can derive a formula for this end correction: $\Delta L = 8H(\alpha)d_a/3\pi$, where $\alpha = (S_a/S_b)^{1/2}$, d_a is the diameter of the inlet pipe, and H is a function defined by:

$$H(\alpha)=\begin{cases}0.875(1-\alpha)(1.371-\alpha), & 0.5\le\alpha\le 1.0\\ 1-1.238\alpha, & \alpha<0.5\end{cases}. \tag{13.40}$$

Note that even with no side branch present, a virtual branch can be added to account for the end correction or near fields.

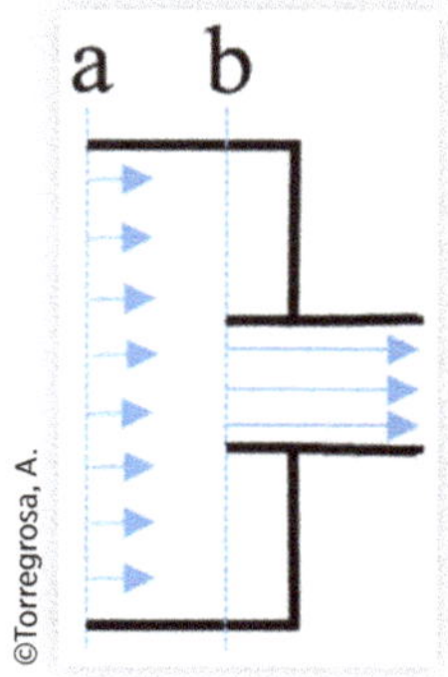

FIGURE 13.11 A sudden area contraction.

13.2.2.4.6 **A Sudden Area Contraction.** The outlet of an expansion chamber silencer creates a sudden area contraction (Figure 13.11). For this case Davies [15] suggests that the losses can be considered much smaller compared to an expansion. One can then apply conservation of energy which, using the same assumptions (quasi-stationarity and incompressible flow) as before, leads to conservation of enthalpy:

$$\frac{p_a}{\rho_0}+U_a u_a=\frac{p_b}{\rho_0}+U_b u_b, \tag{13.41}$$

where "a" and "b" denote cross sections up- and downstream of the contraction where the flow is uniform.

For this case losses are neglected compared to the sudden expansion. In reality a so-called vena contracta will form downstream of the outlet pipe opening creating flow separation and losses. This is neglected here and the cross section "b" is assumed to be at the opening of the outlet pipe.

Applying in addition conservation of mass and following the same steps used for the expansion, one can derive:

$$\begin{pmatrix}p_a\\ q_a\end{pmatrix}=\underbrace{\begin{pmatrix}1-Z_a^o M_a/Z_s & Z_a^o M_a\left((S_a/S_b)^2-1\right)\\ 1/Z_s & 1\end{pmatrix}}_{T_{ac}}\begin{pmatrix}p_b\\ q_b\end{pmatrix}. \tag{13.42}$$

Just as for the expansion, an end correction must be included for this case using the formula given above.

The effect of flow and acoustic near fields on an expansion chamber silencer can now be estimated by combining the transfer matrices (Equations 13.39 and 13.42) for an expansion and contraction with the transfer matrix for a straight duct section (Equation 13.28). The length of the straight duct equals the geometric length of the chamber minus the end corrections at the inlet and outlet.

13.2.2.4.7 **An Arbitrary Flow Constriction.** One can also derive the transfer matrix for an arbitrary constriction simply characterized by its pressure loss coefficient C_L, defined as $\Delta p_s = C_L\frac{1}{2}\rho_0 U_a^2$, where Δp_s is loss in stagnation pressure. The results are as follows:

$$\begin{pmatrix}p_a\\ q_a\end{pmatrix}=\begin{pmatrix}1 & Z_a^o M_a\left(C_L+(S_a/S_b)^2-1\right)\\ 0 & 1\end{pmatrix}\begin{pmatrix}p_b\\ q_b\end{pmatrix}. \tag{13.43}$$

Loss coefficients for typical elements such as orifice plates, bends, etc. can be found in many handbooks; see for instance [49].

13.2.2.4.8 **A Perforated Surface.** This element can be at an opening of a side branch or in the form of a pipe in many mufflers. In general the holes on the surface can be subjected to both grazing and bias flow which affects the impedance. In the literature a number of semiempirical impedance formulas have been presented, and a recent review can be found in [50]. Here one of the more general formulas for the impedance Z_p (=$\Delta p/u_n$, where n denotes normal direction) across a perforate will be presented [27]:

$$\begin{cases} \theta = \mathrm{Re}\left\{\dfrac{jk}{\sigma C_D}\left[\dfrac{t}{F(jK')}+\dfrac{\delta_{re}}{F(jK)}f_{int}\right]\right\}+\left(\dfrac{1-\sigma^2}{\sigma^2 C_D^2}\right)\dfrac{1}{2c}|u_n|+\dfrac{1}{\sigma}\left[1-\dfrac{2J_1(kd)}{kd}\right]+\dfrac{0.5}{\sigma}M_g+\dfrac{1.15}{\sigma C_D}M_b \\ \chi = \mathrm{Im}\left\{\dfrac{jk}{\sigma C_D}\left[\dfrac{t}{F(jK')}+\dfrac{0.5d}{F(jK)}f_{int}\right]\right\}-\left(\dfrac{1-\sigma^2}{\sigma^2 C_D^2}\right)\dfrac{1}{2c}\dfrac{|u_n|}{3}-\dfrac{0.3}{\sigma}M_g \end{cases} \quad (13.44)$$

where $Z_p/Z^o = \theta + j\chi$ and t is the thickness of the surface, d the diameter of the holes, σ the perforation ratio, C_D the orifice discharge coefficient, M_g the grazing Mach number, M_b the bias Mach number, J_n is the Bessel function of the first kind and order n, u_n acoustic velocity in the holes, $K = \sqrt{2}(d/\delta_v)$ (see Equation (13.10)), and K′ is obtained by multiplying kinematic viscosity by 2.179. Furthermore, $\delta_{re} = 0.2(d/1m) + 200(d/1m)^2 + 16{,}000(d/1m)^3$, $f_{int} = 1-1.47\sqrt{\sigma}+0.47\sqrt{\sigma^3}$, and $F(x) = 1-\dfrac{4J_1(x/2)}{xJ_0(x/2)}$.

The formula assumes that the holes are acoustically compact ($kd \ll 1$) and is based on data collected for Mach numbers < 0.3.

As seen from the formula, the effect of a mean flow is to increase the real part or the losses and reduce the imaginary part or the end correction. The losses are related to vortex-sound which for low frequencies always will give a dissipative effect. The formula also describes the effect of high amplitudes which can be important for automotive applications. An example of applying the above formula to model a complex perforated muffler can be found in Ref. [27].

Perforated plates with holes so small that the viscous boundary layer is of the same magnitude as the diameter are called micro-perforates [51]. Traditionally such plates have been developed for use as panel absorbers, e.g., in room acoustics. But as it has been demonstrated, they can also be used to build efficient mufflers [52, 53]. Formulas for the impedance of micro-perforates are similar to the equation above and can be found in Ref. [54]. Unlike traditional perforated silencers, micro-perforated configurations can dissipate sound efficiently also at low or no flow conditions, i.e., are not dependent on a mean flow to work. However, since micro-perforates often have low porosities (<10%), they are sensitive to high amplitudes and nonlinearities. This aspect is discussed and analyzed for instance in the paper by Allam and Åbom [55].

13.2.3 3D Methods

The models presented in the previous sections have been mainly devoted to sound propagation below the onset of 3D waves. As indicated, even before this take place, non-propagating higher-order modes will exist in the form of exponentially decaying non-propagating fields close to discontinuities in order to adapt the acoustic field to the geometry, the corresponding acoustic phenomenon being modeled through end corrections [2, 48, 56, 57, 58]. Once the duct cut-on frequency has been exceeded, however, 3D solutions of the wave equation (13.1) are necessary for an accurate prediction of the acoustic behavior, usually in the expansion form described by Equation (13.3). This may be the case of high-frequency problems or chamber geometries with large transverse dimensions [59, 60], where a number of higher-order modes get cut on even when the inlet and outlet ducts only carry plane waves (except in the region very close to the area changes).

With reference to Figure 13.7, the expansion chamber consists of three ducts and, for each one, the general pressure solution of the wave equation depends on mode shapes, modal wave numbers, and modal amplitudes, as detailed in Equation (13.3). Note that the acoustic velocity is then obtained by using the linearized equation of motion [2, 61], as indicated previously. While mode shapes and wave numbers can be determined for each duct separately, the computation of the modal amplitudes requires the enforcement of suitable compatibility conditions at the area discontinuities between all the ducts involved. In the absence of flow, continuity requirements are that acoustic pressure and axial particle velocity be equal at the junctions, together with zero axial velocity at rigid end plates [2]. When mean flow is present, additional considerations are admitted and some inconsistencies and numerical issues have been found in the bibliography [62, 63, 64, 65]. In particular, for a circular dissipative silencer containing mean flow and a perforated pipe, there exists evidence indicating that the axial kinematic compatibility condition should equate to that chosen for the radial kinematic boundary condition over the interface between the airway and the material. Thus, if the radial compatibility condition is continuity of displacement, then the axial compatibility condition should also be continuity of displacement [64].

Although a number of 3D analytical techniques have been proposed to speed up silencer TL calculations in the last decades [44, 64, 66, 67, 68, 69, 70, 71, 72, 73, 74, 75, 76, 77], they are limited to simple geometries for which analytical solutions are available (e.g., rectangular, circular, and elliptical). In the case of rigid wall ducts and concentric configurations [70, 74], the computational effort is strongly reduced, but for asymmetric geometries the algebra is usually very complicated and cumbersome. Additional problems appear for silencers incorporating a perforated duct and absorbent material. In this latter case, the necessity to include higher-order modes complicates the solution since a transverse governing eigenequation must be solved for the silencer section. The use of iterative methods, such as Newton-Raphson method or the Secant method [64, 78], has a common problem related to the possibility of missing roots, which is further exacerbated in the presence of mean flow. Therefore, an effort has been made in the last years to overcome these limitations without having the computational expenditure of a full 3D numerical approach. The details of this approach are given in Section 13.2.3.1, and, as it will be shown, a simplification is made regarding the silencer cross section, which is assumed axially uniform. For arbitrary geometries that do not meet this condition, the general numerical models briefly described in Section 13.2.3.2 should be considered. As is often the case with fully numerical methods, having a general computational tool usually involves the loss of much of the information related to the modal characteristics of the acoustic system.

13.2.3.1 3D METHODS BASED ON MATCHING THE MODAL EXPANSIONS

For practical purposes, it is especially interesting that there exists axial uniformity in the problem under consideration with a view to reducing the computational expenditure as much as possible when compared to, e.g., a 3D finite element method (FEM) analysis. In this case, the silencer cross section can be modeled as a 2D problem through FEM [79, 80, 81, 82, 83, 84], followed by the application of a modal coupling technique between

FIGURE 13.12 Scheme of a dissipative silencer. It contains a perforated pipe carrying a uniform mean flow. Absorbent material is placed in the outer central chamber: heterogeneities can exist, for example, due to transverse temperature gradients in the material.

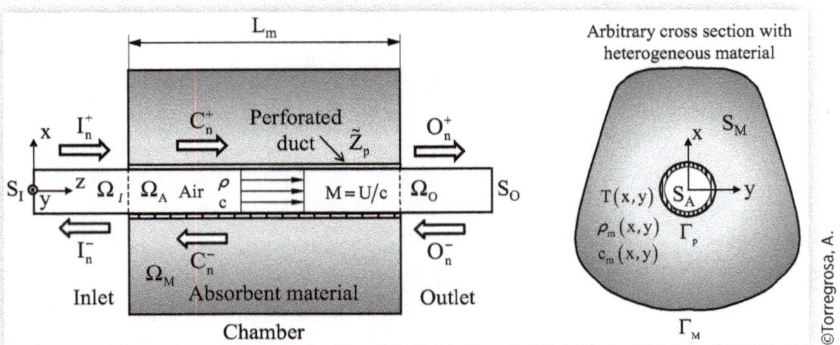

the subdomains involved to obtain the complete acoustic field. Figure 13.12 depicts a silencer scheme used to illustrate the acoustic modeling through this kind of matching approach. The geometry is divided into four subdomains including the inlet and outlet pipes (denoted as Ω_I and Ω_O, respectively), a perforated central duct Ω_A, and an outer chamber Ω_M with absorbent material. As indicated previously, the density and speed of sound in the air ($\Omega_I \cup \Omega_A \cup \Omega_O$) are denoted by ρ and c, whereas ρ_m and c_m are used for the equivalent acoustic properties of the dissipative region [61, 85]. In this first approach, axial uniformity is supposed for the silencer cross section; the properties ρ, c, ρ_m, and c_m; the perforated duct impedance $\tilde{Z}_p$; and the mean flow Mach number M. It is worth noting here that, compared to the impedance equation (13.44), now the perforated duct impedance notation has been modified to indicate the influence of the absorbent material [86]. Uniformity in z means that, for example, any axial temperature gradient is omitted in the mathematical approach (the inlet/outlet temperature average could be used to compute the relevant properties and the silencer performance), while the presence of transverse temperature variations is taken into account. Some procedures to overcome this hypothesis may be, for example, to consider a number of axial segments with piecewise uniform characteristics [69, 71] or to apply a general numerical approach based on FEM (see Section 13.2.3.2).

13.2.3.1.1 **Two-Dimensional Eigenvalue Problem.** First, the cross section is characterized through a transverse eigenvalue problem [81, 83, 84]. Equation (13.2) governs sound propagation in $\Omega_I \cup \Omega_A \cup \Omega_O$ and is repeated here for convenience.

$$\nabla^2 p - M^2 \frac{\partial^2 p}{\partial z^2} - 2jMk\frac{\partial p}{\partial z} + k^2 p = 0. \tag{13.45}$$

In the absence of mean flow, the wave equation for a heterogeneous absorbent material in Ω_M can be written as [87, 88]:

$$\nabla \cdot \left(\frac{1}{\rho_m} \nabla p_m \right) + \frac{k_m^2}{\rho_m} p_m = 0, \tag{13.46}$$

where subscript m is used for the absorbent material. Since axial uniformity along the z axis is assumed for the material properties and the silencer cross section, the solution of the wave equations (13.45) and (13.46) is decomposed into axial and transverse functions by using separation of variables (the treatment of the inlet and outlet pipes is relatively straightforward, and the details are omitted for the sake of brevity). The acoustic pressure in the transverse section of the chamber is expressed in the form:

$$p_C(x,y,z) = \Psi(x,y)e^{-jk_z z} \Rightarrow \Psi(x,y) = \begin{cases} \Psi^A(x,y), & (x,y) \in S_A \\ \Psi^M(x,y), & (x,y) \in S_M \end{cases} \tag{13.47}$$

Substituting Equation (13.47) in Equations (13.45) and (13.46) provides the two-dimensional (2D) governing equations in the cross section:

$$\nabla_{xy}^2 \Psi^A + \left(k^2 - 2Mkk_z - \left(1-M^2\right)k_z^2\right)\Psi^A = 0, \tag{13.48}$$

$$\nabla_{xy} \cdot \left(\frac{1}{\rho_m} \nabla_{xy} \Psi^M\right) + \frac{k_m^2 - k_z^2}{\rho_m} \Psi^M = 0, \tag{13.49}$$

where the subscript xy represents the 2D transverse coordinate system.

A 2D FEM discretization of the chamber cross section is used, together with the Green's theorem and the weighted residual method [89]. This leads to the following equations:

$$-\int_{S_A} \nabla_{xy} N^T \nabla_{xy} N \tilde{\Psi}^A dS + \left(k^2 - 2Mkk_z - \left(1-M^2\right)k_z^2\right) \int_{S_A} N^T N \tilde{\Psi}^A dS = -\int_{\Gamma_p} N^T \frac{\partial \Psi^A}{\partial n} d\Gamma, \tag{13.50}$$

$$-\int_{S_M} \frac{1}{\rho_m} \nabla_{xy} N^T \nabla_{xy} N \tilde{\Psi}^M dS + \int_{S_M} \frac{k_m^2 - k_z^2}{\rho_m} N^T N \tilde{\Psi}^M dS = -\int_{\Gamma_p} \frac{1}{\rho_m} N^T \frac{\partial \Psi^M}{\partial n} d\Gamma, \tag{13.51}$$

where **N** are the usual FEM shape functions, $\tilde{\Psi}^A$ and $\tilde{\Psi}^M$ contain the unknown pressure nodal values, and n is the outward unit vector. Note that the outer wall of the chamber is considered rigid and impervious, and therefore the boundary integrals in the right-hand side of Equations (13.50) and (13.51) are carried out only along the perforated duct line Γ_p. These boundary integrals allow the acoustic coupling between both subdomains through the perforated duct impedance $\tilde{Z}_p$ (defined as the ratio of the pressure jump to the acoustic velocity normal to surface). Here, continuity of the normal acoustic velocity is considered for simplicity, although different possibilities exist in the literature [2, 62, 64, 65, 90].

The procedure yields, from Equation (13.50), the following system of equations:

$$\left(\mathbf{K}^{AA} + k_z \mathbf{D}^{AA} + k_z^2 \mathbf{M}^{AA}\right)\tilde{\boldsymbol{\Psi}}^A + \left(\mathbf{K}^{AM} + k_z \mathbf{D}^{AM}\right)\tilde{\boldsymbol{\Psi}}^M = \mathbf{0}, \tag{13.52}$$

with

$$\begin{gathered} \mathbf{K}^{AA} = -\mathbf{K}_A + k^2 \mathbf{M}_A - \rho j\omega \mathbf{F}_{AA}, \quad \mathbf{D}^{AA} = -2Mk\mathbf{M}_A + \rho jMc\mathbf{F}_{AA}, \\ \mathbf{M}^{AA} = -\left(1-M^2\right)\mathbf{M}_A, \quad \mathbf{K}^{AM} = \rho j\omega \mathbf{F}_{AM}, \quad \mathbf{D}^{AM} = -\rho jMc\mathbf{F}_{AM}, \\ \mathbf{K}_A = \int_{S_A} \nabla_{xy} \mathbf{N}^T \nabla_{xy} \mathbf{N} dS, \quad \mathbf{M}_A = \int_{S_A} \mathbf{N}^T \mathbf{N} dS, \\ \mathbf{F}_{AA} = \int_{\Gamma_p} \frac{1}{\tilde{Z}_p} \mathbf{N}^T \mathbf{N} d\Gamma, \quad \mathbf{F}_{AM} = \int_{\Gamma_p} \frac{1}{\tilde{Z}_p} \mathbf{N}^T \mathbf{N} d\Gamma. \end{gathered} \tag{13.53)-(13.61}$$

CHAPTER 13

Applying a similar FEM procedure, Equation (13.51) provides

$$\left(\mathbf{K}^{MM} + k_z^2\mathbf{M}^{MM}\right)\tilde{\Psi}^M + \mathbf{K}^{MA}\tilde{\Psi}^A = \mathbf{0}, \tag{13.62}$$

where the following matrices have been computed

$$\mathrm{K}^{MM} = -\mathrm{K}_M + \mathrm{M}_{M1} - j\omega \mathrm{F}_{MM}, \quad \mathrm{M}^{MM} = -\mathrm{M}_{M2}, \quad \mathrm{K}^{MA} = j\omega \mathrm{F}_{MA},$$
$$\mathrm{K}_M = \int_{S_M} \frac{1}{\rho_m} \nabla_{xy}\mathrm{N}^T\nabla_{xy}\mathrm{N}dS, \quad \mathrm{M}_{M1} = \int_{S_M} \frac{k_m^2}{\rho_m}\mathrm{N}^T\mathrm{N}dS, \tag{13.63)-(13.70}$$
$$\mathrm{M}_{M2} = \int_{S_M} \frac{1}{\rho_m}\mathrm{N}^T\mathrm{N}dS, \quad \mathrm{F}_{MM} = \int_{\Gamma_P} \frac{1}{\tilde{Z}_p}\mathrm{N}^T\mathrm{N}d\Gamma, \quad \mathrm{F}_{MA} = \int_{\Gamma_P} \frac{1}{\tilde{Z}_p}\mathrm{N}^T\mathrm{N}d\Gamma.$$

Note that the perforated duct impedance $\tilde{Z}_p$ is retained within the integrals to take into account possible variations due to material heterogeneities. Equations (13.52) and (13.62) can be combined, giving the following second-order eigenvalue problem in k_z,

$$\left(\begin{pmatrix} \mathbf{K}^{AA} & \mathbf{K}^{AM} \\ \mathbf{K}^{MA} & \mathbf{K}^{MM} \end{pmatrix} + k_z \begin{pmatrix} \mathbf{D}^{AA} & \mathbf{D}^{AM} \\ \mathbf{0} & \mathbf{0} \end{pmatrix} + k_z^2 \begin{pmatrix} \mathbf{M}^{AA} & \mathbf{0} \\ \mathbf{0} & \mathbf{M}^{MM} \end{pmatrix}\right) \begin{Bmatrix} \tilde{\Psi}^A \\ \tilde{\Psi}^M \end{Bmatrix} = \begin{Bmatrix} \mathbf{0} \\ \mathbf{0} \end{Bmatrix}, \tag{13.71}$$

that is expressed in matrizant form as a standard eigenvalue problem

$$\begin{pmatrix} \mathbf{0} & \mathbf{I} \\ -\mathbf{M}^{-1}\mathbf{K} & -\mathbf{M}^{-1}\mathbf{D} \end{pmatrix} \begin{Bmatrix} \tilde{\Psi} \\ k_z\tilde{\Psi} \end{Bmatrix} = k_z \begin{Bmatrix} \tilde{\Psi} \\ k_z\tilde{\Psi} \end{Bmatrix}, \tag{13.72}$$

I being an identity matrix and **0** a matrix containing zeros. The following matrix definitions have been introduced in Equation (13.72):

$$\mathbf{K} = \begin{pmatrix} \mathbf{K}^{AA} & \mathbf{K}^{AM} \\ \mathbf{K}^{MA} & \mathbf{K}^{MM} \end{pmatrix}, \quad \mathbf{D} = \begin{pmatrix} \mathbf{D}^{AA} & \mathbf{D}^{AM} \\ \mathbf{0} & \mathbf{0} \end{pmatrix},$$
$$\mathbf{M} = \begin{pmatrix} \mathbf{M}^{AA} & \mathbf{0} \\ \mathbf{0} & \mathbf{M}^{MM} \end{pmatrix}, \quad \tilde{\Psi} = \begin{Bmatrix} \tilde{\Psi}^A \\ \tilde{\Psi}^M \end{Bmatrix}. \tag{13.73)-(13.76}$$

Note that, in the absence of flow, matrices $\mathbf{D}^{AA}$ and $\mathbf{D}^{AM}$ are null and therefore the eigenproblem can be solved as a standard one without the need to double the size of the system.

The solution of the problem defined by Equation (13.72) is the basis for the modal expansion used later for the matching approaches based on point collocation (PC) and weighted mode matching (MM). For higher-order modes, a negative imaginary part of the axial wave number k_z is associated with progressive waves, while a positive one is associated with regressive waves. As indicated in the work of Kirby [81], previous to the application of the matching procedure, the eigenvalues and their associated eigenvectors have to be sorted into an ascending order by the modulus of the imaginary part of the incident waves and by the imaginary part of the reflected waves. An example is provided in Table 13.1 for the first ten wave numbers. A silencer with circular cross section has been considered, with chamber radius $r_2 = 0.091875$ m, perforated duct radius $r_1 = 0.0268$ m, thickness $t = 0.001$ m, porosity $\sigma = 10\%$, and hole diameter $d = 0.0035$ mx [91, 92].

TABLE 13.1 Wave numbers of a dissipative configuration with circular chamber cross

Modal number	Progressive wave number	Regressive wave number
n = 1	0.7255 − j0.0940	−0.9818 + j0.1962
n = 2	1.7155 − j11.0641	−1.3829 + j16.4184
n = 3	1.1584 − j58.3534	−1.2854 + j59.6355
n = 4	1.1830 − j104.3578	−1.0014 + j103.7760
n = 5	0.0877 − j141.1106	0.0083 + j144.6966
n = 6	0.5235 − j147.9142	−1.3000 + j154.2342
n = 7	0.7257 − j197.3504	−0.7749 + j197.6001
n = 8	1.0703 − j246.0008	−0.4516 + j244.3934
n = 9	−0.7794 − j257.1017	−0.5206 + j268.4032
n = 10	0.4557 − j292.3685	−0.6624 + j293.1184

The perforated pipe carries a mean flow with Mach number M = 0.1, and it is backed by texturized glass fiber with resistivity R = 4186 Pa s/m^2 [78].

Note that the 2D FEM-based procedure presented here is general and allows the consideration of arbitrary transverse geometry and heterogeneous material properties. Even in the case of very simple geometries (e.g., circular), the current approach can outperform alternative analytical methods from a computational point of view [93]. This is due to the fact that an iterative algorithm to find the roots of a nonlinear characteristic equation is not required (also, the potential risk of having missing modes is strongly reduced). Further details can be found in references [81, 83, 84].

13.2.3.1.2 **Compatibility Conditions.** Acoustic pressure and axial acoustic velocity are assumed to be continuous over $S_I \equiv S_A$ (expansion),

$$p_I(x,y,z=0) = p_A(x,y,z=0), \quad (x,y) \in S_I \equiv S_A, \tag{13.77}$$

$$u_I(x,y,z=0) = u_A(x,y,z=0), \quad (x,y) \in S_I \equiv S_A, \tag{13.78}$$

while the rigid end plate on the left implies zero axial velocity, that is,

$$u_M(x,y,z=0) = 0, \quad (x,y) \in S_M. \tag{13.79}$$

Similarly, the conditions at the contraction can be written as:

$$p_A(x,y,z=L_m) = p_O(x,y,z'=0), \quad (x,y) \in S_O \equiv S_A, \tag{13.80}$$

$$u_A(x,y,z=L_m) = u_O(x,y,z'=0), \quad (x,y) \in S_O \equiv S_A, \tag{13.81}$$

$$u_M(x,y,z=L_m) = 0, \quad (x,y) \in S_M. \tag{13.82}$$

Note that the axial kinematic compatibility condition equates to that chosen for the radial kinematic boundary condition over the perforated duct interface between the airway and the material, that is, velocity is considered. Further details can be found in reference [64].

13.2.3.1.3 **Point Collocation Technique.** Equations (13.77)-(13.82) can now be expressed in terms of modal expansions similar to Equation (13.3), but here the summations start at 1 for convenience. As indicated previously, acoustic velocity is obtained from pressure by using the linearized equation of motion. At the expansion, the continuity of pressure yields

$$\sum_{n=1}^{\infty}\left(I_n^+\Psi_n^{I+}(x,y)+I_n^-\Psi_n^{I-}(x,y)\right)=\sum_{n=1}^{\infty}\left(C_n^+\Psi_n^{A+}(x,y)+C_n^-\Psi_n^{A-}(x,y)\right)\quad (x,y)\in S_I\equiv S_A. \tag{13.83}$$

The continuity of axial velocity can be written as

$$\frac{1}{\rho c}\left(\sum_{n=1}^{\infty}\frac{k_{z,n}^{I+}I_n^+\Psi_n^{I+}(x,y)}{k-Mk_{z,n}^{I+}}+\sum_{n=0}^{\infty}\frac{k_{z,n}^{I-}I_n^-\Psi_n^{I-}(x,y)}{k-Mk_{z,n}^{I-}}\right)$$
$$=\frac{1}{\rho c}\sum_{n=1}^{\infty}\left(\frac{k_{z,n}^{C+}C_n^+\Psi_n^{A+}(x,y)}{k-Mk_{z,n}^{C+}}+\frac{k_{z,n}^{C-}C_n^-\Psi_n^{A-}(x,y)}{k-Mk_{z,n}^{C-}}\right)\quad (x,y)\in S_I\equiv S_A, \tag{13.84}$$

while the rigid end plate provides

$$\frac{1}{\rho_m c_m}\sum_{n=1}^{\infty}\left(\frac{k_{z,n}^{C+}C_n^+\Psi_n^{M+}(x,y)}{k_m}+\frac{k_{z,n}^{C-}C_n^-\Psi_n^{M-}(x,y)}{k_m}\right)=0\quad (x,y)\in S_M. \tag{13.85}$$

In the previous expressions, $k_{z,n}^{I+}$, I_n^+, and $\Psi_n^{I+}(x,y)$ are, respectively, the axial wave number, the unknown pressure modal amplitude, and the corresponding pressure mode for the nth modal term, all of them belonging to the incident wave. Similarly $k_{z,n}^{I-}$, I_n^-, and $\Psi_n^{I-}(x,y)$ denote those terms belonging to the reflected waves. Regarding the chamber, both the central perforated airway and the dissipative region have common axial wave numbers $k_{z,n}^{C\pm}$ and unknown modal amplitudes $C_n^{\pm}$. For consistency with the notation of Equation (13.47), however, the pressure modes of the chamber are classified with superscript A (central pipe) and M (outer chamber with absorbent material). Finally, the previous definitions of the inlet pipe can be also used to describe the acoustic fields in the outlet duct, by suitably replacing the modal terms $k_{z,n}^{I\pm}$, $I_n^{\pm}$, and $\Psi_n^{I\pm}(x,y)$ by $k_{z,n}^{O\pm}$, $O_n^{\pm}$, and $\Psi_n^{O\pm}(x,y)$. It is worth noting that, for the inlet and outlet ducts I and O (with rigid walls), progressive and regressive modes are equal, that is, $\Psi_n^+(x,y)=\Psi_n^-(x,y)$, even in the presence of mean flow [2].

The conditions at the expansion, given by Equations (13.83)-(13.85), are now enforced at collocation points. A suitable choice is associated with the nodes of the transverse FEM mesh used for the computation of the eigenvalue problem [81]. The number of nodes is denoted by N_1 and N_2, corresponding to $S_I \equiv S_A$ and S_M, respectively. The modal expansions are truncated at the number of collocation points to guarantee the same number of equations and unknowns, providing N_1 terms for the inlet duct I and N_1 + N_2 terms for the chamber. Also, to compute the acoustic attenuation, the inlet duct is usually assumed to contain an incident unit plane wave, that is, $I_1^+\Psi_1^{I+}=1$ and $I_n^+=0$ for $n>1$. Thus, Equation (13.83) can be written as

$$1+\sum_{n=1}^{N_1}I_n^-\tilde{\Psi}_{n,q}^{I-}=\sum_{n=1}^{N_1+N_2}\left(C_n^+\tilde{\Psi}_{n,q}^{A+}+C_n^-\tilde{\Psi}_{n,q}^{A-}\right), \tag{13.86}$$

where $q = 1, \ldots, N_1$ and $\tilde{\Psi}^{I-}_{n,q}$ $\tilde{\Psi}^{A\pm}_{n,q}$ refer to the qth nodal value of transverse pressure modes for the inlet pipe and central passage, respectively. Similarly, Equation (13.84) yields, for $q = 1, \ldots, N_1$,

$$1+\sum_{n=1}^{N_1}\frac{k^{I-}_{z,n}I^-_n\tilde{\Psi}^{I-}_{n,q}}{k-Mk^{I-}_{z,n}}=\sum_{n=1}^{N_1+N_2}\left(\frac{k^{C+}_{z,n}C^+_n\tilde{\Psi}^{A+}_{n,q}}{k-Mk^{C+}_{z,n}}+\frac{k^{C-}_{z,n}C^-_n\tilde{\Psi}^{A-}_{n,q}}{k-Mk^{C-}_{z,n}}\right), \tag{13.87}$$

and finally the rigid wall condition given by Equation (13.85) gives, for $q = 1, \ldots, N_2$,

$$\sum_{n=1}^{N_1+N_2}\left(k^{C+}_{z,n}C^+_n\tilde{\Psi}^{M+}_{n,q}+k^{C-}_{z,n}C^-_n\tilde{\Psi}^{M-}_{n,q}\right)=0, \tag{13.88}$$

where $\tilde{\Psi}^{M\pm}_{n,q}$ refers to the qth nodal value of transverse pressure mode for the absorbent material. Equations (13.86)-(13.88) provide $2N_1 + N_2$ relations associated with the expansion.

For the contraction, the application of the same procedure to Equations (13.80)-(13.82) yields the rest of required equations (details can be found in reference [83]). For the computation of the silencer TL, the usual anechoic termination is considered here [2], that is, no outlet reflections are considered ($O^-_n = 0\ \forall n$). The final equations for the contraction are, for $q = 1, \ldots, N_1$,

$$\sum_{n=1}^{N_1+N_2}\left(C^+_n\tilde{\Psi}^{A+}_{n,q}e^{-jk^{C+}_{z,n}L_m}+C^-_n\tilde{\Psi}^{A-}_{n,q}e^{-jk^{C-}_{z,n}L_m}\right)=\sum_{n=1}^{N_1}O^+_n\tilde{\Psi}^{O+}_{n,q}, \tag{13.89}$$

and

$$\sum_{n=1}^{N_1+N_2}\left(\frac{k^{C+}_{z,n}C^+_n\tilde{\Psi}^{A+}_{n,q}e^{-jk^{C+}_{z,n}L_m}}{k-Mk^{C+}_{z,n}}+\frac{k^{C-}_{z,n}C^-_n\tilde{\Psi}^{A-}_{n,q}e^{-jk^{C-}_{z,n}L_m}}{k-Mk^{C-}_{z,n}}\right)=\sum_{n=1}^{N_1}\frac{k^{O+}_{z,n}O^+_n\tilde{\Psi}^{O+}_{n,q}}{k-Mk^{O+}_{z,n}}, \tag{13.90}$$

that are associated with continuity of pressure and velocity. The rigid wall condition on the right gives, for $q = 1, \ldots, N_2$,

$$\sum_{n=1}^{N_1+N_2}\left(k^{C+}_{z,n}C^+_n\tilde{\Psi}^{M+}_{n,q}e^{-jk^{C+}_{z,n}L_m}+k^{C-}_{z,n}C^-_n\tilde{\Psi}^{M-}_{n,q}e^{-jk^{C-}_{z,n}L_m}\right)=0. \tag{13.91}$$

Equations (13.86)-(13.91) provide an algebraic system consisting of $4N_1 + 2N_2$ equations with the same number of unknowns. For each excitation frequency, the unknown modal amplitudes I^-_n, $C^\pm_n$ and O^+_n can now be computed. The TL is finally obtained as [2]

$$TL=-20\log\left|O^+_1\Psi^{O+}_1\right|, \tag{13.92}$$

where it is assumed that the inlet and outlet ducts are long enough to guarantee the rapid decay of higher-order evanescent modes.

As indicated previously, a suitable choice of collocation points is important from a computational point of view, since the rate of convergence of this method is sensitive to the grid chosen [79]. The nodes of the transverse FEM mesh associated with the computation of the eigenproblem have been shown to provide satisfactory results [81]. Note that the presence of the perforated duct avoids the existence of singular points such as the vertices found, for example, in empty expansion chambers. In this latter configuration, acoustic scattering is likely to be pronounced, and a relatively large number of evanescent modes could be required to represent the pressure and velocity fields at particular locations if modal expansions are used, which means that the computational performance of a PC scheme can be considerably reduced. Some possibilities to improve the convergence, such as the consideration of small duct extensions [94] or the exclusion of the vertices at the geometrical discontinuities [95] can be incorporated to the PC scheme. An alternative is presented in the next section, based on satisfying the pressure and velocity compatibility conditions in an integral sense, rather than at discrete points.

13.2.3.1.4 **Weighted Mode Matching Method.** In this section, weighted integration is used to enforce the conditions at the expansion and contraction expressed in Equations (13.77)-(13.82), instead of matching over discrete collocation points on the silencer cross section. A weighting function is chosen for each condition, and then an integration is carried out over the corresponding cross section. The procedure is repeated with a suitable number of weighting functions to form a complete set of algebraic equations. According to the bibliography [61, 64, 70, 96], using the duct pressure modes as weighting functions delivers a convergent system of equations in the absence of mean flow, due to the orthogonality of the duct eigenfunctions. In the presence of mean flow, orthogonality is no longer achieved and nonphysical predictions can occur (e.g., negative TL). Anyway, convergence is still possible without a true orthogonality relation provided that consistent axial and radial kinematic matching conditions are chosen in order to obtain a convergent system of equations [64].

The procedure here proceeds by using the incident pressure modes in the inlet duct as weighting functions for Equation (13.77) and the incident pressure modes in the central chamber as weighting functions for Equations (13.78) and (13.79). For the expansion, the following expression are obtained after integration

$$\int_{S_I} p_I(x,y,z=0)\Psi_n^{I+}(x,y)dS_I = \int_{S_A} p_A(x,y,z=0)\Psi_n^{I+}(x,y)dS_A, \tag{13.93}$$

$$\int_{S_I} u_I(x,y,z=0)\Psi_n^{A+}(x,y)dS_I = \int_{S_A} u_A(x,y,z=0)\Psi_n^{A+}(x,y)dS_A, \tag{13.94}$$

$$\int_{S_M} u_M(x,y,z=0)\Psi_n^{M+}(x,y)dS_M = 0. \tag{13.95}$$

Equations (13.78) and (13.79) corresponding to the contraction are treated in a similar way, giving

$$\int_{S_A} p_A(x,y,z=L_m)\Psi_n^{O+}(x,y)dS_A = \int_{S_O} p_O(x,y,z'=0)\Psi_n^{O+}(x,y)dS_O, \tag{13.96}$$

$$\int_{S_A} u_A(x,y,z=L_M)\Psi_n^{A+}(x,y)dS_A = \int_{S_O} u_O(x,y,z'=0)\Psi_n^{A+}(x,y)dS_O, \tag{13.97}$$

$$\int_{S_M} u_M(x,y,z'=L_m)\Psi_n^{M+}(x,y)dS_M = 0. \tag{13.98}$$

The acoustic fields in the inlet/outlet ducts and chamber are written in terms of modal expansions and truncated at N_1 and $N_1 + N_2$, respectively. Equation (13.93) yields, after considering $q = 1, \ldots, N_1$ weighting functions,

$$\sum_{n=1}^{N_1}\left(I_n^+\int_{S_I}\Psi_n^{I+}(x,y)\Psi_q^{I+}(x,y)dS_I + I_n^-\int_{S_I}\Psi_n^{I-}(x,y)\Psi_q^{I+}(x,y)dS_I\right)$$
$$=\sum_{n=1}^{N_1+N_2}\left(C_n^+\int_{S_A}\Psi_n^{A+}(x,y)\Psi_q^{I+}(x,y)dS_A + C_n^-\int_{S_A}\Psi_n^{A-}(x,y)\Psi_q^{I+}(x,y)dS_A\right), \qquad (13.99)$$

Equations (13.94) and (13.95) are considered with $q = 1, \ldots, N_1 + N_2$ weighted functions. Integration and summation provide

$$\frac{1}{\rho c}\sum_{n=1}^{N_1}\left(\frac{k_{z,n}^{I+}I_n^+}{k-Mk_{z,n}^{I+}}\int_{S_I}\Psi_n^{I+}(x,y)\Psi_q^{A+}(x,y)dS_I + \frac{k_{z,n}^{I-}I_n^-}{k-Mk_{z,n}^{I-}}\int_{S_I}\Psi_n^{I-}(x,y)\Psi_q^{A+}(x,y)dS_I\right)=$$
$$=\frac{1}{\rho c}\sum_{n=1}^{N_1+N_2}\left(\frac{k_{z,n}^{C+}C_n^+}{k-Mk_{z,n}^{C+}}\int_{S_A}\Psi_n^{A+}(x,y)\Psi_q^{A+}(x,y)dS_I + \frac{k_{z,n}^{C-}C_n^-}{k-Mk_{z,n}^{C-}}\int_{S_A}\Psi_n^{A-}(x,y)\Psi_q^{A+}(x,y)dS_I\right)$$
$$+\sum_{n=1}^{N_1+N_2}\left(\frac{k_{z,n}^{C+}C_n^+}{k_m}\int_{S_M}\frac{1}{\rho_m c_m}\Psi_n^{M+}(x,y)\Psi_q^{M+}(x,y)dS_M + \frac{k_{z,n}^{C-}C_n^-}{k_m}\int_{S_M}\frac{1}{\rho_m c_m}\Psi_n^{M-}(x,y)\Psi_q^{M+}(x,y)dS_M\right). \qquad (13.100)$$

For the contraction, the corresponding equations are, in final form

$$\sum_{n=1}^{N_1+N_2}\left(C_n^+e^{-jk_{z,n}^{C+}L_m}\int_{S_A}\Psi_n^{A+}(x,y)\Psi_q^{O+}(x,y)dS_A + C_n^-e^{-jk_{z,n}^{C-}L_m}\int_{S_A}\Psi_n^{A-}(x,y)\Psi_q^{O+}(x,y)dS_A\right)$$
$$=\sum_{n=1}^{N_1}\left(O_n^+\int_{S_O}\Psi_n^{O+}(x,y)\Psi_q^{O+}(x,y)dS_O + O_n^-\int_{S_O}\Psi_n^{O-}(x,y)\Psi_q^{O+}(x,y)dS_O\right) \qquad (13.101)$$

$$\frac{1}{\rho c}\sum_{n=1}^{N_1+N_2}\left(\frac{k_{z,n}^{C+}C_n^+e^{-jk_{z,n}^{C+}L_m}}{k-Mk_{z,n}^{C+}}\int_{S_A}\Psi_n^{A+}(x,y)\Psi_q^{A+}(x,y)dS_I + \frac{k_{z,n}^{C-}C_n^-e^{-jk_{z,n}^{C-}L_m}}{k-Mk_{z,n}^{C-}}\int_{S_A}\Psi_n^{A-}(x,y)\Psi_q^{A+}(x,y)dS_I\right)$$
$$+\sum_{n=1}^{N_1+N_2}\left(\frac{k_{z,n}^{C+}C_n^+e^{-jk_{z,n}^{C+}L_m}}{k_m}\int_{S_M}\frac{1}{\rho_m c_m}\Psi_n^{M+}(x,y)\Psi_q^{M+}(x,y)dS_M\right.$$
$$\left.+\frac{k_{z,n}^{C-}C_n^-e^{-jk_{z,n}^{C-}L_m}}{k_m}\int_{S_M}\frac{1}{\rho_m c_m}\Psi_n^{M-}(x,y)\Psi_q^{M+}(x,y)dS_M\right)$$
$$=\frac{1}{\rho c}\sum_{n=1}^{N_1}\left(\frac{k_{z,n}^{O+}O_n^+}{k-Mk_{z,n}^{O+}}\int_{S_O}\Psi_n^{O+}(x,y)\Psi_q^{A+}(x,y)dS_O + \frac{k_{z,n}^{O-}O_n^-}{k-Mk_{z,n}^{O-}}\int_{S_O}\Psi_n^{O-}(x,y)\Psi_q^{A+}(x,y)dS_O\right). \qquad (13.102)$$

Similar to the PC technique, an incident unit plane wave is assumed in the inlet duct, whereas no outlet reflections are considered. Equations (13.99) and (13.100) form a set of $4N_1 + 2N_2$ equations with the same number of unknowns. The TL is computed through Equation (13.92).

Figure 13.13 shows a comparison between the TL obtained with point collocation (PC) and weighted MM for an axisymmetric perforated dissipative silencer.

FIGURE 13.13 Comparison of TL computed with PC and MM (transversal FEM mesh with 22 nodes and approximate element size of 0.01 m): ooo, PC; ▬▬, MM, six modes.

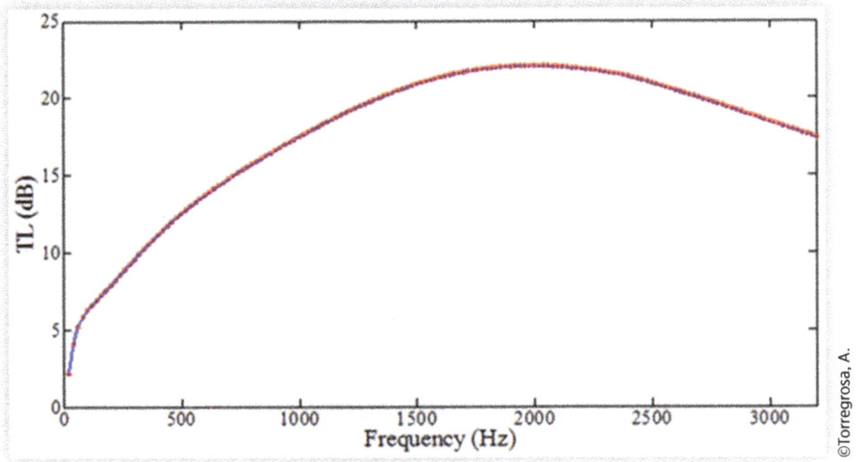

The geometry is defined by a chamber with length $L_m = 0.3$ m and radius $r_2 = 0.091875$ m. The perforated duct is characterized by a radius $r_1 = 0.0268$ m, thickness t = 0.001 m, porosity $\sigma = 10\%$, and hole diameter d = 0.0035 m. The perforated pipe carries a mean flow with Mach number M = 0.1, and it is backed by E fiberglass with a reference resistivity R = 30716 Pa s/m^2 at 25°C. A radial temperature gradient exists within the absorbent material, ranging from 250°C at the inner region (perforated surface with r_1) to 150°C (chamber outer surface with r_2). The reader is referred to Chapter 5 of reference [84] for detailed information about the corresponding acoustic models. As can be seen in the figure, the agreement of the computations is excellent, with undistinguishable curves over all the frequency ranges.

Figures 13.14 and 13.15 present a more detailed comparison of approaches in terms of relative error and computation time, respectively. This can be achieved by comparing

FIGURE 13.14 Comparison of average TL relative error (%) versus number of nodes for different meshes of a perforated dissipative silencer: ▬▬, MM, one mode; ▬▬, MM, three modes; ▬▬, MM, four modes; ▬▬, MM, six modes; ‒ ‒ ‒, PC at nodes; ‒ ‒ ‒, PC at Gauss points.

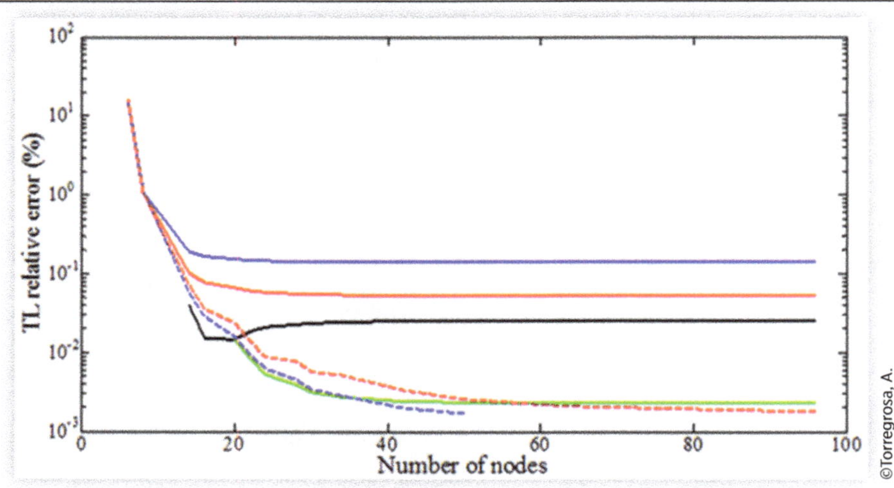

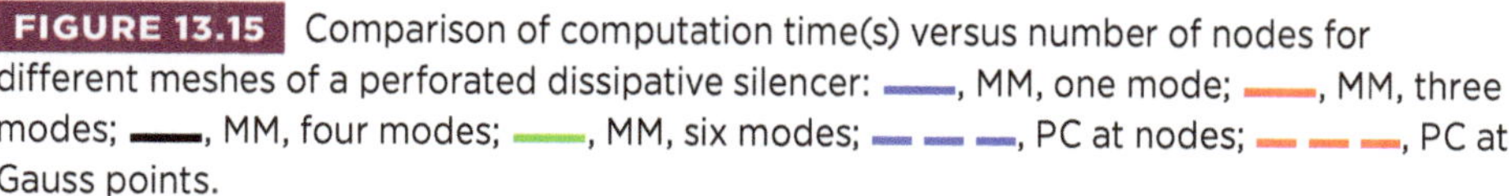

FIGURE 13.15 Comparison of computation time(s) versus number of nodes for different meshes of a perforated dissipative silencer: ——, MM, one mode; ——, MM, three modes; ——, MM, four modes; ——, MM, six modes; – – –, PC at nodes; – – –, PC at Gauss points.

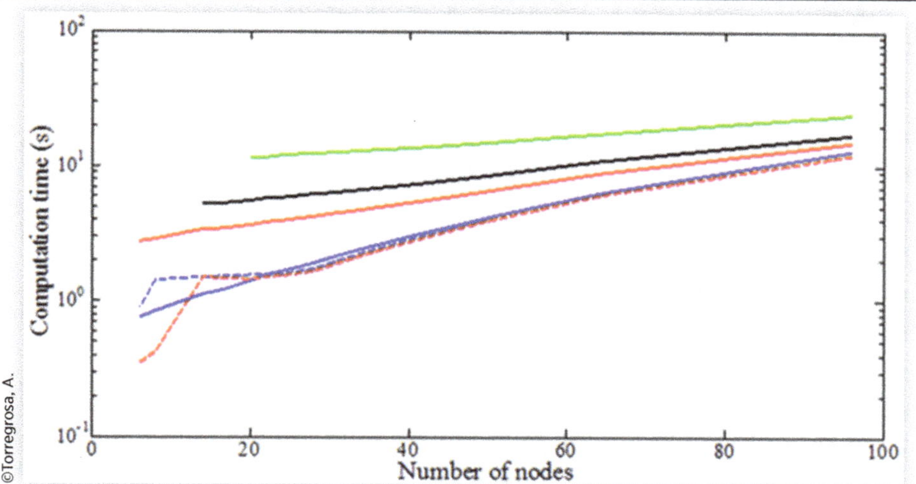

the TL obtained through the PC and the MM techniques with a reference FEM solution [84]. The mesh used in this reference computation is composed of eight-node quadratic quadrilateral elements, their size being 0.001 m, which permits obtaining an accurate solution. This provides about 100 elements per wavelength for the maximum frequency f_{max} = 3200 Hz considered in the computations. The meshes associated with the transversal discretizations of the PC and the MM approaches are composed of 3-node quadratic one-dimensional elements, since the geometry under analysis is axisymmetric, and, therefore, the silencer cross section can be represented by a line. Figure 13.14 represents the average TL relative error [84, 95] versus the number of nodes for several meshes under study. It can be deduced that, from a practical point of view, the average error is negligible if a reasonable number of nodes (more than 10) is used, since its value does not exceed the 0.2% in any case for both the PC and the MM approaches when a mesh of more than 14 nodes is used. It is also observed that for a given number of modes in MM, as the number of nodes increases, the error tends to an asymptotic value. On the other hand, when applying PC, different collocation schemes can be assessed. Here, nodes and Gauss points are compared, the TL relative error for a given FEM mesh being lower for the former point selection (nodes), since the dimension of the system of equations is larger.

Figure 13.15 shows some curves related to the computational effort. As the number of modes considered in the MM technique is higher, the error decreases at the expense of increasing the computation time. It should be noted that the PC and the MM errors are comparable if six modes are considered in the latter technique, but the collocation techniques have a certain advantage in terms of computational expenditure. Regarding PC at nodes and Gauss points, the most time-consuming operation of these methods is related to the assembly of the final system of equations when a standard matrix process is followed. In this case, the PC method at Gauss points requires a lower computational cost. To improve the computation time, the MATLAB function bsxfun has been used to avoid some time-consuming loops in the code when assembling the final system of equations in all the methodologies presented here [84]. This fact supposes that the computation times are quite similar if the MM approach with one mode and the PC approaches (nodes and Gauss points) are compared, although the error of the former is considerably higher (see Figure 13.14).

The computation cost increases with a higher number of degrees of freedom, as well as with a larger number of weighting functions in the case of the MM method. As can be seen, the error is higher when using this latter method for a given computational expenditure.

13.2.3.2 NUMERICAL 3D METHODS

Two methods will be briefly described here: the first is based on the convective wave equation, and the second is based on the LNSE. In both cases the problem can be solved using FEM in the frequency domain, and commercial solvers are available, e.g., COMSOL. The difference is that formulating a wave equation decouples vortex-sound effects from the sound field. Such effects are related to low-frequency dissipation in regions with flow separation but can also lead to amplification of sound in certain frequency ranges. The latter effect is the origin of whistling for instance in a side-branch resonator.

13.2.3.2.1 **The Convective Wave Equation.** In some situations, certain conditions required by the previous models may not be fulfilled, e.g., silencers with large transversal dimensions, high frequencies, nonuniform mean flow, complex geometries, temperature gradients, general boundary conditions, etc. For such cases, the use of a general FEM-based approach [89] can be useful at the expense of a high computational expenditure. A literature review of FEM applications to acoustic problems, with emphasis on the specific case of silencers, is provided in references [2, 84]. In the presence of mean flow, it is necessary to obtain the solution of the convective wave equation, which is usually expressed in terms of an acoustic velocity potential, rather than pressure, unless the mean flow field is uniform [97]. The familiar pressure-based wave equation for stationary medium [2] admits a similar pressure representation in a moving medium if the partial time derivative is replaced by the material one, Equation (13.1), but this is only valid in uniform steady flow. Actually, as shown in [97], no equation of comparable simplicity is known to exist when the flow field varies with the coordinates. In this latter case, the convective wave equation expressed in terms of acoustic pressure also involves derivatives of the acoustic velocities, thus having more than one dependent variable (see Eq. (15) of reference [97]). To avoid this drawback and to retain the simplicity of the wave equation in terms of a single dependent variable in the presence of nonuniform mean flow, an attractive alternative is to consider an acoustic velocity potential. First the mean flow field can be computed by solving the Laplace's equation; then, it is introduced in the convective terms of the wave equation expressed in terms of the acoustic velocity potential [98]. In this case, however, the physical state variables (acoustic pressure and velocity) have to be determined once the acoustic potential has been computed. A FEM approach applied in the convective wave equation for ducts with flow is further described in [2, 98]. Some additional applications of the convective wave equation with inhomogeneous properties and nonuniform moving media are presented in [84, 92] for perforated dissipative silencers, including the influence of temperature gradients.

13.2.3.2.2 **The Linearized Navier-Stokes Equations (LNSE).** The procedure is based on linearizing the fundamental equations of fluid mechanics describing the conservation of mass, momentum, and energy. This means that all field quantities involved are split into a steady part plus a time-varying part and then after substitution only terms linear in the time-varying part is kept. Some of the versions of this approach neglect viscosity and use the Euler equation to formulate the conservation of momentum [99]. However, for numerical stability reasons, dissipation is required and then it makes sense to start from Navier-Stokes to relate dissipation to an existing mechanism. It has also been demonstrated that for low Mach number (<0.3) problems, one can assume adiabatic changes of state and drop the energy equation [100]. Of course to solve the LNSE, one must first determine the steady flow part normally by a separate flow solver, e.g., Reynolds averaged Navier-Stokes (RANS). For accurate results the quality of the flow solution and interpolation from the mesh for the flow part to the acoustic mesh are important to control.

Another choice is if the problem should be done in the time or frequency domain. Moving to the frequency domain has the advantage of removing absolute instabilities only leaving convective instabilities. In addition, formulation of acoustic boundary conditions is simpler in the frequency domain. Of course for problems involving nonsteady-state behavior, e.g., transients or sound generation time-domain formulations can be better. Some recent works on LNSE have demonstrated the importance to also include turbulent damping in the form of eddy viscosity in the model [101, 102].

As an example of the strength of the LNSE method, amplification of sound and whistling will be discussed. Assuming plane waves and two openings, the first step is to compute the 2-port using the LNSE method. The whistling potential or amplification of the element can then be analyzed using the acoustic exergy formulation; see Equation 13.23. As proposed by Auregan and Starobinski [24], the maximum and minimum amplification can be obtained as the eigenvalues for the Hermitian exergy scattering matrix. Furthermore, the boundary conditions surrounding the 2-port will produce reflections on the up- and downstream side. If these reflections are collected into a reflection matrix **R**, one can write down the following equation:

$$\mathrm{p}_- = \mathrm{Sp}_+ \Rightarrow \mathrm{p}_- = \mathrm{Sp}_+ \Rightarrow \mathrm{p}_- = \mathrm{SRp}_- \Rightarrow (\mathrm{E} - \mathrm{SR})\mathrm{p}_- = 0 \qquad (13.103)$$

This equation describes the eigenfrequencies of the subsystem, and amplifying eigenfrequencies correspond to zeroes in the lower half plane. The presence of such zeroes can conveniently be found by using the so-called Nyquist stability criterion [25]. Examples of papers using the above methods to study whistling based on a LNSE model are Kierkegard et al. [103] and Du et al. [104].

13.3 Engine Silencers: Nonlinear (Time Domain) Methods

An essential issue in predictive models for silencers is that, apart from reproducing in a suitable manner the acoustic behavior of the muffler, they may be incorporated into a global exhaust noise prediction, so that it may be guessed in advance if the exhaust line complies with the requirements of the engine manufacturer. In this sense, time-domain modeling is a convenient solution, since any issues related with cylinder discharge and back reaction of the exhaust line on the engine are immediately taken into account. These gas-dynamic methods make use of thermodynamic models for in-cylinder processes, which are coupled by means of suitable boundary conditions at the valve with the flow in the exhaust system. Obviously, the key point in such methods is the possibility to account for muffler performance, which requires the definition of suitable models representing the silencers.

In the rest of this section, several options of different complexity that may allow for a proper description of the muffler behavior are described.

13.3.1 1D Methods

In the context of gas-dynamic time-domain simulation of exhaust systems, the assumption of one-dimensional flow is generally accepted and forms the natural basis for the development of muffler models. However, while flow in the ducts of the exhaust system may be soundly assumed to be one-dimensional, this is not the case in general for the mufflers, which may exhibit multi-dimensional features in their frequency response. Only in simple, even if important, cases such as long symmetric expansion chambers, quarter-wave resonators, or Herschel-Quincke tubes, the formulation of the model is straightforward [105], since wave propagation in those systems is essentially one-dimensional, and the boundary conditions required in order to couple the different one-dimensional elements (most notably sudden area changes and duct junctions) must be available in any case for other purposes. The only additional requirement would be the addition of suitable length corrections at discontinuities than can be borrowed or adapted from linear acoustics. This does not apply, however, to other equally simple systems, such as expansion chambers with offset inlet and/or outlet, reversal chambers, or even short symmetrical chambers. In such cases, and in general, it is necessary to devise a system that, being built up with one-dimensional elements, may behave as a whole in a multidimensional manner.

In order to achieve this goal, a suitable idea is to actually build an equivalent muffler model, whose geometry need not be the same as that of the real muffler (in fact, it might be a totally different one), but with the requirement that the same attenuation mechanisms as in the real case are present. Obviously, the initial requirement that the basic elements considered in the model are themselves one-dimensional (or even zero-dimensional, as it would be the case of an ideal cavity) establishes a certain upper frequency limit for the validity of the model so conceived. This frequency limit should be consistent with the relevant engine orders as specified by the engine manufacturer.

The definition of such an equivalent model should start with the identification of the main geometric characteristics influencing the acoustic behavior of the muffler (most notably, longitudinal or transversal propagation directions and their associated lengths) and of the main attenuation mechanisms acting inside the muffler, for which the use of experimentally validated multidimensional calculations is probably the best option, as they provide a detailed picture of the unsteady flow field inside the silencer (see Section 13.3.2 below). Once these are clear, it is then possible to define the equivalent muffler model, this is, the physical system whose acoustic behavior, for the frequency range specified, is essentially the same as that of the real muffler.

At this point, it should be stressed that one such model is not related to any particular set of equations, save for the general assumption of one-dimensional propagation through its basic elements. Therefore, the model may be used either in the time domain or the frequency domain, as suitable descriptions of the basic elements are available in both cases. However, as it has been shown in previous sections of this chapter, more precise models are available in the frequency domain, and thus the concept described above finds its natural application in the frame of time-domain gas-dynamic simulations.

As an example of the development of such a model, one may consider the short reverse double-chamber geometry shown in Figure 13.16. It can be observed that the working principle of this silencer is the interference between two wave paths: one going through the perforated sections of the inlet and outlet duct and the main chamber and the other one going through the back reverse chamber.

FIGURE 13.16 A short reverse double-chamber silencer.

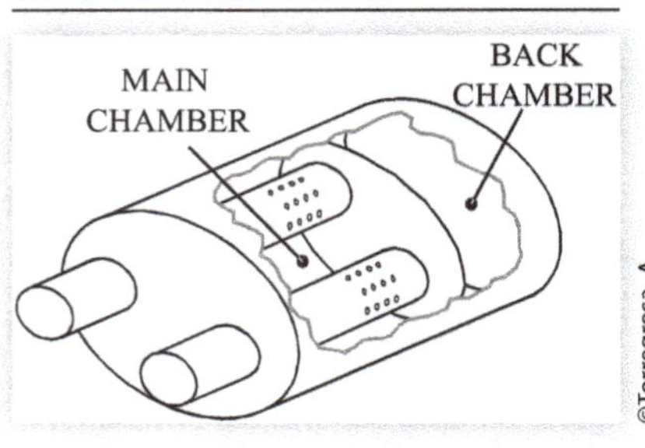

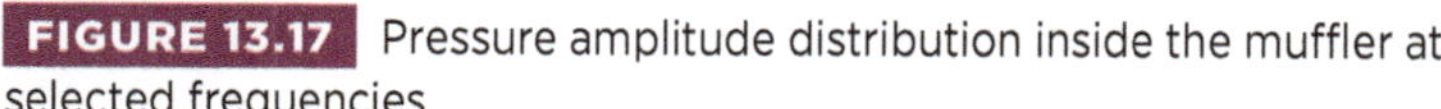
FIGURE 13.17 Pressure amplitude distribution inside the muffler at selected frequencies.

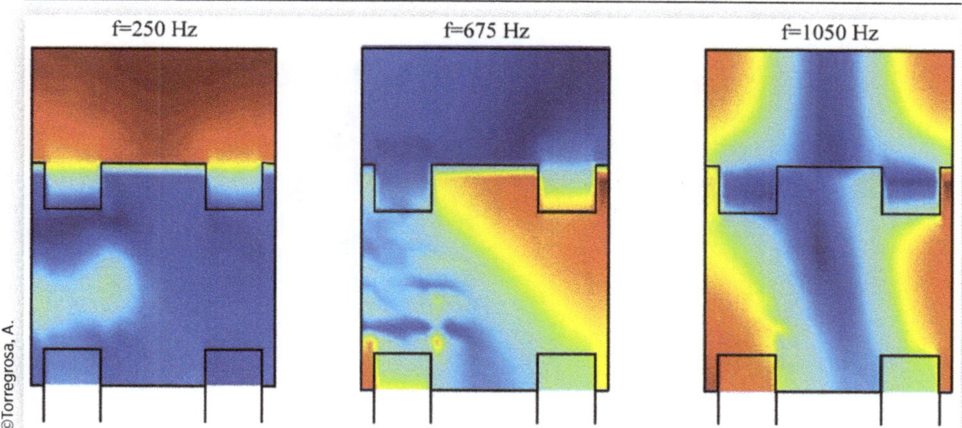

The relatively short dimensions of the two chambers strongly suggest that one should not expect that the main propagation direction corresponds to the silencer axis and thus that a simple translation of the geometry into axial one-dimensional elements may reproduce the associated acoustic behavior. This is confirmed by the results of a 3D computation, shown in Figure 13.17, where the pressure amplitude distribution inside the silencer can be observed at different frequencies at ambient temperature.

It can be readily observed that, at low frequencies, the pressure distribution in both the main chamber and the back chamber is almost uniform, which is consistent with the fact that the chamber dimensions are much smaller than the associated wavelength. Therefore, for this frequency range, the acoustic behavior of the muffler might be represented even by simply considering two ideal cavities connected by means of two straight ducts. On the contrary, at intermediate frequencies it is clearly observed that the main chamber, and to a lesser extent the reverse chamber, exhibits well-defined pressure gradients that are not longitudinal but mostly transversal, thus suggesting that the longest axis of the cross-sectional area is the characteristic dimension of these chambers for wave propagation. In fact, in the case of the main chamber, wave propagation appears to take place more in the diagonal direction than in the transversal one. In general, if the chamber length is shorter than the transversal dimension, it can be assumed that the characteristic dimension of the chamber will be the transversal direction [72]. Finally, at relatively high frequencies, it can be observed that the behavior of the muffler, while still exhibiting an important tendency to transversal propagation, starts to show a clearly 3D behavior that cannot be identified with any characteristic length, thus setting a limit for the possibility of using simple one-dimensional elements.

In view of the picture obtained from the 3D computation, a possible equivalent model for the silencer, valid up to medium frequencies, can be defined as that shown in Figure 13.18. Colors indicate the correspondence between silencer parts and equivalent one-dimensional elements, and junctions are represented with black circles and local pressure losses with grey rectangles.

As it can be observed, the model makes use only of zero-dimensional and one-dimensional basic elements. Transversal propagation in both chambers is modeled making use of a simple duct coupled with two quarter-wave resonators (that computationally are simply ducts terminated with a closed-end boundary condition). The effect of the perforated sections is accounted for by a local pressure drop, and a constant-pressure junction model is used in all the connections. Details on how the dimensions

FIGURE 13.18 Equivalent model for the silencer.

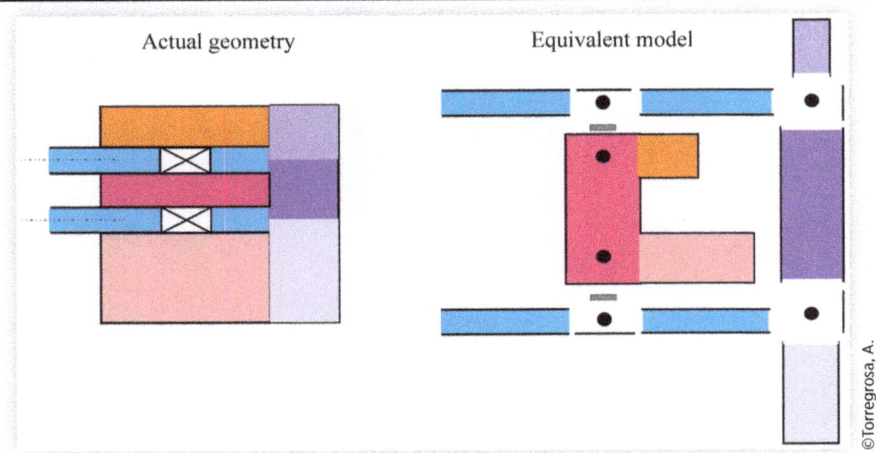

of the equivalent ducts and the value of the pressure drop coefficient at the perforates are chosen can be found in [106].

As an example of the performance of this simple model, the TL computed for a particular geometry is compared in Figure 13.19 with experimental results obtained with an improved version of the impulse method [107]. In fact, the model results were obtained by directly simulating the experimental setup used.

It appears that, consistently with the initial analysis and with the characteristics of the model defined, the results obtained are suitable for any practical purpose for frequencies below those in which, according to the 3D analysis, the use of one-dimensional elements for representing wave propagation inside the silencer would not be advisable.

Of course, a shortcoming of this modeling approach is that a suitable model must be defined for each silencer geometry (or even for the same geometry, depending on the precise values of the geometrical dimensions). However, this may not be a severe limitation from the point of view of exhaust suppliers trying to simplify their portfolio, and the possibility to incorporate these models into global engine simulations in order to actually predict the tailpipe noise (see Section 13.4.2) makes them especially attractive.

FIGURE 13.19 Example of results from equivalent model.

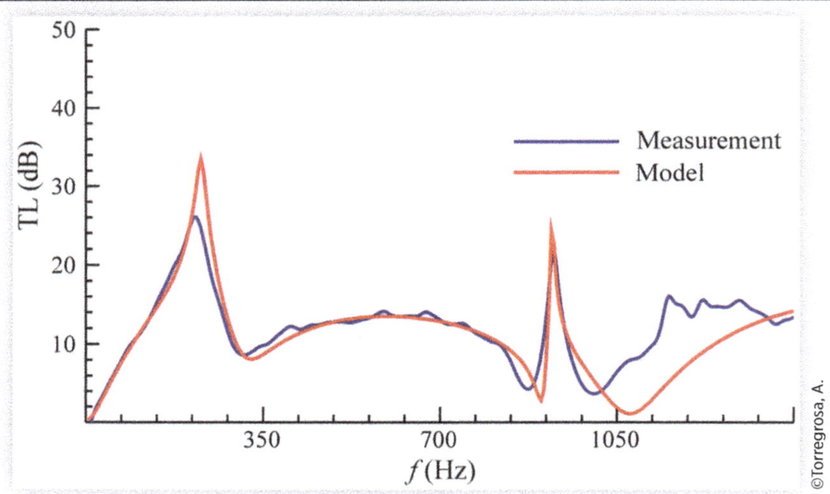

13.3.2 3D Methods

These methods constitute a further step in the direction indicated by the numerical 3D methods discussed in Section 13.2.3.2. Typically, full CFD computations are used, so that the mean flow effects are thus directly considered, since the full flow equations are solved in the time domain. Moreover, the dissipative effects may also be accounted for, the only limitation being the validity of the turbulence model used. CFD has been mostly used to assess the steady pressure drop generated by mufflers, but efforts to extend its use to unsteady flow situations date back to the works of Isshiki and Kim [108, 109]. However, these simulations were validated against linear acoustic results, by setting very small excitations, so that any potential benefits of CFD for the description of dissipative effects were not exploited.

A first attempt at the consideration of such effects was presented in [110], with an approach based on the direct simulation, in the time domain, of a modified impulse method [107] which incorporates a reasonable amount of flow and dissipative effects while keeping the acoustic characteristics essentially dependent on the muffler itself and not, as it would be the case in a severely nonlinear situation, on the particular excitation used. The method was applied to different reactive silencers without perforated elements, and the results were evaluated by comparison with the experimental TL and with the results from a conventional FEM calculation, showing that a better description of the silencer behavior was obtained when the dissipative effects were taken into account.

A similar procedure was later used in references [111] and [112], now applied to different concentric perforated silencers (straight-through and cross-flow). Apart from the ability of these methods to handle the effect of a mean flow, the effect of temperature was also considered. This last effect was precisely the main objective of the work reported in [113], where a complex multichamber muffler was considered. The results indicated that the temperature distribution had an important influence on the TL (in the case of the complex muffler, much more significant than that of the mean flow). It was also determined that the TL computed with the actual temperature field deviated significantly from that obtained with an equivalent uniform temperature field.

This is still an active research topic, and recently several studies with different modifications to the original computational setup have been reported. In [114], it is the difference between a steady computation and the unsteady case with a pulse with superimposed mean flow that is used, together with a plane wave decomposition technique, to produce the input for the attenuation analysis. In [115], the excitation used is a square pulse, and the results are used to optimize an actual intake resonator that is subsequently tested on engine with good results.

As an illustration of the information offered by these techniques, in Figure 13.20 results are shown for the same typology of silencer considered in Section 13.3.1. The measured TL is compared with the results of the full computational approach, and with a reference solution obtained with the FEM.

It is apparent that good qualitative agreement between calculations and measurement is overall achieved in all cases. The CFD prediction of the silencer response is very good in the low frequency range, below 1 kHz, whereas for higher frequencies, the FEM calculation appears to be better suited. This is directly related with the need to keep the mesh size of the CFD method within reasonable limits, what necessarily implies some degradation as frequency increases. Regarding dissipation, the differences between the measurement and the FEM calculation in the two first passbands are evident here, whereas the CFD approach provides very good agreement with the measured results. This is precisely the output expected from such a computation.

FIGURE 13.20 Example of results from 3D CFD model.

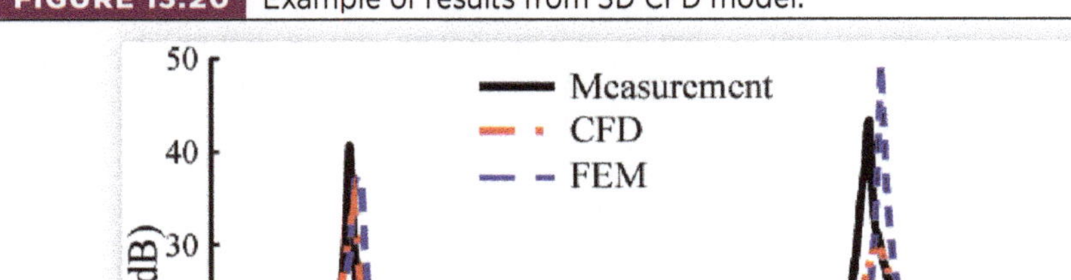

Of course, details of the acoustic pressure distribution inside the silencer are also available with such a method. In fact, those were the 3D results commented in Section 13.3.1 and already shown in Figure 13.17. It was checked in [110] that these results agree very well qualitatively with those provided by the FEM.

However, these methods may also provide additional insight into the actual attenuation mechanisms of a given silencer. As an example, in Figure 13.21 the velocity field inside the silencer is represented at several time instants.

It can be observed that the incoming flow expands and generates a toroidal vortex (non-symmetric, due to the interaction with the wall). The vortex slides along the wall, growing in size only away from the wall. Then, once the pulse reaches the inlet of the second chamber, the flow is deflected toward the wall, and the resulting perturbation in this second chamber is considerably weaker. Part of its remaining energy is dissipated by the central vortex generated in the confined space due to the interaction with the walls. As the flow exits the second chamber, it entrains the air of the first chamber, with the result that the central vortex previously originated by the inlet flow grows further in size, while losing strength. It may thus be interpreted that it is this interaction by the intermediation of a vortex what turns out to be the actual time-domain mechanism that produces the interference effect on which the action of this silencer is based.

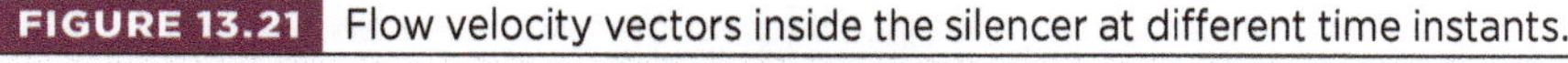

FIGURE 13.21 Flow velocity vectors inside the silencer at different time instants.

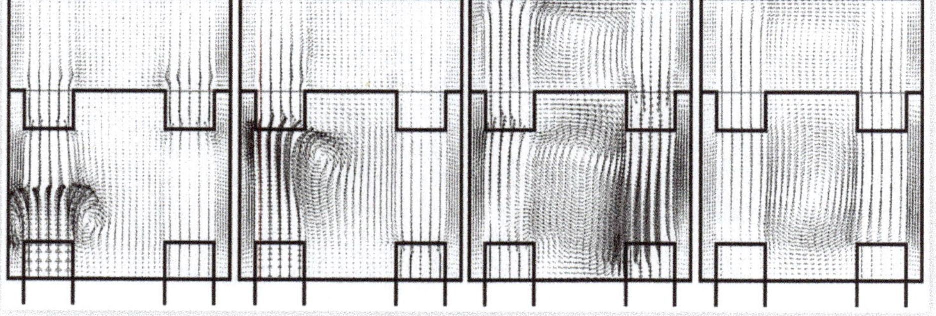

13.4 Tailpipe Noise Prediction

13.4.1 Linear (Frequency Domain) Methods

The linear frequency domain model is described under Sections 13.2.2.1 and 13.2.2.2. To set up the model, one first needs a pre-knowledge of the flow and temperature distribution in the system. In practice this is normally obtained from the 1D codes used to simulate the engine gas dynamics, e.g., BOOST and GT-Power [116, 117]. After this the system is broken down into 2-ports for which the flow and thermodynamic states are known. Normally the details of the temperature variation in each element are not important so that in practice one assumes an average value if the detailed distribution unknown. Finally the approach and models described in Sections 13.2.2.3, 13.2.2.4, and 13.2.3 are applied to compute the 2-ports for all elements. But there is still one piece of the model required to build a complete system not discussed above: the radiation impedance.

13.4.1.1 RADIATION IMPEDANCE

There are two cases of interest an intake and an exhaust pipe. In the classical paper by Davies [15], some formulas are presented for both cases. Here the intake pipe is based on those results, while for the exhaust pipe, more recent work based on the paper by Munt [45] is available.

13.4.1.1.1 **Exhaust Pipe.** Radiation from a straight pipe with no flow was analyzed in a classical paper by Levine and Schwinger [118] using the Wiener-Hopf method. The same problem including a hot outflow was analyzed by Munt [45] who, assuming a simplified jet model, again could solve the problem using the Wiener-Hopf method. A key assumption in the Munt paper is that flow leaves the duct satisfying a Kutta condition, i.e., the streamline at the wall does not bend. Furthermore the jet does not expand after it leaves the pipe implying that the reflection coefficient at the pipe end is completely determined by conditions at the exit. A key aspect here is the Kutta condition that couples sound to vorticity production at the exit. This vorticity will eventually be dissipated thereby absorbing some of the sound power that leaves the pipe [22].

The solution by Munt is complicated but Hofmans [119] has presented a set of formulas which approximate the solution in the plane wave range. The formulas give the reflection coefficient R for a circular unbaffled pipe with radius "a" radiating to free space:

$$R_{Munt} = -|R| \exp(-ik2a\delta), \tag{13.104}$$

where the magnitude $|R|$ and end correction δ are functions of the Helmholtz (ka) and Strouhal numbers ($St = ka/M$): $|R| = (1 + \xi(St))|R_0(ka)|$, $\delta = \delta_0(ka) \cdot F(St)$.

The no-flow part of these formulas are given by:

$$|R_0(ka)| = 1 + 0.01336 \cdot ka - 0.59079 \cdot (ka)^2 + 0.33576 \cdot (ka)^3 - 0.06432 \cdot (ka)^4, \quad ka < 1.8$$

$$\delta_0(ka) = \begin{cases} 0.6133 - 0.1168 \cdot (ka)^2, & ka < 0.5 \\ 0.6393 - 0.1104 \cdot ka, & 0.5 \le ka < 1.8 \end{cases} \tag{13.105}$$

and the mean flow part by:

$$\xi(\mathrm{St})=\begin{cases}\mathrm{St}^2/3, & 0\le \mathrm{St}<1\\ (2\mathrm{St}-1)/3, & 1\le \mathrm{St}<1.85\\ 0.9, & 1.85\le \mathrm{St}\end{cases} \quad \text{and} \quad F(\mathrm{St})=\begin{cases}1, & \mathrm{St}\ge 1\\ 0.33+0.65\cdot \mathrm{St}^2, & 0\le \mathrm{St}<1\end{cases}. \tag{13.106}$$

The Munt solution presented above has been validated by a number of experiments including hot flows [119, 120, 121].

Based on the reflection coefficient, the radiation impedance is obtained from:

$$Z_r/Z^{\circ}=\frac{1+R}{1-R}. \tag{13.107}$$

13.4.1.1.2 **Intake Pipe.** The general formula (13.104) still applies but based on Davies [15] the magnitude $|R|$ and end correction δ only depends on the Mach number (in the flow direction):

$$|R|=|R_0|\frac{(1-\alpha M)^{0.9}}{(1+\alpha M)^{0.9}},\quad \alpha M<0.6 \quad \text{and} \quad \delta=\delta_0\left(1-M^2\right), \tag{13.108}$$

which is valid for $M < 0.4$ and where $\alpha=1+\sqrt{0.4C_L}$ is related to the inflow loss coefficient. The impedance is then obtained from Equation (13.107).

13.4.2 Nonlinear (Time Domain) Methods

Exhaust noise emission is the result of the interaction between the flow fluctuations induced by engine operation at the exhaust mouth and the surrounding atmosphere. The direct solution of this coupled interior-exterior problem may be extremely complicated. While some attempts have been reported, for instance, in the case of the discharge of an impulsive wave [122, 123], the extension of such techniques to an engine exhaust does not seem cost-effective from a practical point of view. Therefore, all the past and present methods used in practice for exhaust noise prediction in the time domain assume that the interior and exterior problems are decoupled and can be solved independently [124]: the flow inside the tailpipe is computed making use of a suitable boundary condition representing the open end, and the resulting flow fluctuation computed at the open end is then used to feed some source model allowing for the prediction of the radiated acoustic field.

For modeling the unsteady exhaust gas flow, it is generally accepted that a sufficiently representative description is achieved by assuming quasi-steady flow at the open end, making the necessary difference between inflow and outflow. In the case of outflow, the pressure drop between the last section of the duct and the atmosphere is not significant, and thus the usual hypothesis consists precisely in assuming that the pressure in the analyzed section is equal to the atmospheric pressure and that the outgoing fluid is laminated at constant pressure until it loses its kinetic energy by friction with the surrounding air. However, the atmospheric pressure is not reached in the outlet section but downstream in the jet. The location of that section is very important, as the real position of the end reflection point affects the calculation of the characteristic frequencies of the system. It is thus usual to extend the length of the duct in the computation, in what would be a rough time-domain equivalent of introducing an end correction.

In the case of inflow, it is usual to assume energy conservation between the external fluid and the conditions at the end section of the duct and a second equation that defines the quality of the expansion process undergone by the fluid. In the general case of open ducts, the entrance is sharp and energy losses occur alongside the *vena contracta* phenomenon, what decreases the velocity of the fluid and the incoming mass flow. This is usually represented by a non-isentropic expansion, in which the generation of entropy is estimated by a coefficient expressing the velocity reduction in comparison with the ideal isentropic case.

This approach has been extensively checked in the time domain, in studies aiming at the prediction of engine performance, but its frequency domain behavior has not been systematically studied. Additionally, the eventual importance of unsteady effects at the discharge was already stressed in [124], where the work of Rudinger [125] was suggested as a promising basis for the development of a more accurate description of the open end flow. Rudinger coupled the frequency response of an open end to the flow calculation making use of the inverse Fourier transform of the termination impedance and performing the convolution of this impulse response with the flow velocity at the open end. This idea was modified and adapted to the exhaust conditions in [126], so that a boundary condition able to account for the influence of the past history of the flow on its current state was finally developed. The improvements brought in by this new boundary conditions were experimentally checked against shock-tube measurements showing a noticeable improvement in the in-duct results. However, the impact of this improved model on the resulting estimation of the radiated noise was not assessed.

In any case, a one-dimensional estimate of the flow fluctuation at the open end can be achieved and used in turn as the source for the noise emission process. Again, this is a rather crude assumption since, despite this fluctuation is likely to be the origin of any noise emission, the emission process itself is extremely complex: the surrounding air is not merely a passive propagation medium, but its own motion, induced by its interaction with the internal flow, plays an active role in the process. It would therefore be necessary to compute the details of the flow at least in the vicinity of the open end (both inside and outside the duct) and, after defining the external region where the actual source is supposed to be located, applying any of the available acoustic analogies in order to predict the radiated sound field. While the application of such techniques has been reported in simple flow configurations [127], no attempt has been found on their application to exhaust noise.

In this situation, more simple representations are needed, even though they might be affected by intrinsic limitations. It is usually admitted that the associated sound radiation is directly proportional to the volume acceleration at the tailpipe end [128]. Then, the resulting interaction with the outside conditions is described in terms of a simple emission model consisting of an isotropic monopole source characterized by the same volume velocity as that existing at the open end [129], so that the pressure at a distance r from the open end would be given by:

$$p(r,t)=\frac{\rho S}{4\pi r}\dot{u}\left(t-\frac{r}{a_0}\right) \tag{13.109}$$

Here, the density used is that of the cold surrounding air. This description is valid only at low Helmholtz numbers and may be convenient for a compact source in a nonzero mean flow, if the source location is known and the measurement position is located at a suitable distance from the source. Awareness of the importance of the influence of the mean flow on the source behavior dates back to the pioneering work of Blair and Coates [130]. Besides, the source is only approximately compact and its location is not known in detail.

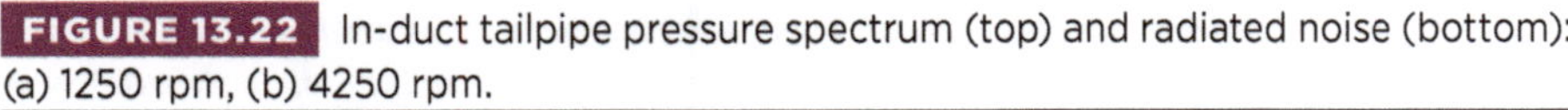

FIGURE 13.22 In-duct tailpipe pressure spectrum (top) and radiated noise (bottom): (a) 1250 rpm, (b) 4250 rpm.

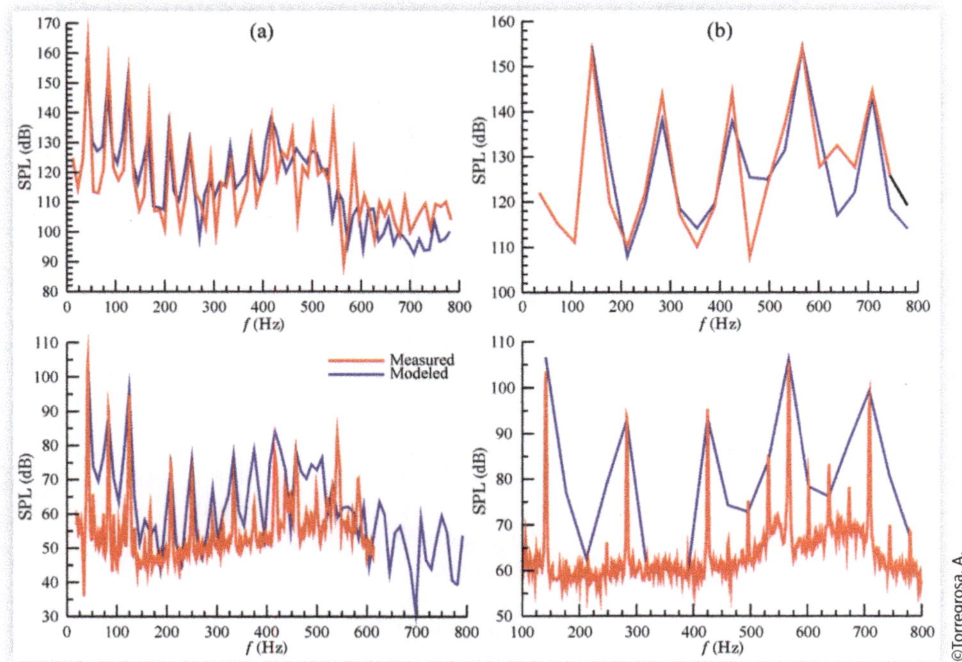

Even with all these limitations, this simple approach has been extensively (and almost exclusively) used. As an example of its performance, results of noise prediction for a four-cylinder four-stroke spark-ignited engine are shown in Figure 13.22. Simple silencer models as that described in Section 13.3.1 were used, and the open end was modeled with the quasi-steady approach described above.

It is apparent that, considering all the simplifications used, the results are acceptably indicative of the real behavior observed, both for the in-duct pressure at the tailpipe, that is essentially affected by the open end boundary condition, and for the radiated noise, that depends on both the boundary condition and the emission model.

In general, one observes that the quality of the noise prediction results depends strongly on the engine speed. This aspect was addressed in [131], where the volume velocity fluctuation at the discharge section was estimated from in-duct pressure measurements, and the results of applying different emission models were checked against noise measurements. The results indicated that a simple monopole model may provide acceptable results for medium and high engine speeds, while radiated noise was clearly overestimated at low engine speeds. It was shown that such influence is related to the corresponding effect on the ratio of the velocity fluctuation to the mean velocity at the open end, so that errors around ±1 dB may be expected when this ratio is lower than 0.5, i.e., when the amplitude of the acoustic pulsation does not dominate over the mean flow velocity and thus there is no inflow at the open end. In the other case, the model overestimates the emitted noise, and thus the acoustic energy available at the open end is not entirely radiated to the free field. It is likely that this is due to wave energy absorption in the unsteady shear layers at the open end, as suggested by the fact that these velocity ratios correspond to acoustic Strouhal numbers $Sr_{ac} < 1$, at which the flow pattern in the vicinity of the open end is defined by both the unsteady separation of the acoustic flow and the formation of vortices.

13.4.3 Hybrid Time-Frequency Methods

As indicated in the previous section, nonlinear one-dimensional unsteady flow calculations provide a good description of the interaction between the source and the duct system but, as according to the contents of Section 13.3.1, their weak point lies in the description of the silencers. Conversely, in Section 13.2 it has been clearly shown how powerful are frequency domain methods for such a description, whereas, as discussed in Section 13.4.1, the limitations of the source representation used constitute the main drawback of frequency domain methods regarding noise prediction.

Therefore, hybrid time-frequency methods have been developed in an attempt to overcome the limitations of both approaches. The simplest of these hybrid methods is that discussed in [132], where the possibility of using a time-domain calculation to determine the instantaneous volume velocity at the source and then applying the linear acoustic description of the piping system in order to compute the volume velocity at the open end was studied. This simple, decoupled solution, presents the obvious limitation that no account is taken for the interaction between the source and the system, since the frequency domain information is introduced as an a posteriori correction to the result of the time-domain computation.

The first method that took into account the coupling between the source and the duct system was that proposed in [133] where the frequency domain description of the discharge system of a compressor was used as a boundary condition for the calculation of the conditions inside the cylinder. The method consisted in a cycle-to-cycle iteration in which in-cylinder conditions were calculated setting the pressure at the valve to that obtained by transforming into the time domain the result of multiplying in the frequency domain the acoustic impedance of the whole discharge system and the spectrum of the volume velocity computed in the previous cycle. Upon convergence of the flow at the valve, the conditions at the open end could be readily calculated. In these two methods, the time-domain side is related only with the source computation, and the frequency domain side with the attached system, so that obviously eventual nonlinearities in pressure wave propagation cannot be accounted for.

A method in which both time- and frequency-domain calculations are performed in the exhaust system is that proposed by Davies and Harrison [134] starting from the work of Rudinger [125] commented above. Davies and Harrison extended the work of Rudinger in the sense that they coupled the frequency response of the silencer(s) placed downstream of the exhaust pipe to the flow computation in the duct making use of the inverse Fourier transform of the filter impedance and performing the convolution of this impulse response with the flow velocity at the inlet of the silencing system. Upon convergence, the acoustic variables at that point are transferred to the open end in the frequency domain and orifice noise is then predicted. The essence of the method is depicted in Figure 13.23.

FIGURE 13.23 Simple hybrid model.

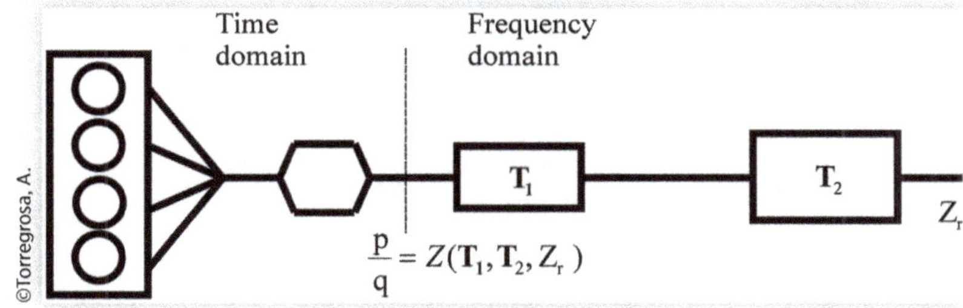

Such an approach allows the use of the frequency domain description of the open end, both through the use of the radiation impedance Z_r in the computation of the exhaust impedance Z and through the use of a frequency domain emission model on the resulting fluctuation at the open end. In this way, the problems related with the time-domain description of the exhaust orifice are avoided. Another relevant contribution was the use of a suitable approximation in order to compute the initial flow conditions along the exhaust pipe, so that the convergence of the procedure is accelerated.

A further step in the development of hybrid methods was that described in [135]. As in the previous case, the method was based on the coupling of the acoustic representation of a certain singularity to a nonlinear unsteady flow calculation, such coupling being performed in such a way that the acoustic representation of the singularity was imposed as a boundary condition for a conventional time-domain computation performed with the method of characteristics, as described in [136]. In this method, the time-domain calculation was performed not only upstream of the singularity considered but also downstream of it so that the flow in the ducts between silencers and in the tailpipe was also computed by means of the method of characteristics. In this way, any silencer was regarded as an acoustic 2-port characterized by its transmission matrix, that is, its transmission and reflection coefficients as schematically represented in Figure 13.24 (R_i is the reflection coefficient as seen from port i, and T_{ij} is the transmission coefficient between ports i and j). The wave components exciting the silencer were obtained in a natural fashion from the perturbations associated with the forward and backward characteristic lines, as detailed in [136].

While this method presents the advantage that the linear description is used only locally, the fact that the tailpipe and the open end are solved in the time domain, with all the associated problems, is certainly a shortcoming.

The methods previously discussed perform the coupling on a cycle-to-cycle basis, so that the inverse Fourier transform of the output from the frequency domain calculation is set as the boundary for the time-domain calculation in the subsequent cycle. Obviously, the convergence of the procedure is quite slow, which represents a serious shortcoming. Subsequent efforts reported in the literature were thus mainly focused on the development of non-iterative methods. Among these, the hybrid boundary proposed in [137] is based on an analytical approach and does not require any starting solution, but its applicability is still limited to periodic excitations. Other solutions avoiding this limitation have been proposed, as those described and reviewed in [138], where both convolution methods and expansion methods were considered. In convolution methods the target value is obtained from the convolution of a certain number of previous values with the corresponding samples of the impulse response or the

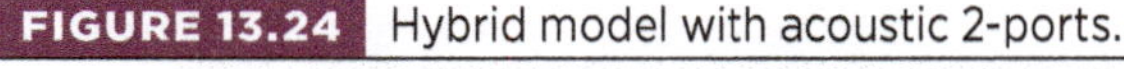
FIGURE 13.24 Hybrid model with acoustic 2-ports.

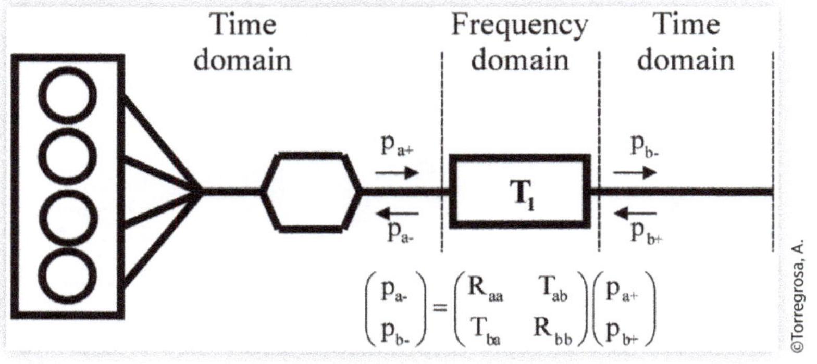

reflection function, which are obtained by the inverse Fourier transform of the impedance or the reflection coefficient, respectively. In expansion methods, the impedance is expanded in powers of iω, that when transformed into the time domain become time derivatives of the same order, so that the acoustic response of the system is described by an equivalent time-domain differential equation. However, depending on the number of terms retained in expansion, the differential equation obtained may be of high order and ill-conditioned, which may lead to numerical instabilities. As an alternative, the use of mechanical analogies has been proposed [139]. With such a method, one obtains second-order differential equations that derive from the classical damped mass-spring system. The numerical solution of this well-posed problem is less prone to show any instabilities.

A recent contribution to the development of hybrid methods proposes the use of finite impulse response (FIR) filters [140] and will be described below. Starting from a passive 2-port model in the frequency domain, e.g., in the form of a scattering matrix, see Figure 13.6 and Equation (13.20), one takes the inverse Fourier-transform to arrive at:

$$\begin{pmatrix} p'_{1-}(t) \\ p'_{2-}(t) \end{pmatrix} = \begin{pmatrix} r_{11}(t) & t_{12}(t) \\ t_{21}(t) & r_{22}(t) \end{pmatrix} \otimes \begin{pmatrix} p'_{1+}(t) \\ p'_{2+}(t) \end{pmatrix} = \begin{pmatrix} r_{11}(t)\otimes p'_{1+}(t) & t_{12}(t)\otimes p'_{2+}(t) \\ t_{21}(t)\otimes p'_{1+}(t) & r_{22}(t)\otimes p'_{2+}(t) \end{pmatrix}, \quad (13.110)$$

where ⊗ denotes a faltung and the matrix elements represent the impulse response of the corresponding scattering-matrix elements. For numerical purposes Equation (13.110) can be written based on points sampled in time: $t_n = n\Delta t, \quad n = \ldots, -3, -2, -1, 0, 1, 2, 3, \ldots$. The sampling frequency $(1/\Delta t)$ must as usual be selected to be at least twice the highest frequency of interest to satisfy the Nyquist criterion. The impulse responses are then assumed to be finite implying that a so-called FIR model from digital signal processing can be used. This leads to the following sampled model:

$$\begin{pmatrix} p'_{1-}(n) \\ p'_{2-}(n) \end{pmatrix} = \begin{pmatrix} \sum_{k=0}^{N} r_{11}(k)p'_{1+}(n-k) & \sum_{k=0}^{N} t_{12}(k)p'_{2+}(n-k) \\ \sum_{k=0}^{N} t_{21}(k)p'_{1+}(n-k) & \sum_{k=0}^{N} r_{22}(k)p'_{2+}(n-k) \end{pmatrix}, \quad (13.111)$$

where the argument denotes the time sample and the impulse response is assumed to be zero after sample N. The model in Equation (13.111) can now be implemented in a nonlinear time step solution of a system by inserting the known incident waves from the previous N steps leaving the four waves in step n as unknowns. The surrounding nonlinear systems can formally be written, assuming that side 1 is the source side and side 2 is source free, on a time sampled form as:

$$\begin{cases} O_1\left[p'_{1+}(n), p'_{1-}(n)\right] = s(n) \\ O_2\left[p'_{2+}(n), p'_{2-}(n)\right] = 0 \end{cases}, \quad (13.112)$$

where

O[...] denotes nonlinear operators

s a source function

CHAPTER 13

Combining Equations (13.111) and (13.112), one can set up a system with four equations for the four unknowns: $p'_{1+}(n), p'_{1-}(n), p'_{2+}(n), p'_{2-}(n)$.

In principle when the FIR model is used, any matrix formulation works and in the work of Knutsson et al. [140] an impedance matrix is chosen. The choice can in practice be dictated by the available options to define a 2-port in commercial codes. However, if a more general system model, such as an IIR model (infinite impulse response) is used, then to guarantee causality a scattering-matrix model must be used.

Definitions, Acronyms, and Abbreviations

Acronyms

1D-2D-3D - One-two-three-dimensional
CFD - Computational fluid dynamics
FEM - Finite element method
FIR - Finite impulse response
LNSE - Linearized Navier-Stokes equations
MM - Weighted mode-matching
PC - Point collocation
RANS - Reynolds averaged Navier-Stokes

Symbols

c - Adiabatic speed of sound
C_D - Orifice discharge coefficient
C_L - Pressure loss coefficient
d - Duct diameter
f - Frequency
G - Passive multi-port matrix
IL - Insertion loss
k - Wave number
L - Length
M - Mach number
m - Acoustic mass
p - Complex valued acoustic pressure
Pr - Prandtl number
q - Acoustic volume flow
r - Radius, distance
R - Reflection coefficient
R - Reflection matrix
S - Cross-sectional area
S - Scattering matrix
Sh - Shear wave number
St - Strouhal number
t - Time
T - Transfer matrix
T - Transmission coefficient
TL - Transmission loss

u - Complex valued acoustic velocity
U - Mean flow velocity
V - Volume
$\mathbf{W^s}$ - Sound power
x,y,z - Spatial dimensions
Y - Mobility or admittance matrix
$\mathbf{X_+}$, $\mathbf{X_-}$ - Acoustic exergy
Z - Impedance matrix
$\mathbf{Z^0}$ - Plane wave impedance
$\mathbf{Z^c}$ - Characteristic impedance

Greeks

α - Kirchoff attenuation coefficient
δ - Boundary layer thickness; end correction
γ - Ratio of specific heats
ν - Kinematic viscosity
ρ - Density
σ - Area ratio; perforation ratio
ω - Angular frequency
Ψ - Mode shape
Ω - Subdomain

Subscripts

+, - - Forward, backwards
b - Bias
g - Grazing
p - Perforate
r - Radiation
th - Thermal
v - Viscous

Superscripts

' - Fluctuating
+, - - Forward, backwards
dis - Dissipative
refl - Reflective

References

1. Goldstein, M.E., *Aeroacoustics* (McGraw-Hill, 1976).
2. Munjal, M.L., *Acoustics of Ducts and Mufflers*, 2nd edn. (Wiley, 2014).
3. Åbom, M., *An Introduction to Flow Acoustics* (2006), www.researchgate.net/publication/296691390_An_Introduction_to_Flow_Acoustics.
4. Ljung, L., *System Identification - Theory for the User*, 2nd edn. (Prentice Hall, 1999).
5. Bodén, H. and Åbom, M., "Modelling of Fluid Machines as Sources of Sound in Duct and Pipe Systems," *Acta Acoustica* 3, no. 6 (1995): 549-560.

6. Sack, S., "Experimental and Numerical Multi-Port Education for Duct Acoustics," Doctoral thesis, KTH-Royal Institute of Technology, TRITA AVE 2017-30, 2017.
7. Sack, S. and Åbom, M., "Investigation of Orifice Aeroacoustics by Means of Multi-Port Methods," *J. Sound Vib.* 407 (2017): 32-45, doi:10.1016/j.jsv.2017.06.026.
8. Sack, S., Shur, M., Åbom, M., Strelets, M. and Travin, A., "Numerical Education of Active Multi-Port Data for In-Duct Obstructions," *J. Sound Vib.* 411 (2017): 328-345, doi:10.1016/j.jsv.2017.09.012.
9. Frid, A., "Fluid Vibration in Piping Systems – A Structural Mechanics Approach. 1. Theory," *J. Sound Vib.* 133, no. 3 (1989): 423-438, doi:10.1016/0022-460X(89)90608-1.
10. Karlsson, M. and Åbom, M., "Quasi-Steady Model of the Acoustic Scattering Properties of a T-Junction," *J. Sound Vib.* 330, no. 21 (2011): 5131-5137, doi:10.1016/j.jsv.2011.05.012.
11. Pierce, A.D., *Acoustics - An Introduction to Its Physical Principles and Applications* (New York: American Institute of Physics, 1989).
12. Lyamshev, L.M., "A Question in Connection with the Principle of Reciprocity in Acoustics," *Sov. Phys. Dokl.* 4 (1959): 405-409.
13. Dokumaci, E., "On Transmission of Sound in Circular and Rectangular Narrow Pipes with Superimposed Mean Flow," *J. Sound Vib.* 210, no. 3 (1998): 375-389, doi:10.1006/jsvi.1997.1336.
14. Allam, S. and Åbom, M., "Modeling and Testing of After-Treatment Devices," *J. Vib. Acoust. - Trans. ASME* 128, no. 3 (2006): 347-356, doi:10.1115/1.2172262.
15. Davies, P.O.A.L., "Practical Flow Duct Acoustics," *J. Sound Vib.* 124, no. 1 (1988): 91-115, doi:0.1016/S0022-460X(88)81407-X.
16. Dokumaci, E., "A Note on Transmission of Sound in a Wide Pipe with Mean Flow and Viscothermal Attenuation," *J. Sound Vib.* 208, no. 4 (1997): 653-655, doi:10.1006/jsvi.1997.1043.
17. Weng, C. and Bake, F., "An Analytical Model for Boundary Layer Attenuation of Acoustic Modes in Rigid Circular Ducts with Uniform Flow," *Acta Acust. United Acust.*102, no. 6 (2016): 1138-1141, doi:10.3813/AAA.919025.
18. Howe, M.S., "The Damping of Sound by Wall Turbulent Sheer Layers," *J. Acoust. Soc. Am.* 98, no. 3 (1995): 1723-1730, doi:10.1121/1.414408.
19. Weng, C., Boij, S., and Hanifi, A., "The Attenuation of Sound by Turbulence in Internal Flows," *J. Acoust. Soc. Am.* 133, no. 6 (2013): 3764-3776, doi:10.1121/1.4802894.
20. Knutsson, M. and Åbom, M., "Sound Propagation in Narrow Tubes Including Effects of Viscothermal and Turbulent Damping with Application to Charge Air Coolers," *J. Sound Vib.* 320, no. 1-2 (2009): 289-321, doi:10.1016/j.jsv.2008.07.006.
21. Fairbrother, R., Bodén, H., and Glav, R., "Linear Acoustic Exhaust System Simulation Using Source Data from Non-Linear Simulation," SAE Technical Paper 2005-01-2358, 2005, doi:10.4271/2005-01-2358.
22. Bechert, D.W., "Sound Absorption Caused by Vorticity Shedding, Demonstrated with a Jet Flow," *J. Sound Vib.* 70, no. 3 (1980): 389-405, doi:10.1016/0022-460X(80)90307-7.
23. Glav, R. and Åbom, M., "A General Formalism for Analyzing Acoustic 2-Port Networks," *J. Sound Vib.* 202, no. 5 (1997): 739-747, doi:10.1006/jsvi.1996.0808.
24. Aurégan, Y. and Starobinski, R., "Determination of Acoustical Dissipation/Production Potentiality from the Acoustical Transfer Function of a Multiport," *Acta Acust. United Acust.* 85, no. 6 (1999): 788-792.

25. Karlsson, M. and Åbom, M., "On the Use of Linear Acoustic Multiports to Predict Whistling in Confined Flows," *Acta Acust. United Acust.* 97, no. 1 (2011): 24-33, doi:10.3813/AAA.918383.

26. Kabral, R., Du, L., and Åbom, M., "Optimum Sound Attenuation in Flow Ducts Based on the "Exact" Cremer Impedance," *Acta Acust. United Acust.* 102, no. 5 (2016): 851-860, doi:10.3813/AAA.918999.

27. Elnady, T., Åbom, M., and Allam, S., "Modeling Perforates in Mufflers Using Two-Ports," *J. Vib. Acoust. - Trans. ASME* 132, no. 6 (2010): 061010, doi:10.1115/1.4001510.

28. Elnady, T. and Åbom, M., "SIDLAB: New 1D Sound Propagation Simulation Software for Complex Duct Networks," *13th International Congress on Sound and Vibration, ICSV 2006*, Vienna, 2006.

29. Hua, X., Jiang, C., Herrin, D.W., and Wu, T.W., "Determination of Transmission and Insertion Loss for Multi-Inlet Mufflers Using Impedance Matrix and Superposition Approaches with Comparisons," *J. Sound Vib.* 333, no. 22 (2014): 5680-5692, doi:10.1016/j.jsv.2014.06.016.

30. Elnady, T., Elsaadany, S., and Åbom, M., "Flow and Pressure Drop Calculation Using Two-Ports," *J. Vib. Acoust. - Trans. ASME* 133, no. 4 (2011): 041016, doi:10.1115/1.4003593.

31. Elsahar, W., Elnady, T., and Åbom, M., "Validation of Two-Port Temperature Models for Heat Dissipation in Exhaust Systems," *J. Acoust. Soc. Am.* 134, no. 5 (2013): 4217.

32. Alenius, E., Åbom, M., and Fuchs, L., "Large Eddy Simulations of Acoustic-Flow Interaction at an Orifice Plate," *J. Sound Vib.* 345 (2015): 162-177, doi:10.1016/j.jsv.2015.02.012.

33. Kabral, R. and Åbom, M., "Investigation of Turbocharger Compressor Surge Inception by Means of an Acoustic Two-Port Model," *J. Sound Vib.* 412 (2018): 270-286, doi:10.1016/j.jsv.2017.10.003.

34. Holmberg, A., Åbom, M., and Bodén, H., "Accurate Experimental Two-Port Analysis of Flow Generated Sound," *J. Sound Vib.* 330, no. 26 (2011): 6336-6354, doi:10.1016/j.jsv.2011.07.041.

35. Nygård, S., "Modeling of Low Frequency Sound in Duct Networks," Licentiate thesis, KTH-Royal Institute of Technology, TRITA-FKT 2000:57, 2000.

36. Nelson, P.A. and Morfey, C.L., "Aerodynamic Sound Prediction in Low Speed Flow Ducts," *J. Sound Vib.* 79, no. 2 (1981): 263-289, doi:10.1016/0022-460X(81)90372-2.

37. Kårekull, O., Efraimsson, G., and Åbom, M., "Prediction Model of Flow Duct Constriction Noise," *Appl. Acoust.* 82 (2014): 45-52, doi:10.1016/j.apacoust.2014.03.001.

38. Kårekull, O., Efraimsson, G., and Åbom, M., "Revisiting the Nelson-Morfey Scaling Law for Flow Noise from Duct Constrictions," *J. Sound Vib.* 357 (2015): 233-244, doi:10.1016/j.jsv.2015.06.019.

39. Abosrea, A. and Elnady, T., "Measurement of Flow-Generated Noise inside Mufflers," SAE Technical Paper 2017-01-1795, 2017, doi:10.4271/2017-01-1795.

40. Miles, J.H., "Acoustic Transmission Matrix of a Variable Area Duct or Nozzle Carrying a Compressible Subsonic Flow," *J. Acoust. Soc. Am.* 69, no. 6 (1981): 1577-1586, doi:10.1121/1.385961.

41. Åbom, M., "An Analytical Model for Reactive Silencers Based on Bragg-Scattering," *J. Sound Vib.* 112, no. 2 (1987): 384-388.

42. Aurégan, Y., "On the Use of a Stress–Impedance Model to Describe Sound Propagation in a Lined Duct with Grazing Flow," *J. Acoust. Soc. Am.* 143, no. 5 (2018): 2975-2979, doi:10.1121/1.5037585.

43. Bauer, B.A., "Impedance Theory and Measurements on Porous Acoustic Liners," *J. Aircraft* 14, no. 8 (1977): 720-728, doi:10.2514/3.58844.

44. Glav, R., Regaud, P.-L., and Åbom, M., "Study of a Folded Resonator Including the Effects of Higher Order Modes," *J. Sound Vib.* 273, no. 4-5 (2004): 777-792, doi:10.1016/j.jsv.2003.05.024.

45. Munt, R.M., "Acoustic Transmission Properties of a Jet Pipe with Subsonic Jet Flow: I. The Cold Jet Reflection Coefficient," *J. Sound Vib.* 142, no. 3 (1990): 413-436, doi:10.1016/0022-460X(90)90659-N.

46. Boij, S. and Nilsson, B., "Reflection of Sound at Area Expansions in a Flow Duct," *J. Sound Vib.* 260, no. 3 (2003): 477-498, doi:10.1016/S0022-460X(02)00950-1.

47. Cummings, A. and Haddad, H., "Sudden Area Changes in Flow Ducts: Further Thoughts," *J. Sound Vib.* 54, no. 4 (1977): 611-612, doi:10.1016/0022-460X(77)90620-4.

48. Karal, F., "The Analogous Acoustical Impedance for Discontinuities and Constrictions of Circular Cross Section," *J. Acoust. Soc. Am.* 25, no. 2 (1953): 327-334, doi:10.1121/1.1907041.

49. Blevins, R.D., *Applied Fluid Mechanics Handbook* (Krieger Publ. Company, 2003).

50. Lahiri, C. and Bak, F., "A Review of Bias Flow Liners for Acoustic Damping in Gas Turbine Combustors," *J. Sound Vib.* 400 (2017): 564-605, doi:10.1016/j.jsv.2017.04.005.

51. Maa, D.Y., "Potential of Microperforated Panel Absorber," *J. Acoust. Soc. Am.* 104, no. 5 (1998): 2861-2866, doi:10.1121/1.423870.

52. Kabral, R., Du, L., Åbom, M., and Knutsson, M., "A Compact Silencer for the Control of Compressor Noise," *SAE Int. J. Engines* 7, no. 3 (2014): 1572-1578, doi:10.4271/2014-01-2060.

53. Bravo, T., Maury, C., and Pinhéde, C., "Optimisation of Microperforated Cylindrical Silencers in Linear and Non-Linear Regimes," *J. Sound Vib.* 363 (2016): 359-379, doi:10.1016/j.jsv.2015.11.011.

54. Guo, Y., Allam, S., and Åbom, M., "Microperforated Plates for Vehicle Applications," *Inter-Noise 2008*, China, October 26-29, 2008.

55. Allam, S. and Åbom, M., "A New Type of Muffler Based on Microperforated Tubes," *J. Vib. Acoust. - Trans. ASME* 133, no. 3 (2011): 031005, doi:10.1115/1.4002956.

56. Peat, K.S., "The Acoustical Impedance at the Junction of an Extended Inlet or Outlet Duct," *J. Sound Vib.* 150, no. 1 (1991): 101-110, doi:10.1016/0022-460X(91)90404-8.

57. Torregrosa, A.J., Broatch, A., Payri, R., and González, F., "Numerical Estimation of End Corrections in Extended-Duct and Perforated-Duct Mufflers," *J. Vib. Acoust. - Trans. ASME* 121, no. 3 (1999): 302-308, doi:10.1115/1.2893980.

58. Selamet, A., Denia, F.D., and Besa, A.J., "Acoustic Behavior of Circular Dual-Chamber Mufflers," *J. Sound Vib.* 265, no. 5 (2003): 967-985, doi:10.1016/S0022-460X(02)01258-0.

59. Kirby, R., Williams, P.T., and Hill, J., "A Three Dimensional Investigation into the Acoustic Performance of Dissipative Splitter Silencers," *J. Acoust. Soc. Am.* 135, no. 5 (2014): 2727-2737, doi:10.1121/1.4869089.

60. Williams, P., Kirby, R., Hill, J., Åbom, M. and Malecki, C., "Reducing Low Frequency Tonal Noise in Large Ducts Using a Hybrid Reactive-Dissipative Silencer," *Appl. Acoust.* 131 (2018): 61-69, doi:10.1016/j.apacoust.2017.10.018.

61. Mechel, F.P., *Formulas of Acoustics* (Springer, 2008).

62. Bento-Coelho, J.L., "Acoustic Characteristics of Perforate Liners in Expansion Chambers," Doctoral thesis, ISVR, University of Southampton, 1983.

63. Cummings, A. and Chang, I.J., "Sound Attenuation of a Finite Length Dissipative Flow Duct Silencer with Internal Mean Flow in the Absorbent," *J. Sound Vib.* 127, no. 1 (1988): 1-17, doi:10.1016/0022-460X(88)90347-1.

64. Kirby, R. and Denia, F.D., "Analytic Mode Matching for a Circular Dissipative Silencer Containing Mean Flow and a Perforated Pipe," *J. Acoust. Soc. Am.* 122, no. 6 (2007): 3471-3482, doi:10.1121/1.2793614.

65. Renou, Y. and Aurégan, Y., "Failure of the Ingard-Myers Boundary Condition for a Lined Duct: An Experimental Investigation," *J. Acoust. Soc. Am.* 130, no. 1 (2011): 52-60, doi:10.1121/1.3586789.

66. Ih, J.G. and Lee, B.H., "Analysis of Higher-Order Mode Effects in the Circular Expansion Chamber with Mean Flow," *J. Acoust. Soc. Am.* 77, no. 4 (1985): 1377-1388, doi:10.1121/1.392029.

67. Ih, J.G. and Lee, B.H., "Theoretical Prediction of the Transmission Loss of Circular Reversing Chamber Mufflers," *J. Sound Vib.* 112, no. 2 (1987): 261-272, doi:10.1016/S0022-460X(87)80194-3.

68. Kim, J. and Soedel, W., "Development of a General Procedure to Formulate Four Pole Parameters by Modal Expansion and Its Application to Three-Dimensional Cavities," *J. Vib. Acoust. - Trans. ASME* 112, no. 4 (1990): 452-459, doi:10.1115/1.2930128.

69. Kim, Y.H., Choi, J.W., and Lim, B.D., "Acoustic Characteristics of an Expansion Chamber with Constant Mass Flow and Steady Temperature Gradient (Theory and Numerical Simulation)," *J. Vib. Acoust. - Trans. ASME* 112, no. 4 (1990): 460-467, doi:10.1115/1.2930129.

70. Åbom, M., "Derivation of Four-Pole Parameters Including Higher Order Mode Effects for Expansion Chamber Mufflers with Extended Inlet and Outlet," *J. Sound Vib.* 137, no. 3 (1990): 403-418, doi:10.1016/0022-460X(90)90807-C.

71. Kim, Y.H. and Choi, J.W., "General Solution of Acoustic Wave Equation for Circular Reversing Chamber with Temperature Gradient," *J. Vib. Acoust. - Trans. ASME* 113, no. 4 (1991): 543-550, doi:10.1115/1.2930219.

72. Selamet, A. and Radavich, P.M., "The Effect of Length on the Acoustic Attenuation Performance of Concentric Expansion Chambers: An Analytical, Computational and Experimental Investigation," *J. Sound Vib.* 201, no. 4 (1997): 407-426, doi:10.1006/jsvi.1996.0720.

73. Selamet, A. and Ji, Z.L., "Acoustic Attenuation Performance of Circular Expansion Chambers with Offset Inlet/Outlet: I. Analytical Approach," *J. Sound Vib.* 213, no. 4 (1998): 601-617, doi:10.1006/jsvi.1998.1514.

74. Selamet, A. and Ji, Z.L., "Acoustic Attenuation Performance of Circular Expansion Chambers with Extended Inlet/Outlet," *J. Sound Vib.* 223, no. 2 (1999): 197-212, doi:10.1006/jsvi.1998.2138.

75. Kirby, R., "Simplified Techniques for Predicting the Transmission Loss of a Circular Dissipative Silencer," *J. Sound Vib.* 243, no. 3 (2001): 403-426, doi:10.1006/jsvi.2000.3425.

76. Denia, F.D., Albelda, J., Fuenmayor, F.J., and Torregrosa, A.J., "Acoustic Behaviour of Elliptical Chamber Mufflers," *J. Sound Vib.* 241, no. 3 (2001): 401-421, doi:10.1006/jsvi.2000.3289.

77. Denia, F.D., Selamet, A., Fuenmayor, F.J., and Kirby, R., "Acoustic Attenuation Performance of Perforated Dissipative Mufflers with Empty Inlet/Outlet Extensions," *J. Sound Vib.* 302, no. 4-5 (2007): 1000-1017, doi:10.1016/j.jsv.2007.01.005.

78. Selamet, A., Xu, M.B., Lee, I.J., and Huff, N.T., "Analytical Approach for Sound Attenuation in Perforated Dissipative Silencers," *J. Acoust. Soc. Am.* 115, no. 5 (2004): 2091-2099, doi:10.1121/1.1694994.

79. Glav, R., "The Point-Matching Method on Dissipative Silencers of Arbitrary Cross-Section," *J. Sound Vib.* 189, no. 1 (1996): 123-135, doi:10.1006/jsvi.1996.0009.

80. Glav, R., "The Transfer Matrix for a Dissipative Silencer of Arbitrary Cross-Section," *J. Sound Vib.* 236, no. 4 (2000): 575-594, doi:10.1006/jsvi.2000.2994.

81. Kirby, R., "Transmission Loss Predictions for Dissipative Silencers of Arbitrary Cross Section in the Presence of Mean Flow," *J. Acoust. Soc. Am.* 114, no. 1 (2003): 200-209, doi:10.1121/1.1582448.

82. Albelda, J., Denia, F.D., Torres, M.I., and Fuenmayor, F.J., "A Transversal Substructuring Mode Matching Method Applied to the Acoustic Analysis of Dissipative Mufflers," *J. Sound Vib.* 303, no. 3-5 (2007): 614-631, doi:10.1016/j.jsv.2007.01.043.

83. Denia, F.D., Sánchez-Orgaz, E.M., Baeza, L., and Kirby, R., "Point Collocation Scheme in Silencers with Temperature Gradient and Mean Flow," *J. Comput. Appl. Math.* 291 (2016): 127-141, doi:10.1016/j.cam.2015.02.007.

84. Sánchez-Orgaz, E.M., "Advanced Numerical Techniques for the Acoustic Modelling of Materials and Noise Control Devices in the Exhaust System of Internal Combustion Engines," Doctoral thesis, Research Centre on Mechanical Engineering-CIIM, Universitat Politècnica de València, 2016.

85. Allard, J.F. and Atalla, N., *Propagation of Sound in Porous Media* (Wiley, 2009).

86. Kirby, R. and Cummings, A., "The Impedance of Perforated Plates Subjected to Grazing Gas Flow and Backed by Porous Media," *J. Sound Vib.* 217, no. 4 (1998): 619-636, doi:10.1006/jsvi.1998.1811.

87. Murphy, J.E. and Chin-Bing, S.A., "A Finite Element Model for Ocean Acoustic Propagation and Scattering," *J. Acoust. Soc. Am.* 86, no. 4 (1989): 1478-1483, doi:10.1121/1.398708.

88. Antebas, A.G., Denia, F.D., Pedrosa, A.M., and Fuenmayor, F.J., "A Finite Element Approach for the Acoustic Modeling of Perforated Dissipative Mufflers with Non-homogeneous Properties," *Math. Comput. Model.* 57, no. 7-8 (2013): 1970-1978, doi:10.1016/j.mcm.2012.01.021.

89. Zienkiewicz, O.C., Taylor, R.L., and Zhu, J.Z., *The Finite Element Method: Its Basis and Fundamentals* (Elsevier Butterworth-Heinemann, 2005).

90. Dokumaci, E., "A Quasi-One-Dimensional Theory of Sound Propagation in Lined Ducts with Mean Flow," *J. Sound Vib.* 419 (2018): 1-17, doi:10.1016/j.jsv.2017.12.039.

91. Lee, S.H. and Ih, J.G., "Empirical Model of the Acoustic Impedance of a Circular Orifice in Grazing Mean Flow," *J. Acoust. Soc. Am.* 114, no. 1 (2003): 98-113, doi:10.1121/1.1581280.

92. Denia, F.D., Sánchez-Orgaz, E.M., Martínez-Casas, J., and Kirby, R. "Finite Element Based Acoustic Analysis of Dissipative Silencers with High Temperature and Thermal-Induced Heterogeneity," *Finite Elem. Anal. Des.* 101 (2015): 46-57, doi:10.1016/j.finel.2015.04.004.

93. Kirby, R., "A Comparison between Analytic and Numerical Methods for Modelling Automotive Dissipative Silencers with Mean Flow," *J. Sound Vib.* 325, no. 3 (2009): 565-582, doi:10.1016/j.jsv.2009.03.032.

94. Kirby, R., "Modeling Sound Propagation in Acoustic Waveguides Using a Hybrid Numerical Method," *J. Acoust. Soc. Am.* 124, no. 4 (2008): 1930-1940, doi:10.1121/1.2967837.

95. Denia, F.D., Martínez-Casas, J., Carballeira, J., Nadal, E. and Fuenmayor, F.J., "Computational Performance of Analytical Methods for the Acoustic Modelling of Automotive Exhaust Devices Incorporating Monoliths," *J. Comput. Appl. Math.* 330 (2018): 995-1006, doi:10.1016/j.cam.2017.03.010.

96. Xu, M.B., Selamet, A., Lee, I.J., and Huff, N.T., "Sound Attenuation in Dissipative Expansion Chambers," *J. Sound Vib.* 272, no. 3-5 (2004): 1125-1133, doi:10.1016/j.jsv.2003.07.025.

97. Pierce, A.D., "Wave-Equation for Sound in Fluids with Unsteady Inhomogeneous Flow," *J. Acoust. Soc. Am.* 87 (1990): 2292-2299, doi:10.1121/1.399073.

98. Peat, K.S., "Evaluation of Four-Pole Parameters for Ducts with Flow by the Finite Element Method," *J. Sound Vib.* 84, no. 3 (1982): 389-395, doi:10.1016/0022-460X(82)90485-0.

99. Gracia, M.M., Korchagin, V., De Roeck, W., and Desmet, W., "Validation of a Surface Based Analogy for the LEE Accounting for Scattering Effects," *22nd AIAA/CEAS Aeroacoustics Conference*, 2717, 2016.

100. Kierkegaard, A., Boij, S., and Efraimsson, G., "A Frequency Domain Linearized Navier-Stokes Equations Approach to Acoustic Propagation in Flow Ducts with Sharp Edges," *J. Acoust. Soc. Am.* 127, no. 2 (2010): 710-719, doi:10.1121/1.3273899.

101. Gikadi, J., Föller, S., and Sattelmayer, T., "Impact of Turbulence on the Prediction of Linear Aeroacoustic Interactions: Acoustic Response of a Turbulent Shear Layer," *J. Sound Vib.* 333, no. 24 (2014): 6548-6559, doi:10.1016/j.jsv.2014.06.033.

102. Holmberg, A., Kierkegaard, A., and Weng, C., "A Frequency Domain Linearized Navier–Stokes Method Including Acoustic Damping by Eddy Viscosity Using RANS," *J. Sound Vib.* 346 (2015): 229-247, doi:10.1016/j.jsv.2015.02.030.

103. Kierkegaard, A., Allam, S., Efraimsson, G., and Åbom, M., "Simulations of Whistling and the Whistling Potentiality of an In-Duct Orifice with Linear Aeroacoustics," *J. Sound Vib.* 331, no. 5 (2012): 1084-1096, doi:10.1016/j.jsv.2011.10.028.

104. Du, L., Abom, M., Karlsson, M., and Knutsson, M., "Modelling of Acoustic Resonators Using the Linearized Navier Stokes Equations," SAE Technical Paper 2016-01-1821, 2016, doi:10.4271/2016-01-1821.

105. Broatch, A., Serrano, J.R., Arnau, F.J., and Moya, D., "Time-Domain Computation of Muffler Frequency Response: Comparison of Different Numerical Schemes," *J. Sound Vib.* 305, no. 1-2 (2007): 333-347, doi:10.1016/j.jsv.2007.04.023.

106. Torregrosa, A.J., Broatch, A., and Payri, R., "The Use of Transfer Matrix for the Design of Interferential Systems in Exhaust Mufflers," SAE Technical Paper 2000-01-0728, 2000, doi:10.4271/2000-01-0728.

107. Payri, F., Desantes, J.M., and Broatch, A., "Modified Impulse Method for the Measurement of the Frequency Response of Acoustic Filters to Weakly Nonlinear Transient Excitations," *J. Acoust. Soc. Am.* 107, no. 2 (2000): 731-738, doi:10.1121/1.428256.

108. Isshiki, Y., Shimamoto, Y., and Wakisaka, T., "Simultaneous Prediction of Pressure Losses and Acoustic Characteristics in Silencers by Numerical Simulation," SAE Technical Paper 960637, 1996, doi:10.4271/960637.

109. Kim, M.H., "Three-Dimensional Numerical Study on the Pulsating Flow Inside Automotive Muffler with Complicated Flow Path," SAE Technical Paper 2001-01-0944, 2001, doi:10.4271/2001-01-0944.

110. Broatch, A., Margot, X., and Gil, A., "A CFD Approach to the Computation of the Acoustic Response of Exhaust Mufflers," *J. Comp. Acoust.* 13, no. 2 (2005): 301-316, doi:10.1142/S0218396X05002682.

111. Ji, Z.L., Xu, H.S., and Kang, Z.X., "Influence of Mean Flow on Acoustic Attenuation Performance of Straight-Through Perforated Tube Reactive Silencers and Resonators," *Noise Control Eng. J.* 58, no. 1 (2010): 12-17, doi:10.3397/1.3244593.

112. Liu, C. and Ji, Z., "Computational Fluid Dynamics-Based Numerical Analysis of Acoustic Attenuation and Flow Resistance Characteristics of Perforated Tube Silencers," *J. Vib. Acoust. - Trans. ASME* 136, no. 2 (2014): 021006, doi:10.1115/1.4026137.

113. Xuan, L.K., Gong, J.F., Ming, P.J., Jin, G.Y. et al., "A Time-Domain Finite Volume Approach for Prediction of Muffler Transmission Loss Including Thermal Effects," *Proc. IMechE Part C: J. Mech. Eng. Sci.* 228, no. 1 (2014): 108-118, doi:10.1177/0954406213481714.

114. Zhu, D.D. and Ji, Z.L., "Transmission Loss Prediction of Reactive Silencers Using 3-D Time Domain CFD Approach and Plane Wave Decomposition Technique," *Appl. Acoust.* 112 (2016): 25-31, doi:10.1016/j.apacoust.2016.05.004.

115. Li, H., Hao, Z.Y., Zheng, X., and Liu, L.Y., "Transmission Loss Simulation Based on Rectangular-Pulse Method and Experimental Evaluation of Acoustic Performance of Perforated Intake Pipe," *J. Cent. South Univ.* 24 (2017): 2605-2612, doi:10.1007/s11771-017-3674-7.

116. www.avl.com/boost/.

117. www.gtisoft.com/gt-suite-applications/propulsion-systems/gt-power-engine-simulation-software/.

118. Levine, H. and Schwinger, J., "On the Radiation of Sound from an Unflanged Circular Pipe," *Phys. Rev.* 73, no. 4 (1980): 383-406, doi:10.1103/PhysRev.73.383.

119. Hofmans, G.C.J., "Vortex Sound in Confined Flows," Doctoral thesis, Eindhoven Technical University, 1998.

120. Allam, S. and Åbom, M., "Investigation of Damping and Radiation Using Full Plane Wave Decomposition in Ducts," *J. Sound Vib.* 292, no. 3-5 (2006): 519-534, doi:10.1016/j.jsv.2005.08.016.

121. Tiikoja, H., Lavrentjev, J., Rämmal, H., and Åbom, M., "Experimental Investigations of Sound Reflection from Hot and Subsonic Flow Duct Termination," *J. Sound Vib.* 333, no. 3 (2014): 788-800, doi:10.1016/j.jsv.2013.09.030.

122. Kim, H.D. and Setoguchi, T., "Study of the Discharge of Weak Shocks from an Open End of a Duct," *J. Sound Vib.* 226, no. 5 (1999): 1011-1028, doi:10.1006/jsvi.1999.2376.

123. Liang, S.M. and Tai, C.S., "Analysis and Prediction of Shock-Induced Near-Field Acoustics from an Exhaust Pipe," *Comp. Fluids* 45 (2011): 222-232, doi:10.1016/j.compfluid.2011.01.031.

124. Jones, A.D., "Modelling of the Exhaust Noise Radiated from Reciprocating Internal Combustion Engines - A Literature Review," *Noise Control Eng. J.* 23, no. 1 (1984): 12-31, doi:10.3397/1.2827635.

125. Rudinger, G., "The Reflection of Pressure Waves of Finite Amplitude from an Open End of a Duct," *J. Fluid Mech.* 3, no. 1 (1957): 48-66, doi:10.1017/S0022112057000476.

126. Piscaglia, F., Montorfano, A., Onorati, A., and Ferrari, G., "Modeling of Pressure Wave Reflection from Open-Ends in I.C.E. Duct Systems," SAE Technical Paper 2010-01-1051, 2010, doi:10.4271/2010-01-1051.

127. Tam, C.K.W., "Computational Aeroacoustics: Issues and Methods," *AIAA J.* 33, no. 10 (1995): 1788-1796, doi:10.2514/3.12728.

128. Dowling, A.P. and Ffowcs-Williams, J.E., *Sound and Sources of Sound* (Ellis Horwood, 1983).

129. Landau, L.D. and Lifshitz, E.M., *Fluid Mechanics*, 2nd edn. (Pergamon Press, 1987).

130. Blair, G.P. and Coates, S.W., "Noise Produced by Unsteady Exhaust Efflux from an Internal Combustion Engine," SAE Technical Paper 730160, 1973, doi:10.4271/730160.

131. Torregrosa, A.J., Broatch, A., Bermúdez, V., and Andrés, I., "Experimental Assessment of Emission Models Used for IC Engine Exhaust Noise Prediction," *Exp. Therm. Fluid Sci.* 30, no. 2 (2005): 97-107, doi:10.1016/j.expthermflusci.2005.05.001.

132. Jones, A.D., Van Moorhem, W.K., and Voland, R.T., "Is a Full Nonlinear Method Necessary for the Prediction of Radiated Engine Exhaust Noise?," *Noise Control Eng. J.* 26, no. 2 (1986): 74-80, doi:10.3397/1.2827665.

133. Singh, R. and Soedel, W., "Mathematical Modelling of Multicylinder Compressor Discharge System Interactions," *J. Sound Vib.* 63, no. 1 (1979): 125-143, doi:10.1016/0022-460X(79)90382-1.

134. Davies, P.O.A.L. and Harrison, M.F., "Hybrid Systems for I.C. Engine Breathing Noise Synthesis," *Proc. I.O.A.* 15, no. 3 (1993): 369-374.

135. Desantes, J.M., Torregrosa, A.J., and Broatch, A., "Hybrid Linear/Nonlinear Method for Exhaust Noise Prediction," SAE Technical Paper 950545, 1995, doi:10.4271/950545.

136. Payri, F., Desantes, J.M., and Torregrosa, A.J., "Acoustic Boundary Condition for Unsteady One-Dimensional Flow Calculations," *J. Sound Vib.* 188, no. 1 (1995): 85-110, doi:10.1006/jsvi.1995.0580.

137. Harrison, M.F. and Perez-Arenas, R., "A Hybrid Boundary for the Prediction of Intake Wave Dynamics in IC Engines," *J. Sound Vib.* 270, no. 1-2 (2004): 111-136, doi:10.1016/S0022-460X(03)00486-3.

138. Albertson, F., Bodén, H., and Gilbert, J., "Comparison of Different Methods to Couple Nonlinear Source Descriptions in the Time Domain to Linear System Descriptions in the Frequency Domain - Application to a Simple Valveless One-Cylinder Cold Engine," *J. Sound Vib.* 291, no. 3-5 (2006): 963-985, 10.1016/j.jsv.2005.07.046.

139. Rodriguez, S., Gibiat, V., Lefebvre, A., and Guilain, S., "Use of a Mechanical Analogy to Couple the Time-Domain of a One-Cylinder Cold-Engine Model with the Input Impedance of Its Intake System," *J. Sound Vib.* 330, no. 23 (2011): 5769-5783, doi:10.1016/j.jsv.2011.07.023.

140. Knutsson, M., Lennblad, J., Bodén, H., and Abom, M., "A Study on Acoustical Time-Domain Two-Ports Based on Digital Filters with Application to Automotive Air Intake Systems," *SAE Int. J. Passeng. Cars - Mech. Syst.* 4, no. 2 (2011): 970-982, doi:10.4271/2011-01-1522.

about the authors

Mats Åbom is Professor in Fluid Acoustics at KTH-Royal Institute of Technology, Stockholm, since 2001. He holds an MSc in engineering physics (1979) and a PhD in engineering acoustics (1989) also from KTH. During 1996-2000 he worked at ABB Corporate Research as main project leader in acoustics. During 2003-2015 he was head of the acoustics group at KTH, and during 2014-2016 vice president for the European Acoustics Association. Since 2015 he is director for the Centre for Sustainable Aviation at KTH.

The main area of research is aeroacoustics of interior flows in particular related to sound generation, propagation, and source characterization in fluid machinery systems. Recently works related to thermoacoustics and metamaterials have come into focus as well as noise prediction for aircrafts. His work has resulted in over 3300 citations (h-index 28, i10-index 59), He is also co-author of ten patents.

Jakob Andert received his Dipl.-Ing. and Dr.-Ing. degrees at the RWTH Aachen University in 2007 and 2012, respectively. From 2012 to 2014, he was a project manager at FEV GmbH in the field of hybrid and electrical drivetrains. Since 2014, he is junior professor for mechatronic systems in combustion engines at the Institute for Combustion Engines (VKA) at RWTH Aachen University. He received his Dr.-Ing. degree for a dissertation about a real-time cycle-to-cycle control of homogeneous charge compression ignition engines. The junior professorship addresses mechatronic systems in combustion engines, with a special focus on the seamless simulation model transfer from model-in-the-loop to software- and hardware-in-the-loop. The real-time co-simulation coupling for heterogeneous virtual/physical validation approaches is of particular interest for him. He is involved as co-organizer of the session "0-D and 1-D Modeling and Numerics" for the WCX SAE World Congress.

Michael Bargende is a full professor at the Institute for Internal Combustion Engines and Automotive Engineering (IVK), University of Stuttgart, and holds the Chair of Automotive Powertrains. In joint appointment, he is also a member of the Board of Management of the Research Institute for Automotive Engineering and Vehicle Engines Stuttgart (FKFS) in Stuttgart, Germany.

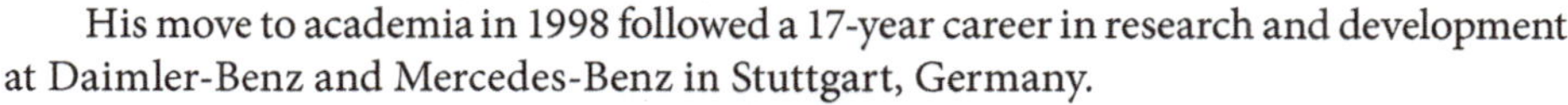

His move to academia in 1998 followed a 17-year career in research and development at Daimler-Benz and Mercedes-Benz in Stuttgart, Germany.

He has been the advisor for more than 60 Ph.D. dissertations on combustion engines and powertrain-related topics. He is the author or co-author of more than 250 scientific publications (35 SAE papers) on the subjects of combustion, thermodynamics, gas exchange, NVH, mechanics, and modeling, simulation, and control of internal combustion engines, as well as hybrid powertrains. He is the editor or co-author of several professional books (reference books, volumes, conference proceedings) on the subject of gasoline and diesel engines, natural gas engines, and hybrid powertrains.

Michael Bargende is a member of various scientific organizations, associations, and councils, and he is an evaluator for several German research associations, such as DFG and AiF. In 2015, he was elected to become an SAE Fellow.

Fabio Bozza is Full Professor of fluid machines, automotive powertrains, and optimization and simulation of internal combustion engines (ICEs) at the Industrial Engineering Department of the University of Naples "Federico II" (Italy). His research interests are related to experimental and numerical studies of internal combustion engines, including turbulence, combustion, knock, radiated noise, turbocharging, emissions, engine design and virtual calibration, and hybrid electric powertrains. He is author of more than 160 scientific publications, is SAE member for more than 15 years, and is member of the Governing Board of SAE-Naples Section. He cooperates to the "Formula SAE" Project of the University of Naples. He received the Oral Presentation Award at SAE World Congress more than twice, and thus he received the Lloyd L. Withrow Distinguished Speaker Award at SAE World Congress 2016. He is involved in various projects with automotive companies and research centers, including the H2020 European project (EAGLE).

Marcello Canova is Associate Professor in mechanical engineering and associate director of the Center for Automotive Research, at The Ohio State University. He earned his *Diploma di Laurea "Summa Cum Laude"* and his PhD in mechanical engineering from the University of Parma (Italy).

Dr. Canova conducts research in the broad areas of powertrain electrification and transportation systems, with emphasis on modeling, optimization, and associated control problems. His research has been funded by, among others, Ford, General Motors, Fiat Chrysler Automobiles, Cummins, the National Science Foundation, the US Department of Energy, and ARPA-E.

Dr. Canova is a 2016 NSF CAREER Award recipient, and he has earned the Kappa Delta Distinguished Faculty Award (2011), the SAE Vincent Bendix Automotive Electronics Engineering Award (2011), the Lumley Interdisciplinary Research Award (2012), the SAE Ralph Teetor Educational Award (2016), and the Michael J. Moran Award for Excellence in Teaching (2017). He has published over 120 articles in journals and conference proceedings.

Marco Chiodi studied Mechanical Engineering and graduated from the Politecnico di Milano, Italy, in 1998. In 1995, he took part in the ERASMUS student exchange program and since then he moved to the University of Stuttgart, Germany.

He completed his doctoral program (summa cum laude) in 2010 at the University of Stuttgart. In 2010, his doctoral thesis has been awarded with the Südwestmetall award for outstanding works of young academic researchers. Since 2003, he is Project Manager for 3D-CFD simulations of internal combustion engines at FKFS (Research Institute of Automotive Engineering and Vehicle Engines Stuttgart) and IVK (Institute for Internal Combustion Engines and Automotive Engineering of the University of Stuttgart). His team is composed of around 6 research assistants and several students.

He has published around 50 papers in international conferences and journals. His research focus is the development of an innovative 3D-CFD-approach toward virtual development of internal combustion engines in all possible application fields: research, mass-production, and motorsports. This software has remarkably contributed to the investigation and improvement of several engine designs and concepts – also in combination with innovative fuels – toward efficiency increase and emissions reduction.

Héctor Climent has been working with internal combustion engines for more than two decades. His expertise lies in the field of internal fluid dynamics phenomena involved in the gases exchange processes in the engine. He is a full professor at Universitat Politècnica de València where he combines both teaching and research activities in the CMT-Motores Térmicos institute. You should find him either in front of his computer dealing with engine models or in the CMT lab persuading the engine to behave as the model predicted.

Vincenzo De Bellis graduated with honors in mechanical engineering at the University of Naples "Federico II" (Italy) in 2005. He received a PhD in mechanical systems engineering at University of Naples "Federico II" in 2012, discussing a thesis concerning the 1D simulation of steady and unsteady operations of turbochargers for automotive applications. He is a researcher at the Industrial Engineering Department of the University of Naples "Federico II," where he teaches fluid machines, automotive powertrains, and optimization and simulation of ICEs. Research interests include experimental and numerical studies of turbochargers, internal combustion engines, and hybrid electric powertrains. He is author of more than 50 scientific publications and an SAE member for more than 10 years.

Francisco D. Denia has a doctoral thesis from UPV, Spain, since 2004. He works as a Professor of Mechanical Engineering at UPV, dedicated to teaching and research activities as well as to industry-sponsored research projects in the fields of acoustics and vibration. His research activity in the last two decades has focused on the acoustic design of automotive exhaust silencers and aftertreatment devices, including the development of efficient computational models for sound wave propagation analysis.

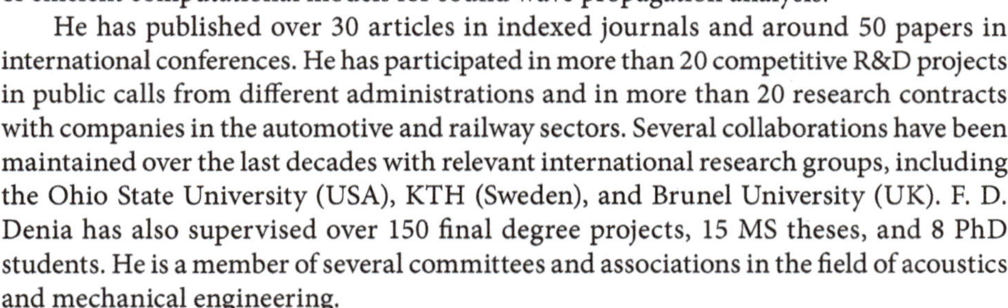

He has published over 30 articles in indexed journals and around 50 papers in international conferences. He has participated in more than 20 competitive R&D projects in public calls from different administrations and in more than 20 research contracts with companies in the automotive and railway sectors. Several collaborations have been maintained over the last decades with relevant international research groups, including the Ohio State University (USA), KTH (Sweden), and Brunel University (UK). F. D. Denia has also supervised over 150 final degree projects, 15 MS theses, and 8 PhD students. He is a member of several committees and associations in the field of acoustics and mechanical engineering.

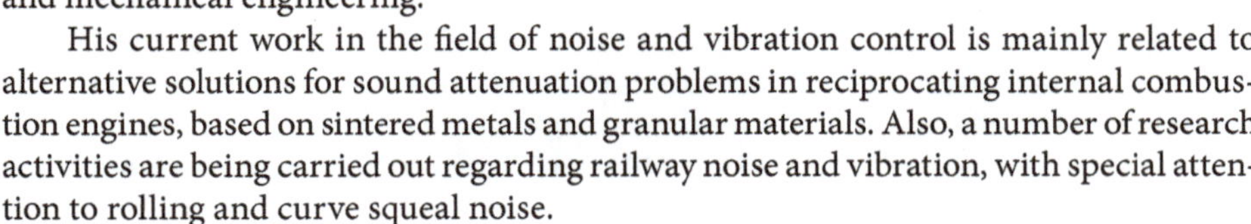

His current work in the field of noise and vibration control is mainly related to alternative solutions for sound attenuation problems in reciprocating internal combustion engines, based on sintered metals and granular materials. Also, a number of research activities are being carried out regarding railway noise and vibration, with special attention to rolling and curve squeal noise.

Alessio Dulbecco is a senior research engineer at IFP Energies nouvelles (IFPEN) French Research Institute (France). His research interests are related to 0D to 3D computational approaches for turbulence, spray, turbulent combustion, and pollutant emissions, applied to automotive and aeronautics domains. He obtained a double master degree in mechanical engineering at the Politecnico di Torino (Italy) and at the Université de Compiègne (France) in 2005. He received a PhD in energetics and transfer at the Ecole Polytechnique of Toulouse (France) in 2009, Research Laboratory CERFACS. He worked as research associate at the Imperial College London (United Kingdom) in 2010. Since 2013 he has been part-time Associate Professor of internal combustion engines at the Université Paris 13 (France). He is the author of scientific publications and participated in international scientific conferences.

Robert Fairbrother graduated from Imperial College (London) in 1994 with a Ph.D. in Mechanical Engineering. He started his professional career at Ford Motor Company (Dunton, UK) before working for Ricardo (Shoreham, UK, and Chicago, USA). He became a senior development engineer in the Advanced Simulation Technologies department at AVL List GmbH (Graz, Austria) in 2001. He is now a Project Manager for Software Development of Multiphysical Systems at AVL. His responsibilities include the development of engine performance prediction software, gas exchange and combustion analysis, as well as linear and non-linear acoustics. He has authored and co-authored a number of research papers on these topics. He is a Chartered Engineer (CEng) and a member of the IMechE, the IIAV, and the SAE.

José Galindo is Professor of internal combustion engine at CMT-Motores Térmicos of the Universitat Politècnica de València. He has been the leader of the activities in engine gas exchange process and turbocharging since 2000. His research has always been related with 1D wave dynamics modeling in engines, in which he has contributed in topics such as exhaust manifold junctions, two-stroke engines, sequential turbocharging, compressor surge, and others. He has published more than 60 papers in research journals and contributed to a similar number of conferences.

Michael Grill studied Mechanical Engineering and graduated from the University of Stuttgart, Germany, in 2003. He completed his doctoral program in 2006. Till 2015, he was head of the 0D/1D simulation department of FKFS (Research Institute of Automotive Engineering and Vehicle Engines Stuttgart). There are around 15 research assistants and lots of students in his 0D/1D simulation department. He gives lectures regarding modeling and simulation of internal combustion engines at the University of Stuttgart.

He has published around 80 papers in international conferences and journals. The research focus of his department is on the development of better submodels for 0D/1D simulation and a deeper understanding of the physics and chemistry in the combustion chamber. The commercial focus of his department is on concept studies regarding possible future engine concepts.

Santhosh Gundlapally received his PhD in chemical engineering from University of Houston in 2011. Santhosh worked with Professor Vemuri Balakotaiah for his doctorate and later on with Professor William Epling as a research assistant. His PhD research was on the mathematical modeling and analysis of catalytic monoliths and packed bed reactors. Specifically, he has investigated the effect of various design and operating conditions on the steady state and dynamic performance of aftertreatment reactors using nonlinear bifurcation analysis and numerical simulations. He joined Gamma Technologies in 2012 and since then he has been actively involved in the development of advanced mathematical models and simulation tools for aftertreatment reactors, combustion, knock, and in-cylinder emissions. His research interests are in heterogeneous catalysis, chemical kinetics, and control-oriented models.

Dr. Onoufrios Haralampous is an Associate Professor at the University of Thessaly. He is specialized in exhaust after-treatment technology for internal combustion engines and his research interests extend to heat and mass transport phenomena, filtration, catalysis and numerical methods. After receiving his PhD from the Department of Mechanical Engineering at Aristotle University of Thessaloniki (AUTh) in 2005, he worked at the Laboratory of Applied Thermodynamics (LAT) and participated in the foundation of spinoff company Exothermia SA. Since 2010, he has been pursuing a career in academia and has served as a principal investigator in three research projects for the automotive industry. He has coauthored more than 30 scientific papers in international journals and conferences in the field of exhaust after-treatment modeling, one of which has been selected by SAE for the "Arch T. Colwell Merit Award".

Mahir Tim Keskin studied Automotive Engineering at the Politecnico di Torino (Italy) and at the University of Stuttgart (Germany), from where he graduated in 2012. Having completed his doctoral program in 2015 at the same institution, he now works in the 0D/1D engine simulation department of the Research Institute of Automotive Engineering and Vehicle Engines Stuttgart (FKFS).

His research activity so far has focused on 0D combustion modeling mainly, noticeably with the development of a fast, predictive burn rate model for gasoline homogeneous charge compression ignition (gHCCI). Other publications include work on natural gas engines and on virtual engine development for RDE (Real Driving Emissions).

Apart from scientific work, his current activities extend to joint projects with the automotive industry, where he is actively involved in concept studies that investigate future combustion technologies such as lean combustion, water injection, and prechamber spark plugs.

Dr. Grigorios Koltsakis received his diploma and PhD from the Department of Mechanical Engineering of Aristotle University of Thessaloniki. He is a professor at the Laboratory of Applied Thermodynamics of Aristotle University, specializing in internal combustion engines and exhaust aftertreatment technology. He has authored more than 100 papers in scientific journals and international conferences and holds two patents. Dr. Koltsakis has received the Arch T. Colwell Merit Award as wells as two oral presentation awards from SAE International. He has also received the first innovation prize from the Research Committee of Aristotle University. Dr. Koltsakis leads a research group active in the simulation and evaluation of aftertreatment technologies in close collaboration with the international automotive industry. He is a co-founder and scientific director of Exothermia SA, a university spinoff company developing emission control system simulation software.

Tommaso Lucchini is an associate professor at the Energy Department of Politecnico di Milano, where he graduated in Mechanical Engineering in 2002 and achieved PhD in Energetics in 2006. His main research activities are focused on CFD modelling of in-cylinder flows and combustion in internal combustion engines. He develops the Lib-ICE code, which is based on the OpenFOAM® technology. In his career he has dealt with different CFD simulation projects including diesel and spark-ignition engines focused on efficiency improvement and fuel consumption reduction. In 2017, he received the SAE John Johnson Award for Outstanding Research in Diesel Engines.

Jan Macek graduated at master level at the Czech Technical University (CTU) in Prague in 1972. He achieved his PhD level in the National Institute for Machine Research in 1980 with thesis focused on thermodynamic simulations of turbocharged diesels and Dr. Sc. techn. at the CTU in 1993. He obtained full professorship at the same university in 1994. He acted at this university as a chairman of Automobile, Internal Combustion Engine and Rail Vehicle Department between 1993 and 2015. Before it, he was employed in CKD Praha Heavy Industries Research from 1972 to 1990, and he was active in medium speed and locomotive diesel and gas engine research.

Federico Millo received his master degree in mechanical engineering at the Politecnico di Torino in 1989.

In 1991 he joined the faculty at Politecnico di Torino as a researcher assistant, and in 1998 he was appointed as associate professor, being then elevated to the role of full professor in 2016.

Since 1998 he has been giving lectures in the graduate course *Automotive Internal Combustion Engines and since 2004 in the Pollutant Emissions Control graduate course.*

His research activity has been entirely focused on internal combustion engines, in particular on the analysis and on the diagnostic of the combustion process, on the use of alternative fuels, on pollutant emissions control in SI and diesel engines, on engine modeling, and on the development of engine control strategies for conventional as well as for hybrid powertrains.

He has been responsible and principal investigator for several research projects with major OEMs such as General Motors, Honda, FCA, and Ferrari and the coordinator for several research projects funded by national and regional Italian public institutions.

F. Millo has published over 100 articles, most of which on international journals.

He has been nominated SAE (Society of Automotive Engineers) Fellow in 2015, being the first Italian from the academia to be elevated to the role of Fellow.

Gianluca Montenegro achieved the MSc degree in mechanical engineering at the Politecnico di Milano in 1999 and the PhD in energy engineering in 2002, in the same university. He became permanent staff in 2006 working as assistant professor, and he is associate professor at Politecnico di Milano, Department of Energy, since 2015.

He chairs the courses of fluid machines and modeling techniques for fluid machines at Politecnico di Milano for energy engineering and mechanical engineering students. He works as member of the Internal Combustion Engine (ICE) Group of the Energy Department at Politecnico di Milano coordinating the research activity on 1D and 3D modeling of intake and exhaust systems. The main topics of his research are the development and application of 1D and 3D models for the simulation of unsteady reacting flows in IC engine duct systems and aftertreatment devices; the development and application of 1D-3D coupling techniques, and the development of quasi-3D models for the acoustic and fluid dynamic simulation of intake and exhaust systems for IC. He is the author of more than 60 indexed publications.

He is associate editor for the *SAE Int. Journal of Engines*. In April 2015 he received the Lloyd L. Withrow Distinguished Speaker SAE Award.

Roberto Navarro was born in Valencia (Spain) in 1986. He obtained his PhD degree in transport propulsion systems in 2014 at Universitat Politècnica de València, where he is currently employed as a senior lecturer. Particularly, he is a member of CMT-Motores Térmicos research institute, and his area of interest is focused on engine air management, including the numerical and experimental analysis of EGR flow phenomena. Navarro has published 18 scientific papers in JCR journals and 11 contributions to conferences and has participated in 7 competitive public-funded research programs and 18 private-funded research collaborations. He has received several awards, including an outstanding PhD thesis prize by Universitat Politècnica de València.

Angelo Onorati graduated in 1989 in mechanical engineering at the Politecnico di Milano (PoliMi), then he achieved a PhD in Energy Engineering in 1993. He became a lecturer at the Department of Energy of Politecnico di Milano in 1993 (in the field of "fluid machines"). From 1998 to 2003, he was an associate professor in the same university, then since the beginning of 2004, he become a full professor. His main research subjects are 1D and CFD modeling of unsteady reacting flows in IC engine duct systems and aftertreatment devices; modeling of SI and CI engine combustion process, prediction of emissions; and prediction of tailpipe noise and silencer modelling. He is the author of more than 120 publications.

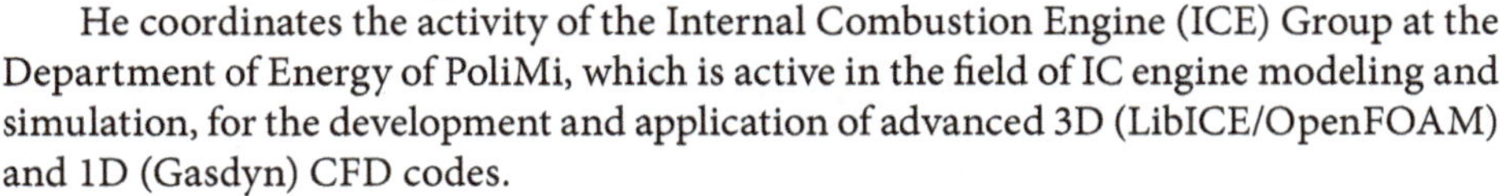

He coordinates the activity of the Internal Combustion Engine (ICE) Group at the Department of Energy of PoliMi, which is active in the field of IC engine modeling and simulation, for the development and application of advanced 3D (LibICE/OpenFOAM) and 1D (Gasdyn) CFD codes.

He is involved in the editorial board of the IJER (*International Journal of Engine Research*) and of the *SAE Int. Journal of Engines*. In April 2011 he received the Lloyd L. Withrow Distinguished Speaker SAE Award. He is involved as co-organizer of the session "0-D and 1-D Modeling and Numerics" for the WCX SAE World Congress.

Christof Schernus (Dipl.-Ing.), Vice President, Business Development Research and Innovation at FEV Europe GmbH.

Christof formerly worked as Department Manager for Innovation and Research and as Senior Project Manager in gasoline engines. He joined FEV after receiving his diploma in mechanical engineering from RWTH Aachen University in 1989. In the almost 30 years of his career, he dealt with a wide range of different analytical and simulation projects including energy management in electrified powertrains, heat pumps, acoustics, 1D- and 3D-CFD, and internal combustion engines. As an expert in the field of simulations, he is one of the organizers of the session 0D and 1D modeling and numerics for over ten years at the SAE World Congress and since 2015 in the ICE Conference by SAE Naples section. He has been author and co-author of more than 60 papers and presentations. Since 2015, he coordinates FEV's European research and innovation activities covering the entire portfolio from internal combustion engines, electric vehicles, transmissions, and autonomous driving, with simulations playing an important role in all these fields, too.

Stephanie Stockar is an assistant professor in the Department of Mechanical Engineering (ME) at Penn State University. Dr. Stockar conducts research in the areas of modeling and optimization of nonlinear dynamical systems, with focus on automotive applications and energy systems. Her research approach hinges upon the multidisciplinary integration of thermo-fluid sciences with dynamic systems, modeling, optimization, and control. Dr. Stockar's work has been funded by Ford Motor Company, Fiat Chrysler Automobiles, the National Science Foundation, the US Department of Energy, and ARPA-E.

Before joining the ME Department at Penn State University in 2016, she was a research associate at The Ohio State University Center for Automotive Research (OSU-CAR). She earned her BS and MS in mechanical engineering from the Swiss Federal Institute of Technology (ETH) Zurich in 2007 and 2010, respectively, and obtained her PhD in mechanical engineering from The Ohio State University in 2013.

Augusto Della Torre graduated in Mechanical Engineering at Politecnico di Milano in 2009. He achieved Ph.D. in Energy and Nuclear Science and Technology in 2013, discussing a thesis entitled "Multi-Scale CFD Modeling of Intake and Exhaust Systems for Internal Combustion Engines." As part of the Ph.D., in 2012, he spent a research period at the University of Exeter (UK), working in the research group of prof. Gavin Tabor. From 2015, he is part of the faculty staff as a researcher at the Department of Energy of Politecnico di Milano. He is currently lecturer for the course of "Fluid Machines and Energy System" and "Modeling Techniques for Fluid Machines."

His research activity is focused on thermo-fluid dynamic modeling of internal combustion engines, with particular interest in the simulation of unsteady flows in intake and exhaust systems. He developed models for the simulation and the optimization of the acoustic properties of silencers at different detail levels, ranging from fast quasi-3D approaches to detailed CFD models. Moreover, his research activity includes the CFD modeling of reacting flows in the after-treatment systems, the study and the optimization of innovative substrates for catalytic devices, the CFD modeling of gas exchange process, and the simulation of the engine thermal management.

Antonio J. Torregrosa has been full professor at CMT-Motores Térmicos, Universitat Politècnica de València (UPV), Spain, since 2002. He obtained his PhD at the same university in 1993. His research activity has focused on the thermo- and fluid dynamics of reciprocating internal combustion engines and in particular in those aspects related to noise control and thermal management of engines.

He has published around 50 articles in indexed journals and around 30 papers in international conferences. He has participated in more than 30 competitive R&D projects subsidized in public calls by different administrations and in more than 50 research contracts with companies in the automotive sector.

His current work in the field of noise control is mostly related with intake systems of turbocharged engines, with special interest on elements with flexible and/or porous walls. In the field of thermal management, the focus is on the problems associated with heat flows in propulsion plants of hybrid vehicles.

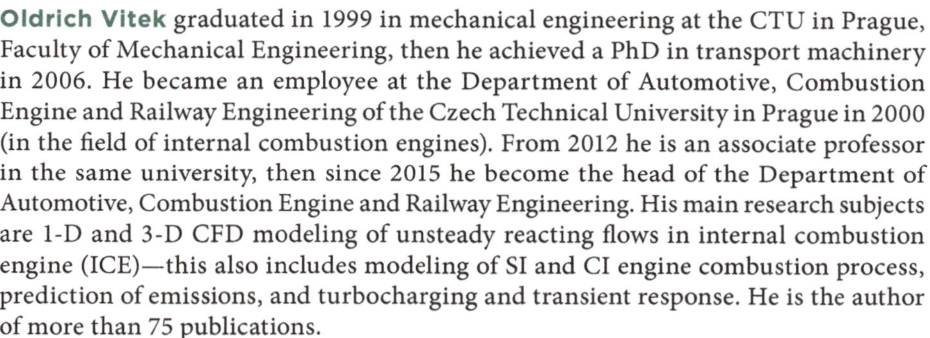

Oldrich Vitek graduated in 1999 in mechanical engineering at the CTU in Prague, Faculty of Mechanical Engineering, then he achieved a PhD in transport machinery in 2006. He became an employee at the Department of Automotive, Combustion Engine and Railway Engineering of the Czech Technical University in Prague in 2000 (in the field of internal combustion engines). From 2012 he is an associate professor in the same university, then since 2015 he become the head of the Department of Automotive, Combustion Engine and Railway Engineering. His main research subjects are 1-D and 3-D CFD modeling of unsteady reacting flows in internal combustion engine (ICE)—this also includes modeling of SI and CI engine combustion process, prediction of emissions, and turbocharging and transient response. He is the author of more than 75 publications.

He coordinates the activity of the whole Department of Automotive, Combustion Engine and Railway Engineering while focusing on ICE Group, which is active in the field of ICE experiments, modeling, and simulation.

He is involved in the editorial board of the MECCA (*Journal of Middle European Construction and Design of Cars*). He is also involved as co-organizer of the session "0-D and 1-D Modeling and Numerics" for the WCX SAE World Congress.

Syed Wahiduzzaman received his PhD in mechanical engineering from Purdue University in 1983 with a concentration in turbulent fluid flow, combustion, and heat transfer. Following which he joined Integral Technologies (ITI) Incorporated as a senior engineer until 1987 where he played a key role in the development of IRIS for engine in-cylinder processes simulation. During his ITI days and subsequently in Ricardo North America, he developed models for simulation of combustion in gasoline and diesel engines, for predictions of phenomena such as engine knock, in-cylinder heat transfer, and engine-out emissions. He played a key role in integrating the two ITI codes (WAVE and IRIS) in order to create a comprehensive engine simulation code that models all the relevant processes that occur from the air cleaner to the tailpipe. He left Ricardo North America in 1995 after serving as Manager of Research and Development and thereafter proceeded to co-found Gamma Technologies, a company dedicated in developing a state-of-the-art multi-physics CAE system simulation software. From 1995 to 2014, he was a director and principal engineer at Gamma, where he was involved in the development of simulation platform for modeling advanced combustion concepts for both spark-ignited and compression-ignited engines. Additionally, he led the research and development of advanced modeling tools for simulation of exhaust aftertreatment components and systems. He is a Vice President at Gamma and continues to be active in in-cylinder processes and aftertreatment modeling methodologies. Syed Wahiduzzaman has published over 50 peer-reviewed articles.

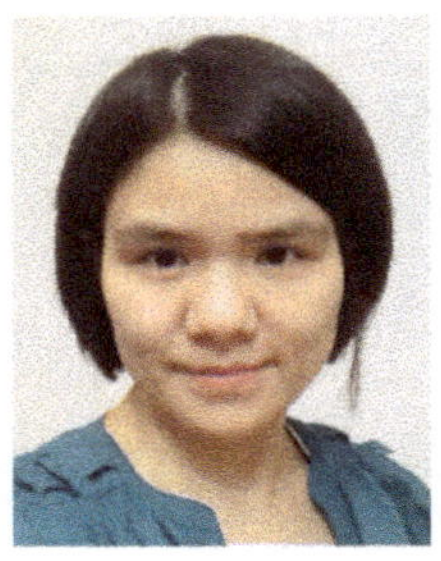

Dr. Wen Wang is a staff engineer at Gamma Technologies working on aftertreatment system model development. She has been working on one-dimensional and multidimensional computational models and code development for exhaust aftertreatment devices including DOC, TWC, SCR, DPF, GPF, etc. Dr. Wang received her PhD in mechanical engineering from Iowa State University in 2008. Her research work was focused on turbulent flow and heat transfer modeling, and she has developed large eddy simulation codes for the cooling flow of nuclear reactors.

Yuri Wright is a senior IT engineer and head of the IC Engine Computational Reactive Fluid Dynamics group at the Aerothermochemistry and Combustion Systems Laboratory of ETH Zurich since 2006. He received his diploma in mechanical engineering in 2000 and Dr. sc. techn. in 2005 from ETH Zurich in the field of turbulent spray combustion modelling. His research interests include single- and two-phase flows; heat transfer; premixed, non-premixed, and dual fuel combustion; and pre-chamber ignition systems in the broader IC engines context, as well as after-treatment systems. He has authored over 60 Web of Science indexed publications and acts as a reviewer for 12 journals in his field. From 2018 to 2024, he has been appointed associate editor for the Proceedings of the Combustion Institute. Since 2014, he is also the leader for the Gas Engines collaborative task in the Technology Collaboration Programme on "Energy Conservation and Emissions Reduction in Combustion" of the International Energy Agency. He is also part-time employed as a project manager at the ETH spin-off company Combustion and Flow Solutions GmbH in Zurich.

Feihong Xia received her bachelor's degree in mechanical engineering and her master's degree in automotive engineering and transport at RWTH Aachen University in Germany. Since 2015 she works as a PhD candidate at the junior professorship for mechatronic systems for combustion engines at Institute for Combustion Engines under the supervision of Professor Jakob Andert. Her work focusses on the 1D CFD simulation of internal combustion engines, not only for engine concept study but also for X-in-the-loop applications like function development, system testing, and virtual calibration.

Zdenek Zak graduated in mechanical engineering from CTU in Prague in 2009. He received a PhD in mechanical engineering from CTU in Prague in 2018. He works as a research engineer at CTU from 2009. Zdenek Zak's teaching and research interests lie in the areas of internal combustion engines and turbochargers. He is a member of SAE.

index

www.ingramcontent.com/pod-product-compliance
Lightning Source LLC
Chambersburg PA
CBHW050104220326
41598CB00043B/7376

* 9 7 8 0 7 6 8 0 9 3 5 2 0 *